Benzimidazoles and Congeneric Tricyclic Compounds

IN TWO PARTS

PART ONE

This is the fortieth volume in the series

THE CHEMISTRY OF HETEROCYCLIC COMPOUNDS

THE CHEMISTRY OF HETEROCYCLIC COMPOUNDS

A SERIES OF MONOGRAPHS

ARNOLD WEISSBERGER and EDWARD C. TAYLOR

Editors

BENZIMIDAZOLES AND CONGENERIC TRICYCLIC COMPOUNDS

PART 1

Edited by

P. N. PRESTON

DEPARTMENT OF CHEMISTRY,
HERIOT–WATT UNIVERSITY,
EDINBURGH, SCOTLAND

With contributions by

D. M. SMITH

DEPARTMENT OF CHEMISTRY,
UNIVERSITY OF ST. ANDREWS,
ST. ANDREWS,
SCOTLAND

G. TENNANT

DEPARTMENT OF CHEMISTRY,
UNIVERSITY OF EDINBURGH,
EDINBURGH,
SCOTLAND

AN INTERSCIENCE ® PUBLICATION

JOHN WILEY & SONS

New York . Chichester . Brisbane . Toronto

An Interscience ® Publication
Copyright © 1981 by John Wiley & Sons, Inc.

Library of Congress Cataloging in Publication Data:

Main entry under title:

Benzimidazoles and congeneric tricyclic compounds.

　(The Chemistry of heterocyclic compounds;
-v. 40, pt. 1　　ISSN 0069-3154)
　"An Interscience publication."
　Includes index
　1. Benzimidazoles. I. Preston, P. N.
QD401.B46　　547'.593　　80-17383
ISBN 0-471-03792-3 (v. 1)
ISBN 0-471-08189-2 (v. 2)

The Chemistry of Heterocyclic Compounds

The chemistry of heterocyclic compounds is one of the most complex branches of organic chemistry. It is equally interesting for its theoretical implications, for the diversity of its synthetic procedures, and for the physiological and industrial significance of heterocyclic compounds.

A field of such importance and intrinsic difficulty should be made as readily accessible as possible, and the lack of a modern detailed and comprehensive presentation of heterocyclic chemistry is therefore keenly felt. It is the intention of the present series to fill this gap by expert presentations of the various branches of heterocyclic chemistry. The subdivisions have been designed to cover the field in its entirety by monographs which reflect the importance and the interrelations of the various compounds, and accommodate the specific interests of the authors.

In order to continue to make heterocyclic chemistry as readily accessible as possible, new editions are planned for those areas where the respective volumes in the first edition have become obsolete by overwhelming progress. If, however, the changes are not too great so that the first editions can be brought up to date by supplementary volumes, supplements to the respective volumes will be published in the first edition.

ARNOLD WEISSBERGER

Research Laboratories
Eastman Kodak Company
Rochester, New York

EDWARD C. TAYLOR

Princeton University
Princeton, New Jersey

Preface to Part 1

More than 25 years have elapsed since the publication in this series of
Imidazole and Its Derivatives by Klaus Hofmann. In updating this work,
Leroy Townsend has undertaken the task of editing a volume on monocyclic
imidazoles, and the present book covers the chemistry of benzimidazole and
its dihydro derivatives, as well as congeneric tricyclic compounds that
contain a condensed benzimidazole moiety. Because many ring systems are
covered, it has proved necessary to divide the volume into Part 1 (Chapters
1 to 5) and Part 2 (Chapters 6 to 10).

Chapters 1 to 3 on benzimidazoles, benzimidazole *N*-oxides, and dihydro
derivatives update the book of Hofmann through Volume 87 of *Chemical
Abstracts*. The chemistry of tricyclic compounds containing a condensed
benzimidazole moiety is covered comprehensively from early literature
through the same Volume 87 of *Chemical Abstracts*.

Chapters 4 to 9 on the condensed ring systems are organized in terms of
the position and size of the ring fused to the benzimidazole skeleton
(denoted "6-5"). Thus Chapters 4 through 8 are concerned with compounds
in which fusion of the third ring is at the benzo and imidazole rings
respectively.

Chapter 9 deals with the chemistry of tricyclic compounds in which a
benzimidazole moiety may be considered to be formally annulated from N-1
to C-7.

The growth of benzimidazole chemistry in the past 25 years has paralleled
that of purines and stems from the determination of the partial structures of
nucleic acids in the early 1950s. Benzimidazoles and congeneric compounds
are substrates that might act as inhibitors in nucleic acid biosynthesis, and
their relative ease of preparation and low cost make them attractive as
potential pharmacological agents. The variety of marketed products described in chapter 10 bears witness to the large commitment to benzimidazole chemistry. I hope that this book will stimulate further research,
particularly on the synthesis of new tricyclic derivatives and related condensed analogs.

I am indebted to a number of friends and colleagues who have contrib-
uted to this book. It has been a pleasure to collaborate with David Smith
and with Malcolm Stevens and George Tennant, and I thank them for their
large collective contribution. Information on commercially marketed products is difficult to obtain, but my task was simplified with the generous
assistance of Colin C. Beard, Gerald Farrow, Janet M. Shether, Brian K.
Snell, and Ian S. Swanson. I also thank my wife, Veronica, who carried out
an initial estimate of the magnitude of literature on benzimidazoles and

congeneric tricyclic compounds. Thanks are due also to Susan Bobby who typed part of the manuscript, Anthony F. Fell who translated a number of documents from Russian, and my former research students Alex Davidson and Ian E. P. Murray who helped to check the manuscript. Finally, I express my appreciation of the help and enthusiasm of the Series Editors, Edward C. Taylor and Arnold Weissberger, of Stanley F. Kudzin, and of the staff of John Wiley and Sons, Inc.

P. N. PRESTON

Edinburgh, Scotland
January 1981

Contents

PART ONE

PART TWO

Benzimidazoles and Congeneric Tricyclic Compounds

IN TWO PARTS

PART ONE

This is the fortieth volume in the series

THE CHEMISTRY OF HETEROCYCLIC COMPOUNDS

CHAPTER 1

Benzimidazoles

P. N. PRESTON

1

4   Benzimidazoles

1.1. INTRODUCTION: LITERATURE COVERAGE AND ORGANIZATION OF THE BOOK

In 1953 the extent of the literature was such that Klaus Hofmann was able to comprehensively cover the entire chemistry of monocyclic imidazoles (**1.1**) and benzimidazoles (**1.2**) in *Imidazole and Its Derivatives, Part I.*[1]

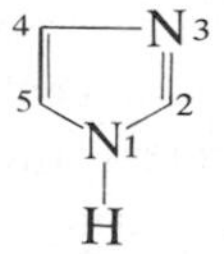

(**1.1**) 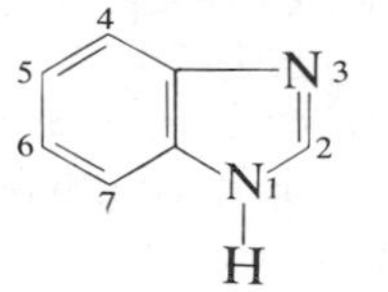   (**1.2**) 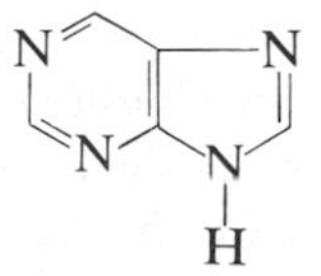   (**1.3**)

The early 1950s was an important period regarding the biological significance of benzimidazoles and the closely related purines (**1.3**);[2] the vital role of purines in biological systems was established,[2] and it was discovered that 5,6-dimethyl-1-(α-D-ribofuranosyl)benzimidazole is an integral part of the structure of vitamin B_{12}.[3] These findings stimulated great interest in the chemistry of imidazoles and related compounds, and considerable commercial success has accrued from these studies: a new antibacterial agent [azomycin (**1.4**)],[4] a trichomonacide [metronidazole (**1.5**)],[5] and a variety of benzimidazole derivatives of use as anthelmintic agents [e.g., thiabendazole (**1.6**)] and fungicides [e.g., benomyl (**1.7**)] are well-established marketed products (see also Chapter 10).

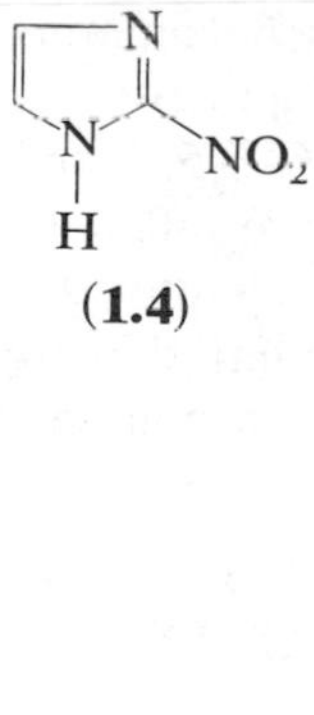

(**1.4**) 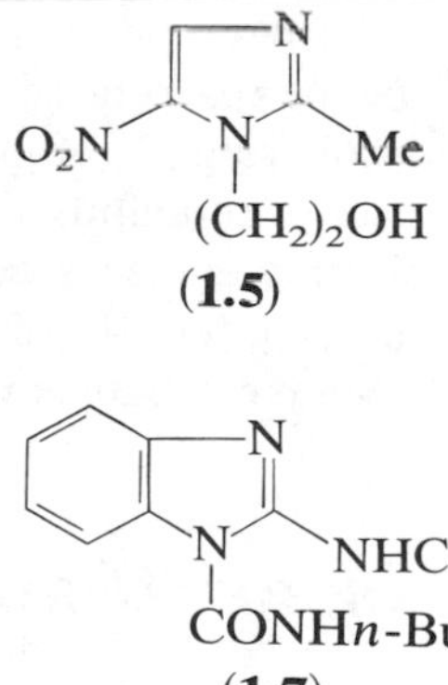   (**1.5**) 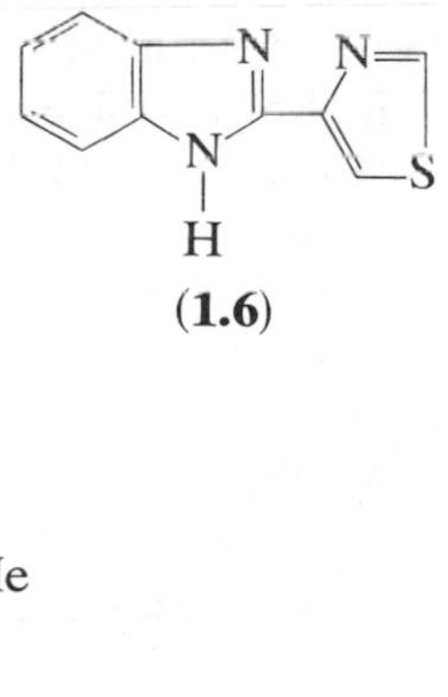   (**1.6**)

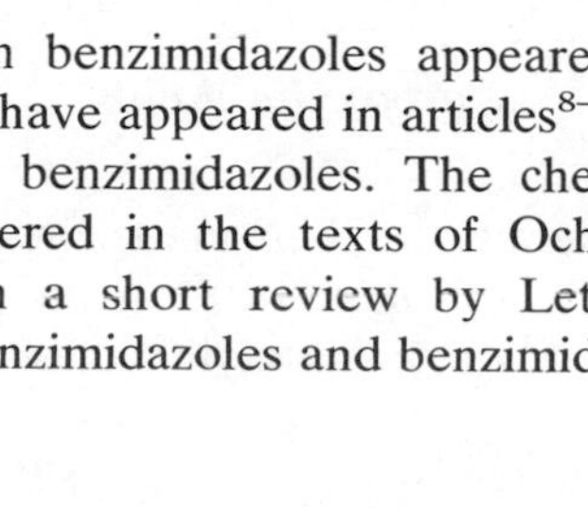

(**1.7**)

Systematic reviews on benzimidazoles appeared in 1951,[6] 1953,[1] and 1974,[7] although surveys have appeared in articles[8–10] covering the chemistry of both imidazoles and benzimidazoles. The chemistry of benzimidazole *N*-oxides has been covered in the texts of Ochiai[11] and Katritzky and Lagowski[12] and also in a short review by Lettau.[13] Specific synthetic procedures leading to benzimidazoles and benzimidazole *N*-oxides based on the use of *ortho*-substituted nitrobenzene derivatives have been described,[14] and methods employing *ortho*-substituted *tert*-alkylanilines have been reviewed.[15]

The magnitude of the literature on benzimidazoles presents a difficult problem for a reviewer. In the twenty-five-year period since publication of Hofmann's book[1] there have been approximately 25,000 *Chemical Abstracts* compound citations on benzimidazoles; by comparison it is interesting to note that Lister's text[2] in this series refers to 3000 citations on purines from the origin of such studies through 1970. Accordingly no attempt has been made in this chapter to provide comprehensive physicochemical data (e.g., mp, bp) on all cited benzimidazole derivatives. Nevertheless, the chapter is comprehensive in the sense that relevant publications appearing in *Chemical Abstracts* subsections on imidazoles are included (Volumes 48 through 78); material has also been selectively abstracted from the subject index, but the chemistry of benzimidazole nucleosides[16] and transition metal organometallic complexes of benzimidazoles is excluded.

The subject matter in Part 1 (Chapters 1 to 5) and Part 2 (Chapters 6 to 10) is organized in terms of synthesis, physicochemical properties, and reactions; synthetic methods are described in terms of the types of starting materials employed, and the sections on reactions are categorized on a mechanistic basis rather than on product type. The appended tables of data at the end of this chapter contain comprehensive lists of benzimidazole derivatives categorized according to functional group type.

Benzimidazole *N*-oxides (Chapter 2) and dihydrobenzimidazoles (Chapter 3) are treated separately, and condensed benzimidazoles are organized according to the size and position of fusion of the condensed ring. Using a 6-5 notation for benzimidazole, chapters on condensed compounds are included in which fusion is on the benzo ring [5-6-5 and 6-6-5 systems (Chapters 4 and 5)] or on the imidazole ring [6-5-5, 6-5-6, 6-5-7, and higher (Chapters 6 to 8)]; the chemistry of compounds that are annulated from positions 1 to 7 of a benzimidazole ring is discussed in Chapter 9.

An attempt has been made in Chapter 10 to present data on benzimidazoles (including dihydro derivatives) that have been marketed commercially in the last decade.

1.2. SYNTHESIS OF BENZIMIDAZOLES

1.2.1. From Reactions of *o*-Arylenediamines with:

Carboxylic Acids

Synthetic methods leading to benzimidazoles (Table 1.1)[17-110] and bisbenzimidazoles[28,111-129] (Table 1.2) from the reaction of *o*-arylenediamines and mono- or dicarboxylic acids are widely applicable. In

TABLE 1.1. SYNTHESIS OF BENZIMIDAZOLES BY THE REACTION OF o-ARYLENEDIAMINES WITH CARBOXYLIC ACIDS

Carboxylic acid	Ref.
$H-CO_2H$	17–35
$H-{}^{14}CO_2H$	36
CH_3CO_2H	18, 22, 23, 25, 27
	32–35, 37, 38, 39
$R-CO_2H$	40[b]
(R = alkyl, adamantyl)	
$ClCH_2-CO_2H$	41, 42
Cl_2CH-CO_2H	43–45
Cl_3C-CO_2H	44, 46
F_3C-CO_2H	47–53, 55
$F(CF_2)_n-CO_2H$ (n = 1, 2)	51, 53, 54
$HSCH_2CO_2H$	56
$HS(CH_2)_2CO_2H$	57
$HO(CR^1R^2)CO_2H$ (R^1, R^2 = H, alkyl, aryl)	58, 59
α-Amino acids	60–65
cyclohexyl-$(CH_2)_n CO_2H$ (n = 2, 3)	66
$HO_2C(CH_2)_2CO_2H^a$	67
$PhSO_2NHCH(R)(CH_2)_n CO_2H$ (R = CO_2H, etc.)	68
$RSCH_2CO_2H$ (R = alkyl)	69
$ArOCH_2CO_2H$ (Ar = aryl)	70, 71
$RCH_2CH(NH_2)CO_2H$ (R = aryl, hetaryl, etc.)	72
(ferrocenyl)$(CH_2)_n CO_2H$ (n = 1–5)	73
$PhCH_2CO_2H$	18, 22, 27, 74 76
$p-IC_6H_4CO_2H$	77a
$p-H_2NO_2SC_6H_4(CH_2)_n CO_2H$ (n = 1, 2)	77b
$ArCH(OH)CO_2H$	78
(2-benzimidazolyl)$(CH_2)_2CO_2H$	79
(1- and 2-benzotriazolyl)CH_2CO_2H	80
(3-indolyl)CH_2CO_2H	81
(2-oxo-3-benzoxazolyl)$(CH_2)_n CO_2H$ (n = 1–3)	82
$(CH_3)_2C\!=\!CHCO_2H$	83
$CH_3C(Cl)-CHCH_2CO_2H$	84
$CH_3C(Cl)\!=\!CHCH_2CH_2CO_2H$	85–87
$ArCO_2H$ (Ar = aryl)	22, 23, 25, 27,
	88–100
3-Fluoronaphthyl-2-CO_2H	101
3-(ArSO$_2$NH)naphthyl-2-CO_2H	100
$R-CO_2H$ (R = 5-membered ring containing N and S)	102, 103
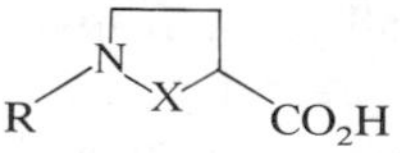 [X = CH_2, $(CH_2)_2$; R = H, alk, etc.]	104
$R-CO_2H$ (R = 2-furyl, 2-pyridyl, etc.)	105
(2-pyridyl)-CO_2H	106
(4-pyridyl)-CO_2H	107
(2-quinolinyl)-CO_2H	108
(5-benzimidazolyl)carboxylic acid derivatives	109, 110

[a] The product is a carboxylalkylbenzimidazole.
[b] See text.

7

TABLE 1.2. SYNTHESIS OF BISBENZIMIDAZOLES BY
THE REACTION OF o-ARYLENEDIAMINES[a]
WITH DICARBOXYLIC ACIDS

X in HO_2CXCO_2H	Ref.
$(CH_2)_2$	28, 111, 112
$(CH_2)_4$	113
$CH{=}CH$	114
$CH_2CH(OH)$	114
$[CH(OH)]_2$	115
$(CH_2)_8$	116
$(CF_2)_n$ $(n = 1, 2)$	117
$(CH_2)_n$ $(n = 2, 4, 6, 8)$	
$[CH(OH)]_2$	118
$(CH_2)_2S(CH_2)_2$	
$[CH(OH)]_4$	119
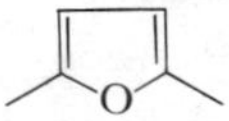	120
$CH(OH)CH_2$	121
$(CH_2)_n$ $(n = 2{-}4, 6)$	
$[CH(OH)]_n$ $(n = 2, 4)$	
$CH(CH_3)CH_2$	122
$(CH_2)_2S(CH_2)_2$	
$CH(OH)CH_2$	
$CH(OH)CH(R)$ $(R = H, OH)$	123
$(CH_2)_n$ $(n = 1{-}8, 10)$	124
$[(CH_2)_nCR_2(CH_2)_m]_2SO_2$	125
$(n = 0, 1; m = 0{-}4; R = H, alk)$	
	126
 (in reactions with tetra amino derivatives)	127 ($cf.$ 128, 129)

[a] The nature of substituents in these aryl rings is not indicated.

general the Phillips synthesis, in which the carboxylic acid is heated with the diamine in aqueous hydrochloric acid,[6] is used. This conventional procedure works satisfactorily for the preparation of most 2-alkyl derivatives but frequently fails or gives poor yields when applied to 2-aryl analogs. For example, the Phillips synthesis of 2-phenylbenzimidazole gives only a trace amount of the desired product,[130] and it is necessary to effect this transformation in a sealed tube at 180°.[131] Nevertheless, incorporation of either electron-withdrawing or -attracting groups into the aryl ring of the diamine results in only fair to poor yields of benzimidazole derivatives.[132] The discovery that polyphosphoric acid is an effective, convenient, and general catalyst for effecting such reactions is thus an important one.[94] Using this procedure, 2-arylbenzimidazoles are obtained in moderate to good yields (generally 50–85%) and high working pressures are not required. Alternative procedures for 2-aryl derivatives include the use of phosphorous pentoxide as a dehydrating agent[99] or azeotropic dehydration from kerosine[97] or xylene[99] solvent.

As mentioned earlier, most aliphatic carboxylic acids react normally in the Phillips process, but acids containing bulky substituents, such as adamantane 1-carboxylic acid[133] and 2,2-dimethylpropionic acid,[134,135] either do not react or give low yields of benzimidazoles. This poor reactivity has been attributed[134] to a combined effect of steric hindrance and a diminished electrophilic reactivity of the carboxyl group. Reactions in which ionic intermediates are formed from neutral reactants are normally accelerated in polar solvents at very high pressures,[136] and with this in mind Holan and coworkers[40] have been successful in converting hindered aliphatic acids into benzimidazoles at pressures of up to 8 kbar (1 kbar = 10^5 N m^{-2} ≡ 986.1 atm); yields of 2-R-substituted derivatives are, for example, R = t-Bu (66%), CMe$_2$CH$_2$Cl (83%), and 1-adamantyl (48%). Reactions of the Phillips type probably[137] occur via intermediate o-aminoanilides, and while these were not isolated in the high-pressure reactions[40] it is interesting to note the formation of 1,3,4,5-tetrahydro-3,3-dimethyl-1,5-benzodiazepin-2-one (1.8) from the reaction of o-phenylenediamine with 3-fluoro-2,2-dimethylpropionic acid. When the reaction is carried out in the presence of an equivalent amount of hydrochloric acid, the appropriate benzimidazole derivative (1.9) is formed, but this is not converted into the benzodiazepinone (1.8) under the original conditions. Evidently the strong acid suppresses nucleophilic displacement of fluoride and favors benzimidazole formation.

(1.8)

(1.9)

An additional modification to the original procudure is the use of formic acid in the presence of a mixture of hydrochloric acid and an acid resin [Dowex-50W-X8 (strong sulfonic acid)];[138] the reactions are carried out at room temperature, hence operating conditions are somewhat milder than in the normal Phillips method (4 M HCl under reflux). An example of the utility of this method is the high-yield (96%) synthesis of the 5-R derivative (R = CH=CHCO$_2$H), which cannot be synthesized by the Phillips method.[138] The technique has been used to synthesize other benzimidazole-5-R derivatives (R = CO$_2$H, 62%; Cl, 30%; Me, 30%)[139] and 5,5'-bisbenzimidazole (75%), but the scope of the reaction in the context of synthesis of a variety of 2-substituted benzimidazoles has not been evaluated.

Carboxylic Acid Derivatives (Esters, Amides, Anhydrides, and Chlorides)

Routine procedures involving the condensation of o-arylenediamines and carboxylic acid derivatives are listed in Table 1.3.[140–170] A recent variant of the conventional procedure using esters is the method using thioesters.[154] In a typical reaction the diamine is treated with the thioester in aqueous ethanol at room temperature with the pH adjusted to ~8; hydrogen sulfide is aspirated, and the reaction is monitored by disappearance of the red color of the thioester. Yields are ~90%, and the method has been used for the synthesis of 2-carbamic acid ester derivatives.[155]

The reaction of o-phenylenediamine with fluoren-9-ylidenecyanoacetic ester (see Table 1.3) proceeds normally to give the desired benzimidazole (**1.10**), but 1-naphthylidene analogs give rise to 1-naphthylbenzimidazole (**1.11**) in good yield.[149] The 1-naphthyl analog of (**1.10**) is easily accessible, however, by allowing 1-cyanomethylbenzimidazole to react with 1-naphthaldehyde.[149]

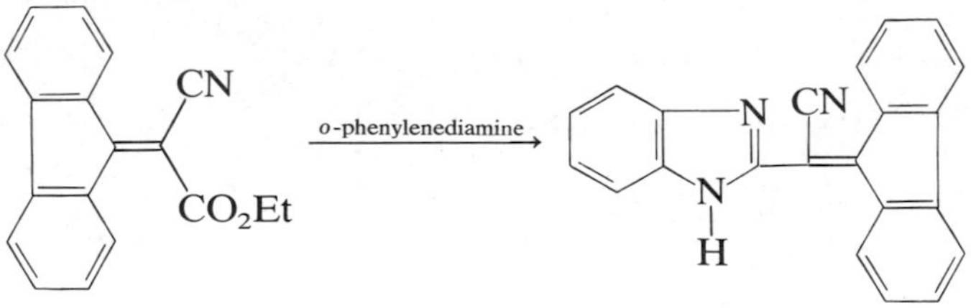

(**1.10**)

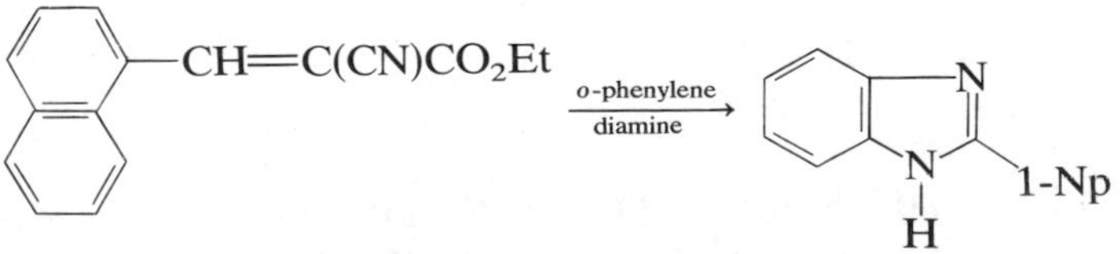

(**1.11**)

(1-Np = 1-naphthyl)

TABLE 1.3. SYNTHESIS OF BENZIMIDAZOLES BY THE
REACTION OF *o*-ARYLENEDIAMINES WITH
CARBOXYLIC ACID ESTERS, AMIDES,
ANHYDRIDES, AND CHLORIDES

Carboxylic acid derivative	Ref.
$H_2NCH_2CO_2Et$	140
$NCCH_2CO_2Et$	141–143
$NCCH(R)CO_2Et$ (R = alkyl)	144
$CF_3(CF_2)_2CO_2Ph$	145
$BrCH_2COCO_2Et$	146, 147
$PhCO_2Ph$	148

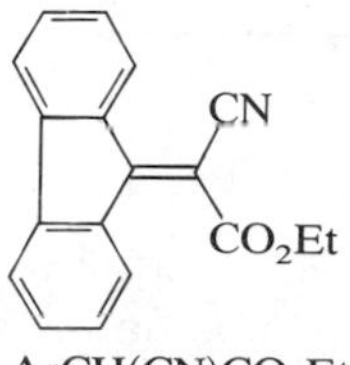

149[a]

| $ArCH(CN)CO_2Et$ (Ar = aryl) | 150 |

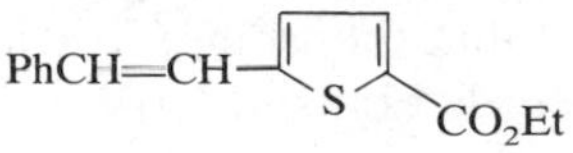

151

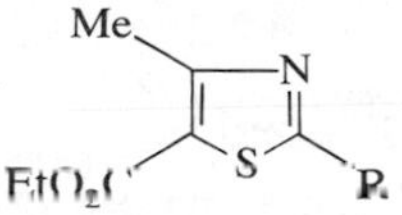

(R = halogen, NH-alkyl)

152

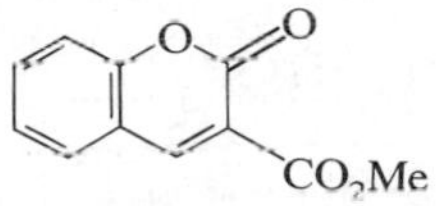

153

$R-CS_2R^1$ (R = Ph, Me; R^1 = Et, CH_2CO_2H)	154[a]
RO_2CNHCS_2R (R = alkyl)	155[a]
$EtO_2CCH_2CO_2Et$	

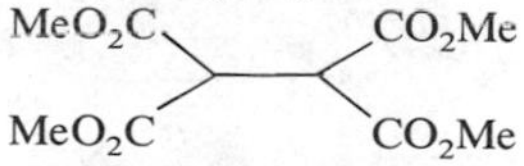

156

$RCONHCH(R)CONH_2$ (R = alkyl or aryl)	157
$H_2NCO(CH_2)_nCONH_2$ (n = 0, 2)	158
$RNHCO(CH_2)_nCONHR$ (n = 0, 4, 8, R = $HOCH_2CH_2$)	
$NCCH_2CONH_2$	159
$PhCONH_2$	160
$(CH_3CO)_2O$	161–163
pyromellitic anhydride	164, 165[a]

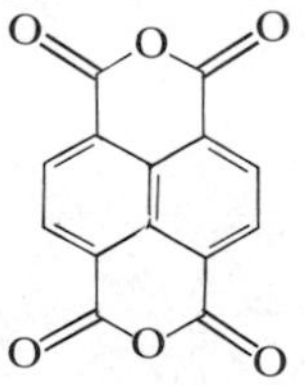

166

11

TABLE 1.3.   (*Continued*)

Carboxylic acid derivative	Ref.
CH_3COCl	167, 168
$o\text{-}ClC_6H_4COCl$	169

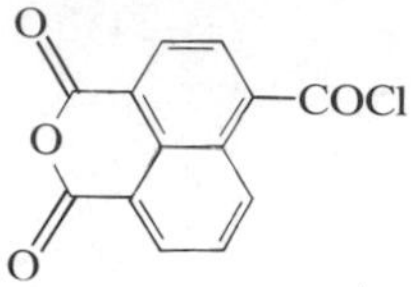

170

a See text.

Reactions[164,165] of pyromellitic anhydride (**1.12**) and the analogous ether (**1.13**) with *o*-phenylenediamine are of interest as model studies of the synthesis and properties of polybenzimidazoles.[171] In this manner substrates (**1.14, 1.15**) for the synthesis of complex model compounds (**1.16, 1.17**) have been prepared, although it may be noted that the isomer distributions in mixtures of (**1.14**) and (**1.15**) have not been determined.

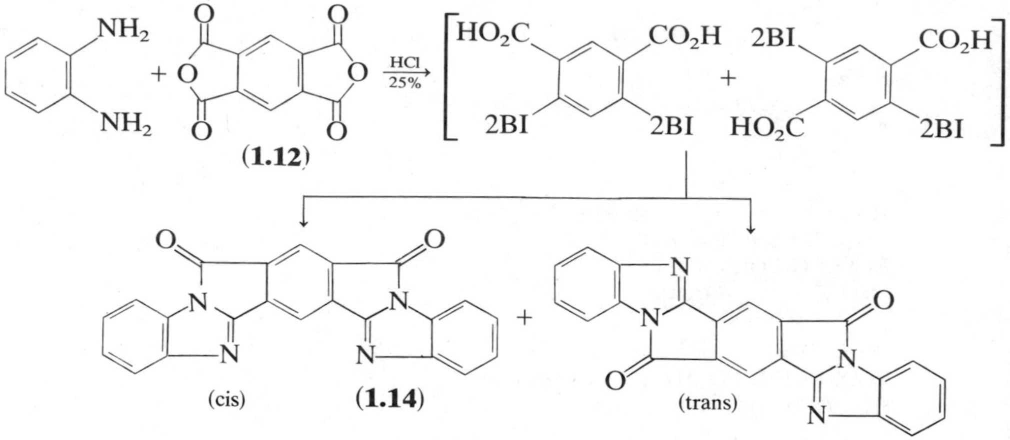

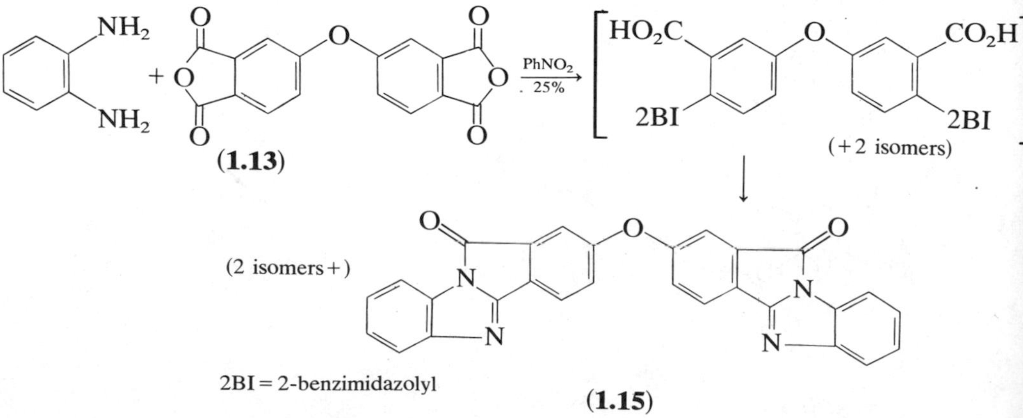

(cis)   (**1.14**)   (trans)

(2 isomers+)

2BI = 2-benzimidazolyl

(**1.15**)

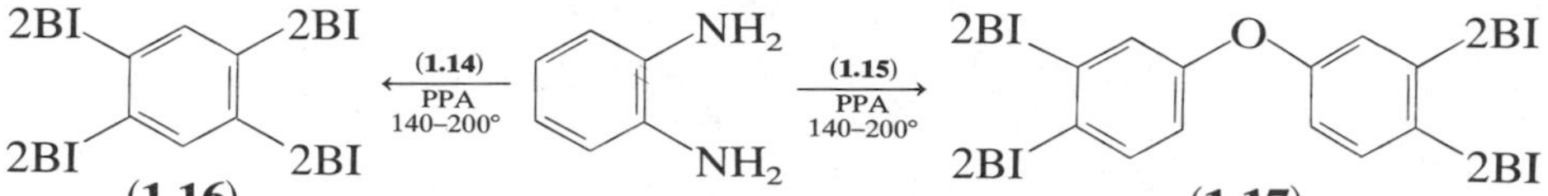

2BI = 2-benzimidazolyl
PPA = polyphosphoric acid

Imino Ethers (Imidates)

Examples of the synthesis of benzimidazoles from reactions of *o*-arylenediamines with imino ethers (imidates) are summarized in Table 1.4.[172–199] The scope of the imidate[200] procedure was assessed in early work by Acheson and King.[201] One problem concerning the Phillips reaction is that the diamine often competes successfully for the proton of the acid catalyst, hence inhibiting nucleophilic addition to the carbonyl group; this difficulty is surmounted by replacing the carbonyl group by the more basic imino group, and very often the imidate method is superior to the conventional Phillips approach. For example, Phillips reaction of *o*-phenylenediamine with 2,4-dinitrophenylacetic acid takes place only under drastic conditions, and considerable resinification occurs;[188] conversely, reaction of the diamine with the hydrochloride of ethyl 2,4-dinitrophenylacetimidate under reflux gives the desired 2-(2,4-dinitrobenzyl)-benzimidazole in 84% yield.[188]

The synthesis of 2-(trichloromethyl)benzimidazole presents similar difficulties by the Phillips approach: a complex mixture is obtained from the reaction of *o*-phenylenediamine and trichloroacetic acid, from which 2,2′-bibenzimidazolyl (**1.18**) and the anilide (**1.19**) are isolated.[202] Since these

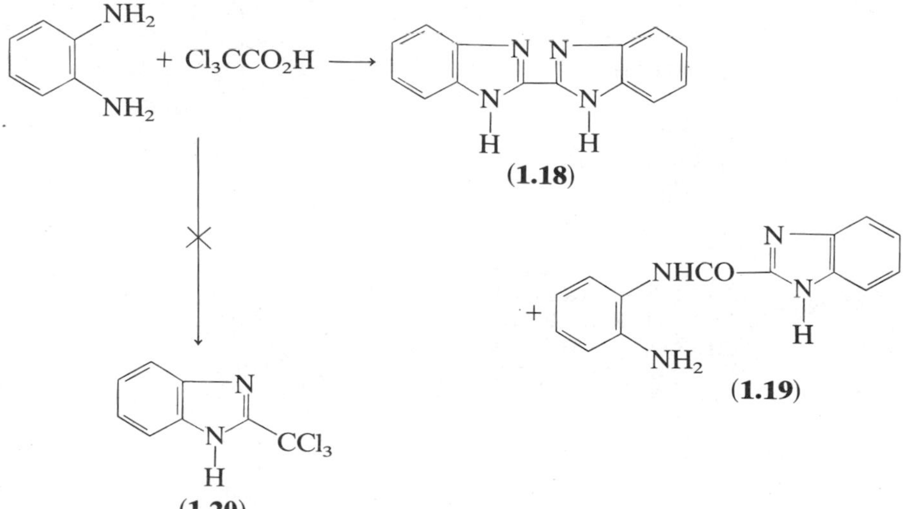

TABLE 1.4. SYNTHESIS OF BENZIMIDAZOLES BY THE REACTION OF o-ARYLENE-DIAMINES WITH IMINO ETHERS (IMIDATES)

Substituent (R^1) in the imino ether: $R^1 \diagup NH$ OR^2 (R^2 = Me or Et)	Ref.
CCl_3	172–175
H, alkyl, hydroxylalkyl, etc.	176
H, alkyl, aralkyl, etc.	177
Polyhaloalkyl, alkenyl, alkyl-S-	178
$CH_2CH_2CMe_2$ $\vert$ NO_2	179
$(CH_2)_2C(NO_2)_2R$ ($R = NO_2$, $CH_2CH_2CO_2Me$)	180
CH_2CO_2Et	181, 182
CH_2CN	182
$(CH_2)_n Br$ ($n = 3–5$)	183
$CH_2CH_2NR^1R^2$ (NR^1R^2 = piperidino, etc.)	184
$HN \diagdown$ $\diagup$ CH_2 EtO	185
$HN \diagdown$ $\diagup (CH_2)_n$ ($n = 4, 8$) EtO	186
$ArCH_2$ (Ar = aryl)	187–189
$ArNHCH_2$	190
$PhCH_2OCONHC$ $\overset{R_1}{\underset{R^2}{\vert\;\vert}}$ (R^1, R^2 = H or alkyl)	191
$RO—\!\!\langle\;\rangle\!\!—$  ($R = H$, alkyl)	192
$ArNHCH_2CH_2$	193
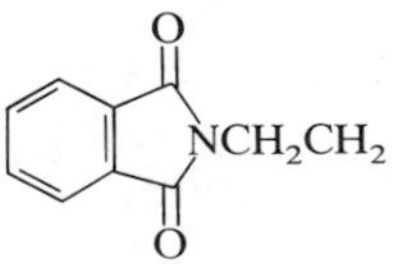 phthalimido-NCH_2CH_2	194
2-Furyl	195
CH_2CH_2-2-benzimidazolyl	196
$(CH_2)_2PO(OR)_2$ (R = alkyl)	197
$(CH_2)_2OP(OR)Me$ (R = alkyl)	198
$(CH_2)_2C(NO_2)_2(CH_2)_2C(OEt)\!\!=\!\!NH$	199

two products (**1.18** and **1.19**) are also formed by the reaction of 2-trichloro-methylbenzimidazole (**1.20**) with o-phenylenediamine, the former is evidently an intermediate. Reaction of the appropriate imidate ester with o-phenylenediamine or its disalt gives the bibenzimidazolyl (**1.18**) in 90% yield, and indeed this is the synthetic method of choice for this compound and substituted analogs. However, by using o-phenylenediamine monosalt as such, or by generating it in situ, the reaction with methyl trichloroacetimidate gives 95% of the 2-trichloro derivative (**1.20**) even at room temperature.[202] Suppression of the reaction giving rise to 2,2'-bibenzimidazolyl (**1.18**) requires a rapid and complete initial reaction of the diamine with the imidate to give an intermediate (**1.21**), and this can be achieved at intermediate acidities (pH ~ 4).

(**1.20**)   (**1.21**)

Aldehydes

A summary of procedures leading to 2-substituted benzimidazoles by the reaction of o-arylenediamines with aldehydes is presented in Table 1.5.[203–225] In this method, the reactants are condensed in the presence of an oxidant[1,6] such as cupric acetate (Weidenhagen procedure),[226] mercuric oxide (for 2-NHCO$_2$Me)[204] or chloranil (for 2-furyl).[219b]

An improvement on the conventional method is the use of the sodium bisulfite addition adduct of the aldehyde.[227] The reactions are carried out in boiling ethanol, yields are good [e.g., 2-Ph (90%); 2(3-pyridyl) (97%)], and there is little risk of decomposition of labile substituents. Evidently, the aldehyde route is suitable for the preparation of 2-hetaryl derivatives, and improved yields are also obtained by using the bisulfite addition complex of the aldehyde (for 2-furyl derivatives).[219c]

The superficially simple reaction of o-phenylenediamine and its derivatives with benzaldehyde has been studied in detail.[210] Benzaldehyde and o-phenylenediamine react rapidly at −20° to produce the imine (**1.21A**), and when the mixture is warmed to room temperature, 2-phenylbenzimidazole

TABLE 1.5.   SYNTHESIS OF BENZIMIDAZOLES
BY THE REACTION OF o-ARYL-
ENEDIAMINES WITH ALDE-
HYDES

Aldehyde[a]	Ref.
RCHO (R = alkyl, aryl, aralkyl, hetaryl)	203
MeO$_2$CNHCHO	204
CH$_2$O[b]	205
ArCHO (Ar = aryl)	206–214
2-Furyl-CHO	215–219
(2-Furyl)CH=CHCHO	220
(3-thienyl)CHO	221
(2-pyridyl)CHO	222
(1-isoquinolinyl)CHO (2-quinolinyl)CHO	223
(3-pyridyl)CHO (3-quinolinyl)CHO	219a
(3-phthalimido)CHCHO (R = H, alkyl) ⎪ R	224

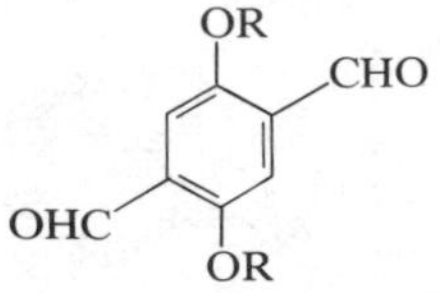

225

---

[a] Substituents in the aldehyde and diamine are not shown.
[b] The products are 1-methylbenzimidazoles.

(**1.22**) is formed in 19% yield together with a molecular complex of (**1.23**) and (**1.24**). 2-Phenylbenzimidazole (**1.22**) must arise by oxidation of precursors (**1.21** or **1.25**), and it is suggested that the compensating reduction is that of the imine (**1.21**) to the diamine (**1.26**).[210] In support of this it has been shown that (**1.26**) reacts rapidly with benzaldehyde to give the dihydro compound (**1.24**), as the molecular complex with (**1.23**), together with 1,3-dibenzyl-2-phenylbenzimidazoline (**1.27**); NMR evidence has been adduced for the intermediacy of an imine in this reaction, and an oxidation–reduction step similar to (**1.21** → **1.26**) presumably occurs. A brief study of substituent effects has been carried out.[214] 3-Methyl-o-phenylenediamine reacts with benzaldehyde in acetic acid to give 2-phenyl-4-methylbenzimidazole (**1.28**) and the 1-benzyl derivative (**1.29**) but not the isomer (**1.30**).

The reaction of o-arylenediamines with formaldehyde gives rise to 1-methylbenzimidazoles,[228,229] and this type of process has been studied in detail.[205] It is possible to use the more easily accessible N-monoacetyl o-phenylenediamine derivatives, and in this manner a variety of 1-methyl derivatives can be prepared in yields of 42 to 59%.[205]

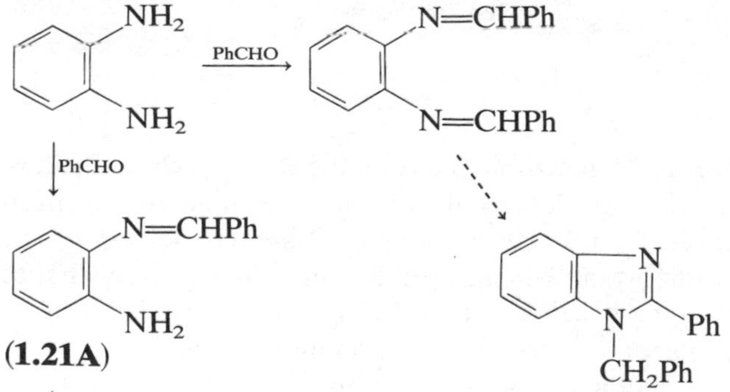

(1.21A)   **(1.23)**

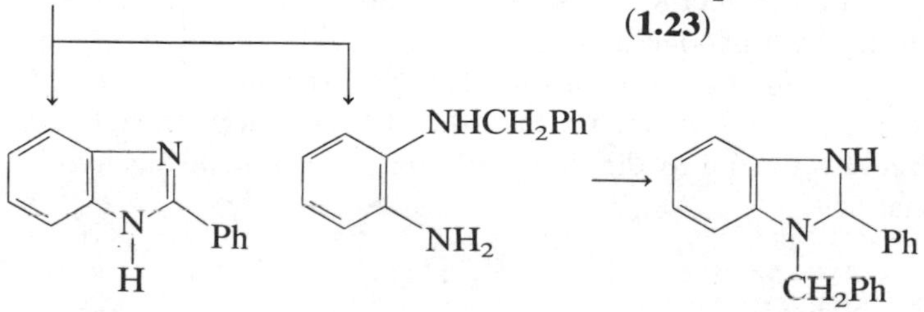

(1.22)   **(1.26)**   **(1.24)**

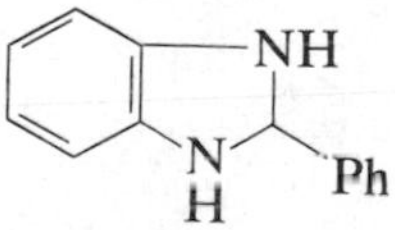

(1.25)

(**1.26**) $\xrightarrow{\text{PhCHO}}$ $\xrightarrow{(1.24)}$ + (**1.23**)

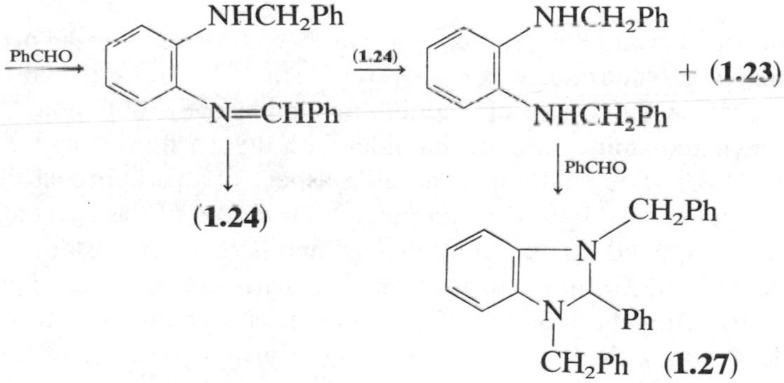

(1.24)

(1.27)

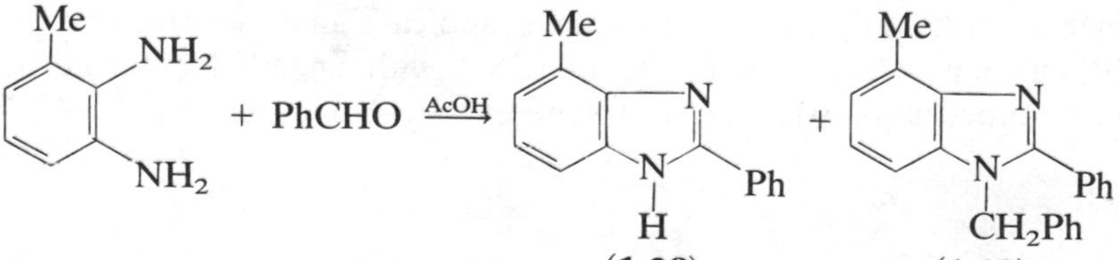

(1.28)   **(1.29)**

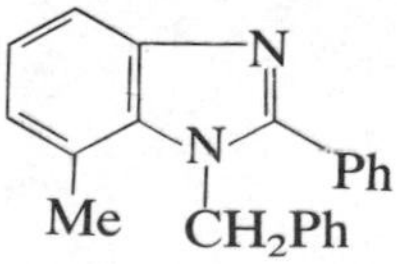

(1.30)

17

Cyano Derivatives

The well-established[1,6] benzimidazole synthesis in which a nitrile is heated with *o*-phenylenediamine hydrochloride has been used to prepare 2-benzylbenzimidazole[230] and also a series of 2-hetaryl derivatives including the commercially important 2-(4-thiazolyl) compounds[231] (see Chapter 10).

The standard route[1,6] using cyanogen bromide has also been applied routinely to prepare a variety of 2-aminobenzimidazoles.[232–235] The cyanogen bromide method has been extended from diamines to closely related *o*-acylhydrazidoanilines, and this provides a route to synthetically valuable 1,2-diaminobenzimidazole (**1.31**)[236] (see section 7.3.1 in Chapter 7). Subsequently it was shown that the diamino derivative (**1.31**) can be synthesized by the more direct amination procedure using hydroxylamine *O*-sulfonic acid.[237]

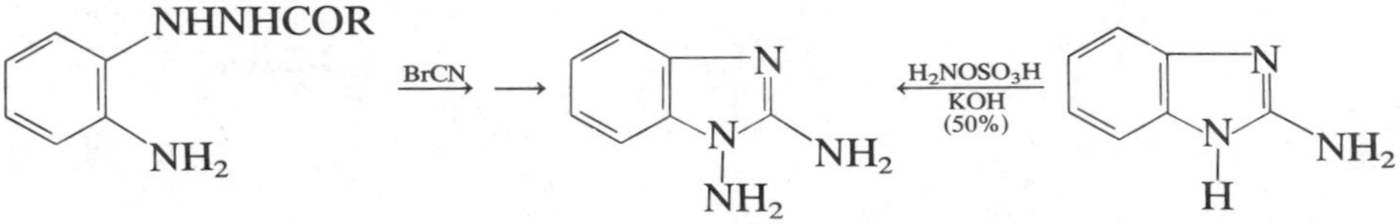

(**1.31**)

The use of cyanamide, rather than cyanogen bromide, for the preparation of 2-aminobenzimidazoles was discovered[238] in 1961 and was elaborated upon later.[239,240] A 92% yield of 2-aminobenzimidazole is obtained[239] by heating *o*-phenylenediamine and cyanamide with hydrochloric acid followed by sodium hydroxide solution; a notable aspect of this approach is the ready commercial availability of cyanamide.[241] A variant of the cyanamide method has been applied to the synthesis of industrially important 2-carbamate derivatives (**1.32**). In this procedure cyanocarbamates are used as such[242,243] or *in situ* from the reaction of cyanamide with chloroformate esters;[244–248] condensation with the diamine is effected in the presence of bases such as sodium hydroxide[244] or triethylamine,[245] or at elevated temperature.[248] An alternative method of generation of the cyano carbamate involves treatment of the calcium salt of cyanamide ($CaCN_2$) with an alkyl chloroformate; ensuing condensation with the *o*-arylenediamine is effected at pH ~4.[249–253]

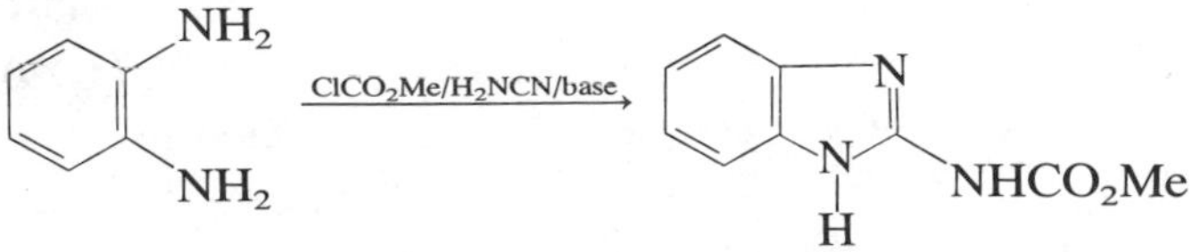

(**1.32**)

Alternatively, the barium salts of cyanamides may be used, and this technique has been put to good effect for the synthesis of a radioactive ^{14}C-labeled fungicide[254] [benlate (**1.33** and **1.34**); see Chapter 10].

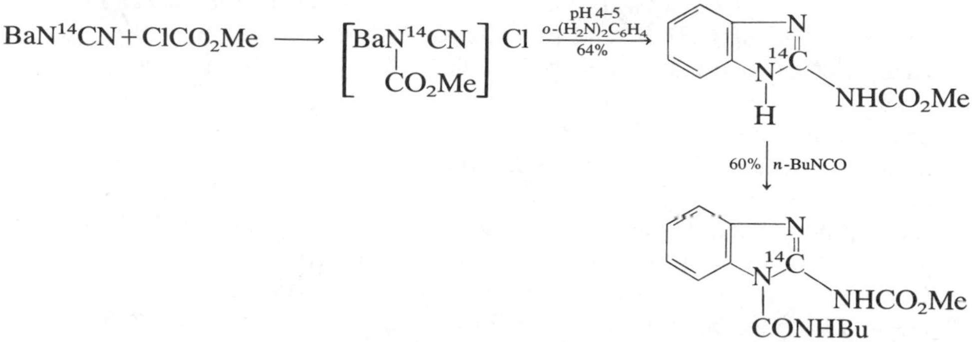

$$BaN^{14}CN + ClCO_2Me \longrightarrow \left[\begin{matrix} BaN^{14}CN \\ | \\ CO_2Me \end{matrix} \right] Cl \xrightarrow[64\%]{\substack{pH\ 4-5 \\ o-(H_2N)_2C_6H_4}}$$

(benzimidazole structure, ^{14}C, NHCO$_2$Me)

$$60\% \Big| n\text{-BuNCO}$$

(benzimidazole structure, ^{14}C, NHCO$_2$Me, CONHBu)

(**1.33**)

$$^{14}CH_3OH + COCl_2 \longrightarrow ClCO_2{}^{14}CH_3 \xrightarrow{\text{BaNCN}} \left[\begin{matrix} BaNCN \\ | \\ CO_2\ {}^{14}CH_3 \end{matrix} \right] Cl$$

$$15\% \Big| o\text{-}(H_2N)_2C_6H_4,\ pH\ 4-5$$

(benzimidazole structure, NHCO$_2$^{14}CH$_3$) $\xleftarrow[68\%]{n\text{BuNCO}}$ (benzimidazole structure, NHCO$_2$^{14}CH$_3$, CONHBu)

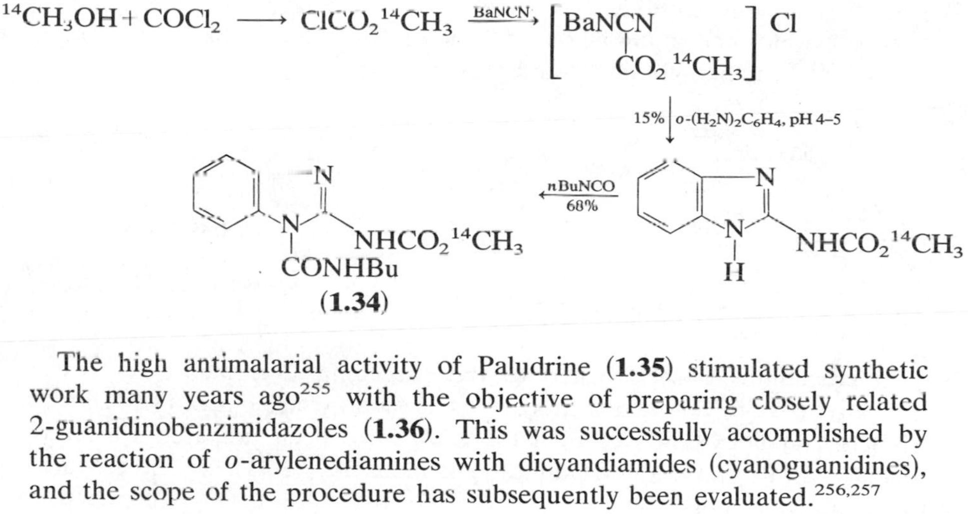

(**1.34**)

The high antimalarial activity of Paludrine (**1.35**) stimulated synthetic work many years ago[255] with the objective of preparing closely related 2-guanidinobenzimidazoles (**1.36**). This was successfully accomplished by the reaction of o-arylenediamines with dicyandiamides (cyanoguanidines), and the scope of the procedure has subsequently been evaluated.[256,257]

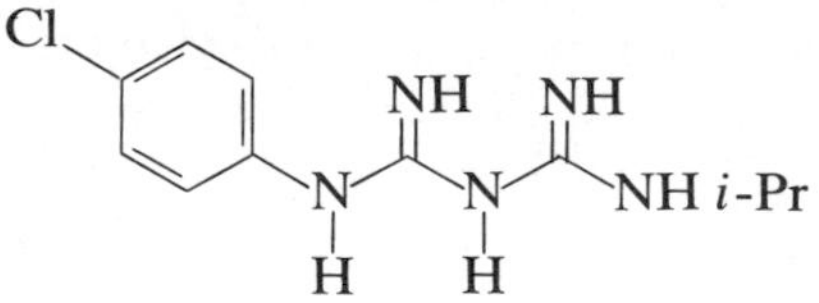

(structure of Paludrine with Cl, NH, NH, NHi-Pr)

(**1.35**)

(o-phenylenediamine structure) + NCNH (structure NHR, NH) $\xrightarrow[66\%\ (\text{for } R=H)]{HCl/H_2O}$ (benzimidazole structure NHR, NH)

(R = H, alk)

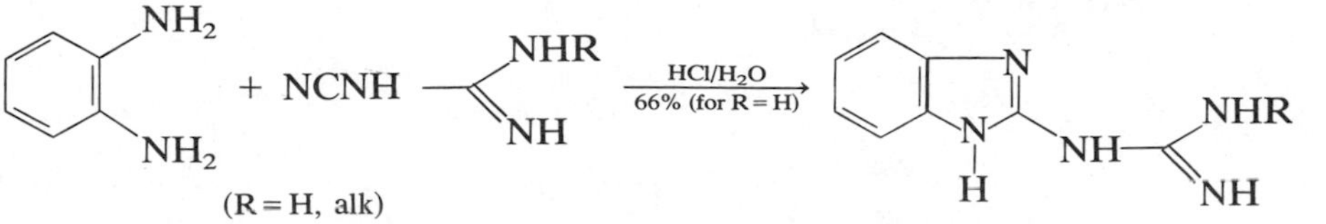

(**1.36**)

Isothioureas and Related Compounds

The synthesis of alkylbenzimidazole 2-carbamates (**1.38**) by the reaction of o-phenylenediamine with 3-alkoxycarbonyl-2-methylisothioureas (**1.37**) was described in 1961,[258] and has subsequently been used to synthesize a variety of 5-substituted analogs including styryl,[259] sulfonic acid phenyl ester,[260] aryloxy and arylthio,[261–263] and (2-pyridinyl)carbonyl[264] derivatives. The isothiourea derivatives (**1.37**) are generated[264a] *in situ* by treating S-methylisothiourea with an alkyl chloroformate, and subsequent condensation with the diamine is effected at pH ~ 6; yields are moderate to good for this synthesis (e.g., 35–83% for 5-styryl derivatives).[259] An equally effective synthesis involves the use of 1,3-bis(alkoxycarbonyl)-2-methylisothioureas (**1.39**) in a procedure that is operated at near the same pH.[265] The latter method has also been used extensively to prepare anthelmintic compounds bearing 5-substituents such as alkyl,[266–268] halo,[266] hydroxyalkyl,[269] benzoyl,[270] (2-thienyl)carbonyl (as ketal),[271] (hetaryl)alkyloxy and -thio,[272] fluoroalkylthio and fluoroalkoxy,[273] (hetaryl)oxy and -thio,[274,278] alkylsulfinyl,[275] (2-naphthyl)thio,[276] and alkenylthio.[277]

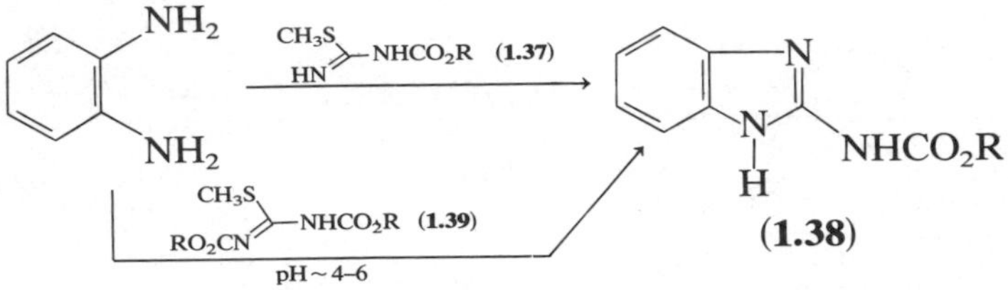

In procedures closely related to the preceding, 2-carbamic acid ester derivatives (**1.38**) can be prepared by using O-alkyl analogs of (**1.37**).[279,280] Thioalkyl analogs (**1.42**) of benlate (**1.33**) have been prepared via intermediate 2-(alkylthiocarbonylamino)benzimidazoles (**1.41**), which are also prepared using appropriate isothiourea derivatives (**1.40**).[281,282] The reaction of acyl- and aroyl halides with S-methylisothiourea gives rise to acyl analogs of (**1.37**), which are converted by o-phenylenediamine into 2-acylamino- and -aroylaminobenzimidazoles.[283]

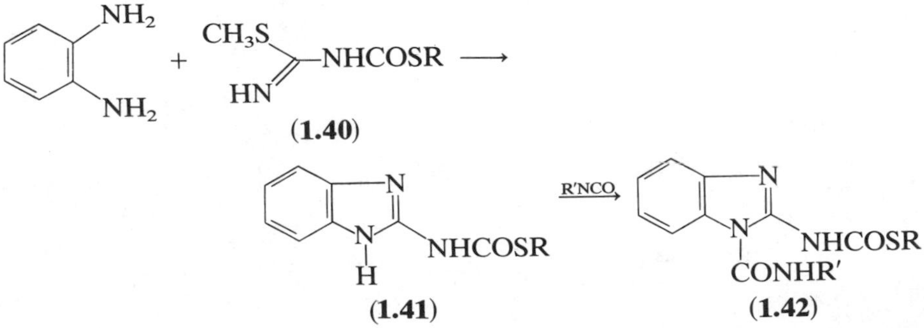

Miscellaneous Compounds

Benzimidazoles have been synthesized by a variety of routes involving the reaction of o-arylenediamines with a number of substrates (see Table 1.6).[284–314] Many of these reactions are limited in scope but occasionally are valuable for the synthesis of specific types of benzimidazole derivatives; some of these are discussed below in the order in which they appear in Table 1.6.

The use of lactones and lactams provides a good method for the synthesis of 2-hydroxyalkyl[285–287] and 2-aminoalkyl derivatives.[288] Reasonable yields are obtained, and the products (**1.43**) are useful intermediates for the synthesis of pyrrolo[1,2-a]benzimidazoles (**1.44**)[286] (see section 6.1.1 in Chapter 6). The reaction of diamines with dichloroalkenes (**1.45**) has proved to be effective for the preparation of 2-nitromethylbenzimidazoles (**1.46**) under very mild conditions.[291] A more recent variant of this method involves the use of 1-nitro-2,2-bismethylmercaptoethylene (**1.47**).[293]

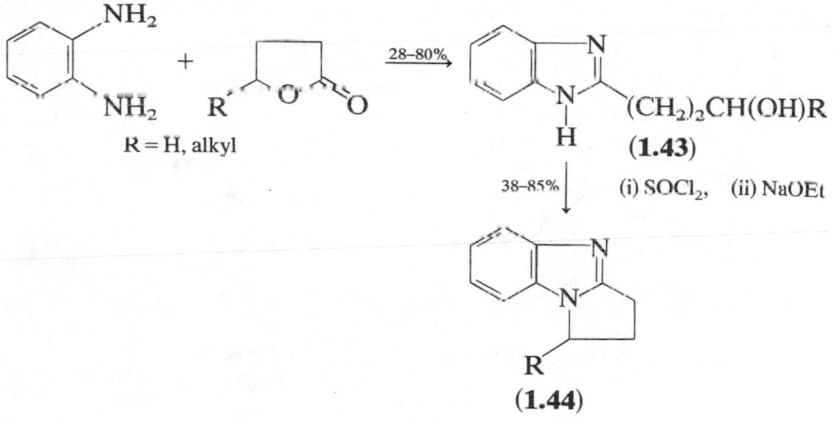

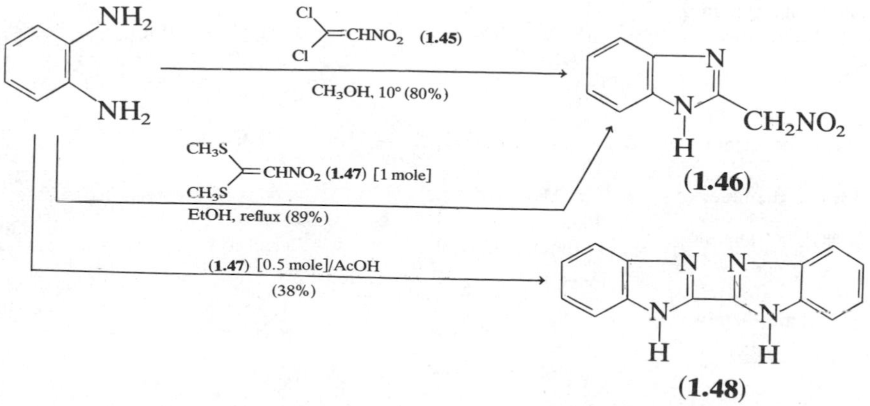

TABLE 1.6. SYNTHESIS OF BENZIMIDAZOLES BY THE REACTION OF *o*-ARYLENEDIAMINES WITH MISCELLANEOUS COMPOUNDS

Type of diamine derivative	Type of reagent	Type of benzimidazole product	Ref.
o-Phenylene diamine	Lactone	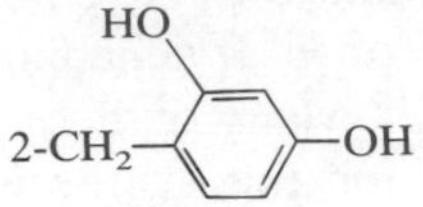 2-CH$_2$— (HO...—OH structure)	284
o-Phenylene diamine	Lactone	2-(CH$_2$)$_3$OH	285[a]
o-Phenylene diamine	Lactone	2-(CH$_2$)$_2$CH(OH)R (R = H, alkyl)	286[a]
o-Arylene diamine	Lactone	2-(CH$_2$)$_n$CH(OH)R (R = H, CCl$_3$, alk; n = 1, 2)	287[a]
o-Arylene diamine	Lactam	2-CH$_2$CR$_2$(CH$_2$)$_n$NHR^1 (R = H, Me; R^1 = H, Me, Ph; n = 0–3, 9)	288[a]
o-Phenylene diamine	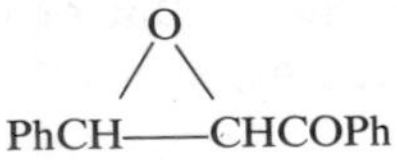 PhCH—CHCOPh (epoxide O)	2-Ph	289
o-Arylene diamine	Cl$_2$C=CHSO$_2$Ph	2-CH$_2$SO$_2$Ph	290
o-Arylene diamine	Cl$_2$C=CHNO$_2$	2-CH$_2$NO$_2$	291[a]
o-Arylene diamine	Cl$_2$C=C(Cl)NO$_2$	2-CHClNO$_2$	292[a]
o-Phenylene diamine	(MeS)$_2$C=CHNO$_2$	2-CH$_2$NO$_2$	293[a]
o-Phenylene diamine	ArCH=C(CN)$_2$	2-Aryl	294
o-Phenylene diamine	Cl$_2$C=NCO$_2$R	2-NHCO$_2$R (R = alkyl, aralkyl)	295
o-Phenylene diamine	(MeS)$_2$C=NCO$_2$Me	2-NHCO$_2$Me	296
o-Phenylene diamine⎫ N-Alkyl derivatives⎭	PhN=C=NPh or PhN=C(NHMe)(SMe)	1-Alkyl-2-NHPh	297
o-Phenylene diamine	Cl$_2$C=NCO$_2$CH$_2$CH=CH$_2$	2-NHCO$_2$CH$_2$CH=CH$_2$	298
o-Arylene diamines	ArCH=NAr	2-Aryl	299
o-Phenylene diamine	PhCCl$_3$ (hetaryl)CCl$_3$ ⎫⎭	2-Ph 2-Hetaryl	300
o-Phenylene diamine	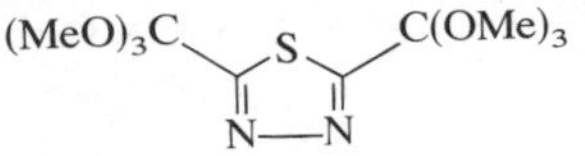 (MeO)$_3$C—[S,N—N ring]—C(OMe)$_3$	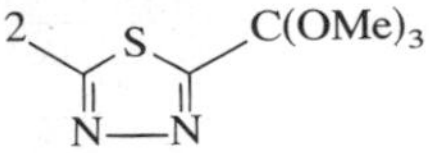 2-[S,N—N ring]—C(OMe)$_3$	301

22

TABLE 1.6.   (*Continued*)

Type of diamine derivative	Type of reagent	Type of benzimidazole product	Ref.
o-Arylene diamine	$(AcO)_2CHCONH(CH_2)_n Ar$	$2-CONH(CH_2)_n Ar$	302
o-(N-Alkylamino)arylamines			
o-(N,N-Dialkylamino)-arylamines	MnO_2 or CF_3CO_3H	1-Alkyl; 1,2-dialkyl; and tricyclic derivatives	303, 304
o-(2-Aminoaryl)cycloalkyl-amines			
N,N-Dimethyl-o-phenylene diamine	SO_2Cl_2	$1-Me-4,5,6,7-Cl_4$	305[a]
N,N-Dimethyl o-phenylene diamine	Cl_3CCHO/H_2NOH/aq. HCl	1-Me-2-CH=NOH	
N-(2-Aminoaryl)piperidine		1-(5-chloro-n-pentyl)-2-CH=NOH	306[a]
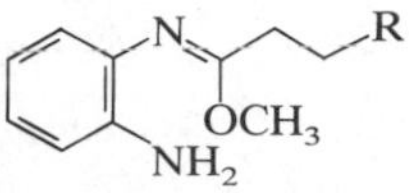 (R = Ph, 2-furyl)	Heat	$2-CH_2CH_2R$	307[a]
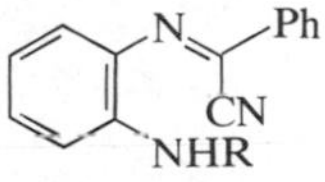 (R = H, Ph)	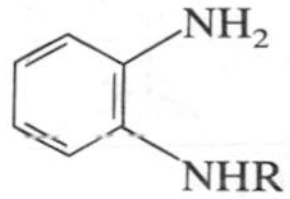	1-R-2-Ph	308[a]
o-Phenylene diamine	$CF_3CF=CF_2$	$2-CHFCF_3$	309
o-Phenylene diamine		2-Aryl	308
o-Phenylene diamine	$RCSNHCO_2Et$	2-R (R = alkyl, aryl, hetaryl)	310[a]
o-Arylene diamines	α-Picoline/S	2-Pyridyl	106[a]
o-Phenylene diamines	R-Me/S	2-R (R = hetaryl)	311[a]
o-Arylene diamines	β-Keto esters	2-Alkyl and substituted alkyl	312[a], 313
o-Phenylene diamine	2-Bromocyclobutanone	2-Cyclopropyl	314[a]

[a] See text.

When a 2:1 mole ratio of diamine to (**1.47**) is used, 2,2'-bisbenzimidazole (**1.48**) is formed, presumably via 2-nitromethylbenzimidazole (**1.46**); the latter can also be use to prepare pyrido[1,2-a]benzimidazoles[293] (see section 7.1.1 in Chapter 7).

The formation of 1-methylbenzimidazole 2-aldoxime from the reaction of N,N-dimethyl-o-phenylene diamine with choral hydrate and hydroxylamine was first discovered by Petrov and coworkers,[306a,b] who were actually intending to prepare o-(N,N-dimethylamino)isonitrosoacetanilide (**1.49**) [cf. the behavior of aniline].[315] Subsequently, a series of related oximes (**1.50a**) have been prepared in this manner in ~60% yield from appropriate

N-(2-aminoaryl)piperidine.[306c] Derivatives (**1.50b,c**) other than 2-oximes can be isolated by use of the analogous reagents, semicarbazide and phenylhydrazine, respectively. The mechanism[306c] probably involves an intermediate chlorimine (**1.51a**) rather than an intermediate isonitrosoacetanilide (**1.51b**), since the latter is not converted into the benzimidazole derivative under the reaction conditions. The chloroalkyl derivatives (**1.50a**) are valuable intermediates for the synthesis of imidazo[4,5,1-*jk*]benzazepines[306c] (see section 9.1.1.2 in Chapter 9).

(**1.49**)

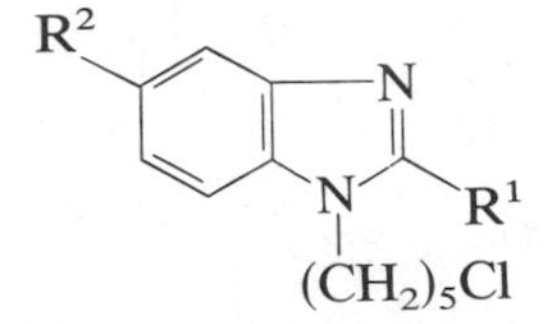

(**1.50**)

a, $R^1 = CH = NOH$; $R^2 = H$, Cl, NO_2
b, $R^1 = CH = NNHCONH_2$; $R^2 = H$
c, $R^1 = CH = NNHPh$; $R^2 = H$.

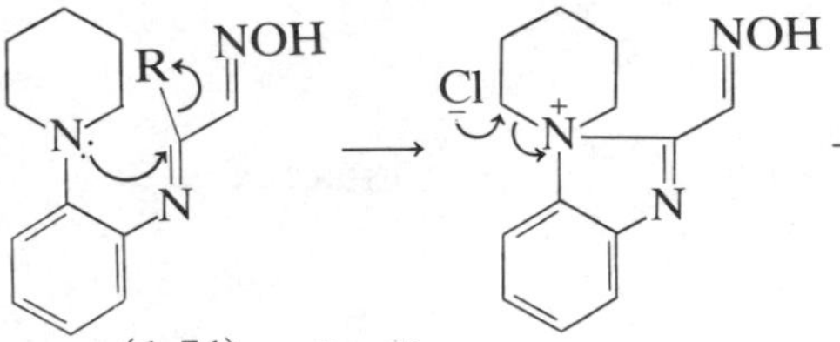

(**1.51**) a, R = Cl
           b, R = OH

When *N,N*-dimethyl-*o*-phenylenediamine is allowed to react with sulfuryl chloride at room temperature, a reasonable yield of 1-methyl-4,5,6,7-tetrachlorobenzimidazole (**1.52**) is isolated;[305] it is important to note that prior to this work,[305] a four-stage synthesis of this material was necessary from 1,2,3-trichlorobenzene.[316] The reaction leading to (**1.52**) is believed[305] to follow two distinct pathways of electrophilic substitution and sulfonylamine (**1.53** or **1.54**) formation; ensuing benzimidazole formation via a nitrene intermediate (**1.55**) or a sulfinic acid (**1.56**) can be envisaged.

The synthesis of 1- and 1,2-disubstituted benzimidazoles from 1-methoxy-3-substituted propylidene *o*-aminoanilines[307] and cyanoanils[308] both involve simple addition–elimination mechanisms, and a comparable route[308] can be

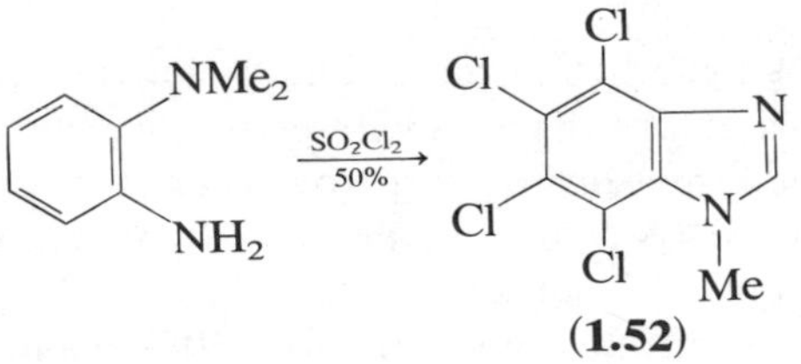

(**1.52**)

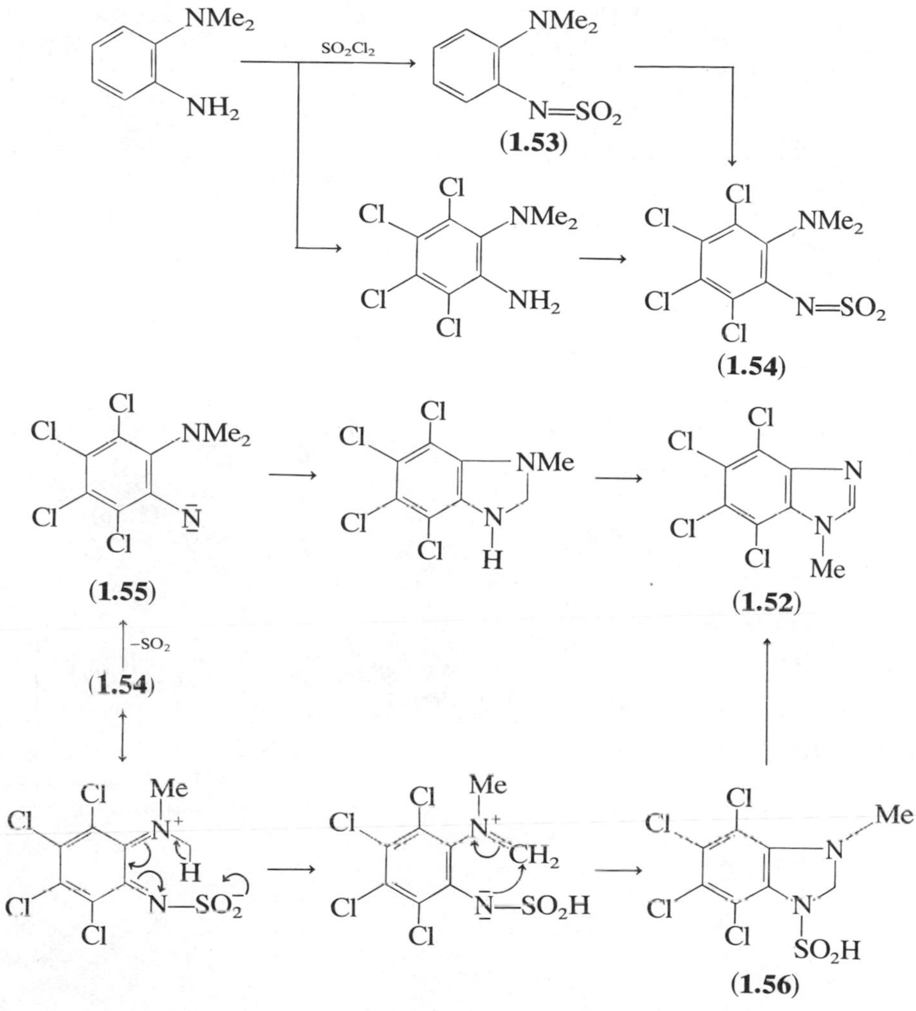

envisaged for 2-arylbenzimidazole formation from the reaction of *o*-phenylenediamine with aryl-*N*-(*p*-dimethylaminophenyl)nitrones (cf. **1.57** → **1.58**); yields in both these types of reactions are good (50 to 90%).

The synthesis of a number of heterocycles from the reaction of *N*-ethoxycarbonylthioamides (**1.59**) with 1,2-dinucleophilic reagents has been carried out;[310] using *o*-phenylenediamine, excellent yields of 2-substituted benzimidazoles (**1.60**) are obtained, and the method circumvents the use of acidic catalysts. The reagents (**1.59**) are easily obtained in one step from ethoxycarbonyl isothiocyanate and simple aromatic or heterocyclic compounds or alkylmagnesium halides.[317,318]

A number of benzazole derivatives (e.g., **1.62**) have been synthesized by the reaction of *o*-phenylenediamine and related compounds with heterocyclic compounds containing reactive methyl groups (e.g., **1.61**); the reactions

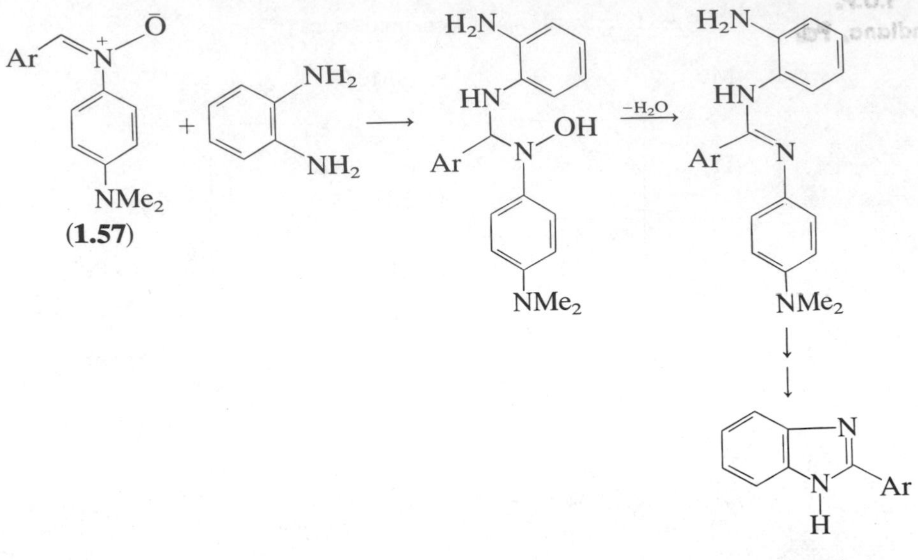
(1.57)
$-H_2O$
(1.58)

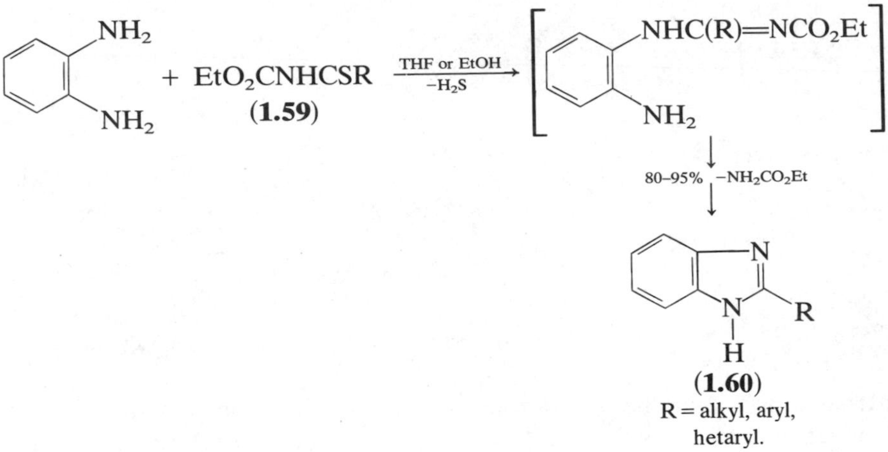
$EtO_2CNHCSR$
(1.59)
THF or EtOH
$-H_2S$
80–95%
$-NH_2CO_2Et$
(1.60)
R = alkyl, aryl, hetaryl.

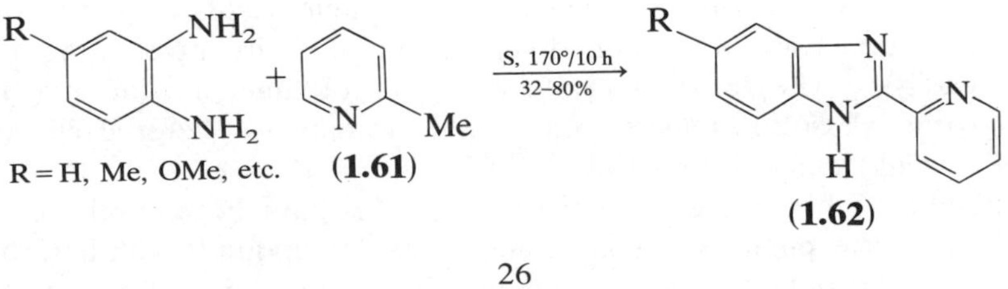
R = H, Me, OMe, etc.
(1.61)
S, 170°/10 h
32–80%
(1.62)

are conducted in the presence of sulfur at elevated temperatures.[319] Yields obtained in this type of synthesis compare favorably with those obtained from reaction of the o-arylenediamine with picolinic acid,[106] and the procedure can be used to prepare a variety of 2-hetaryl derivatives including 2-quinolinyl, 2-imidazolyl, and 1-methyl-3H-imidazo[4,5-b]pyridin-2-yl.[311]

A detailed study[312a] has been made of the conditions necessary for formation[312b] of benzimidazoles and benzimidazolones from the reaction of β-keto esters with o-phenylenediamine. Under neutral conditions in xylene a mixture of a benzimidazolone (**1.63**) and a benzo-1,4-diazepinone (**1.64**) are formed, but under acidic conditions, 2-methylbenzimidazole (**1.65**) is produced in good yields; extension of this type of procedure to 2-carbethoxycyclohexanone provides a synthesis of ethyl 6-(benzimidazol-2-yl) caproate (**1.66**). A possible route[312a] to the benzimidazole products (**1.63** and **1.65**) via an aminocrotonate intermediate (**1.67**) is outlined. Subsequently it has been shown that such intermediates (**1.67**) can be isolated from the reaction of o-arylenediamines and β-keto esters.[313]

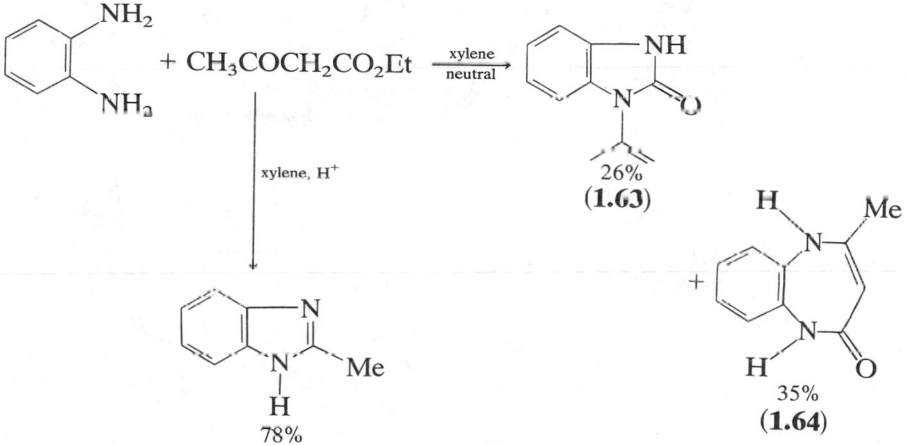

A compound thought to be 1,2,2a,3-tetrahydrocyclobuta[b]quinoxaline (**1.68**)[320] was isolated from the reaction of o-phenylenediamine with 2-bromocyclobutanone, but this was subsequently shown to be 2-cyclopropylbenzimidazole (**1.70**);[314] a spiro cyclobutane intermediate (**1.69**) is envisaged.[314]

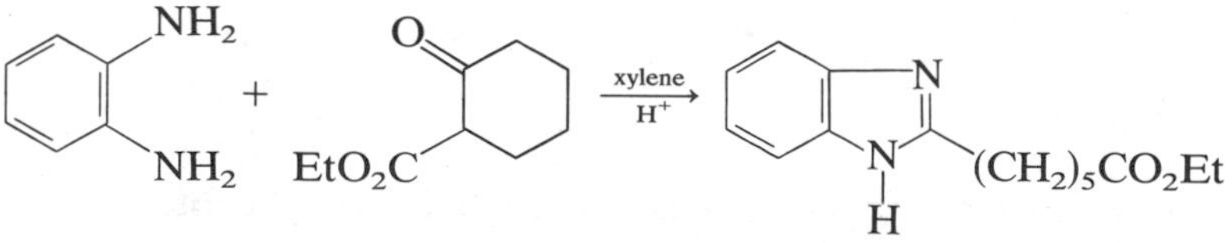

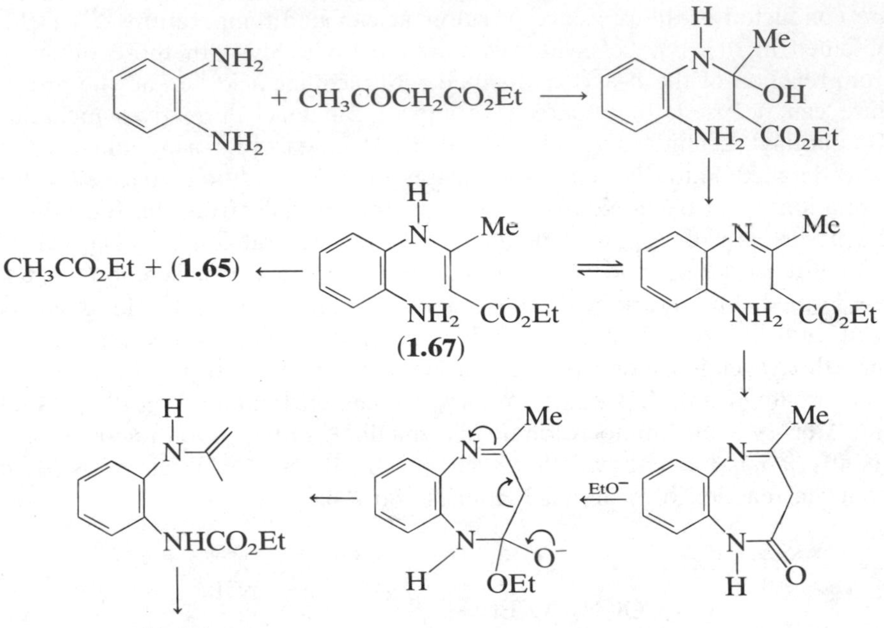

1.2.2. From o-(N-Acylamino- and -Aroylamino)arylamines and -Nitroarenes

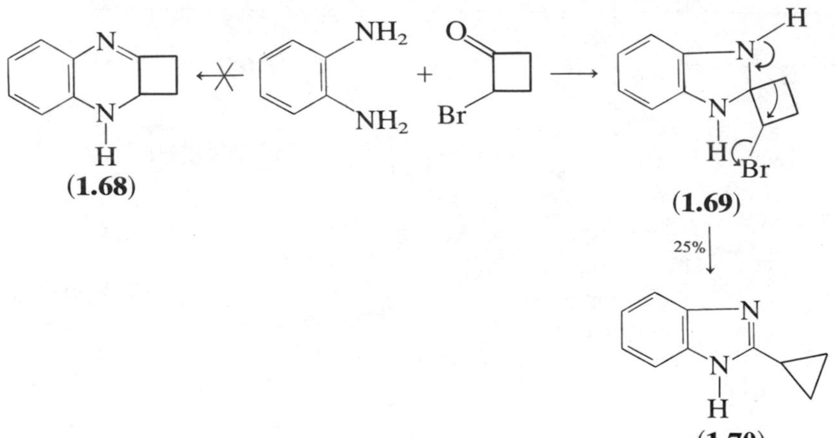

The formation of benzimidazoles (**1.71**) by the reaction of o-arylenediamines with carboxylic acids and related compounds[137] is presumed to involve the (monoacyl-) or (monoaroylamino)arylamines (**1.70**), and a number of compounds have been synthesized by cyclization of such intermediates under acid-catalyzed conditions or in uncatalyzed thermal reactions (see Table 1.7).[321–338] The o-arylenediamine can also be generated

TABLE 1.7. SYNTHESIS OF BENZIMIDAZOLES FROM o-(N-ACYLAMINO AND -AROYLAMINO) ARYLAMINES

Starting material[a]

R^1	R^2	R^3	Reaction conditions	Ref.
H	Me	Dodecyl (dialkylamino)alkyl hydroxyalkyl	HCl/EtOH/room temp.	321
H	Me	Aralkyl	aq. HCl/heat	322
H	Me	COPh	Heat in a melt	323
H	Me	$(CH_2)_2Cl$	aq. HCl	324
H	Me	H	Conc. HCl	325
H	i-Pr	H	Heat	326
H	$CH_2CH(OH)CCl_3$	H	Heat in xylene	327
H	Chloroalkyl	Chloroacyl	Heat	328
H	(Dialkylamino)alkyl	Aryl	aq. HCl/heat	329
H	$(CF_2)_n CONH_2$	H	heat	330
H	Aryl	H	$4 M$ HCl or $2 M H_2SO_4$	331
H	Aryl	H	aq. HCl or H_2SO_4/heat	332
H, Ac, Et, $Et_2NCH_2CH_2$	Aryl	H	aq. HCl/heat	333
H	p-$O_2NC_6H_4$	aryl	$SnCl_2/HCl^b$	334
H	p-$H_2NC_6H_4$	H	aq. HCl	335
H	4-Thiazoyl	H	H_3PO_4/heat	336
H	2-Halo-4-thiazolyl	H	aq. HCl/heat	337
H	2-Benzofuryl	H	Heat	338

[a] Substituents in the aryl ring are not included.
[b] The product is a 2-(p-aminophenyl)benzimidazole.

in situ from the appropriate o-nitroarylamine (**1.72**) by using a variety of reductants (see Table 1.8).[339–345]

(**1.70**)   (**1.71**)   (**1.72**)

R^1–R^3 = H, alkyl, aryl, etc.
R^4 = H, CF_3

TABLE 1.8. SYNTHESIS OF BENZIMIDAZOLES FROM
o-[N-ACYLAMINO AND -AROYL-
AMINO]NITROBENZENES

Starting material[a]

R¹
|
NCOR²

NO₂

R¹	R²	Reaction conditions	Ref.
H	Me	Sn/AcOH/heat	339
H	Me	$Na_2S_2O_4$/heat	340
H	CF₃	H_2/Pd/C	341
H	CF₃	Raney Ni	342
H, Me	Aryl	SnCl₂/HCl/heat	343
H	4-Thiazolyl	(i) H_2/Pd/C (ii) aq. HCl/heat	344
H	4-Thiazolyl	Fe/AcOH[b]	345

[a] Substituents in the aryl ring are not included.
[b] Various reductants have been employed depending on the nature of the aryl ring substituent; e.g. $(NH_4)_2$S/aq. DMF or PtO_2/H_2 for an isopropyloxycarbonylamino substituent and Sn/HCl, HCO_2NH_4/$HCONH_2$/$NaHSO_3$, Zn/alc. NH_4OH, Raney Ni/DMF, or PtO_2/DMSO for an aryl substituent.

In general, yields for the reactions described in Tables 1.7 and 1.8 are good, and occasionally the method is superior to conventional procedures. For example, reduction of N-(2-nitro-5-trifluoromethylphenyl)trifluoro-acetamide with Raney nickel gives 2,5-bistrifluoromethylbenzimidazole in 60% yield,[342] which compares with a 48% yield of this product from reaction of the appropriate diamine with trifluoroacetic acid.[346] It is interesting to note, however, that the nitroarene (**1.72**, R¹ = H; R² = R⁴ = CF₃) is reduced by Raney nickel to the amino compound but is not cyclized to the benzimidazole derivative; a possible explanation for this is that the hydrogen atoms of the amino group are strongly hydrogen-bonded to the trifluoromethyl group.[342]

The procedure just described (cf. **1.70** → **1.71**) has been successfully applied to the synthesis of bisbenzimidazoles (**1.73**)[347,348] in good yields, and related compounds (**1.74**, **1.75**) have been synthesized by similar methods.[347]

The ring-opening aminolysis of lactams (**1.76**) has been developed as a convenient route to five- and six-membered 2-(ω-aminoalkyl)-substituted heterocycles including benzimidazoles (**1.78**)[349] (see also Ref. 288 in section 1.2.1). Intermediate anilides (**1.77**) are envisaged in these processes in which the reactants are heated in the presence of a catalytic amount of p-toluenesulfonic acid or 85% phosphoric acid; milder reaction conditions are

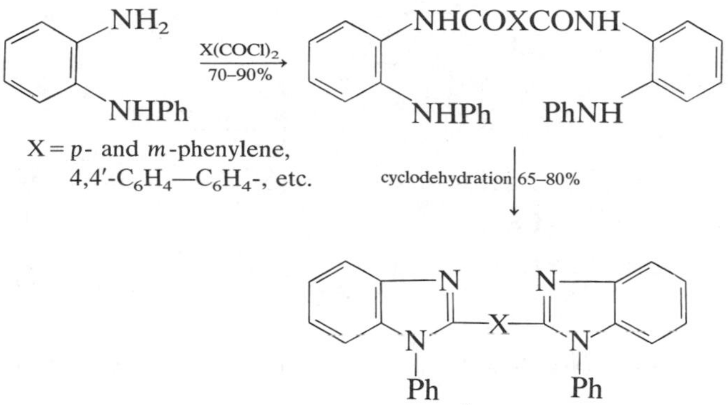
X(COCl)2
70–90%
cyclodehydration
65–80%
X = p- and m-phenylene,
4,4′-C6H4—C6H4-, etc.

(**1.73**)

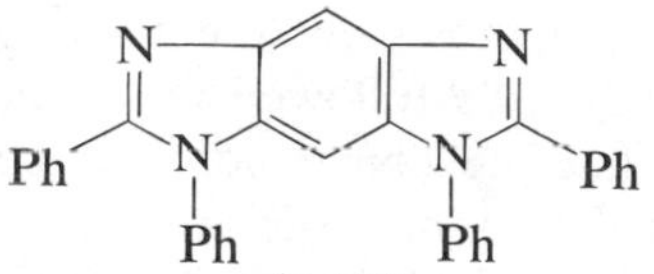

(**1.74**)

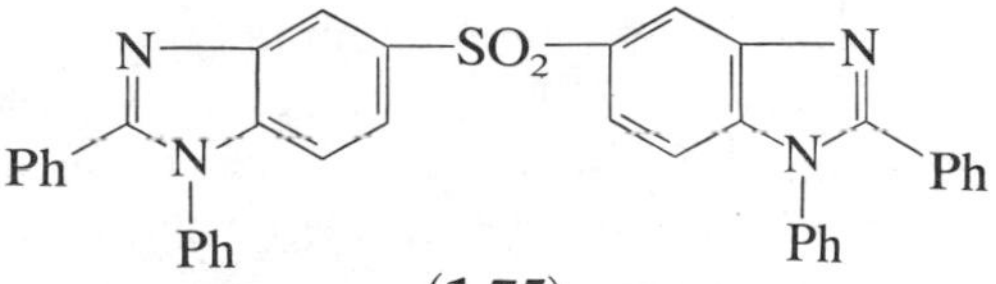

(**1.75**)

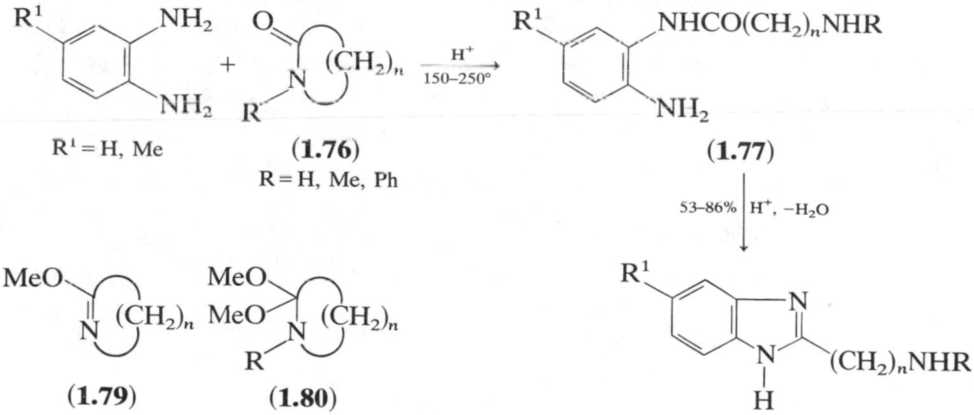
H+
150–250°
53–86%
H+, −H2O
R1 = H, Me
R = H, Me, Ph

(**1.76**) (**1.77**) (**1.78**) (**1.79**) (**1.80**)

required if lactim ethers (e.g., **1.79**) or lactam acetals (e.g., **1.80**) are used instead of free lactams.[349]

Synthesis of the condensed benzimidazoles (**1.82**) from the acylamino derivatives (**1.81**) is a more interesting example of cyclization in this group in that a skeletal rearrangement occurs;[350] a more detailed discussion of such reactions is presented in Chapter 9, section 9.1.1.

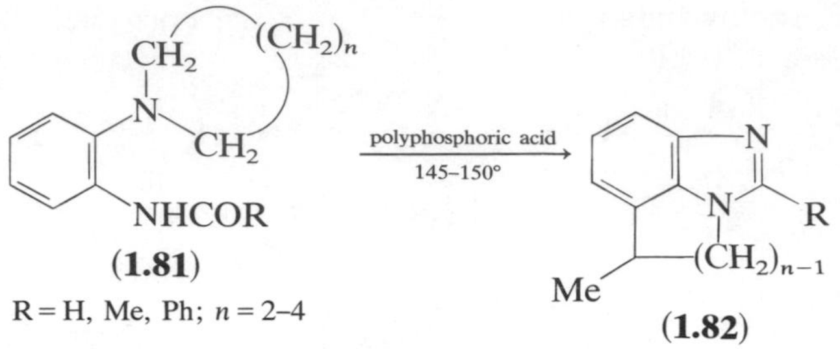

(1.81)

R = H, Me, Ph; $n = 2-4$

(1.82)

1.2.3. From *N-o-*Aminoaryl-*N′-*Arylthioureas and Related Compounds

The synthesis of 2-phenylaminobenzimidazole (**1.84**, R = Ph) from *N-o-*aminophenyl-*N′*-phenylthiourea (**1.83**, R = Ph) was first achieved by heating these reactants with 5% ethanolic potassium hydroxide followed by acidic work-up.[351] Subsequently, it has been shown that the conversion can be effected by mercuric chloride,[352] mercuric oxide,[353] or mercuric acetate,[353] the best conditions being the use of ~3-mole equivalents of mercuric chloride in chloroform under reflux.[353] An alternative cyclodesulfurization method entails heating the urea derivatives (**1.83**) with 8 equivalents of an alkyl halide in ethanol.[354–356] The oxidative cyclization presumably involves carbodiimide intermediates, but these cannot[353] be isolated, while the alkyl halide-induced cyclization probably involves the intermediacy of *S*-alkyl isothiuronium compounds; intermediates of the latter type can also be generated by the use of 2-mole equivalents of dimethylsulfate in ethanol.[357,358]

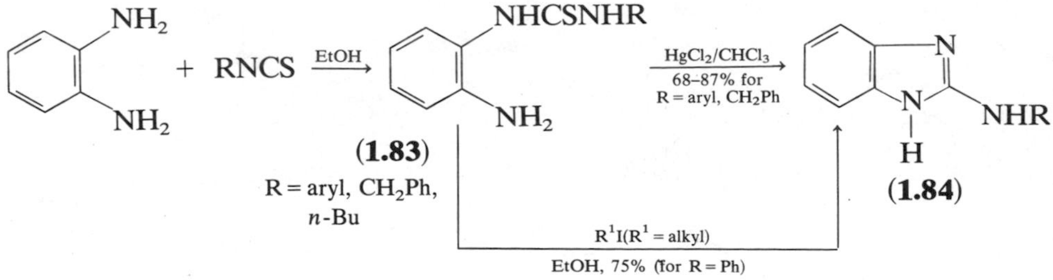

Reactions of the type described in the preceding discussion are useful since they offer an alternative synthesis of 2-amino derivatives, which are normally prepared from the reaction of 2-chloro derivatives with amines, often under forcing conditions (see section 1.4.3).

In reactions closely related to the process described previously, the mono- (**1.85a**) and bis- (**1.85b**) thioallophanic acid esters can be converted into the commercially important benzimidazole-2-carbamic acid ester derivatives (**1.86**) under a variety of conditions including the use of strongly protic solvents,[359] triethylamine,[360] sodium hydroxide/methyl iodide,[361] sodium hydroxide/dimethylsulfate,[362] heating in water,[363] or heating the calcium salt.[364]

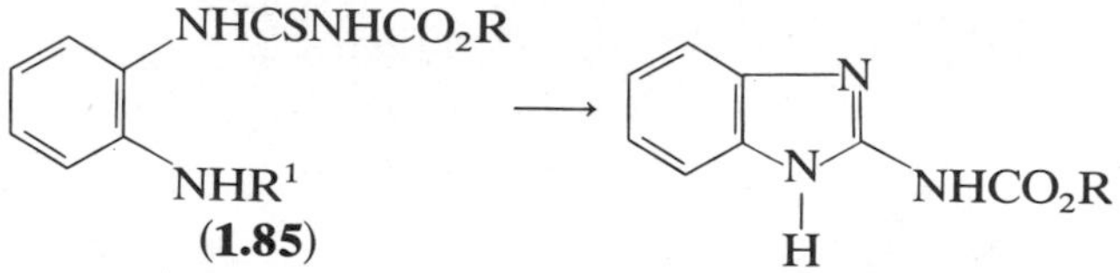

R = alkyl
a, R^1 = H
b, R^1 = CSNHCO$_2$R

Little attention has been paid to urea derivatives that are analogous to (**1.83**), but it has been shown that N-o-aminophenyl-N' benzoylurea (**1.87**) is converted into 2-benzamidobenzimidazole (**1.88**) by heating it in toluene in the presence of an acidic catalyst.[365]

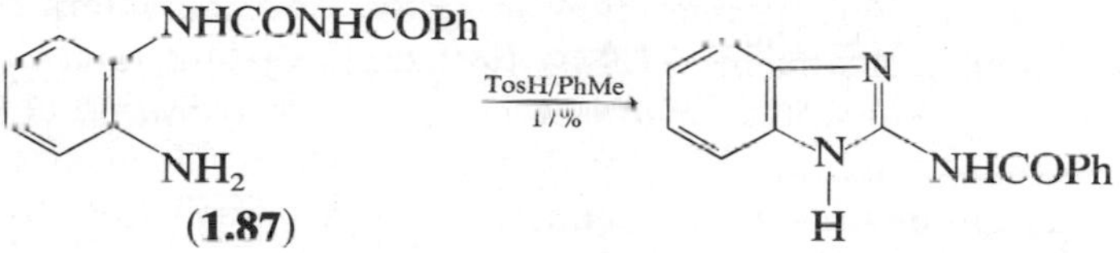

1.2.4. From N-Substituted N'-(o-Amino- and -Nitro)arylhydrazines

Conversion of N-formyl-N'-(o-aminophenyl)hydrazines (**1.89**, R = H) into 1-aminobenzimidazole (**1.90**) was first achieved by the use of m-nitrobenzenesulfonic acid,[366] but better yields of 1-amino derivatives can be obtained via 1-acylamino derivatives (**1.91**), which in turn are prepared from the reaction of the hydrazines (**1.89**) with aliphatic acids.[367] In a subsequent detailed study,[368] it has been shown that cyclization of the formylhydrazine (**1.89**, R = H) with acetic or propionic acid results in the exclusive formation of the acylaminobenzimidazoles, (**1.92**) and (**1.93**), respectively. Similarly, the reaction of o-acetylhydrazinoaniline (**1.89**, R = Me) with propionic acid gives 2-methyl-1-propionylaminobenzimidazole (**1.94**) as the only isolable product. However, when the acetyl hydrazine (**1.89**, R = Me) or the analogous propionyl derivative (**1.89**, R = Et) were treated with formic acid, mixed products were obtained in which either the alkyl group of the original acylhydrazino moiety or the formic acid hydrogen had become attached to C-2 of the resulting aminobenzimidazole product.

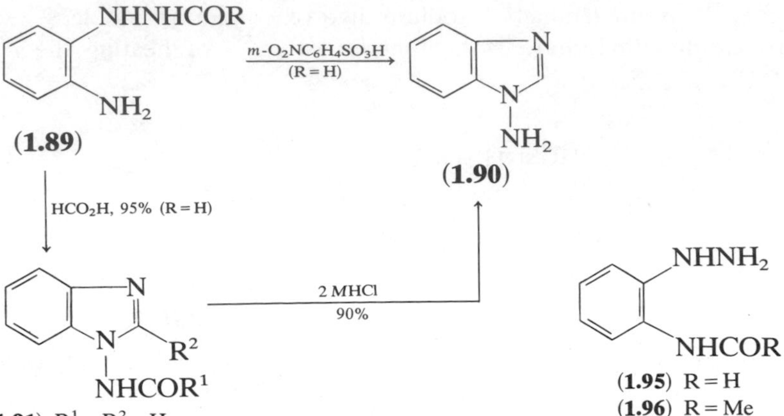

NHNHCOR

$m\text{-}O_2NC_6H_4SO_3H$
(R = H)

(1.89)

NH$_2$

(1.90)

NH$_2$

HCO$_2$H, 95% (R = H)

N

N

R^2

NHCOR^1

2 MHCl
90%

NHNH$_2$

NHCOR

(1.95) R = H
(1.96) R = Me

(1.91) R^1 = R^2 = H
(1.92) R^1 = Me, R^2 = H
(1.93) R^1 = Et, R^2 = H
(1.94) R^1 = Et, R^2 = Me

The exclusive formation of compounds (1.92) and (1.93) in the reactions described previously supports the original suggestion[366] that cyclization proceeds via formyl group migration in (1.89) and the intermediacy of o-formylaminophenylhydrazine (1.95); the corresponding intermediate in the synthesis of (1.94) would be o-acetylaminophenylhydrazine (1.96). The formation of mixed products[368] is thus presumed to result from a reduced rate of acyl group migration and competitive formylation of the free amino group (cf. 1.89 → 1.97 + 1.98).

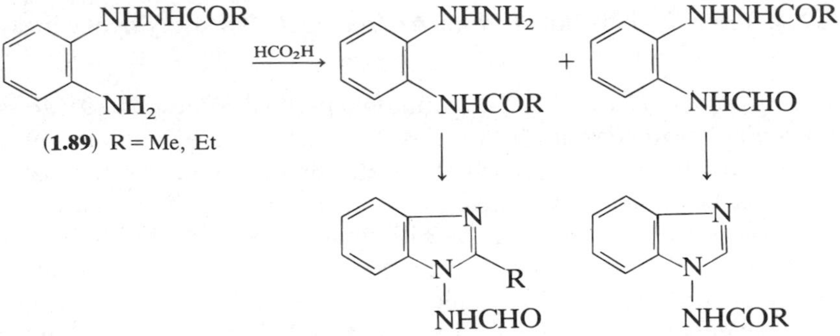

NHNHCOR

HCO$_2$H

NHNH$_2$

NHCOR

+

NHNHCOR

NHCHO

NH$_2$

(1.89) R = Me, Et

N

N

R

NHCHO

(1.97)

N

N

NHCOR

(1.98)

A more complicated mechanism is clearly involved in the conversion of o-nitroarylhydrazine derivatives (cf. 1.99) into 1-aminobenzimidazoles (cf. 1.100) under the influence of hydrochloric acid.[369] Reactions of this type give good yields of 1-amino derivatives except for nitro derivatives (e.g., 1.99, R^2 = NO$_2$), where the formation of benzotriazoles and benzotriazole

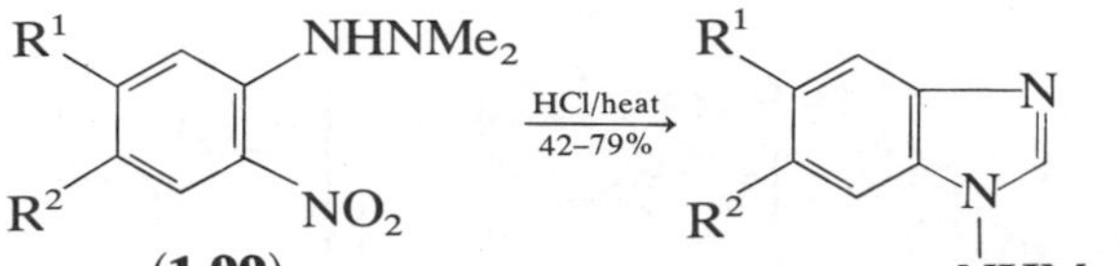

$R^1 = H, Cl$
$R^2 = H, NO_2, CF_3, CO_2Et$

N-oxides competes successfully with benzimidazole formation. A mechanism involving a [1,5]sigmatropic rearrangement[369] is envisaged for this type of process (see path *a* in Scheme 1.1), and this is considered to have precedent[369] in the thermal uncatalyzed cyclization of N-cyclohexyl-*o*-nitroaniline (**1.101**) into the tricyclic compound (**1.102**).[370] The former method has also been used for the synthesis of [1,2]diazepino [1,7-*a*]benzimidazoles[369] (see section 8.2.1 in Chapter 8).

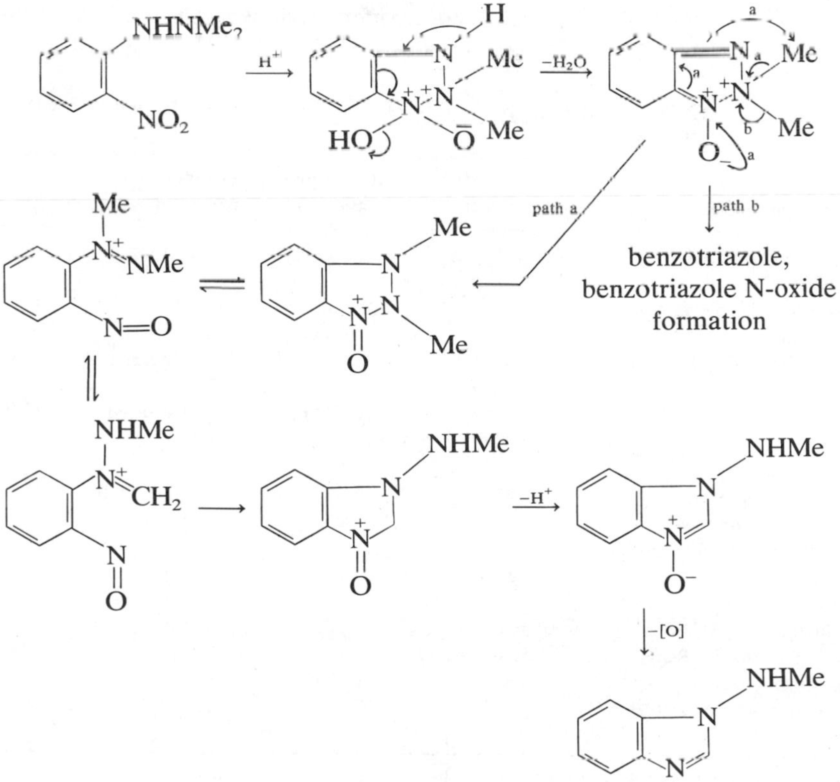

Scheme 1.1

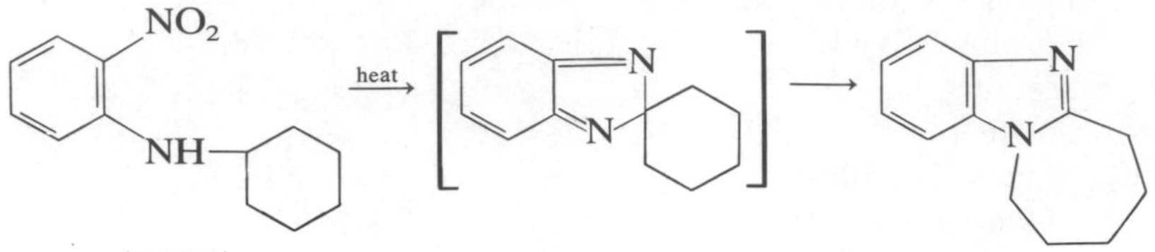

(1.101)   **(1.102)**

It is pertinent to mention that the method of choice for the synthesis
1-aminobenzimidazoles derivatives is the direct amination of benzimidazole
by hydroxylamine O-sulfonic acid.[237]

1.2.5. From o-Nitroarylamines and o-Dinitroarenes

A number of methods have been developed in which benzimidazole
syntheses are accomplished in a single step from an o-nitroarylamine or an
o-dinitroarene; some examples of these direct methods are shown in Table
1.9.[371-380] Of particular interest from a commercial aspect is the formation of

TABLE 1.9.   SYNTHESIS OF BENZIMIDAZOLES FROM o-NITROARYLAMINES
             AND o-DINITROARENES

Starting materials	Reductant or catalyst	Reaction conditions	Product benzimidazole	Ref.
o-Nitroaniline/HCONH$_2$/ HCO$_2$NH$_4$	NaHSO$_3$	150°/1 hr	Unsubstituted	371
o-Nitroaniline/benzaldehyde derivative		Xylene/ reflux	2-Aryl	372
2,4-Dinitro-6-methoxy-N- methylaniline	BaSO$_4$/Pd/H$_2$ or Sn/HCl		1-Me-5-NH$_2$ 7-MeO	373
o-Nitroaniline/HCO$_2$H	Zn	100°/3 hr	Unsubstituted	374a
2-Nitro-4,5-dimethylaniline/ HCO$_2$H	Na$_2$S$_2$O$_5$		5,6-Me$_2$	374b
2-Nitro-S^a-isopropyloxy- carbonylamino-N- (4-thiazolylmethyl)aniline	Na$_2$SO$_3$	aq. H$_2$SO$_4$/ reflux	2-(4-Thiazolyl)- isopropyloxy- carbonylamino	375
o-Nitroaniline/ROH^b	Al$_2$O$_3$	320°	1-Alkyl and 1,2-dialkyl	376
o-Nitroaniline/ROH^b	V or Cu/Al$_2$O$_3$	320°	2-Alkyl	377
o-Dinitrobenzene/ROH^b	V or Cu/Al$_2$O$_3$	320°	2-Alkyl	378
o-Dinitrobenzene/CF$_3$CO$_2$H	Sn/HCl	60–70°	2-CF$_3$	379
2,3-Dinitroanisole/PhCH$_2$NH$_2$		Xylene/ reflux	2-Ph-4-MeO	380

^{a} S denotes that position of substituent in the aryl ring is not stated.
^{b} See text.

benzimidazoles by thermolysis of nitroarene–alcohol mixtures in the gas phase. Reasonable yields are obtained in these processes, but when *o*-nitroanilines are used a secondary reaction converts the 2-alkyl-benzimidazoles into 1,2-dialkyl derivatives;[376] the products from reaction of *o*-nitroaniline with methanol, ethanol, and *n*-propanol at 320° over alumina are thus 1-methyl (67%), 1-ethyl-2-methyl (57%), and 1-*n*-propyl-2-ethyl-benzimidazole (22%), respectively.[376] *o*-Dinitroarenes behave in similar fashion with alcohols over alumina-copper or vanadium, but in this case the products are exclusively 2-alkylbenzimidazoles. Yields are approximately the same for both types of catalyst ($\sim 50\%$), but the activity of the vanadium lasts longer than the alumina-copper catalyst.[378]

The direct reductive method from the carbanilic acid derivatives (**1.103**) leading to alkyl esters of 2-amino-1-benzimidazole carboxylic acids (**1.104**) is also of commercial value, since products of the latter type are transformed thermally into alkyl esters of benzimidazole-2-carbamic acids (**1.105**)[381] (see Chapter 10).

| **(1.103)** | **(1.104)** | **(1.105)** |

Benzimidazoles have also been obtained in thermal uncatalyzed cycliza-tion reactions of *N*-substituted *o*-nitroaniline derivatives (cf. **1.106** → **1.107**),[382] and the scope of the synthesis for condensed imidazoles has been evaluated.[383] A reaction of the type (**1.106** → **1.107**) is probably also involved in the direct conversion of *o*-nitrochloroarenes into benzimidazoles by benzylamine and sodium acetate at 200°.[384] Thermolysis of *o*-nitro- and 2,4-dinitrophenyl derivatives of α-amino acids (cf. **1.108**) also affords benzimidazoles, although with the exception of *o*-nitrophenylalanine (which leads to **1.109** and **1.110**), yields are poor.[385] Cyclization reactions of the type (**1.106** → **1.107**) are presumed[382,383] to involve an *aci*-nitro mechanism of the type depicted, and it is interesting to note the formation of 2-phenylbenzimidazole *N*-oxide by pyrolysis of the phenyl analog of (**1.108**). The isolation of benzimidazolones and bisbenzimidazoles (cf. **1.110**) in reactions of this type is also consistent with the intermediacy of benzimidazole *N*-oxides.[386,387]

N,N-Disubstituted *o*-nitroanilines have been converted into condensed benzimidazoles photochemically in acid solution[388] (see section 7.1.1 in Chapter 7) and also by heating with zinc chloride and acetic anhydride. The latter type of cyclization was originally reported by Van Romburgh and coworkers,[389] but

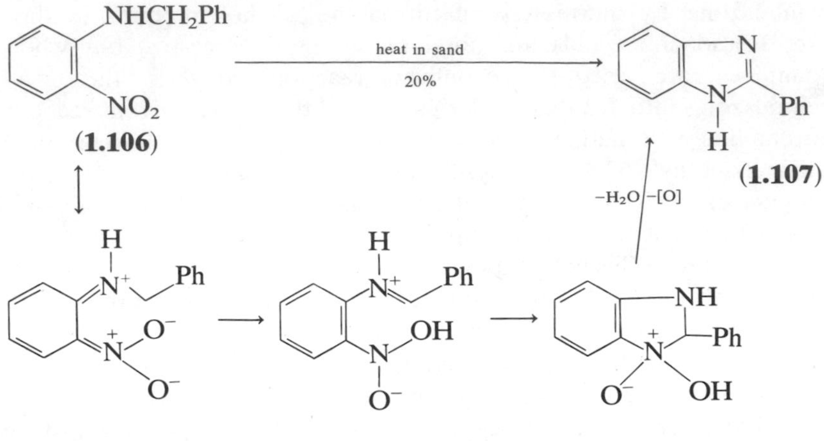

(1.106) → (1.107)

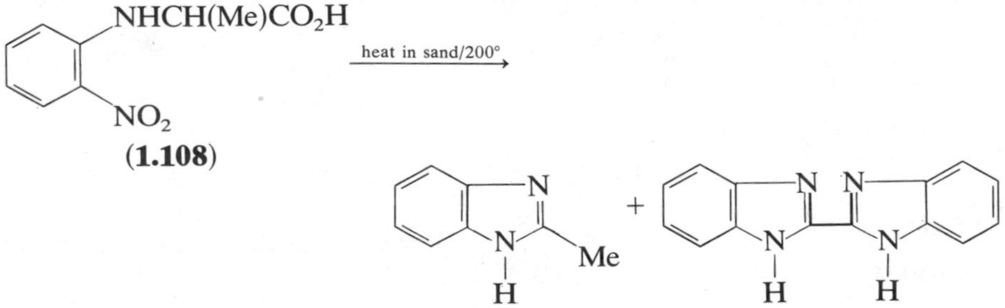

heat in sand
20%

−H₂O / −[O]

(1.109) (1.110)

has subsequently been clarified by Grantham and Meth-Cohn.[390] *N,N*-Dimethyl *o*-nitroanilines give rise to benzimidazolone derivatives (see section 3.2.1.3 in Chapter 3), but *N,N*-diethyl-2,4-dinitroaniline (1.111) is converted by zinc chloride-acetic anhydride into the 1-ethylbenzimidazole derivative (1.113) and not the quinoxalinone derivative (1.112), as reported by Van Romburgh and coworkers.[389]

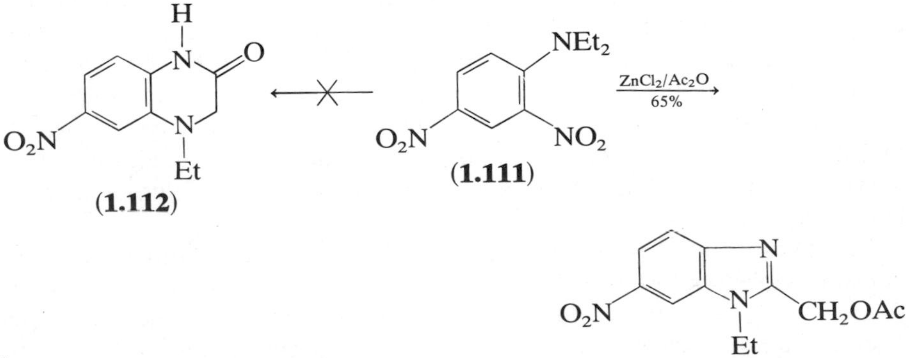

(1.112) ← ✕ — (1.111) → (1.113)

ZnCl₂/Ac₂O
65%

1.2.6. From *ortho*-Substituted N-Benzylideneaniline Derivatives

Nitro Compounds

The reduction of aromatic nitro compounds by triethyl phosphite and related reagents has been widely used as a simple, effective route to a variety of nitrogen-containing heterocycles.[391] N-Benzylidene-2-nitroaniline derivatives (**1.114**) are converted[392] into 2-phenylbenzimidazoles (**1.115**) in this manner, and it is interesting to note that the yields (33, 47%) are slightly higher than those obtained by use of the classical Weidenhagen procedure (see section 1.2.1).

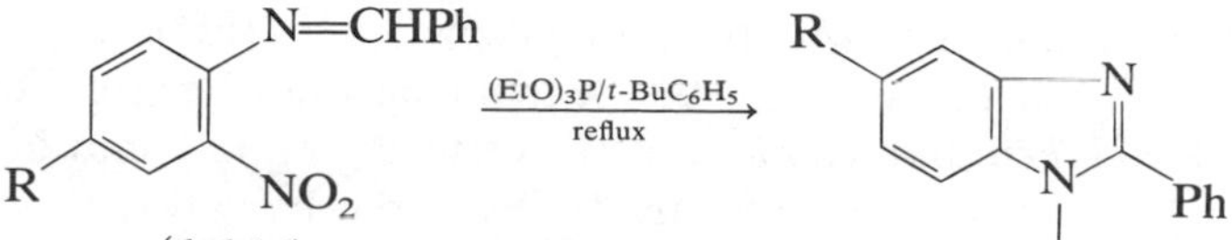

$(EtO)_3P/t-BuC_6H_5$, reflux

(**1.114**)

R = H, 47%
R = Me, 33%

(**1.115**)

Azido Compounds

Good yields of benzimidazoles (**1.117**) can be obtained by heating N-benzylidene-2-azidoanilines (**1.116**) in 1,2-dichlorobenzene[393] or dimethylformamide,[394] and the method can be used to prepare compounds with a heterocyclic group in the 2-position (see Table 1.10). Reactions of this type

TABLE 1.10. SYNTHESIS OF BENZIMIDAZOLES FROM ANILS OF *o*-AZIDOANILINE

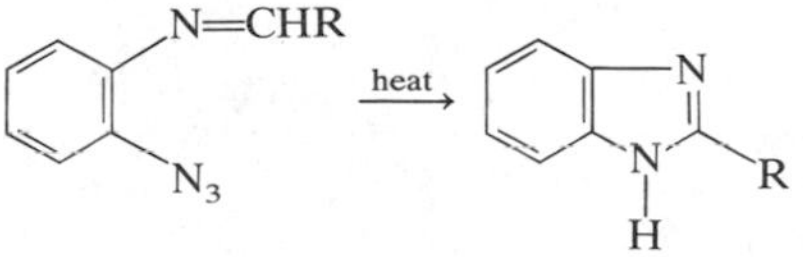

(**1.116**)   (**1.117**)

(R) in benzimidazole derivative (**1.117**)	Pyrolysis temperature (°C)	Yield of (**1.117**) %	Ref.
$p\text{-}O_2NC_6H_4$	130	48	394
$p\text{-}NCC_6H_4$	130	56	394
o-Pyridyl	130	52	394
2-Quinoxalinyl	130	71	394
$p\text{-}ClC_6H_4$	140	96	393
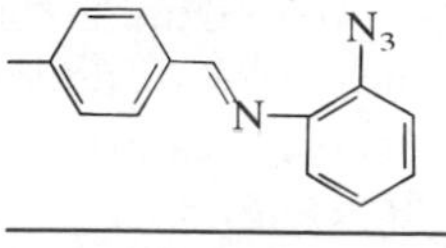	140–145	85[a]	393

[a] The product is a bisbenzimidazole.

may proceed via aryl nitrenes, but it is notable that the decomposition temperatures (ca. 130°) are about 30° below that of phenyl azide and a concerted loss of nitrogen with concomitant cyclization cannot be ruled out.

Amino Compounds

The oxidative heteroaromatization of N-benzylidene-o-phenylenediamines (1.118) to benzimidazoles (1.119) is well documented and can be effected by, for example, cupric salts,[395] lead tetracetate,[396] active manganese dioxide,[397] and nickel peroxide[398] (see Table 1.11). More recently, 2-phenyl-benzimidazole has been prepared in 89% yield by oxidation of N-benzyl-idene o-phenylenediamine with plumbophosphates {$H_2[Pb(H_2PO_4)_2(HPO_4)_2 + Pb(H_2PO_4)_2]$}, but the scope of the reaction has not been established.[399] The bis anils, (1.120 and 1.122), are converted[400] rapidly into bisbenzimi-dazoles, (1.121 and 1.123), respectively, by aerial oxidation in solution at 60° but surprisingly are not cyclized by irradiation with ultraviolet light (cf. the photolytic conversion of N-benzylidene o-phenylenediamine into 2-phenylbenzimidazole[401]).

TABLE 1.11. SYNTHESIS OF BENZIMIDAZOLES
            BY OXIDATION OF
            N-BENZYLIDENE-o-PHENYLENE
            DIAMINES

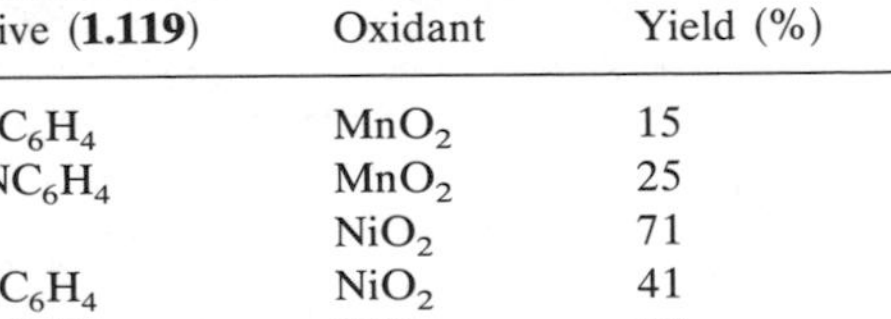

(1.118)   (1.119)

R in benzimidazole derivative (1.119)	Oxidant	Yield (%)	Ref.
p-$O_2NC_6H_4$	MnO_2	15	397
m-$O_2NC_6H_4$	MnO_2	25	397
Ph	NiO_2	71	398
o-$O_2NC_6H_4$	NiO_2	41	398
p-$O_2NC_6H_4$	NiO_2	57	398

Anils derived from o-arylenediamines (1.124) are probably intermediates in the formation of 2-[2-(2,4-dimethyl-3-oxopentyl)]benzimidazoles (1.125) from the acid-catalyzed reaction of o-arylenediamines with dimethylketene dimer.[402] Conversion of the ketones (1.125) into pyrido[1,2-a]benzimidazoles is described in section 7.1.1 of Chapter 7.

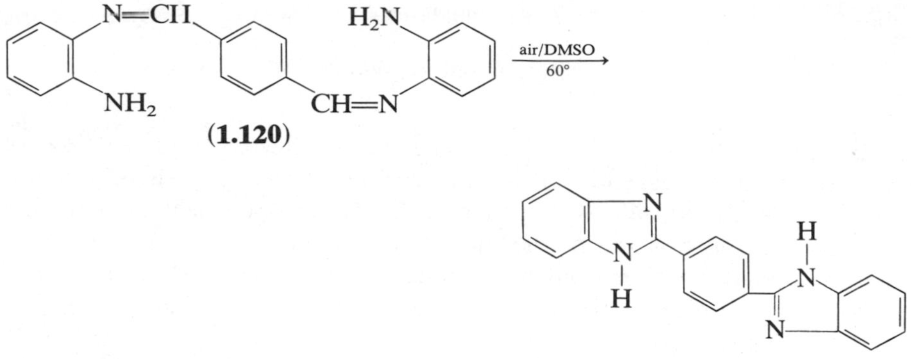

(1.120)

air/DMSO
60°
→

(1.121)

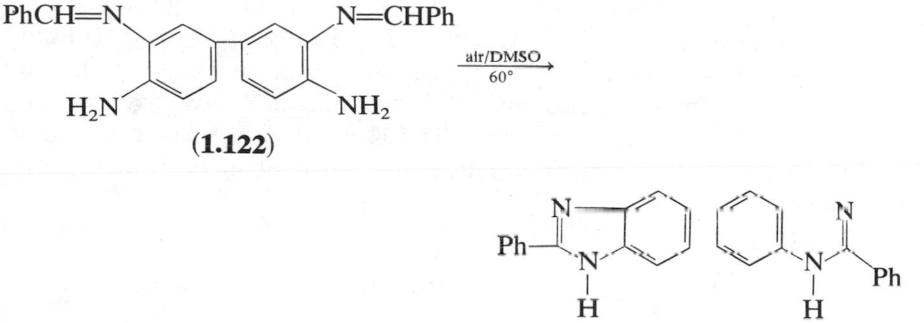

PhCH=N $\quad$ N=CHPh

$H_2N \quad$ NH$_2$

(1.122)

air/DMSO
60°
→

Ph— $\quad$ Ph

H $\qquad$ H

(1.123)

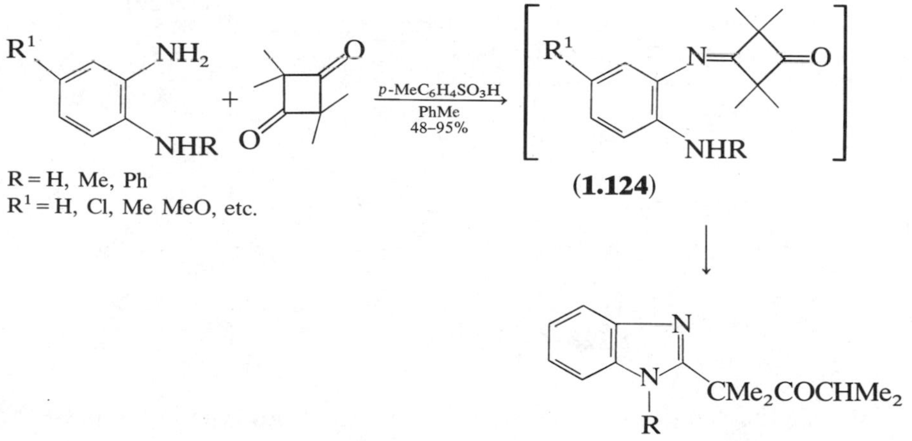

R^1 — NH$_2$

  +

NHR O

p-MeC$_6$H$_4$SO$_3$H
PhMe
48–95%
→

$\left[R^1 - \quad N= \quad =O \quad NHR \right]$

(1.124)

R = H, Me, Ph
R^1 = H, Cl, Me MeO, etc.

↓

CMe$_2$COCHMe$_2$
R

(1.125)

41

1.2.7. From Amidines

In general, the synthesis of benzimidazoles is effected from reactions of
o-phenylenediamines and compounds derived therefrom. Their preparation[403] from N'-aryl-N-hydroxyamidines (**1.126a**) is unusual in the sense
that both nitrogen atoms of the imidazole ring arise from one side chain.
The reactions are carried out under mild conditions using benzenesulfonyl
chloride in pyridine or triethylamine; generally, yields are good (>60%),
and the method has been used to prepare a variety of compounds with
substituents in the aryl ring.[403] Subsequently, it has been shown that such
products (**1.127**) can be obtained from the parent amidines (**1.126b**) by
oxidation with sodium hypochlorite under basic conditions.[404] Using this
more direct procedure, very high yields of benzimidazoles are obtained
(70–98%), and the method can be extended to prepare imidazopyridines (cf.
1.128) and triazolopyridines (cf. **1.129**). N-Chloro derivatives (**1.126c**) are
intermediates in these reactions, but the mechanism of their conversion into
benzimidazoles is in doubt: two possibilities considered[404] involve either the
intermediacy of a discrete iminonitrene (**1.130**) (path a in Scheme 1.2) or a
concerted process of dehydrochlorination with concomitant cyclization (path
b in Scheme 1.2).

$$\xrightarrow[\substack{<10° \\ >60\% \text{ yield}}]{\text{PhSO}_2\text{Cl/Et}_3\text{N or pyridine (for \textbf{1.126a})}}$$

(**1.126**)

(**1.127**)

a, R^1 = H, alkyl, alkoxy; R^2 = Ph; R^3 = OH
b, R^1 = H; R^2 = 4-thiazolyl, Ph, Et; R^3 = H
c, R^1 = H; R^2 = 4-thiazolyl, Ph, Et; R^3 = Cl

(**1.128**)

(**1.129**)

It would be interesting to carry out reactions of this type in the presence
of 1,3-dipolarophiles with a view to trapping the intermediate nitrene;
alternatively, competition experiments involving intramolecular cycloaddition or insertion reactions with a suitable *ortho* side chain could be informa-
tive.

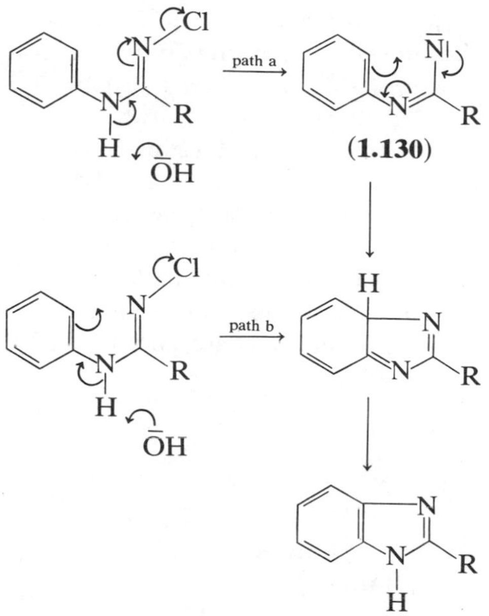

Scheme 1.2

Iminonitrenes (cf. **1.130**) are probably also intermediates in the photolytic and thermal transformation of N-(N-arylimidoyl)sulfimides (**1.131**) into benzimidazoles[405] (**1.132**). Good yields (30–96%) of benzimidazoles are obtained in the photochemical reactions, but thermal reactions lead to poor yields of benzimidazoles (cf. **1.133**) together with higher yields of quinazolines (cf. **1.134**) and 1,3,5-triazines (cf. **1.135**).

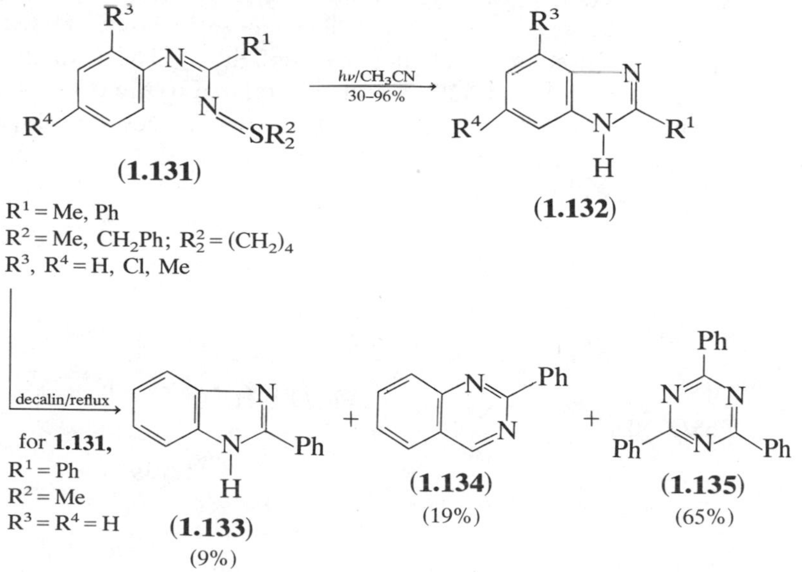

(**1.131**)

$R^1 = Me$, Ph
$R^2 = Me$, CH_2Ph; $R_2^2 = (CH_2)_4$
R^3, $R^4 = H$, Cl, Me

$h\nu/CH_3CN$
30–96%

(**1.132**)

decalin/reflux

for **1.131**,
$R^1 = Ph$
$R^2 = Me$
$R^3 = R^4 = H$

(**1.133**)
(9%)

(**1.134**)
(19%)

(**1.135**)
(65%)

The methods of Partridge and Turner[403] and Grenda and coworkers[404] have been subsequently employed for the preparation of 2-aryl,[406] 2-aralkyl,[406] and 2-hetarylbenzimidazoles.[407] Alternative cyclization procedures have been developed,[408] including manganese dioxide and lead tetraacetate oxidation of unsubstituted amidines (cf. **1.126b**); N-chloro (cf. **1.126c**), sulfonylmethyl, and p-tosyl derivatives (cf. **1.126**; $R^3 = SO_2Me$, p-$MeC_6H_4SO_2$) have been cyclized by thermal and photochemical reactions using benzoyl peroxide as an initiator or by the use of ferric chloride in methanol.[408] The method has also been extended to O-acylated 1-phenyl 2-hydroxy-3,3-dialkylguanidines (**1.136**), but yields of 2-aminobenzimidazoles (**1.137**) from these pyrolytic reactions are poor (13–27%).[409]

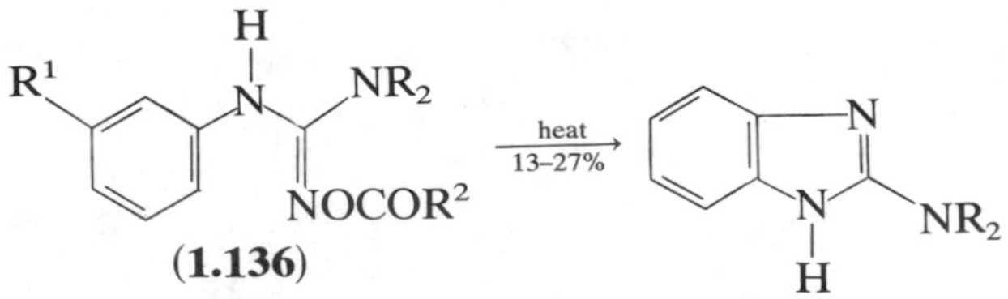

(1.136)   **(1.137)**

R = Me, n-Pr or NR_2 = morpholino
R^1 = H, Cl
R^2 = Ph, NHPh, NHMe

1.2.8. From Quinone Derivatives

A benzimidazole derivative (**1.138**, <5%) and a pyrido[1,2-a]benzimidazole (**1.139**, 23%) are formed together with dibenzenesulfonamides (**1.140a, b**) when the quinone dibenzenesulfonimides (**1.141a, b**, respectively) are exposed to irradiation by sunlight.[410] The formation of these benzimidazoles (**1.138**, **1.139**) is probably related to the conversion of analogous benzoquinones into benzoxazolines[411] but in addition must involve elimination of benzenesulfonic acid.

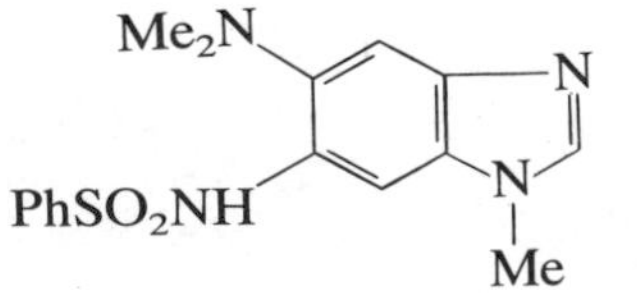

(1.138) 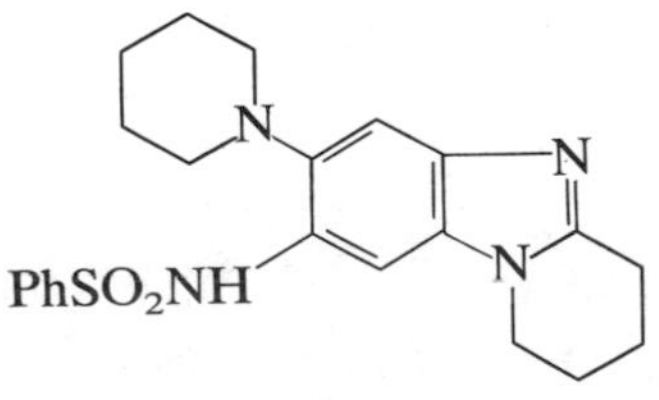   **(1.139)**

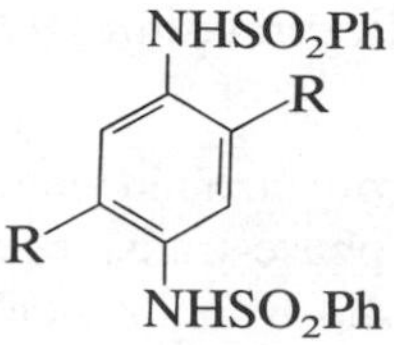

NHSO$_2$Ph

R

R

NHSO$_2$Ph

(**1.140**)

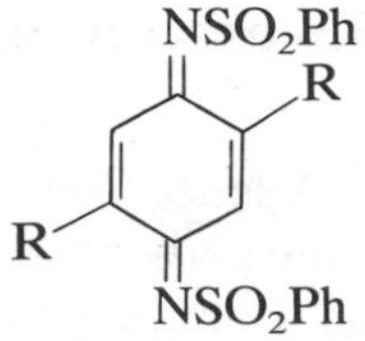

NSO$_2$Ph

R

R

NSO$_2$Ph

(**1.141**)

a, R = NMe$_2$
b, R = 1-piperidyl

A benzimidazole derivative (**1.143**) has also been isolated[412] from reductive cyclization of *o*-benzoquinonedibenzimide (**1.142**) by triphenylphosphine, a possible [412] mechanism for which is outlined in Scheme 1.3.

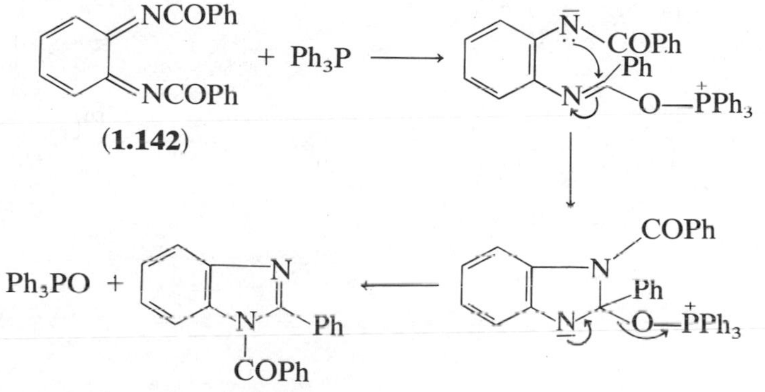

NCOPh

NCOPh

(**1.142**)

+ Ph$_3$P ⟶

N

COPh

Ph

N

O—PPh$_3$

Ph$_3$PO +

N

N

Ph

COPh

(**1.143**)

⟵

COPh

N

Ph

N

O—PPh$_3$

Scheme 1.3

1.2.9. From Heterocyclic Compounds

A number of methods have been devised in which benzimidazoles are synthesized via other heterocyclic compounds; in most cases stable heterocycles are used as starting materials, but in the first example in the following section an unstable diazirine is generated. Some of the procedures described in this section may appear somewhat esoteric but nevertheless may have synthetic value; in many cases neutral conditions are used, hence advantages over the normal *o*-arylenediamine procedures can be envisaged. Applications in this area include routes to commercially important alkylbenzimidazol-2-yl-carbamates and 2-(4-thiazolyl) derivatives.

Three-Membered Ring Compounds: (Intermediate) Diazirines

Dimethylvinylidenecarbene (**1.144**) reacts with azobenzene under conditions in which the carbene is generated by the phase-transfer technique[413] to give 1-phenyl-2-isobutenylbenzimidazole (**1.146**) in poor yield (10%);[83] an unstable intermediate diazirine (**1.145**) is envisaged in this process (see Scheme 1.4).

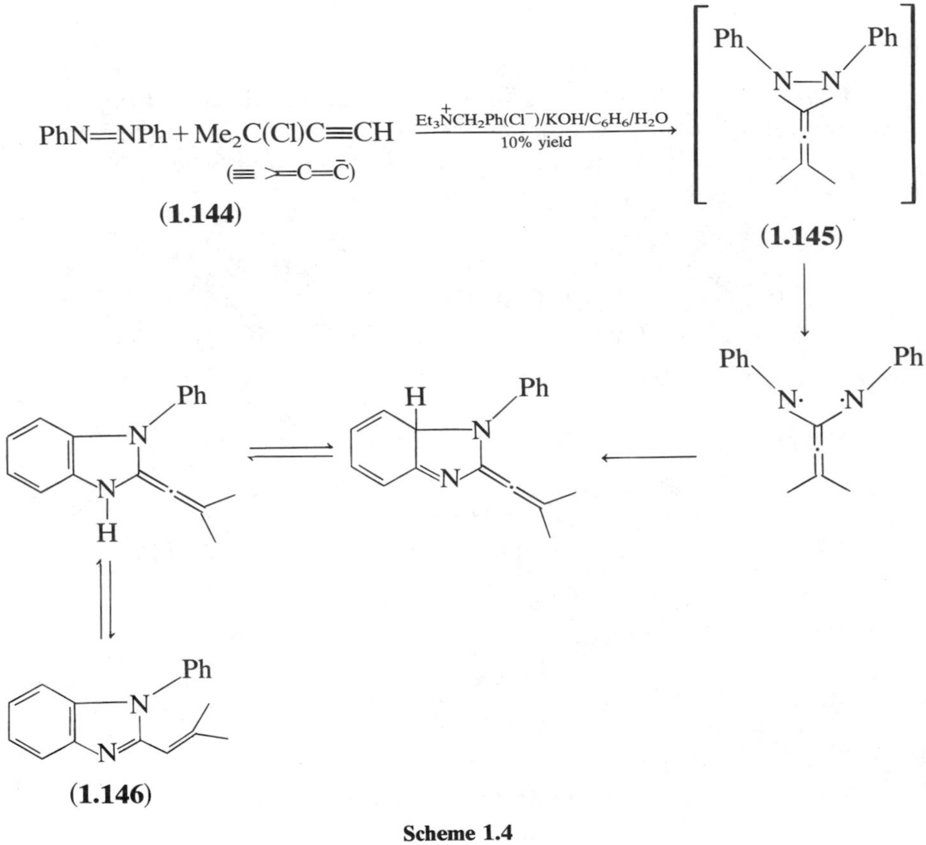

Scheme 1.4

Five-Membered Ring Compounds

IMIDAZOLES. Benzimidazoles are conventionally prepared by ultimate construction of the imidazole ring via o-arylenediamines (see section 1.2.1). The alternative approach in which preformed, suitably functionalized imidazoles, are used as starting materials has recently been described. Imidazole-5-carboxaldehyde derivatives (**1.147**) react with the Grignard reagents (**1.148**) to yield the alcohols (**1.149**); treatment of the latter with acetic acid/sodium acetate under reflux effects a single-step annulation to yield 1,2,6-trialkylbenzimidazoles (**1.150**) via hydrolysis of the acetal, electrophilic substitution, and aromatization.[414] In a simple modified procedure, the alcohols (**1.149**) can be converted via ketones (**1.151**) into secondary alcohols (**1.152**), which in turn are cyclized to 1,2,7-trialkylbenzimidazoles (**1.153**).[414]

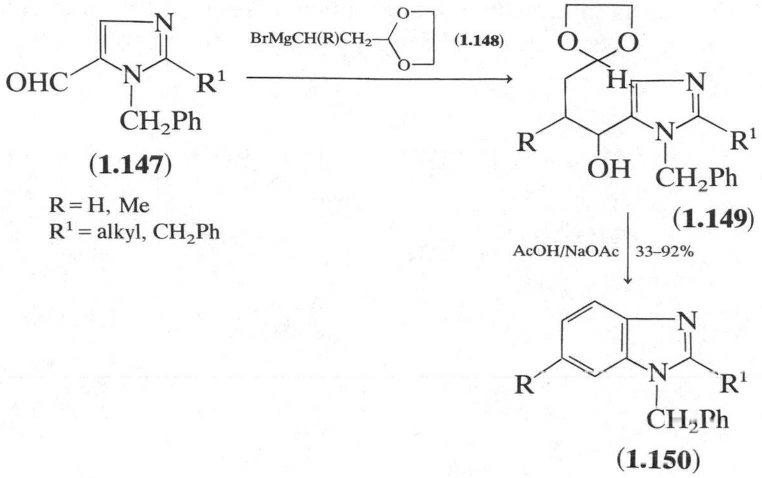

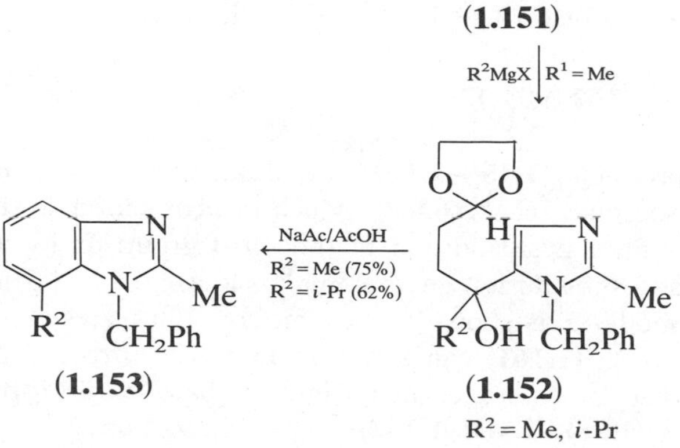

Sulfur extrusion from intermediate imidazothiepins is probably involved in the base-catalyzed reaction of 1,2-disubstituted imidazole-4,5-dicarboxaldehydes (**1.154**) with sulfides (**1.155a–c**) and sulfinyl derivatives (**1.155d**).[415] Poor yields (10–15%) of the benzimidazole (**1.156a**) are obtained from the sulfide (**1.155a**), but in contrast, yields of the dibenzoyl (**1.156b**, 80%) and dipivaloyl derivatives (**1.156c**, 79%) are high. Annulation by use of the sulfinyl derivative (**1.155d**) also proceeds in high yield (75%), whereas an intermediate imidazothiepin is isolated from the reaction of the analogous sulfone (**1.155e**) (see "Seven-Membered Ring Compounds" below).

(**1.154**)   (**1.155**)   (**1.156**)

a, R = OMe, X = S   a, R = OMe
b, R = Ph, X = S   b, R = Ph
c, R = CH₂COtBu, X = S   c, R = CH₂COtBu
d, R = OMe, X = SO
e, R = OMe, X = SO₂

A complex series of reactions (i–iii) occurs during ultraviolet photolysis of (dinitro-2′,4′-phenyl) 1-imidazoles (**1.157a–c**) in ethanol; the products contain benzimidazoles (**1.158**, **1.160**), benzimidazole *N*-oxides (**1.159**, **1.161**), or a benzimidazolone (**1.162**), depending on the structure of the imidazole derivative (**1.157**).[416]

i. **1.157a** $\xrightarrow{h\nu}$ **1.158** (50%)

ii. **1.157b** $\xrightarrow{h\nu}$ **1.159** (30%) + **1.160** (10%) + **1.161** (20%)

iii. **1.157c** $\xrightarrow{h\nu}$ **1.162** (90%).

The photoproducts (**1.158–1.162**) are thought to originate from intermediate nitrosoarenes (cf. **1.163**),[416] which in turn can arise from either an intramolecular photocycloaddition of the nitro group or by a more direct process of photosubstitution in the imidazole ring. [1,5] Rupture of the primary intermediate as shown in Scheme 1.5 illustrates a mechanism by which the *N*-oxide (**1.161**) can arise; an isomeric nitrosoarene formed by [1,2] rupture can be envisaged as an intermediate in the formation of the other products (**1.158–1.160**, **1.162**).

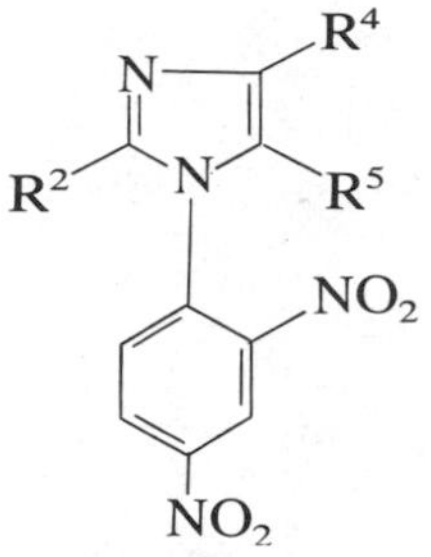

(1.157)

a, $R^4 = Ph$; $R^2 - R^5 = H$
b, $R^4 = R^5 = Ph$; $R^2 = H$
c, $R^2 = Ph$; $R^4 = R^5 = H$

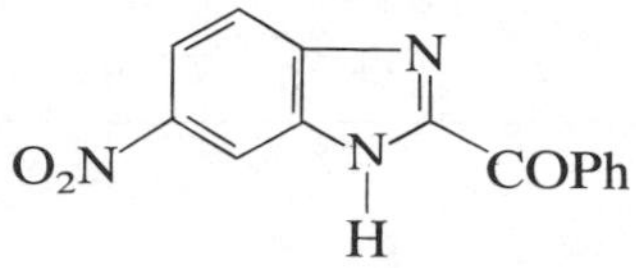

(1.158)

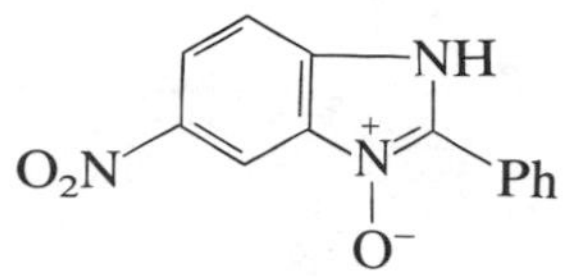

(1.159)

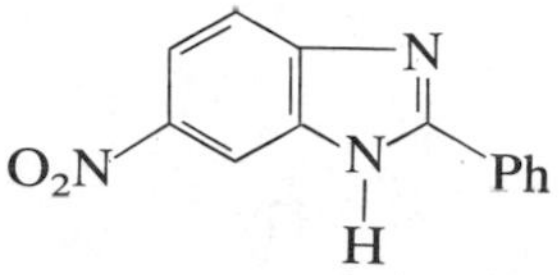

(1.160)

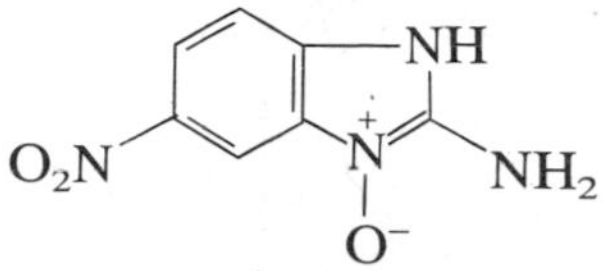

(1.161)

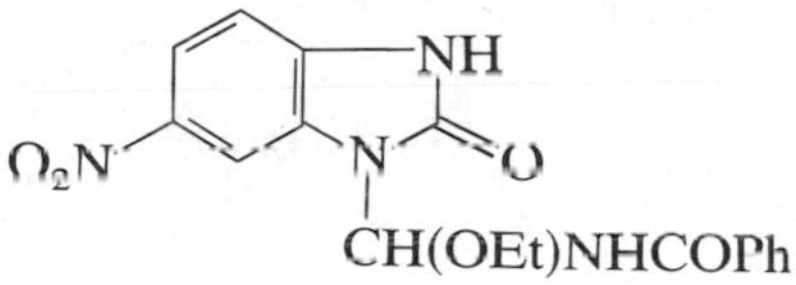

(1.162)

INDAZOLES. Benzimidazoles are among the products formed by the photolysis of indazoles, but the course of such reactions is markedly dependent on the position of the substituent in the heterocyclic ring.[417] Thus in the absence of a substituent, indazole (1.164) is converted into benzimidazole in poor yield together with trace quantities of 2-aminobenzonitrile (1.165); $N(2)$-alkylated indazoles (1.166) with fixed quinonoid chromophores are converted in very good yield into 1-alkylbenzimidazoles (1.167); and $N(1)$-alkylated indazoles (1.168), which have fixed benzenoid chromophores, by opening of the N—N bond and migration of hydrogen from C-3 to N-1 give 2-alkylaminobenzonitriles (1.169) in variable yield (10–60%).[417] The detailed mechanims of these indazole photorearrangement processes has not been elucidated, and it has proved impossible to detect intermediates in such reactions by photolysis at low temperatures[418] (cf identification of 2-hydroxyphenylisocyanide as an intermediate during the analogous low-temperature photolytic transformation of indoxazene into benzoxazole).[418]

OXADIAZOLES. 3,4-Diaryl-1,2,4-oxadiazol-5-ones (1.170) are converted[419] by pyrolysis at ~190–260° into 2-arylbenzimidazoles (1.171) in generally very good yields (see Table 1.12);[419b] reactions of this type are useful, since

(1.157)

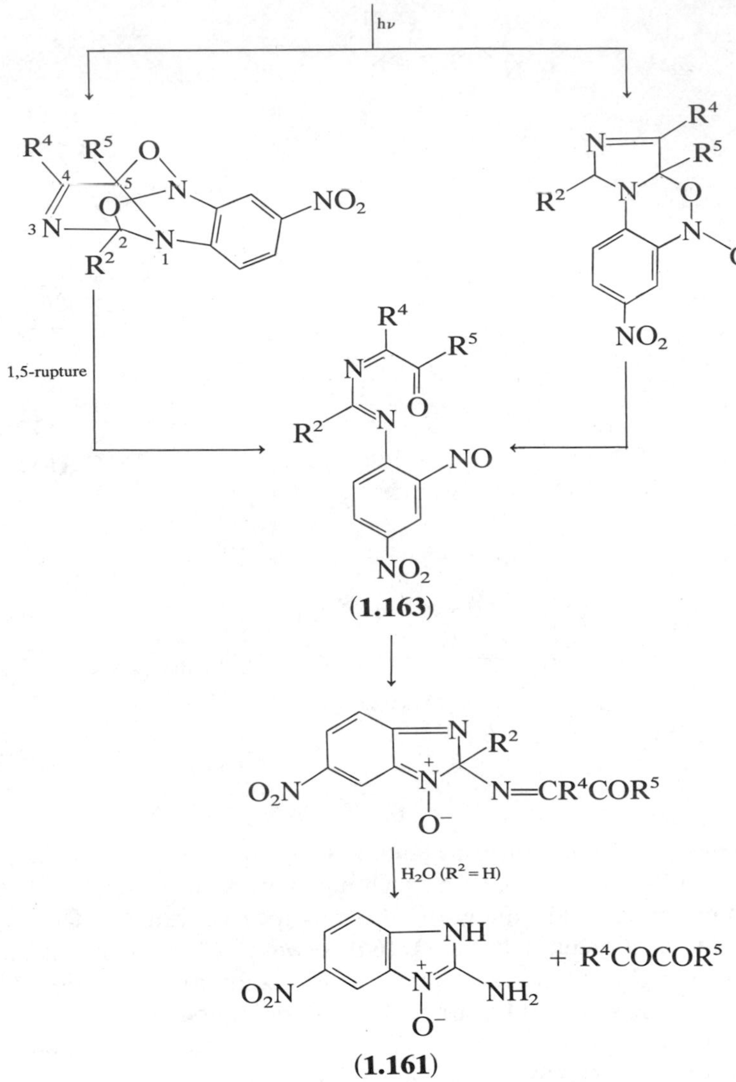

$h\nu$

R^4 R^5 O ... N ... NO$_2$

1,5-rupture

R^4 R^5 N O N—O·

NO$_2$

R^4 R^5 N O R^2 N

NO

NO$_2$

(1.163)

$$O_2N \quad \overset{N}{\underset{\underset{O^-}{N^+}}{\big|}} \overset{R^2}{\underset{}{}} N{=}CR^4COR^5$$

H$_2$O (R^2 = H)

$$O_2N \quad \overset{NH}{\underset{\underset{O^-}{N^+}}{\big|}} NH_2 \quad + \; R^4COCOR^5$$

(1.161)

Scheme 1.5

the starting materials (**1.170**) are easily synthesized from the reaction of N-hydroxy-N'-arylamidines with ethyl chloroformate.[419b] Subsequent modifications of the method (cf. **1.170** → **1.171**) include peroxide-initiated thermolysis[420] and photolysis[420,421] of oxadiazolones (**1.170**) or analogous thiones in dioxan solution. Procedures of this type, which presumably[421] occur via intermediate iminonitrenes, have been used successfully for the synthesis of a number of commercially important 2-(4-thiazolyl)-5-alkoxycarbonylamino["cambendazole"] derivatives[420] (cf. Chapter 10).

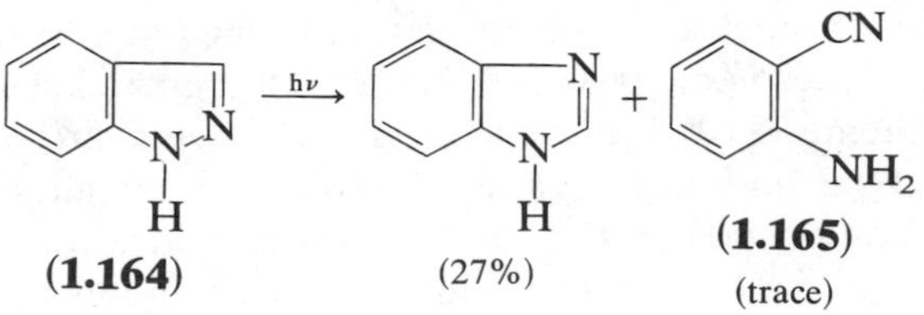

(1.164) → (1.165)

(1.164) (27%) (1.165)
(trace)

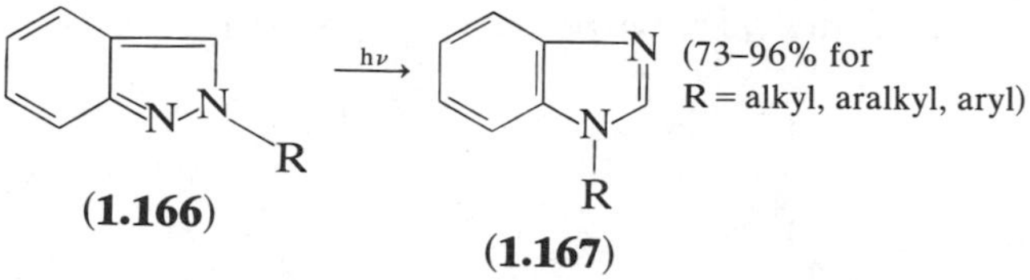

(1.166) → (1.167) (73–96% for R = alkyl, aralkyl, aryl)

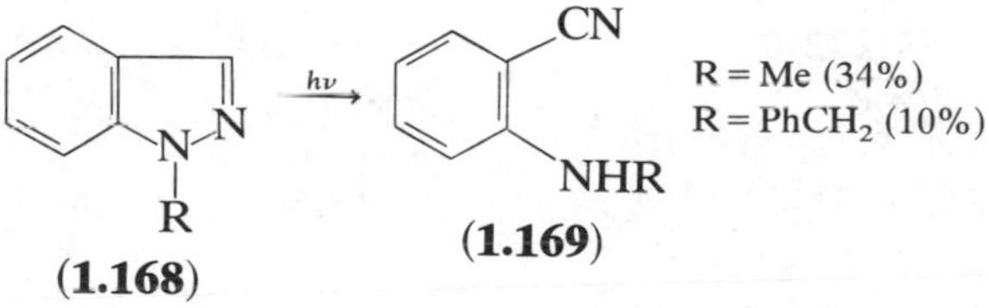

(1.168) → (1.169) R = Me (34%)
R = PhCH₂ (10%)

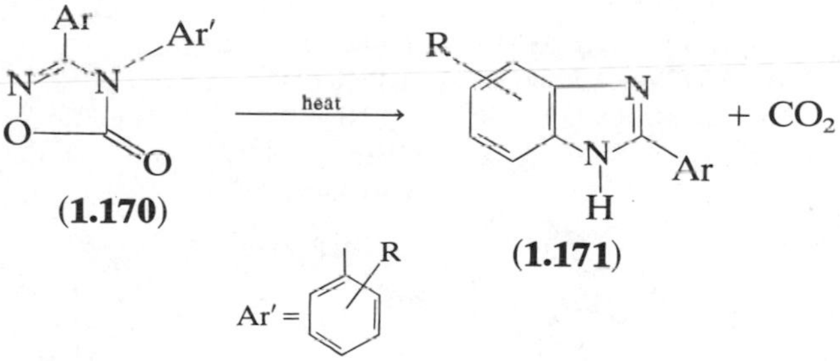

(1.170) → (1.171) + CO₂

Ar' =

TABLE 1.12. BENZIMIDAZOLES FROM PYROLYSIS OF 3,4-DIARYL-1,2,4-OXADIAZOL-5-ONES[419b]

Starting material (1.170)		Product benz-imidazole (1.171)	Decomposition temp. (°C)	Yield of (1.171) %
Ar	Ar'			
Ph	o-MeC$_6$H$_4$	2-Ph-4-Me	220	91
Ph	o-O$_2$NC$_6$H$_4$	2-Ph-4-NO$_2$	255	78
Ph	p-O$_2$NC$_6$H$_4$	2-Ph-5-NO$_2$	230	75
p-O$_2$NC$_6$H$_4$	Ph	2-(p-O$_2$NC$_6$H$_4$)	220	74
p-MeOC$_6$H$_4$	Ph	2-(p-MeOC$_6$H$_4$)	190	14

TRIAZOLES. Iminonitrenes are probably also intermediates in the photolytic transformation of mesoionic triazoles (**1.172**) into benzimidazoles (**1.173**) [cf. (i)–(iii) in Scheme 1.6], but only moderate yields are obtained in reactions of this type and little synthetic value can be anticipated; the formation of azo compounds (**1.174**) and carbamic acid esters (**1.175**) are competing processes.[422] The suggested mechanism[422] involving isomerization of a bicyclic intermediate (cf. **1.176**) to a triazolone (cf. **1.177**) accounts satisfactorily for all the reaction products (see Scheme 1.7).

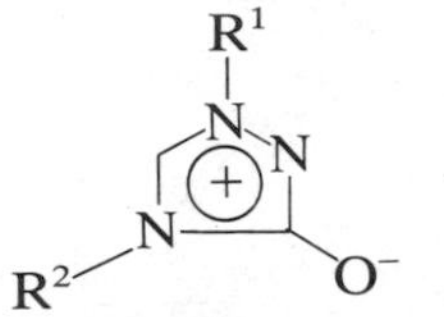

(1.172)

a, $R^1 = R^2 = Ph$
b, $R^1 = Ph$; $R^2 = p\text{-MeC}_6\text{H}_4$
c, $R^1 = p\text{-MeC}_6\text{H}_4$; $R^2 = Ph$

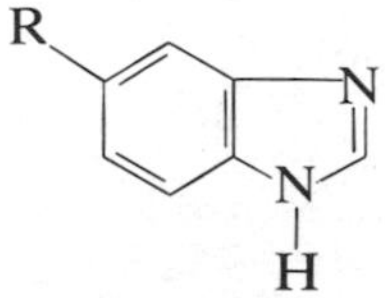

(1.173)

a, R = H
b, R = Me

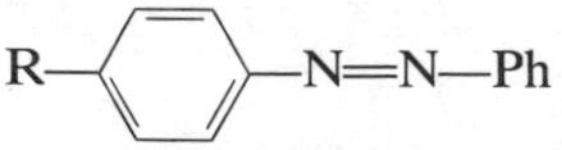

(1.174)

a, R = H
b, R = Me

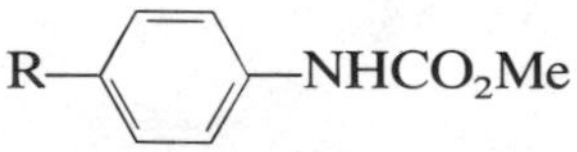

(1.175)

a, R = H
b, R = Me

(i)   **1.172a** → **1.173a** (18%) + **1.174a** (4%) + **1.175a** (25%)
(ii) **1.172b** → **1.173a** (32%) + **1.174b** (23%) + **1.175b** (44%)
(iii) **1.172c** → **1.173b** (44%) + **1.174b** (10%) + **1.175a** (32%)

Scheme 1.6

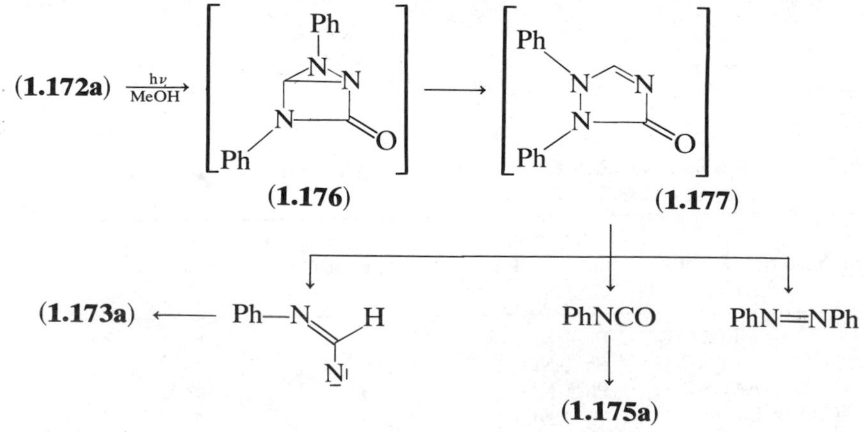

(1.172a) $\xrightarrow[\text{MeOH}]{h\nu}$

(1.176)   **(1.177)**

(1.173a) ⟵ Ph—N   H

PhNCO   PhN=NPh

(1.175a)

Scheme 1.7

TETRAZOLES. 2-Substituted benzimidazoles (**1.179**) are formed, albeit as minor products (8 to 19%), accompanying carbodiimides (**1.180**) from the pyrolysis of 1,5-diaryltetrazoles (**1.178**).[423,424] The maximum yield (19%) of 2-arylbenzimidazole is provided by the *p*-chloro derivative (**1.178a**), and this is thought[424] to imply that this substituent has a retarding effect on the competing carbodiimide reaction (cf. its behavior[425] in the analogous Beckmann rearrangement). This type of reaction has been extended[423] using 1-(1-naphthyl)-5-phenyltetrazole (**1.178**: R = H; Ar = 1-naphthyl), which provides the imidazonaphthalene derivative (**1.181**) in 25% yield but unfortunately cannot be applied to heterocyclic analogs (e.g., **1.178**; 2-pyridyl or 2-quinolyl for Ar). An investigation of the effect of catalysts show that metallic copper lowers the temperature required for the decomposition of 1,5-diphenyltetrazole by ~60°, but a complex mixture is formed from which no pure compounds can be isolated.[424]

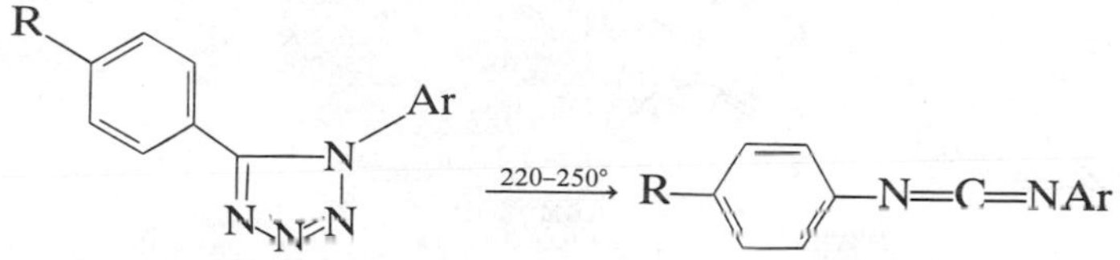

(**1.178**)

a, R = Cl
b, R = NO$_2$

$\xrightarrow{220-250°}$ R—⟨ ⟩—N=C=NAr

(**1.180**)

(70% for R = H)

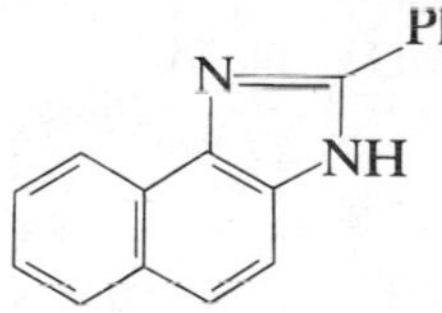

(**1.181**)

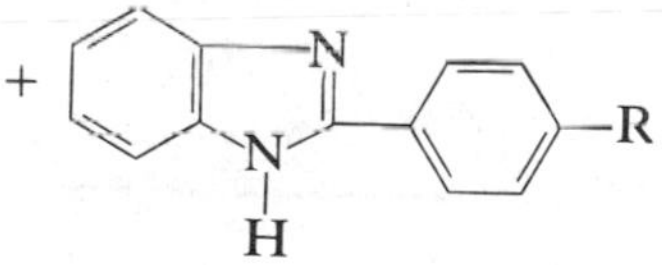

+ (**1.179**)

(14% for R = H)

N-Arylbenzimidoyl nitrenes are probably intermediates in these tetrazole pyrolyses, and the situation for arenes with blocked *ortho* positions is intriguing from a mechanistic viewpoint. The pathway for thermal decomposition of 1-xylyl-5-phenyltetrazole is outlined in Scheme 1.8, in which a series of consecutive sigmatropic migration processes[425a] is envisaged for the formation of isomeric 2-phenylbenzimidazole derivatives and the cyclopenta[*d*]pyrimidine; the major product is again the carbodiimide derivative.[425a]

The claim[426] that the 2-alkenylbenzimidazole derivative (**1.183**) is produced by pyrolysis of the tetrazole (**1.182**) at 160° is surprising, since formation of the isomeric carbodiimide (**1.184**) would be anticipated.

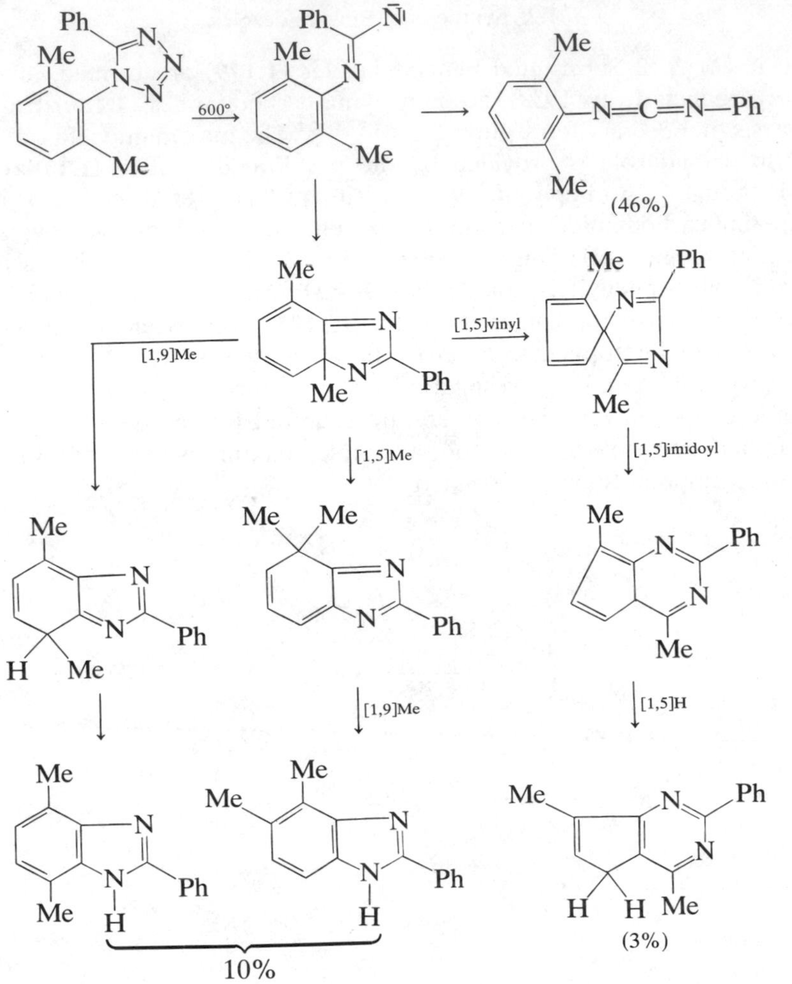

Scheme 1.8

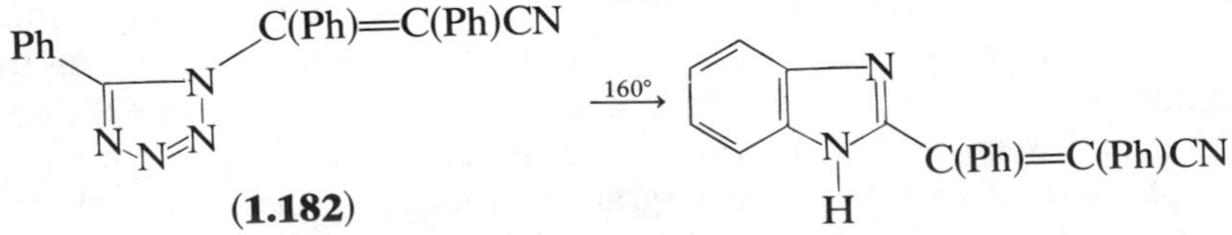

(**1.182**)

(**1.183**)

Ph—N=C=N—C(Ph)=C(Ph)CN

(**1.184**)

The formation of benzimidazoles by ultraviolet photolysis of tetrazoles was first noted by Moriarty and Kliegman,[427] who obtained 2-phenylbenzimidazole in 42% yield by photolysis of 1,5-diphenyltetrazole (**1.185a**). Analogous behavior, namely formation of 2-phenoxy-benzimidazole, was reported[428] for the photolysis of 5-phenoxy-1-phenyl-1H-tetrazole (**1.185b**) in acetonitrile but four compounds (**1.186–1.189**) are formed by photolysis of the tetrazole (**1.185b**) in benzene.[429] The formation of these products (**1.186–1.189**) can be accounted for by the mechanism shown in Scheme 1.9. Initial loss of nitrogen is followed by partial secondary photochemical decomposition of 2-phenoxybenzimidazole (**1.186**), and it has been confirmed[129] that the latter is transformed photochemically into a mixture of (**1.187**) and (**1.188**). It is noteworthy that this is the first

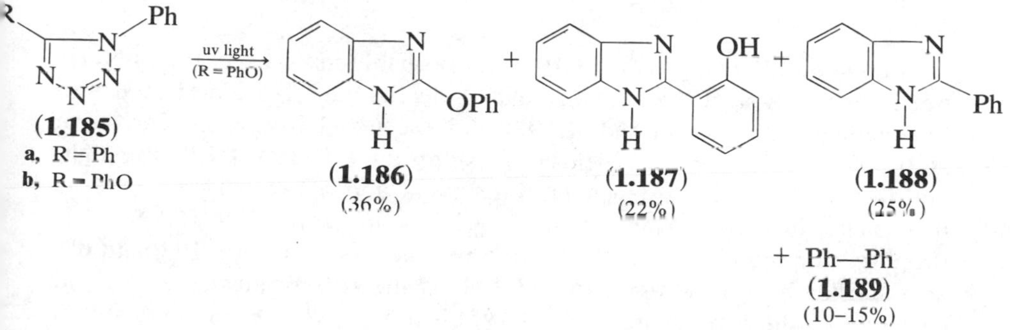

example of a photo-Fries rearrangement involving migration of a heterocyclic moiety, and extensions of this type of reaction to other heterocycles will be of interest from both a synthetic and a mechanistic viewpoint.

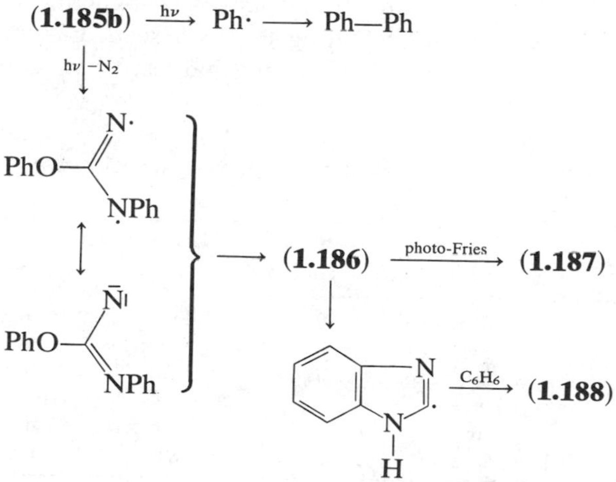

Scheme 1.9

Six-Membered Ring Compounds

QUINOXALINES. 2,3-Diphenylquinoxaline-1-oxide (**1.190**) is converted by ultraviolet photolysis into the oxazirino[2,3-*a*]quinoxaline derivative (**1.191**); in boiling aqueous methanol, the latter derivative (**1.191**) undergoes facile ring contraction to give 1-benzoyl-2-phenylbenzimidazole (**1.192**).[430]

(**1.190**)   (**1.191**)   (**1.192**)

Intermediates analogous to (**1.191**) are probably involved in the photolytic transformation of methylquinoxalin-2-ylcarbamate 1-oxide (**1.193**) into the methylcarbamates (**1.194, 1.195**) and the novel isocyanide derivative (**1.196**).[431] Conversions and yields of these products (**1.194–1.196**) are high, and the yield of each is dependent on solvent and pH (see Table 1.13). A mechanism involving an intermediate 3,1,5-benzoxadiazepine (**1.197**)[431] is proposed for the formation of all three products (see Scheme 1.10). In dry acetonitrile the formyl derivative (**1.194**) is the sole product, but a rapid solvolysis to the parent carbamate (**1.195**) is though to occur in methanol; in methylene chloride, which is known to contain hydrochloric acid, or in methanol containing sulfuric acid, ring-opening of an intermediate protonated species (**1.198**) is envisaged.

It is interesting that when the fungitoxic *N*-oxide (**1.193**) is applied to the roots of cucumbers the compound is translocated into the leaves without decomposition, but on exposure of the leaves to light the benzimidazole derivative (**1.195**) is present.[432] During use as a seed dressing against soil and seedborne fungal pathogens and in postharvest treatments, compound

TABLE 1.13. PHOTOLYSIS OF METHYLQUINOXALIN-2-YLCARBAMATE 1-OXIDE[431]

Solvent	Light Source	Time (hr)	Yield (%) of products[a]		
			(**1.196**)	(**1.194**)	(**1.195**)
CH$_2$Cl$_2$	Tungsten	50	110	0	0
CH$_2$Cl$_2$	uv	3	95	0	0
CH$_2$Cl$_2$/Et$_3$N	uv	3	0	0	103
MeCN	uv	5	0	105	0
MeOH	uv	3	0	0	105
10^{-3} M H$_2$SO$_4$/MeOH	uv	3	81	0	0

[a] Estimated from uv spectra; accuracy ±10%.

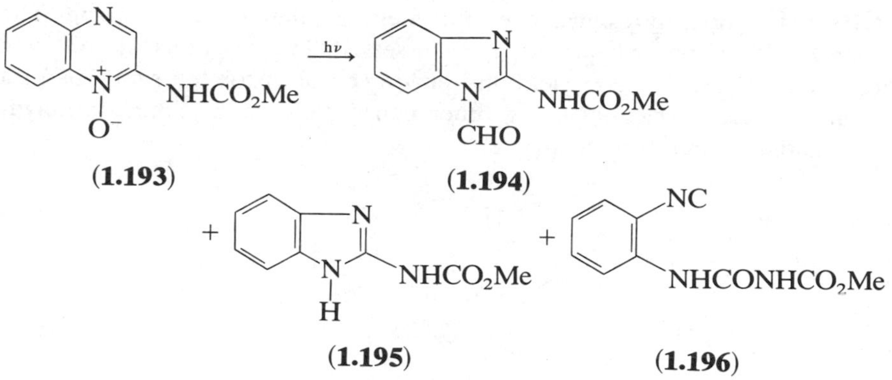

(**1.193**)

(**1.194**)

+

(**1.195**)

+

(**1.196**)

(**1.193**) was found to be inactive in the absence of light, which implies that the activity is due to the production of methylbenzimidazol-2-yl carbamate (**1.195**) [cf. the fungitoxic agent benomyl (**1.195**; N-CONHBu for N—H) described in Chapter 10].

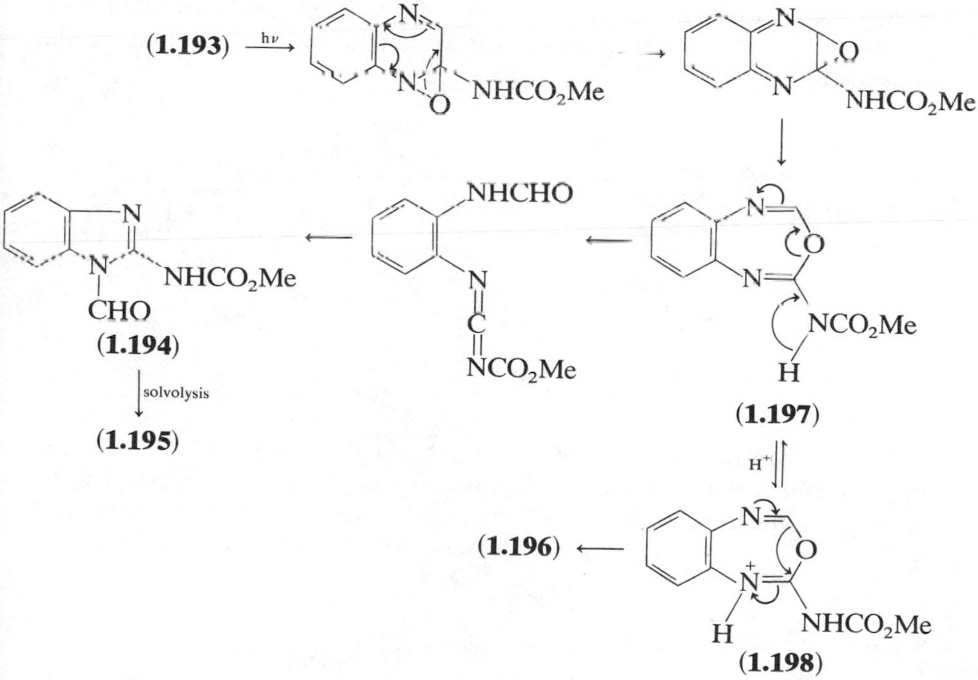

(**1.193**) $\xrightarrow{h\nu}$

(**1.194**)

↓ solvolysis

(**1.195**)

(**1.197**)

NHCHO

(**1.196**) ←

H^+

(**1.198**)

Scheme 1.10

s-TRIAZINES. Benzimidazole and its 1-methyl derivative (**1.200a, b**) have been obtained[433] in 100 and 50% yields, respectively, by allowing o-phenylenediamine or N-methyl-o-phenylenediamine to react with s-triazine

(**1.199**) at temperatures just over the melting point of the diamine. The scope of this approach has not been assessed, but reactions of this type generally proceed in good yield and have been used for the synthesis of a variety of heterocycles including other benzazoles, imidazolines, tetrahydropyrimidines, and purines.

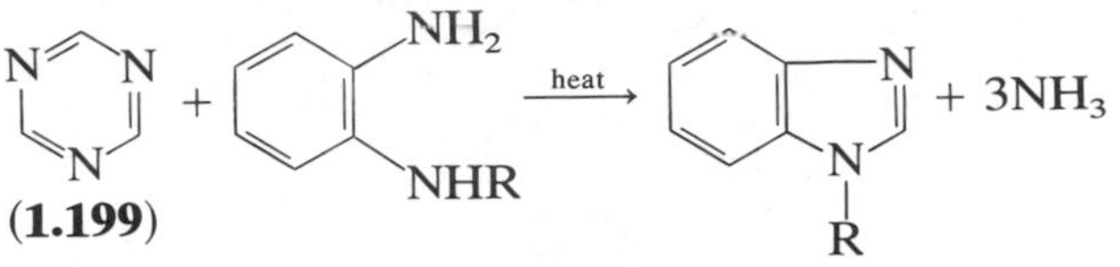

(**1.200**)

a, R = H (100%)
b, R = Me (50%)

BENZO-1,2,4-TRIAZINES. Anthelmintic 2-(4-thiazolyl)benzimidazoles (**1.202**) are formed by reduction of the benzotriazine-1-oxides (**1.201**) using a number of methods including zinc/acetic acid or platinum oxide in ethanol.[434] Reactions of this type presumably occur via intermediate o-aminoaryl-substituted amidines which would arise by reductive ring-opening of the triazine ring.

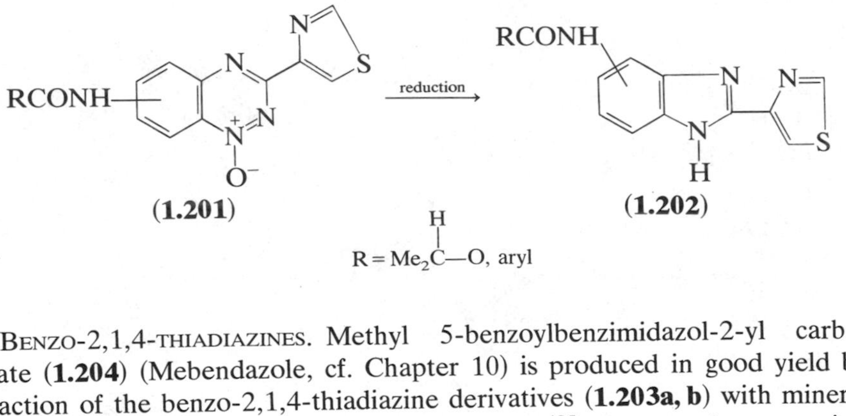

(**1.201**)

(**1.202**)

$$R = Me_2\overset{\overset{\displaystyle H}{|}}{C}-O, \text{ aryl}$$

BENZO-2,1,4-THIADIAZINES. Methyl 5-benzoylbenzimidazol-2-yl carbamate (**1.204**) (Mebendazole, cf. Chapter 10) is produced in good yield by reaction of the benzo-2,1,4-thiadiazine derivatives (**1.203a, b**) with mineral acid or with triphenylphosphine, respectively.[435] The starting materials

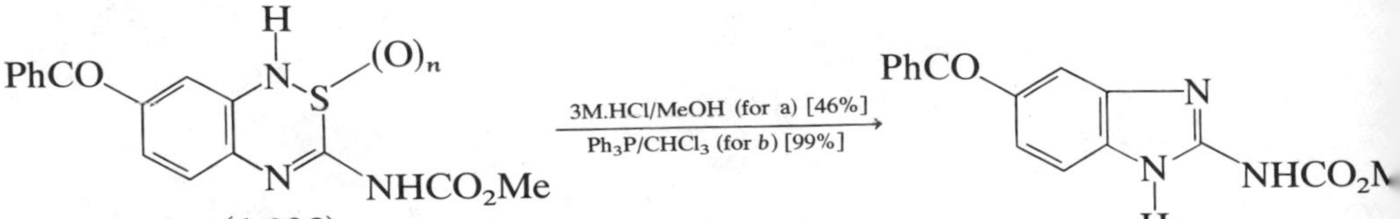

(**1.203**)

a, n = 1
b, n = 0

(**1.204**)

(**1.203**) are easily accessible from sodium dithionite reduction of *o*-nitroarylthiocarbamoylcarbamates (e.g., **1.205**),[435] and the method has been subsequently applied to the synthesis of Mebendazole analogs (**1.206a**[436] and **1.206b**[437]).

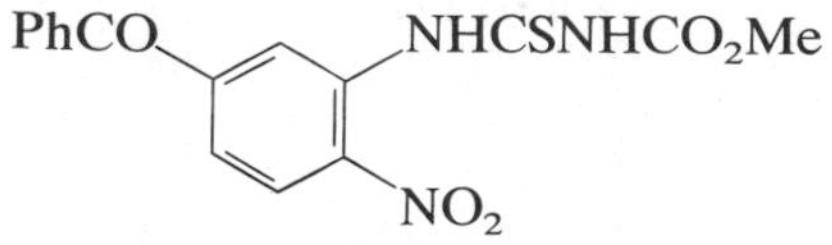

PhCO NHCSNHCO$_2$Me

NO$_2$

(**1.205**)

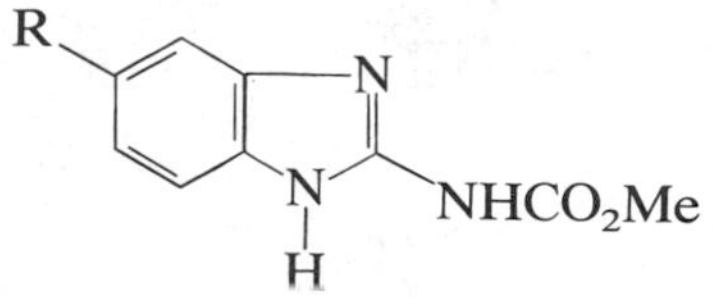

R

N

N
H

NHCO$_2$Me

(**1.206**)

a, R = ArX (X = O, S)
b, R = ArX (X = OSO$_2$, SO$_2$O)

Seven-Membered Ring Compounds

THIEPINO[5,4-*d*]IMIDAZOLES. In contrast to the behavior[415] of sulfide and sulfinyl analogs (see preceding section, "Five-Membered Ring Compounds"), the sulfone derivative (**1.207**) reacts with the 1,2-disubstituted imidazole-4,5 dialdehyde (**1.154**) to give an isolable thiepino[5,4-*d*]imidazole (**1.208**); the latter compound (**1.208**) is slowly decomposed under reflux in sulfolan to give the benzimidazole (**1.209**) in moderate yields.[415]

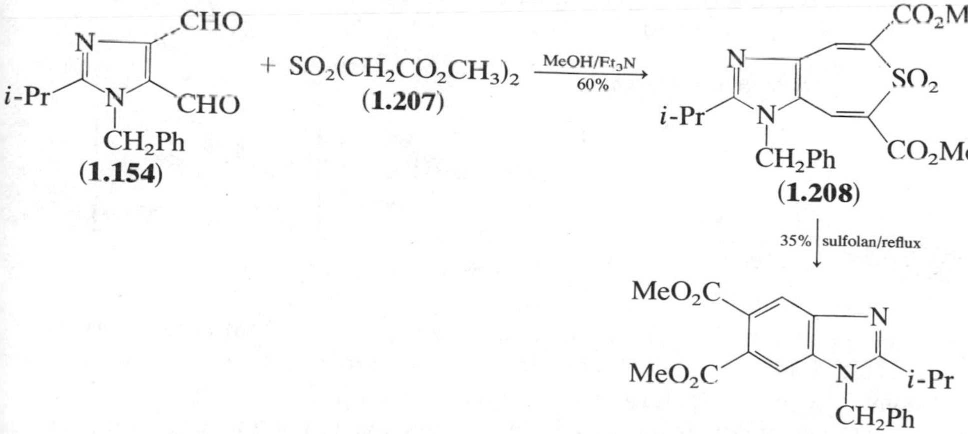

i-Pr

N

N
CH$_2$Ph

CHO

CHO

(**1.154**)

+ SO$_2$(CH$_2$CO$_2$CH$_3$)$_2$
(**1.207**)

$\xrightarrow[60\%]{\text{MeOH/Et}_3\text{N}}$

i-Pr

N

N
CH$_2$Ph

CO$_2$Me

SO$_2$

CO$_2$Me

(**1.208**)

35% | sulfolan/reflux

MeO$_2$C

MeO$_2$C

N

N
CH$_2$Ph

i-Pr

(**1.209**)

BENZO[2,3][1,4]DIAZEPINES. The diazepin derivative (**1.210**) is converted in a single step reaction into the α-(2-benzimidazolyl)valerolactone derivative (**1.212**) in good yield;[438] evidently, hydrolytic opening of the azepin ring is followed by cyclization of an intermediate amide (**1.211**).

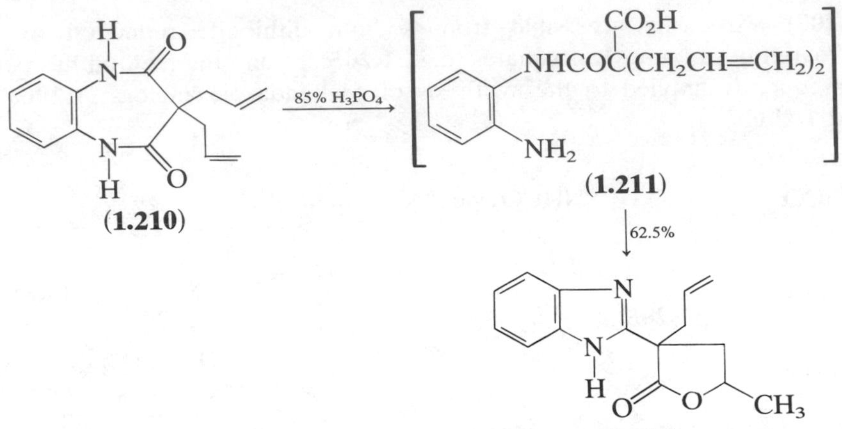

(1.210) $\xrightarrow{\text{85\% H}_3\text{PO}_4}$ **(1.211)**

62.5%

(1.212)

1.3. PHYSICOCHEMICAL STUDIES

1.3.1. Crystal and Molecular Structure

The crystal and molecular structures of benzimidazole[439] and its 2-ethyl[440] and 2-(4-thiazolyl)[441] derivatives have been determined. The structures of the molecular complexes, 1,3-dimethylbenzimidazolium (tetracyanoquinodimethane) **(1.213)**[442] and benzimidazole benzimidazolium tetrafluoroborate **(1.214)**,[443] as well as the organometallic complex iron cyclopentadienyldicarbonyl benzimidazolium tetraphenylborate **(1.215)**,[444] have also been elucidated.

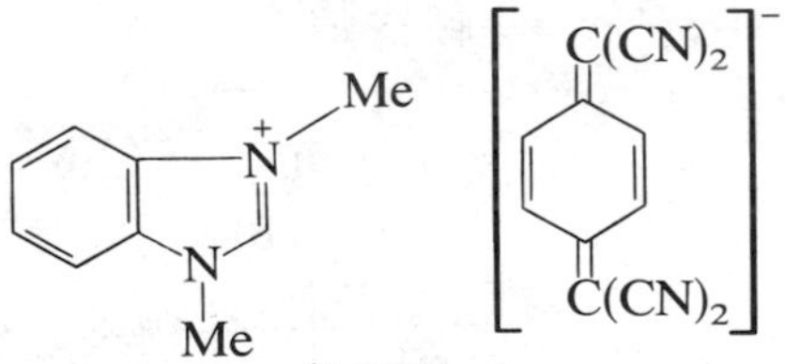

$[BI(H)BI]^+BF_4^-$
BI = benzimidazole
(1.214)

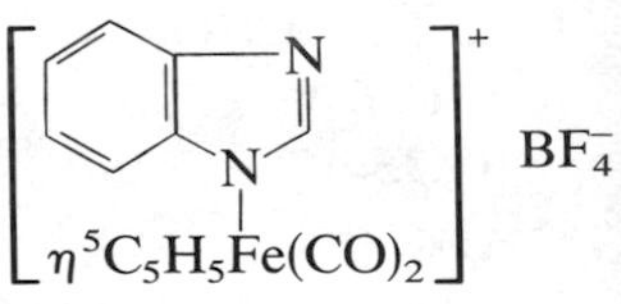

(1.213)   **(1.215)**

The detailed molecular structure of benzimidazole **(1.216)** is summarized in Table 1.14, and crystal data for benzimidazole and its derivatives is summarized in Table 1.15. In benzimidazole **(1.216)**,[439] each molecule is connected to two neighbors, generated by a twofold screw axis along the x-axis, via a hydrogen bond of 2.00(4) Å from H6 to N2. A similar crystal structure is also evident in 2-ethylbenzimidazole,[440] for which there are two independent molecules in the unit cell. The NH⋯⋯N bond angle for hydrogen bonding is 157.7° for molecule 1 and 156.4° for molecule 2. This is comparable to other benzimidazoles,[439,441] but is slightly more bent than in imidazole.[445]

TABLE 1.14. MOLECULAR STRUCTURE OF BENZIMIDAZOLE[439]

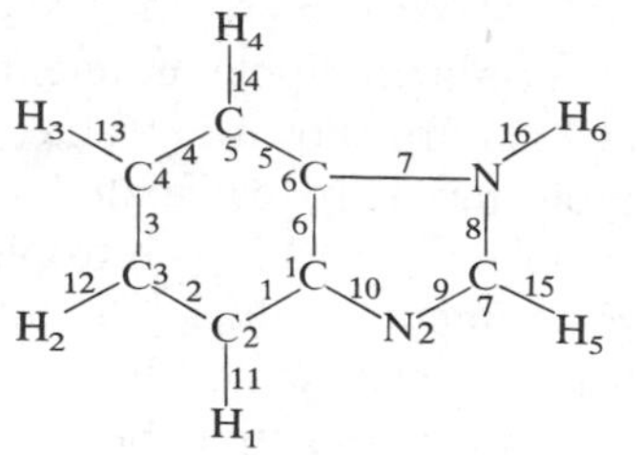

(1.216)

Bond distance (Å)

1 = 1.389(4)	6 = 1.392(4)	11 = 0.94(4)
2 = 1.386(4)	7 = 1.372(4)	12 = 1.02(4)
3 = 1.401(5)	8 = 1.346(4)	13 = 1.07(4)
4 = 1.378(5)	9 = 1.311(5)	14 = 0.98(4)
5 = 1.401(4)	10 = 1.395(3)	15 = 1.03(4)
		16 = 0.90(4)

Angles (°)

1, 2 = 117.8(3)	5, 7 = 131.9(3)	1, 11 = 117(2)	4, 14 = 121(2)
1, 6 = 120.6(3)	6, 7 = 105.8(2)	2, 11 = 126(2)	5, 14 = 123(2)
1, 10 = 130.0(2)	6, 10 = 109.5(2)	2, 12 = 117(2)	7, 16 = 126(2)
2, 3 = 120.9(3)	7, 8 = 106.6(3)	3, 12 = 122(2)	8, 16 = 127(2)
3, 4 = 122.3(3)	8, 9 = 114.0(3)	3, 13 = 111(3)	8, 15 = 121(2)
4, 5 = 116.1(3)	9, 10 = 104.2(2)	4, 13 = 125(3)	9, 15 = 125(2)
5, 6 = 122.4(3)			

TABLE 1.15. CRYSTAL DATA FOR BENZIMIDAZOLE AND ITS DERIVATIVES

Compound	Crystal type	a (Å)	b (Å)	c (Å)	Dm (g cm^{-3})	Z	Space group	R	Ref.
Benzimidazole (1.216)	Monoclinic	6.940	13.498	6.808		4	$P2_1nb$	0.04	439
2-Ethylbenz-imidazole	Monocyclic						$P2_1/c$	0.072	440
2-(4-Thiazolyl)-benzimidazole	Orthorhombic	17.052	10.998	10.030	1.414		$Pbca$	0.066	441
(1.213)	Triclinic	8.726	13.266	7.870	1.303		$P\bar{1}$	0.042	442
(1.214)	Monoclinic	5.797	9.231	27.480	1.45	4	$P2_1/n$	0.050	443
(1.215)	Monoclinic	13.626	25.410	10.088	1.28	4	$P2_1/n$	0.079	444

In the benzimidazolium derivative (1.214),[443] the planar rings are inclined at 7.6° to each other and linked by a hydrogen bond [2.787(3) Å] with the proton in an asymmetric single minimum. The asymmetric siting of the hydrogen atom produces characterizing effects on the bond lengths and angles to which it is attached.

1.3.2. Dipole Moments

The experimentally determined dipole moments of a number of benzimidazole derivatives (**1.217**) are shown in Table 1.16.[446-450] It is interesting to note that the dipole moments calculated vectorially for the planar conformations of 2-*o*-nitrophenyl- and 2-*m*-nitrophenylbenzimidazole differ considerably from the experimental values.[449] This is easily rationalized in terms of a steric effect for the *ortho* isomer, but it is apparent that there is a low stabilization energy from conjugation between the rings through the C_2-aryl bond for the *meta* compound. It is suggested that this low-resonance energy is due to counter effects arising as a result of the electron-withdrawing nature of both the imidazole and nitro-aryl rings.[449]

It may be noted from Table 1.16 that values of dipole moments in dioxan are higher than values measured in benzene solution (cf. the cases of 1-propyl,[448] 2-methylthio,[450] and 1-methyl-2-methylthio[450] derivatives). In dioxan solution the solute forms hydrogen bonds to solvent molecules, and this reduces any tendency to solute association. The increment in dipole moment from such complex formation (ca. 0.5D for 2-methyl-thiobenzimidazole)[450] may be compared with values in the range, 0.1–0.4D, for substituted pyrroles.[451]

TABLE 1.16.   DIPOLE MOMENTS OF
BENZIMIDAZOLE DERIVATIVES

(**1.217**)

Compound (**1.217**)		Dipole		
R^1	R^2	moment (D)	Solvent	Ref.
H	H	4.03	—	446
H	H	3.99	Dioxan	447
Me	H	4.04	—	446
Pr	H	3.72	C_6H_6	448
Pr	H	3.95	Dioxan	448
CH_2Ph	H	3.37	C_6H_6	448
Ph	H	3.37	C_6H_6	448
$(O_2N)_2C_6H_3$	H	3.17	C_6H_6	448
m-$O_2NC_6H_4$	H	3.45	Dioxan	449
o-$O_2NC_6H_4$	H	5.91	Dioxan	449
H	m-$O_2NC_6H_4$	3.24	Dioxan	449
Et	m-$O_2NC_6H_4$	4.59	Dioxan	449
H	SCH_3	2.50	C_6H_6	450
H	SCH_3	2.99	Dioxan	450
CH_3	SCH_3	2.51	C_6H_6	450
CH_3	SCH_3	2.58	Dioxan	450

Theoretical calculations of the dipole moments of benzimidazole and its derivatives have been reported.[452-454] It has been shown that an additive valence-optical scheme can be used to calculate the polarizability tensors and dipole moments of benzimidazole.[447]

1.3.3. Spectroscopic Properties

Infrared Spectra

Systematic studies of the infrared spectra of benzimidazoles have been carried out by Morgan[455] and by Rabiger and Joullié.[456,457]

Spectra of simple alkyl and perfluoroalkyl benzimidazoles in the solid phase are characterized by a series of strong, broad bands in the region, 2400–3200 cm^{-1};[455] they have no band in the region normally associated with N—H stretching frequency. On the other hand, in solution this broad absorption is replaced by a single sharp band near 3400 cm^{-1} which may be ascribed to the N—H stretching vibration. By examination of 1-deuterobenzimidazole in the solid phase, it has been shown that the strong bands near 2400–3200 cm^{-1} are best assigned to strong hydrogen bonds of the type N—H·····N.

Characteristic bands in the solution and solid phase spectra of selected derivatives are shown in Table 1.17. By reference to these data and also to extensive studies of Rabiger and Joullié,[456,457] a number of assignments can be made: these are summarized in Table 1.18. The 1650–1500 cm^{-1} region is a particularly characteristic one for benzimidazoles. All substituted derivatives have bands in this region that vary in position and intensity with the nature and position of the substituent.

TABLE 1.17. CHARACTERISTIC BANDS IN THE INFRARED SPECTRA OF BENZIMIDAZOLES

Benzimidazole derivative	N—H band (in CH₂Cl₂)	Other bands (nujol or fluorolub mull)								
Unsubstituted	3460	1621 w	1605 w 1591 m	—	1500 w	1351 w	—ᵃ	—ᵃ	—ᵃ	—ᵃ
Me	3460	1628 m	1592 w	1562 m	1512 w	—	—	—	—	—
Me	3460	1613 w	1595 mw	—	—	—	—	—	—	—
Me	3460	1628 w	1590 w	—	—	1345 w	—	—	—	—
-Me₂	3460	—	—	—	—	—	—	—	—	—
CF₃	3430	1625 w	1597 mw	1555 m	1501 mw	1329 ms	1192	1172	1143 1132	—
CF₃	3440	1633 mw	1604 w	—	1501 w	1335 s	—	1175	1125	1100
Me-4-CF₃	3450	1624 w	1608 mw	1550 m	1510 m	1340 ms	1195	1165	1127	1119
Me-2-CF₃	3440	1625 mw	1610 mw	1551 m	1514 mw	1325 s	1182	1170	1142	—

ᵃ Only strong bands in this region are quoted.

TABLE 1.18. ASSIGNMENTS OF CHARACTERISTIC
BANDS IN THE INFRARED SPECTRA OF
BENZIMIDAZOLES[455-457]

Region of Absorption (cm^{-1})	Assignment
2400–3200	Intramolecular hydrogen bonding
3390–3460	N—H stretch
1650–1500	C=C and C=N stretch
1400–1500	Skeletal in-plane vibrations
1000, 960	Benzenoid ring breathing modes[a]
880, 760	Heterocyclic ring breathing modes

[a] see also D. G. O'Sullivan, *Spectrochim. Acta*, **16,** 764 (1960).

In general, groups attached to the rings of benzimidazole show their normal characteristic bands. However, some of these vibrations may be modified by the imidazole ring if a strong electronic interaction occurs. For example, in 2-acetylbenzimidazole the carbonyl absorption at 1664 cm^{-1} [457] is typical of $\alpha\beta$-unsaturated ketones.

Infrared spectroscopy has been used specifically for a number of problems in benzimidazole chemistry, particularly in relation to the phenomenon of hydrogen bonding.[458-461] By a detailed study of the intensities of bands due to free and associated NH groups of a series of 2-trifluoromethyl derivatives it has been concluded that there is no intramolecular hydrogen bonding to the trifluoromethyl moiety.[459] The nature of intermolecular hydrogen bonds between benzimidazole and a number of proton acceptors such as benzoyl peroxide and furan has been studied. From absorbance measurements on the N—H band at 3480 cm^{-1}, a good correlation was found between diminution of the frequency of the absorption band and an enthalpy value corrected to refer only to the formation of the hydrogen bond.[460] Infrared data has shown that 2-benzimidazolyl hydrazones and formazans are associated by intermolecular hydrogen bonding in solution[462] and that the sodium salts of 1-alkyl-2-aminobenzimidazoles exist in the amino form in the crystalline state.[463]

The far-infrared spectrum of benzimidazole has been measured by means of the internal reflection technique, but no structural information has been adduced from this study.[464]

Ultraviolet and Fluorescence Spectra

ULTRAVIOLET SPECTRA. Detailed studies of the ultraviolet spectra of benzimidazoles have been made,[457,465,466] and selected data are collected in Table 1.19. From these and related spectra[457,465,466] it is evident that the absorption pattern of benzimidazole resembles that of a substituted benzene derivative; the short- and long-wavelength absorption bands correspond to transitions in the imidazole and aryl rings, respectively.

TABLE 1.19. CHARACTERISTIC BANDS IN THE ULTRAVIOLET SPECTRA OF BENZIMIDAZOLES[457,466]

(1.218)

Compound (1.218)

R	R¹	R²	R³	R⁴	R⁵	λ_{max}^{EtOH}(nm); log ε	λ_{max}^{HCl}(nm); log ε^a	λ_{max}^{NaOH}(nm); log ε^a
H	H	H	H	H	H	280 (3.89) 272 (3.91) 243 (3.80)	274 (3.91) 268 (3.92) 235 (3.61)	277 (3.75) 271 (3.74) 240 (3.63)
H	H	H	H	i-Pr	H	280 (3.84) 272 (3.86) 241 (3.73)	274 (3.85) 267 (3.87)	278 (3.65) 270 (3.63) 240 (3.49)
H	H	H	H	COCH₃	H	300 (3.92) 235 (3.65)	300 (3.76) 275 (3.83) 267 (3.77) 234 (3.64)	320 (3.97) 237 (3.94)
H	H	H	H	H	Et	281 (3.64) 274 (3.66) 266 (3.62) 254 (3.78) 249 (3.78)	281 (3.67)ᵇ 275 (3.74) 267 (3.74) 254 (3.84) 248 (3.83)	—
H	OEt	H	H	H	H	290 (3.71) 286 (3.74) 243 (3.58)	283 (3.84)	283 (3.76) 241 (3.49)
H	NO₂	H	H	H	H	306 (3.89) 234–237 (3.99) 286–289 (3.79) 243–246 (3.57)	284 (3.93)	286–289 (3.71)
OMe	H	H	OMe	H	H	251 (3.77)	263–266 (3.79)	2.49 (3.58)

ᵃ The hydrochloric acid and sodium hydroxide were 0.01 N.
ᵇ Refers to the hydrobromide salt.

The spectra of 2-alkyl derivatives are similar to benzimidazole, but substitution of alkyl groups in the aryl ring causes small bathochromic shifts of all bands. An electron-withdrawing group such as acetyl in the 2-position modifies the spectrum considerably: an intense band appears at 300 nm, and the 240-nm band shows a hypsochromic shift. 5,6-Dimethoxy derivatives show loss of fine structure, but 4- and 4,7-dimethoxy substituents cause drastic changes such that the characteristic benzimidazole spectrum is lost. In the spectra of 4- and 5-nitro derivatives a new band appears near 300 nm and a regression of the shorter wavelength band occurs. In contrast to the effects of other substituents (e.g., methyl[467]), a larger bathochromic shift is noted for the 4-isomer. Such behavior is rationalized by Leandri and coworkers[465] by consideration of the compounds as a nitrobenzene system slightly modified by different substitution patterns (cf. an earlier suggestion[468] that chelation might be involved).

In dilute hydrochloric acid solution the benzimidazole ring is protonated to form the appropriate benzimidazolium ion, and this is usually manifested by small hypsochromic shifts in spectra compared to those in ethanol; an exception to this general rule is noted for 4,7-dimethoxy derivatives.[457] Conversely, in dilute basic solution where salt formation should occur, it might be expected that bathochromic shifts would be observed. This is certainly the case for 4- and 5-nitro- and 2-acetylbenzimidazole, but in most cases only very small shifts occur, although some loss of fine structure is observed.

The ultraviolet spectra have been recorded and discussed in detail for the following compounds: 2-hydroxyaryl,[469] 2-aryl,[470] 1-vinyl,[471] 5-(5-nitrofurfurylideneamino),[472] 2-furyl,[218a] and 2-(phenyl)furyl[218b] derivatives.

The vapor absorption spectrum of benzimidazole near 2850 Å has been analyzed.[473] This investigation indicates that the strongest band is the origin at 36,023 cm^{-1}. By comparison of the spectrum with those of closely related heterocycles and with aromatic compounds, it has been concluded[473] that the 2850 Å transitions in benzimidazole are localized within the aryl ring.

FLUORESCENCE SPECTRA. Fluorescence spectra of benzimidazole and some of its derivatives have been recorded in neutral and acidic solution (see Table 1.20).[474–478]

Neutral molecules of benzimidazole show an emission band near 290 nm, whereas the cation gives rise to a broad structureless band with a maximum near 360 nm.[476] Derivatives of benzimidazole such as 2-alkyl compounds also show an emission maximum near 300 nm, but a more complex behavior

TABLE 1.20. CHARACTERISTIC FLUORESCENCE SPECTRA OF
            BENZIMIDAZOLES

Benzimidazole derivative	pK$_a$ of fluorescence activation	pK$_a$	Maximum fluorescence quantum efficiency	Fluorescence emission maximum (nm)	Ref.
Unsubstituted	5.3, 12.4	5.5, 12.3	0.10 (pH 2) 0.16 (pH 9)	360 (pH 2) 290 (pH 9)	476
2-Me	—	—	—	280, 360 (cation) 293 (neutral molecule)	477
2-Et	—	—	—	365 (pH 2–5) 295 (pH 8–10)	478
5-Me	—	—	—	365 (pH 2–4) 303 (pH 8–10)	478
5-Cl	—	—	—	295 (cation) 304 (neutral molecule)	477

TABLE 1.21.   CLASSIFICATION OF FLUORESCENCE
SPECTRA OF BENZIMIDAZOLIUM
CATIONS[477]

Benzimidazolium cation derivatives	pK_a	Emission band		Class[a]
		ca 290 nm	ca 360 nm	
Unsubstituted	5.5	−	+	A
2-Me	6.1	+	+	B
5-Me	5.7	−	+	A
5-Cl	4.9	+	−	C

[a] A denotes an emmision spectrum at higher wavelength.
B denotes an emission spectrum consisting of two bands.
C denotes an emission spectrum at lower wavelength.

is evident in acid solution where benzimidazolium ion formation occurs; for these cases emission bands near 290 and 360 nm may or may not occur depending on the type and position of the substituent[477] (see Table 1.21). From polarization measurements it has been concluded that for benzimidazole, fluorescence occurs from the ^1L_a state;[475] the observed spectra described previously are thus critically dependent on the separation between the ^1L_a and ^1L_b states.

The fluorescence lifetime $(0.55 \pm 0.05$ nsec) of a benzimidazolium carbocyanine dye (**1.219**) has been measured in methanol,[479] and the phosphorescence lifetime of benzimidazole is 2.3 sec on an emission maximum at 406 nm.[480] Phosphorescence polarization measurements indicate that the emission vector is perpendicular to that of absorption.[475]

(**1.219**)

Nuclear Magnetic Resonance Spectra

^1H SPECTRA. Following early proton magnetic resonance studies of benzimidazole,[481] detailed analyses of a variety of derivatives have been presented;[482–488] selected data are collected in Table 1.22. Because of the tautomeric equilibrium in the heterocyclic ring of benzimidazole, the aromatic ring proton resonances appear as AA′BB′ multiplets in NMR spectra.[481] Only small differences in chemical shift values are observed when the 1- and 2-positions of benzimidazole are substituted by methyl[482] and alkylsulfonyl[483] substituents, respectively. From an extensive study of 1-methyl-2-chlorobenzimidazoles bearing aryl ring substituents, it is concluded that

TABLE 1.22. CHARACTERISTIC PROTON MAGNETIC RESONANCE SPECTRA OF BENZIMIDAZOLES

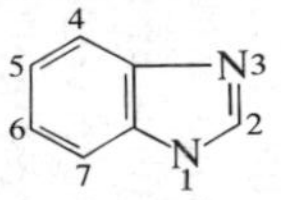

Compound (1.220)

N_3, N_1 (1.220) — benzimidazole, positions 4, 5, 6, 7 and 2.

Compound (1.220)						Solvent	Chemical shifts (δ-values from TMS)						Coupling constants (hertz)	Ref.
1	2	4	5	6	7		1	2	4	5	6	7		
H	H	H	H	H	H	Acetone-d_6	—		7.70	7.26	7.26	7.70	$J_{45}=J_{67}=8.2$ $J_{56}=7.1$ $J_{46}=J_{57}=1.4$ $J_{47}=0.7$	481
H	H	H	H	H	H	DMSO-d_6	—	8.29	7.65	7.22	7.22	7.65	$J_{45}=J_{67}=8.0$ $J_{56}=7.0$ $J_{46}=J_{57}=1.1$ $J_{47}=0.7$	482
CH$_3$	H	H	H	H	H	Acetone-d_6	3.82	7.98	7.67	7.20	7.24	7.45	$J_{45}=J_{67}=8.1$ $J_{56}=7.2$ $J_{46}=J_{57}=1.1$ $J_{47}=0.7$	482

CH$_3^a$	Cl	H	H	H	CDCl$_3$	—	7.67	7.17	7.24	7.32	$J_{45} = 7.99$ $J_{67} = 7.93$ $J_{56} = 7.48$ $J_{57} = 1.19$ $J_{46} = 1.16$ $J_{47} = 0.67$	484
CH$_3^b$	Cl	NO$_2$	H	H	CDCl$_3$	—	—	8.12	7.41	7.64	$J_{67} = 8.30$ $J_{56} = 8.29$ $J_{57} = 0.95$	484
CH$_3^c$	Cl	H	OCH$_3$	H	CDCl$_3$	—	7.14	—	6.91	7.11	$J_{67} = 8.85$ $J_{46} = 2.46$ $J_{47} = 0.49$	484
CH$_3^d$	Cl	H	H	CH$_3$	CDCl$_3$	—	7.53	7.07	—	6.98	$J_{45} = 7.90$ $J_{57} = 1.30$ $J_{47} = 0.56$	484
CH$_3^e$	Cl	H	H	OCH$_3$	CDCl$_3$	—	7.26	7.10	6.65	—	$J_{45} = 8.20$ $J_{56} = 8.14$ $J_{46} = 0.80$	484

[a,b,c,d,e] The NMe resonances are at $\delta 3.74$, 3.94, 3.72, 3.65, and 3.95, respectively.

conjugation effects are present between substituents and the nitrogen atoms of the imidazole ring.[484] Proton–proton coupling constants are rationalized in terms of additive substituent effects when there are no strong mesomeric influences between substituents and the heterocyclic ring.[484]

Prototropic rearrangement rates in azoles[489] are normally very rapid even at low temperature, but it has proved possible to study the tautomeric equilibrium in 2-chloro-5-methoxybenzimidazole by NMR spectroscopy; the coalescence temperature is −35°C, and free energy of activation, ΔG^*, is 45.6 KJ/mole.[485] A similar value of ΔG^* (44 KJ/mole) is calculated from low-temperature (−75°C) NMR spectra of the 5,6-dimethyl derivative (**1.221**), but in this case it is concluded that intramolecular hydrogen bonding is responsible for a slowing of the proton exchange rate[486] (cf. **1.221a** ⇌ **1.221** ⇌ **1.221b**).

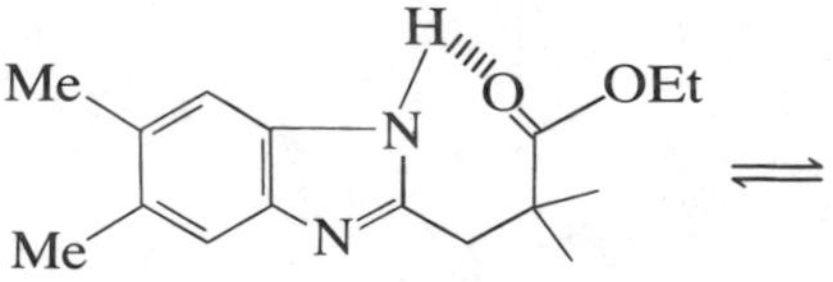

(**1.221a**)

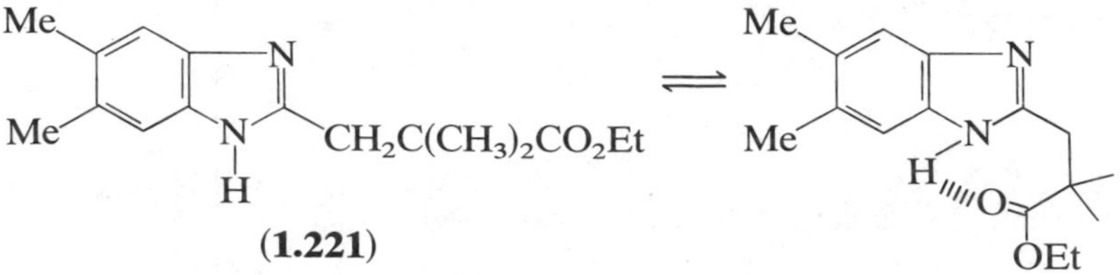

(**1.221**)

(**1.221b**)

The spectra of 1-H, —CH$_3$, and —COCH$_3$ derivatives with nitro substituents in the aryl ring have been analyzed;[487,488] for the 1-acetyl derivatives it is inferred that configurational isomer (**1.122a**) predominates in the equilibrium, (**1.122a** ⇌ **1.122b**).[488]

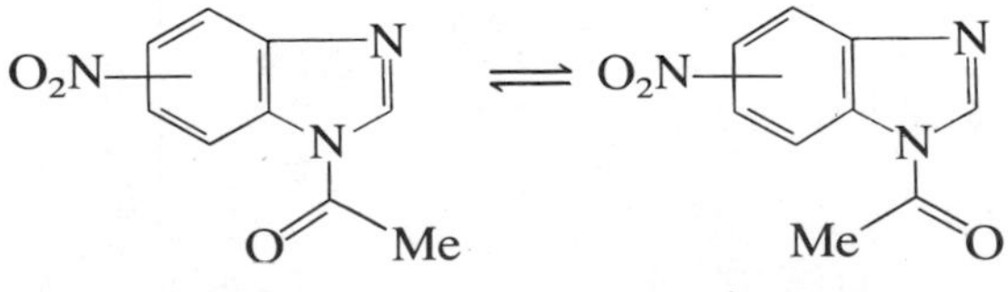

(**1.122a**)   (**1.122b**)

In dilute sulfuric acid, benzimidazole is protonated to form the benzimidazolium ion, but a rapid proton exchange with solvent occurs;[490] in concentrated sulfuric acid this process is considerably decelerated, and in this medium H-2 appears as a triplet with $J_{12} = J_{23} = 2.5$–2.6 Hz.[490,491] NMR data for benzimidazole in acidic solutions, together with comparative data for the free base, are shown in Table 1.23. The low-field shift observed passing from the free base to the benzimidazolium ion[492] is characteristic for other heterocyclic compounds; conversely, small high-field shifts are noted in the NMR spectra of solutions of the sodium salts of benzimidazole and its 2-methyl derivative.[493]

Proton chemical shifts have been used to obtain estimates of the electron distribution in benzimidazole.[494]

TABLE 1.23. PROTON MAGNETIC RESONANCE SPECTRA OF BENZIMIDAZOLE IN ACID SOLUTION

(1.223)

Compound	Solvent	Chemical shifts (δ values from TMS)				Coupling constants (Hz)	Ref.
		4H, 7H	5H, 6H	2H	1H, 3H		
1.223	CD$_3$OD	7.61	7.24	8.12	—	—	492
1.223 (HCl)	CD$_3$OD	7.88	7.66	9.50	—	—	492
1.223	H$_2$SO$_4$	—	—	9.01	11.84	$J_{12} = J_{23} = 2.6$	491

Two brief studies of the use of paramagnetic shift reagents have been reported. In 1,5- and 1,6-dimethylbenzimidazole derivatives, the effect of bis(acetylacetonato)nickel(II) is to cause relatively large contact shifts (ca. 0.4 ppm) to the low field of H-7.[495] Coordination presumably occurs to N-3, and the shifts are thus rationalized in terms of inter-ring spin delocalization through the σ-skeleton in a characteristic zigzag or "W-plan" arrangement.[495] On the other hand, the lanthanide shift reagent [Eu(dpm)$_3$] causes the greatest aromatic pseudocontact shift on H-4, and this method has proved to be of value in interpretation of the aromatic region of the NMR spectra of 1-methylbenzimidazoles bearing aryl ring substituents;[205] examples of the magnitudes of such induced shifts are shown in Table 1.24.

A contact shift is also induced in the C—H protons of benzimidazole by the stable organic free radical, di-t-butyl nitroxide (1.225), and this new procedure allows a subtle observation of the tautomeric behavior of the heterocyclic ring (cf. 1.226).[496] The free radical associates with the substrate by hydrogen bonding. In the configuration shown in (1.226) the largest shift is induced at H-4 via a W-plan contact interaction; the shift at H-7 is

TABLE 1.24. EXAMPLES OF THE EFFECT OF A LANTHANIDE
SHIFT REAGENT ON NMR SPECTRA OF
SUBSTITUTED 1-METHYLBENZIMIDAZOLES[205]

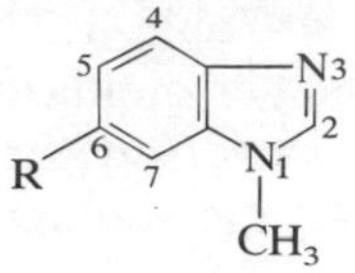

(1.224)

Compound (1.224) R	Chemical shifts (δ values from TMS in $CDCl_3$)						Coupling constants (Hz)
	N-1	2	4	5	6	7	
CH_3	3.36	7.01	← 7.38–6.80^a →				$J_{45} = 9.0$ $J_{57} = 0.8$
CH_3/Eu(dpm)$_3^b$	4.40	9.88	8.50	7.54	—	7.70	—
NO_2	3.95	← 8.40–7.75 →					$J_{45} = 3.0$ $J_{57} = 2.5$
NO_2/Eu(dpm)$_3^c$	7.34	15.44	11.80	10.72	—	11.54	—

^{a} The 6-Me resonance is at δ 2.35 and is shifted to δ 2.74.
^{b} With 0.066 mol of reagent/mole.
^{c} With 0.674 mol of reagent/mole.

identical. It is interesting to note that H-2 experiences an upfield shift. In
this case, the slightly acidic H-2 proton serves as the substrate for hydrogen
bonding and senses an upfield contact interaction.

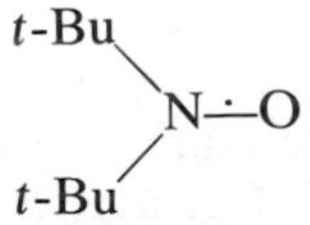

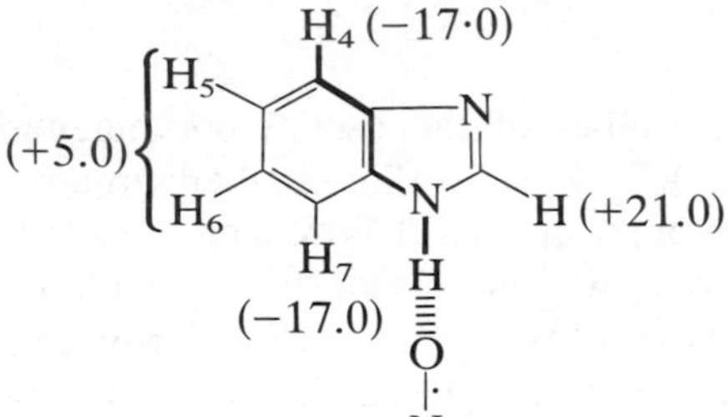

Shifts are in Hz at 220 MHz.
Downfield is negative.

¹³C SPECTRA. ¹³C NMR spectra have been interpreted for benzimidazole
and its anion and cation[497] (see Table 1.25), and for benzimidazolium
derivatives;[498] molecular-orbital methods [CNDO–SCF (complete neglect of
differential overlap–self-consistent field)] have been used to rationalize
chemical-shift data.[497,498] Other selected values of ¹³C chemical shifts are

TABLE 1.25. CHARACTERISTIC ^{13}C NMR SPECTRA OF
BENZIMIDAZOLE AND ITS ANION AND
CATION[497]

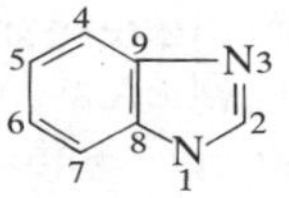

(1.227)

Carbon atom number in (1.227)	Chemical shifts (ppm downfield from TMS)		
	Benzimidazole[a]	Benzimidazole anion[b]	Benzimidazole cation[c]
2	141.46	150.45	139.58
4, 7	115.41	116.41	114.44
5, 6	122.87	120.10	127.29
8, 9	137.92	143.88	129.79

[a] In EtOH
[b] K salt in EtOH
[c] HCl salt in H_2O

illustrated within formulas (**1.228** and **1.229**). As might be expected, the largest effects upon chemical-shift values on protonation of, or proton abstraction from, benzimidazole are experienced by carbon atoms of the imidazole ring. It is significant that the chemical shift change of C-2 to higher field with decreasing pH is paralleled by the behavior of imidazole;[497] in this context the relatively low-field chemical shift of C-2 in the benzimidazolium derivative (**1.228**)[498] is surprising. The room-temperature ^{13}C NMR data illustrated in structure (**1.229**) illustrate the effect of an intramolecular hydrogen bond. It may be noted that the aromatic region is unsymmetrical. If a trace of acid is added, C-8 and C-9 coalesce as do C-5 and C-6, and carbons 4 and 7 are greatly broadened.

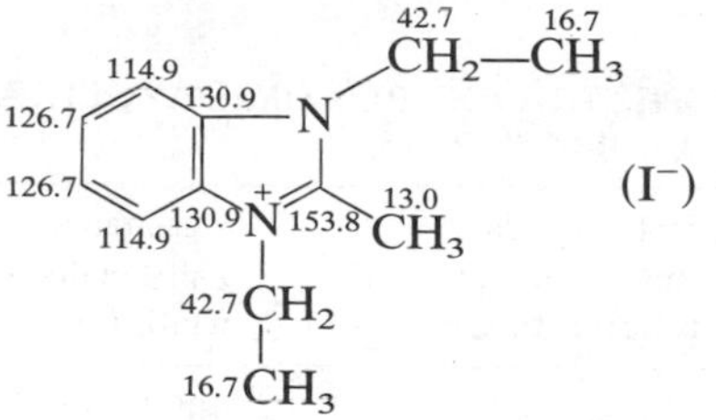

(1.228)

(DMSO solvent)[498]

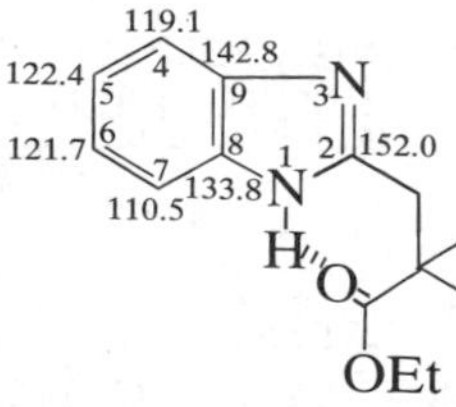

(1.229)

(in CDCl$_3$)[486]

^{19}F SPECTRA. ^{19}F NMR spectroscopy has been employed to evaluate the cis/trans ratios in a series of 1-fluoroalkenylbenzimidazoles.[499] Chemical-shift data and coupling constants for a typical example are illustrated with respect to structure (**1.230**). Figures shown are for trans and cis isomers, respectively, and are quoted as ppm with respect to external trifluoroacetic acid. It is interesting to note that the CF_3^a resonance of the trans isomer is to the low field of the cis isomer, but this situation is reversed for the CF_3^b resonance; it is also notable that there is no observable coupling between CF_3^a and CF_3^b nuclei in (**1.230**), whereas a coupling constant of 3.5 Hz is evident in the spectrum of a closely related compound (**1.230**; PhS for F^1 and F^2).

$$CF_3^b(t, -13.25; c, -13.86)$$

$$(t, +34.9; c, +15.86) \; F^1C$$

$$CF_3^a(t, -8.25; c, -7.4)$$

$$F^2(t, +84.8; c, +72.62)$$

$$\text{(t)} \quad \text{(c)}$$
$$J_{12} = 134 \quad 28 \text{ Hz}$$
$$J_{1a} = 21 \quad 9.5$$
$$J_{2a} = 11 \quad 12$$
$$J_{1b} = 4.5 \quad 2$$
$$J_{2b} = 3 \quad <1$$

(**1.230**)
t = trans
c = cis

^{14}N Spectra. Nitrogen-14 NMR spectra of benzimidazole and 1-methylbenzimidazole have been measured by the direct method[500] and also indirectly by the ^{1}H-[^{14}N] heteronuclear double-resonance technique[501] (see Table 1.26). Because of the tautomeric nature of benzimidazole the ^{14}N resonance appears as a single peak, but in the 1-methyl derivative two peaks are observed; the mean of the latter two chemical-shift values approximates to the ^{14}N chemical shift of benzimidazole.

^{14}N shifts of imidazole and benzimidazole are modified slightly depending upon whether a proton donor or proton acceptor solvent is used.[501] For example, a marked upfield shift in the nitrogen resonance of benzimidazole is observed in passing from acetone to dimethylsulfoxide solvent.[501]

TABLE 1.26.   NITROGEN-14 NMR SPECTRA OF BENZIMIDAZOLE AND 1-METHYLBENZIMIDAZOLE[500,501]

Compound	Solvent	^{14}N chemical shift (ppm) referred to $CH_3NO_2^a$	Resonance half-height width (Hz)	Technique used[b]
Benzimidazole	$(CH_3)_2CO$	$+185 \pm 5$	400 ± 20	A
Benzimidazole	CH_3OH	$+192 \pm 5$	750 ± 50	A
Benzimidazole	DMSO	$+234 \pm 4$	—	B
1-Methylbenzimidazole	$(CH_3)_2CO$	$+130 \pm 8(N)$ $+228 \pm 5(NMe)$	350 ± 50 300 ± 50	A
1-Methylbenzimidazole	CCl_4	$+124 \pm 6(N)$ $+234 \pm 5(NMe)$	—	B

^{a} Positive sign denotes high-field direction.
^{b} A indicates measurement by the direct procedure at 4.33 MHz.[500] B indicates measurement by the double resonance method.[501]

The ^{14}N spectra show a linear relationship between chemical shifts and SCF–PPP–MO (self-consistent field–Pariser-Parr-Pople–molecular orbital) π-charge densities.[500]

^3H Spectra. Triton magnetic resonance spectra of benzimidazole and 1-methylbenzimidazole indicate that triton and proton chemical shifts in respect to the 2-position are virtually identical.[502]

Electron Spin Resonance Spectra

Spectra of Neutral Free Radicals. No structural information was adduced from the electron spin resonance spectrum of the hydrazyl radical (1.231),[503] but a well-resolved 41-line spectrum of the nitroxide derivative (1.232a) is obtained by lead tetraacetate oxidation of the 1-hydroxybenzimidazole 3-oxide derivative (1.233a) in benzene solution.[504] The spectrum is analyzed in terms of $a_{N1} = a_{N3} = 0.42$ mT; $a_{CH_3} = 0.28$ mT; and $a_{4H} = a_{5H} = a_{6H} = a_{7H} = 0.07$ mT. The spectrum of the 2-phenyl analog (1.232b) is less well resolved, but a five-line pattern with $a_{N1} = a_{N3} = 0.43$ mT is evident.

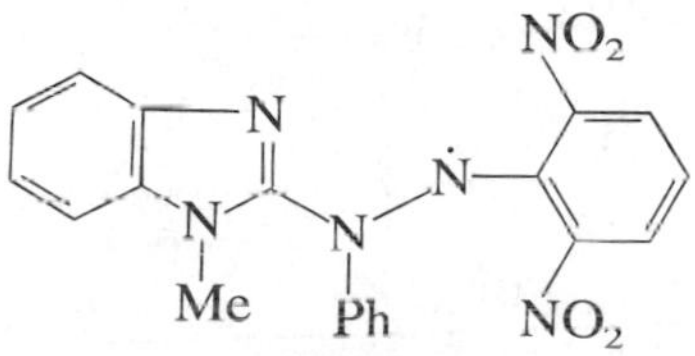

(1.231)

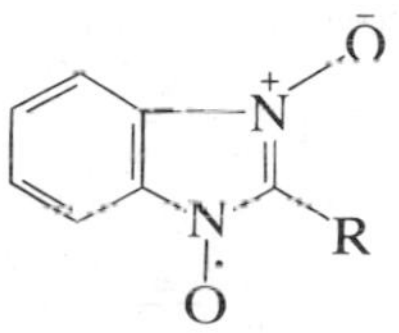

(1.232)

a, R = Me
b, R = Ph

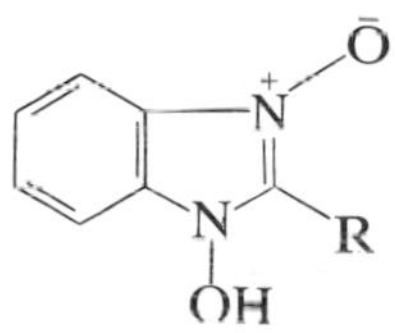

(1.233)

a, R = Me
b, R = Ph

Neutral free radicals are also produced by X-irradiation of crystalline imidazole and benzimidazole and their 2-methyl homologs.[505] Significantly, the radicals produced from imidazole and benzimidazole are almost identical and consist of a 1:2:1 triplet ($a = 4.5$ mT) with each line showing further poorly resolved hyperfine splitting. The radical produced from imidazole is thus assigned structure (1.234), in which coupling occurs with two equivalent protons (H_A). On this basis structure (1.235), which was earlier assigned to the same species produced from imidazole by γ-irradiation,[506] is unacceptable.

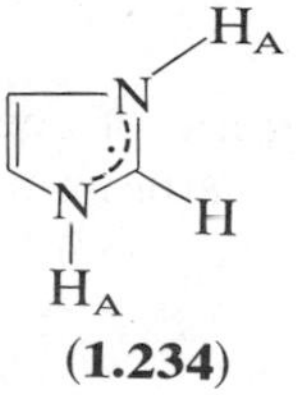

(1.234)

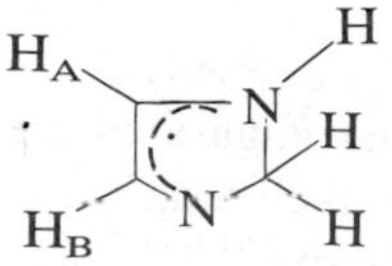

(1.235)

SPECTRA OF RADICAL ANIONS. Free-radical dianions (e.g., **1.237**, **1.239**) are formed by treating imidazoquinones (e.g., **1.236**) or 5,6-dihydroxybenzimidazoles (e.g., **1.238**) with potassium *t*-butoxide in dimethylsulfoxide.[507] An interesting feature of the spectral analyses is the relative magnitude of hyperfine splitting of H-2 in these compounds (0.05 and 0.504 mT); these values are confirmed by replacement of H-2 by methyl groups for both (**1.237**) and (**1.239**); for the latter compounds, methyl hfs constants of 0.069 and 0.502 mT are evident. These data are used to refute an earlier suggestion[508] that the paramagnetic species formed by electrolytic reduction of (**1.236**) was a radical anion rather than a radical dianion.

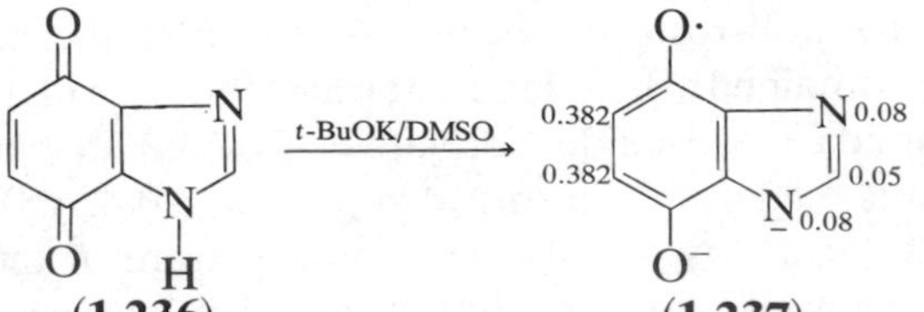

(1.236)   **(1.237)**

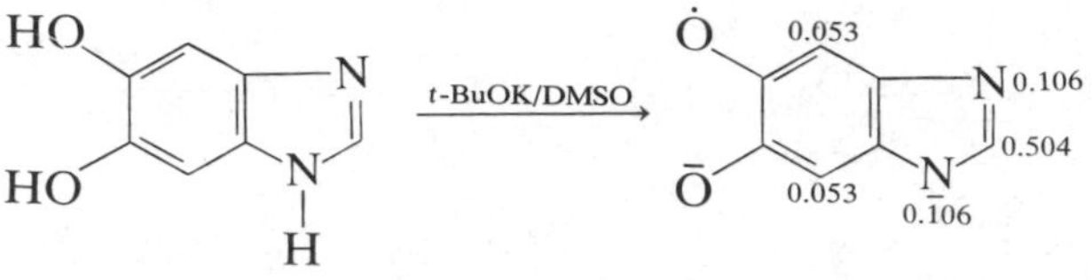

(1.238)   **(1.239)**

Nuclear Quadrupole Resonance Spectra

Nuclear quadrupole resonance (NQR) of ^{14}N has been observed in benzimidazole by the use of a double-resonance technique.[509] Values of ν_+, ν_- (where measured), ν_0, e^2qQ, and η for N_1-H and N_3 are: 2.590, —, 0.475, 3.14, 0.303; and 1.855, 1.855, 1.100, 0.770, 1.96, 0.786, respectively. These data have proven valuable as an aid in assigning the NQR of ^{14}N in purines.[509]

Mass Spectra

A number of investigations on the mass spectra of benzimidazoles have been carried out including benzimidazolium barbiturates,[510] alkyl,[511] 1-skatyl,[512] 1,2-diphenyl,[513] 2-phenoxy,[514] 2-(4-thiazolyl),[515] and N-oxide[516] derivatives.

A systematic study of 22 benzimidazole derivatives has been reported by Bowie and coworkers,[517] from which it appears that the fragmentation pathways of simple benzimidazoles are similar to those of imidazoles.[518] The mass spectrum of benzimidazole exhibits the molecular ion as the base peak and indicates an ensuing sequential loss of two molecules of hydrogen cyanide; N-d_1-benzimidazole shows loss of both HCN and DCN from the molecular ion, which indicates that initial loss of HCN is a nonspecific process (cf. the M-HCN fragmentation mode in imidazole[518]). The detailed mechanism of the initial HCN loss has been evaluated by a study of the mass spectrum of benzimidazole-2-^{13}C.[519] Evidently, the [M-HCN]$^{+}$ ion arises from two competitive reactions: ions at m/e 91 and 92 can be rationalized[519] by an initial isomerization of the molecular ion to o-aminobenzonitrile radical cation, which can fragment by two different modes of HCN loss (see Scheme 1.11). Analogous loss of RCN from 2-alkylbenzimidazoles is a less

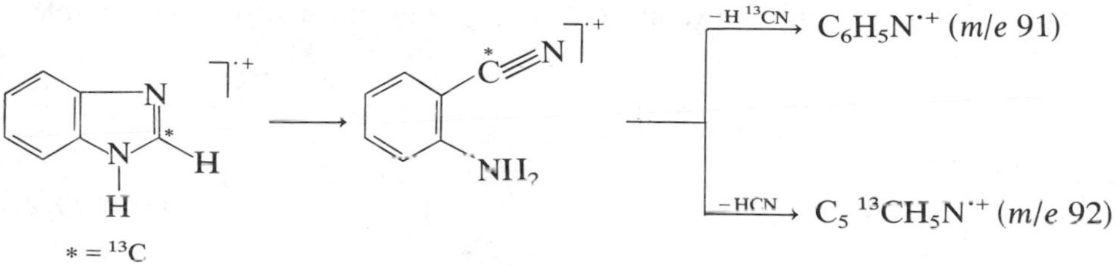

$$-\text{H}\,^{13}\text{CN} \longrightarrow C_6H_5N^{\cdot+} \ (m/e\ 91)$$

$$-\text{HCN} \longrightarrow C_5\,^{13}CH_5N^{\cdot+} \ (m/e\ 92)$$

$* = {}^{13}C$

Scheme 1.11

important fragmentation process. 2-Methylbenzimidazole molecular ion loses a hydrogen atom to give an ion that has been formulated as a protonated quinoxalinium ion (see Scheme 1.12).[520] Evidently, for related 2-alkyl derivatives (e.g., 2-n-Pr and 2-n-Bu) β-scission to the heterocyclic ring occurs, since the m/e 131 ion is also observed in these cases; concomitant hydrogen migration must occur, however, because the m/e 131 ion arises from an ion at m/e 132, as evidenced by a metastable ion at m/e 130. A McLafferty rearrangement has been invoked to explain the origin of the ion at m/e 132 (see Scheme 1.13).[517,520]

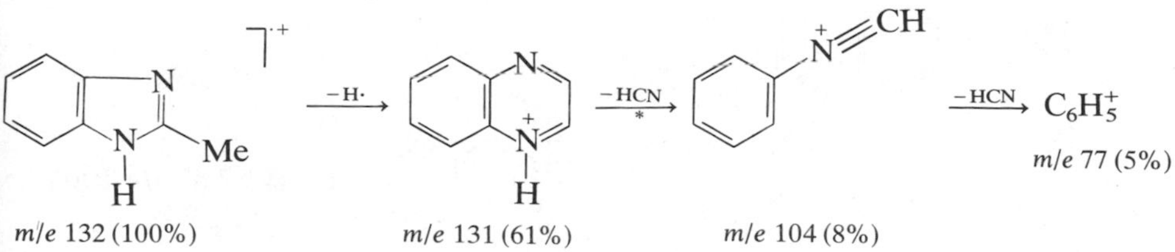

$$-\text{H}\cdot \qquad -\text{HCN} \qquad -\text{HCN} \longrightarrow C_6H_5^+$$

m/e 132 (100%)　　m/e 131 (61%)　　m/e 104 (8%)　　m/e 77 (5%)

Scheme 1.12

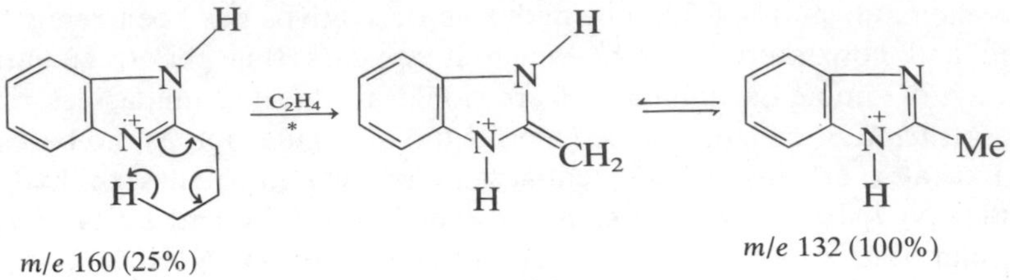

m/e 160 (25%)   *m/e* 132 (100%)

Scheme 1.13

A quinoxalinium structure has also been proposed to rationalize the mass spectrum of 1-ethylbenzimidazole, and evidence in this context has been adduced from studies of 1-ethylbenzimidazole-2-^{13}C and 1-ethyl-d_5-benzimidazole (see Scheme 1.14).[521]

As with 2-acylthiophenes,[522] 2-acyl- and 2-benzoylbenzimidazoles are characterized[517] by loss of carbon monoxide from the molecular ion, and since this fragmentation mode is not apparent in the spectra of acylbenzenes,[523] it is a diagnostic transition for acyl substituents in the imidazole ring.

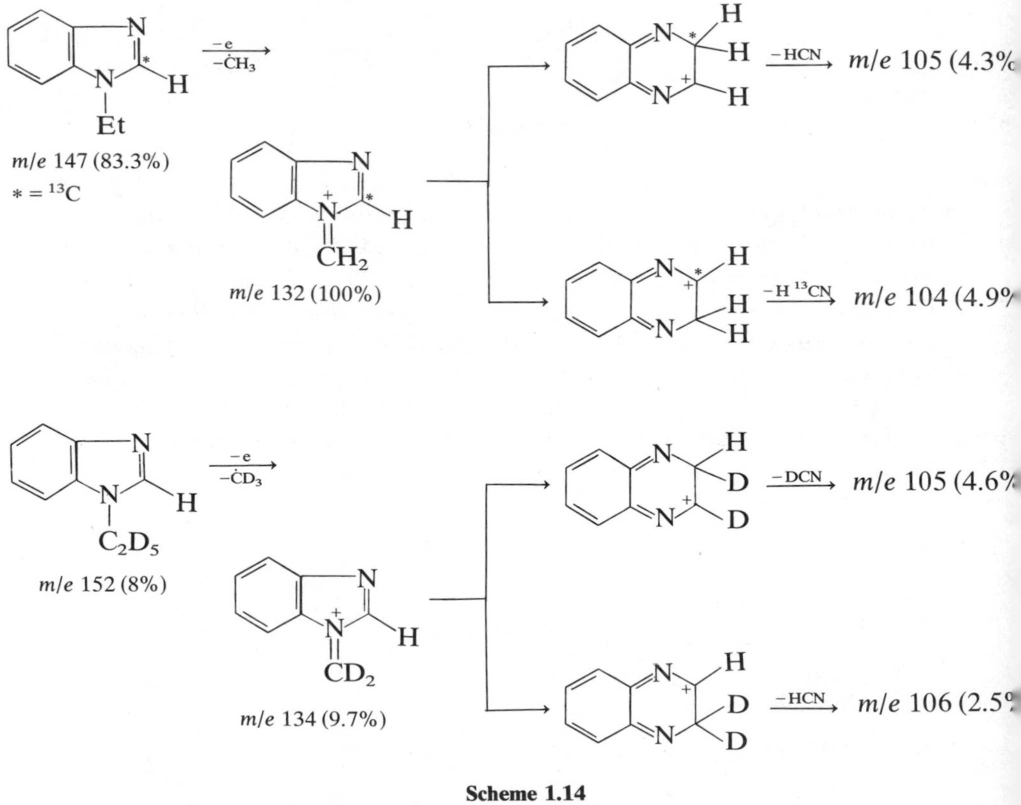

m/e 147 (83.3%)
* = ^{13}C

m/e 132 (100%)

m/e 152 (8%)

m/e 134 (9.7%)

Scheme 1.14

Two important features emerge from the series of secondary alcohols (**1.240**).[517] Combined loss of the substituent and water from the molecular ion gives rise to an ion at m/e 205 [represented as (**1.241**)] for (**1.240a**), but not for the methyl homolog (**1.240b**); the presence of an ion at m/e 205 can thus be used to detect the presence of a 1H substituent.

Similarly, loss of water from the molecular ion occurs only for the 1-H and not the 1-Me derivatives.

(**1.240**)

a, $R^2 = H$
b, $R^2 = Me$

(**1.241**)

1.3.4. General Studies

Acidity Constants

The acidity of benzimidazole and its derivatives is reflected in the ability of such compounds to form metallic salts; details of the early literature on salt formation have been summarized by Hofmann.[1] Subsequently, a pK_a value of 12.57 has been estimated for benzimidazole from hydrolysis constants[524] (cf. a value of 14.2 for imidazole,[524] and a value of 12.3 for benzimidazole from electrometric titration[525]). Aqueous thermodynamic pK_a values have been obtained spectrophotometrically for a series of 5-substituted derivatives.[526] It has been demonstrated[526] that there is a good correlation of ΔpK_a (acidity) values with ΔpK^+ (basicity) values. A correlation has also been found of pK_a values (in acetonitrile) of 2-substituted derivatives with the magnitude of the electrostatic energy of interaction (ε) of the electron pair of the nitrogen atom being protonated, with the π-system;[527] the ε values were calculated by the MO–SCF (molecular orbital–self-consistent field) method with the Pariser-Parr-Pople approximation. Some reservations have been expressed concerning this notion, however: pK_a values in this series also correlate well with σm-constants, and it is proposed that the effect of a 2-substituent on the reaction center is predominantly inductive.[528] pK_a data for 2-pyridyl-5-substituted benzimidazoles can be correlated by means of the Hammett equation with σp-values; a ρ-value of 1.3 is observed.[106]

Imidazole is surprisingly basic ($pK_a \sim 7.0$; cf. pyridine, $pKa \sim 5.2$), presumably because of stabilization as depicted in the canonical structures

TABLE 1.27. ACIDITY CONSTANTS OF
            BENZIMIDAZOLES

Compound	pKa as base		pK_a as acid, solvent (ref.)
	ln H_2O^a	ln EtOH^a	
Unsubstituted	5.48	4.98	13.25, CH_3CN (527)
2-CH_3	6.19	5.48	14.20, CH_3CN (527)
2-NH_2	—	7.39	15.95, CH_3CN (527)
4-CH_3	5.67	—	—
5-CH_3	5.81	—	13.0, H_2O (526)
5-NO_2	—	2.68	10.86, H_2O (526)
5-NH_2	6.11	—	13.13, H_2O (526)

^{a} From Ref. 1, p. 251; cf. M. T. Davies, P. Mamalis, V. Petrov,
and B. Sturgeon, *J. Pharm. Pharmacol.*, **3**, 420 (1951).

(**1.242**). Annulation of a benzo ring reduces the basicity of the pyridine-like
nitrogen and results in a p$K_a \sim 5.5$ for benzimidazole. A selection of pK_a
values for benzimidazoles as acids and bases is collected in Table 1.27. A
discussion of the influence of substituents on the basicity of benzimidazoles
has been presented by Hofmann.[1]

(**1.242a**)   (**1.242b**)

Formation of Molecular Complexes

A number of molecular complexes containing benzimidazole and its
derivatives have been reported; some examples of substrates that have been
used are methyl trans-cinnamate,[529] napthalene,[529] tetracyanoquinodimethane,[530,531] tetracyanoethylene,[531] organohalostannanes,[532]
trinitroanisole,[533] iodine,[534] and boron trihalides.[535] Stability constants of
the complexes are usually evaluated from infrared and electronic absorption
spectral data, but measurements of electron spin resonance spectra[530,531]
and dipole moments[531,534] have also been carried out. It has been concluded[534] that complexes of iodine with benzimidazole are of the *nσ* type
with localization of the coordinate link on the pyridine-type nitrogen.

Benzimidazole forms two complexes of 1:1 and 3:1 mole ratio, respectively, with the haloketone $(ClF_2C)_2CO$, and complexes of these types are of
interest as fungicides.[535] The utility of a 1:1 benzimidazole–sulfur dioxide
complex is claimed as a curing agent for epoxide resins.[536]

Molecular-Orbital Calculations

The semiempirical Pariser-Parr-Pople SCF method has been used to calculate the electronic spectrum of benzimidazole,[537,538] and a good correlation of calculated with experimental data is evident.

A semiempirical SCF–MO procedure has been used to estimate a resonance energy of 129.3 KJ/m for benzimidazole[539] (cf values of 87.6 KJ/m for pyridine and 83.8 KJ/m for benzene). Benzimidazole thus has a relatively high resonance energy, and this is reflected in the good thermal stability of polymers containing the benzimidazole moiety.[540]

The semiempirical Pariser-Parr-Pople SCF method and also the CNDO/2 procedure have provided data on π-electron distribution, bond orders, dipole moments, and energies associated with the highest molecular orbitals.[538,541a,542] In the most recent calculations[541a] both CNDO/2 and CNDO/S methods with configuration interaction have been used. The results are in agreement with those for other heterocycles but are not in accord with earlier calculations for benzimidazole.[541b] However, it should be noted that there are distinct differences in the results obtained by the two procedures. The charge distributions on nitrogen atoms N1 and N3 calculated by the two methods are:

	CNDO/2		CNDO/S	
	$\sigma + \pi$	π	$\sigma + \pi$	π
N1	5.1282	1.6726	5.0977	1.7226
N3	5.2250	1.2110	5.2369	1.2106

Thus N1 is apparently more of a π-donor from the CNDO/2 than from the CNDO/S model, although little differences are observed for N3. The CNDO/S results provide, as might be expected, reasonable predictions of the electronic spectrum, and calculated values of dipole moments (3.90 by CNDO/S; 3.48 by CNDO/2)[541a] agree satisfactorily with the experimental value of 4.03D (cf. section 1.3.2).

A theoretical interpretation of the relative reactivities of the 2-position in imidazoles and benzimidazoles has been attempted.[543] The difference in behavior of electrophilic and nucleophilic reagents at the 2-position is considered in terms of charge densities at this position; it is concluded that such differences can be ascribed to the electron acceptor properties of the carbocyclic ring.

The ionization potential of benzimidazole has been calculated by use of the PPP–SCF–MO method and Del Re's procedure. A good correlation is obtained by plotting this (9.67) and values for other nitrogen heterocycles against pK_a.[544]

Analytical Studies

CHROMATOGRAPHIC ANALYSIS. Thin-layer chromatographic (TLC) analysis of benzimidazoles using both silica gel[545] and alumina[546] as stationary phases has been described. Silica gel containing a fluorescent indicator is a useful phase for TLC analysis of the systemic fungicide, 1-(butylcarbamoyl)-2-benzimidazolecarbamic acid methyl ester[benomyl], and its metabolites;[547] the separation and determination were achieved by the use of a two-dimensional technique, and an efficient spray reagent (N-2,5-trichloro-p-benzoquinoneimine in cyclohexane) was found for 2-benzimidazolecarbamic acid methyl ester. A variety of solvent combinations have been used to evaluate the TLC behavior of a series of 2-alkyl and -aryl, -4-thiazolyl, 5-hydroxy, and bisbenzimidazole derivatives.[548] The solvent systems, ethyl acetate–toluene–ammonia (60:40:2) and toluene–ethyl acetate–ethanol–ammonia (60:10:10:2) are particularly useful for separation purposes.

Gas–liquid chromatographic (GLC) analysis of benzimidazoles has been studied using silicone greases on celite as the stationary phases.[548,549] It appears that the silylation procedure is of little value in the GLC analysis of 2-alkyl and 2-aryl derivatives, but is a useful method for the analysis of 5-hydroxy derivatives.[548]

The use of high-pressure liquid chromatography for the analysis of benzimidazole[550] and a number of derivatives including benomyl and thiabendazole[551] has been evaluated. Successful methods have been achieved by the use of both reverse phase and adsorption systems.

THERMAL ANALYSIS. The melting properties (T_m, ΔH, ΔS) of benzimidazole, 2-phenylbenzimidazole, and a number of bisbenzimidazoles have been investigated by differential scanning calorimetry.[552] By comparison with polyphenyls it appears that bisbenzimidazoles have higher values of T_m, and it is suggested that hydrogen bonding in the latter examples is a factor in a relative lowering of ΔS values.[552] A combination of the techniques of differential thermal analysis, differential thermogravimetric analysis, and differential enthalpic analysis has been employed to study the melting and thermal decomposition of a series of 1-(4-substituted aryl)-5-aminobenzimidazole derivatives.[553] All compounds in this series form stable hydrates, and the detailed mode of dehydration and melting of one such derivative has been studied in detail. An interesting feature of the dehydration process that occurs at 60° for 1-(4-bromophenyl)-5-aminobenzimidazole is that it is a two-stage process. It is concluded that the water is bound in the crystal lattice by hydrogen bonds to two (separate) types of nitrogens of the imidazole rings.[553]

1.4. REACTIONS OF BENZIMIDAZOLES

1.4.1. Reactions with Electrophilic Reagents

Substitution in the Carbocyclic Ring

Examples of electrophilic aromatic substitution reactions in the carbocyclic ring of benzimidazole are collected in Table 1.28. Unsubstituted benzimidazoles are attacked preferentially by electrophiles at the 5-position. If a 5-substituent does not influence the course of subsequent substitution (e.g., methyl[554] and methoxy[555] groups), the second substituent normally enters the 6-position. However, if the 5-substituent is powerfully electron-releasing, for example, amino or hydroxyl groups, the second substituent enters at the 4-position.[556] On the other hand, an electron-withdrawing substituent at the 5-position directs the entering electrophile to the 6-position and to a lesser extent the 7-position.[557]

Particular attention has been paid to nitration in relation to substituent effects:[565,566] benzimidazole is nitrated exclusively in the 5-position,[564] and on further nitration a mixture of the 5,6-dinitro (54%) and 4,6-dinitro (21%) derivatives is obtained. It is interesting to note, however, that 4-fluorobenzimidazole is nitrated at the 7-position under relatively mild conditions, and ensuing nitration under more vigorous conditions gives the 4-fluoro-5,7-dinitro derivative.[565] Evidently, the electronic influence of the fluorine atom takes precedence over that of the imidazole ring in this case. It is apparent also that steric effects can play a vital role in the course of aromatic nitration: 1,2-dimethyl-4-fluorobenzimidazole is nitrated exclusively at the 5-position, presumably because of steric repulsion in regard to nitration at the 7-position.[565] A 2-trifluoromethyl substituent appears to have a considerable deactivating influence since only mononitration (at the 5-position) can be achieved under conditions where 5-nitrobenzimidazole is converted into a mixture of 5,6- and 4,6-dinitro derivatives.[565]

Interesting data on the relative reactivity of 2-aminobenzimidazoles toward diazonium coupling has been adduced from a study of the reaction of 2-aminobenzimidazoles (1.244) with nitrosylsulfuric acid.[577] By introducing arenes into the reaction medium, two competitive processes occur: self-coupling (cf. 1.245) and coupling to the arene (cf. 1.246). Poor yields of arene-coupled products (1.246) are obtained using benzene and toluene, but appreciable yields are obtained with mesitylene. It appears that the reactivity of the 5- and 6-positions in the (protonated) 2-aminobenzimidazoles (cf. 1.244) is comparable to that of mesitylene, while a methyl group in the 5-position (1.244b) lowers the yield of arene-coupled product (1.246b) by

TABLE 1.28. ELECTROPHILIC AROMATIC SUBSTITUTION REACTIONS OF BENZIMIDAZOLES

Types of process	Substituents in the benzimidazole derivative	Reagent	Product	Ref.
Sulfonation	None	H_2SO_4	$5\text{-}SO_3H$	559
	None, 2-aryl	H_2SO_4	$2\text{-}H(R)\text{-}5\text{-}SO_3H$	560
	None, 2-alkyl, 2-aryl	$ClSO_3H$	$5\text{-}SO_3H$ and/or $5\text{-}SO_2Cl$ derivatives	561
Nitration	2-Me-4-AcNH	H_2SO_4/HNO_3	$2\text{-}Me\text{-}4\text{-}AcNH\text{-}7\text{-}(and\ 5\text{-})NO_2$	562
	4-NHCHO	H_2SO_4/HNO_3	$4\text{-}NH_2\text{-}5\text{-}(and\ 7\text{-})NO_2$	563
	2-CH$_2$Ph	H_2SO_4/HNO_3	$5\text{-}NO_2\text{-}4'\text{-}NO_2$	
	5-NO$_2$	H_2SO_4/HNO_3	$5,6\text{-}(NO_2)_2 + 4,6\text{-}(NO_2)_2$	564
	2-Me-5-NO$_2$	H_2SO_4/HNO_3	$2\text{-}Me\text{-}5,6\text{-}(NO_2)_2 + 2\text{-}Me\text{-}4,6\text{-}(NO_2)_2$	
	4-F	H_2SO_4/HNO_3	$4\text{-}F\text{-}7\text{-}NO_2$	
	2-CF$_3$-4-F	H_2SO_4/HNO_3	$2\text{-}CF_3\text{-}4\text{-}F\text{-}7\text{-}NO_2$	
	2-Me-4-F	H_2SO_4/HNO_3	$2\text{-}Me\text{-}4\text{-}F\text{-}5,7\text{-}(NO_2)_2$	565^c
	1,2-Me$_2$-4-F	H_2SO_4/HNO_3	$1,2\text{-}Me_2\text{-}4\text{-}F\text{-}5\text{-}NO_2$	
	1,2-Me$_2$-4-F-5-NO$_2$	H_2SO_4/HNO_3	$1,2\text{-}Me_2\text{-}4\text{-}F\text{-}5,7\text{-}(and\ 5,6\text{-})(NO_2)_2$	
	2-CF$_3$	H_2SO_4/HNO_3	$2\text{-}CF_3\text{-}5\text{-}NO_2$	
	2-CF$_3$-5,6-Cl$_2$	H_2SO_4/HNO_3	$2\text{-}CF_3\text{-}4\text{-}NO_2\text{-}5,6\text{-}Cl_2$	
	2-CF$_3$-4,6-Br$_2$	H_2SO_4/HNO_3	$2\text{-}CF_3\text{-}4\text{-}NO_2\text{-}5,7\text{-}Br_2$	
	2-CF$_3$-5,6,7-Br$_3$	H_2SO_4/HNO_3	$2\text{-}CF_3\text{-}4\text{-}NO_2\text{-}5,6,7\text{-}Br_3$	
	2-CF$_3$-5-Cl	H_2SO_4/HNO_3	$2\text{-}CF_3\text{-}4\text{-}NO_2\text{-}5\text{-}Cl + 2\text{-}CF_3\text{-}5\text{-}Cl\text{-}6\text{-}NO_2$	566
	2-CF$_3$-5-Br	H_2SO_4/HNO_3	$2\text{-}CF_3\text{-}4\text{-}NO_2\text{-}5\text{-}Br + 2\text{-}CF_3\text{-}5\text{-}Br\text{-}6\text{-}NO_2$	
	2-i-Pr-5-CF$_3$	H_2SO_4/HNO_3	$2\text{-}i\text{-}Pr\text{-}5\text{-}CF_3\text{-}6\text{-}NO_2$	326
	1-alkyl-5-CF$_3$-6-R derivatives (R = e.g., Cl, MeO)	H_2SO_4/HNO_3	$1\text{-}Alkyl\text{-}5\text{-}CF_3\text{-}6\text{-}R\text{-}7\text{-}NO_2$	176
	2-(2-pyridyl)	H_2SO_4/HNO_3	$2\text{-}(2\text{-}pyridyl)\text{-}5\text{-}NO_2$	106
	1-Me-2-NH$_2$	H_2SO_4/KNO_3	$1\text{-}Me\text{-}2\text{-}NH_2\text{-}5\text{-}(and\ 6\text{-})NO_2$	

Reaction	Substrate	Reagent	Product	No.
	1-Me-2-NH$_2$	H$_2$SO$_4$/xsKNO$_3$	1-Me-2-NH$_2$-5,6-(NO$_2$)$_2$	567
	1,3-Me$_2$-2-(=NH)(nitrate)	H$_2$SO$_4$	1,3-Me$_2$-2-(=NH)-5-NO$_2$	
	1,3-Me$_2$-2-(=NH)-5-NO$_2$	H$_2$SO$_4$/HNO$_3$	1,3-Me$_2$-2-(=NH)-5,6(NO$_2$)$_2$	
Halogenation	None	Cl$_2$	4,5,6-Cl$_3$	568
	2-Me	Cl$_2$	2-Me-4,5,6-Cl$_3$	
	2-Me	Br$_2$	2-Me-4,5-Br$_2$	
	2-CHO	Cl$_2$	2-CHO-4,5-Cl$_2$	
	2-Me	Br$_2$/CHCl$_3$	2-Me-5-Br + 2-Me-4,6-Br$_2$	
	2-CO$_2$H	Br$_2$/CHCl$_3$	2-CO$_2$H-5-Br	569
	2-CF$_3$	Br$_2$/AcOH	2-CF$_3$-5-Br	
	2-CF$_3$	xsBr$_2$/AcOH	2-CF$_3$-4,5,6-Br$_3$ + 4,5,6,7-Br$_4$	566
	2-CF$_3$-5-Cl	Br$_2$/AcOH	2-CF$_3$-5-Cl-6-Br	570[a]
	2-SH-4,5,6,7-Cl$_4$	Cl$_2$/HCl/AcOH	2,4,5,6,7-Cl$_5$	
	2-Cl-5-(4'-chlorophenyl-sulfonylamino)	Cl$_2$/FeCl$_3$	2,4,6,7-Cl$_4$-5-(4'-chlorophenylsulfonylamino)	
	2-CCl$_2$NO$_2$	Cl$_2$/AcOH	2-CCl$_2$NO$_2$-4,5,6,7-Cl$_4$	571
	2-CCl$_2$NO$_2$-5-Me	Cl$_2$/AcOH	2-CCl$_2$NO$_2$-5-Me-4,6,7-Cl$_3$	
Alkylation	None	Phthalic Anhydride/AlCl$_3$	5-(2-carboxybenzoyl) derivatives	572
	2-OH-6-Me			
	2-OH-1,3-Me$_2$			
	2-NMe$_2$			
	2-NHCO$_2$Me	γ-Butyrolactone/AlCl$_3$	2-NH$_2$-5-CHMeCH$_2$CO$_2$H	573
	2-NHCO$_2$Me	CH$_3$CHClCH$_2$CO$_2$H/AlCl$_3$	2-NHCO$_2$Me-5-CMe$_2$CO$_2$H	
	2-NHCO$_2$Me	γ-Valerolactone/AlCl$_3$	2-NHCO$_2$Me-5-CHMe(CH$_2$)$_2$CO$_2$H	574
	2-NHCO$_2$Me	γ-Valerolactone/xs AlCl$_3$	2-NH$_2$-5-CHMe(CH$_2$)$_2$CO$_2$H	
	5-OH	sec-amines/CH$_2$O	4-Aminomethyl-5-OH	575
	1-Ph-5-OH	p-HO$_3$SC$_6$H$_4$N$_2^+$	1-Ph-4-N=NAr-5-OH	576
Diazonium coupling	1-Me-2-NH$_2$-5-Br	ONOSO$_3$H	1-Me-2-NH$_2$-5-Br-6-N=N—BI[b]	577[c]
	1-Me-2-NH$_2$-6-Br	ONOSO$_3$H	1-Me-2-NH$_2$-5-N=N-BI[b]-6-Br	

[a] Thirtysix halogeno derivatives are described in this patent

[b] BI = appropriate 2-benzimidazolyl-N=N moiety

[c] See text

facilitating self-coupling. Diazonium coupling at the 5- and 6-positions also occurs when the 5,6-dibromo derivative (**1.244c**) is treated with nitrosylsulfuric acid in concentrated sulfuric acid. The latter reaction constitutes one of the few examples of displacement of bromonium ion by a diazonium ion in an electrophilic substitution process.

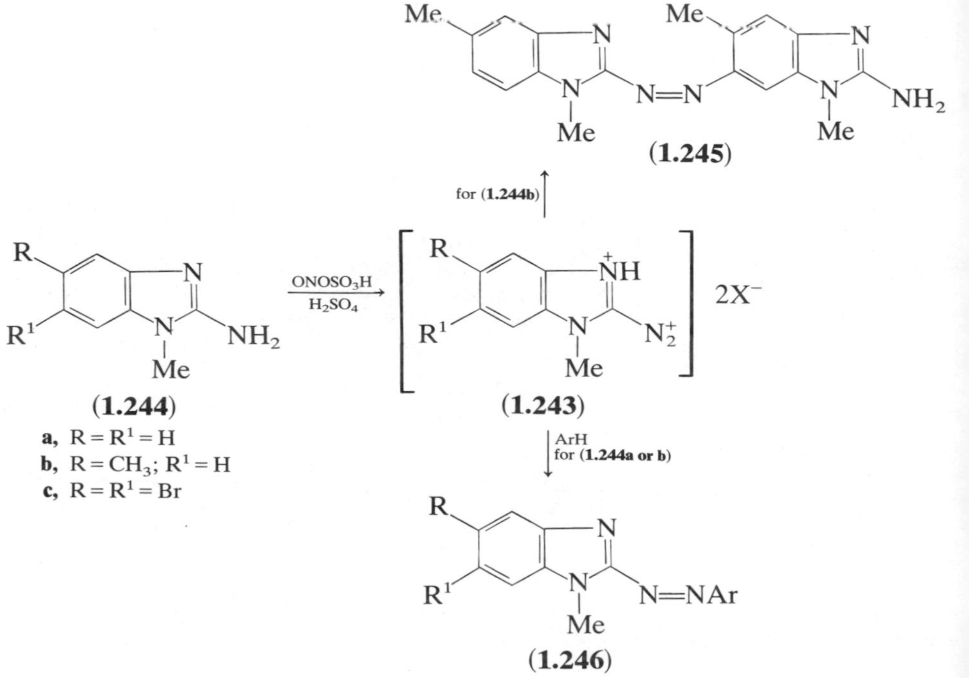

(**1.245**)

for (**1.244b**)

(**1.244**)

a, R = R¹ = H
b, R = CH₃; R¹ = H
c, R = R¹ = Br

(**1.243**)

(**1.246**)

Molecular orbital calculations were carried out many years ago in an attempt to rationalize the pattern of electrophilic aromatic substitution in benzimidazole.[558] From this work it was concluded that substitution should occur at the 4-position in the benzimidazolium ion and at the 5-position in the neutral molecule. This is evidently not the case, and de la Mare and Ridd[578] believe that such calculations overestimate the stability of transition states for substitution *ortho* to a group conjugated with the aromatic ring.

Substitution in the Imidazole Ring

Electrophilic substitution does not occur in the imidazole ring of benzimidazoles. Benzimidazole is quantitatively iodinated when treated with iodine in aqueous sodium hydroxide solution[579] to give a product that was believed to be 2-iodobenzimidazole.[579] Iodine is lost very rapidly from this product, however, and following reservations expressed concerning its formulation,[578] it was subsequently shown to be 1-iodobenzimidazole.[580]

Benzimidazole and 1-methylbenzimidazole react with bromine in chloroform at room temperature to give a dicoordinate complex and an n-donor complex that are formulated as (**1.247**) and (**1.248**), respectively.[581]

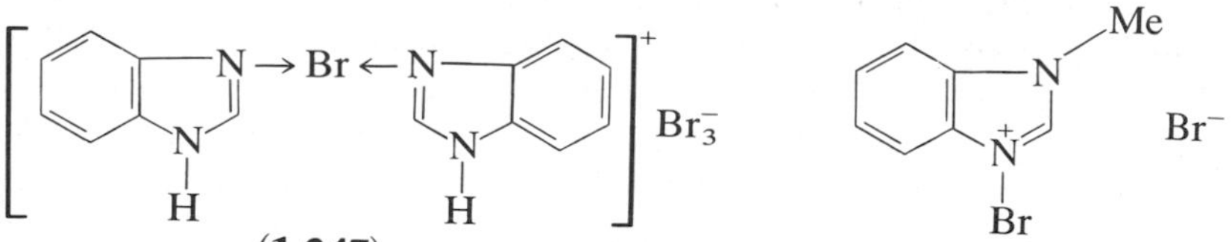

(1.247)   **(1.248)**

Electrophilic Attack at the 1-(or 3-) Position: Alkylation and Related Reactions

GENERAL COMMENTS. The mechanism of alkylation of imidazoles and benzimidazoles has been discussed in detail by Simonov and coworkers.[9] Four distinct mechanims of alkylation of the imidazole ring are possible, depending on whether the substrate is alkylated as a base (S_E2 and S_E2' mechanisms), as an anion (S_E2 CB mechanism), or as the conjugate acid (S_E2 CA mechanism). Alkylation of neutral imidazoles[582,584] and benzimidazole[583,584] by alkyl halides actually occurs by the S_E2' mechanism in which electrophilic attack is directed at the pyridine-like nitrogen. Under neutral conditions, the ensuing benzimidazolium intermediate (**1.249**) reacts with unchanged benzimidazole (**1.250**) with the consequence that yields are restricted to around 50%. This problem is best alleviated by alkylation of benzimidazoles in the presence of bases, and satsfactory yields of alkylated products can be obtained by using 2 moles of alkyl halide and 1.5 moles of base per mole of benzimidazole.

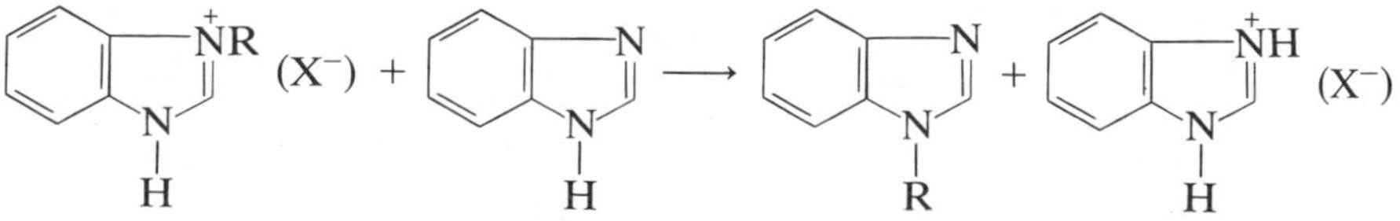

(1.249)   **(1.250)**

Because of the tautomeric equilibrium in benzimidazoles, unsymmetrically substituted derivatives possess two possible sites for alkylation. By analogy with early work of Phillips,[585] it was considered that treatment of 2-methyl-5-nitrobenzimidazole with diethylsulfate had provided 1-ethyl-2-methyl-6-nitrobenzimidazole as the most abundant isomer.[586] Subsequently, however,

it was shown that methylation of 2-methyl-5-nitrobenzimidazole with dimethyl sulfate in the absence of base gives a mixture containing an approximately equal proportion of 1,2,5- and 1,2,6-isomers.[583] Alkylation of 2-
chloro-5-substituted derivatives with alkyl halides in the presence of base
also gives a mixture of 1,2,5- and 1,2,6-isomers, but alkylation of 2-chloro-
4-substituted derivatives under similar conditions gives predominantly the
1,2,4- rather than the 1,2,7-isomer.[587] Isomer ratios observed during alkylation of the free base are thought to be a reflection of the position of the
tautomeric equilibrium;[583] in basic solution, the preferential formation of
the 1,2,4-isomer is ascribed to the different steric requirements of substituents in positions 4 and 7 to the entering alkyl group.[587] (For a related
example of regioselective alkylation, see Ref. 163.)

ALKYLATION BY ALKYL HALIDES AND RELATED COMPOUNDS. Routine procedures for the alkylation of benzimidazole derivatives are summarised in
Table 1.29. As described previously, the best procedure is alkylation in
the presence of a base, but yields vary depending on the substituent in the
carbocyclic ring. Some examples[598] of the utility of direct alkylation of
2-(α-hydroxybenzyl)benzimidazole are shown in Table 1.30 from which it is
interesting to note that yields are comparable or higher by using the direct
process than by synthesis from the appropriate o-nitrochlorobenzene derivative.[626]

Important differences in the mode of reactivity in alkylation are illustrated
with regard to 2-aminobenzimidazole (**1.252**).[603] 1-(3'-Propynyl)-2-
aminobenzimidazole (**1.253**) is formed by treating (**1.252**) with equimolar
quantities of propargyl bromide and sodium amide in liquid ammonia, but
alkylation by propargyl bromide in ethanol affords the benzimidazolium
derivative (**1.254**). The chemistry of these valuable intermediates (**1.253**,
1.254) is described in section 6.2.1 in Chapter 6 in relation to the synthesis
of imidazo[1,2-a]benzimidazoles.

TABLE 1.29. ALKYLATION OF BENZIMIDAZOLES[a] BY
ALKYL HALIDES

Organic halide used R in RX or X—R—X	Base	Ref.
—CH$_2$CN	Et$_3$N	143
—CH$_2$CHR^1(OH) (R^1 = alkyl)		
—CH$_2$CO$_2$R^1 } (R^1 = H, alkyl,	RONa	588
—CH$_2$COR^1 ∫ aryl)		
—CH$_2$CO$_2$Et	NaHCO$_3$	589
—CH$_2$R^1	NaOH	590
(R^1 = CH$_2$OH, CO-aryl, CO$_2$H etc.)		
—CH(CCl$_3$)NHCOR^1	Et$_3$N	591
(R^1 = H, alkyl, alkenyl, etc.)		
—CH$_2$Ph	—	592
C$_{1-20}$ alkyl)	KOH	593
—n-Pr	NaH	163
—CH$_3$, CH$_2$Ph	K$_2$CO$_3$	587b
—n-C$_6$H$_{13}$		
—CH$_2$—	NaH	594[b]
—CH$_2$Ar		
—Bu	NaOMe	595
—(CH$_2$)$_2$—	KOH	596
—(CH$_2$)$_2$SEt	NaH	597
—R	NaOEt	598[b]
(R = alkyl, benzyl, crotyl, allyl)		
—Hydroxyalkyl	NaH	599
—CH$_2$CH=CH$_2$	NaOEt, NaOMe	600, 601
—CH$_2$—C=CH$_2$	NaOEt	600
CH$_3$		
—CH$_2$CH—CH$_2$	NaH	601
—(CH$_2$)$_2$CHMe(CH$_2$)$_2$CH=CMe$_2$	NaH	602
—CH$_2$C≡CH	NaNH$_2$	603[b]
—(CH$_2$)$_2$NR$_2$ (R = alkyl)	NaNH$_2$	604–607
—(CH$_2$)$_2$NEt$_2$	NaOEt	608, 609
—CH$_2$CHMeNR$_2$ (R = Me, Et)	NaNH$_2$	610[b]
—(CH$_2$)$_2$-1-piperidyl	Na/dioxan	611
—CH$_2$NR$_2$	KOH	612
(R = alkyl or R$_2$ = morpholinyl, etc.)		
—(CH$_2$)$_n$NR$_2$	NaOH	613, 614
(n = 2, 3; R = alkyl or R$_2$ = morpholino, etc.)		
—CH$_2$COR (R = alkyl, aryl)	NaOMe	615
—CH$_2$CO$_2$R	K$_2$CO$_3$	616
—CH$_2$Ar	Na/dioxan	617
—Z-aryl (Z = C$_{1-4}$ alkylene)	—	618
—(CH$_2$)$_3$NMe(CH$_2$)$_2$Ar	NaOH	619
—CH$_2$CH=CHPh	NaH	620
—CO$_2$Me	NaOEt	621
—CO$_2$R (R = alkyl)	NaOEt	622
—CO$_2$CH$_2$Ph		
—CO$_2$CH$_2$CH=CH$_2$	K$_2$CO$_3$	623
—CSOMe	NaHCO$_3$	624
—CS$_2$Et	Et$_3$N	623
—(2,4-Dinitrophenyl)	NaOAc	625

[a] Other substituents in the benzimidazole are not shown.
[b] See text.

TABLE 1.30. COMPARISON OF YIELDS OF 1-
ALKYL-2-(α-HYDROXYBENZYL)-
BENZIMIDAZOLE (**1.251**) BY
DIRECT ALKYLATION[a] AND
INDIRECT SYNTHESIS[b]

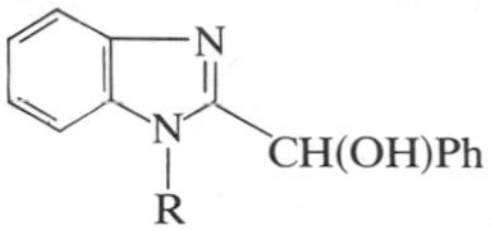

(**1.251**)

R	Reactant	Yield, (%, direct)	Yield, (%, indirect)
^{n}-Pr	RI	51	52
Bu	RI	69	40
CH$_2$Ph	RBr	30	17
CH$_2$CH=CH$_2$	RI	27	—

[a] Reaction conditions were: 2-(α-hydroxybenzyl)benz-
imidazole/RX/NaOEt/PhMe under reflux.
[b] From appropriate o-chloronitrobenzene derivative.[626]

Two investigations in relation to commercially important benzimidazoles are of interest: analgesically active[627] 1-(2-dialkylaminopropyl)-2-benzylbenzimidazoles (**1.255a, c**) have been prepared by alkylation of 2-benzylbenzimidazoles with 2-chloro-1-dialkylaminopropanes.[610] Alkylation by the diethylamino compound gives rise to one product (**1.255c**), but two compounds (**1.255a, b**) are formed in a 7:1 ratio by alkylation with 2-chloro-1-dimethylaminopropane. Preferential ring opening of an inter-mediate ethyleneimonium ion (cf. **1.256**) at the least hindered CH$_2$ position presumably gives rise to a predominance of isomer (**1.255a**).

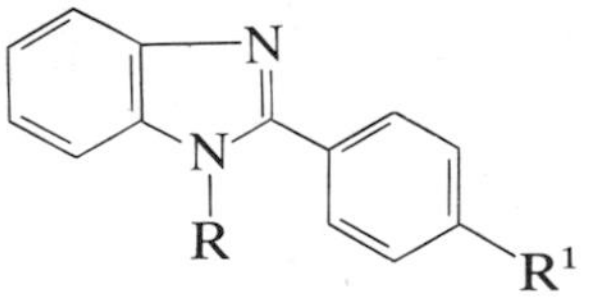

(**1.255**)

a, R = CH$_2$CHMeNMe$_2$; R^1 = H
b, R = CHMeCH$_2$NMe$_2$; R^1 = H
c, R = CH$_2$CHMeNEt$_2$; R^1 = OEt

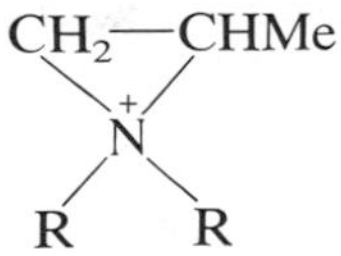

(**1.256**)

R = Me, Et

The work of Maynard and coworkers[594] on the alkylation of 2-(4-thiazolyl)benzimidazole (thiabendazole) constitutes one of the few detailed studies of the chemical reactions of this important anthelmintic agent (cf. Chapter 10). Alkylation of thiabendazole proceeds in normal fashion with n-hexyl and benzyl halides in the presence of sodium hydride.[628] Analogous procedures with $\alpha\omega$-dihalides give rise to a series of bisbenzimidazolylalkanes (**1.257**), but in the absence of base the bromo derivative (**1.258**) is converted into a condensed thiazolium salt (**1.259**) in 24% yield; compounds of the latter type (**1.259**) are valuable intermediates for the synthesis of derivatives containing the [1,4]diazepino[1,2-*a*]benzimidazole ring system (see section 8.2.1 in Chapter 8).

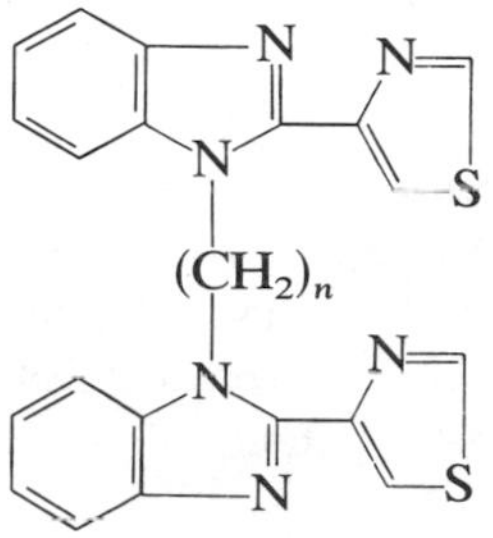

(1.257)

$n = 1, 3, 6, 12$ (50–60% yields)

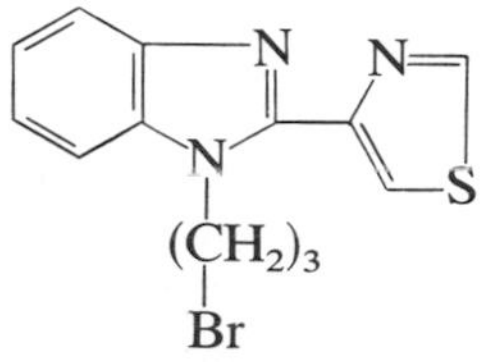

(1.258)

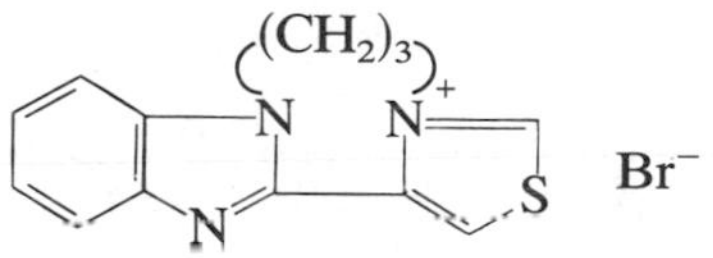

(1.259)

Dialkyl sulfates have been used for the alkylation of benzimidazoles,[188,232,586,629,630] and good yields can be obtained from this type of reaction. For example, 2-(2,4-dinitrobenzyl)benzimidazole is monoalkylated in 90% yield by treatment with dimethylsulfate and sodium bicarbonate in aqueous acetone. Methylation of 2-aminobenzimidazole by dimethylsulfate gives rise to a trimethyl derivative that is formulated as a dihydrobenzimidazole (**1.260**) rather than a dimethylamino derivative (**1.261**) on the basis of ultraviolet absorption spectral data.[630]

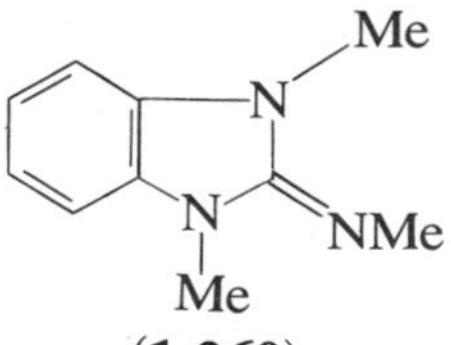

(1.260) 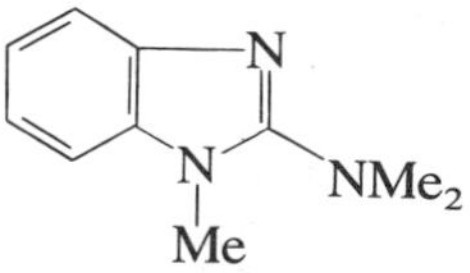   **(1.261)**

Phosphorylation reactions of benzimidazole have not been widely studied: a number of 1-phosphinyl Cambendazole analogs (e.g., **1.262**) (cf. Chapter 10) have been prepared[631] and the properties of simple phosphoryl derivatives (cf. **1.263**) as plant-growth stimulators have been discovered.[632] It is interesting to note that, whereas phosphorylation of benzimidazole with

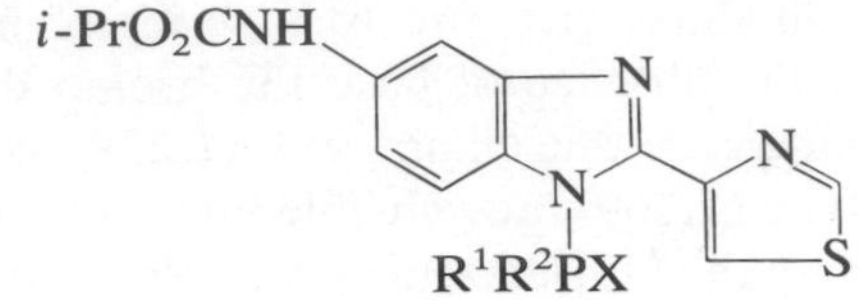

i-PrO$_2$CNH

(1.262)

X = O, S
R^1, R^2 = O-alkyl etc.

phosphorus trichloride, phosphoryl chloride, and thiophosphoryl chloride gives rise to a series of trisbenzimidazoles (**1.263**),[632] reaction with 5,6-dimethylbenzimidazole affords, after hydrolysis, bisbenzimidazoles (**1.264**) under comparable phosphorylation conditions;[633] formation of the bis analogs in the latter cases is ascribed[633] to a steric effect.

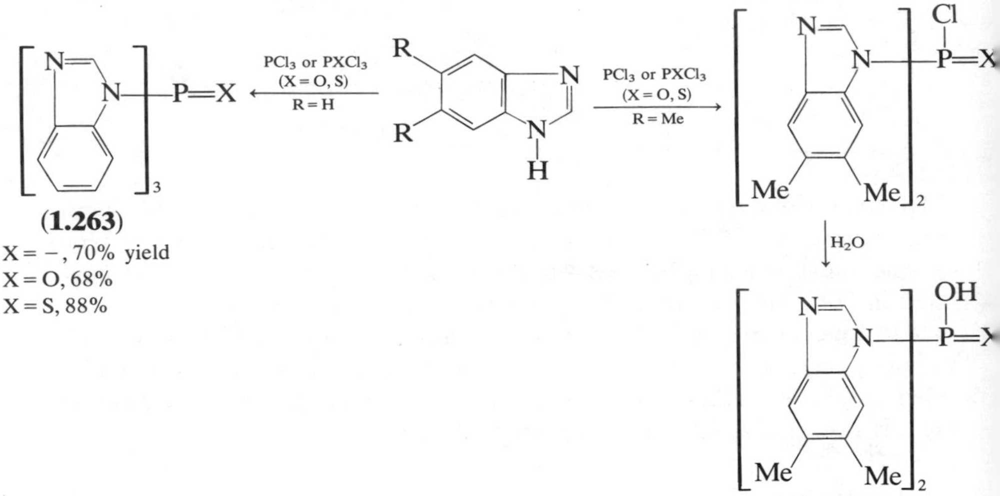

PCl$_3$ or PXCl$_3$
(X = O, S)
R = H

PCl$_3$ or PXCl$_3$
(X = O, S)
R = Me

(1.263)

X = −, 70% yield
X = O, 68%
X = S, 88%

H$_2$O

(1.264)

X = −, 62% yield
X = O, 88% yield
X = S, 92% yield

In reactions related to these procedures, the synthesis of 1-(halogenoalkyl)sulfenyl derivatives of alkyl-2-benzimidazolyl carbamates[634] and 4-mercaptobenzimidazole[635] have been described; sulfenylation is effected by the use of sulfenyl chlorides in the presence of triethylamine and aqueous sodium hydroxide, respectively.

RING OPENING OF EPOXIDES AND LACTONES. Table 1.31 illustrates reactions in which substituted 1-alkylbenzimidazole derivatives are formed by ring

TABLE 1.31. ALKYLATION OF BENZIMIDAZOLES BY EPOXIDE RING-
OPENING

Reactant	Reaction conditions	Type of benzimidazole product[a]	Ref.
Ethylene oxide	EtOH/trace pyridine/heat	1-CH$_2$CH$_2$OH	636
Ethylene oxide	AcOH/HCl	1-CH$_2$CH$_2$OH	637
Propylene oxide	EtOH/trace pyridine/65°	1-CH$_2$CH(OH)CH$_3$	638
ArOCH$_2$CH—CH$_2$ (epoxide)	EtOH/trace pyridine/heat	1-CH$_2$—CHCH$_2$OAr | OH	639
RCH—CH$_2$ (epoxide) (R = H, Me, CH$_2$OH)	EtOH/trace pyridine	1-CH$_2$CH(OH)R	640
o-ClC$_6$H$_4$OCH$_2$—CH$_2$ (epoxide)	Pyridine(cat)/100°	1-CH$_2$CH(OH)CH$_2$OC$_6$H$_4$Cl-o	641
RCH CH$_2$ (epoxide) (R – alkyl, aryl)	EtOH/trace NaOH	1-CH$_2$CH(OH)R	642[b]
Styrene oxide	NaH/HCONMe$_2$	1-CH=CHPh-2-CO$_2$H[c]	643[b]

[a] Other substituents in the benzimidazole derivative are not shown.
[b] See text.
[c] The starting material is ethyl benzimidazole-2-carboxylate.

opening of epoxides. The method of choice involves heating the benzimidazole derivative with the epoxide in the presence of a catalytic quantity
of base. Good yields are obtained in this type of reaction (e.g., 44 to 92%
from ring opening of akyl- and aryl-substituted epoxides by 2-
chlorobenzimidazole derivatives[642]), and ring opening provides the β-
hydroxyethyl isomer as might be expected.

An unusual synthesis of 1-styrylbenzimidazole-2-carboxylic acid (**1.267**) was
achieved during an attempt to alkylate ethyl benzimidazole-2-carboxylate
(**1.265**) with styrene oxide.[643] Small quantities (~2%) of the cyclic lactone
(**1.266**) are also isolated, and a mechanism involving the intermediacy of
such a compound has been suggested (see Scheme 1.15).[643] This type of
procedure has been extended to prepare derivatives of other heterocycles
such as pyrazoles,[643] indoles,[643] and pyrroles.[643,644]

1-(Carboxyalkyl)benzimidazoles can be prepared by heating the parent
compound with a lactone in dimethylformamide in the presence of sodium
hydride. One example of this type of transformation is conversion of the
erythronolactone (**1.268**) into the D-erythronic acid derivative (**1.269**).[645]

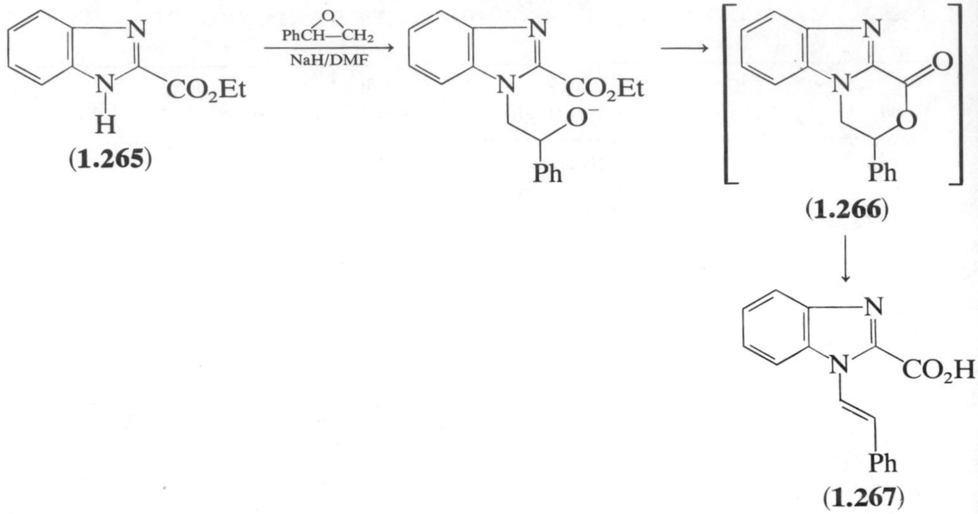

(**1.265**) (**1.266**) (**1.267**) 57% yield

Scheme 1.15

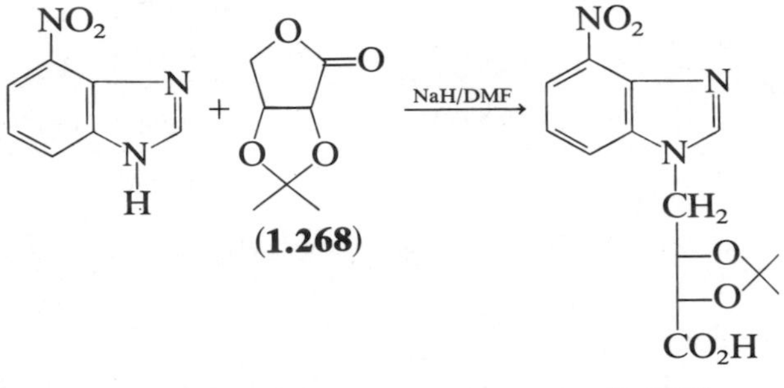

(**1.268**) (**1.269**)

ALKYLATION BY MANNICH REACTIONS. The synthesis of 1-(aminomethyl)benzimidazoles (cf. **1.270**) by the use of the Mannich procedure was first described by Bachman and Heisey.[646] This efficient method has been subsequently applied to the preparation of a variety of benzimidazole derivatives including Benlate analogs (cf. **1.271**)[652] (see Table 1.32).

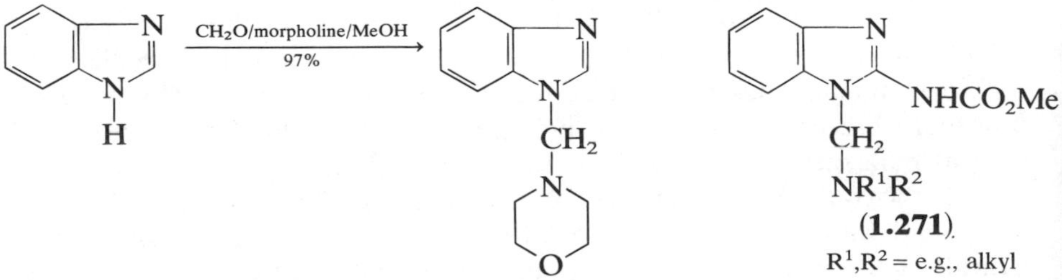

(**1.270**) (**1.271**) R^1,R^2 = e.g., alkyl

TABLE 1.32. ALKYLATION OF BENZIMIDAZOLES BY USE OF THE MANNICH PROCEDURE

Reactants	Reaction conditions	Type of benzimidazole product[a]	Ref.
$H_2N(CH_2)_2$-2-pyridyl/CH_2O Et_2NH/CH_2O	EtOH/reflux	1-$CH_2NH(CH_2)_2$-2-pyridyl 1-CH_2R	647
Morpholine/CH_2O Piperidine/CH_2O	EtOH/HCl/reflux	R = NEt_2, 1-morpholinyl, 1-piperidinyl	648
$HN(CH_2CH_2OH)_2$/CH_2O	EtOH/reflux	1-$CH_2N(CH_2CH_2OH)_2$	649
Morpholine/CH_2O	MeOH	1-CH_2(-1-morpholinyl)	650
p-$H_2NC_6H_4CO_2Et$/CH_2O	EtOH/0°	1-$CH_2NHC_6H_4CO_2Et(p)$	651
R^1R^2NH/CH_2O (R^1R^2 = alkyl, benzyl, etc.)	CH_2Cl_2/38°	1-$CH_2NR^1R^2$	652
$ArNH_2/CH_2O$ (Ar = aryl)	EtOH/reflux	1-CH_2NHAr	653
R_2NH/CH_2O (R = Me or R_2 = piperidinyl and piperazinyl derivatives)	MeOH	1-CH_2NR_2	654

[a] Other substituents in the benzimidazole derivative are not shown.

The amine exchange reaction[655] of Mannich bases has been successfully applied to the synthesis of 1-alkylated benzimidazoles containing a γ-carbonyl function (e.g., **1.272**[656] and **1.273**[657]).

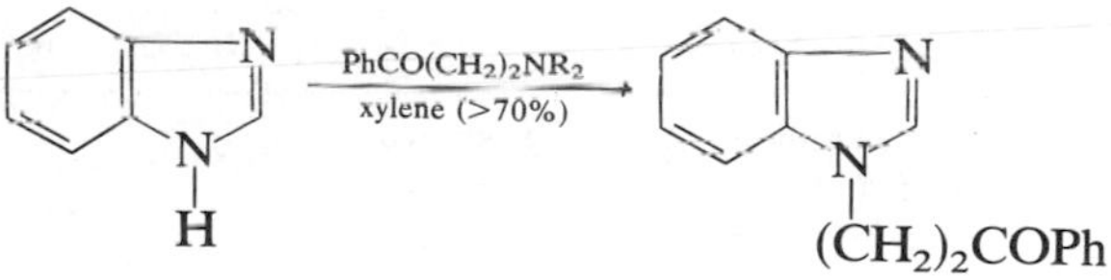

(**1.272**)

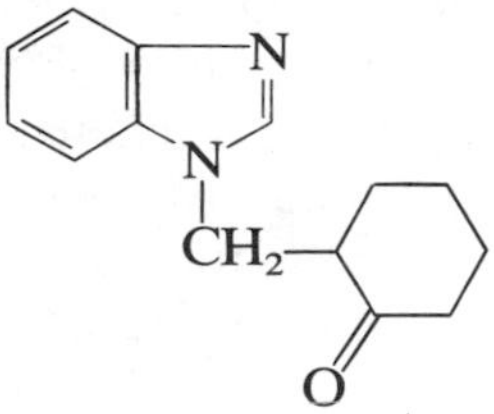

(**1.273**)

ALKYLATION BY ACTIVATED ALKENES. Benzimidazole derivatives have been alkylated by activated alkenes in thermal reactions and also in acid- and base-catalyzed processes (see Table 1.33); moderate to good yields are obtained in reactions of this type (e.g., 33 and 77% in alkylation by 4-vinylpyridine[672] and N-phenylmaleimide,[668] respectively).

TABLE 1.33. ALKYLATION OF BENZIMIDAZOLES BY ACTIVATED ALKENES

Reactant	Reaction conditions	Type of benzimidazole product[a]	Ref.
CH_2=CHCN	aq. dioxan/ $Et_3\overset{+}{N}HCH_2Ph(OH^-)/$ 45–60[ob]	1-CH_2CH_2CN	658–660
CH_3CH=$CHCO_2H$	Pyridine catalyst	1-$CHMeCH_2CO_2H$	661
CH_2=$C(R)CO_2H$ R = H, Me	150–200°	1-CH_2CHRCO_2H	662
CH=$CHCONH_2$	$Et_3\overset{+}{N}CH_2Ph(OH^-)$/pyridine	1-$CH_2CH_2CONH_2$	663
CH_2=$CHNO_2$	EtOH/room temp.	1-$CH_2CH_2NO_2$	664
HO_2CCH=$CHCO_2Na$	aq. NaOH/heat	1-$CH(CO_2H)CH_2CO_2Na$	665
HO_2CCH=CHCONHR (R = H, alkyl, etc.)	(i) DMF/110° (ii) polyphosphoric acid/110°		666
(R = H, alkyl etc.)	Dioxan or acetonitrile/ heat		667, 668
CH_2=CHOCO-2-furyl	Heat in H_2O	1-CH_2CH_2OCO-2-furyl	669
CH_2=CH-2-pyridyl	AcOH/120°	1-CH_2CH_2-2-pyridyl	670, 671
CH_2=CH-4-pyridyl	i-PrOH/reflux	1-CH_2CH_2-4-pyridyl	672

[a] Other substituents in the benzimidazole are not shown.
[b] Reaction conditions of Ref. 658.

Treatment of benzimidazole with 1 mole of diphenylketene affords 1-(diphenylacetyl)benzimidazole (**1.274**), but with 3 moles of this reagent a tricyclic compound is produced that was originally formulated as (**1.275**).[673] From a more recent detailed study of the reaction of a variety of benzazoles with diphenylketene, it has been concluded that the tricyclic product has the [1,3]oxazino[3,4-*a*]benzimidazol-3(4*H*)-one structure (**1.276**) rather than (**1.275**)[674]; a mechanism involving a zwitterionic intermediate is envisaged[674] (see structure (**1.277**) and also section 7.2.1 in Chapter 7).

ACYLATION AND SULFONATION BY ACID HALIDES AND SULFONYL HALIDES. Benzimidazoles can be acylated or sulfonated by treating them with acid halides or sulfonyl halides, respectively, in the presence of bases such as sodium hydride or triethylamine (see Table 1.34). Particular attention has been focused on the synthesis of analogs of commercially useful 2-(4-thiazolyl)- and alkylbenzimidazole 2-carbamate derivatives (cf. Chapter

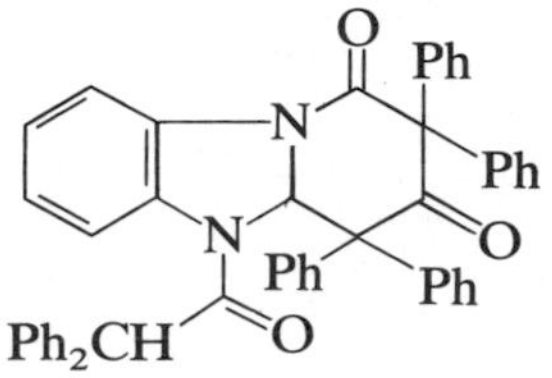

(1.275)

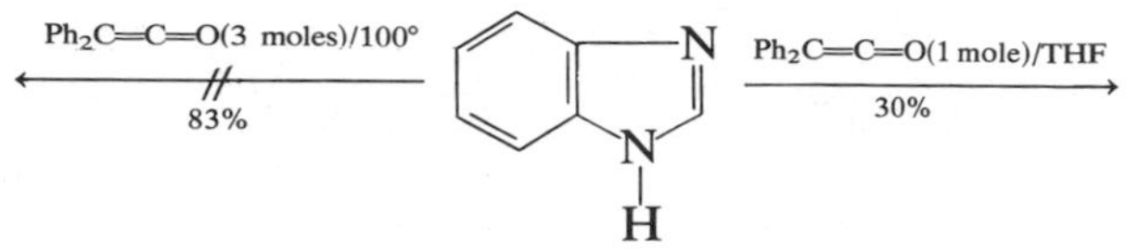

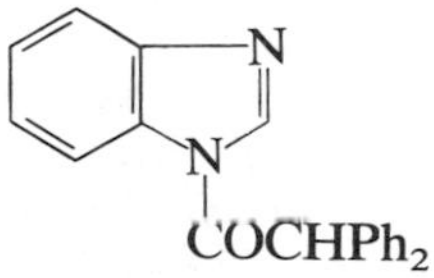

(1.274)

10), and good yields of acylated derivatives have been obtained (e.g., ca. 68% for acylation of alkylbenzimidazole-2-carbamates[677]). The synthesis of tri-N-alkoxycarbonyl derivatives (**1.278** and **1.279**) has been achieved by acylation reactions of 2-aminobenzimidazole and its mono- and di-(N-alkoxycarbonyl) derivatives.[684] The di-N-alkoxycarbonyl compounds (**1.280**) are accessible in moderate yields (33 to 49%) by hydrolysis of the trisubstituted derivatives (**1.278**) with sodium hydroxide in ethanol.[684]

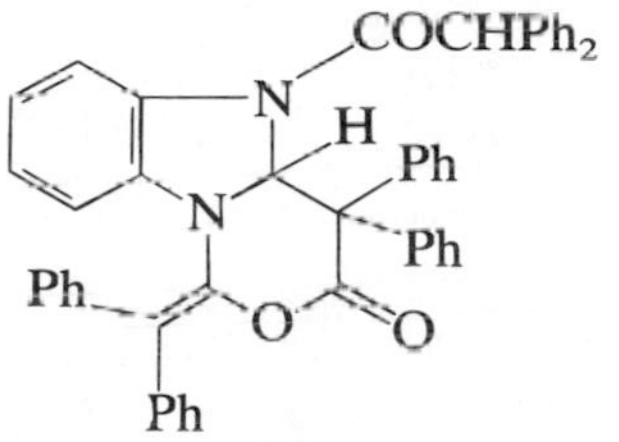

(1.276)

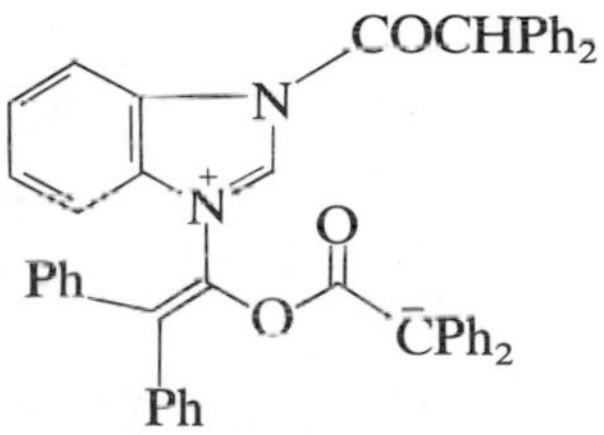

(1.277)

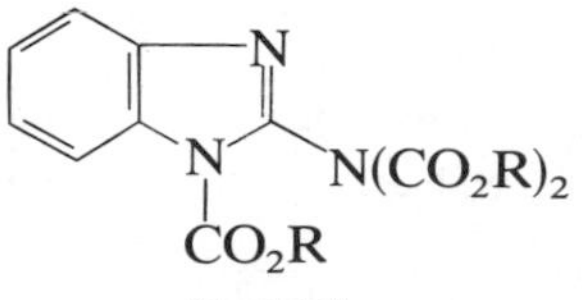

(1.278)

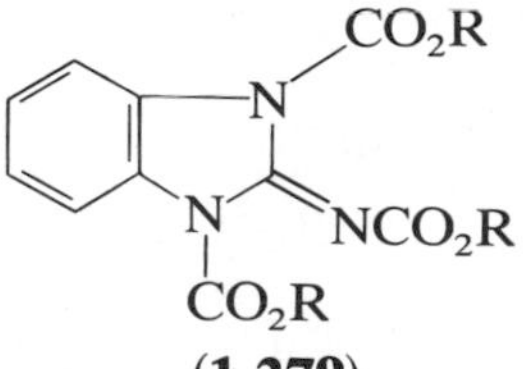

(1.279)

(R = alkyl)

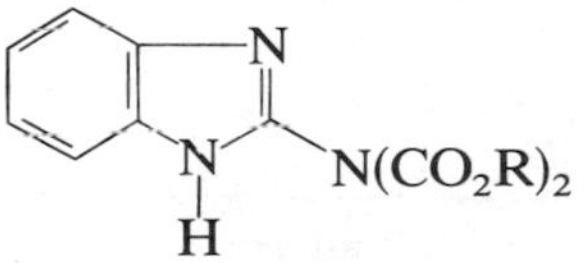

(1.280)

The kinetics of benzoylation of monocylic imidazoles and also benzimidazole and two of its derivatives have been studied.[710,711] Benzoylation

TABLE 1.34. N(1)-ACYLATION AND -SULFONATION OF BENZIMIDAZOLES[a]

Reactant	Reaction conditions	Ref.
RCOCl (R = alkyl, aryl)	NaH/PhMe/60°	675, 676
RCOCl (R = alkyl, aryl, alkenyl, etc.)	Et$_3$N/CHCl$_3$/room temp.	677
RCOCl (R = alkyl, aryl, CH$_2$OPh, O-alkyl)	Et$_3$N or pyridine/room temp.	678
RO(CH$_2$CH$_2$O)$_n$COCl (R = alkyl, CH$_2$Ph; n = 1, 2, 4)	Pyridine	679
RCOCl (R = substituted cyclopropyl or cyclopentenyl)	CHCl$_3$/Et$_3$N/>38°	680
RCOCl (R = p-Me$_2$CHCH$_2$C$_6$H$_4$CHMe)	Et$_3$N	681
RCOCl (R = PhCH=CH)	NaH	682
RCOCl (R = aryl)	C$_6$H$_6$/Et$_3$N/heat	683
ClCO$_2$R (R = alkyl)	Pyridine/20°	684
ClCO$_2$CH$_3$	Et$_3$N[b]	685–687
ClCO$_2$i-Pr	NaoMe/Me$_2$CO	688
ClCO$_2$R (R = alkyl, aryl, alkenyl, etc.)	NaOEt/Me$_2$CO[c]	689–691
ClCO$_2$(CH$_2$)$_n$Ph (n = 1, 2)	Heat	692
ClCO$_2$R (R = CH$_2$Ph,) CH$_2$CH=CH$_2$)	K$_2$CO$_3$/Me$_2$CO	693
ClCO$_2$N=$\overset{R^1}{\underset{R^2}{<}}$ (R^1, R^2 = alkyl, aryl, etc.)	Me$_2$CO/CHCl$_3$/K$_2$CO$_3$/0–14°[d]	694–697
ClCO$_2$-(4-thiazolyl)	CHCl$_3$/NaHCO$_3$/reflux	698
ClCO$_2$XCO$_2$CH$_2$CH$_2$COAr (X = C$_{5-8}$ alkylene chain)	CH$_2$Cl$_2$/pyridine	699
ClCON(CH$_2$CH$_2$Cl)$_2$	NaH/DMF/reflux	700
RSO$_2$Cl (R = alkyl, aryl)	base/Me$_2$CO (base = e.g., Et$_3$N, NaHCO$_3$ or NaH)	701–707
RSO$_2$Cl (R = alkyl, aryl, dialkylamino)	Et$_3$N/CHCl$_3$/25–30°[e]	708, 709

[a] Other substituents in the benzimidazole derivative are not shown.
[b] Conditions of Ref. 685.
[c] Conditions of Ref. 689.
[d] Conditions of Ref. 694.
[e] Conditions of Ref. 708.

TABLE 1.35.   CARBAMOYLATION OF ALKYLBENZIMIDAZOLE-2-CARBAMATES
WITH ISOCYANATE DERIVATIVES

Type of isocyanate (R in RNCO)	Ref.	Type of isocyanate (R in RNCO)	Ref.
n-Bu	712	R^1CO_2X	726
Alkyl, Ph, CH_2CO_2Et	713	(R^1 = Me, Bu;	
$Cl(CH_2)_6$	714	X = alkenyl, aralkylene,	
Bromoalkyl		chloroalkylene)	
Bromo(alkenyl)	715	$R^1(CH_2)_n$	727
Bromo(cyclopentyl)		(R^1 = $PhCH_2S$, COS-alkyl,	
$NC(CH_2)_n$	716	SCOAr, 3-phenyl-1,2,4-	
(n = 5, 11)		oxadiazol-5-yl)	
$NCCR^1R^2$	717	R^1SCH_2CO	728
(R^1, R^2 = alkyl, aryl;		(R^1 = Et, aryl, CH_2Ph)	
$R^1R^2 = (CH_2)_5$)		CH_2	
(Haloalkyl)-OCH_2	718		
$ArOCH_2CO$	719	SR^1	729
(Ar = aryl)			
R^1XCO	720	(R^1 = alkyl, phenyl)	
(R^1 = alkyl, aryl;		R^1SO_2	730
X = O, S)		(R^1 = chloroalkyl)	
$R^1S(CH_2)_2$	721	$R^1SO_2N(R^2)SO_2$	731
(R^1 = alkyl)		(R^1 = alk, aryl,	
R^1SX	722	R^2 = alk, cycloalk)	
(X = $(CH_2)_2$;		$ArOSO_2$	732
$R^1 = SCCl_2$, $CHCl_2$)		(Ar = aryl)	
$R^1S(CH_2)_n$	723	Tetrahydropyranyl,	733
(n = 1, 2;		dihydropyranyl,	
R^1 = MeS, Ar, (alkenyl)S)		tetrahydrofuryl	
$R^1R^2N(CH_2)_n$	724		
(R^1, R^2 = alkyl;		X	734
NR^1R^2 = morpholino, etc.;		O	
n = 1–10)		(X = O, S)	
Chloroaryl	725^a	R^1CO	735
		(R^1 = aryloxyalkyl)	

^{a} Refers to reactions of benzimidazole rather than 2-carbamate derivatives.

of benzimidazole with benzoyl chloride in benzene at 25° is slower (k = $1.02 \pm 0.04 \, \text{lm}^{-1} \, \text{sec}^{-1}$) than imidazole ($20.2 \pm 0.6$) but is faster than 2-ethylbenzimidazole (0.022 ± 0.001); the latter result is rationalized in terms of a steric effect operating in a mechanism of the S_E2' type.[711]

CARBAMOYLATION BY ISOCYANATE DERIVATIVES. The commercial success of the DuPont fungicide Benlate (1.281) (cf. Chapter 10) has encouraged work on the synthesis of a variety of analogs in which the benzimidazole 1-substitutent is modified. Such compounds are prepared routinely in good yield by allowing the alkylbenzimidazole 2-carbamate to react with the appropriate isocyanate or diisocyanate (see Tables 1.35 and 1.36, respectively). The preparation of 1-carbamoyl derivatives of 2-(4-thiazolyl)[740] and 2-acylaminobenzimidazoles[741] has also been reported.

TABLE 1.36.   SYNTHESIS OF BIS-
                BENZIMIDAZOLES BY
                CARBAMOYLATION
                OF BENZIMIDAZOLES
                USING DIISOCYANATES

Type of diisocyanate (X in ONCXNCO)	Ref.
$(CH_2)_n$ ($n = 4$–12)	736, 737
$(CH_2)_6$ arylene-CH_2-arylene	738
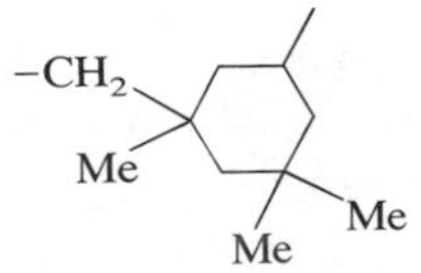 $-CH_2$ Me Me Me	739
$CH_2CMe_2CH_2CHMe(CH_2)_2$ -arylene- -arylene-CH_2-arylene	

Analogous thiocarbamates[742] are obtained by the use of isothiocyanates, and haloakyl derivatives of the latter type have been used[743–746] for the synthesis of 1-thiazolinyl- and 1-thiazinylbenzimidazole (cf. **1.282**); antiviral activity is claimed[746] for such compounds.

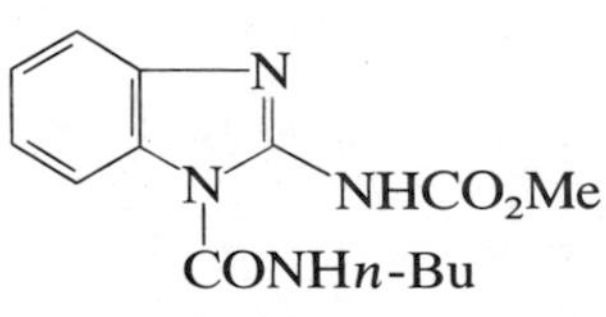

(**1.281**)

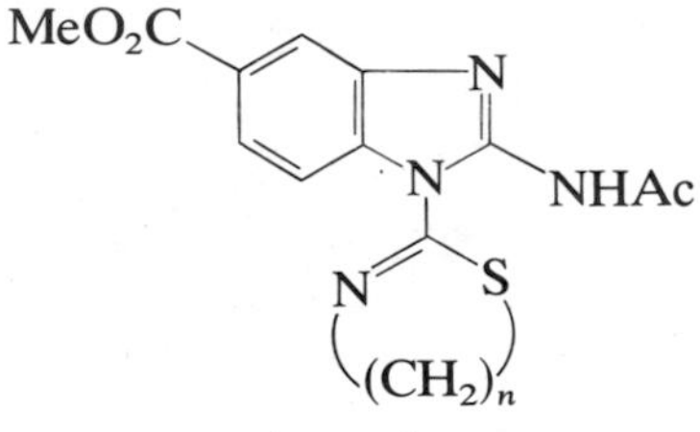

(**1.282**)

$n = 2, 3$

Benlate analogs (cf. **1.281**) have also been prepared by reaction of alkylbenzimidazole-2-carbamates with phosgene in the presence of the appropriate amine[747–750] or N-hydroxyformamidate derivative.[751]

REACTIONS WITH ACTIVATED ALKYNES. 1-Vinylbenzimidazole is formed by nucleophilic addition of benzimidazole to acetylene in aqueous dioxan at 160°,[752] and analogous products are produced (cf. **1.283a, b**) from benzimidazoles and hexafluorobut-2-yne in methanol and tetrahydrofuran, respectively.[753] When dimethylacetylene dicarboxylate (DMAD) is treated

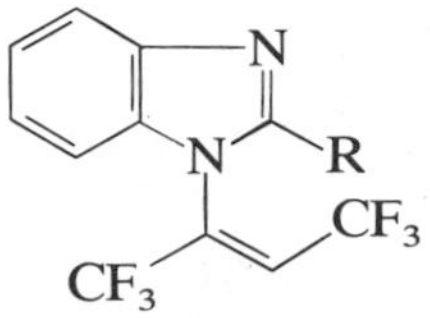

(1.283)

a, R = H (70%)
b, R = SCH₂CN (23%)

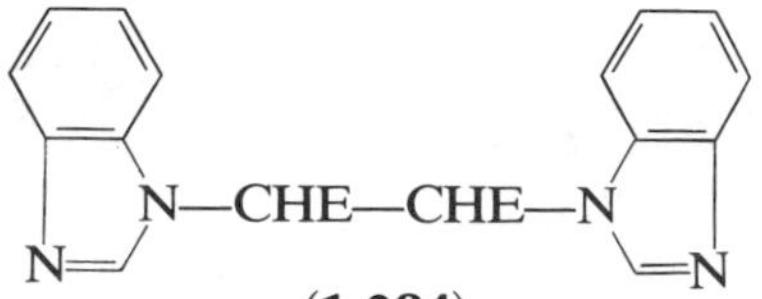

(1.284)

E = CO₂Me

with an excess of benzimidazole in benzene, the dimethyl succinate derivative (**1.284**) is obtained,[754] presumably by addition to the intermediate
(**1.285**). On the other hand, when an excess of DMAD is used, a pyrido[1,2-
a]benzimidazole is formed (see section 7.1.1 in Chapter 7) together with the
dihydro derivative (**1.287a**) as a minor product. The yield of this product
increases when methanol is added to the reactants, and an analog (**1.287b**) is
obtained from diethylacetylene dicarboxylate and ethanol; the intermediacy
of a benzimidazolium derivative (**1.286**) is suggested.[754] Compounds analogous to (**1.285**) are formed by treating 2-benzylbenzimidazole with DMAD,[755]
and also by treating a series of 2-alkyl, -aryl, and -aralkyl derivatives with
methyl propiolate.[756]

It may be noted that the reactions of benzimidazole derivatives with
DMAD and related compounds have proved to be of value for the syntheses
of condensed benzimidazoles of the 6-5-5, 6-5-6, and 6-5-7 types (see
Chapters 6, 7, and 8, respectively).

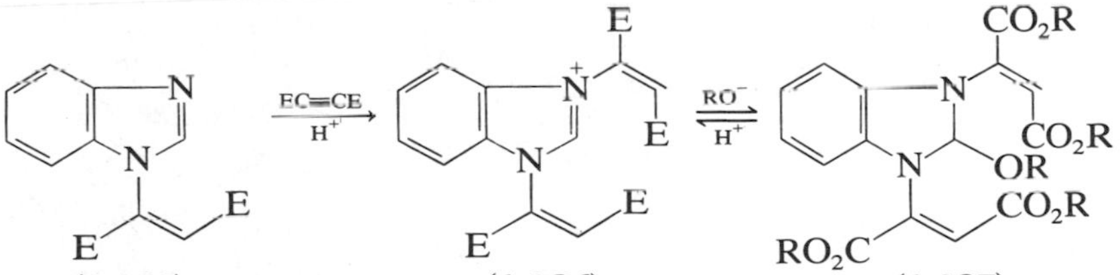

(1.285)

E = CO₂Me or CO₂Et

(1.286)

(1.287)

a, R = Me
b, R = Et

INTRAMOLECULAR ALKYLATION AND ACYLATION. Reactions in this category
form the basis of the preparation of a variety of tricyclic derivatives
containing a condensed benzimidazole moiety. Such processes are discussed
in detail in later chapters, and this section is restricted to reactions leading to
condensed polycyclic systems containing at least four rings.

Treatment of 2-(3,4-dimethoxybenzyl)benzimidazole (**1.288**) with an
anhydride in the presence of a catalytic quantity of perchloric acid leads to
acylation products of the veratrole ring (cf. **1.289**), but if an equimolar
quantity of perchloric acid is used the products are benzimidazo[3,2-
b]isoquinolinium perchlorates (cf. **1.290**).[757] Other compounds in this ring

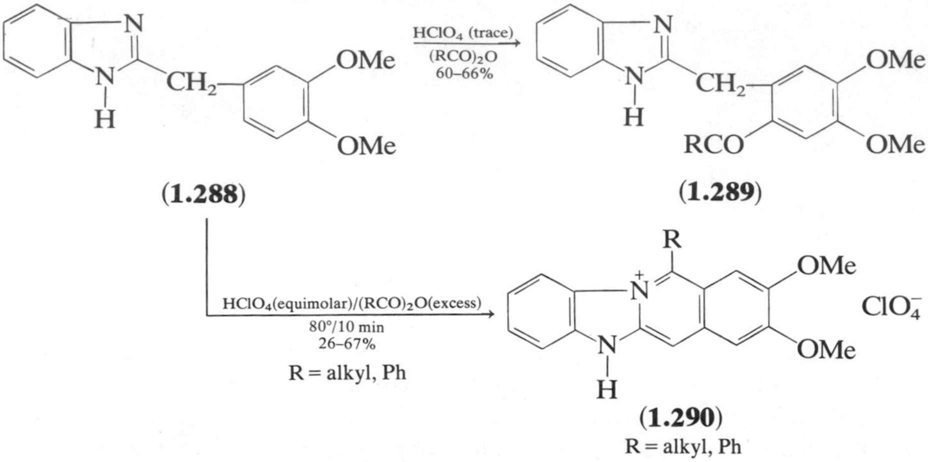

(**1.288**)   (**1.289**)

(**1.290**)

R = alkyl, Ph

system have subsequently been prepared by related intramolecular cyclization reactions of 2-(2-carboxybenzyl)- and 2-(2-carboxybenzoyl)benzimi-
dazoles (see Scheme 1.16).[758].

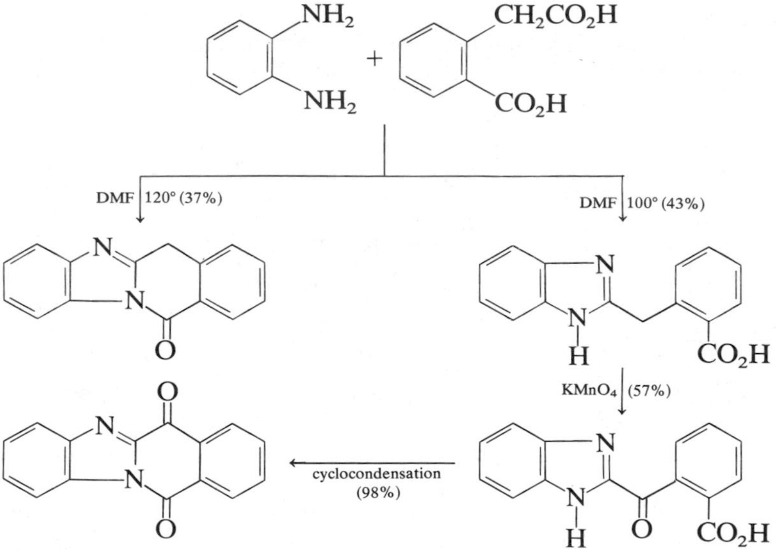

Scheme 1.16

Tetracyclic   derivatives   (**1.291a, b**)   are   also   formed   when   2-
aminobenzimidazoles are treated with phthaloyl chloride in pyridine.[759] The
recent definitive formulation[759] of such products as benzimidazo[1,2-*b*]
[2,4]benzodiazepin derivatives is important, since one such product (**1.291a**)
has been erroneously described as the isomeric 2-phthalimidobenzimidazole
(**1.292**).[760]

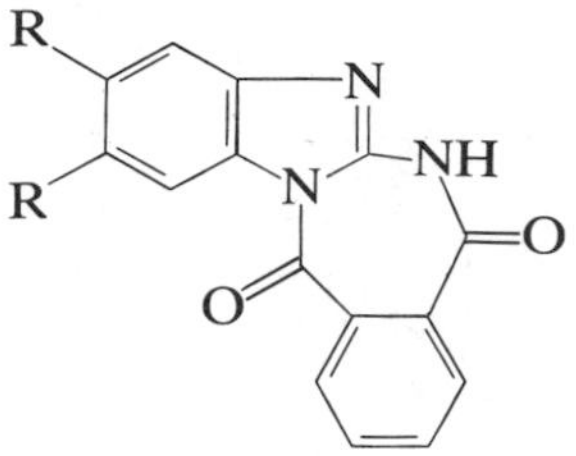

(1.291)

a, R = H (66% yield)
b, R = Me (58% yield)

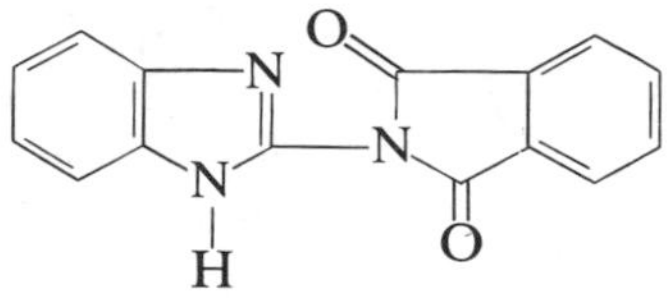

(1.292)

A combination of inter- and intramolecular alkylation reactions of the pyridinium salts (**1.293a, b**) makes accessible the bisbenzimidazo[1,2-*a*; 1′,2′-*e*][1,5]diazocine derivatives (**1.294a** and **b**) in good yield;[761] the molecular structure of (**1.294a**) has been verified by an X-ray diffraction study.[761]

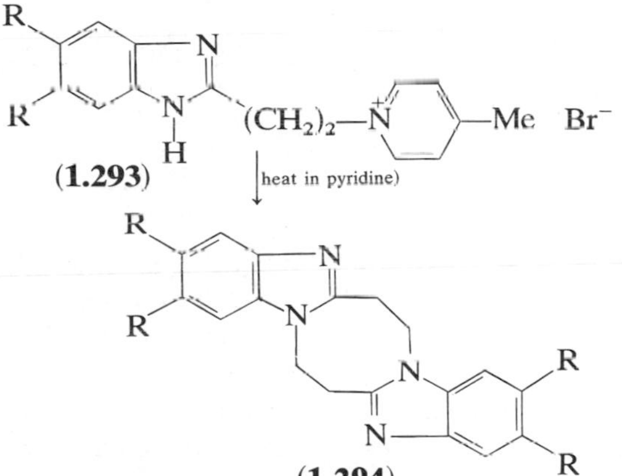

(1.293)

heat in pyridine

(1.294)

a, R = H (42% yield)
b, R = Me (95%)

An interesting dearomatization reaction of the imidazole ring of the bis amides (**1.295a** and **b**) occurs when they are heated under reduced pressure;[762] products analogous to (**1.296**) are also formed in related reactions of 4-quinazolone and with acyclic amidines.[762]

SYNTHESIS OF BENZIMIDAZOLIUM COMPOUNDS. 1-Alkyl- or 1-arylbenzimidazoles are alkylated at the 3-position when they are heated with an alkyl halide in methanol under pressure at temperatures in the range, 110–150°. Benzimidazolium salts that have been routinely synthesized in this manner or by related procedures are collected in the Appendix of Derivatives, Table 1.74.

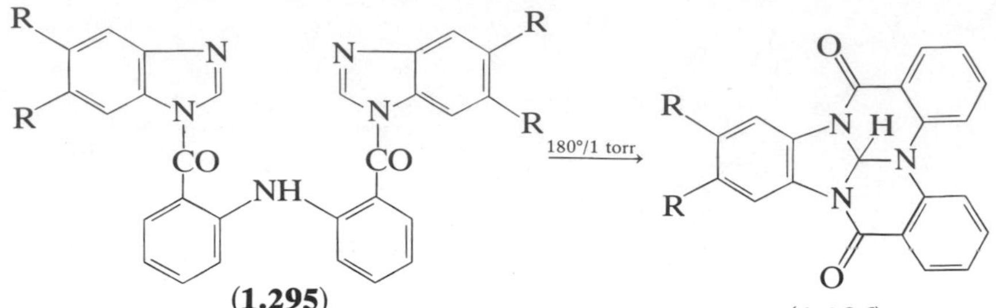

(1.295)

(1.296)

a, R = H (74% yield)
b, R = Me (60%)

Direct *N*-amination of 1-alkylbenzimidazoles has been effected[763] using *O*-mesitylenesulfonylhydroxylamine (MSH)[764] (cf. **1.297** → **1.298**), but it has been subsequently shown that aminations of this type can also be carried out using a cheaper reagent, *O*-*p*-tolylsulfonylhydroxylamine (TSH).[765]

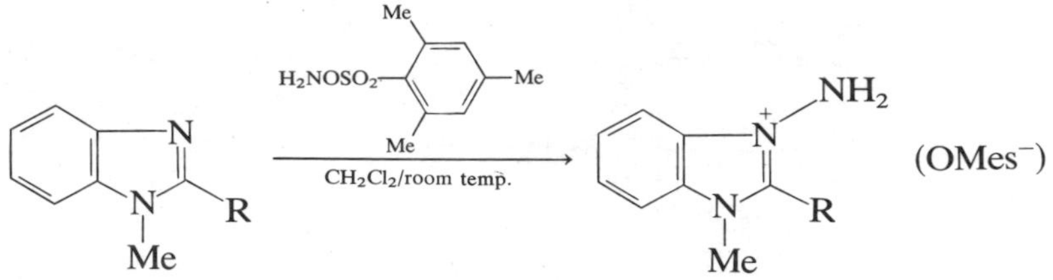

(1.297)

(1.298)

a, R = H (96% yield)
b, R = CH$_3$ (86%)

A 1,2-shift in an intermediate benzimidazolium compound (cf. **1.299**) is envisaged following the reaction of benzimidazoles with phthaloyl chloride;[766] this procedure constitutes an unusual synthesis of benzimidazo[1,2-*b*]isoquinolin-5,10-diones (**1.300**) albeit in poor yields (cf. analogous 1,2-shifts observed during the reaction of 1-alkylimidazoles with acyl halides).[766]

MISCELLANEOUS REACTIONS. 1-Cyanobenzimidazoles are formed when cyanogen bromide is allowed to react with 1-unsubstituted benzimidazoles in the presence of sodium hydride;[767] an alternative method is use of the silver salt of the benzimidazole derivative.[768]

An unusual transformation has been observed when the thiocyano derivatives (**1.301a** and **b**) are treated with carbon disulfide.[769] The anticipated tricyclic derivatives (**1.302**) are not produced, and the products are formulated[769] as [1,3,5]thiadiazino[3,2-*a*: 5,6-*a'*]bisbenzimidazoles (**1.303a** and **b**). The scope and mechanism of this interesting reaction have not been evaluated.

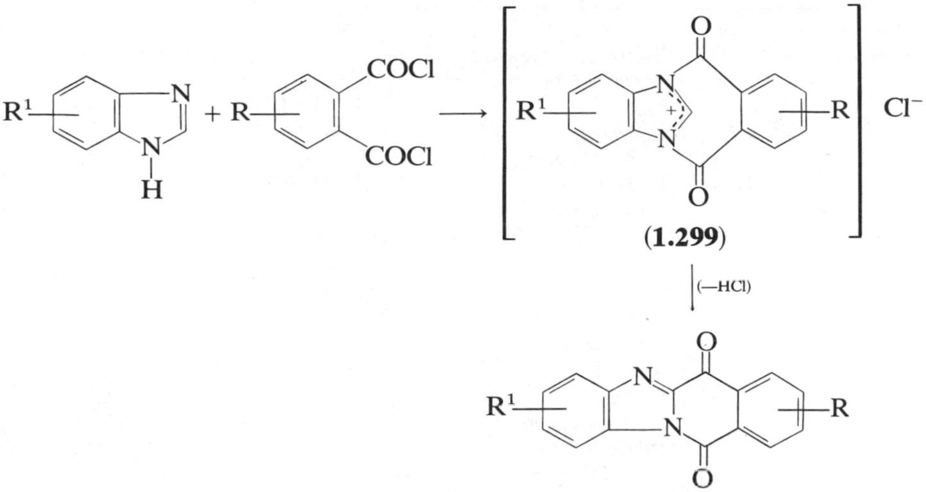

(1.299)

$\downarrow$ (—HCl)

(1.300)

R^1 = H,hal,NO$_2$,CF$_3$
R = H,hal,NO$_2$,MeO $\Big\}$ 2.5–30% yield

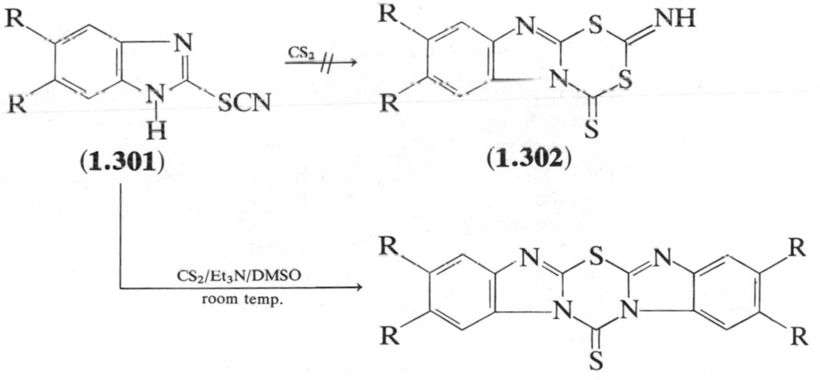

(1.301) $\xrightarrow{\text{CS}_2} \not\rightarrow$ **(1.302)**

$\downarrow$ CS$_2$/Et$_3$N/DMSO
room temp.

(1.303)
a, R = H (86% yield)
b, R = Me

Electrophilic Attack at Side-Chain Substituents

Reactions in this category include substitution and addition processes and entail to a large extent reactions of the 2-amino function. These, and also modifications of other substituents by electrophilic reagents are listed in Table 1.37, and selected reactions in the table are discussed here.

TABLE 1.37. REACTIONS OF BENZIMIDAZOLE SIDE-CHAIN SUBSTITUENTS WITH ELECTROPHILIC REAGENTS

Position and type of substituent nucleophile	Electrophilic reagent	Type of benzimidazole product[a]	Ref.
1-CH$_2$CH$_2$OH	Ph$_2$CHCl	1-CH$_2$CH$_2$OCHPh$_2$	770
1-CH$_2$CH$_2$OH	RCOCl (R = alkyl, aryl)	1-CH$_2$CH$_2$OCOR	771
1-CH$_2$CONHNH$_2$	RCHO (R = aryl, 2-furyl)	1-CH$_2$CONHN=CHR	589
2-NH$_2$	Alkyl halides	2-Alkylamino 2-N,N-dialkylamino derivatives	772[e]
2-NH$_2$	RCOCl (R = aryl, hetaryl)	2-NHCOR	773
2-NH$_2$	RCOCl (RCO)$_2$O (R = alkyl, aryl)	2-NHCOR	774
2-NH$_2$	RCOCl (R = Me, Et)	2-NHCOR	775
2-NH$_2$	RCOCl (R = 2-furyl, 2-thienyl)	2-NHCOR	776
2-NH$_2$	RCOCl (R = 2-(5-nitrofuryl))	2-NHCOR	777
2-NH$_2$	RCOCl R=O⟨⟩O 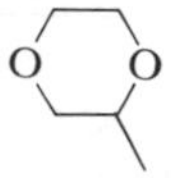	2-NHCOR	778
2-NH$_2$	ClCO$_2$Et	2-NHCO$_2$Et	779
2-NH$_2$	ClCO$_2$R (R = alkyl)	2-NHCO$_2$R 2-N(CO$_2$R)$_2$ derivatives	780
2-NH$_2$	ClCO$_2$Ph	2-NHCO$_2$Ph	781
2-NH$_2$	ClCO$_2$R (R = 2-thienyl, 4-thiazolyl)	2-NHCO$_2$R	782
2-NHCO$_2$R (R = Me, Et)	R^{1}COCl (R^{1}CO)$_2$O (R^1 = alkyl)	2-N(COR^1)CO$_2$R	783
2-NHPh	ClCO$_2$Me	2-N(Ph)CO$_2$Me	784
2-NH$_2$	(RO)$_2$CO (R = Me, Et, Bu)	2-NHCO$_2$R	785
2-NH$_2$	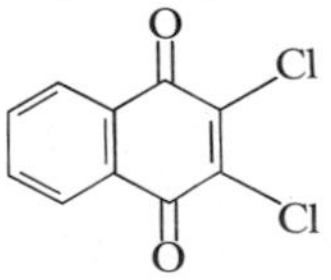	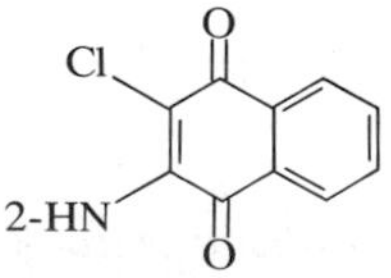 2-HN	788
2-NH$_2$	ArCHO	2-N=CHAr	789–791
2-NH$_2$	RCHO (R = 2-furyl)	2-N=CHR	792
2-NH$_2$	RCHO (R = 2-furyl, 2-thienyl)	2-N=CHR	793

106

TABLE 1.37 (Continued)

Position and type of substituent nucleophile	Electrophilic reagent	Type of benzimidazole product[a]	Ref.
2-NH₂	RNCO (R = alkyl, cycloalkyl)	2-NHCONHR	794
2-NH₂	O=⟨▢⟩=O (X = alkenyl alkylene phenylene etc.)	2-NHCOXCO₂H	795
2-NH₂	Isopentyl nitrite	2-NHNO	796[e]
2-NH₂	NaNO₂/H₃PO₄	2-NHNO	797
2-NH₂	Isoamyl nitrite	2-NHNO	798[e]
2-CH₂CH₂NH₂	CS₂/NH₂NH₂	2-CH₂CH₂NHCSNHNH₂	799
2-⟨thiazole⟩NH₂	ArNCO		800
2-NHNO	H₂SO₄	2-BI—NH—N=N-2-BI[b]	801[e]
2-CHMeNH-4-thiazolyl	ClCO₂Me	2-CHMeN(CO₂Me)-4-thiazolyl	802
2-(4-pyridyl)	RX (R = alkyl, alkenyl, allyl, etc.)	2-(4-pyridinium salts)	803
2-NHNH₂	R¹COR (R,R¹ = H, Me, aryl, etc.)	2-NHN=CRR¹	804
2-NHNH₂	MeCOCH₂COMe		805
2-XCSNH₂ (X = bond, CH₂, (CH₂)₂, CH=CH)	BrCH₂COR (R = Me, Ph)		806
2-SCH₂CONHNH₂	RCHO (R = aryl, styryl)	2-SCH₂CONHN=CHR	807
2-SCH₂CONHNH₂	RNCS (R = aryl, allyl)	2-SCH₂CONHNHCSNHR	808
2-Me	ArCHO	2-CH=CHAr	809
2-Me	CH₂O	2-CH₂CH₂OH	810
2-Me	(ClCH=N⁺Me₂)Cl⁻	2-C—CH=N⁺Me₂(Cl⁻) ‖ CHNMe₂	811
2-Me	PrONO₂	2-CH₂NO₂	812
2-CH₂CN	PhCH₂Cl	2-CH(CN)CH₂Ph	813
2-CH₂CN	Aldehydes Nitroso derivatives	2-C(CN)=CRR¹ (R = H, Me; R¹ = alkyl, aryl, hetaryl; 2-C(CN)=NAr)	814
2-CH₂CN	ArNCX (X = O, S)	2-CH(CN)CXNHAr	815

TABLE 1.37   (*Continued*)

Position and type of substituent nucleophile	Electrophilic reagent	Type of benzimidazole product[a]	Ref.
2-CH$_2$Ar	Ar'CHO	2-C(Ar)=CHAr'	816
2-COCH$_3$	RCHO (R = aryl, hetaryl)	2-COCH=CHAr	817–819
2-COCH$_3$	Br$_2$	2-COCHBr$_2$ 2-COCH$_2$Br	820
2-NHN=CHR	ArN$_2^+$ 2-BIN$_2^{+b}$	2-NHN=C(R)N$_2$R (R = aryl, 2-BI^b)	821–826
2-CH$_2$SH	ClCH$_2$CONHAr	2-CH$_2$SCH$_2$CONHAr	827
2-CH$_2$SH	ZX$_2$ (Z = alkylene, alkenyl, arylene etc., X = Cl, Br)	[2-BI—CH$_2$—S]$_2$Z^b	828
2-CH$_2$SH	ArCOCH$_2$X (X = Cl, Br)	2-CH$_2$SCH$_2$COAr	829
2-CH$_2$SH	ArCH$_2$X (X = Cl, Br)	2-CH$_2$SCH$_2$Ar	830
2-CH$_2$SH	ArCOCl	2-CH$_2$SCOAr	831
2-(CH$_2$)$_x$CO$_2$Na (x = 0–2)	Cl(CH$_2$)NR$_2$ (R = Me, Et)	2-(CH$_2$)$_x$CO$_2$(CH$_2$)$_2$NR$_2$	832
2-CH=NOH	ClCON[(CH$_2$)$_2$Cl]$_2$	2-CH=NOCON[(CH$_2$)$_2$Cl]$_2$	833
3-C̄HCOAr^c	RNCS (R = alkyl, aryl, etc.)	3-C=(O$^-$)Ar^d $\mid$ CSNHR	834
4-NH$_2^f$	CH$_3$CH$_2$CO$_2$H	4-NHCOCH$_2$CH$_3$	163
5-NH$_2$	ArCN	5-NH—C(NH)(Ar)	835
5-NH$_2$	CSCl$_2$	5-NCS	836, 837

[a] Other benzimidazole substituents are not shown.
[b] 2BI represents the 2-benzimidazolyl moiety.
[c] The reactants are benzimidazolium ylides.
[d] The products are benzimidazolium betaines.
[e] See text.
[f] This regiospecific acylation is carried out on 1-unsubstituted derivatives.

Alkylation of the *N*-anion of 1-ethyl-2-aminobenzimidazole in liquid ammonia leads to a mixture of products resulting from mono- and dialkylation.[772] The relative yield of the dialkyl derivative increases in passing from bromides to iodides and also increases in passing from high molecular weight and branched alkyl groups (e.g., *t*-Bu, *i*-Pr) to smaller groups (e.g., methyl). In contrast to the benzylation of the *N*-anion of 4-aminoquinoline with benzyl chloride,[838] benzylation of 1-ethyl-2-aminobenzimidazole gives a relatively high proportion of monoalkylated product. It may be noted that mixtures of mono- and dialkylated products are easily separated, and this method is thus a useful one.

Nitrosation of 2-aminobenzimidazoles has been effected[796,798] by the Bamberger procedure, in which the sodium salt of the heterocycle (cf. **1.304**) is treated with isoamyl nitrite. Isolable intermediate diazotates (cf. **1.305**) can then be converted into stable yellow crystalline nitrosamines (cf. **1.306**) at acidities in the pH region 5.5–7.5. When such nitrosamines (cf. **1.306**) are treated in organic solvents with a catalytic quantity of sulfuric acid, they are converted into 1,3-di(2-benzimidazolyl)triazenes (cf. **1.307**), perhaps by way of intermediate N-nitrosotriazenes.[839]

(1.304)   **(1.305)**   **(1.306)**

a, R = CH$_3$ (74% yield)
b, R = CH$_2$Ph (60%)

(1.307)
a, R = Me (74%)
b, R = CH$_2$Ph (75%)

2-Methylbenzimidazoles are of interest, since the weakly acidic 2-methyl substituent has a reactivity comparable with that of 2-methylpyridine and 2-methylquinolines; accordingly, condensations with aldehydes[809,810] and other electrophilic reagents can be achieved (cf. Ref. 811). The synthesis of 2-nitromethylbenzimidazoles by nitration of 2-methyl-1-phenyl-benzimidazole with propyl nitrate[812] is a valuable application of this method, since such derivatives cannot be prepared by conventional synthetic procedures based on cyclization of diamines.

Treatment of 2-methyl- or 2-benzylbenzimidazole (**1.308**) with 2 moles of n-butyllithium at 0° results in abstraction of the heterocyclic N–H as well as the α-hydrogen of the 2-alkyl substituent.[840] Treatment of the ensuing dianion (e.g., **1.309**) with alkyl halides or with aldehydes and ketones causes a selective reaction at the side-chain carbanion center to yield 2-alkyl (e.g., **1.310**) and 2-(2-hydroxyalkyl) derivatives (e.g., **1.311**), respectively. It should be noted that this procedure[840] entails relatively mild reaction conditions so that carbinol derivatives (cf. **1.311**) rather than styryl derivatives are isolated. A restriction is that the method cannot be applied to (saturated) alkyl homologs (cf. **1.308**, R = alkyl), but from a more recent study of 1-methyl analogs of (**1.308**) it has been shown that carbinols can be isolated by treating (**1.308**) (N–Me for NH, R = Et) with sodium (or lithium) naphthalenide followed by benzophenone in tetrahydrofuran at 0°.[841]

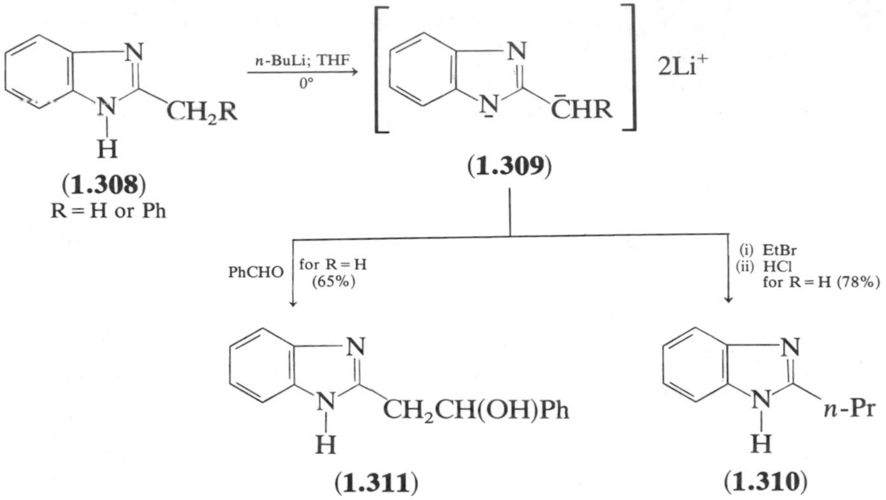

(**1.308**)
R = H or Ph

(**1.309**)

(**1.311**)

(**1.310**)

1.4.2. Metalation Reactions

1-Methylbenzimidazole is metalated by n-butyllithium in dry diethyl ether at −60°;[842] reaction of the ensuing 2-lithio derivative with carbon dioxide affords 1-methylbenzimidazole-2-carboxylic acid in 45% yield. If the metalation is carried out at room temperature with an excess of n-butyllithium, a mixture containing a 3:1 molar ratio of the bisbenzimidazole derivatives (**1.312** and **1.313**) is formed;[842] a simple addition mechanism presumably operates (see Scheme 1.17).

Metalation of 1-alkylbenzimidazole can also be achieved by the use of phenyl sodium at low temperature,[843,844] although for 1-arylbenzimidazoles this procedure can give rise to mixtures of the desired 2-sodio derivative together with 3-sodio addition products; however, metalation of 1-phenylbenzimidazole can be effected by phenyllithium.[845] An alternative method of metalation[846,847] of 1-alkyl derivatives involves the use of sodium or potassium in a mixture of benzene and isoamyl alcohol, although application of this type of reaction at room temperature gives rise to dimer formation[847] (cf. Ref. 842 and compound **1.313**).

2-Lithiobenzimidazoles,[848] and 2-magnesium derivatives[849] prepared in similar fashion by metalation with di-n-butylmagnesium, are useful intermediates for the synthesis of 2-bromo- and 2-iodobenzimidazoles. 1-Alkyl-2-sodio derivatives react in normal fashion with aldehydes to give secondary alcohols.[850]

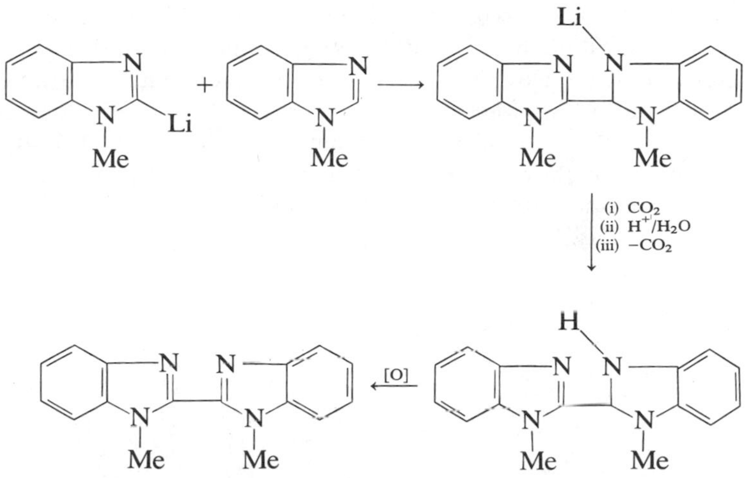

(i) CO_2
(ii) H^+/H_2O
(iii) $-CO_2$

(1.313)   (1.312)

53.5% yield

Scheme 1.17

1.4.3. Reactions with Nucleophilic Reagents

Substitution in the Carbocyclic Ring

Very little use has been made of nucleophilic displacement reactions in the carbocyclic ring of benzimidazole. A 4-chloro substituent in 2-trifluoromethyl-7-nitro derivatives can be displaced by amines and other nucleophiles (e.g., RS⁻, RO⁻) providing the 5-position has a nitro-[851–853] or trifluoromethyl[854] substitutent.

Only one quantitative study has been carried out:[565] apparently a condensed imidazole ring has a marked activating effect on the rate of nucleophilic aromatic substitution compared with 2,4-dinitrofluorobenzene (cf. Table 1.38). Relative retardations due to methyl substituents at the 1 and/or

TABLE 1.38. RELATIVE RATES OF REACTIONS OF 2,4-DINITROFLUOROBENZENE AND 4-FLUOROBENZIMIDAZOLE WITH ALANYLGLYCINE AT pH 7.95

Compound	Relative rate
2,4-Dinitrofluorobenzene	1
4-Fluoro-5,7-dinitrobenzimidazole	84.2
4-Fluoro-2-methyl-5,7-dinitrobenzimidazole	37.3
4-Fluoro-1,2-dimethyl-5,7-dinitrobenzimidazole	22.5

2-positions are in accord with the behavior expected from inductive effects while the overall rate enhancements in the benzimidazole series are ascribed to electronic activation by the imidazole ring rather than to intramolecular interactions within the Meisenheimer complex. Such activation may be visualized as resulting from an additional canonical form (**1.314**) for the anion hybrid. As might be expected, 4-fluoro-1,2-dimethyl-5,6-dinitrobenzimidazole is considerably less reactive toward nucleophiles than the 5,7-dinitro analog.

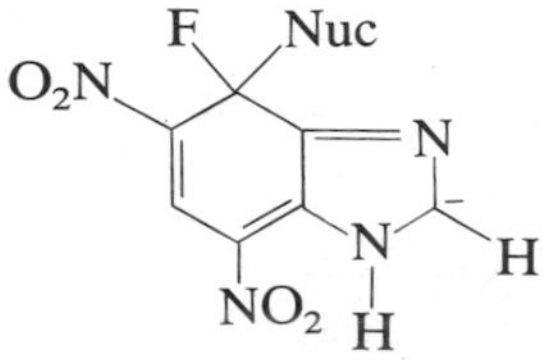

(1.314)

Nuc = alanylglycine moiety.

Substitution in the Imidazole Ring

THE CHICHIBABIN REACTION. The Chichibabin reaction (cf. **1.315 → 1.316**) has been widely used by Simonov and coworkers for the synthesis of a variety of 2-aminobenzimidazole derivatives (see Table 1.39). In this procedure the benzimidazole derivative is heated in xylene with sodium amide, although it may be noted from Table 1.39 that the reaction cannot be used for derivatives with no substituent at position 1; for compounds of this type, formation of the anion of the heterocycle prohibits nucleophilic attack at the 2-position. The reaction is also sensitive to the nature of the substituent at the 5-position: thus 5-alkyl, -alkoxy, and -thioalkyl derivatives are successfully aminated, whereas 5-hydroxy, -halogeno, -nitro, and -carboxy derivatives do not enter into the Chichibabin reaction.[859] 1-Benzyl-4,7-dimethoxybenzimidazole is aminated, but it is interesting to note that 5,6-dimethoxy-1-methylbenzimidazole does not react with sodium amide either neat or at

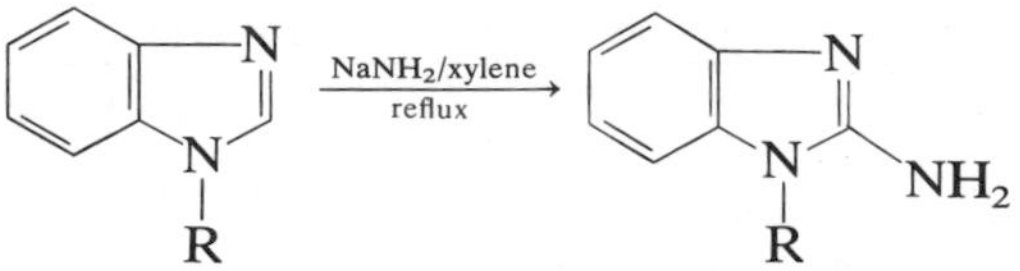

(1.315)   **(1.316)**

(R = alkyl, aryl, aralkyl)

TABLE 1.39.   SYNTHESIS OF 2-AMINOBENZ-
             IMIDAZOLES BY THE
             CHICHIBABIN REACTION[a]

Benzimidazole derivative	Ref.
1-Me	855
1-Me-5-MeO	855, 856
1-Me-5-MeS	857
1,5-$(Me)_2$	858
1-Me-5-EtO	859
1 iPr	860[c]
1-Et-5-MeO	859, 861
1-cyclohexyl	862
1-phenyl	
1-R-5-R^1[b]	863
1-R (R = t-Bu, n-C_9H_{19}, n-$C_{11}H_{23}$)	864[c]
1-[4-$MeOC_6H_4CH_2$]	865[c]
1-$(CH_2)_n$-1-benzimidazolyl (n = 3–5)	866[c]

[a] The reactions are usually carried out using sodium
amide under reflux in either xylene or N,N-dimethyl-
aniline.
[b] Substituent (R, R^1) combinations are 1-Me, -Et,
-$(CH_2)_2NEt_2$ and -$(CH_2)_3NEt_2$, and 5-OEt, -OCH$_2$Ph
and -H.
[c] See text.

180° as a dispersion in xylene or mineral oil;[865] this inhibiting "dimethoxy
effect"[865] is also manifested in 1-aryl derivatives in which the methoxy
substituents are in the 1-aryl substituent. In the former case the 5,6-
dimethoxy moiety is converted by sodium amide into an o-methoxyaryloxy
anion, which inhibits amination at the 2-position.

A slight retarding effect of bulky 1-substituents has been noted, but this
probelm can be alleviated by prolonged heating and by the use of a large
excess of sodium amide. In this manner the yield of 2-amino-1-
isopropylbenzimidazole can be raised from 39 to 64% by using a 2.5-fold
molar excess of sodium amide;[860] the very hindered 1-t-butyl analog can be
prepared in this way, albeit in poor yield (21%).[864] An inhibition of the
Chichibabin reaction has also been observed during attempts to prepare a
series of bis benzimidazoles (cf. **1.317** → **1.318** in Scheme 1.18).[866] Of the
series of compounds (**1.317**), only **c**, **d**, and **e** are aminated, while the
remainder do not react with sodium amide in xylene. The mechanism of the
Chichibabin reaction is believed to involve an initial step in which the
pyridine-like nitrogen coordinates to Na$^+$ ions of sodium amide. Compounds
(**1.317c**, **d**, and **e**) are thus thought[866] to behave like 1-alkylbenzimidazoles,
but in the remainder, the electron-withdrawing effect of the 1-substituent
lowers the pK_a below a critical value at which coordination can occur.

$$1\text{-BI}\!-\!X\!-\!1\text{-BI} \xrightarrow{\text{NaNH}_2}$$

(1.317)

1-BI = 1-benzimidazolyl

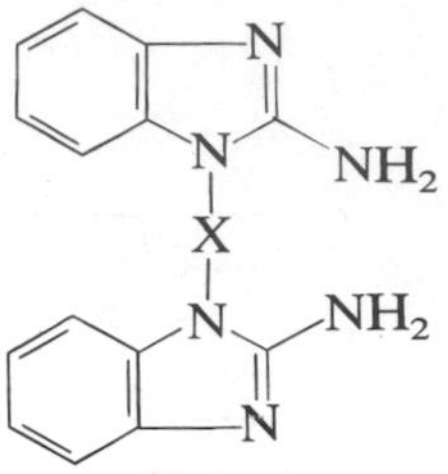

X	$pK_a^{25°}(\text{H}_2\text{O/EtOH})$
a CH_2	4.22
b $(CH_2)_2$	4.56
c $(CH_2)_3$	5.02
d $(CH_2)_4$	5.13
e $(CH_2)_5$	5.17
f $CH_2C_6H_4CH_2(p)$	4.80
g CH_2OCH_2	4.38

(1.318)

Scheme 1.18

DISPLACEMENT OF CHLORIDE AND OTHER ANIONS. Unusual nucleophilic displacements at the 2-position of the imidazole ring are those of a hydrazino group by thionyl chloride,[867] a nitro group by ethoxide ion,[868] and a phenyl substituent by amide ion;[869] a sulfonic acid group at the 2-position can be readily displaced by alkoxy,[870] hydrazino,[871] and alkylamino substituents.[872] Other displacements at the 2-position involve a chloro substituent by a variety of nucleophiles (see Table 1.40), some examples of which are discussed below.

Reaction of 2-chlorobenzimidazole with pyridine gives rise to the pyridinium salt (1.319),[886] which is a precursor to an interesting analog (1.320)[886] of pyridinium cyclopentadienide.[895] The chemistry of these novel betaines (cf. 1.320) has not been explored, but new compounds have been synthesized in this category in which the benzo and pyridyl rings contain dinitro and methyl substituents, respectively.[887]

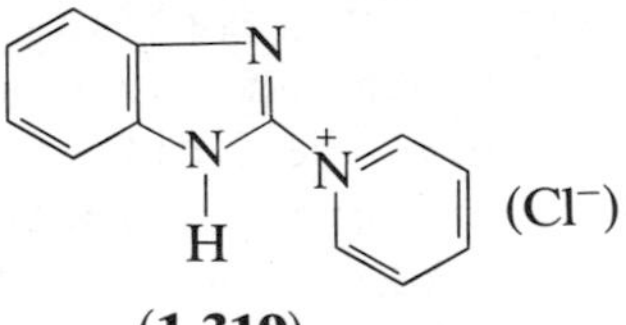

(1.319) 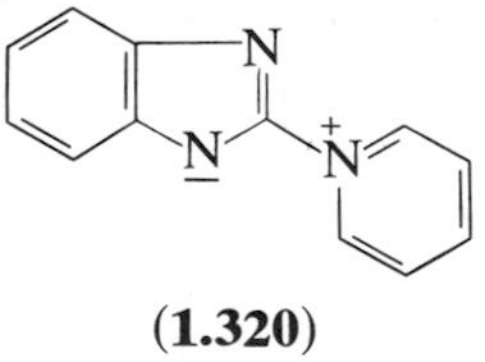   (1.320)

An extensive kinetic study has been made of the nucleophilic displacement of chloride ion from 2-chlorobenzimidazole and its derivatives by piperidine and by methoxide ion. Apparently, the reactivity of the 2-chloro derivative is comparable to that of 2-chloro-1-methylbenzimidazole but is lower than that of 2-chlorobenzothiazole and 2-chlorobenzoxazole.[882] As might be anticipated, such reactions are facilitated by electron-withdrawing substituents (e.g., NO_2, Cl) in the aryl ring.[883–885]

TABLE 1.40. NUCLEOPHILIC SUBSTITUTION REACTIONS INVOLVING 2-CHLOROBENZIMIDAZOLES[a]

Nucleophile	Ref.
NH$_4$OH, morpholine, PhCH$_2$NH$_2$	873
Ethylene diamine derivatives	874
Hydrazine	875, 876
Alkylamines, dialkylamines, arylalkylamines, etc.	876
m-ClC$_6$H$_4$NH$_2$	877
Piperazine derivatives	878
Aminothiazole, aminothiadiazole	879
Triazole derivatives	880
Triazole, 2-methylthiazole	881
Piperidine[b]	882–885
Pyridine derivatives[b]	886, 887
Alkoxy[b]	882, 883, 885, 888, 889
(Amino)alkyloxy	890
(Amino)alkylthio	
Aryloxy	891
Alkylthio	892
Arylthio	893, 894
Thiourea,[c] benzimidazolin-2-thione	894

[a] Other substituents in the benzimidazole derivative are not shown.
[b] See text.
[c] The product is benzimidazolin-2-thione which is presumably formed via an intermediate isothiourea derivative.

An important general principle in connection with the design of nucleophilic substitution reactions has been pointed out by Harrison and Ralph:[888] for unsubstituted 2-halogenobenzimidazoles a competition exists between proton abstraction by the nucleophile at the 1-position with concomitant retardation of 2-substitution, and nucleophilic substitution at the 2-position. Accordingly, chloride ion is not displaced from 2-chlorobenzimidazole by the powerful nucleophiles RO$^-$ (R = Me, Et, or t-Bu), whereas 2-chloro-1-methylbenzimidazole reacts readily with sodium methoxide or -ethoxide; even the latter process is susceptible to steric effects, since 2-chloro-1-isopropylbenzimidazole reacts only slowly with sodium methoxide or -ethoxide, and 2-chloro-1-methylbenzimidazole gives no alkoxy derivative with sodium t-butoxide. Despite these limitations, 2-alkoxybenzimidazoles are accessible via nucleophilic displacement of chloride ion from 2-chloro-1-isopropenyl derivatives by alkoxide ion followed by oxidative cleavage of the isopropenyl group.[889]

Substitution Within Side-Chain Substituents

Reactions in which benzimidazole side-chain substituents undergo substitution reactions with nucleophilic reagents are collected in Table 1.41. Most of these transformations are routine procedures on 2-chloromethyl-benzimidazoles, but the studies of Holan and coworkers[925,926,932-934] on the mode of reactivity of 2-trichloromethylbenzimidazole with nucleophiles are especially notable.

There is considerable evidence[935] that the benzimidazole nucleus is significant in relation to purine antimetabolite behavior, which suggests that benzimidazole may act biologically by competitive inhibition in nucleic acid synthesis. With this in mind, Holan and coworkers[925,926,932-934] have investigated 2-trichloromethyl derivatives as model systems with a view to providing substrates which might react irreversibly at nucleophilic enzyme sites.

Reaction of 2-trichloromethylbenzimidazole (**1.321a**) with aqueous ammonia gives a mixture of 2-cyanobenzimidazole (**1.321b**, 50%) and the pyrazine diimine (**1.322a**, 30%), although with anhydrous ammonia the cyano derivative is the sole product in 86% yield. With primary alkyl and arylamines the trichloro compound (**1.321a**) is converted into amidine derivatives (**1.321c**), although in the absence of an additional base or in an acidic medium the reaction of arylamines produces amides (**1.321d**); with secondary amines the appropriate anilides (**1.321e**) are formed.

(**1.321**)   (**1.322**)

a, R = CCl$_3$ a, = X = NH
b, R = CN   b, = X = O

c, R = (structure with =NR^1 and NHR^1)   (R^1 = alkyl or aryl)

d, R = CONHAr
e, R = CONR^{1}R^2 (R^1 = alkyl, R^2 = alkyl or aryl)

Interestingly, 1-methyl-2-trichloromethylbenzimidazole is considerably less susceptible to nucleophilic attack, which suggests[926] that the anion (**1.323**) is a probable intermediate in transformations of the trichloromethyl derivative (**1.321a** (cf. Scheme 1.19 and also the reactivity[936] of analogous trichloromethylpurine derivatives). An alternative route[926] to the nitrile (**1.321b**) could involve the intermediacy of a ketenimine (**1.325**), and indeed

Type of benzimidazole side-chain[b]	Nucleophilic reagent	Type of benzimidazole product[b]	Ref.
1-CH_2Cl	NH_3	1-CH_2NH_2	896
1-CH_2Cl	RO^-	1-CH_2OR	897, 898
1-CH_2CH_2Cl	Cyclic amines	1-$CH_2CH_2N(CH_2)_n$	899
1-$(CH_2)_nX$ ($n = 2,3$; X = halogeno)	Acyclic and cyclic amines	1-$(CH_2)_nN(CH_2)_n$ 1-$(CH_2)_nNR_2$	900
2-CH_2Cl	$AcNH\bar{C}(CO_2Et)_2$	2-$CH_2CH(NH_2)CO_2H$	901
2-CH_2Cl	Dialkylamines cyclic amines	2-CH_2NR_2 2-$CH_2N(CH_2)_n$	654, 902– 906
2-CH_2Cl	$HN(CH_2CH_2OH)_2$	2-$CH_2N(CH_2CH_2OH)_2$	904, 907, 908
2-CH_2Cl	Trialkylamines	Quaternary salts	909, 910
2-CH_2Cl	Arylamines	2-CH_2NHAr	911
2-CH_2Cl	p-$H_2NC_6H_4COG^c$	2-$CH_2NHC_6H_4COG$-$(p)^c$	912
2-CH_2Cl	Pyridine Isoquinoline	Pyridinium Isoquinolinium } Salts	913
2-CH_2Cl	$HO(CH_2)_2O^-$ $Et_2N(CH_2)_2O^-$	2-$CH_2O(CH_2)_2OH$ 2-$CH_2O(CH_2)_2NEt_2$	914
2-CH_2Cl	RO^-	2-CH_2OR	915
2-CH_2Cl	Salicylamide derivatives	2-$CH_2OC_6H_4CONHCH_2$-$NHCONHAr(-o)$	916
2-CH_2Cl	8-hydroxyquinoline	2-CH_2O(8-quinolinyl)	917
2-CH_2Cl	SCN^-	2-CH_2SCN	918
2-CH_2Cl	$CS(NH_2)_2$	[2-$BICH_2S]_2^d$	919
2-CH_2Cl	$CS(SK)_2$	[2-$BICH_2S]CS^d$	920
2-CH_2Cl	p-ClC_6H_4SH	2-$CH_2SC_6H_4Cl(p)$	921
2-$CHCl_2$	H_2O	2-CHO	922
2-$CHClPh$	R^1NH_2 (R^1 = H, alkyl, aryl, benzyl)	2-$CH(NHR^1)Ph$	923
2-CCl_3	NH_3	2-CN^e	924-926
2-$COCH_2Cl$	$H_2NCS_2^-$	2-(2-mercapto-4-thiazolyl)	927
2-$CH(R^1)OCOCH_2Cl$ (R^1 = H, alkyl)	SCN^-	2-$CH(R^1)OCOCH_2SCN$	928
2-SO_2Cl	$H_2N(CH_2)_nNH_2$ ($n = 2$–10, 12)	2-BI-$SO_2NH(CH_2)_nNHO_2S$-2-BI^d	929
2-(5-halogenofuryl)	NO_2^-	2(5-nitrofuryl)	930
4-SO_2Cl	Dialkylamines Cyclic amines	4-SO_2NR_2 4-$SO_2N(CH_2)_n$	931

[a] Other benzimidazole substituents are not shown.
[b] R and Ar denote alkyl and aryl groups, respectively.
[c] G = —$NHCH(CO_2Et)CH_2CH_2CO_2Et$.
[d] 2-BI denotes an unsubstituted[919,929] or substituted[920] 2-benzimidazolyl moiety.
[e] See text for a detailed discussion of the reaction of 2-trichloromethylbenzimidazoles with nucleophiles.

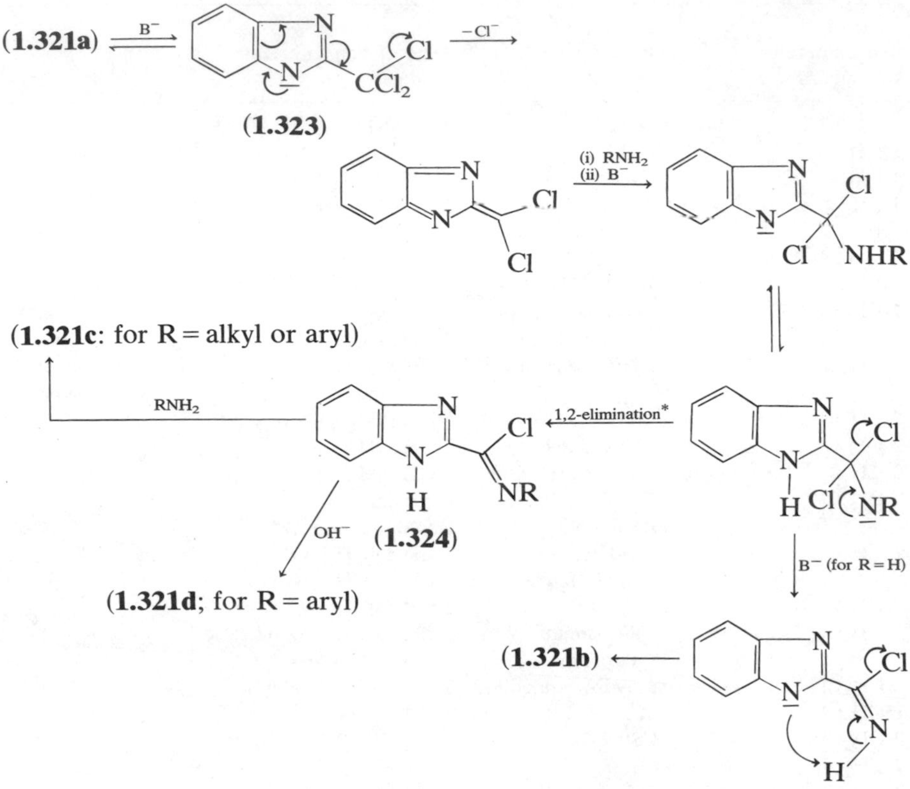

(1.321a) ⇌ (1.323)

(1.321c: for R = alkyl or aryl)

(1.324)

(1.321d; for R = aryl)

(1.321b)

* An alternative 1,6-elimination mechanism[926] can be envisaged for formation of the imidoyl chloride.

Scheme 1.19

such a pathway is an attractive one in relation to formation of the pyrazine diimine (1.322a) (see Scheme 1.20). Although the mechanisms depicted in Schemes 1.19 and 1.20 are speculative, it is noteworthy that reaction of the trichloromethyl derivative (1.321a) with 2-amino-5-chloropyridine provides an isolable imidoyl chloride intermediate (cf. 1.324).[926]

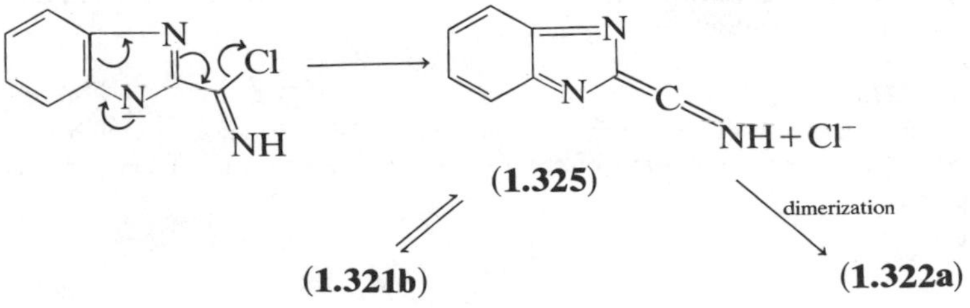

(1.325)

(1.321b)

(1.322a)

Scheme 1.20

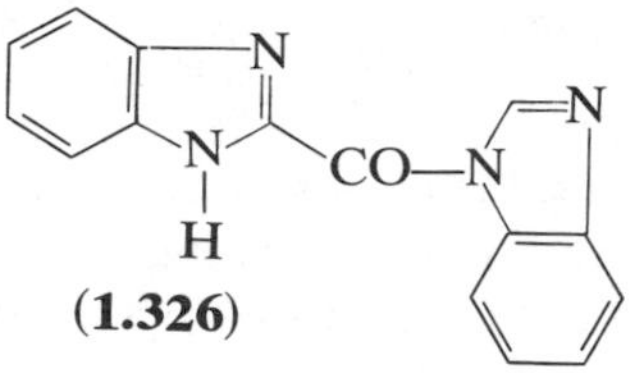

(1.326)

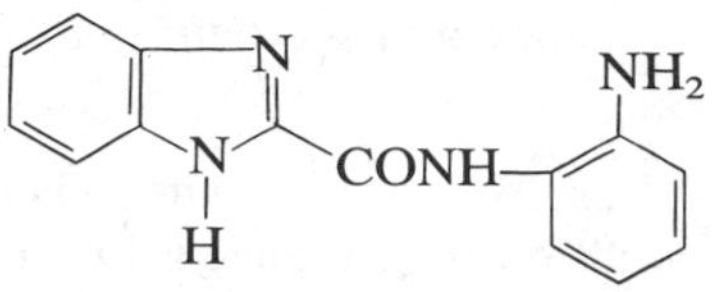

(1.327)

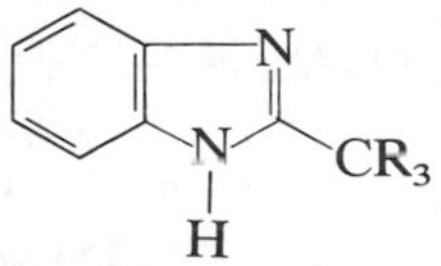

(1.328)

a, R = OMe
b, R = SMe
c, SC₆H₄Cl(p)

A wider investigation of this type of reaction using a variety of mono- and difunctional nucleophiles indicates considerable synthetic potential.[933,934] Carefully controlled basic hydrolysis[933] of (1.321a) can lead to the pyrazine dione (1.322b) or its decomposition products (e.g., 1.326 and 1.327), while reactions with methanol methanethiol, and p-chlorothiophenol under basic conditions lead to the ortho esters (1.328a–c), respectively. Reactions with difunctional nucleophiles[934,937,938] are especially versatile and afford an effective method for the synthesis of benzimidazoles containing a variety of heterocyclic substituents in the 2-position; some examples of this type of process are shown in Scheme 1.21.[934]

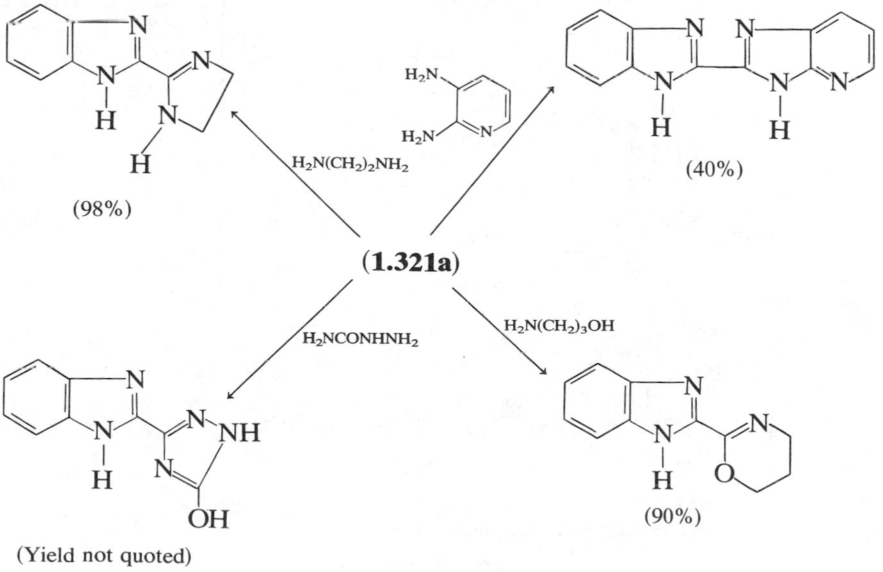

(98%)

(40%)

(1.321a)

(Yield not quoted)

(90%)

Scheme 1.21

Addition Reactions of Nucelophilic Reagents

To the Imidazole Ring. The reactivity of the 2-position toward nucleophilic addition is enhanced by the presence of a strongly electron-withdrawing substituent in the 1-position. The preparative value of this behavior has been demonstrated[939] by allowing the readily accessible *N*-(1,1,2-trifluorochloroethyl)benzimidazole (**1.329**) to react with a variety of nucleophiles (see Scheme 1.22). Opening of the imidazole ring followed by ring closure via the very reactive difluoromethylene group can produce 2-(chlorofluoromethyl) benzimidazole (**1.330**), which may undergo subsequent transformations (cf. **1.331–1.333**). Using hydrazine as the nucleophile, the preferred pathway involves triazole formation followed by Wollf–Kishner reduction to give a 3-methyltriazole derivative ((**1.334**) in 73% yield in a single-step process.

The conversion[940] of benzimidazole into the butadiene derivative (**1.336**) by reaction with ethoxymethylenemalononitrile is considered[940] to be another example of a process in which the 2-position of benzimidazole is activated toward nucleophilic addition; in this case, however, the adduct (**1.335**) is thought to rearrange as shown in Scheme 1.23.

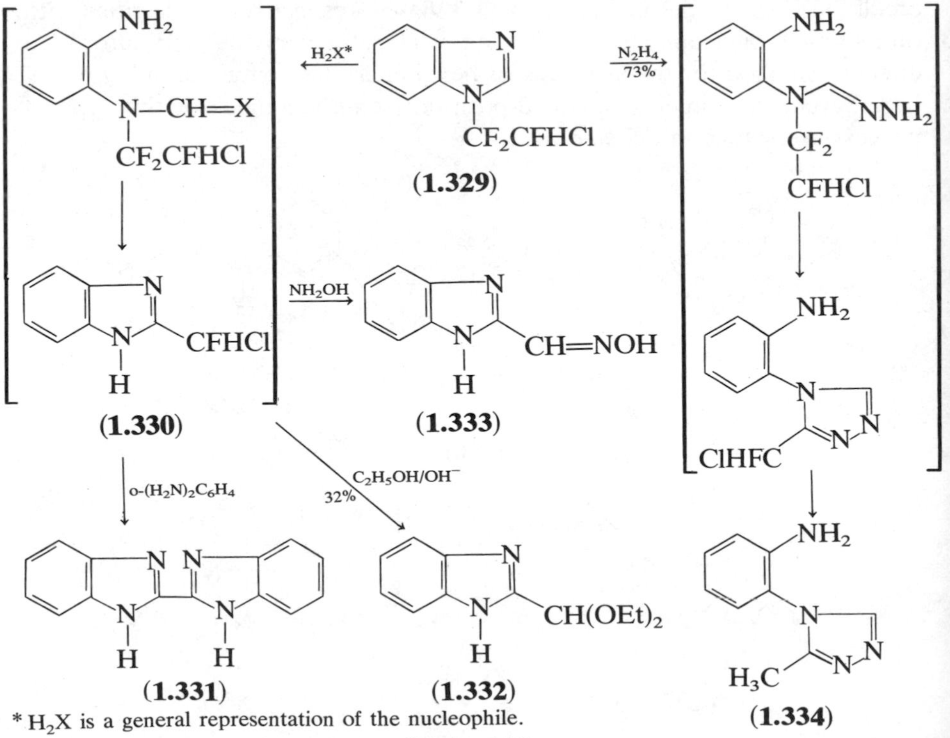

* H_2X is a general representation of the nucleophile.

Scheme 1.22

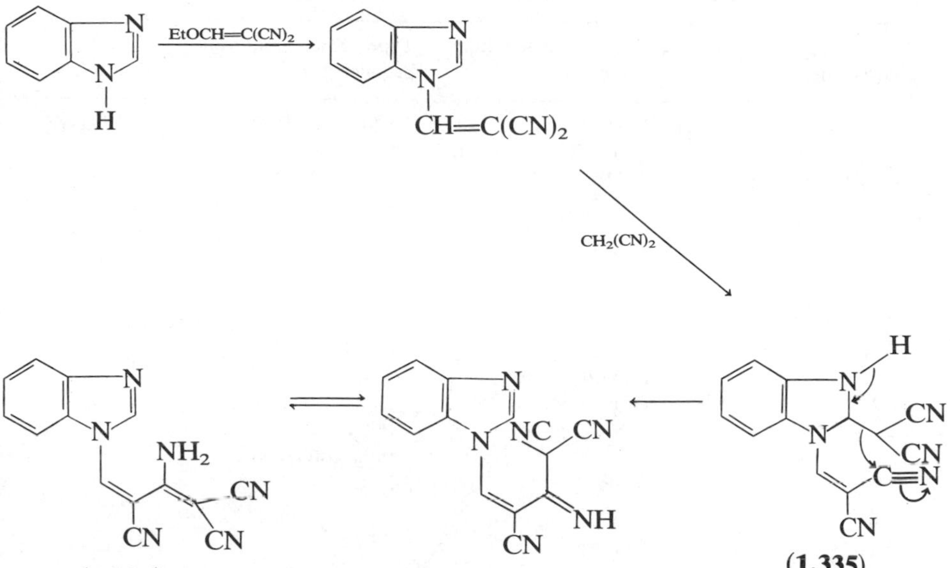

Scheme 1.23

These investigations[939,940] highlight an area where a number of novel compounds should be accessible by careful choice of the alkylating or acylating agent, and the entering nucleophile.

TO SIDE-CHAIN SUBSTITUENTS. Table 1.42 shows reactions in which nucleophilic addition to benzimidazole side-chain substituents occurs. Most of these transformations have been effected upon groups at the 2-position and commonly relate to the synthesis of benzimidazole derivatives which contain a heterocyclic ring at that position [cf. the commercial success 2-(4-thiazolyl)- and 2-(2-furyl)benzimidazole described in Chapter 10].

It is interesting to note that the reactivity of an acetylenic group in the 2-position is modified by the benzimidazole nucleus and behaves in the same manner as diacetylenes[967] in the context of conjugate addition reactions (cf. **1.336A → 1.337**). The intramolecular process depicted in (**1.336A → 1.338**)[968] is also analogous to the behavior of diynes.[967]

A kinetic study of intramolecular general base catalysis of hydrolysis of 4-acetoxymethylbenzimidazole (**1.339**) has been carried out.[969] Compared with the monocyclic derivative (**1.340**), a relative rate of 4.1 is obtained after correction for basicity difference and a polar substituent effect. The value of 4.1 is an expected[969] magnitude due to freezing of an internal rotation. Introduction of a 5-methyl substituent into (**1.339**) enhances this relative rate by a factor of 1.4, which is an indication of the restriction imposed upon the second internal rotation.

TABLE 1.42. NUCLEOPHILIC ADDITION REACTIONS INVOLVING
BENZIMIDAZOLE SIDE-CHAIN SUBSTITUENTS[a,b]

Type of benzimidazole side-chain	Type of nucleophilic reagent	Type of benzimidazole product	Ref.
1-CN	R^1R^2NH (R^1, R^2 = H, alkyl)	1-C(=NH)NR^{1}R^2	791
1-CH$_2$CO$_2$Et	N$_2$H$_4$	1-CH$_2$CONHNH$_2$	589
1-CH(CO$_2$H)CH$_2$CO$_2$H	RNH$_2$		941
1-CONH(CH$_2$)$_6$NCO	RCH	1-CONH(CH$_2$)$_6$NHCO$_2$R	738
2-CN	MeCSNH$_2$	2-CSNH$_2$	942
2-CN	H$_2$NOH	2-C(NH$_2$)=NOH	942
2-CN	Difunctional nucleophiles	2-Hetaryl [e.g., 2-(1-Et-imidazolin-2-yl)]	942
2-NHCN	ROH	2-NHCO$_2$R	943
2-NHCN	Ethylene diamine derivatives	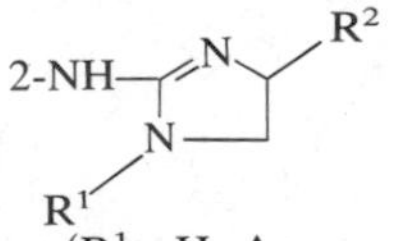 (R^1 = H, Ac R^2 = H, Me)	257
2-SCH$_2$CN	H$_2$NOH	2-SCH$_2$C(=NH)NHOH	944
2-CO$_2$H	ROH	2-CO$_2$R	945
2-(CH$_2$)$_2$CO$_2$H	EtOH	2-(CH$_2$)$_2$CO$_2$Et	946
2-C$_6$H$_4$COCl(p)	p-OHC$_6$H$_4$- (2-benzoxazolyl)	2-C$_6$H$_4$CO$_2$C$_6$H$_4$- (2-benzoxazolyl)	947
2-NHCO$_2$Ph	ROH (R = alkyl, allyl)	2-NHCO$_2$R	948
2-NHCO$_2$CHR^{3}CHR^{4}OH (R^3, R^4 = H, alkyl, Ph)	ROH (R = C$_{1-4}$alkyl)	2-NHCO$_2$R	949
2-CHO	R-Me (R = aryl, hetaryl)	2CH=CHR	950
2-CHO	RMgBr (R = hetaryl)	2-CH(OH)R	951
2-CHO	RCH$_2$NO$_2$/NaOH (R = H, Ph)	2-CH(OH)CHRNO$_2$	952
2-CHO	H$_2$NOH	2-CH=NOH	952
2-CHO	Hydantoin/piperidine	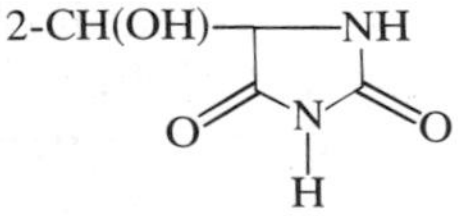	953
2-CHO	Thiazolidinone derivatives	(R^1 = H, Ph, CH$_2$Ph; R^2 = O, S, NH R = NPh, NCH$_2$Ph, O, S)	954
2-CHO	Ph$_3$$\overset{+}{P}\bar{C}$(Br)CO$_2$R	2-CH=C(Br)CO$_2$R	955–957

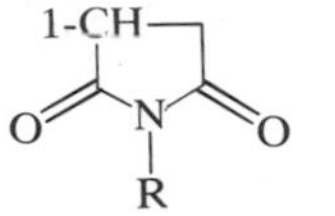

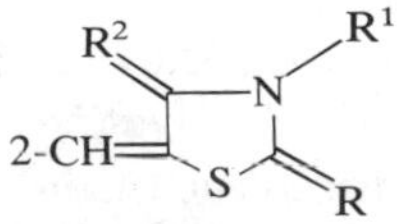

122

TABLE 1.42 *(Continued)*

Type of benzimidazole side-chain	Type of nucleophilic reagent	Type of benzimidazole product	Ref.
2-COPh	RLi RMgBr (R = alkyl, alkenyl, Ph, etc.)	2-CR(OH)Ph	817, 951
2-CH$_2$COCH$_2$CH$_2$R (R = NR$_2^1$, piperidinyl, etc.)	R^{2}NHCSNHNH$_2$ (R^2 = H, Ph)	2-CH$_2$C=NNHCSNHR^2	958
2-CH$_2$CO(CH$_2$)$_2$NR$_2^1$ (NR$_2^1$ = piperidino, etc.)	ArMgBr	2-CH$_2$C(OH)Ar(CH$_2$)$_2$NR$_2^1$ $\;$ CH$_2$CH$_2$R	959
2-(CH$_2$)$_n$COCHO (n = 0, 1)	RNH$_2$ (R = aryl, 2-thiazolyl)	2-(CH$_2$)$_n$COCH=NR	960
2-CHR(CH$_2$)$_2$COMe (R = H, Me)	PhNHNH$_2$	2-CH(R)CH$_2$– [2-methylindole ring, N–H] (Me)	85
2-COCH=CHAr	PhMgBr	2-COCH$_2$CH(Ar)Ph	817
2-COCH=CHPh	CS(NH$_2$)$_2$	2-C [dihydropyrimidine-2-thione ring with Ph; N, NH, S]	817
2-COCH=CHAr	NCCH$_2$CO$_2$Et/NH$_4$Ac	2-C [dihydropyridinone ring bearing Ar, CN, =O, N–H] (and related compounds)	961
2-COCH=CHAr	H$_2$NNH$_2$	2-C [pyrazoline ring with Ar; N, N, N–H]	962
2-COCH=CHAr	H$_2$NOH	2-C [isoxazoline ring with Ar; N, O]	962
2-COCH=CHR (R = aryl, hetaryl)	PhNHNH$_2$	2-C [pyrazoline ring with R; N, N–Ph]	963
2-C≡CH	ROH	2-CH=CHOR	964, 965
5-(and 6-)NCS	N$_3^-$	5- (and 6-)—N [thiadiazole ring; S, N, N, NH]	966
7-N$_2^+$	2-BI—NHN=CHR^c (R = Me, Ph)	7-N=N—C(R)=NNH-2BI^c	967

[a] Other benzimidazole side-chain substituents are not shown.
[b] R and Ar denote alkyl and aryl substituents, respectively, unless otherwise stated.
[c] 2-BI denotes 1-benzyl-2-benzimidazolyl moiety.

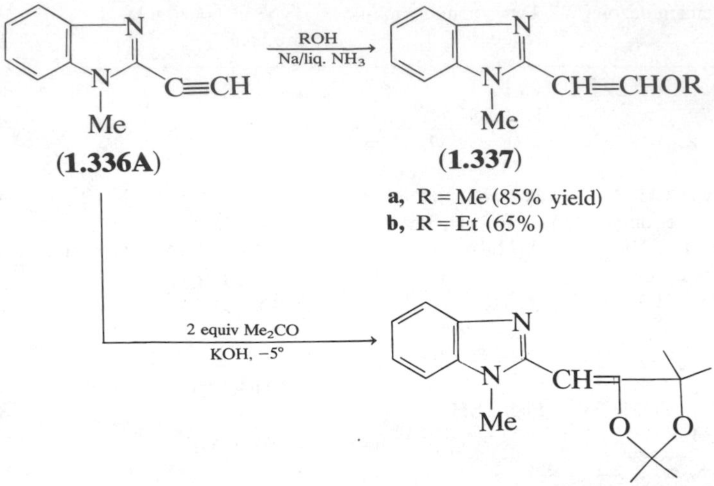

(1.336A)   **(1.337)**

a, R = Me (85% yield)
b, R = Et (65%)

(1.338)

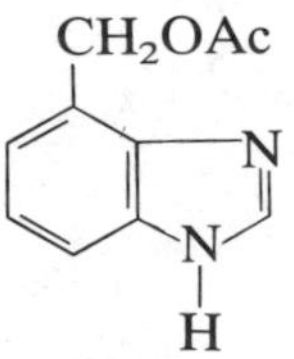

(1.339) 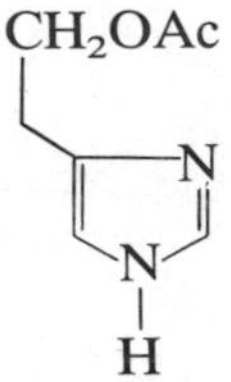   **(1.340)**

Reactions of Benzimidazolium Derivatives

NUCLEOPHILIC SUBSTITUTION IN THE IMIDAZOLE RING. 1-Ethyl-2-chloro-substituted benzimidazolium tetrafluoroborates undergo nucleophilic substitution by reagents such as phenols, thiophenols, and N-alkylanilines (cf. **1.341 → 1.342**).[970] A more complicated reaction is probably involved in the

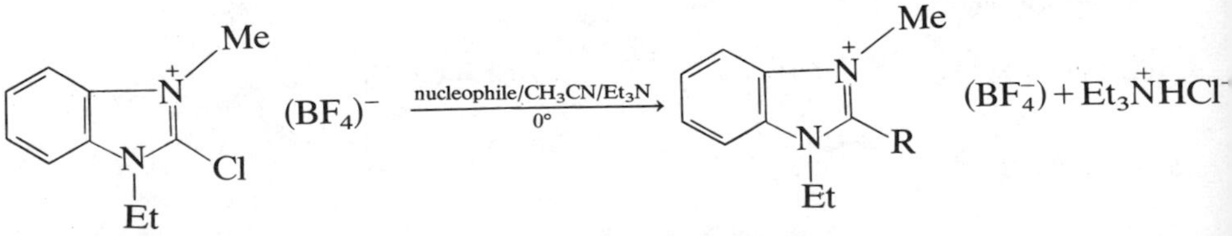

(1.341)   **(1.342)**

R	Yield (%)
a, N(Me)Ph	71
b, OPh	59
c, O-2-Naphthyl	59
d, SPh	65

reactions of 3-methoxy-1-methylbenzimidazolium iodide with nucleophiles in which 2-substituted benzimidazoles are formed (see **1.343** → **1.344** for examples of these transformations).[971]

(1.343)   **(1.344)** $+ CH_3OH$

R	Yield (%)
a, CN	100
b, OMe	95
c, NHPh	85
d, CH(CN)$_2$	75
e, H	100

In general, yields are very high for this type of reaction, and in contrast to the case for pyridine analogs they can be used to synthesize 2-hydroxy and 2-alkylamino derivatives. The mechanism of these processes is probably different from the simple addition–elimination reactions (cf. Path A in Scheme 1.24) responsible for reactions in the pyridine series.[972] The NMR spectrum of the benzimidazolium derivative (**1.343**) indicates that the 2-hydrogen is considerably deshielded ($\delta = 11.00$ in CDCl$_3$) and is, therefore, unusually acidic. A mechanism involving the formation of a carbenic intermediate (**1.345**) is thus envisaged[971] [cf. Path B and formation of the bisbenzimidazole-N-oxide (**1.346**) illustrated in Scheme 1.24, and also related reactions in the chemistry of imidazoles[973] and thiazoles[974]].

Formation of the dimer (**1.348**) from the benzimidazolium tetrafluoroborate (**1.347a**) in basic media is probably a closely related process, and indeed the intermediacy of carbenes in this type of reaction is convincingly demonstrated by their interception in reactions with ketene diethylthioketal (cf. **1.347a** → **1.349** and **1.347b** → **1.350**).[975]

INTRAMOLECULAR NUCLEOPHILIC ADDITION. Anhydro bases of merocyanines containing a 1,3-dimethylbenzimidazolium moiety (**1.351**) may be converted into spiro(benzimidazolin-2,2'-[2H]chromene) derivatives (**1.352**) by intramolecular nucleophilic attack at the 2-position of the imidazole ring.[976] It may be noted that cyclizations of this type do not occur unless the benzo ring of (**1.351**) contains an electron-withdrawing substituent.[977]

BASE-INDUCED CLEAVAGE OF THE IMIDAZOLE RING. Under the influence of hot alkali, 1,3-dialkylbenzimidazolium salts undergo cleavage of the imidazole ring with the formation of N,N'-dialkyl-o-phenylenediamines and a

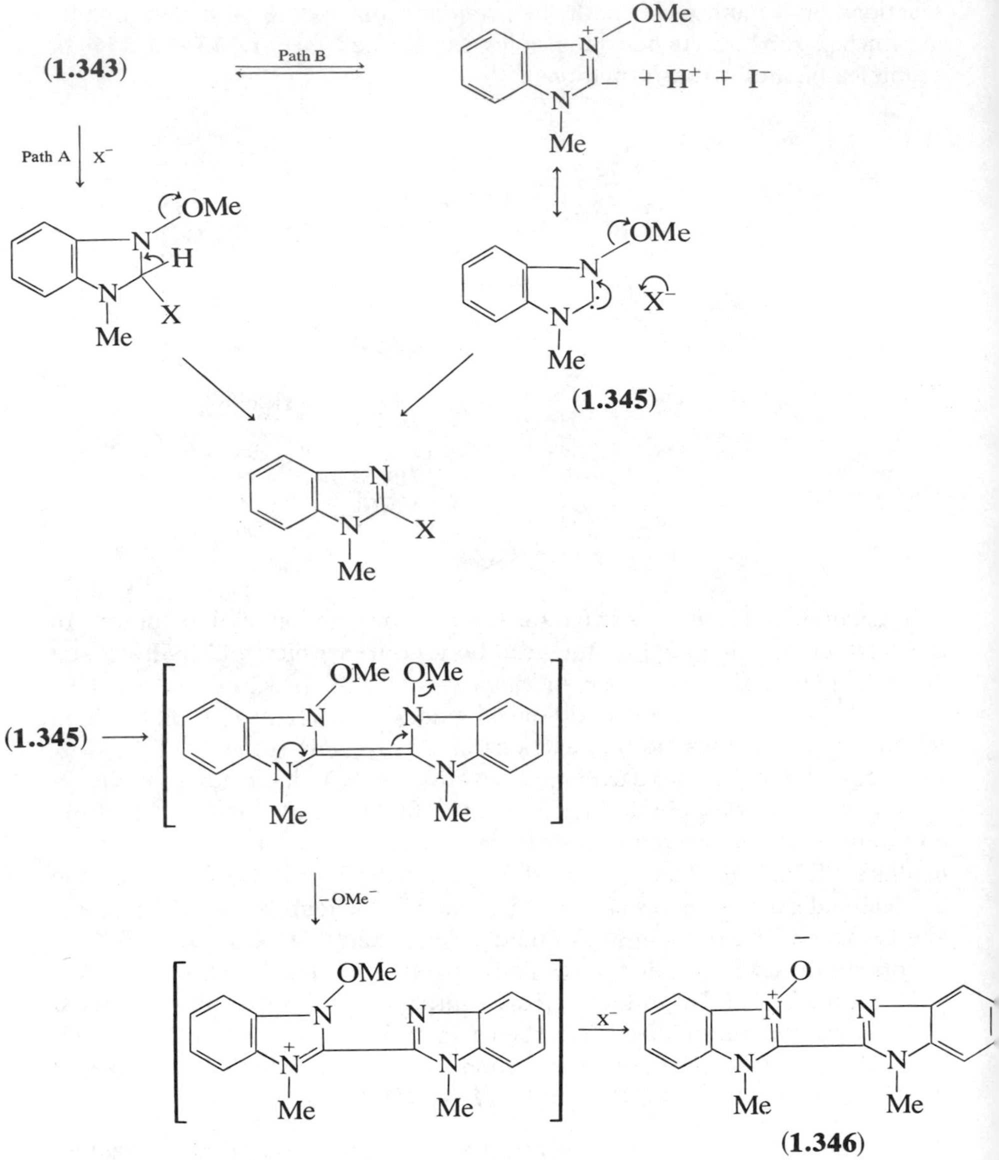

(1.343) $\overset{\text{Path B}}{\rightleftharpoons}$

$-$ + H$^+$ + I$^-$

Path A $\downarrow$ X$^-$

(1.345)

(1.345) $\longrightarrow$

$\downarrow$ $-$OMe$^-$

$\overset{\text{X}^-}{\longrightarrow}$

(1.346)

Scheme 1.24

molecule of carboxylic acid; the 2-unsubstituted, -methyl, -ethyl, and -phenyl derivatives yield formic, acetic, propionic, and benzoic acids, respectively. Ring opening is particularly facile when a nitrogen atom of the imidazole ring has a 2,4-dinitroaryl substituent, and for these cases ring-opening can be effected by weak bases such as aniline and pyridine.[978,979]

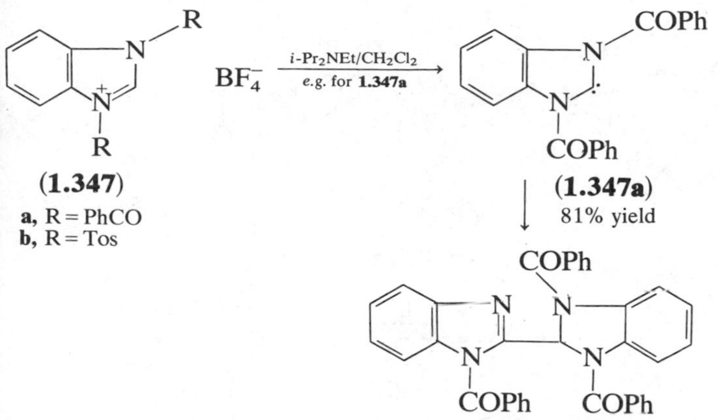

(1.347)

a, R = PhCO
b, R = Tos

(1.347a) 81% yield

(1.348)

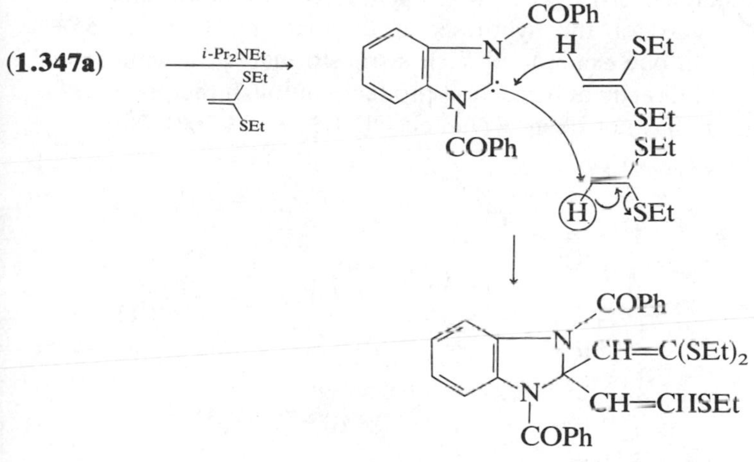

(1.349)

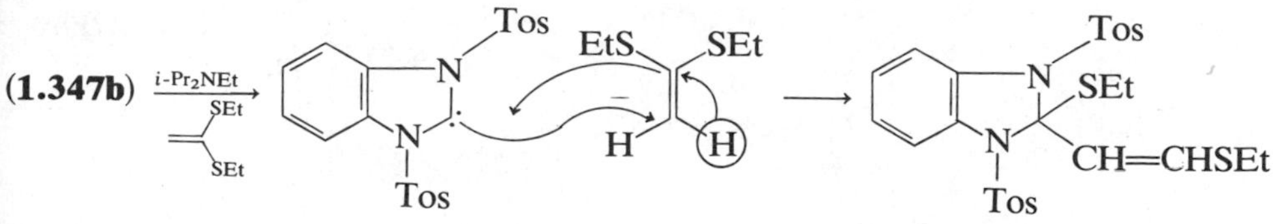

(1.350)

Treatment of 2-methyl-1-phenylbenzimidazole (**1.353**) with benzoyl chloride also results in the formation of an *o*-phenylenediamine derivative (**1.354**), but in this case an enol ester (**1.355**) is also formed. By analogy with comparable reactions of condensed benzimidazoles[980] a mechanism of the type shown in Scheme 1.25 appears likely.

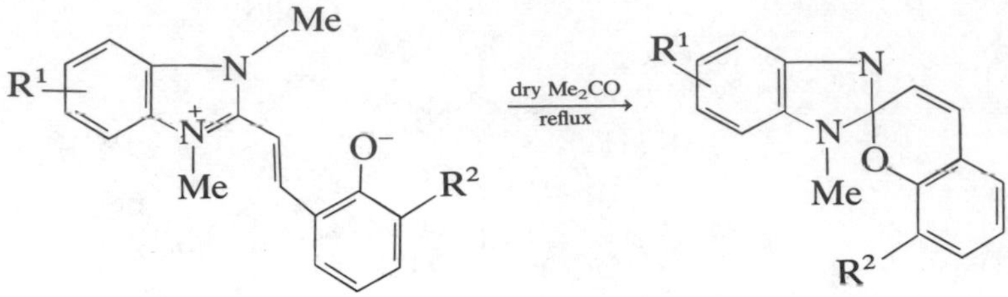

(1.351)   (1.352)

	R¹	R²	Yield (%)
a,	5-NO₂	H	20
b,	5,6-(NO₂)₂	H	43
c,	5,6-(NO₂)₂	OMe	32

The *o*-arylenediamine products of ring-opening of benzimidazolium derivatives are of value for the synthesis of quinoxalines (e.g., **1.356** →
1.357 → **1.358**);[981] in one example of such a transformation, a quinoxaline
derivative is isolated directly as a reaction product, although the precise role
of dimethylsulfoxide has not been evaluated (cf. **1.359** → **1.360**).[982]

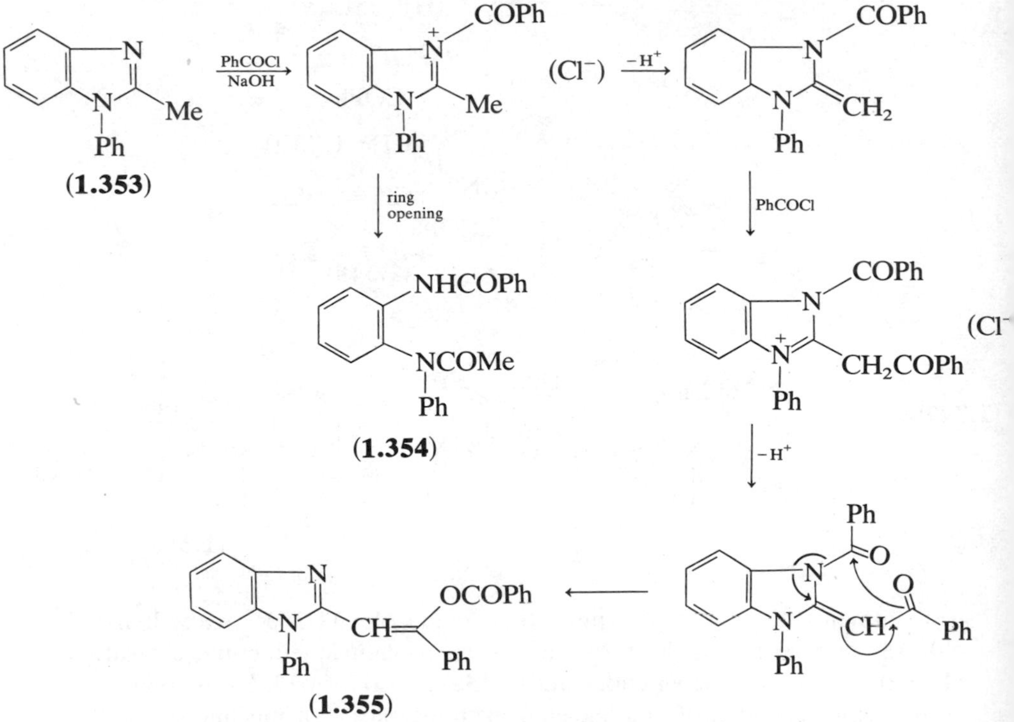

(1.353)

(1.354)

(1.355)

Scheme 1.25

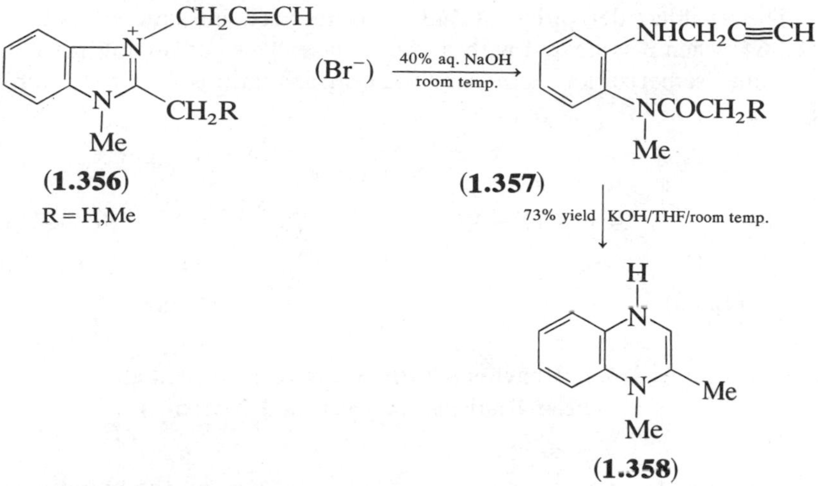
(1.356)
R = H,Me
(Br^-)
40% aq. NaOH
room temp.
(1.357)
73% yield
KOH/THF/room temp.
(1.358)

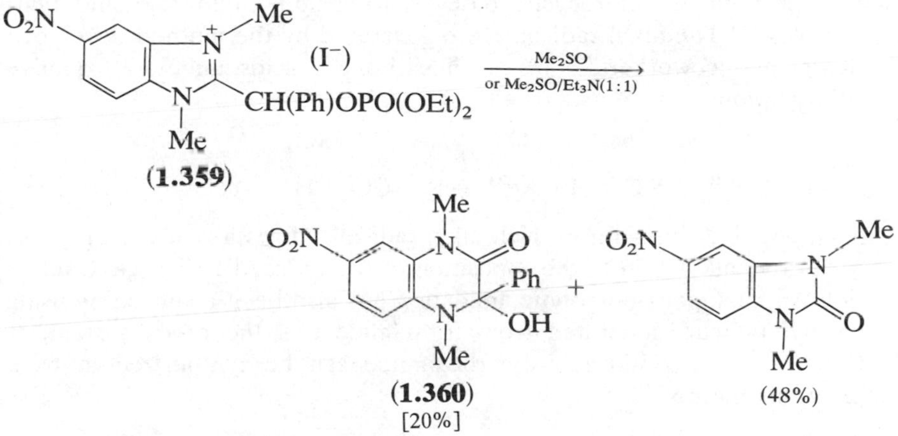
(1.359)
(I^-)
Me_2SO
or $Me_2SO/Et_3N(1:1)$
(1.360)
[20%]
(48%)

Miscellaneous Reactions. A series of labile nitrile oxides (**1.362**) is formed when benzimidazolyl-2-methylhydroxamoyl chlorides (**1.361**) are allowed to react with triethylamine.[983] The most stable of these derivatives in solution is the 1-isopropyl compound as evidenced by infrared analysis. The nitrile oxides (**1.362**) have not been used synthetically.

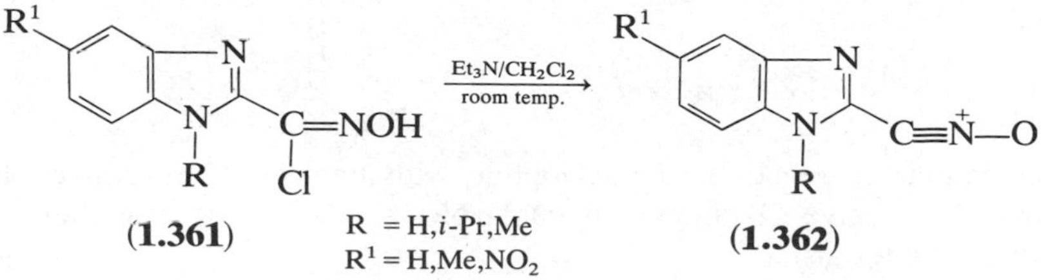
Et_3N/CH_2Cl_2
room temp.
(1.361)
R = H,i-Pr,Me
R^1 = H,Me,NO_2
(1.362)

The allylthio derivative (**1.363**) is isomerized into the propenyl isomer (**1.364**) when it is heated with a strong base. This conversion is followed by cis/trans isomerization: after 4 hr the cis/trans ratio is $4:3$, but after 12 hr it is $1:2$.[984]

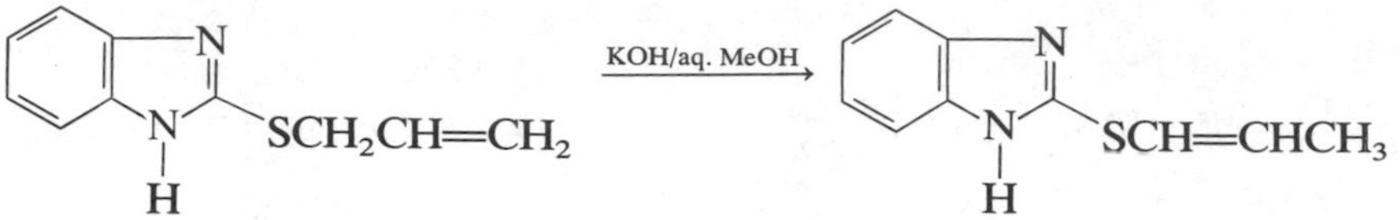

(**1.363**)   (**1.364**)

1.4.4. Reactions with Reactive Intermediates
(Free Radicals, Arynes, and Nitrenes)

Minisci and coworkers have extended their work on the homolytic alkylation of protonated heterocyclic bases[985] to include imidazole and benzimidazoles.[986] The alkyl radicals were generated by the method developed by Kochi and coworkers[987] in which carboxylic acids undergo oxidative decarboxylation:

$$2Ag^+ + S_2O_8^{2-} \rightarrow 2Ag^{2+} + 2SO_4^{2-}$$

$$RCO_2H + Ag^{2+} \rightarrow R\cdot + CO_2 + H^+ + Ag^+$$

The nucleophilic character of such alkyl radicals allows a ready, high yield, regioselective alkylation at the 2-position of the imidazole ring (cf. **1.365** → **1.366**). An analogous homolytic amidation has also been achieved by using carbamoyl radicals generated from formamide and the redox system: t-BuOOH/Fe^{2+}; benzimidazole-2-carboxamide can be synthesized in 60% yield by this method.[986]

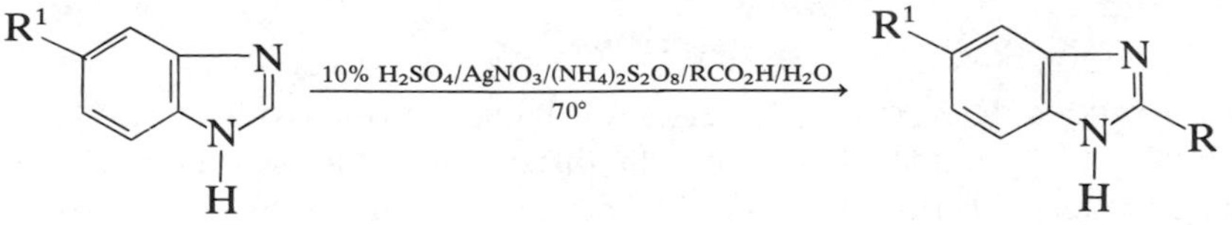

(**1.365**)   (**1.366**)

	R^1	R	Yield (%)
a,	H	PhOCH$_2$	93
b,	H	cyclohexyl	70
c,	H	t-Bu	68
d,	H	n-Pr	50
e,	H	i-Pr	54
f,	Cl	i-Pr	78

Benzimidazole reacts as a nucleophile with benzyne[988] in conventional fashion[989] to give 1-phenylbenzimidazole (29% yield) together with polymeric materials.

Nitrene or nitrene-like intermediates generated from azides and from haloimines have been used to good effect for the final step in the synthesis of the anthelmintic agent Cambendazole (**1.368**)[990] [cf Chapter 10]. It is interesting to note that the carbocyclic ring is substituted selectively at the 5-position in a manner characteristic of attack by electrophilic reagents (cf. section 1.4.1).

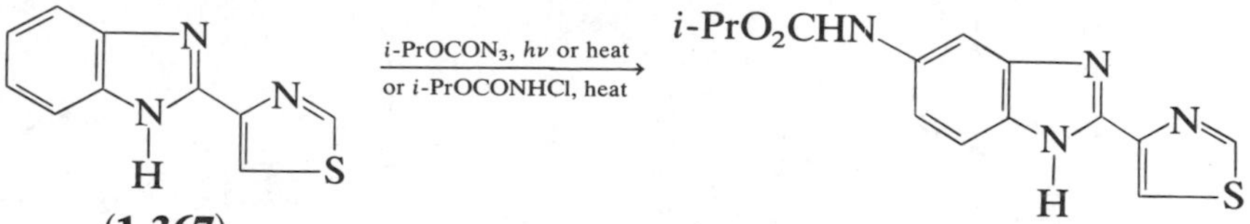
i-PrOCON3, hν or heat
or i-PrOCONHCl, heat
i-PrO2CHN

(**1.367**) (**1.368**)

1.4.5. Thermal and Photochemical Reactions

Thermal Reactions

The high thermal stability of the benzimidazole nucleus has been described in section 1.3.4. Typical decomposition temperatures are benzimidazole, 405°; 2-methylbenzimidazole, 382°; 2-phenylbenzimidazole, 393°; and 2,2′-*p*-phenylenebisbenzimidazole, 550°.[992] The reactions illustrated in this section are thus confined to transformations within substituent groups, and are collected in Table 1.43 and also in formulae (**1.369**→**1.370**;[1000] **1.371**→**1.372**[365,1001] and **1.373**→**1.374**[1002]). The most important

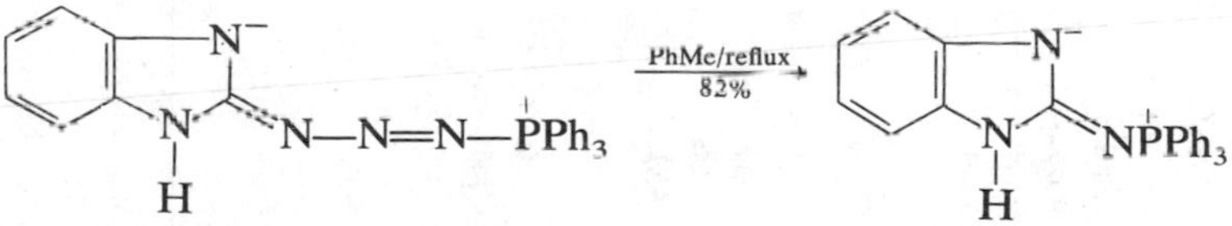
PhMe/reflux
82%
N—N=N—PPh3
NPPh3

(**1.369**) (**1.370**)

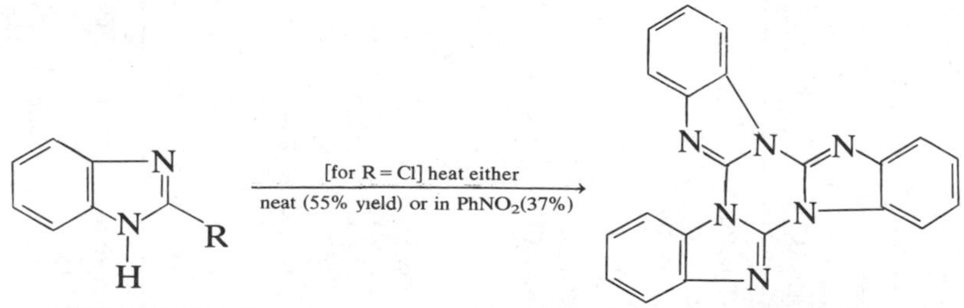
[for R = Cl] heat either
neat (55% yield) or in PhNO2(37%)

R = Cl[1001] NHCOPh[365]

(**1.371**) (**1.372**)

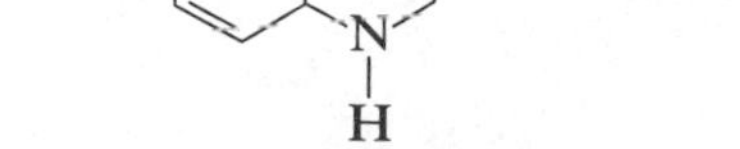
O2N
quinoline/600°

"benzimidazolyl quinoline"

(**1.374**)

(**1.373**)

TABLE 1.43. THERMAL REACTIONS OF BENZIMIDAZOLE DERIVATIVES

Functional group(s) transformed	Other substituents	Reaction conditions	Reaction product	Ref.
1-CONHn-Bu	2-NHCO$_2$Me	CHCl$_3$/room temp. or heat	2-NHCO$_2$Me	993
1-CONHn-Bu	2-NHCO$_2$Me	dimethylsulfoxide/ 140–170°	2-NHCO$_2$Me	994
1-CH$_2$OH	2-CH$_2$CH$_2$OH	Xylene/reflux	2-CH$_2$CH$_2$OH	810
1-CONHPh	none	60° [c]	benzimidazole	995
2-CH=NOCON(CH$_2$CH$_2$Cl)$_2$	none	Heat beyond m.p.	2-CN	833
2-NH$_2$	1-Me	heat with 1-Me-2NH$_2 \cdot$ HCl 250°/5 hr	2-BI-NH-2-BI[a]	996
2-NHCOMe	1-Me	240–245°/5 hr	2-BI-NH-2-BI[a]	996
1-CO$_2$R-2-NH$_2$ (R = alkyl)	Unsubstituted, or alkyl in carbocyclic ring	heat at m.p. for 5 hr or mesitylene/reflux	2-NHCO$_2$R[b]	381, 997, 998
1-CO$_2$N=CRR^1-2-NH$_2$ (R = Me, Et; R^1 = Me, Ph, or RR^1 = cyclohexylidene)	5-H or Me	heat under reflux	2-NHCO$_2$N=CRR^{1} [b]	999

[a] 2-BI denotes 1-methyl-2-benzimidazolyl substituent.
[b] See text.
[c] The amide is considerably dissociated at this temperature as evidenced by infrared spectroscopy.

132

type of thermal reaction from a commercial viewpoint is the alkoxycarbonyl migration (cf. **1.375**→**1.376**).[381,997,998] This reaction provides an alternative procedure for the synthesis of alkylbenzimidazole-2-carbamates of value as anthelmintic agents and fungicides (cf. Chapter 10).

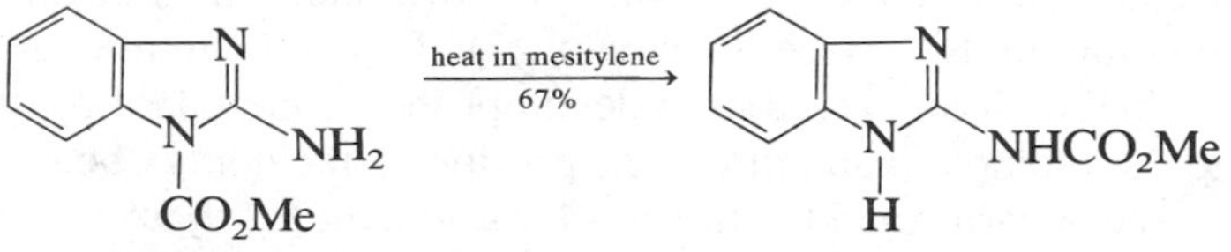

(**1.375**) (**1.376**)

Thermolysis of the benzimidazolium betaine (**1.377a**) without solvent at 20–30° higher than the melting point gives the 2-benzamido derivative (**1.378**), but the 3-ethoxycarbonyl analog (**1.377b**) is not converted into an alkylbenzimidazole-2-carbamate[1003] [cf. the photolytic behavior of (**1.377b**) described below under "Photochemical Reactions"].[1003] A mechanism involving an intermediate diaziridine (**1.379**) has been suggested[1003] (see Scheme 1.26).

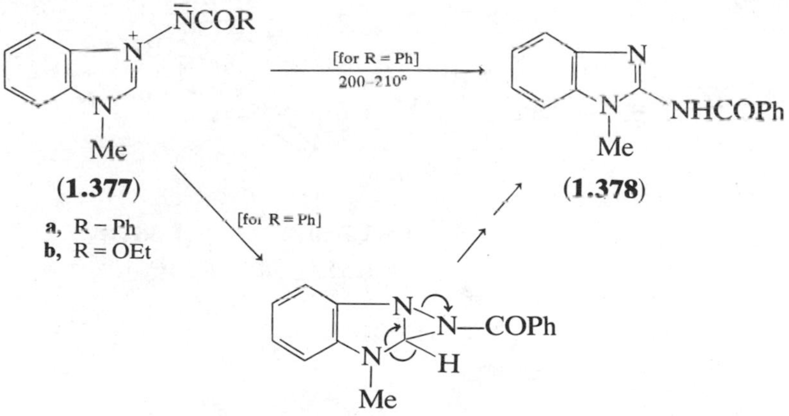

(**1.377**) (**1.378**)

a, R = Ph
b, R = OEt

(**1.379**)

Scheme 1.26

Photochemical Reactions

Benzimidazole is transformed by ultraviolet irradiation into a mixture of dehydro dimers (**1.380** and **1.381**);[1004] it is interesting to note that neither of these is the symmetrical compound. The two dimers must be formed by homolytic substitution at the 4- and 5-positions of benzimidazole by a

2-benzimidazolyl free radical, but it not known whether the latter species is formed directly by photolysis or whether it arises by isomerization of a 1-benzimidazolyl radical [cf. other studies[1005] (section 1.4.6) in which the existence of 2-benzimidazolyl radicals is suggested].

The fungicide "Benlate" (**1.382**) is converted photochemically into methyl benzimidazol-2-ylcarbamate, and this in turn can be transformed into guanidino compounds[1006] (see Scheme 1.27). The anthelmintic and fungicidal agent, thiabendazole (**1.383**), is decomposed in methanol by ultraviolet light into a complex mixture of products including benzimidazole, benzimidazole-2-carboxamide, thiazol-4-carboxamide, thiazol-4-ylamidine, and methyl thiazole-4-carboxylate.[1007,1008] Of more preparative value is the ultraviolet photolysis of 2-(1-benzotriazolyl)benzimidazoles (**1.384**), which are converted into benzimidazo[1,2-*a*]benzimidazoles (**1.385**);[1009] the yields of products (**1.385a** and **b**) have subsequently been raised to 80 and 50%, respectively, by the use of a low-pressure lamp.[1010]

The photo-Fries rearrangement of 2-aryloxybenzimidazoles (**1.386**) to 2-(hydroxyaryl)benzimidazoles (**1.387** and **1.388**) has been described in section 1.2.9; rather poor yields are obtained in this type of reaction.[1011] The quantum yield (0.15) of product formation of 2-(4-methylphenoxy)benzimidazole does not change with concentration and irradiation

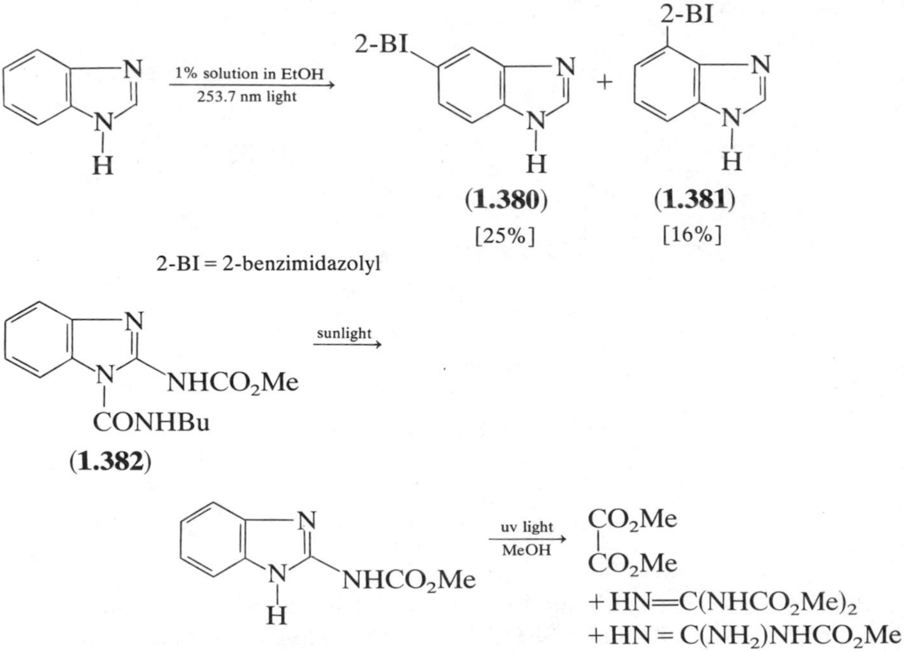

2-BI = 2-benzimidazolyl

Scheme 1.27

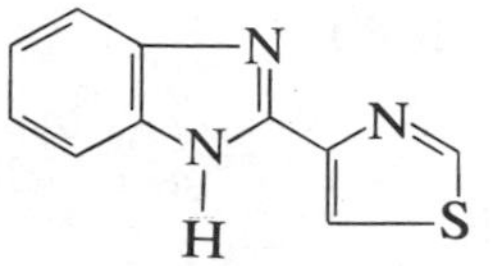

(1.383)

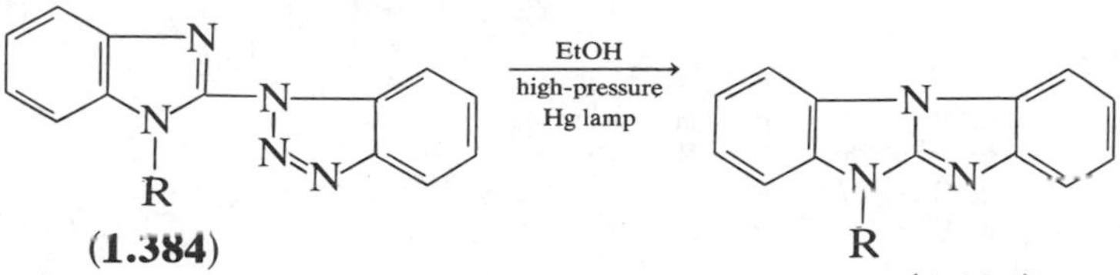

EtOH
high-pressure
Hg lamp

(1.384)

(1.385)

	R	Yield (%)
a,	H	25
b,	Me	30
c,	CH₂Ph	30

time in the presence of piperylene as a triplet quencher; therefore, it has been concluded[1011] that the reaction follows the normal photo-Fries mechanism involving an intramolecular process via the lowest excited singlet state.

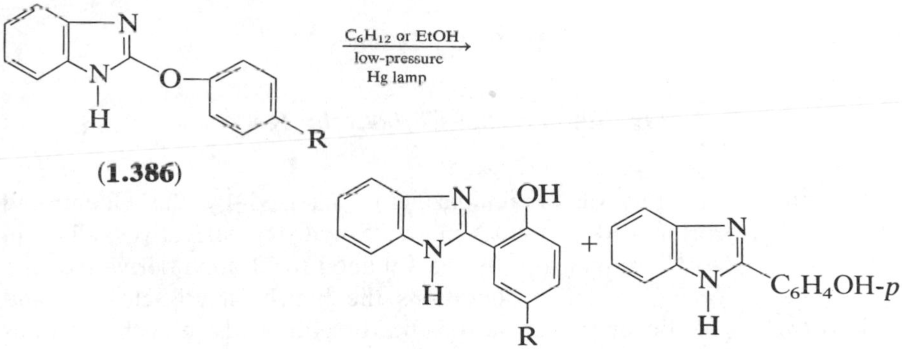

C₆H₁₂ or EtOH
low-pressure
Hg lamp

(1.386)

(1.387)
a, R = H [15%]
b, R = Me [11%]

(for R = H)
(1.388)
[21%]

The nature of photoproducts from the benzimidazolium betaines (1.389) is markedly dependent on the wavelength of incident light and the nature of the imino substituent[1003] (see Scheme 1.28). The benzoyl derivatives (1.389a, b) are not decomposed by light of wavelength >300 nm, but in a quartz apparatus, N–N bond scission occurs. In contrast, irradiation of the 3-ethoxycarbonyl imines (1.389c, d) in a Pyrex vessel gives the ethylbenzimidazole-2-carbamate derivatives (1.390); intermediate di-aziridines have been suggested to rationalize this type of process.[1003]

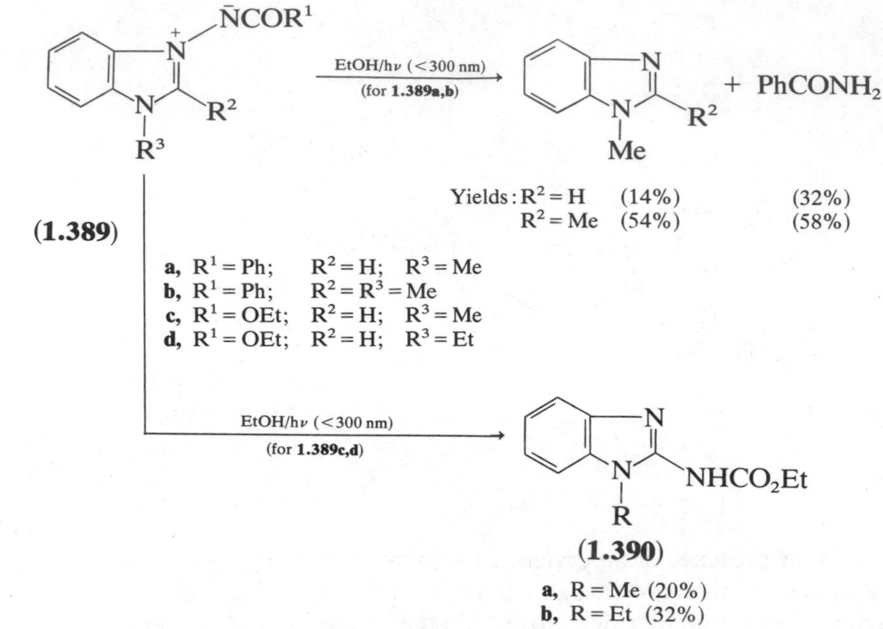

ṄCOR[1]
EtOH/hν (<300 nm)
(for 1.389a,b)
+ PhCONH2
Me
Yields: R2 = H (14%) (32%)
R2 = Me (54%) (58%)
(1.389)
a, R1 = Ph; R2 = H; R3 = Me
b, R1 = Ph; R2 = R3 = Me
c, R1 = OEt; R2 = H; R3 = Me
d, R1 = OEt; R2 = H; R3 = Et
EtOH/hν (<300 nm)
(for 1.389c,d)
NHCO2Et
(1.390)
a, R = Me (20%)
b, R = Et (32%)

Scheme 1.28

1.4.6. Oxidation

Oxidation of the Carbocyclic Ring

The carbocyclic ring of benzimidazole is oxidized by the Udenfriend method[1012] (ascorbic acid–Fe^{2+}–O_2) to give 5-hydroxy derivatives, albeit in poor yield (16%);[1013] similar results are obtained for 2-alkyl derivatives.[1013] Under more forcing oxidative conditions the benzo ring is cleaved, and indeed the formation of imidazole 4,5-dicarboxylic acids in such reactions was used by Bamberger[1014] in early structural work in imidazole chemistry. Examples of oxidants that have been used are chromic acid,[1015] hydrogen peroxide,[124,1016,1017] and ozone;[1018] practicable yields of the carboxylic acids can be isolated (e.g., 30–85% yields of tetracarboxylic acids from hydrogen peroxide oxidation of a series of bisbenzimidazoles[124]).

Oxidation of the Imidazole Ring

The synthesis of benzimidazole *N*-oxides by direct oxidation procedures has not yet been achieved (cf. Refs. 1045 to 1047 and Chapter 2). Thus hydrogen peroxide or peracid oxidation of benzimidazole derivatives containing 4-thiazolyl and 2- and 3-pyridyl substituents give rise to benzimidazole-substituted heteroaromatic *N*-oxides.[1041–1044]

Oxidation of benzimidazole with lead dioxide provides $\Delta^{2,2'}$-biisobenzimidazolylidine (**1.391**) in poor yield;[1005] the reaction is slow and inefficient, and the product is more easily accessible by a similar oxidation of 2,2'-bisbenzimidazole (**1.392**).[1005]

The mechanism of formation of this tetraazadibenzofulvene analog (**1.391**) is unclear, but Hill[1005] has tentatively suggested that an intermediate 2,2'-bibenzimidazole is formed by dimerization of a 2-benzimidazole free radical. The latter could arise by isomerization of a 1-benzimidazolyl radical; a rearrangement of this type could have precedent in the thermally induced isomerization of 1-trityl- to 2-tritylimidazole derivatives.[1048]

II   PbO₂/C₆H₆/reflux, 240 h, 8.5% → (**1.391**)

PbO₂/C₆H₆/reflux 24 h, 65%

(**1.392**)

Oxidation of Substituents

A variety of oxidants has been used for the oxidation of side-chain substituents; these are listed in Table 1.44, and selected examples are discussed in the following in the order in which they appear in the table.

Difficulties have been encountered in the synthesis of benzimidazole carboxaldehyde[1026] by use of the direct Vilsmeier–Haack procedure,[1049] and also by chromium trioxide[1050] or -persulfate[1051] oxidation of methylbenzimidazoles. Oxidation of the latter can be accomplished by ceric ammonium nitrate in sulfuric acid, but side reactions occur and yields are generally moderate (25 to 60%).[1026] It is interesting to note that a 5-methyl substituent is selectively oxidized to a formyl group in the presence of 2- and/or 6-methyl substituents. The resistance of a 2-methyl substituent toward oxidation by ceric ammonium nitrate is in accord with the reported stability of this group to oxidation by dichromate–sulfuric acid mixture,[1015] but it may be noted that this transformation can be achieved by the use of selenium dioxide.[1025]

Interest concerning the oxidation of 1-amino- and 2-aminobenzimidazoles has stemmed from the commercial success of an azo compound (**1.393**) with neuromuscular blocking activity[1052] on the one hand, and an antibacterial agent (**1.394**)[1053] and a trichomonacide (**1.395**)[1054] on the other. Oxidation of 1-amino-3-methylbenzimidazolium derivatives with aqueous bromine[368]

TABLE 1.44.   OXIDATION OF BENZIMIDAZOLE DERIVATIVES

Substituent(s) oxidised[a]	Oxidant	Product(s)	Ref.
4,7-Dihydroxy derivatives[b]	FeCl$_3$, K$_2$Cr$_2$O$_7$, or Na$_2$CrO$_3$	4,7-Quinones	1019, 1020
6,7-Dihydroxy derivatives	FeCl$_3$	6,7-Quinones	1021
5,6-Dihydroxy derivatives	Ag$_2$O	5,6-Quinones	1022
5,6-(OMe)$_2$	HNO$_2$	5,6-Quinone	1023
1,5-Me$_2$	aq. KMnO$_4$	1-Me-5-CO$_2$H	1024
1,2-Me$_2$[c]	SeO$_2$	1-Me-2-CHO	1025
5-Me[c]	Ce(NH$_4$)$_2$(NO$_3$)$_3$/H$_2$SO$_4$	5-CHO	1026
1-Me-2-CH$_2$OH	AgNO$_3$/H$_2$SO$_4$/K$_2$S$_2$O$_8$	1-Me-2-CHO	1025
5-CH$_2$OH	Jones reagent	5-CHO	1027
2-CH$_2$OH-5-Br	KMnO$_4$/Na$_2$CO$_3$/H$_2$O	2-CO$_2$H-5-Br	569
1—(CH$_2$)$_3$CH(OH)C$_6$H$_4$F-p-2-SR 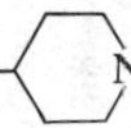	MnO$_2$	1—(CH$_2$)$_3$COC$_6$H$_4$F-p-2-SR 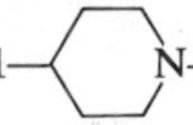	1028
1-CH$_2$CH$_2$NEt$_2$-2-CH(OH) aralkyl	various oxidants, e.g., CrO$_3$, Pb(OAc)$_4$	1-CH$_2$CH$_2$NEt$_2$-2-CO-aralkyl	1029, 1030
2-CH$_2$CHPh$_2$	SeO$_2$	2-COCHPh$_2$	1030
2-COCH$_3$	SeO$_2$	2-COCHO	960
2-CH$_2$COCH$_3$	SeO$_2$	2-CH$_2$COCHO	960
1-Me-2-C≡CH	Cu$_2$Cl$_2$/pyridine	R—C≡C—C≡C—R (R = 1-Me-2-benzimidazolyl)	1031
2-SCH$_2$CH$_2$—N(X) (X = CH$_2$, O) 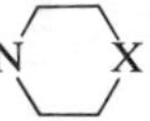	H$_2$O$_2$/AcOH	2-SOCH$_2$CH$_2$—N(X) 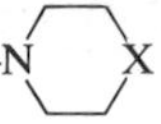	1032

Starting material	Reagent	Product	Ref.
2-NHCO$_2$R-5-SPh (R = alkyl)	H$_2$O$_2$/AcOH	2-NHCO$_2$R-5-SOPh	1033, 1034
1-(CH$_2$)$_2$SEt	m-ClC$_6$H$_4$CO$_3$H	1-(CH$_2$)$_2$SO$_2$Et	597
2-CH$_2$SR (R = Me, Et)	m-ClC$_6$H$_4$CO$_3$H	2-CH$_2$SO$_2$R	915
1-R-2-NH$_2$ (R = alkyl, aryl)	aq. NaOCl	2,2'-Azobenzimidazole derivatives	1035
1-NH$_2$ benzimidazolium derivatives	aq. Br$_2$	1,1'-azobenzimidazolium derivatives	368
2-NHNH$_2$	MeCH=NNHAr/air	2-N=NC(Me)=NNHAr	1036, 1037
1-CH$_2$Ph-2-NH$_2$[c]	(i) Na (2 moles)/NH$_3$ (ii) autoxidation	2-NH$_2$	1038
		2,2'-Azobenzimidazole	1039
1-CH$_2$Ph-2-NH$_2$[c]	(i) Na (4 moles)/NH$_3$ (ii) autoxidation	2-NO$_2$	1038
		2,2'-Azobenzimidazole	1039
1-R-2-NH$_2$[c] (R=H, Me)	(i) NaNH$_2$ or KNH$_2$/NH$_3$ (ii) autoxidation or KMnO$_4$ or K$_2$S$_2$O$_8$ or I$_2$	2,2'-Azobenzimidazoles	1040
2-(4-thiazolyl)[c]	CF$_3$CO$_3$H or Cl$_2$CHCO$_3$H	2-(3-oxo-4-thiazolyl)	1041, 1042
1-(CH$_2$)$_2$-pyridyl[c]	H$_2$O$_2$/AcOH	1-[CH$_2$CH$_2$-(1-oxo-2-py:idyl)]	1043
2-(2- and 3-pyridyl)[c]	Peroxyacids	Pyridine N-oxide derivatives	1044

[a] All benzimidazole substituents not shown in every example.
[b] Type of oxidant used depends on the nature of other substituents.
[c] See text.

139

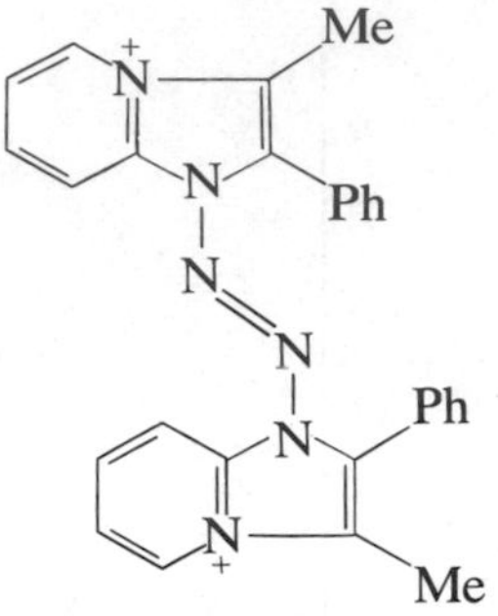

(1.393)   **(1.394)**   **(1.395)**

$2X^-$ (X = Cl,Br)

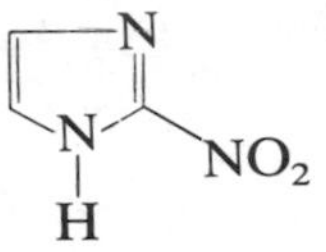

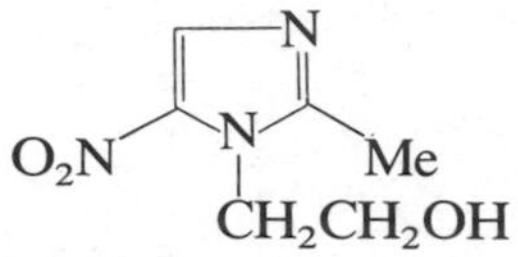

affords 3,3′-dimethyl-1,1′-azobenzimidazolium salts that are structurally related to the azo derivative (**1.393**).

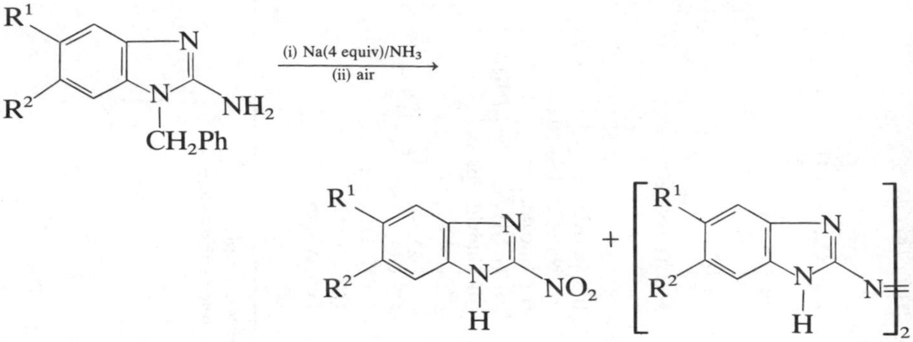

| | Yields (%) | |
	2-nitro	azo
$R^1 = R^2 = H$:	55	43
$R^1 = Me$; $R^2 = H$:	61	35
$R^1 = MeO$; $R^2 = H$:	61	35
$R^1 = R^2 = Me$:	81	17

Scheme 1.29

2-Amino-1-benzylbenzimidazoles are oxidized by air in the presence of sodium amide in liquid ammonia to give good yields of azo and nitro compounds[1038,1039] (see Scheme 1.29). Oxidation in the presence of only 2 molar equivalents of sodium, or oxidation of 1-alkyl derivatives, produces only azo compounds. It has been concluded that the nitro- and azobenzimidazoles are formed via trianions and dianions, respectively (see Scheme 1.30),[1038,1039] and it has been subsequently shown that intermediate dianions can also be oxidized by potassium permanganate and potassium persulfate.[1040]

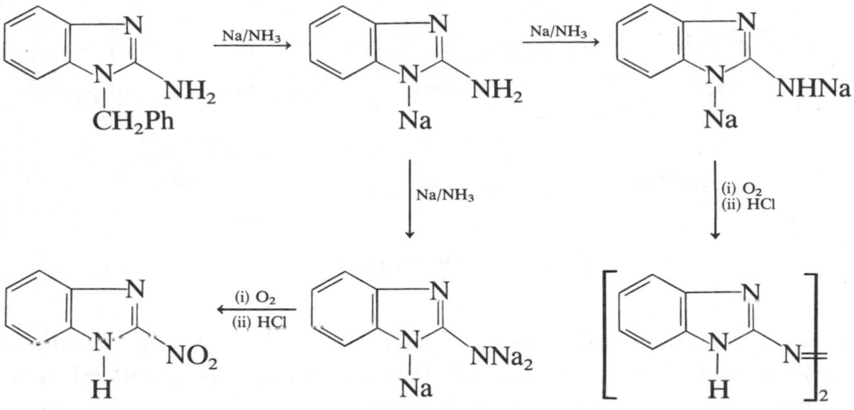

Scheme 1.30

1.4.7. Reduction

Reduction of the Carbocyclic Ring

The standard method for the synthesis of 4,5,6,7-tetrahydrobenzimidazoles is hydrogenation of benzimidazole in acetic acid in the presence of a platinum catalyst (cf. Ref. 6), although more recently palladium,[1055] platinum oxide,[1056] and rhodium catalysts[1057,1058] have been used. The optimum conditions for benzimidazole reduction using a rhodium catalyst are indictated in the following (cf. **1.396 → 1.397**),[1059] but it may be noted that 5,6-dimethylbenzimidazole is not hydrogenated using this catalyst at 100° under hydrogen pressures up to 280 atm.[1059]

Reduction of the Imidazole Ring

The synthesis of 1,3-dihydro derivatives from benzimidazole is discussed in detail in Chapter 3 and is mentioned only briefly here. Reduction can be effected by lithium aluminium hydride,[1060] and reductive acetylation is achieved by catalytic hydrogenation in acetic anhydride (cf. **1.398 → 1.399**).[1061]

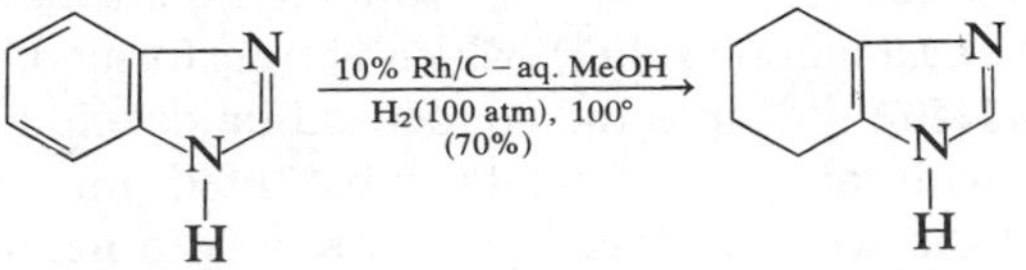

(1.396)   **(1.397)**

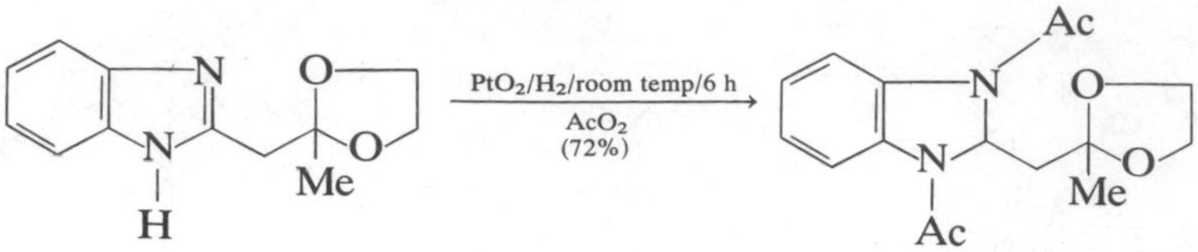

(**1.398**)   (**1.399**)

Reduction of Substituents

Routine procedures involving the reduction of benzimidazole substituents are collected in Table 1.45, and additional examples are discussed in the following.

TABLE 1.45.   REDUCTION OF BENZIMIDAZOLE DERIVATIVES

Compound reduced	Reductant	Product	Ref.
2-C(CN)=R (R = 9-fluorenylidene, CH-1-naphthyl)	NaBH$_4$/i-PrOH	2-CH(CN)CHR (R = 9-fluorenyl, 2-CH(CN)CH$_2$-1-naphthyl)	149
2-C(CN)=CHAr (Ar = p-MeC$_6$H$_4$, o-HOC$_6$H$_4$)	NaBH$_4$/PrOH	2-CH(CN)CH$_2$Ar	294
2,6-Me$_2$-4-NHCOEt	B$_2$H$_6$/THF	2,6-Me$_2$-4NHnPr	163
1-R-2-COCH=CHR^1 (R = H, CH$_2$CH$_2$OH; R^1 = aryl, hetaryl)	NaBH$_4$/pyridine/EtOH	1-R-2-CH(OH)CH$_2$CH$_2$R^1	1062
1-Et-2-Me-5-CF$_3$-6- MeO-7-NO$_2$	H$_2$/PtO$_2$/EtOAc	1-Et-2-Me-5-CF$_3$-6- MeO-7-NH$_2$	176
2-(5-R-2-furyl)-5-NO$_2$ (R = H, hal, CO$_2$Me)	SnCl$_2$/HCl	2-(5-R-2-furyl)-5-NH$_2$	218a
2-[3,5-(O$_2$N)$_2$C$_6$H$_3$]	Raney nickel	2-[3,5-(H$_2$N)$_2$C$_6$H$_3$]	96
2-Pyridinium salts	NaBH$_4$/C$_{1-3}$ alcohols or DMF	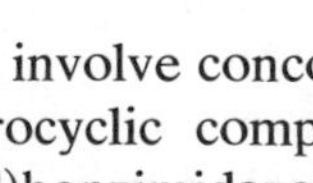	1063

(R = alkyl, allyl, CH$_2$Ph, etc.)

Three interesting cases of substituent reduction involve concomitant cyclization and lead ultimately to condensed heterocyclic compounds. Thus electrochemical reduction[169] of 2-(2-chlorophenyl)benzimidazole (**1.400**) initially gives the radical anion (**1.401**), which is transformed, probably by way of the σ-radical (**1.402**),[1064] into the condensed imidazole (**1.403**). A similar reaction occurs with an appropriately substituted monocyclic imidazole derivative,[169] but the scope of this type of process in heterocyclic synthesis has not been assessed.

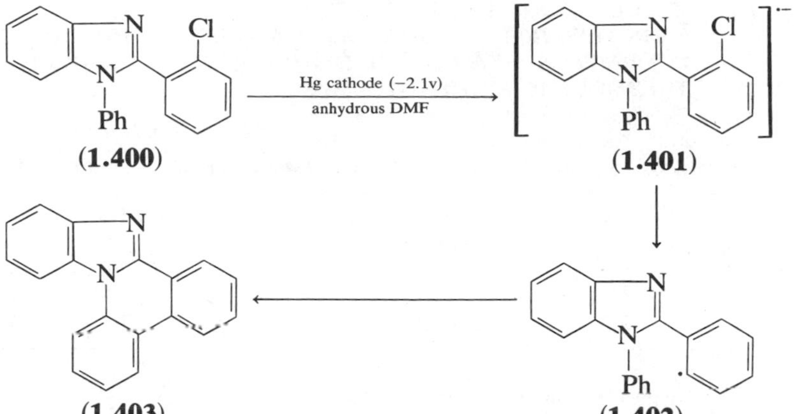

(1.400)

(1.401)

(1.403)

(1.402)

Triethylphosphite reduction of 1-methyl-2-(2-nitrophenyl)benzimidazole (**1.404**, R = H) probably generates a nitrenophenyl intermediate (**1.405**, R = H) that cyclizes to give the condensed imidazole (**1.406**, R = H) in good yield.[1065] The intermediate nitrene has also been generated by thermolysis and photolysis of analogous azides, and its mode of reactivity has been evaluated by using 1-methyl and -isopropyl derivatives.[1065] It appears that the intermediate generated from phosphite reduction behaves in a similar manner to the nitrene generated by azide thermolysis (singlet route) but in a different manner from the nitrene generated by azide photolysis (triplet route)[1066] (see Scheme 1.31 and Table 1.46).

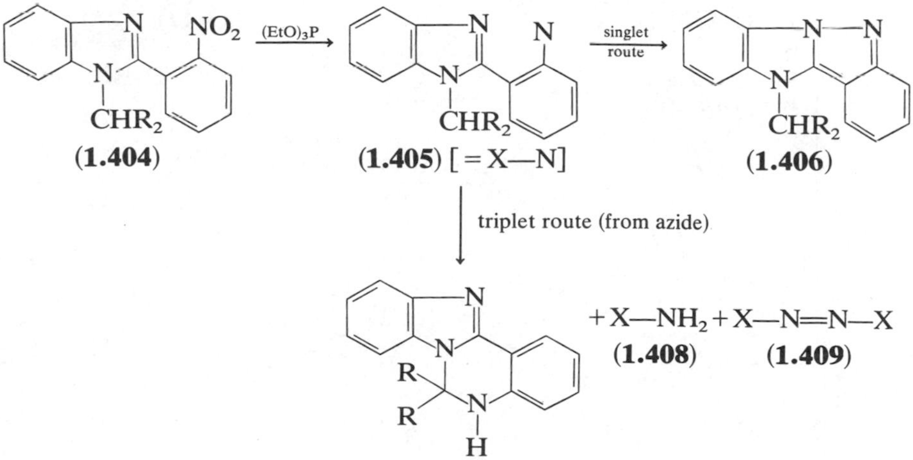

(1.404)

(1.405) [= X—N]

(1.406)

triplet route (from azide)

+X—NH$_2$+X—N=N—X
(1.408)   (1.409)

(1.407)

Scheme 1.31

TABLE 1.46.   REDUCTION OF 1-METHYL-2-(2-
NITROPHENYL)BENZIMIDAZOLE BY TRIETHYLPHOSPHITE.
COMPARISON OF PRODUCT DISTRIBUTION WITH AZIDE
THERMOLYSIS AND PHOTOLYSIS

R in (**1.404**)	Source of (**1.405**)	Reagent or conditions	Solvent	Products (%)			
				(**1.406**)	(**1.407**)	(**1.408**)	(**1.409**)
H	Nitro	$(EtO)_3P$	Cumene	83	0	Trace	0
H	Azido	Δ	Cumene	96	0	4	0
H	Azido	$h\nu$	PhCOMe[a]	15	0	26	8
Me	Azido	Δ	PhBr[b]	66	0	14	14
Me	Azido	$h\nu$	PhCOMe[a]	0	59	15	0

[a] Triplet sensitizer.
[b] Triplet promotor (heavy atom effect).

Lithium aluminum hydride reduction of *N*-cycloalkyl derivatives of the
1-benzimidazolyl succinimide derivatives (cf. **1.410**) affords the expected
pyrrolidine derivatives (cf. **1.411**), but a more complex reaction occurs for
the *N*-aryl derivatives whereby pyrrolo[2,3-*b*]quinoxaline derivatives (cf.
1.412) are obtained (see Scheme 1.32).[668] Reduction of the *N*-phenyl
derivative with lithium aluminum tetradeuteride affords the pentadeuterio

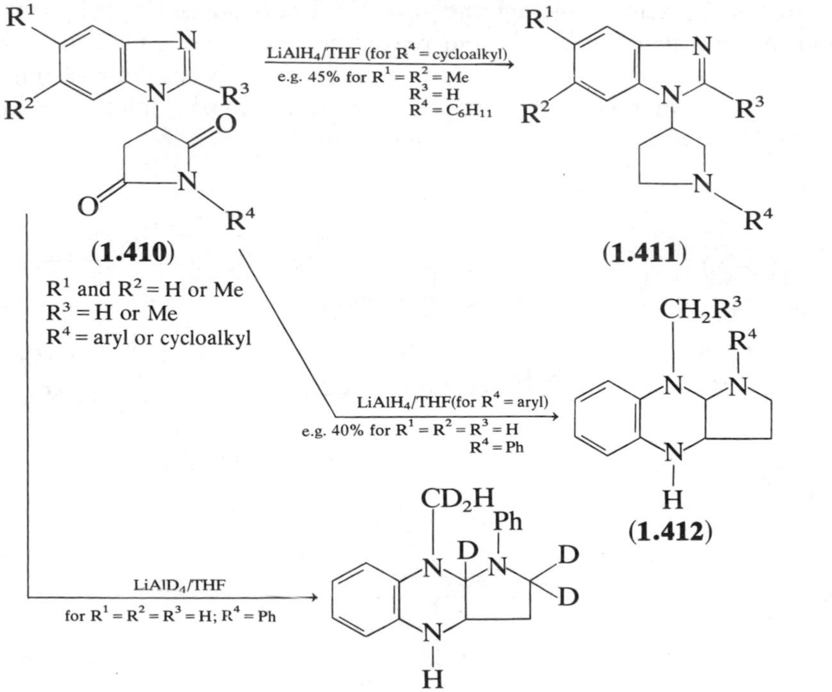

Scheme 1.32

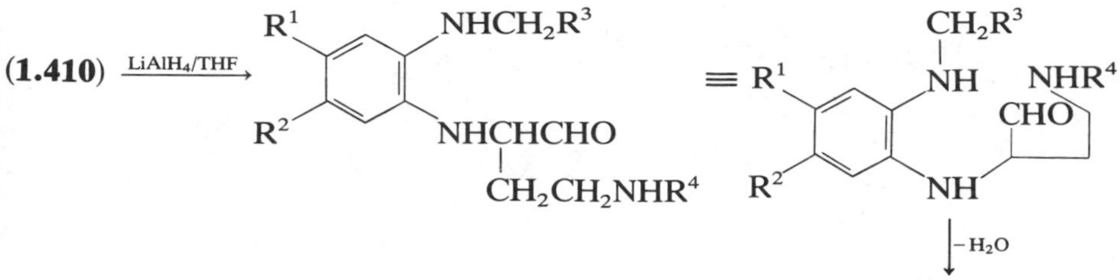

(1.410) $\xrightarrow{\text{LiAlH}_4/\text{THF}}$

Scheme 1.33

compound (1.413), and on this basis a mechanism of the type shown in Scheme 1.33 has been proposed.[668] The different mode of reactivity of aryl from cycloalkyl derivatives is rationalized[668] on the grounds of stabilization in the former cases of intermediates that would arise by opening of the succinimide ring (see Scheme 1.34).

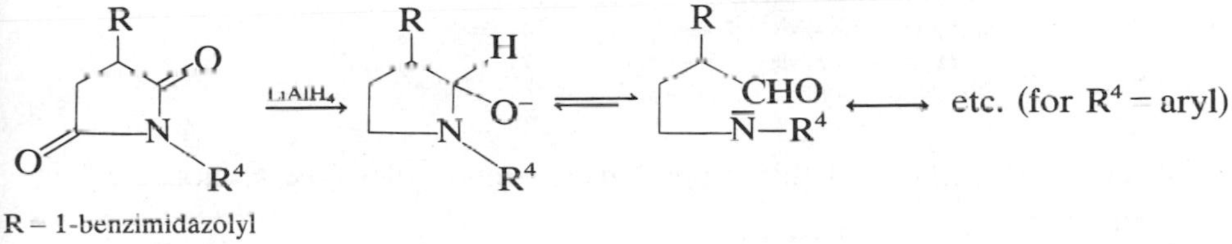

R — 1-benzimidazolyl

Scheme 1.34

1.4.8. Electrocyclic Reactions

Benzimidazoles

2-(Benzimidazolyl)-*N*-phenylnitrone (1.414) reacts[1067] with methyl acrylate and styrene in conventional fashion[1068,1069] to afford isoxazolidines (1.415a and b) in moderate yield (40 and 50%, respectively); the types of isomers formed are in accord with previous reports[1068a] of nitrone-alkene 1,3-dipolar cycloaddition reactions. A series of triazoles (1.417) has been synthesized by the reaction of 2-azido-1-methylbenzimidazole (1.416) with activated acetylenes.[880]

An electrocyclic reaction of the "ene" type[1070] rather than a Michael process has been used to explain the formation of adducts (1.419) of the esters (1.418) with diethylazodicarboxylate, and such products have proved to be valuable intermediates for the synthesis of as-triazino[4,5-*a*]-benzimidazoles[1071] (see section 7.3.1 in Chapter 7). Also described later is the Claisen rearrangement of 2-propargylthiobenzimidazoles, which affords

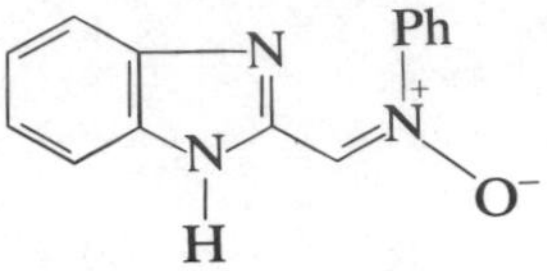

(1.414)

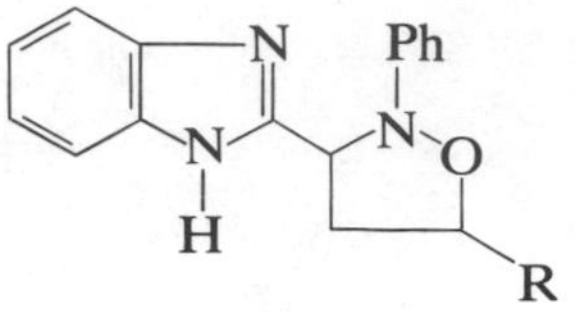

(1.415)

a, R = CO₂Me
b, R = Ph

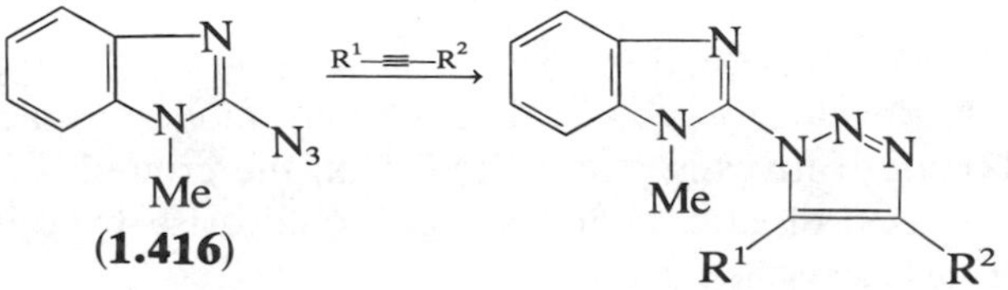

(1.416)

(1.417)

R¹	R²	Solvent	Yield (%)
CO₂Me	CO₂Me	C₆H₆	4
H	CO₂Me	MeCN	37
Et₂N	Ph	Dioxan	86

an efficient synthesis of thiazolo[3,2-*a*]benzimidazoles (see section 6.2.1 in Chapter 6).[1072]

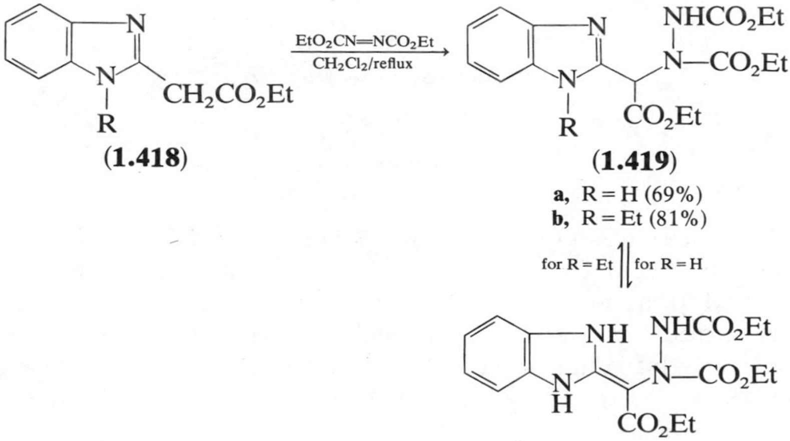

(1.418)

(1.419)

a, R = H (69%)
b, R = Et (81%)

for R = Et   for R = H

Benzimidazolium Compounds

Electrocyclic reactions of benzimidazolium ylides (e.g. **1.420**) with reactive acetylenes have been used for the synthesis of pyrrolo[1,2-*a*]benzimidazoles[1073,1074] (see section 6.1.1 in Chapter 6). In analogous

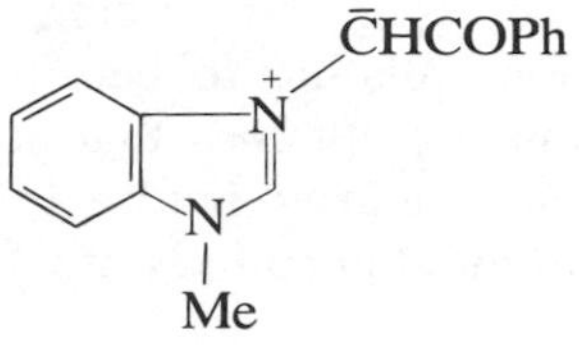

$$\overset{\overset{\displaystyle \bar{C}HCOPh}{|}}{\text{(structure 1.420)}}$$

(1.420)

processes with benzimidazolium imines (1.421), ring-opening of inter-mediate pyrazolo[1,5-a]benzimidazoles occurs, and the products isolated are pyrazole derivatives[1075] (see Scheme 1.35).

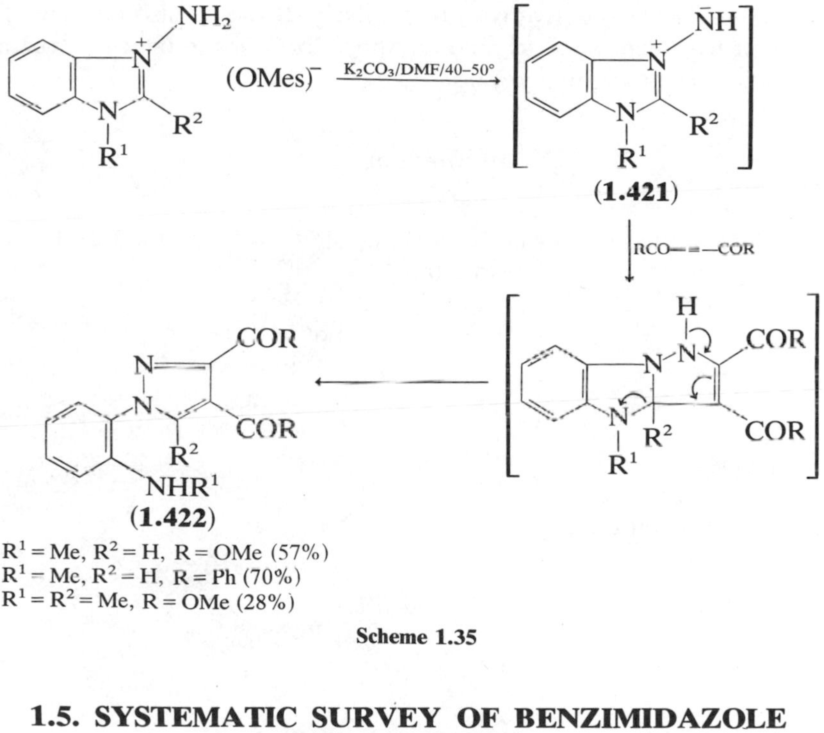

$R^1 = Me, R^2 = H, R = OMe$ (57%)
$R^1 = Me, R^2 = H, R = Ph$ (70%)
$R^1 = R^2 = Me, R = OMe$ (28%)

Scheme 1.35

1.5. SYSTEMATIC SURVEY OF BENZIMIDAZOLE DERIVATIVES

Scope of the Survey

This section contains two separate noncritical surveys of benzimidazole derivatives to cover the period from 1919 through 1950 and 1950 through 1977. In view of the very large number of benzimidazoles cited in *Chemical Abstracts* during the latter period, detailed physicochemical data such as melting point or boiling point have not been included; rather, the information provided directs the user to groups of compounds containing a common functionality. It has been necessary also to omit from the survey in Appendix A (1950–1977) citations relating to benzimidazole nucleosides,[16] organometallic derivatives, and compounds that are reduced in the aromatic ring. Compounds containing a reduced imidazole ring and condensed tricyclic analogs are described in the Appendixes of Chapter 3 and Chapters 4–9, respectively.

In view of the limited availability of the earlier volume in this series, *Imidazole and Its Derivatives, Part I* by K. Hofmann,[1] it has been decided to reprint the original survey of benzimidazole derivatives as Appendix B (1919–1950) in the present volume. The chemical nonenclature and content of Appendix B have not been changed from the original version. Thus 1-unsubstituted compounds bearing a 4- or 5-substituent are described as 4-(or 7-) and 5-(or 6-) derivatives, respectively. It should also be noted that no attempt has been made to modify Appendix B in the light of structural revisions in the intervening period.

Abbreviations

D   refers to the appropriate *o*-arylenediamine (cf. section 1.2.1)
R   alkyl (unless otherwise stated)
Me   methyl
Et   ethyl
Pr   propyl
Bu   butyl
Am   amyl
Ph   phenyl
Ar   aryl
Tos   *p*-toluene sulfonyl

Appendix A. Benzimidazole Derivatives Cited During the Period 1950-1977

TABLE 1.47. ALKYL AND CYCLOALKYL DERIVATIVES

Substituent	Other substituents	Method of preparation	Ref.
1-Me	None	o-$O_2NC_6H_4NMe_2$/ ferrous oxalate	1084
1-Me	None	s-Triazine/o-$H_2NC_6H_4NHMe$	433
1-Me[a]	Various substituents at 5- (or 6-)	D[b] (or mono Ac derivative)/CH_2O	205
1-Me	None 2-Me	o-$H_2NC_6H_4NHMe$/MnO_2	303
1-Me	2-CH_2Ph, CHMePh	2-CH_2Ph/Na/NH_3/MeI	755
1-Me	2-CN, OH, Me, SO_3H, OMe, Oi-Pr, $NHNH_2$, NH_2, NHMe, NMe_2, NHPh, $CH(CN)CO_2Me$, $CH(CN)_2$	1-Me-3-MeO benzimidazolium iodide/nucleophiles	971
1-Me, Et	2-Alkyl	2-Alkyl/RX/$NaNH_2$	754
1-Me, Et	2-NO_2	2-NO_2/MeI or CH_2N_2	868
1-Et	2-Me, various combinations of 5, 6, and 7 substituents	D[b]/carboxylic acid or derivative	37
1-Et	None, 2-Me	o-$H_2NC_6H_4NHMe$/MnO_2	303
1-Et	None	N-Phenylamidine derivative/ NaOCl/base	404
1-n-Pr	2,5-Me_2-7-NH_2 2,6-Me_2-4-NH_2	D[b]/Ac_2O	163
1-Alkyl	5-, 6-, and 7-Me	N(2)-Alkylindazoles/ photolysis	418
1-Alkyl	2-Alkyl	o-Nitroarylamine derivative/ Al_2O_3/aliphatic alcohol/320°	376
2-Me	None	D[b]/$MeCS_2Et$	154
2-Me	5-CO_2H	Cyclization of o-[N-acetylamino]-nitroarene	339
2-Me	5-Alkyl, RO, CN, COR, etc.	Cyclization of o-N-[acylamino]nitroarene	340
2-Me	None 5-Cl, NO_2 4,6-X_2 (X = Cl, Br)	D[b]/$EtCOCH(R)CO_2Et$ R = Et, n-Pr	313
2-Me, Et	None	D[b]/carboxylic acid	28
2-R (R = Me, Et, i-Pr, t-Bu)	None 6-Me 7-Me, i-Pr	Annulation of functionalised imidazoles	414
2-Et	None	D[b]/$EtCSNHCO_2Et$	310
2-n-Pr	None	2-Me/ (i) n-BuLi (ii) EtX (X = hal)	840

149

TABLE 1.47 (*Continued*)

Substituent	Other substituents	Method of preparation	Ref.
2-*n*-Pr 2-*i*Pr	None 5-Cl	Homolytic alkylation by R· (from RCO_2H/peroxydi- sulfate/Ag^+)	986
2-*i*-Pr	None 1-Me 5 (or 6)-NO_2, CF_3	(i) o-$H_2NC_6H_4NHCO$*i*-Pr (and CF_3 derivatives)/heat (ii) nitration by HNO_3/H_2SO_4	326
2-*i*-Pr	1-CH_2Ph-5,6-R_2 R = CO_2Me, COPh	1-CH_2Ph-2-*i*-Pr-imidazole- 4,5-dialdehyde/$(RCH_2)_2$X/ MeONa/MeOH [X = S (for R = CO_2Me, COPh), SO (for R = CO_2Me)]	415
2-CH_2CHMe_2	None	2-Me/ (i) *n*-BuLi (ii) *i*-PrX (X = hal)	840
2-*t*-Bu	None	Homolytic alkylation by R· (from RCO_2H/peroxydi- sulfate/Ag^+	986
2-*t*-Bu	None 5-Cl	D^b/carboxylic acid/high pressure	40
2-Alkyl	None	o-$H_2NC_6H_4NO_2$/V/Cu/Al_2O_3/ 320°/ROH	377
2-Alkyl	None	N (1 or 2) unsubstituted indiazoles/photolysis	418
5-Me	6-Me	D^b + 97% HCO_2H	1105
5-Me	6-Me 2,6-Me_2	o-Nitroarylamine/amide/ $NaHSO_3$	371
2-Cyclopropyl	None	D^b/2-bromocyclobutanone or cyclopropane carboxylic acid	314
2-Cyclopentyl	None	D^b/carboxylic acid	382
2-Cyclohexyl	None	Homolytic alkylation by R· (from RCO_2H/peroxydi- sulfate/Ag^+)	988
2-(1-adamantyl)	None 5-Cl, Me; 5,6-Me_2	D^b/carboxylic acid/high pressure	40
2-$(CH_2)_n$-cyclohexyl (n = 2,3)	5-Cl, Me	D^b/carboxylic acid	66

[a] Isomer mixtures (5 + 6 derivatives) are formed in such cases.
[b] D = appropriate o-arylenediamine derivative.

TABLE 1.48. ARALKYL DERIVATIVES

Substituent	Other substituents	Method of preparation	Ref.
1-CH$_2$Ph	4,7-(MeO)$_2$	D^a/formic acid	865
1-CH$_2$Ph	2-alkyl, aryl, 5-Cl, NO$_2$, etc. 6-Me	D^a/carboxylic acid or aldehyde	74
1-CH$_2$Ph	2-NO$_2$	2-NO$_2$/PhCH$_2$$\overset{+}{N}Me_2$Ph(Cl$^-$)	868
1-CH$_2$C$_6$H$_4$Cl-p	2-Alkyl	1-Alkylation by ArCH$_2$Br/Na	617
1-CH$_2$C$_6$H$_4$R-o R = NO$_2$, NH$_2$	2-Me	(i) Cyclization of o-[N-acylamino]aryl-amine (for R = NO$_2$) (ii) Reduction/H$_2$NNH$_2$·H$_2$O (for R = NH$_2$)	322
1-CH$_2$C$_6$H$_4$OMe-p	None 2-NH$_2$	(i) Benzimidazole/ KOH/aralkyl chloride (ii) Chichibabin	865
1-CH$_2$Ar	2-CH$_2$Cl	D^a carboxylic acid	203
1-[3,4(MeO)$_2$C$_6$H$_3$CH$_2$]	None	benzimidazole/KOH/ aralkyl chloride	865
1-Z-aryl (Z = C$_{1-4}$ alkylene, straight or branched)	2-NH$_2$, various benzimidazole aryl substituents	Alkylation by ArZBr	618
2-CH$_2$Ph	None	D^a/carboxylic acid	28
2-CH$_2$Ph	None	D^a/nitrile	230
2-CH$_2$Ph	None	N^1-phenyl-N-hydroxy-amidine derivative/ PhSO$_2$Cl/OH$^-$	403
2-CH$_2$Ph	None, 5-Cl	N-arylamidines/ NaOCl/base	406
2-CH$_2$Ph	4-NO$_2$ 5-NO$_2$	D^a/imidate	563
2-CH$_2$Ph	5-NH$_2$	D^a/carboxylic acid	75
2-CH$_2$C$_6$H$_4$NO$_2$-p	(a) None (b) 5-NO$_2$	D^a/imidate [for (a)] 2-CH$_2$Ph/conc. H$_2$SO$_4$/ conc. HNO$_3$ [for (b)]	563
2-CH$_2$C$_6$H$_4$I-p	None	D^a/carboxylic acid	77
2-CH$_2$C$_6$H$_4$CO$_2$H-o	None	D^a/carboxylic acid/ DMF/100°	758
2-CH$_2$Ar (Ar = 2,4(O$_2$N)$_2$C$_6$H$_3$)	1-Me	2-CH$_2$Ar/Me$_2$SO$_4$/ NaHCO$_3$	188
2-CH$_2$Ar (Ar = 2,4-(O$_2$N)$_2$C$_6$H$_3$)	5-NO$_2$	2-CH$_2$Ar/H$_2$SO$_4$/HNO$_3$	188
2-CH(R)Ar (R = H, Me; Ar = 2,4-(O$_2$N)$_2$C$_6$H$_3$)	Cl, Br, Me, MeO in aryl ring	D^a/imidate	188
2-CH$_2$-[2,4-(HO)$_2$C$_6$H$_4$]	None	D^a/lactone	284
2-CH$_2$-[2-COR-4,5-(OMe)$_2$C$_6$H$_2$] (R = Me, Et, Pr, Ph)	None	2-CH$_2$-[3,4-(OMe)$_2$C$_6$H$_3$]/ (RCO)$_2$O/trace HClO$_4$	757
2-C(n-Bu)(Et)Ph	None	2-CH(Et)Ph/ (i) n-BuLi (ii) n-BuBr	840

151

TABLE 1.48 (*Continued*)

Substituent	Other substituents	Method of preparation	Ref.
2-CH$_2$(1-naphthyl)	None	D^a/carboxylic acid	105
2-CHPh$_2$	None	o-N$_3$C$_6$H$_4$N=CCHPh$_2$/ thermolysis	394
2-CPh$_3$	None	2-CCl$_3$/AlCl$_3$/C$_6$H$_6$	933
2-(CH$_2$)$_2$Ph	None	2-Me/ (i) n-BuLi (ii) PhCH$_2$X (X = hal)	840
2-(CH$_2$)$_2$Ph	None	1-MeO-3-Ph propylidene-o-aminoaniline/heat	307
2-CH$_2$CHPh$_2$	None 1-Me	(i) D^a/carboxylic acid (ii) 1-methylation by MeI/NaH	674
2-(CH$_2$)$_n$C$_6$H$_4$SO$_2$NH$_2$-p (n = 1,2)	5-Cl, Me, MeO, CO$_2$H, etc.	D^a/carboxylic acid	77b

^{a} D = appropriate o-arylenediamine derivative.

TABLE 1.49. ALKENYL DERIVATIVES

Substituents	Other substituents	Method of preparation	Ref.
1-CH=CH$_2$	None	Benzimidazole/Br(CH$_2$)$_2$Br/KOH	596
1-CH=CH$_2$	None	Benzimidazole/KOH/C$_2$H$_2$	752
1-C(Me)=CH$_2$	2-Cl	2-OH derivative/POCl$_3$	889
1-C(Me)=CH$_2$	2-OR (R = Me, Et, i-Pr)	2-Cl derivative/RO$^-$	889
1-CH=CHPh	2-CO$_2$H	2-CO$_2$Et/NaH/styrene oxide	643
1-CH=CHCO$_2$Me	None 2-alkyl, CH$_2$Ph, aryl	Conjugate addition to HC≡CCO$_2$Me	756
1-CE=CHE (E = CO$_2$Me)	2-CH$_2$Ph	2-CH$_2$Ph/EC≡CE	755
1-CF=CClF	2-Me	2-Me (K salt)/CF$_2$=CClF	499
1-CF=CF$_2$	2-Me	2-Me (K salt)/CF$_2$=CF$_2$	499
1-CF=CFCF$_3$	2-Me, CF$_3$	2-Me or CF$_3$ (K salt)/ CF$_2$=CFCF$_3$	499
1-C(SPh)=C(SPh)CF$_3$	2-Me, CF$_3$	1-CF=CFCF$_3$ derivative/ PhSH/Li	499
1-CH=C(CN)—C- (NH$_2$)=C(CN)$_2$	None 5,6-Me$_2$, Cl$_2$	Benzimidazole/EtOCH=C(CN)$_2$/ CH$_2$(CN)$_2$	940
1-CH=C(CO$_2$Et)—C- (NH$_2$)=C(CN)CO$_2$Et	None 5,6-Me$_2$	Benzimidazole/NCCH$_2$CO$_2$Et/ EtOCH=C(CN)CO$_2$Et	940
1-CH$_2$CH=CH$_2$	None	1-Alkylation by allyl bromide/ NaOMe	601
1-CH$_2$CH=CH$_2$	2-Furyl	2-Furyl/CH$_2$=CHCH$_2$Br/NaOEt	600
1-CH$_2$C(Me)=CH$_2$	2-CH(OH)Ph, CH$_2$Ph	1-Alkylation by β-methallyl chloride/NaOEt	600
1-CH$_2$CH=CHMe	2-CH(OH)Ph, CH$_2$Ph	1-Alkylation by crotyl chloride/NaOEt	600

TABLE 1.49 (*Continued*)

Substituents	Other substituents	Method of preparation	Ref.
1-CH$_2$CH=CHPh	2-Me, hydroxyalkyl, aryl, aminoalkyl	Alkylation by PhCH=CHCH$_2$Cl/ NaH	620
1-(CH$_2$)$_2$CHMe(CH$_2$)$_2$-CH=CMe$_2$	2-Ph	Alkylation by citronellyl bromide/NaH	602
2-CH=CH$_2$	None	2-CH$_2$$\overset{+}{P}Ph_3$(Cl$^-$)/CH$_2$O	1130
2-CR=CH$_2$ (R = alkyl)	None	2-CHRCH$_2$OH/Ac$_2$O	144
2-CH=CHCN	1-H, Me	Dehydration of amido derivatives	1098
2-CH=CHOR (R = alkyl)	1-Me 5-Me, CH$_2$OH	2-C≡CH derivative/ROH/Na/ liq NH$_3$	964, 965
2-CH=C(R)NO$_2$ (R = H, Ph)	1-Me	1-Me-2-CH(OH)CH(R)NO$_2$/Ac$_2$O	952
2-CH=CHCH$_3$	None	D^a/CH$_3$CHClCH$_2$CO$_2$H/H$_3$PO$_4$	328
2-CH=CHC$_6$H$_4$I-*p*	None	D^a/carboxylic acid	77
2-CH=CHC$_6$H$_4$NO$_2$-*o*	None	*o*-N$_3$C$_6$H$_4$CH=CHC$_6$H$_4$NO$_2$-*o*/ thermolysis	394
2-C=CHAr | R (R = H, Ph)	None 1-Me	2-Me benzimidazole derivative/ArCHO	92
2-CH=CH-(2-furyl)	Cl, Br in furyl ring	D^a/aldehyde or 2-Me/furfural	220
2-CH=CH-[2-(4-R-thiazolyl)] (R = Me, Ph)	None 1-Me, (CH$_2$)$_2$CSNH$_2$	2-CH=CHCSNH$_2$/BrCH$_2$COR	806
2-CH=CH-(2-benzoxazolyl)	1-Me	2-CHO/2-Me benzoxazole	950
2-CH=CHR (R = aryl, 2-thienyl)	None 5-Cl, NO$_2$	2-Me derivative/RCHO	809
2-CH=CHR (R = aryl, hetaryl)	None 1-Me	2-CHO/RMe	950
2-CH=CHCOAr	None	2-CHO/ArCOMe	950
2-CH=CHCOPh	1-Me	1-Me-2-CHO/PhCOMe	1025
2-CH=CHCO$_2$R (R = Me, Et)	None 1-Me 5-Me, MeO	2-CHO derivative/Ph$_3$$\overset{+}{P}\overset{-}{C}HCO_2$R	956
2-CH=CMe$_2$	1-Ph	D^a/carboxylic acid or PhN$_2$Ph/ dimethylvinylidene carbene	83
2-C(=CPh$_2$)OCOCHPh$_2$	1-Me, CH$_2$Ph, Ph	1-Substituted derivative/ Ph$_2$C=C=O	674
2-CH=C(Br)CO$_2$Me	1-Me-5-R (R = H, MeO, Me)	2-CHO derivative/ Ph$_3$$\overset{+}{P}\overset{-}{C}$(Br)CO$_2$Me	955, 957
2-CH=	1-Me	1-Me-2-C≡CH/Me$_2$CO/KOH	968
2-CH=	1-Me	1-Me-2-CHO/hydantoin/ AcONa/AcOH	953

TABLE 1.49   (*Continued*)

Substituents	Other substituents	Method of preparation	Ref.
2-CH=⟨structure: R^2, S, N—R^1, R⟩ R = PhN, PhCH$_2$N, O, S; R^1 = H, Ph, PhCH$_2$; R^2 = O, S, NH)	None	2-CHO/thiazolidinone derivative	954
2-C(Ar)=CHAr	None	2-CH$_2$Ar/ArCHO/Ac$_2$O	816
2-C(CN)=CHAr	None	2-CH$_2$CN/ArCHO	294
2-C(CN)=CH-1-naphthyl	None	2-CH$_2$CN/OHC-1-naphthyl	149
2-C(CN)=CHNHCOPh	1-Me	3-Benzoylamino-1-methyl-benzimidazolium betaine/ NCCH=CHCN	116
2-C(CN)=CR^{1}R (R^1 = H, Me; R = alkyl, aryl, hetaryl)	None 1-Me 5-, 6- or 7-Me	2-CH$_2$CN derivative/ RCHO or RCOMe	814
2-C(CO$_2$Me)= CHNHCOR	1-Me	3-Acyl-1-methylbenzimi-dazolium betaines/ HC≡CCO$_2$Me	116
2-C(CH=$\overset{+}{N}$Me$_2$)= CHNMe$_2$(Cl$^-$)	1-Ph-6-Cl	1-Ph-2-Me-6-Cl/ClCH=$\overset{+}{N}$Me$_2$/ (Cl$^-$)	811
2-C(CN)=⟨fluorenylidene structure⟩	None	D^a/ester	149
2-CH$_2$CH=CClCH$_3$	None	D^a/carboxylic acid	84
2-(CH$_2$)$_2$CH=CH$_2$	None	2-Me/(i) *n*-BuLi (ii) CH$_2$=CHCH$_2$X (X = hal)	840
2-(CH$_2$)$_2$CH=C(Cl)CH$_3$	None	2-Me/(i) *n*-BuLi (ii) CH$_3$C(Cl)=CHCH$_2$X (X = hal)	840
2-CH(R)CH$_2$CH=C-(Cl)Me (R = H, Me)	None	D^a/carboxylic acid	85–

^{a} D = appropriate *o*-arylenediamine derivative

TABLE 1.50. ALKYNYL DERIVATIVES

Substituents	Other substituents	Method of preparation	Ref.
1-CH$_2$C≡CH	2-NH$_2$	2-NH$_2$/NaNH$_2$/BrCH$_2$C≡CH	603
2-C≡CR (R = Me, Et)	1-Me-5-R (R = H, Me, MeO)	2-C≡CH derivative/RX/Na/NH$_3$ (X = hal)	955
2-C≡CCO$_2$H	1-Me-5-R (R = H, MeO, Me)	2-CH=C(Br)CO$_2$Me derivative/KOH	957
2-C≡C—C≡CPh	1-Me-5-R (R = H, Me)	2-C≡CH derivative/BrC≡CPh/ n-BuNH$_2$/H$_2$NOH·HCl/CuCl	955
2-C≡C—C≡CR (R = 1-methyl-2-benzimidazolyl)	1-Me	1-Me-2-C≡CH/O$_2$/Cu$_2$Cl$_2$/pyridine	1031
2-C≡C—CH=C(OMe)R (R = 1-methyl-2-benzimidazolyl)	1-Me	1-Me-2-C≡CH/O$_2$/Cu$_2$Cl$_2$/ MeOH/pyridine	1031

TABLE 1.51. ARYL DERIVATIVES

Substituents	Other substituents	Method of preparation	Ref
1-Ph	None	D^c/carboxylic acid	674
1 Ph	None	benzimidazole/benzyne	988
1-Ph	None 5-NO$_2$	Pyrolysis of aryl derivatives of α-amino acids	385
1-C$_6$H$_4$OR-p (R = H, Me)	5-NO$_2$	D^c/HCO$_2$H	31
1-C$_6$H$_4$OR-p (R = H, Me)	5-NH$_2$	1-aryl-5-NO$_2$/Na$_2$S	31
1-[2,4-(O$_2$N)$_2$C$_6$H$_3$]	None 2-Me	1-arylation by aryl halide/NaAc or fusion	625
1-(MeO)$_2$C$_6$H$_3$ (Methoxy substituents are 3,4-, 2,4-, and 2,5-)	None	deamination of 5-NH$_2$ derivatives via diazonium salts	865
2-Ph	None	Cyclization of o-H$_2$NC$_6$H$_4$NHCOPh	332
2-Ph	None	o-H$_2$NC$_6$H$_4$N=CHPh/ H$_2$[Pb(H$_2$PO$_4$)$_2$(HPO$_4$)$_2$]/ Pb(H$_4$PO$_4$)$_2$	399
2-Ph	None	D^c/PhCS$_2$CH$_2$CO$_2$H	154
2-Ph	None	D^c/PhCCl$_3$	300
2-Ph	None	N-phenylamidine derivative/NaOCl/base	404
2-Ph	None	1,5-diphenyltetrazole/ pyrolysis or photolysis	423, 427
2-Ph	4-Me	D^c/aldehyde	214
2-Ph	None 4-Me	o-chloronitroarene/PhCH$_2$NH$_2$	384
2-Ph^a	1-CH$_2$Ph	D^c/aldehyde	210

TABLE 1.51 (*Continued*)

Substituents	Other substituents	Method of preparation	F
2-Ph	None 1-phenethyl	Pyrolysis of *N*-benzyl- or *N*-phenethyl-*o*- nitroaniline with or without ferrous oxalate	3
2-Ph	4-MeO	2,3-$(O_2N)_2C_6H_4OMe/PhCH_2NH_2$	3
2-Ph	None 5-Me, Cl	D^c/aldehyde or *o*-$O_2NC_6H_4N$=CHPh (or derivative)/(EtO)$_3$P	3
2-Ph	4-Cl, Me 5-Cl	Photolysis of ArN=C(Ph)N=SR$_2$ (R = Me, CH$_2$Ph or Ph)	4
2-Ph	4-Cl, Me, NO$_2$	D^c/aldehyde	2
2-Ph	Me, Me$_2$, MeO, and Cl in aryl ring	N^1-Aryl-*N*-hydroxyamidines/ PhSO$_2$Cl/OH$^-$	4
2-Ph	1-Me, CH$_2$Ph, allyl, cinnamyl 5-Me, MeO 6-Me	D^c/carboxylic acid (or derivative)	9
2-Ph	1-COPh	*o*-Benzoquinonedi- benzimide/Ph$_3$P	4
2-C$_6$H$_4$R-*p* (R = H, Me)	5-Cl, Me	*N*-arylamidine/NaOCl/base	4
2-C$_6$H$_4$R-*p* (R = Me, MeO)	None	D^c/ArCSNHCO$_2$Et	3
2-C$_6$H$_4$R-*p* (R = H, Me, MeO)	H, Me, MeO in (benzimidazole)aryl ring	*o*-Nitroarylamine/ArCHO	3
2-C$_6$H$_4$R-*p* (R = H, Me, CO$_2$Me)	5-Me, CO$_2$Me	D^c/imidate	1
2-C$_6$H$_4$R-*p* (R = H, Me, Cl, NO$_2$)	None	*o*-$N_3C_6H_4$=CHAr /thermolysis	3
2-C$_6$H$_4$OH-*o*	None	D^c/*o*-MeOC$_6$H$_4$CO$_2$H	9
2-C$_6$H$_4$OH-*o*	None	D^c/amide	1
2-C$_6$H$_4$OH^b	1-alkyl	D^c/aldehyde	2
2-C$_6$H$_4$R-*o* (R = H, OH)	None	5-PhO-1-Ph-1-*H*-tetrazole/ photolysis	4
2-C$_6$H$_4$OH (*o*- and *p*-)	None	Photo-Fries rearrangement of 2-aryloxy derivatives	8
2-C$_6$H$_4$OH-*m*	None	D^c/carboxylic acid	6
2-C$_6$H$_4$OH-*p*	None	D^c/aldehyde	6
2-(2-OH-5-Me-C$_6$H$_3$)	None	Photo-Fries rearrangement of 2-aryloxy derivatives	8
2-[3,5-R$_2$-4-OH-C$_6$H$_2$] (R = CH$_2$—X; X = piperidino, pyrrolidino, 2,6- dimethylmorpholino, etc.)	None	2-(4-HOC$_6$H$_4$) derivatives/ CH$_2$O/amine	6
2-C$_6$H$_4$NH$_2$-*m*	5-NH$_2$	D^c/carboxylic acid	9
2-C$_6$H$_4$NH$_2$-*p*	5-NH$_2$	D^c/carboxylic acid	9
2-C$_6$H$_4$NH$_2$-*p*	5-NH$_2$	*o*-H$_2$NC$_6$H$_4$NHCOC$_6$H$_4$NH$_2$-*p*/ HCl	3

TABLE 1.51   (*Continued*)

Substituent	Other substituents	Method of preparation	Ref.
2-$C_6H_4NH_2$-*p*	1-Ar-5-NH_2	*o*-ArNHC$_6$H$_4$NHCOC$_6$H$_4$NO$_2$-*p*/ SnCl$_2$/HCl	334
2-C_6H_4R (R = H, *m*- and *p*-NH$_2$	None	*o*-H$_2$NC$_6$H$_4$NHCOAr/heat	331
2-[3,4-(H$_2$N)$_2$C$_6$H$_3$]	None	D^c/carboxylic acid	109
2-[3,5-(H$_2$N)$_2$C$_6$H$_3$]	None	Reduction of dinitro analog by Raney Ni	96
2-C_6H_4R-*p* (R = H, NO$_2$)	None	D^c/carboxylic acid	28
2-$C_6H_4NO_2$-*p*	1-Me-5-NO$_2$	Reductive cyclization of *o*-[*N*-aroylamino]- nitroarene	343
2-C_6H_4R (R = H, *o*-, *m*- and *p*-NO$_2$)	None	*o*-H$_2$NC$_6$H$_4$NH== CHAr/NiO$_2$	398
2-C_6H_4R (R = *o*- and *p*-NO$_2$, *p*-CN)	None	*o*-N$_3$C$_6$H$_4$N==CHAr/ thermolysis	394
2-C_6H_4R-*p* (R = H, NO$_2$, MeO)	4-Me, NO$_2$ 5-Me, NO$_2$, Cl	3,4-diaryl-1,2,4-oxadiazol- 5-ones/pyrolysis	419
2-C_6H_4F-*o*	None 1-Ac, Et, (CH$_2$)$_2$NEt$_2$	Cyclization of *o*-[N- aroylamino]arylamine derivative	333
2-C_6H_4Cl-*o*	1-Ph	D^c/acid chloride	169
2-C_6H_4R-*p* (R = H, Cl)	None	1,5-diaryltetrazoles/ pyrolysis	424
2-C_6H_4R (R = *o*-, *m*-, and *p*-halogeno	5-NO$_2$	D^c/carboxylic acid	88
2-C_6H_4R (R = Cl, NH$_2$)	None	D^c/carboxylic acid	95
2-C_6H_4R (R = *p*-Me, Br, *o*-Br, *m*-Cl)	None	D^c/ArCH==C(R)CN (R = CN, CO$_2$H, CO$_2$Et, CONH$_2$)	294
2-C_6H_4R (R = e.g. halogeno, MeO, etc.)	None 1-Me, Ph	D^c/carboxylic acid or aldehyde	92
2-C_6H_4OMe-*p*	None	D^c/aldehyde	208
2-C_6H_4R (R = *p*-MeO, *m*-NO$_2$)	None	D^c/aldehyde	213
2-$C_6H_4O(CH_2)_2NMe_2$	None 5-NO$_2$	D^c/aldehyde	809
2-$C_6H_4CO_2H$-*p*	5-, 6-(NH$_2$)$_2$	(i) 1,2-(H$_2$N)$_2$-4,5- (O$_2$N)C$_6$H$_2$/ *p*-MeO$_2$CC$_6$H$_4$COCl (ii) hydrolysis (iii) reduction	1151
2-*p*-C$_6$H$_4$CO$_2$C$_6$H$_4$-(2- benzoxazolyl)-*p*	None	2-C$_6$H$_4$COCl-*p*/*p*-HOC$_6$H$_4$- (2-benzoxazolyl)	947
2-$C_6H_4SO_2R$-*p* (R = NH$_2$, NH alkyl, N,N-dialkyl)	None	D^c/carboxylic acid	89

TABLE 1.51 *(Continued)*

Substituents	Other substituents	Method of preparation	F
2-C$_6$H$_4$R-o (R = p-MeC$_6$H$_4$SO$_2$NH)	None 5-Cl, Me, NO$_2$	D^c/carboxylic acid	1
2-aryl	None	D^c/carboxylic acid or derivative	9
2-aryl, 39 derivatives, (e.g., 2-(2-OH-5-Cl) 2-(2-MeO-3,5-Br$_2$) 2-(2-SH), etc.)	None	D^c/carboxylic acid	9
2-Aryl	None 1-CH$_2$Ar	D^c/ArCHO	2
2-Ar	None 1-Ph 5,6-Cl$_2$	D^c/aldehyde	2
2-(1-Naphthyl)	None	D^c/carboxylic acid	1
2-(1-Naphthyl)	None	D^c/NC(R)C=CH-1-naphthyl (R = CO$_2$Et, CONH$_2$, CO$_2$H, CN)	1
2-(2-Naphthyl)	None 5-Cl	N-arylamidine/NaOCl/base	4
2-(3-Fluoro-2-naphthyl)	1-Me, CH$_2$Ph 5-C$_6$H$_4$F-p	D^c/carboxylic acid (or derivative) or aldehyde	1
2-[2-(3-NHR-naphthyl)] (R = p-MeC$_6$H$_4$SO$_2$)	None 5-Cl, Me	D^c/carboxylic acid	1
2-(2-OH-3-naphthyl)	None	D^c/carboxylic acid	1
2- (structure)	Me, MeO, EtO, NO$_2$ in benzimidazole aryl ring	D^c/acid chloride	1

^{a} This compound is formed as a 1:1 complex with the analogous dihydro compound.
^{b} These derivatives have been used to prepare light-sensitive N-alkyl-2-(benzimidazolylaryl) naphthoquinone diazide sulfonate esters.
^{c} D = appropriate o-arylenediamine derivative.

TABLE 1.52. HETARYL DERIVATIVES

Substituent	Other Substituents	Method of preparation	Ref.
2-(2-Furyl)	None	D^a/aldehyde	203, 221
2-(2-Furyl)	None	D^a/carboxylic acid	105
2-(2-Furyl)	None	D^a/aldehyde NaHSO$_3$ addition product	219c
2-(2-Furyl)	5-NO$_2$	D^a/aldehyde	217
2-Furyl	Combinations of 5- and 6-H, Cl, Me, NO$_2$	D^a/imidate or aldehyde	195
2-(5-Halo-2-furyl)	1-Alkyl	(i) D^a/aldehyde (ii) alkylation by Me$_2$SO$_4$ or RI	215
2-[2-(5-R-furyl)] (R = 2,4-Cl$_2$C$_6$H$_3$, 3,4-Cl$_2$C$_6$H$_3$)	1-Me, Ph	D^a/aldehyde	219b
2-(5-nitro-2-furyl)	1-Me, CH$_2$Ph	2-(5-halogenofuryl) derivative/ NaNO$_2$/AcOH	930
2-(5-R-2-furyl) (R − NO$_2$, hal, CO$_2$Me)	5-NO$_2$, NH$_2$	(i) D^a/aldehyde (ii) 5-NO$_2$ → 5-NH$_2$ by SnCl$_2$	218a
2-[2-(5-furyl)] (R = 3-pyridyl, 3-quinolinyl)	None 5-Cl	D^a/aldehyde/heat	219
1-(CH$_2$)$_2$OCO-2-furyl	None	1-Alkylation by CH$_2$=CHCO-2-furyl	669
2-(CH$_2$)$_2$-2-furyl	None	1-MeO-3-furyl propylidene-o-aminoaniline/heat	307
2-(Tetrahydrofuryl)	None	D^a/aldehyde	195
[lactone structure, 2-CH$_2$CH=CH$_2$, O=C–O–Me]	None	[benzimidazole-2,4-dione structure with CH$_2$CH=CH$_2$ substituents] /H$_3$PO$_4$	438
2-Thienyl	None	D^a/carboxylic acid or imidate	195
2-(2-Thienyl)	None	D^a/aldehyde	203
2-(2-Thienyl)	None	D^a/(2-thienyl)CSNHCO$_2$Et	310
2-(2-Thienyl)	5-Cl	D^a/aldehyde	809
2-(5-R-2-thienyl) (R = CH=CHPh)	None	D^a/ester	151
2-(3-thienyl)	5-Me 5,6-Me$_2$	D^a/aldehyde	221
2-Pyrrolyl	None	D^a/aldehyde	195
2-(2-Pyrrolyl)	None	D^a/(2-pyrrolyl)CSNHCO$_2$Et	310
2-(3-Pyrrolyl)	None	D^a/aldehyde	221
1-(1-R-pyrrolidin-3-yl) (R = p-MeC$_6$H$_4$, cycloalkyl)	None 5,6-Me$_2$	reduction of 1-succinimido derivatives/LiAlH$_4$	668
[succinimide structure, 1-NR]	None 2-Me 5-Me, NO$_2$ 6-Me	1-Alkylation by maleimide derivative	668

(R = alkyl, aryl, cycloalkyl)

159

TABLE 1.52 (*Continued*)

Substituents	Other substituents	Method of preparation	Ref.
1-[imide structure] O, N—R, O (R = H, alkyl, aryl etc.)	None 2-Alkyl, haloalkyl, etc. Aryl substituents include alkyl, hal, NO_2, etc.	Alkylation by maleamic acid derivatives	666, 6
1-[structure] O, N, O, R (R = H, alkyl, O-alkyl, Ar, etc.)	None 2-Alkyl, hydroxyalkyl, etc.	1-$CH(CO_2H)CH_2CO_2H$ derivatives/ RNH_2	941
2-[ring] X, NR (X = CH_2, $(CH_2)_2$ R = H, alkyl etc.)	None	D^a/carboxylic acid	104
2-[ring] O, S, $S(CH_2)_2OH$	None	2-CCl_3/$HO(CH_2)_2SH$	934
2-(Oxazol-2-yl)	None	2-CCl_3/$HO(CH_2)_2NH_2$	937
2-(1-Et-2-oxazolin-2-yl)	None	2-CN/2-amino-1-butanol	942
2-[structure] N—CR_2, $(CH_2)_n$, O (R = H, Me n = 1,2)	None	2-CCl_3/amino alcohols	934
2-[ring] R, N, X (X = O, NH, NAc, NMe, $(CH_2)_2OH$; R = H, Me)	None 1-Me 5-Me, Cl, NO_2	2-CCl_3/diamine or amino alcohol	934
2-[ring] COR, N, O, Me (R = Me, Ph)	1-H, Me 5-H, Cl, NO_2	2-$C{\equiv}\overset{+}{N}{-}\bar{O}$/$MeCOCH_2COR$	1076
2-$\overset{+}{N}$[ring] S (Cl^-), R (R = H, Me)	Various aryl substituents	2-Cl derivative/thiazole derivative	881
2-$(CH_2)_n$—$\overset{+}{N}$[ring] S (Cl^-), R (n = 1–4; R = H, Me)	Various aryl substituents	2-Haloalkyl derivative/thiazole derivative	881

TABLE 1.52   (*Continued*)

Substituents	Other substituents	Method of preparation	Ref.
1—⫶S N ⫯(CH₂)ₙ (*n* = 2,3)	2-Ar-5-(and 6-)NCS	(i) 2-Ar-5-NO₂/Cl(CH₂)ₙNCS (ii) reduction (iii) CSCl₂	744
1—⫶S N ⫯(CH₂)ₙ (*n* = 2,3)	2-NHR-5-(and 6-)CO₂R¹ (R = H, CHO, COMe, COEt; R¹ = alkyl, cycloalkyl, etc.)	Alkylation by X(CH₂)ₙNCS (X = hal) and ensuing cyclization	746
1—N S—(CHR)ₙ (R = H,Me; *n* = 1,2)	2-NH₂-5-(and 6-) H, Me, Cl, NO₂	2-NH₂-5-(and 6-)-substituted derivative/Cl(CHR)ₙNCS	743
2-[2(4-methylthiazolyl)]	None	2-CSNH₂/RCOCH₂Br	942
2-(2- and 4-thiazolyl)	1-H, Me, etc. 5-RO, MeS, etc.	*o*-Nitroarylamine/Pd/C/H₂	344
2-(2- and 4-thiazolyl)⎱ 2-(4-Me-2-thiazolyl)⎰	5-, 6-substituents including hal, RS, ArO, ArS, Ar, hetero- cyclic ring	N¹-Aryl-N-hydroxy amidine/ MeSO₂Cl/base	407
2-(CH₂)ₙ-[2(4-R- thiazolyl)] (R = Me, Ph; *n* = 0–2)	None 1 Me, (CH₂)₂CSNH₂	2-(CH₂)ₙCSNH₂ derivative/ BrCH₂COR	806
2-[4-(2-X-thiazolyl)] (X = Cl, Br)	None	*o*-H₂NC₆H₄NHCO-4-thiazolyl derivatives/HCl	337
2-(4-Thiazolyl)	None	N-phenylamidine derivative/ NaOCl/base	404
2-(4-Thiazolyl)	None	*o*-H₂NC₆H₄NHCO-4-thiazolyl/ H₃PO₄	336
2-(4-Thiazolyl)	None 1-alkyl, alkenyl	Dᵃ/carboxylic acid (or derivative)	102
2-(4-Thiazolyl)	None 1-Bu 5-NO₂	Dᵃ/nitrile	231
2-(4-Thiazolyl)	1-Me, CH₂Ph, *n*-C₆H₁₃	2-(4-Thiazolyl)/RI/NaH	594
2-(4-Thiazolyl)	1-(CH₂)₃Br	2-(4-Thiazolyl)/Br(CH₂)₃Br	594
2-(4-Thiazolyl)	1-Hydroxyalkyl	2-(4-Thiazolyl)/bromoalcohol/NaH	599
2-(4-Thiazolyl)	1-Acyl	Dᵃ/carboxylic acid (or derivative)	103
2-(4-Thiazolyl)	1-COR (R = Me, Ph)	Acylation by RCOCl	675
2-(4-Thiazolyl)	1-COR (R = O-alkyl, SCH₂Ph, S-alkyl etc.)	Acylation by RCOCl	691
2-(4-Thiazolyl)	1-CO₂(CH₂)ₙPh (*n* = 1, 2)	Acylation by chloroformate ester	692
2-(4-Thiazolyl)	1-CO(OCH₂CH₂)ₙOR (R = Me, Bu, CH₂Ph; *n* = 1, 2, 4)	Acylation by chloroformate esters	679
2-(4-Thiazolyl)	5-*i*-PrOCONH, *p*-FC₆H₄CONH	2-(4-thiazolyl)/RCON₃/h*ν* or Δ or/ RCONHCl (R = *i*-PrO or *p*-FC₆H₄CO)	990

161

TABLE 1.52 (*Continued*)

Substituents	Other substituents	Method of preparation	Ref.
2-(4-Thiazolyl)	5-NHCOR (R = lower alkoxy, Ar)	Cyclization of o-[N-acylamino]-arylamines or -nitroarenes	345
2-(4-Thiazolyl)	NHCOR in aryl ring (R = i-PrO, Ar)	o-Nitroarylamine derivative/reduction	375
2-(4-Thiazolyl)	5-NHCOR (R = alkoxy, aryl)	Cyclization of N-arylamidines	408
2-(4-Thiazolyl)	5-NHCOR (R = lower alkoxy, aryl)	1,2,4-Oxadiazole derivatives/pyrolysis or photolysis	420
2-(4-Thiazolyl)	5-NHCOR (R = lower alkoxy, aryl)	Reduction of benzotriazine or benzimidazole N-oxide derivatives	434
2-$(CH_2)_n$-4-(2-R-thiazolyl) (R = H, NH_2, CO_2Et; n = 0–2)	None 1-alkyl, carboxyalkyl, etc. 5-CO_2H, NH_2, etc.	(4-Thiazolyl)$(CH_2)_n$(C=NH)NHAr/ NaOCl	407
2-CH(Me)N(CO_2Me)-4-thiazolyl	1-Me	1-Me-2-CH(Me)NH-4-thiazolyl/ $NaHCO_3$/$ClCO_2$Me	802
2-(2-mercapto-5-R-4-thiazolyl) (R = H, Me)	None 1-Ac, COPh 1-CH_2Ph-5,6-Cl_2 5-Me	2-COCH(R)Cl derivative/H_2NCS_2Na	927
2-(3-Oxy-4-thiazolyl) ⎫ 2-(3-Oxy-2-thiazolyl) ⎬	5-NHCOR	2-(thiazolyl)/CF_3CO_3H	104
2-(3-Oxythiazol-4-yl) ⎫ 2-(3-Oxythiazol-2-yl) ⎬	None 1-alkyl, aryl, etc. Various benzimidazole aryl substituents	2-(4- or 2-thiazolyl) derivatives/organic peracid	104
2-(5-Thiazolyl)	None	D^a/carboxylic acid	675
2-(5-Thiazolyl)	1-acyl 5,6-dialkyl	D^a/carboxylic acid (or derivative)	103
2-(5-Thiazolyl)	Me, halogeno, NHAlkyl in thiazole ring	D^a/ester	152
2-(4-Isothiazolyl)	None 1-Ac	D^a/carboxylic acid	675
2-(4-Isothiazolyl)	1-acyl 5,6-dialkyl	D^a/carboxylic acid (or derivative)	103
2-(4-Isothiazolyl)	5-MeO, MeS etc.	o-Nitroarylamine/Pd/C/H_2	344
2-(4-Isothiazolyl)	5-, 6- substituents indluding hal, RS, ArO, ArS, Ar, hetero-cyclic ring	N^1Aryl-N-hydroxyamidine/ $MeSO_2$Cl/base	407
2-(4-Isothiazolyl) ⎫ 2-(3-Me-5-isothiazolyl) ⎬	None 1-alkyl, alkenyl	D^a/carboxylic acid (or derivative)	102
2-(3-Me-5-isothiazolyl)	1-acyl 5,6-dialkyl	D^a/carboxylic acid (or derivative)	103
2-(1-Pyrazolyl)(Cl, Me, Me_2, alkoxy substituents in pyrazole ring)	None 1-Bu, CO_2R	(i) 2-$NHNH_2$/[(MeO)$_2$CH]$_2CH_2$ (ii) 1-Alkylation or acylation	595
2-[1-(3,5-dimethyl-pyrazolyl)]	None 4-Me 5-Me	2-$NHNH_2$ derivative/MeCOCH$_2$COMe	805

TABLE 1.52 (*Continued*)

Substituents	Other substituents	Method of preparation	Ref.
2-(1-phenyl-4,5-dihydro-5-R-pyrazol-3-yl) (R = aryl, hetaryl)	None	2-COCH=CHR/PhNHNH$_2$	963
2-(1-imidazolyl)	None 1-Me	2-SO$_3$H derivative/imidazole	872
2-(2-imidazolyl)	None	D^a/RMe/S (R = 2-imidazolyl)	311
2-(1-Et-2-imidazolin-2-yl)	None	2-CN/N-ethyl ethylene diamine	942
2-(1-R-imidazolin-2-yl) (R = CH$_2$Ph, HO(CH$_2$)$_2$, NC(CH$_2$)$_2$, etc.)	None	2-CCl$_3$/RNH(CH$_2$)$_2$NH$_2$	1126
(X = NH, n = 3,4; X = S, n = 2,3)	None 1-Me	2-CCl$_3$/diamine or mercaptoalkylamine	934
2-[4-(1-Me-5-R-imidazolyl)] (R = H, CO$_2$H)	None	D^a/1-Me-4-CCl$_3$-5-R-imidazole	300
	1-Me	/Na./Hg	953
	1-Me	1-Me-2-CHO/hydantoin/piperidine (cat)	953
	None	D^a/carboxylic acid	62
2-(5-R-1,2,4-oxadiazol-3-yl) (R = CF$_3$, Ph)	None	2-C(NH$_2$)=NOH/(RCO)$_2$O	942
1-(CH$_2$)$_2$[5-phenyl-1,2,4-oxadiazol-3-yl]	None 2-(4-thiazolyl) 5-(and 6-)NO$_2$, NCO	1-(CH$_2$)$_2$C(NHOH)=NH derivatives/ PhCOCl	660
2-(1,2,3-thiadiazol-4-yl)	None	D^a/carboxylic acid (or derivative)	675
2-(1,2,3-Thiadiazol-4-yl)	1-Acyl 5,6-dialkyl	D^a/carboxylic acid (or derivatives)	103
2-(1,2,3-Thiadiazol-4-yl)	5, 6 Substituents including hal, RS, ArO, ArS, Ar, heterocyclic ring	N^1-aryl-N-hydroxyamidine/ MeSO$_2$Cl/base	407

TABLE 1.52   (*Continued*)

Substituents	Other substituents	Method of preparation	Ref.
1,3,4-thiadiazole ring (2-, S, 2-position and R) (R = C(OMe)$_3$, CO$_2$Me)	5-NO$_2$, Cl, etc.	(i) D^a/2,5-bis[(MeO)$_3$C]-1,3,4-thiadiazole (ii) H$_2$SO$_4$ (for CO$_2$Me derivative)	301
2-(CH$_2$)$_2$NH— thiadiazole ring with S, SH	None	2-(CH$_2$)$_2$NHCSNHNH$_2$/CS$_2$/DMF	799
2-(1,2,5-thiadiazol-3-yl)	None 1-COEt	D^a/carboxylic acid	675
2-(1,2,5-thiadiazol-3-yl)	1-Acyl 5,6-dialkyl	D^a/carboxylic acid (or derivative)	103
2-[3-(1,2,5-Thiadiazolyl)]	5-MeO, SPh	o-Nitroarylamine/Pd/C/H$_2$	344
2-N—N triazole ring with R, R^1 (R = H, CO$_2$Me; R^1 = CO$_2$Me; R = NEt$_2$; R^1 = Ph)	1-Me	1-Me-2-N$_3$/RC≡CR^1	880
2-N—N triazole ring with R, CO$_2$Me (R = H, CO$_2$Me)	1-Me	1-Me-2-Cl/triazole derivative/Et$_3$N	880
2- triazole ring N—NH, OH	None	2-CCl$_3$/semicarbazide	934
2-phenyl triazole ring with Ph, NH	None	D^a/2-Ph-5H-s-triazolo[5,1-a]isoindol-5-one	1083
5- (and 6-)—N triazole-thione ring S, NH	1-Me 2-Me, Ph	5- (and 6-)NCS/NaN$_3$	966
2-(tetrahydropyranyl)	None	D^a/aldehyde	195
2-CH$_2$R(Cl$^-$) (R = 1-pyridinium 1-quinolinium)	None	2-CH$_2$Cl/heterocycle	913
2-(CH$_2$)$_2$-1-pyridinium (Br$^-$)	None 5,6-Me$_2$	D^a/carboxylic acid	761

TABLE 1.52 (*Continued*)

Substituents	Other substituents	Method of preparation	Ref.
2—(CHR²)ₙ-N⁺ ring with R, R¹, R substituents (ClO₄)⁻ (R = Ph,Me; R¹ = Ph,H,Me; R² = H,Me; n = 1,2)	None	2-(CHR₂)ₙNH₂/pyrilium perchlorates	1145
5-N⁺ ring with R, R substituents —R (ClO₄⁻) (R = Me,Ph)	1-Me	1-Me-5-NH₂/pyrilium salts	1145
2-N ring X [X = CR=CR (R = H,Me), (CH₂)₂, phenylene, etc.]	Cl, Me, OMe in benzimidazole aryl ring	2-NHCOXCO₂H/Ac₂O/NaAc	795
2-(1-Piperidyl)	None 1-Me	2-SO₃H derivative/piperidine	872
2-(2-Pyridyl)	None	Dᵃ/RMe/S (R = 2-pyridyl)	311
2-(2-Pyridyl)	None	Dᵃ/2-(trichloromethyl)pyridine	300
2-(2- and 4-Pyridyl)	None	Dᵃ/nitrile	231
2-(2-Pyridyl)	None	Dᵃ/carboxylic acid	105
2-Pyridyl	None	Dᵃ/aldehyde	203
2-(2- and 4-Pyridyl)	None	o-N₃C₆H₄N=CH-pyridyl/thermolysis	394
2-(2-Pyridyl)	1-(CH₂)₂OH	Dᵃ/aldehyde	222
2-Pyridyl	1-(CH₂)₂R (R = CN, 2-pyridyl)	2-pyridyl/CH₂=CHR	670
2-(2-Pyridyl)	5-NO₂	2-(2-pyridyl)/H₂SO₄/HNO₃	106
2-(2-Pyridyl)	None 5-Me, MeO, Cl	Dᵃ/2-methylpyridine/S	106
Ar ring with CO₂Et, 2-, N, NH₂	None	2-COCH=CHAr/NCCH₂CO₂Et/NH₄Ac	961
1-(CH₂)₂-(2-pyridyl)	None 2-alkyl, CH₂Ph, Ph	Benzimidazole derivative/CH₂=CH-(2-pyridyl)	671

TABLE 1.52 (*Continued*)

Substituents	Other substituents	Method of preparation	Ref.
1-(CH$_2$)$_2$R (R = 2- and 4-pyridyl)	None	Alkylation by CH$_2$=CHR/ PhCH$_2$N$^+$Me$_3$(OH$^-$)	672
2-(1-oxy-6-R- pyrid-2-yl) and 2-(1-oxypyrid- 3-yl) (R = H, alkyl)	NCS, NO$_2$, alkyl, etc., in benzimidazole aryl ring	2-(2- or 3-pyridyl) derivatives/organic peracid or H$_2$O$_2$	104
1-(CH$_2$)$_2$-(1-oxy-2- pyridyl)	None	1-(CH$_2$)$_2$(2-pyridyl)/H$_2$O$_2$/AcOH	671
2-(CH$_2$)$_2$-(1- oxypyrid-2-yl)	None	2-(CH$_2$)$_2$-(2-pyridyl)/H$_2$O$_2$/AcOH	104
(pyridinone structure: Ar at 4-position, CN at 3-position, 2-substituted, N-H, C=O)	None	2-COCH=CHAr/NCCH$_2$CO$_2$Et/ NH$_4$Ac	961
(pyridinone structure: Ar at 4-position, COMe at 3-position, 2-substituted, N-H, C=O)	None	(pyridinone structure: Ar, CN)/MeMgBr	961
(pyridinone structure: Ar at 4-position, pyrazole ring with Me and N-N-Ph at 3-position, 2-substituted, N-H, C=O)	None	(pyridinone structure: Ar, COCH$_2$COCH$_3$)/PhNHNH$_2$	961
(tetrahydropyridine structure: 2-substituted, NR) (R = C$_{1-16}$ alkyl, allyl, CH$_2$Ph, etc.)	None	2-Pyridinium salts/NaBH$_4$	106
(piperidinone structure: Ar at 4-position, CN at 3-position, 2-substituted, N-H, C=O)	None	2-COCH=CHAr/NCCH$_2$CO$_2$Et/ NH$_4$Ac	961
2-(4-pyridyl)	None 5,6-Me$_2$	D^a/carboxylic acid	107
1-(CH$_2$)$_2$-(4-pyridyl)	None 2-Me	Benzimidazole derivative/CH$_2$=CH- (4-pyridyl)	671
(2-pyridinium structure: 2-substituted, N$^+$-R (X$^-$)) (X = Cl,Br,F)	None	2-(4-pyridyl) benzimidazole/RX	803

166

TABLE 1.52 (*Continued*)

Substituents	Other substituents	Method of preparation	Ref.
1—[structure] N—(CH$_2$)$_3$-COC$_6$H$_4$F-*p*	2-SR (R = H, alkyl, aryl, etc.)	Oxidation of alcohol derivatives by MnO$_2$	1028
1-(2-Thiazolin-2-yl)	2-NH$_2$-5-COR (R = H, alkyl, hetaryl)	2-NH$_2$-5-COR/Cl(CH$_2$)$_2$NCS	745
2-(1-Me-1,4,5,6-tetrahydro-2-pyrimidinyl)	None	2-CN/N-methyl-1,3-diaminopropane	942
2-[structure with Ph] N N S	None	2-COCH=CHPh/(H$_2$N)$_2$CS	817
2-N[structure]N—R (R = H, alkyl, aryl etc.)	None 1-alkyl, acyl, etc. various benzimidazole aryl substituents	2-Cl derivative/piperazine derivative	878
2-CH$_2$N[structure]NCH$_2$COR (R = e.g. NH alkyl, NEt$_2$, etc,)	1-(CH$_2$)$_2$COR^1 (R^1 = aryl, hetaryl)	2-CH$_2$Cl derivative/Na$_2$CO$_3$/heterocyclic base	1123
2-CH$_2$N[structure]N—CH$_2$ CH=CHPh	1-(CH$_2$)$_2$COR (R = 2-furyl, 2-thienyl)	2-CH$_2$Cl derivative/amine	906
2 (3-benzofuryl)	None 1-alkyl Cl, MeO, Me, etc., in benzimidazole aryl ring. Alkyl, alkoxy in benzofuran ring	Cyclization of N-acyl-*o*-phenylene diamines	338
2-CH(R)(CH$_2$)$_n$ – (2-Me-indol-3-yl) (R = H, Me; n = 0, 1)	None	2-CH(R)(CH$_2$)$_n$COMe/PhNHNH$_2$/ H$_2$SO$_4$ (n = 1, 2)	81, 85
2-[structure] N O O	None	6H-benzimidazo[1,2-*b*][2,4]benzo-diazepin-7,12-dione/thermolysis	1140
2-[structure] N O O	5,6-Me$_2$	2-NH$_2$-5,6-Me$_2$/phthalic anhydride	1140
2-(CH$_2$)$_n$—N[structure] O O R (R = H,O$_2$C-alkyl)	None	D^a/carboxylic acid	82

TABLE 1.52 (*Continued*)

Substituents	Other substituents	Method of preparation	Ref
2-R (R = 2-benzoxazolyl 2-benzothiazolyl)	None	2-CCl₃/o-aminophenol or o-aminothiophenol	934
2-(5-chloro-2-benzoxazolyl)	None	2-CN/2-amino-4-chlorophenol	942
2-(CHR)₂—[structure: Q—A benzoxazole ring] (Q = N, A = O, S, NH or NHCO; R = H, OH)	Benzimidazole aryl substituents, e.g., MeO, (MeO)₂, etc.	D^a/imidate	196
2-(2-Benzothiazolyl)	None	2-CH₂NO₂/o-H₂NC₆H₄SH	293
5-(2-Benzothiazolyl)	1- and 2-alkyl, aryl	5-CO₂H/o-H₂NC₆H₄SH	168
2-(4,5,6,7-Tetrahydro-2-benzimidazolyl)	None	2-CN/1,2-diaminocyclohexane	942
2-CH₂R (R = benzotriazol-1-yl, benzotriazol-2-yl)	1-(dialkylamino)alkyl	D/carboxylic acid	80
2—[structure: imidazo-pyridine with NH]	None	2-CCl₃/2,3-diaminopyridine	934
2-(1-Me-3-H-imidazo [4,5-b (and c)]- pyridin-2-yl)	None	D^a/RMe/S (R = 2-hetaryl moiety)	311
2—[structure: triazine with S, O, Ar, N]	None	2-Amino-4-(2-benzimidazolyl)- thiazole/ArNCO	800
2-(3-Coumarinyl)	None 4-Me 5-Cl, Br	D^a/ester	153
2-(2-Quinolinyl)	None	D^a/RMe/S (R = 2-quinolinyl)	311
2-(2-Quinolinyl)	None	D^a/nitrile	231
2-(2-Quinolinyl)	None	D^a/carboxylic acid	105
2-(2-Quinolinyl)	Cl, MeO in quinoline ring	D^a/aldehyde	223
2-(3-Isoquinolinyl)	None 1-(CH₂)₂OH	D^a/aldehyde	222
2-(2-Quinoxalinyl)	None	o-N₃C₆H₄N=CH- (2-quinoxalinyl)/thermolysis	394
2—[structure: pyrano-pyridine with Ar, O, Me]	None	[structure: pyridinone with Ar, COCH₂COCH₃]/EtOH/HCl	961
2-CH(OH)R (R = hetaryl)	None	2-CHO/RMgBr	951

^{a} D = appropriate o-arylenediamine derivative.

TABLE 1.53. NITROSO AND NITRO DERIVATIVES

Substituent	Other substituents	Method of preparation	Ref.
1-NO	2-Alkyl	2-Alkyl/$NaNO_2$/HCl	1096
2-NHNO	1-Me, CH_2Ph 5-Me, NO_2 6-CH_2Ph	2-NH_2 derivative/$NaNO_2$/H_3PO_4	797
2-NHNO	1-Me, Ph, CH_2Ph 5-Me, OR 6-Me	2-NH_2 derivative/ (i) $NaNH_2$/excess i-AmONO (ii) H^+	796, 798
2-NO_2	None	1-CH_2Ph-2-NH_2/Na/NH_3	1038
2-NO_2	None 5-Me, MeO 5,6-Me_2	1-CH_2Ph-2-NH_2 derivative/ (i) Na/NH_3 (ii) air	1039
4-NO_2	2,6-Me_2 2-Me-6-Cl 6-Me	D^a/carboxylic acid or anhydride	161
4-NO_2	6-$CONH_2$	D^a/HC(OEt)$_3$	21a
4-NO_2	6-NH_2	4-NO_2-6-$CONH_2$/Hoffman	21a
4-NO_2	6-SO_2NH_2	D^a/HCO_2H	21b
5-NO_2	2-Me	2-Me/H_2SO_4/HNO_3	22
5-NO_2	2-Me-6-NO_2	2-Me-5-NO_2/H_2SO_4/HNO_3	22
5-NO_2	1-Me-2-NH_2	1-Me-2-NH_2/H_2SO_4/KNO_3	30
5-NO_2	4,7-$(MeO)_2$	4,7-$(MeO)_2$/AcOH/H_2SO_4/HNO_3	32
5-NO_2	1-Me-2-NH_2 1-Me-2-NH_2-6-NO_2 (i) 1,3-Me_2-2-imino	Nitration with conc. H_2SO_4/ HNO_3; or for (i), the nitrate derivative/H_2SO_4/KNO_3	567
5-+6-NO_2	1,2-Me_2	1,2-Me_2/nitration	629
5- and 7-NO_2	2-Me-4-AcNH 4-NH_2	Nitration of 2-Me-4-AcNH or 4-NHCHO derivatives	562
6-NO_2	4-OH	4-MeO-6-NO_2/HBr	24
6-NO_2	4-MeO	D^a/HCO_2H	24
6-NO_2	1-Me-2-NH_2	1-Me-2-NH_2/H_2SO_4/KNO_3	30
6-NO_2	4-CO_2H 2-Me-4-CO_2H	D^a/carboxylic acid or anhydride	161
6-NO_2	1,3-Me_2-2-(=NH)-5-NO_2	Nitration by H_2SO_4/KNO_3	567
2-CH_2NO_2	None	D^a/$(MeS)_2C$=$CHNO_2$	293
2-CH_2NO_2	4- and 5-H, alkyl	D^a/Cl_2C=$CHNO_2$	291
2-CH_2NO_2	1-Me, Ph	2-Me derivative/$NaNH_2$/PrONO	812
2-CH(Cl)NO_2	None 5-Me	D^a/Cl_2C=C(Cl)NO_2	292
2-CCl_2NO_2	4,6,7-Cl_3-5-Me 4,5,6,7-Cl_4	2-$CHClNO_2$ derivatives/Cl_2/ AcOH	571
1-$CHMeCH_2NO_2$	None 2-Me	Alkylation by MeCH=$CHNO_2$	664
2-$(CH_2)_2CMe_2NO_2$	None	D^a/imidate	179
2-$(CH_2)_2C(NO_2)_2R$ (R = NO_2, $(CH_2)_2CO_2Me$)	None	D^a/imidate	180

^{a} D = appropriate o-arylenediamine derivative.

169

TABLE 1.54. AMINO DERIVATIVES

Substituent	Other substituents	Method of preparation	Ref.
1-NH₂	None; 2-Me, Et	1-NHCOR derivative/2 M HCl (R = H, Me, Et)	368
1-NH₂	None	N-Acyl-N'-o-aminophenyl hydrazines/sodium m-nitrobenzene sulfonate	366
1-NH₂	2-Me, Ph; 2-Alkyl, aralkyl	1-NHCOR derivatives/hydrolysis	367
1-NH₂	2-NH₂	1-NHCOR-2-NH₂/hydrolysis (R = H, alkyl, phenyl)	236
1-NH₂	2-NH₂	(i) o-Acylhydrazidoanilines/BrCN (ii) Hydrolysis of ensuing 1-acylamino derivative	1079
1-NH₂	2-NH₂-5-Me, CF₃, Cl; 2-NH₂-5,6-Me₂	(a) Hydrolysis of 1-NHCOMe derivative or (b) 2-NH₂ derivative/H₂NOSO₃H	237
1-NH₂	2-NH₂-5-R-6-R¹ (R = H, Me) (R¹ = H, Cl, Me, CF₃)	(i) o-Acylhydrazidoanilines or (ii) Amination of 2-NH₂ derivatives by H₂NOSO₃H/KOH (for R,R¹ = H, Me)	1078
1-NHCOR (R = H, Me, Et)	None	Cyclization of acylated hydrazino aniline derivatives	368
1-NHCOR (R = H, Me, Et)	2-Me, Et	o-H₂NC₆H₄NHNHCOR/R¹CO₂H R = R¹ = H, Me, Et	367
1-NHCOR (R = H, Me, Et, Ph)	2-NH₂	o-H₂NC₆H₅NHNHCOR/BrCN	236
1-CH₂NH₂	None	1-CH₂Cl/NH₄OH	1125
1-CH₂NHC₆H₄R (R = H, Me, halogeno, alkoxy)	None	Mannich reaction	653
1-CH₂-(1-piperidyl)	2-Me	2-CH₂Ph/CH₂O/piperidine	650
1-CH₂NR₂ (R = Et or NR₂ = 1-piperidyl, 1-morpholinyl)	2-CH₂Ph	Benzimidazole/CH₂O/amine	1125, 648
1-CH₂NR₂ (R = Me;	None; 5-Cl	Mannich reaction	654

1-(CH₂)ₙNR¹R² (R¹, R² = alkyl, allyl; NR¹R² = morpholinyl, etc.; (n = 1-3))	2-SC₆H₄OEt-p	Cl(CH₂)ₙNR¹R²/Na/dioxan	611
1-CH₂NH(CH₂)₂-2-pyridyl	None	Benzimidazole/CH₂O/H₂N(CH₂)-2-pyridyl or 1-CH₂OH/H₂N(CH₂)₂-2-pyridyl	647
1-(CH₂)₂NH₂	2-Me, Ph; 5-OMe	Dᵇ (protected by phthalimido)/ carboxylic acid	586
1-(CH₂)₂NMe₂	2-Ph, CH₂Ph; 7-MeO	Dᵇ/carboxylic acid or PhCHO	29
1-(CH₂)₂NEt₂	2-CH(OAc)Ph	Dᵇ + PhCH(OAc)COCl	1121
1-(CH₂)₂NEt₂	2-CH₂NHAr; 5-NO₂	Dᵇ/imidate	190
1-(CH₂)₂NR¹R² (R¹, R² = Me, Et)	2-CH₂C₆H₄OR-p (R = Et, Me)	1-Alkylation by (amino)alkyl halide/NaNH₂	604
1-(CH₂)₂NEt₂	2-CH(CONH₂)Ar	1-Alkylation by Cl(CH₂)₂NEt₂/NaNH₂	605
1-(CH₂)₂NR₂ (R = Et or NR₂ = morpholine, etc.)	2-aryl, -(γ-pyridyl), -(α-furyl)	2-Aryl, etc./KOH/(dialkylamino)alkyl chloride	612
1-(CH₂)₂NR¹R² (R¹, R² = alkyl; or R¹, R² = morpholino, etc.)	2-CH(R)Ar (R = alkyl, CONH₂ etc.); 2-(CH₂)₂Ar; 2-CH₂-2'-pyridyl; NO₂, NH₂ in aryl ring	Dᵇ/imidate	187
1-(CH₂)₂NEt₂	2-CH₂Ar-5-(and 6-)NO₂	1-Alkylation by Cl(CH₂)₂NEt₂/NaOEt	608
1-(CH₂)₂NR₂ (R = Me, Et)	2-(Methoxyaryl)	1-Alkylation by R₂N(CH₂)₂Cl/NaNH₂	604
1-(CH₂)ₙNR₂ (n = 2, 3; R = Me, Et)	2-Me-5-NHCOMe	1-Alkylation by (dialkylamino)alkyl chloride/NaOEt	609
1-(CH₂)ₙNR₂ (n = 2, 3; R = Et or NR₂ = piperidino, etc.)	None; 2-aryl, hetaryl	Alkylation by (dialkylamino)alkyl chloride/NaOH	613
1-(CH₂)₂-R (R = (substituted) 1-pyrrolidinyl, 1-piperidinyl, and 1-piperazinyl)	2-Me	1-(CH₂)₂Cl-2-Me/amine	899

171

TABLE 1.54 (Continued)

Substituent	Other substituents	Method of preparation	Ref.
1-CHMeCH₂NR₂ (R = Me, Et)	2-Me-5-NHCOMe	1-Alkylation by (dialkylamino)alkyl chloride/NaOEt	609
1-CH₂CH(R)NEt₂ (R¹ = H, Me)	2-C₆H₃R¹-p-NO₂ (R¹ = H, OEt)	Alkylation of 1-unsubstituted derivatives	563
1-CH₂CHMeNR₂ (R = Me, Et)	2-CH₂Ar	Dᵇ/imidate or 1-alkylation	610
1-(CH₂)ₙ NR¹R² (R¹, R² = alkyl, hydroxyalkyl; NR¹R² = heterocyclic ring; n = 2, 3)	2-CH=CHAr	1-(CH₂)ₙX/R¹R²NH (X = hal)	900
1-(CH₂)₃NR¹R² (R¹, R² = Me or NR¹R² = piperidino, etc.)	2-NHCOR (R = aryl, CH=CHPh, 2-furyl etc.) 5-Me, CF₃, etc; 6-Me, Cl	(i) Dᵇ/BrCN (ii) Acylation	235
1-(CH₂)₃N(Me)(CH₂)₂C₆H₄Rₙ (Rₙ = e.g. 4-Cl, 2,4-Cl₂, (OMe)Cl etc.)	None	Alkylation by RCl/NaOH	619
2-NH₂	None	Dᵇ/H₂NCN	239
2-NH₂	None	2-SO₃H/NH₄OH	1164
2-NH₂	1-Me	1-Me-2-Ph/NaNH₂/250°	869
2-NH₂	1-Me	1-Me-2-N₃/HBr	1103
2-NH₂	None, alkyl in carbocyclic ring	o-O₂NC₆H₄NHCN/Raney Ni	381
2-NH₂	1-CH₂C≡CH	2-NH₂/NaNH₂/BrCH₂C≡CH	603
2-NH₂	1-C(R¹R²N)=NH (R¹, R² = H, alkyl, CH₂Ph or R¹, R² = (CH₂)₄, (CH₂)₂O(CH₂)₂)	1-CN-2-NH₂/R¹R²NH	791
2-NH₂	1-C(OPh)=NH	2-NH₂/PhOCN	791
2-NH₂	5-CH(Me)(CH₂)₂CO₂H	2-NHCO₂Me/γ-valerolacetone/AlCl₃	574
2-NH₂	None 1-Me, CH₂Ph 5-Me	2-N₃ derivative/H₂/Raney nickel	875

172

	Substituents	Method	Ref.
2-NH$_2$	1,5-Me$_2$	Chichibabin	858
2-NH$_2$	1-Pr, i-Pr, -Bu	Chichibabin	596
2-NH$_2$	1-R (R = t-Bu, nonyl, undecyl)	Chichibabin	864
2-NH$_2$	1-cyclohexyl, Ph	Chichibabin	862
2-NH$_2$	4-CF$_3$	D[b]/BrCN	342
2-NH$_2$	5-CF$_3$		
2-NH$_2$	1-(CH$_2$)$_n$NR$_2$ (n = 2, 3; R = Me, Et)	2-NH$_2$/R$_2$N(CH$_2$)$_n$Cl/Na	1077
2-NH$_2$	1-CH$_2$CH(OEt)$_2$	2-NH$_2$/BrCH$_2$CH(OEt)$_2$/Na	1077
2-NH$_2$	1-CH$_2$COC$_6$H$_4$R-p (R = H, Cl, Br)	2-NH$_2$/BrCH$_2$COAr	1077
2-NH$_2$	5-SH	2-NH$_2$-5-SMe/Na/NH$_3$	857
2-NH$_2$	5-SMe	Chichibabin	857
2-NH$_2$	1-Me-5-OH	5-OEt derivative/HBr	859
2-NH$_2$	1-Et-5-OMe	Chichibabin	861
2-NH$_2$	1-Me-6-Br	1-MeO-2-NH$_2$/Br$_2$/AcOH	1103
2-NH$_2$	1-Me-5-NH$_2$ / 1-Me-6-NH$_2$ } picrates	Reduction of 5- and 6-azo derivatives	1119
2-NH$_2$	1-Me	Chichibabin	790
2-NH$_2$	1-Me-5-MeO / 1-(CH$_2$)$_3$NEt$_2$-5-MeO / 1-Me-5-OEt / 1-Et-5-OEt / 1-Me-5-OCOPh	Chichibabin	859
2-NH$_2$	1-alkyl (or diethylaminoalkyl)-5-OR	Chichibabin	863
2-NH$_2$	1-Me-5-CO$_2$H	2-Cl derivative/NH$_4$OH	873
2-NH$_2$	5,6-R$_2$ (R = CF$_3$, NO$_2$)	D[b]/BrCN	232
2-NH$_2$	5,6-(OR)$_2$ (R = C$_{1-4}$ alkyl)	D[b]/BrCN	234
2-NH$_2$	Halogeno, C$_{1-4}$ alkyl, etc., in aryl ring	D[b]/aq. NCNH$_2$	240

173

TABLE 1.54 (Continued)

Substituent	Other substituents	Method of preparation	Ref.
2-NH$_2$	1-CO$_2$R (R = alkyl) alkyl in carbocyclic ring	o-O$_2$NC$_6$H$_4$N(CO$_2$R)CN/Raney Ni	381
2-NR^{1}R^2 (R^1, R^2 = H, alkyl, aryl)	1-alkyl, CH$_2$Ph, (CH$_2$)$_2$NR^{1}R^2 (R^1, R^2 = alkyl or NR^{1}R^2 = piperidino) 5-NO$_2$, Me	2-Cl derivative/R^{1}R^{2}NH or R^{1}R^{2}NLi	876
2-NR^{1}R^2 (R^1, R^2 = H, alkyl, aryl)	1-H, alkyl, aminoalkyl	2-Cl/R^{1}R^{2}NH or D^b/carbodiimide or S-methylthiourea	297
2-NHR	5-NO$_2$, Cl, Me		357
	None	2-SO$_3$H derivative/RNH$_2$	872
(R = H, CH$_2$Ph, (CH$_2$)$_2$OH, Ph, 1-ethynyl-2-benzimidazolyl, 4-phenyl-2-thiazolyl)	1-Me		
2-NHMe	1-Me	1-Me-2-NH$_2$-3-COPh(Cl$^-$)/(i) MeI (ii) hydrolysis	1131
2-NHR (R = i-Pr, Ph)	1-Me-5-MeO	o-H$_2$NC$_6$H$_4$N(Me)C(NHR)=NPh/heat or o-H$_2$NC$_6$H$_4$N(Me)CSNHR/MeI/MeOH	356
2-NHR (R = n-Bu, CH$_2$Ph, Ar)	None	o-H$_2$NC$_6$H$_4$NHC(SMe)=NR/HgCl$_2$	358
2-NHR (R = n-Bu, CH$_2$Ph, Ar)	None	o-H$_2$NC$_6$H$_4$NHCSNHR/alkyl iodide or Me$_2$SO$_4$ or HgCl$_2$	352–354, 357
2-NR^{1}R^2 (R^1 = H; R^2 = alkyl or CH$_2$Ph; R^1 = R^2 = alkyl or CH$_2$Ph)	1-Et	1-Et-2-NH$_2$/NaNH$_2$/alkyl halide	772
2-NHCH$_2$Ph	1-Me	1-Me-2-NO$_2$/PhCH$_2$NH$_2$	868
2-NHCH$_2$Ph	1-Ph	Mesoionic thiazolo benzimidazole/PhCH$_2$NH$_2$	1117
2-NHCH$_2$Ph	1-Me-5-CO$_2$H	2-Cl derivative/PhCH$_2$NH$_2$	858
2-NHCH$_2$Ph	1-Me-5-CONHCH$_2$Ph	1-Me-2-Cl-5-CO$_2$H/PhCH$_2$NH$_2$	873
2-NHCH$_2$CH(OH)Ph	1-Me	(a) 1-Me-2-N=C(COPh)NEt$_2$/NaBH$_4$ or (b) 2-NHCOCH(OH)Ph/LiAlH$_4$	880
2-NHCPh$_2$CO$_2$Et	1-Me	Hydrolysis of an imidazo[1,2-a]benzimidazole	880

174

Compound	Substituents	Reagents / Method	Ref.
2-NHPh	1-Ph	o-H$_2$NC$_6$H$_4$NHPh/PhN=CCl$_2$	1164
2-NHC$_6$H$_4$Me-p	1-Me	1-Me-2-Cl/p-toluidine	1103
2-NHC$_6$H$_4$Cl-m	None	2-Cl/m-ClC$_6$H$_4$NH$_2$	877
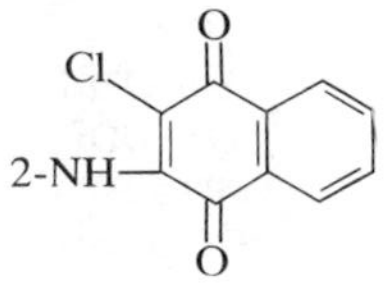 [naphthoquinone structure: Cl, O, O, 2-NH]	1-Et, CH$_2$Ph	2-NH$_2$ derivative/2,3-dichloro-1,4-naphthoquinone	788
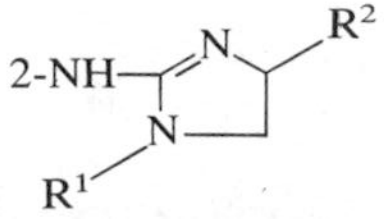 [imidazoline structure: 2-NH, R^2, N, N, R^1] (R^1 = H,Ac; R^2 = H,Me)	1-CH$_2$Ph; 5-Br, SO$_3$H, Cl, Me etc.	2-NHCN derivative/ (i) ethylene diamine derivative (ii) electrophilic aromatic substitution	257
2-NHR (R = COMe, COEt)	Various aryl substituents, e.g., alkyl, MeO, CF$_3$ etc.	(a) 2-NH$_2$ derivative/acylation or (b) D^b/MeSC(NHR)=NH.HI	775
2-N(R)Ac (R = H, Ac)	1-Me-6-OAc	1-Me-2-N$_3$/Ac$_2$O	1103
2-NHCOR (R = alkyl, aryl, CH$_2$Ar)	None; Me, alkoxy in benzimidazolyl aryl ring	2-NH$_2$ derivative/(RCO)$_2$O or RCOCl	774
2-NR^{1}R^2 (R^1 = Ac, Me, COEt; R^2 = H, Ac)	1-Ac; 5,6-(MeO)$_2$; 5,6,7-(MeO)$_3$	(i) D^b/BrCN (ii) acylation	233
2-NHCOPh	None	o-H$_2$NC$_6$H$_4$NHCONHCOPh, p-MeC$_6$H$_4$SO$_3$H/heat	365
2-NHCOPh	1-Me	1-Me-2-NH$_2$,-3-COPh(Cl$^-$)/base	1131
2-NHCOR (R = aryl, CH$_2$OAr, CH(NH$_2$)Ph, 2,4-dioxo-1,2,3,4-tetrahydro-pyrimidin-6-yl)	None	D^b/RCONHC(SMe)=NH	283
2-NHCO-[2-(5-nitrofuryl)]	1-CONHR (R = alkyl, aryl, cyclohexyl)	2-Acylamino derivative/RNCO	777
2-NHCOR (R = 2-furyl, 2-thienyl)	None	2-NH$_2$/2-furoyl chloride or 2-thienoyl chloride	776

TABLE 1.54 (Continued)

Substituent	Other substituents	Method of preparation	Ref.
2-NHCO 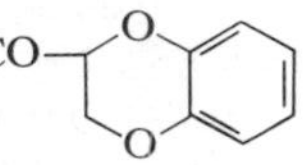 (benzodioxane structure)	None	2-NH$_2$/acid chloride	778
2-NHCOR (R = aryl, hetaryl)	Cl, Me in benzimidazole aryl ring	Acylation of 2-amino derivative	773
2-NHCOCH$_2$-5(2-thiono-3-benzyl-4-oxo-thiazolidinyl]	None	2-NHCOCH=CHCO$_2$H/PhCH$_2$NH$_2$/CS$_2$	795
2-NMe$_2$	1-Me	1-Me-2-Cl/Me$_2$NH	1164
2-NMe$_2$	1-Me-5,6-Cl$_2$	2-NMe$_2$-5,6-Cl$_2$/Me$_2$SO$_4$	232
2-NR$_2$	None	Thermolysis of 0-acylated 1-aryl-2-hydroxy-3,3-dialkyl guanidines	409
2-(1-piperidyl)	None 1-Me	2-Cl derivative/piperidine	882
2-(piperidyl)	1-Me-5-CO$_2$H	2-Cl derivative/piperidine	873
2-N(Me)Ph	1-Me	1-Me-2-Cl/PhNHMe/BuLi	1164
2-N(Me)-[2-(4-methylthiazolyl)]	1-CO$_2$Me	Acylation by ClCO$_2$Me	687
2-N(CH$_2$Ar)(CH$_2$)$_n$ NR^{1}R^2 (n = 2, 3; R^1, R^2 = alkyl or NR^{1}R^2 = morpholino)	None 5-Me, Cl, etc. 5,6-Me$_2$	2-Cl derivative/amine	874
2-N(CO$_2$R)$_2$	None	1-CO$_2$R-2-N(CO$_2$R)$_2$/basic hydrolysis	684
2-(N(CO$_2$R)$_2$ (R = alkyl)	1-CO$_2$R	(a) 1-CO$_2$R-2-NHCO$_2$R/ClCO$_2$R or (ROCO)$_2$O or (b) 2-NHCO$_2$R/ClCO$_2$R or (ROCO)$_2$O	684
2-CH$_2$NH$_2$	1-Me	D^b/carboxylic acid	140
2-CHRNH$_2$ (R = H, alkyl, CH$_2$Ph)	None	2-CHRNHCOPh/hydrolysis	64
2-(CH$_2$)$_n$NH$_2$ (n = 1-5)	Various aryl ring substituents	D^b/carboxylic acid	63
2-CH(NH$_2$)CH$_2$R (R = aminoalkyl)	None	D^b/carboxylic acid	72

2-C(NHR^1)(R^2)Ph R^1 = H, alkyl, aryl, C$_6$H$_{11}$, CH$_2$Ph; (R^2 = H, Ph)	None	2-C(Cl)(R^2)Ph/R^{1}NH$_2$	923
2-CH$_2$N(R)Me (R = H, Me)	1-Me	2-CH$_2$Cl/amine	140
2-CR^{1}R^{2}NHR^3 (R^1, R^2 = H, alkyl, Ph; R^3 = CO$_2$CH$_2$Ph, Tos)	None	D^b/imidate	191
2-CHRNHR^1 (R = H, alkyl, CH$_2$Ph; R^1 = COPh)	None	D^b/carboxylic acid	64
2-CH$_2$NHCH$_2$Ar	None	2-CH$_2$N=CHAr/NaBH$_4$	789
2-CH$_2$NHC$_6$H$_4$R (R = CO$_2$H, SO$_2$NH$_2$)	None	2-CH$_2$Cl/ArNH$_2$	911
2-C(R^1)(R^2)NHAr (R^1, R^2 = H, Me, Ph; Ar = p-C$_6$H$_4$CONHCH(CO$_2$Et)(CH$_2$)$_2$-CO$_2$Et)	Various aryl substituents	2-C(R^1)(R^2)Cl/diethyl p-aminobenzyl glutamate	58
2-CH(NHR)CH(R^1)NO$_2$ (R = p-tolyl; R^1 = H, Ph)	1-Me	1-Me-2-CH=CH(R^1)NO$_2$/p-MeC$_6$H$_4$NH$_2$	952
2-CR^{1}R^{2}NHC$_6$H$_4$CONHCH(CO$_2$Et)(CH$_2$)$_2$-CO$_2$Et (R^1, R^2 = H, Me, Ph)	5-Cl, Me, NO$_2$	2-CR^{1}R^{2}Cl derivative/amino acid diester	912
2-CHRNHCOR R = alkyl, aryl	Various aryl substituents e.g. alkyl, NO$_2$, Cl, etc.	D^b/amide	157
2-(CH$_2$)$_n$NHCOCH$_2$NR$_2$ (n = 1, 2; R = aminoalkyl)	None	D^b/carboxylic acid	72
 2-(CH$_2$)$_n$NHCH$_2$CON◯NR n = 1,2 R = aminoalkyl	None	D^b/carboxylic acid	72
2-CH$_2$NHCOPh	None	D^b/hippuric acid	28

TABLE 1.54 (Continued)

Substituent	Other substituents	Method of preparation	Ref.
2-CH₂NEt₂	1-Me	1-Me-2-CH₂Cl/amine	905
2-CH₂NR₂ (R = alkyl or NRR = piperidinyl)	1-(CH₂)₂COAr	2-CH₂Cl derivative/R₂NH	1124
2-CH₂-(1-pyrrolidinyl)	1-CH₂Ar	2-CH₂Cl derivative/pyrrolidine	902
2-CH₂(1morpholinyl)	1-Alkyl	2-CH₂Cl derivative/amine	905
2-CH₂N(CH₂CO₂H)₂	None	2-CH₂NH₂/ClCH₂CO₂H or	140
	1-Me	1-Me-2-Cl/HN(CH₂CO₂H)₂	
2-CH₂N(CH₂CH₂OH)₂	None	2-CH₂Cl/HN(CH₂CH₂OH)₂	908
2-CH₂N(CH₂CH₂OH)₂	1-Et, Pr	2-CH₂Cl derivative/amine	905
2-CH₂N(CH₂CH₂X)₂ (X = OH, Cl)	None	(i) 2-CH₂Cl/HN(CH₂CH₂OH)₂	41, 42
	1-CH₂CH₂X	(ii) SOCl₂	
	5-Cl		
	5,6-Cl₂		
2-CH₂N(CH₂CH₂Cl)₂	None	1-CH₂CH₂OH-2-CH₂N(CH₂CH₂OH)₂/SOCl₂	908
	1-CH₂CH₂Cl		
2-CH(Ph)N(R¹)CO₂R² [(a) R¹ = C₆H₁₁; R² = CH₂Ph; or (b) R¹ = p-MeC₆H₄; R² = OMe]	None	1-Phenyl-3,4-dioxo-1,2,3,4-tetrahydro-pyrazino[1,2-a]benzimidazole/PhCH₂NH₂ or MeOH	923
2-CH₂N⁺Me₃(X⁻) (X = Cl, Br)	1-aryl, aralkyl	2-CH₂X derivative/Me₃N	909
2-(CH₂)₂NH₂	None	Dᵇ/imidate	194
2-(CH₂)₂NHAr	None	Dᵇ/imidate	193
2-(CH₂)₂NR¹R² NR¹R² = piperidino, morpholino, etc.	None 5-NO₂	Dᵇ/imidate	184
2-CMe₂CH₂NR¹R² (R¹, R² = alkyl or NR¹R² = piperidino)	1-Aryl NO₂, halogeno in benzimidazole aryl ring	Cyclization of o-[N-acylamino]arylamine	329
2-CH(R)CH₂NHCOR¹ (R = CN, CO₂Me; R¹ = OEt, Ph)	1-Me	3-Acylamino-1-Me benzimidazolium betaines/CH₂=CHR	1162

178

Substituent	Substituent(s)	Method	Ref.
2-CH$_2$CR$_2$(CH$_2$)$_n$NHR^1 (R = H, Me; R^1 = H, Me, Ph; n = 0–3, 9)	None; 5-Me	D^b/lactam derivatives	288
2-(CH$_2$)$_n$NHR (n = 3–5; R = H, Me, Ph, and branched chain alkyl analogs)	None; 1-Ph; 5-Me	Cyclization of o-R^{1}HNC$_6$H$_4$NHCO(CH$_2$)$_n$NHR derivatives (generated in situ), and branched chain analogs. (R^1 = H, Ph)	349
2-CHPh(CH$_2$)$_2$NEt$_2$	None	D^b/imidate	607
2-(CH$_2$)$_3$R (R = morpholino, piperidino)	None	2-(CH$_2$)$_3$OH/amine/Ni	285
2-CH$_2$CO(CH$_2$)$_2$NR$_2$^{a} (R = Et or NR$_2$ = pyrrolidino, etc.)	None	2-CH$_2$COCH$_3$/CH$_2$O/R$_2$NH	958
4-NH$_2$	6-NH$_2$	D^b/HCO$_2$H	21a
4-NH$_2$	6-Cl	4-NO$_2$-6-Cl/reduction	21a
4-NH$_2$	6-SO$_3$H, SH, SMe	D^b/HCO$_2$H	21b
4-NH$_2$	1-Ph-5-OH	1-Ph-4-N=N-C$_6$H$_4$SO$_3$H-p-5-OH/Na$_2$S$_2$O$_4$	1094
4-NH-n-Pr	2,6-Me$_2$	2,6-Me$_2$-4-NHCOEt/B$_2$H$_6$	163
4-NHR (R = H, CHO, COMe)	2-H, Me, Ph, CH$_2$Ph	D^b/carboxylic acid	25
4-CH$_2$NH$_2$	None	4-CN/H$_2$/Raney Ni	1058
4-CH$_2$NR^{1}R^2 (R = R^1 = Me, Et or NR^{1}R^2 = piperidino, etc.)	5-OH	5-OH/CH$_2$O/R^{1}R^{2}NH	575
5-NH$_2$	None; 2-Me, CH$_2$Ph, phenethyl Ph	D^b/carboxylic acid	76
5-NH$_2$	2-Me-6-NH$_2$	2-Me-5,6-(NO$_2$)$_2$/SnCl$_2$/HCl	22
5-NH$_2$	4,7-(MeO)$_2$	4,7-(MeO)$_2$-5-NO$_2$/Pd/H$_2$	32
(R, S)-5-CH(OH)CH$_2$NH$_2$	None	(R, S)-1-(3-NO$_2$-4-formamidophenyl)-2-nitroethanol/H$_2$/Pd/C	1080
5-CH$_2$CH(Me)NH$_2$	2-Me	D^b/carboxylic acid	35
6-NH$_2$	1-Et-2-Me	6-NO$_2$ derivative/H$_2$/Raney Ni	586
6-NHMe	1-Et-2-Me	6-NHCHO derivative/LiAlH$_4$	586

^{a} These derivatives have been converted into thiosemicarbazones.
^{b} D = appropriate o-arylenediamine derivative.

179

TABLE 1.55. ALKYL- AND ARYL BENZIMIDAZOLE-2-CARBAMATES[a]

R in 2-NHCO$_2$R	Other substituents	Method of preparation	Ref.
Me	None	D^c/NCNHCO$_2$Me (prepared in situ)	250
Me	None	D^c/CaNCN/ClCO$_2$Me	251
Me	None	D^c/NCNH$_2$/ClCO$_2$Me	245
Me	None	D^c/NCN(CO$_2$Me)$_2$	243
Me	None	D^c/(MeS)$_2$C=NCO$_2$Me	296
Me	None	1-CO$_2$Me-2-NH$_2$/heat in mesitylene	998
Me	None	o-H$_2$NC$_6$H$_4$NHCSNHCO$_2$Me/H$^+$	362
Me	None	Photolysis of methyl quinoxalin-2-ylcarbamate-1-oxide	431
Me	1-CHO	D^c/MeO$_2$CNHCHO/HgO	204
Me	5-Bu	(i) D^c/H$_2$NC(SMe)=NH/ClCO$_2$Me; (ii) acetylation, Ac$_2$O/pyridine	269
Me	5-(CH$_2$)$_3$OAc	2-NHCO$_2$Me/γ-valerolacetone/AlCl$_3$	574
Me	5-CH(Me)(CH$_2$)$_2$CO$_2$H	D^c/MeO$_2$CNHC(SMe)=NH	261
Me	5-OC$_6$H$_4$R (R=NMe$_2$, NEt$_2$)	D^c/NCNH$_2$/ClCO$_2$Me	248
Me	5-SPr	D^c/MeO$_2$CNHCN	1147
Me	5-X-(CH$_2$)$_n$R (X=O, S; n=0-2; R=cycloalkyl)	D^c/MeO$_2$CNHC(SMe)=NH	262
Me	5-X-C$_6$H$_3$RR^1 R=H, Cl, Me; R^1=NH$_2$; X=O, S		
Me	5-X-C$_6$H$_4$OCH$_2$R (X=O, S, CO; R=1-piperidyl, dialkylamino, etc.)	D^c/MeO$_2$CNHC(SMe)=NH	263
Me	5-R [R=S-(2-naphthyl), SO$_2$(2-naphthyl), OC$_6$H$_4$COMe-p, OC$_6$H$_4$SO$_n$Me-p (n=0-2)]	(i) D^c/MeO$_2$CNHC(SMe)=NCO$_2$Me; (ii) Side-chain oxidation (for S derivatives)	276

$$5\text{-}S\text{-}S\overset{X}{\underset{Y}{\diagdown N}}$$

Me	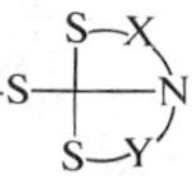 [X,Y (independently) C—C, C═C, C—C—C, C—C═C—C]	D^c/MeO$_2$CNHC(SMe)=NCO$_2$Me	278
Me	5-SO(CH$_2$)$_3$Cl, SCH═CHEt, O(CH$_2$)$_2$OCH$_2$Ph, SCH$_2$C(Cl)═CH$_2$, etc.	(i) D^c/MeO$_2$CNHC(SMe)=NCO$_2$Me (ii) Oxidation by m-ClC$_6$H$_4$CO$_3$H (for sulfoxide derivative)	277
Me	5-SO$_n$Ar (n = 1, 2)	D^c/Cl$_2$C=NCO$_2$Me	1146
Me	5-X-R (X = O, S, SO; R = hetaryl e.g., 2-pyridyl, 2-pyrimidyl, etc.)	D^c/MeO$_2$CNHC(SMe)=NCO$_2$Me	274
Me	5-X(CH$_2$)$_n$R (X = O, S; R = hetaryl, e.g., furyl, pyridyl	D^c/MeO$_2$CNHC(SMe)=NCO$_2$Me	272
Me	5-SCN, SR, alkylsufonyl. SO$_2$R, OR, etc.	D^c/MeO$_2$CNH(SMe)=NCO$_2$R	275
Me	5-SO$_3$C$_6$H$_4$R (R = H, alkyl, alkoxy, etc.)	D^c/MeO$_2$CNHC(SMe)=NH	260
Me	5-COPh	D^c/MeO$_2$CNHC(SMe)=NCO$_2$Me	270
Me	5-COPh	3-NHCO$_2$Me-2,1,4-benzothiadiazine derivatives/H$^+$ or Ph$_3$P	435
Me	5-CR^2(OR^1)OR (RR^1 = (CH$_2$)$_2$, CHMeCH$_2$, etc.; R^2 = aryl, 2-thienyl)	D^c/MeO$_2$CNHC(SMe)=NCO$_2$Me	271
Me	1-COR (R = alkyl, aryl, alkenyl, etc.)	2-NHCO$_2$Me/RCOCl/Et$_3$N	677
Me	1-COR (R = (substituted)cyclopropyl, 2-(dichloromethyl)-3-cyclopenten-1-yl)	2-NHCO$_2$Me/RCOCl/Et$_3$N	680
Me	1-COR (R = alkyl, aryl, CH$_2$OPh, alkoxy)	2-NHCO$_2$Me/RCOCl	678

181

TABLE 1.55 (Continued)

R in 2-NHCO$_2$R	Other substituents	Method of preparation	Ref.
Me	1-CO-[2-(5-nitrofuryl)]	2-NHCO$_2$Me/5-nitro-2-furoyl chloride	777
Me	1-CO$_2$CH$_2$CO$_2$R (R = H, Et)	Acylation by chloroformate ester derivative	696
Me	1-CO$_2$N=C(SMe)Me 1-CO$_2$N=X (X = CMe$_2$, C(Me)Et, cyclohexylidene)	2-NHCO$_2$Me/ClCO$_2$N=X	678
Me	1-CO$_2$-4-thiazolyl	Acylation by ClCO$_2$-4-thiazolyl	698
Me	1-CSOMe	2-NHCO$_2$Me/ClCSOMe/NaHCO$_3$	624
Me	1-CONHBu	2-NHCO$_2$Me/COCl$_2$/BuNH$_2$	749
Me	1-CONHR-5-S-Ph (R = alkyl, (alkoxy)alkyl, (alkoxycarbonyl)alkyl, aryl, etc.)	Acylation by RNCO or RNHCOX (X = hal)	736
Me	1-CONHR-5-SOR^1 (R = alkyl, Ph, CH$_2$CO$_2$Et; R^1 = Ph, CH$_2$OMe)	2-NHCO$_2$Me 5-SOR^1/RNCO	713
Me	1-CONH(CH$_2$)$_6$Cl	2-NHCO$_2$Me/Cl(CH$_2$)$_6$NCO	714
Me	1-CONHR (R = bromoalkyl, bromoallyl, bromocycloalkyl)	2-NHCO$_2$Me/RNCO	715
Me	1-CONHCH(R)CH$_2$Br (R = Ph, CHClCH$_2$Cl, C(Cl)=CH$_2$)	2-NHCO$_2$Me/BrCH$_2$CH(R)NCO	715
Me	1-CONHR (R = CH$_2$O(CH$_2$)$_2$Cl, CH$_2$OCH$_2$CH(Cl)Me, CH$_2$OCH(Me)CH$_2$Cl, etc.) 4- (and 5-)Me	Acylation by isocynate derivatives	718
Me	1-CONHC(R^{1}R^2)CN (R^1 = alkyl, Ph; R^2 = Me; R^{1}R^2 = (CH$_2$)$_5$)	2-NHCO$_2$Me/R^{1}R^{2}C(CN)NCO	717
Me	1-CONHQCO$_2$R (R = Me, Bu; Q = e.g., CHMeCH$_2$, CMe$_2$CH$_2$,)	1-Alkylation by RCO$_2$QNCO	726

182

183

Me	1-CONH(CH$_2$)$_2$SR (R = SCCl$_3$, CHCl$_2$)	2-NHCO$_2$Me/RS(CH$_2$)$_2$NCO	722
Me	1-CONHR (R = (CH$_2$)$_2$SSCCl$_2$CHCl$_2$, (CH$_2$)$_2$SC$_6$H$_4$Cl-p, CH$_2$SMe, (CH$_2$)$_2$SSC(Cl)=CCl$_2$)	2-NHCO$_2$Me/RNCO	723
Me	1-CONHR (R = tetrahydropyranyl, 5,6-dihydro-3-pyranyl, tetrahydro-2-furyl, COC(Cl)=CCl$_2$)	2-NHCO$_2$Me/RNCO/ 1,4-diazabicyclo[2,2,2]octane	733
Me	1-CONHCOCH$_2$SR (R = Et, CH$_2$Ph, Ar)	2-NHCO$_2$Me/isocyanate derivative	728
Me	1-CONHSO$_2$R R = (CH$_2$)$_2$Cl, CHClCH$_2$Cl, CH$_2$CHClCH$_2$CHCl$_2$	2-NHCO$_2$Me/RSO$_2$NCO	730
Me	1-CONHSO$_2$N(R)SO$_2$R^1 (R = alkyl, cyclohexyl; R^1 = alkyl, aryl)	2-NHCO$_2$Me/R^{1}SO$_2$N(R)SO$_2$NCO	731
Me	1-CONHSO$_3$Ar	2-NHCO$_2$Me/ArSO$_3$NCO	732
Me	1-CONHCOR (R = O-aryl, O-alkyl, S-alkyl, O-cyclohexyl)	2-NHCO$_2$Me/isocyanate derivative	720
Me	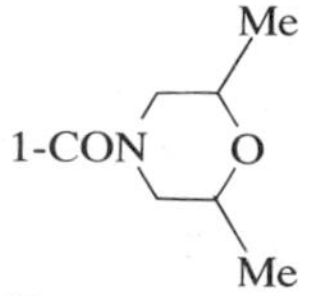 1-CON[morpholine, Me, O, Me]	2-NHCO$_2$Me/COCl$_2$/morpholine derivative	750
Et	None	o-H$_2$NC$_6$H$_4$NHCSNHCO$_2$Et/heat in protic solvent	359
Me, Et	None	D^c/RO$_2$CNHC(OMe)=NH	280
Me, Et	None	D^c/H$_2$NC(SMe)=NCO$_2$R	265
Me, Et	None	D^c/H$_2$NCSNH$_2$/ClCO$_2$R	264a
Me, Et	5-Bu	2-NH$_2$-5-Bu/ClCO$_2$R	779
Me, Et	5-Cl, Me	Cyclization of o-H$_2$NC$_6$H$_4$NHCSNHCO$_2$Me and derivatives	363

TABLE 1.55 (Continued)

R in 2-NHCO$_2$R	Other substituents	Method of preparation	Ref.
Me, Et	None	o-R^{1}NHC$_6$H$_4$NHCSNHCO$_2$R/H$_2$O$_2$/Et$_3$N (R^1 = H, CSNHCO$_2$Et)	360
Me, Et	None	2-NHNa derivative/(RO)$_2$CO	787
	5-Cl, alkyl, SPh		
	6-Me		
Et	5-CH=CHR	D^c/EtO$_2$CN=C(SMe)NH$_2$	259
	(R = aryl, 2-thienyl)		
Me, Bu	5-ArO, PhS	D^c/Cl$_2$C=NCO$_2$R	1152
	6-Cl		
Me, Bu	5-SPh, OAr	D^c/RO$_2$CN=C(SMe)$_2$	1163
	6-Cl		
Me, Bu	5-XAr	2-NH$_2$-5-XAr/MeONa/(RO)$_2$CO	785
	(X = O, S)		
Me, i-Pr	5-SOAr	5-SAr derivative/H$_2$O$_2$/AcOH	1033, 1034
Me, Et	5-COR^1	D^c/NCNHCO$_2$R	242
	(R^1 = Me, Ar)		
Me, Et	5-COR	D^c/MeO$_2$CNH—C(SMe)=NH	264
	(R = 2- and 4-pyridyl, 4-thiazolyl,		
	5-CH$_2$-2-pyridyl)		
Me, Et	1-Na, K	2-NHCO$_2$R/R^{1}ONa (or K) (R^1 = H, Et, t-Bu)	1137
Me, Et	1-CONHR^1	2-NHCO$_2$R/COCl$_2$/R^{1}NH$_2$	748
	(R^1 = alkyl, aryl, cyclohexyl)		
Me, Et	1-CONH(CH$_2$)$_n$CN	2-NHCO$_2$R/NC(CH$_2$)$_n$NCO	716
	(n = 5, 11)		
Alkyl	None	o-R^{1}C$_6$H$_4$NHCSNHCO$_2$R/base (R^1 = NH$_2$, NHCSNH$_2$, NHCSNHCO$_2$R)	361
C$_{1-8}$ Alkyl	None	D^c/CaCN$_2$/ClCO$_2$R/HCl	252
Alkyl	None	o-(RO$_2$CNHCSNH)$_2$C$_6$H$_4$ (Ca salt)/heat	364
Alkyl	None	Thermolysis of 1-CO$_2$R-2-NH$_2$	997

R (substituent)	Ring substituents	Reagents/conditions	Ref.
Alkyl	None	D^c/H_2NCN (or salts)/$ClCO_2Me$	247
Lower alkyl	None	D^c/$H_2NCSNHCO_2R$/Me_2SO_4	264a
Alkyl, aryl	None	D^c/$Cl_2C=NCO_2R$	295
Alkyl	None	D^c/$Cl_2C=NCO_2,CH_2CH=CH_2$	298
Alkyl, aryl	Cl, Me in benzimidazole aryl ring	D^c/$RO_2CNHCSO$-alkyl	155
Alkyl	Alkyl in carbocyclic ring	1-CO_2R-2-NH_2/heat	381
Alkyl	None, Alkyl in carbocyclic ring	Dialkyl ester of 2-imino-1,3-benzimidazoline dicarboxylic acid/base	381
Alkyl, allyl	None	D^c/H_2NCN/$ClCO_2R$	246
Alkyl, CH$_2$Ph	None	2-NHCN derivative/ROH/H$^+$	943
Lower alkyl	Cl, Me in benzimidazole aryl ring; Lower alkyl, halo in carbocyclic ring	D^c/$RO_2CN=C(OR^1)NH_2$ (R^1 = lower alkyl)	279
Alkyl, allyl	None; 5-Me	(a) 2-$NHCO_2Ph$ derivative/pyridine/ROH or (b) 2-NH_2 derivative/$ROCO_2N=C(Me)R$ (R = H, alkyl, aryl)	948
Alkyl	5-CMe_2CO_2H	2-$NHCO_2Me$/$CH_3CHClCH_2CO_2H$/$AlCl_3$	573
Alkyl, allyl	1-$CH(CCl_3)NHCOR^1$ (R^1 = H, alkyl, aryl, etc.); Various benzimidazole aryl substituents	Alkylation by a (substituted) alkyl halide	591
Alkyl	5-X-Ar (X = O, S)	3-$NHCO_2Me$-2,1,4-benzothiadiazine derivatives/H$^+$ or Ph$_3$P	436
Alkyl	5-$XC_6H_4O(CH_2)_n$ NR^1_2 (R^1 = alkyl, n = 2, 3; X = O, S, CO)	2-NH_2 derivative/$(RO)_2CO$/EtONa	786
C$_{1-4}$ Alkyl	C_{1-4} alkyl, C_{1-4} alkoxy, hal, NO_2, etc., in aryl ring	2-$NHCO_2CH(R^1)CH(R^2)OH$ derivative/C_{1-4} alcohols/base (R^1, R^2 = H, alkyl, Ph)	949
Alkyl	5-OSO_2Ar	2-$NHCO_2R$-5-OH/$ArSO_2Cl$	1143
Alkyl	5-XC_6H_4R X = OSO_2, SO_2O (R = H, halogeno, MeO, etc.)	3-$NHCO_2Me$-2,1,4-benzothiadiazine derivatives/H$^+$ or Ph$_3$P	437
Alkyl	1-COR^1 (R^1 = Alkyl)	2-$NHCO_2R$/R^1COX (X = hal) or $(R^1CO)_2O$	676

185

TABLE 1.55 (Continued)

R in 2-NHCO$_2$R	Other substituents	Method of preparation	Ref.
Alkyl	1-CO$_2$N=CMe$_2$	2-NHCO$_2$R/ClCO$_2$N=CMe$_2$	695
C$_{1-4}$ Alkyl	1-CONHR	(i) D'/CaCN$_2$/ClCO$_2$R (ii) RNCO	249
Alkyl	1-CONH(CH$_2$)$_n$CN (n = 5, 11)	2-NHCO$_2$R/isocyanate derivative	716
Alkyl, aryl, hetaryl, etc.	1-CONH(CH$_2$)$_n$NR^{1}R^2 (n = 1–10; R^1, R^2 = alkyl or NR^{1}R^2 = piperidino, etc.)	Acylation by (a) R^{1}R^2(CH$_2$)$_n$NCO or (b) R^{1}R^{2}N(CH$_2$)$_n$NHCOCl or (c) R^{1}R^{2}N(CH$_2$)$_n$NH$_2$/COCl$_2$	724
Alkyl	1-CONH(CH$_2$)$_n$R^2 (R^2 = SCH$_2$Ph, 3-Ph-1,2,4-oxadiazol-5-yl, COSEt, COSPh, SCOAr)	2-NHCO$_2$R/isocyanate derivative	727
Alkyl	1-CONHCOCH$_2$OAr various benzimidazole aryl substituents	2-NHCO$_2$R/ArOCH$_2$CONCO	719
Alkyl	1-COR^1, CONHCOR^1 (R^1 = (aryloxy)alkyl. hal, NO$_2$, alkyl, alkoxy, etc., in benzimidazole aryl ring)	Acylation by R^{1}COCl or R^{1}CONCO	735
Alkyl	1-CONHCH$_2$— 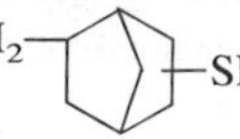 [SR^1 bicyclic structure] (R^1 = alkyl, Ph. Cl, alkyl, OPh etc. in benzimidazole aryl ring)	2-NHCO$_2$R/isocyanate derivative	729
Alkyl	1-CONH 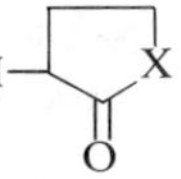 [ring structure, X=O,S]	2-NHCO$_2$R/isocyanate derivatives	734
Alkyl	1-CSNHCOR^1 (R^1 = OMe, SMe)	2-NHCO$_2$R/R^{1}OCNCS	742

(CH₂)₂NEt₂	1-CONHn-Bu	None	Acylation by (a) n-BuNCO or (b) COCl₂/n-BuNH₂	712
NR₂[b] (R = alkyl or NR₂ = piperidyl, etc.)	1-alkyl, CH₂Ph	5-Me, Cl, OEt, etc.	D[c]/H₂NC(SMe)=NH/ClCO₂(CH₂)$_n$NR₂	266
N=CRR¹ (R = Me, Et; R¹ = Me, Ph; or RR¹ = cyclohexylidene)	None	5-Me	1-CO₂N=CRR¹-2-NH₂ derivative/thermolysis	999
N=C(R¹)R² (R¹ = Me, Et; R² = Me; or CR¹R² = cyclohexylidene)	1-COR (R = OMe, ON=CMe₂ NH(CH₂)₅CN, NHBu, etc.)		Acylation by RCOCl	697
Ph		5-Bu	2-NH₂-5-Bu/ClCO₂Ph	781
Ph		5-Bu	D[c]/NCNH₂/ClCO₂Ph	244
Ph		5-Bu	D[c]/MeSC(=NH)NH₂/ClCO₂Ph	267
4-Thiazolyl		5-Bu	D[c]/H₂NC(SMe)=NH/ClCO₂-4-thiazolyl	268
4-Thiazolyl, 2-thienyl etc.		5-Bu	2-NH₂-5-Bu/ClCO₂R (R = hetaryl)	782
4-Thiazolyl	1-CO₂Me-5-Bu		1-Acylation by COCl₂/MeOH	1153
4-Thiazolyl	1-CONHBu		Acylation by (a) COCl₂/BuNH₂ or (b) BuNCO	747

[a] Examples of N-substituted alkyl- and aryl benzimidazole 2-carbamates are c_ted in Ref. 634, 685, 780, 783, and 784. Thio-analogs of alkyl benzimidazole 2-carbamates (viz 2-NHCOSR types) are described in Ref. 281, 282, and 721.

[b] This substituent is of the type 2-NH(CH₂)$_n$CO₂R, n = 2, 3.

[c] D = appropriate o-arylene diamine derivative.

187

TABLE 1.56. MISCELLANEOUS NITROGEN-CONTAINING GROUPS: ANILS, UREAS, THIOUREAS, THIOSEMICARBAZIDES, AMIDINES, AND GUANIDINES

Substituent	Other substituents	Method of preparation
2-N=CHC$_6$H$_4$NO$_2$-p	None	2-NH$_2$/ArCHO
2-N=CHC$_6$H$_4$NO$_2$-p	1,5-Me$_2$	2-NH$_2$ derivative/ArCHO
2-N=CHC$_6$H$_4$NO$_2$-p	1-Et-5-OMe	2-NH$_2$ derivative/ArCHO
2-N=CHAr	None	2-NH$_2$/ArCHO
2-N=CHAr	1-C(OPh)=NH	2-NH$_2$ derivative/ArCHO
2-N=CHR [R = 2-furyl and 2-(5-substituted derivatives)]	1-Alkyl, Ph	2-NH$_2$ derivative/furfurals
2-N=CHR (R = furyl, 5-nitrofuryl)	1-i-Pr	1-i-Pr-2-NH$_2$/furfural or derivative
2-N=CHR (R = 2-furyl, 2-thienyl)	None 2-thiazin-2-yl 5-NO$_2$	2-NH$_2$ derivative/RCHO
2-N=C(COPh)NEt$_2$	1-Me	Pyrolysis of a (1-Me-2-benzimidazolyl) 1H-1,2,3-triazole derivative
2-N=CPh$_2$	1-Me	1-Me-2-NH$_2$/Ph$_2$CO or 1-Me-2-N$_3$/Ph$_2$CS
2-CH$_2$N=CHR (R = aryl, pyridyl)	None	2-CH$_2$NH$_2$/ArCHO
2-C(CN)=NAr	None 1-Me 5-, 6- or 7-Me	2-CH$_2$CN derivative/ArNO
2-C(R)=NAr (R = Me, Ph; Ar = Ph, p-MeC$_6$H$_4$)	None	2-COMe or 2-COPh/arylamine
2-NHCONHn-Bu	None	1,2,3,4-tetrahydro-3-butyl-2,4-dioxo-s-triazino[a]-benzimidazole/NaOH
2-NHCONHR R = alkyl, cyclopropyl	None 5-Bu	2-NH$_2$ derivative/RNCO
5-NHCXNHR (X = O, S; R = alkyl, Ph, Ph(CH$_2$)$_2$)	1-Alkyl, CH$_2$Ph 2-Alkyl	5-NH$_2$ derivative/RNCX
1-(CH$_2$)$_2$NHCONHCOR^1 (R^1 = C$_{1-5}$ alkyl, aryl, CH$_2$Ph, CH=CHPh)	5- and 6-NO$_2$	Alkylation by R^{1}COCl
2-(CH$_2$)$_2$NHCSNH$_2$	None	2-(CH$_2$)$_2$NH$_2$/CS$_2$/N$_2$H$_4$·H$_2$O
2-(CH$_2$)$_2$NHCSNHN=CHR (R = aryl, hetaryl)	None	Thiosemicarbazide/carbonyl derivative
5-NHC(Ar)=NH	None	5-Ammonium benzene sulfonates/ArCN
1-C(R^{1}R^{2}N)=NH (R^1, R^2 = H, alkyl, CH$_2$Ph; R^{1}R^2 = (CH$_2$)$_4$, (CH$_2$)$_2$O(CH$_2$)$_2$)	2-NH$_2$	1-CN-2-NH$_2$/R^{1}R^{2}NH
2-C(=NR^1)NHR^2 (R^1 = R^2 = alkyl, aryl, 2-furyl	None	2-CCl$_3$/amine
2-C(NH$_2$)=NOH	None	2-CN/NH$_2$OH
2-SCH$_2$C(NHOH)=NH	None 5-NO$_2$	2-SCH$_2$CN derivative/H$_2$NOH/Et$_3$N

ostituent	Other substituents	Method of preparation	Ref.

NHC(NH$_2$)=NH	None
	1-Me, CH$_2$Ph
	benzimidazole aryl
	    substituents include
	    Cl, alkyl, NO$_2$, etc.
		D^a/NCNHC(NH$_2$)=NH	257

NHC(=NH)NHR R = alkyl)	5-Cl 5,6-Cl$_2$	D^a/dicyandiamides	256

• = appropriate *o*-arylenediamine derivative.

BLE 1.57.   HYDRAZINES, HYDRAZONES AND FORMAZANS

ostituent	Other substituents	Method of preparation	Ref.
NHNH$_2$	None 1-Me, CH$_2$Ph 5-Me	2-Cl derivative/N$_2$H$_4$	875
NHNH$_2$	1-Me	1-Me-3-oxide/H$_2$NNH$_2$	1104
NHNH$_2$	5-NO$_2$	2-Cl-5-NO$_2$/N$_2$H$_4$	876
NHNHCOR R = aryl, 4-pyridyl)	1-CH$_2$CO$_2$Et 1-CH$_2$CO$_2$Et	1-CH$_2$CO$_2$Et-2-Cl/H$_2$NNHCOR	589
NHN—CHPh	1-Me 5-Me, OMe 6-Me	2-NHNO derivative/(i) Zn/AcOH (ii) PhCHO	796
NHN=CRR^1 R = H, Me; R^1 = aryl, hetaryl, styryl; CRR^1 = 3-isatinylidene; RR^1 = (CH$_2$)$_5$, (CH$_2$)$_4$)	None	2-NHNH$_2$/RR^{1}CO	804
NHN=C(CO$_2$Et)$_2$	1,2-Me$_2$-5-NH$_2$	5-NO$_2$ derivative/H$_2$/Ni	1111
NHN=CHN=NAr	1-CH$_2$Ph	1-CH$_2$Ph-2-NHNH$_2$/ArNHN=CHNO$_2$/ pyridine	1138
NH—N=C(R)N$_2$Ar R = Me, Ph)	1-Me, CH$_2$Ph 1,5,6-Me$_3$	2-hydrazone derivative/ArN$_2^+$ salt	821–823
NHN=C(CH$_2$OH)N=NAr	None	2-NHN=CHCH$_2$OH/ArN$_2^+$	826
NHN=C(R)N=N-[.-(1-benzylbenzimidazolyl)] R = Me, Ph)	1-CH$_2$Ph	2-NHN=CHR/1-CH$_2$Ph-2-NHNH$_2$/[O] in pyridine	1037

NHN=C(Me)N=NR
R = 2-benzoxazolyl,
    2-benzothiazolyl)
	1-CH$_2$Ph	1-CH$_2$Ph-2-NHNH$_2$/MeCH=NNHR/[O]
		    in pyridine
			1037

NHN=C(R)N=N-2- 1-benzylbenzimidazolyl) R = CH=CH$_2$, CH=CHMe)	None	2-NHN=CHR/1-CH$_2$Ph-2-N$_2^+$	826
N=NC(R)=NNHR^1 R = Me, Ph, 2-quinoxalinyl; R^1 = aryl, hetaryl]	1-Me, CH$_2$Ph	2-NHNH$_2$ derivative/hydrazone derivative/air/pyridine	1036
or 7-)-N=N—C(R^1)=NNHR^2 R^1 = Ph, Me; R^2 = 2-(1-benzylbenzimi- dazolyl)	None 1-CH$_2$Ph	4-(or 7-)N$_2^+$/hydrazone derivatives	967

TABLE 1.58. AZO COMPOUNDS AND TRIAZENES

Substituent	Other substituents	Method of preparation	Ref.
2-N=NR [R = (a) 1-Me-2-NH$_2$-5-Br-6-benzimidazolyl; (b) 1-Me-2-NH$_2$-6-Br-5-benzimidazolyl]	In (a) 1-Me-5-Br In (b) 1-Me-6-Br	1-Me-2-NH$_2$-5-(or 6)Br derivatives/ONOSO$_3$H	577
2-N=NR [R = 5- and 6-(1-methyl-2-aminobenzimidazolyl)]	1-Me	1-Me-2-NHNO/conc.H$_2$SO$_4$	796
2-N=NR [R = 5- and 6-(1-R^1-2-NH$_2$ benzimidazolyl); R^1 = Me, Et]	1-Me, Et	Diazotization of 1-Me-2-NH$_2$	1119
4-N=N—Ar	None	Diazonium coupling of 4-NH$_2$	1108
5-N=N-aryl	None 2-Ph 1-Me, cyclohexyl	Diazonium coupling of 5-NH$_2$	1108
7-N=NPh	4-NH$_2$ 2-Me-4-NH$_2$	4-NH$_2$ derivative/PhN$_2^+$Cl$^-$	562
1-C$_6$H$_4$N=NAr-p	None	1-C$_6$H$_4$NH$_2$-p/diazonium coupling	1108
2-C$_6$H$_4$—N=NAr-p	None	2-C$_6$H$_4$NH$_2$-p/diazonium coupling	1108
2-N=NNHR [R = 2-(1-alkyl and benzyl benzimidazolyl)]	1-Alkyl, benzyl	1-alkyl(or benzyl)-2-NHNO/H$_2$SO$_4$ or 1-Me-2-N$_2^+$ (BF$_4^-$)/1-Me-2-NH$_2$	801
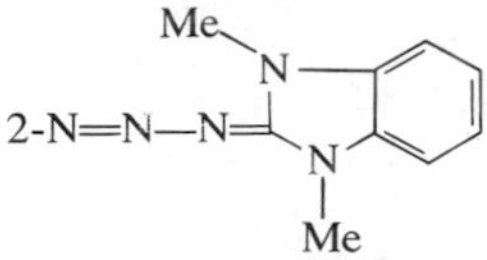 2-N=N—N= (benzimidazoline with Me, N, N-Me groups)	1-Me	1-Me-2-N$_2^+$ (BF$_4^-$)/1,3-Me$_2$-1-iminobenzimidazoline	801

TABLE 1.59. NITRILES AND NITRILE OXIDES

Substituent	Other substituents	Method of preparation	Ref.
1-CN	2-CF$_3$-4,5,7-Cl$_3$ 2-CN-4,5,7-Cl$_3$	1-Unsubstituted derivative/ BrCN/NaH	767
1-CN	2-Alkyl, aryl, CH$_2$Ph, CN, etc.	2-Substituted derivative (or Ag salt)/BrCN	768
1-CH$_2$CN	None 2-alkyl, Ph	Alkylation by ClCH$_2$CN	143
1-(CH$_2$)$_2$CN	5-Me 2,5-Me$_2$ 2,5,6-Me$_3$	Alkylation by CH$_2$=CHCN	658
1-(CH$_2$)$_2$CN	2-(CH$_2$)$_2$CO$_2$H	Alkylation by CH$_2$=CHCN	659

TABLE 1.59 *(Continued)*

Substituent	Other substituents	Method of preparation	Ref.
2-CN	None	2-CCl$_3$ 2-CFCl$_2$ }/NH$_4$OH or NH$_3$ 2-CF$_2$Cl	926
2-CN	1-Me	1-Me-3-oxide/KCN	1104
2-CN	Various aryl substituents, e.g., hal, NO$_2$	2-CCl$_3$ or 2-CF$_2$Cl or 2-CCl$_2$F/NH$_3$	925
2-CN	None 1-Me, CO$_2$Et. halogen substituents in aryl ring	2-CCl$_3$/NH$_3$ or dehydration of 2-CH=NOH	924
2-CN	5-C$_6$H$_4$NH$_2$-*p*	2-NH$_2$-5-Ar/(i) HNO$_2$, (ii) NaCN	1142
2-CH$_2$CN	None	D^a/imidate	182
2-CH$_2$CN	None	D^a/amide	159
2-CH$_2$CN	None	D^a/ester	142
2-CH$_2$CN	None 1-Me	D^a/amide	1165
2-CH$_2$CN	None 1-Me	D^a/EtO$_2$CCH$_2$CN	143
2-CHRCN (R = H, Me, Ph)	None 1-Me	D^a/ester	141
2-CH(R)CN (R = alkyl)	None	D^a/ester	144
2-CH(R)CN (R = alkyl)	None	D^a/ester	1095
2-CH(CN)CH$_2$Ar	None	2-C(CN)=CHAr/NaBH$_4$	294
2-CH(CN)CH$_2$Ph	None 1-CH$_2$Ph	2-CH$_2$CN derivative/PhCH$_2$Cl/NaH	813
2-CH(CN)CO$_2$Me	1-Me	1-Me-3-oxide/NCCH$_2$CO$_2$Me	1104
2-CH(CN)CXNHAr (X = O, S)	None 1-Me 5-Me	2-CH$_2$CN derivative/ArNCX	815
2-(CH$_2$)$_n$CN (n = 2–4)	1-H, Me, (CH$_2$)$_2$CN 5-H, Br, NO$_2$	dehydration of amido derivatives	1098
2-NHCN	None 1-Me, CH$_2$Ph benzimidazole aryl substituents include Cl, alkyl, NO$_2$ etc.	2-NHC(NH$_2$)=NH derivative/HNO$_2$	257
2-NHCOCH$_2$CN	None	2-NH$_2$/NCCH$_2$CO$_2$Et	1118
4-CN	None	4-NH$_2$/Sandmeyer(CuCN)	1058
2-C≡N̄—O̅	None 1-Me, *i*-Pr 5-Me, NO$_2$	2-C(Cl)=NOH derivative/Et$_3$N	983

^{a} D = appropriate *o* = arylenediamine derivative.

191

TABLE 1.60. AZIDO DERIVATIVES

Substituent	Other substituents	Method of preparation	Ref.
2-N_3	None 1-Me, CH_2Ph 5-Me	2-$NHNH_2$ derivatives/$NaNO_2$/AcOH	875
2-N_3	None 1-$(CH_2)_2OH$	2-$NHNH_2$ derivative/HNO_2	111(
2-N_3	1-$CONHCH_2CO_2Et$	2-N_3/EtO_2CCH_2NCO	110(
2-N_3	1-$CONHCH_2CO_2R$ (R = Bu, i-Bu, i-Pr)	2-N_3 derivative/$ONCCH_2CO_2R$	100(

TABLE 1.61. THIOCYANATES AND ISOTHIOCYANATES

Substituent	Other substituents	Method of preparation	Ref.
2-CH_2SCN	None 5-Cl	2-CH_2Cl/NH_4SCN	115!
2-CH(R)SCN (R = H, Me)	None 1-Me 5-Cl, NO_2	2-CH(R)Cl/NH_4SCN	918
2-CH(R)$OCOCH_2SCN$ (R = H, Me)	None 1-Me, CH_2Ph	2-CH(R)$OCOCH_2Cl$ derivative/KSCN	928
5-(and 6-)NCS	1-Me 2-Me, Ph	5-(and 6-)NH_2 derivatives/$CSCl_2$	966
5-NCS	None 1-$(CH_2)_2NMe_2$, 2-thiazolin- 2-yl, 2-hetaryl, cycloalkyl	5-NH_2 derivatives/CS_2	114:
5-NCS 2-(CH=CH)$_n$Ar^a ($n = 0, 1$)	1-Me 6-Cl	5-NH_2 derivative/$CSCl_2$	837
5-NCS	2-(Z)$_n$-aryl^a Z = CH=CH, $(CH_2)_2$ ($n = 0, 1$)	5-NH_2 derivative/$CSCl_2$	836

^{a} Also NCS substituent in this aryl ring.

TABLE 1.62. HALOGENO DERIVATIVES

Substituent	Other substituents	Method of preparation	Ref.
4-F	None 2-CF$_3$	D^b/carboxylic acid	50
4-F	2-Me 1,2-Me$_2$	o-aminoacetanilide derivative/ MeCO$_2$H	50
4-F	None 2-Me, CF$_3$ 1,2-Me$_2$	D^b/carboxylic acid	50
4-F	7-NO$_2$ 1,2-Me$_2$-5-NO$_2$ 2-CF$_3$-7-NO$_2$ 5,7-(NO$_2$)$_2$ 2-Me-5,7-(NO$_2$)$_2$ 1,2-Me$_2$-5,7-(NO$_2$)$_2$	Nitration of 4-F derivatives/HNO$_2$/H$_2$SO$_4$	50
4-F	None	4-NH$_2$/Balz-Schiemann reaction	1120
5-F	None	5-NH$_2$/Balz-Schiemann reaction	1120
2-CHF$_2$	4-CF$_3$	D^b/carboxylic acid	342
2-CHF$_2$	5-SO$_2$CF$_2$R (R = H, CHFCl, CHF$_2$)	D^b/carboxylic acid (or ester)	52
2-CF$_3$	None 4,7-Me$_2$ 5-F	D^b/carboxylic acid	33
2-CF$_3$	5 Br	(a) D^b/carboxylic acid or (b) 2-CF$_3$/Br$_2$/AcOH	566
2-CF$_3$	4,6-Br$_2$	D^b/carboxylic aid	566
2-CF$_3$	5-Cl-6-Br	2-CF$_3$-5-Cl/Br$_2$/AcOH	566
2-CF$_3$	4,5,6,7-Br$_4$ + 4,5,6-Br$_3$	2-CF$_3$/excess Br$_2$/AcOH/H$_2$O	566
2-CF$_3$	1-CO$_2$-i-Pr 4,6-I$_2$	Acylation by ClCO$_2$-i-Pr/Na$_2$CO$_3$	688
2-CF$_3$	1-Alkyl, aryl, etc. (Cl, CN, NO$_2$, etc., in aryl ring	D^b/imidate	178
2-CF$_3$	1-Me 5-Me 5-NO$_2$ 4,7-(OEt)$_2$	D^b/carboxylic acid or nitration (for 5-NO$_2$ derivative)	1085
2-CF$_3$	4-Me, NO$_2$ 5-Me 6-CF$_3$	D^b/carboxylic acid	49
2-CF$_3$	None 5-NO$_2$, Cl 5,6-Cl$_2$ 7-NO$_2$	D^b/carboxylic acid	172
2-CF$_3$	None 1-CO$_2$-alkyl 4-SO$_2$NR^{1}R^2 (R^{1}R^2 = H, alkyl, aryl) 5-, 6- and 7-hal, SO$_2$NR$_2$ (R = alkyl)	4-SO$_2$Cl derivatives/R^{1}R^{2}NH	931, 1127
2-CF$_3$	5,6-Cl$_2$ 4,5-Cl$_2$ 6-SO$_2$Me	o-dinitroarene/CF$_3$CO$_2$H/Sn/HCl (or other reductants)	379

TABLE 1.62   (*Continued*)

Substituent	Other substituents	Method of preparation	
2-CF$_3$	5-Alkylsulfonyl + various aryl substituents	D^b/carboxylic acid	
2-CF$_3$	4-NR^{1}R^2-5,7-(NO$_2$)$_2$ (R^1, R^2 = H, alkyl, fluoroalkyl, etc.; NR^{1}R^2 = piperidino, etc.)	2-CF$_3$-4-Cl-5,7-(NO$_2$)$_2$/R^{1}R^{2}NH	
2-CF$_3$	5-NO$_2$ 4-NO$_2$-5,6-Cl$_2$ 4-NO$_2$-5,7-Br$_2$ 4-NO$_2$-5,6,7-Br$_3$ 4-NO$_2$-5-Cl + 5-Cl-6-NO$_2$ 4-NO$_2$-5-Br + 5-Br-6-NO$_2$	Nitration of appropriate 2-CF$_3$ derivative by H$_2$SO$_4$/HNO$_3$	
4-CF$_3$	None	D^b/HCO$_2$H	
5-CF$_3$	None	D^b/HCO$_2$H	
5-CF$_3$	1-, 2-, and 6-alkyl	D^b/acid chloride	
5-CF$_3$	7-NH$_2$ and various substituents at positions 2- and 6-, e.g., alkyl, Cl, MeO	D^b/imidate	
2-CF$_3$, C$_2$F$_5$, C$_3$F$_7$	1-Me	2-perfluoroalkyl/CH$_2$N$_2$	
2-R (R = CF$_3$, C$_2$F$_5$)	Cl, Br in aryl ring	o-O$_2$NC$_6$H$_4$NHCOR/H$_2$/Pd	
2-CF$_3$, C$_2$F$_5$	4-Alkyl-6-NO$_2$	D^b/carboxylic acid	
2-CF$_2$R (R = F, CF$_3$)	None 1-Ac, Me 4,5,7-Cl$_3$ 4,5,6,7-Cl$_4$	D^b/carboxylic acid	
2-(CF$_2$)$_n$F (n = 1, 2)	1-H, Et various aryl ring substituents	D^b/carboxylic acid	
2-C$_n$F$_{2n+2}$ (n = 1–3)	Various aryl substituents, e.g., CF$_3$, NO$_2$, Cl	(i) D^b/carboxylic acid or (ii) reduction of N-(o-nitroaryl)-perfluoroalkanamides	
2-CMe$_2$CH$_2$F	None	D^b/carboxylic acid (high pressure)	
2-CF$_2$CHCl$_2$	1-Alkyl, aryl, etc. Cl, CN, NO$_2$, etc., in aryl ring	D^b/imidate	
2-CF$_2$CHCl$_2$	5-SO$_2$CF$_2$R (R = H, CHFCl, CHF$_2$)	D^b/carboxylic acid (or ester)	
2-CHFCH$_3$	None	D^b/CF$_3$CF=CF$_2$	
1-CF$_2$CHFCl	None	Benzimidazole/CF$_2$=CFCl/130–140°	
2-C$_2$F$_5$	None 4,7-Me$_2$ 5-F	D^b/carboxylic acid	
2-C$_2$F$_5$	5-SO$_2$CF$_2$R (R = H, CHFCl, CHF$_2$)	D^b/carboxylic acid (or ester)	
2-C$_3$F$_7$	None	D^b/carboxylic acid	
2-Cl	1-Me	1-Me-3-oxide/PCl$_3$	
2-Cl	1,5-Me$_2$	2-OH derivative/POCl$_3$	
2-Cl	1-C(Me)=CH$_2$	2-OH derivative/POCl$_3$	
2-Cl	1-i-Pr 5,6-(NO$_2$)$_2$	2-OH derivative/POCl$_3$	

TABLE 1.62   (*Continued*)

Substituent	Other substituents	Method of preparation	Ref.
2-Cl	None 5-Me 2-Cl-5-Me 2-Cl-5-NO$_2$	2-OH derivative/POCl$_3$	580
2-Cl	1-Me 1-Me-5-CO$_2$H 1-CH$_2$Ph-5-NO$_2$	2-NHNH$_2$ derivative/SOCl$_2$	867
2-Cl	1,4-Me$_2$ 1,7-Me$_2$ 1-Me-4-R 1-Me-7-R (R = Cl, NO$_2$)	2-OH derivative/POCl$_3$	883
2-Cl[a]	4,6,7-Cl$_3$-5-(4-Chlorophenyl- sulfonylamino)	2-Cl-5-substituted derivative/ Cl$_2$/FeCl$_3$	570
2-Cl	4- and 5-Cl, Me, NO$_2$, MeO	2-OH derivative/POCl$_3$	884
4-Cl	6-NO$_2$	4-NH$_2$-6-NO$_2$/Sandmeyer	21a
4-Cl	6-NH$_2$	4-Cl-6-NO$_2$/SnCl$_2$/HCl	21a
4-Cl	6-Cl	4-NH$_2$-6-Cl/Sandmeyer	21a
4-Cl, Br	2-Alkyl, aralkyl	D[b]/carboxylic acid (or derivative)	160
4,5,6-Cl$_3$	None 2-Me	Cl$_2$/DMF	568
4,5,6,7 Cl$_4$	1-Me	*o*-H$_2$NC$_6$H$_4$NMe$_2$/SO$_2$Cl$_2$	305
5-Cl	2-*n*-Pr, *i*-Pr, CF$_3$, cyclopropyl	D[b]/carboxylic acid	55
2-CH$_2$Cl	1-Me	D[b]/carboxylic acid	141
2-CH$_2$Cl	1-CH$_2$Ar	D[b]/carboxylic acid	902
2-CH$_2$Cl	None 5-Cl 5,6-Cl$_2$	D[b]/carboxylic acid	41, 42
2-CHCl$_2$	None	D[b]/carboxylic acid	43, 45
2-C(R^1)(R^2)Cl (R^1, R^2 = H, Me, Ph)	Various aryl ring substituents	2-C(R^1)(R^2)OH/SOCl$_2$	58
2-C(R^1)(R^2)Cl (R^1, R^2 = H, Me, Cl)	5-Cl, Me, NO$_2$	2-C(R^1)(R^2)OH·HCl derivative/ heat or 2-CR^{1}R^{2}OH/SOCl$_2$	912
2-CClF$_2$	None 5-NO$_2$	D[b]/carboxylic acid	172
2-CCl$_2$F	None	D[b]/carboxylic acid	172
2-CCl$_3$	None	D[b]/imidate	175
2-CCl$_3$	5-Me 5,6-Cl$_2$, Me$_2$	D[b]/Cl$_3$CC(=NH)OMe	173
2-CCl$_3$	4,5-Cl$_2$	D[b]/Cl$_3$CCO$_2$H/POCl$_3$	46
2-CCl$_3$	1-Alkyl various aryl substituents, e.g., Cl, Me, NO$_2$	D[b]/imidate	172
1-(CH$_2$)$_2$Cl	2,6-Me$_2$-4-N(CH$_2$CH$_2$Cl)$_2$	1-(CH$_2$)$_2$OH-2,6-Me$_2$-4-NH$_2$/ (i) ethylene oxide; (ii) chlorination	324

TABLE 1.62 *(Continued)*

Substituent	Other substituents	Method of preparation	Ref.
2-CMe$_2$CH$_2$Cl	None 5-Cl 5,6-Cl$_2$, etc.	D^b/carboxylic acid/high pressure	40
2-Br, I	None	1-CH$_2$OR/(i) Bu$_2$Mg, (ii) Br$_2$ or I$_2$, (iii) HBr/AcOH (R = CH$_2$OMe, CH$_2$OCH$_2$Ph)	849
5-Br	2-Me	2-Me/Br$_2$/CHCl$_3$ or NBS/CCl$_4$	569
5-Br	2-Me	D^b/Ac$_2$O	583
5-Br	1-Et 1-Et-6-NO$_2$	5-NH$_2$ derivative/Sandmeyer	1087
5-Br	2-CO$_2$H	(a) 2-CO$_2$H/Br$_2$/AcOH (b) 2-CH$_2$OH-5-Br/KMnO$_4$	569
4,5-Br$_2$	2-Me	2-Me/Br$_2$/DMF	568
4,6-Br$_2$	2-Me	2-Me/Br$_2$/CHCl$_3$ or NBS/CCl$_4$	569
6-Br	1-Et-5-NO$_2$	6-NH$_2$ derivative/Sandmeyer	1087
1-CH$_2$CH$_2$Br	None	Benzimidazole/Br(CH$_2$)$_2$Br/KOH	596
2-(CH$_2$)$_n$Br (n = 3–5)	None	D^b/imidate	183
1-I	None 2-Me	Benzimidazole derivative/I$_2$/NaOH	580
2-X (X = I, Br)	1-R (R = Me, Ph, CH$_2$OMe)	1-R/(i) BuLi, (ii) X$_2$	848
4-I	None	D^b/carboxylic acid	1092

[a] A variety of polyhalogenobenzimidazoles is described in this patent.
[b] D = appropriate *o*-arylenediamine derivative.

TABLE 1.63. HYDROXY AND MERCAPTO DERIVATIVES[a]

Substituent	Other substituents	Method of preparation	Ref.
4-OH	2-R-7-OH (R = H, Me)	2-R-4,7-(MeO)$_2$/HBr	32
4-OH	2-Me-6-NO$_2$	2-Me-4-MeO-6-NO$_2$/HBr	38
4-OH	2-Me-6-NH$_2$	2-Me-4-OH-6-NO$_2$/H$_2$/Pd	38
4,7-(OH)$_2$	None 1-Me 2-Ph 1-Me-2-Ph 1-Me-2-Cl 1-Me-2-NMe$_2$	4,7-(MeO)$_2$ derivatives/HBr	1020
4,7-(OH)$_2$	None 2-Me, (CH$_2$)$_2$CO$_2$Me	4,7-(MeO)$_2$ derivative/H$^+$	1019
5-OH	None 2-Me, Et	Oxidation by ascorbic acid/Fe^{2+}/O$_2$	1013

196

TABLE 1.63 (*Continued*)

Substituent	Other substituents	Method of preparation	Ref.
5-OH(+6-OH)	1-Me	Oxidation by ascorbic acid/Fe^{2+}/O_2	1013
5-OH	6-OH	5,6-$(MeO)_2$/HBr	32
5,6-$(OH)_2$	None	5,6-$(MeO)_2$/HBr	1019
6-OH	4-NO_2	4-NO_2-6-MeO/HBr	38
6-OH	4-NH_2	4-NO_2-6-OH/H_2/Pd	38
1-CH_2OH	None	Benzimidazole/CH_2O	647
1-CH_2OH	(a) 2-Me (b) 2-$(CH_2)_2OH$	2-Me/CH_2O (1:1 molar for a and 1:10 molar for b)	810
4-CH_2OR (R = H, Ac)	None 1-Ac	4-CO_2Me (i) $LiAlH_4$ (ii) Ac_2O	969
4-CH_2OH	5-Me	D^b/formic acid	969
2-CH(OH)R (R = cyclohexyl, aryl)	1-Me 4-Et 5-Me, CF_3, MeO, NO_2	D^b/carboxylic acid	951
2-CH(OH)Ph	2-$(CH_2)_2NEt_2$	1-$(CH_2)_2NEt_2$-2-CH(OAc)Ph/NaOH	1121
2-CH(OH)C_6H_4Cl-*o*	None	D^b/carboxylic acid	78
2-CH(OH)Ar	1-R (R = Me, Pr)	1-R/ (i) PhNa (ii) ArCHO	850
2-CRR^1OH (R, R^1 = H, Me, Ph)	4-NH_2, NO_2 5-NO_2, Cl, Me	D^b/hydroxy acid	912
2-CH(OH)$(CH_2)_2R$ (R = aryl, hetaryl)	None 1-$(CH_2)_2OH$	2-COCH=CHR derivative/$NaBH_4$/pyridine	1062
2-C(OH)(R)Ph (R = alkyl, alkenyl, Ph, CH_2CO_2Et, $CH_2CO_2NH_4$)	5-CF_3	2-COPh/RLi or RMgBr	951
2-CH(OR)Ph (R = H, Me)	1-Alkyl, CH_2Ph, allyl, crotyl	2-CH(OR)Ph/alkyl halide etc/NaOEt	598
2-$C(R^1)(R^2)OH$ (R^1, R^2 = H, Me, Ph)	Various aryl ring substituents	D^b/carboxylic acid	58
2-CH(OH)R (R = H, Me, Ph)	None 5-OMe, OEt	D^b/carboxylic acid	1099
2-$C(R^1)(R^2)OH$ (R^1, R^2 = alkyl, aryl)	None	2-COMe or 2-COPh/Grignard	817
2-CH(OH)CH(R)NO_2 (R = H, Ph)	1-Me	1-Me 2-CHO/RCH_2NO_2/NaOH	952
2-α- and β-hydroxylalkyl	Various aryl substituents	D^b/carboxylic acid	59
1-CH_2R (R = CH_2OH, CH(OH)Me, CH(OH)Ar)	2-Me, SO_2alkyl, SCH_2Ar	Alkylation by α-haloketone, etc.	588
1-$(CH_2)_2OH$	2-Aryl	2-Aryl/ethylene oxide/pyridine catalyst	636a,c
1-$(CH_2)_2OH$	2-C_6H_4OH	2-Aryl/ethylene oxide	636b
1-$(CH_2)_2OH$	2-$COCH_3$	2-$COCH_3$/ethylene oxide	819
1-$CH_2CH(OH)Me$	2-Alkyl, aryl	2-Alk, aryl/propylene oxide/pyridine catalyst	638

TABLE 1.63   (*Continued*)

Substituents	Other substituents	Method of preparation	Ref.
1-CH$_2$CH(OH)R (R = H, Me)	2-CH$_2$Ph-5-NO$_2$	D^b/imidate	563
2-CH$_2$CH(OH)CCl$_3$	None	o-H$_2$NC$_6$H$_4$NHCOCH$_2$- CH(OH)CCl$_3$/heat in xylene	327

2-CH$_2$C(OH)R^{1}R^2
(R^1 = n-C$_6$H$_{13}$, R^2 = H;
R^1 = Ar, R^2 = H, Ph;
R^1 = PhCH=CH, R^2 = H)
	None
	5-Cl
		2-Me derivative/
		  (i) n-BuLi,
		  (ii) aldehyde derivative
			840

1-CH$_2$CH(R)OH (R = H, Me, CH$_2$OH)	2-R^1 (R^1 = alkyl, aralkyl)	Alkylation by epoxides	640
1-CH$_2$CH(OH)R (R = Me, Ph, p-O$_2$NC$_6$H$_4$)	2-Cl 2-Cl-5,6-Me$_2$	Alkylation by epoxides	642
1-CH$_2$CH(OH)CH$_2$OC$_6$H$_4$Cl-o	None	Alkylation by epoxide	641
2-CH(R)C(OH)Ph$_2$ (R = H, Me, Et)	1-Me	1-Me-2-CH(M)R/Ph$_2$CO (M = Li or Na)	841

1-CH(CH$_2$OH)(CH$_2$)$_2$NHPh	5,6-Me$_2$	Reduction of the
		  1-succinimido derivative/
		  LiAlH$_4$
			668

2-(CH$_2$)$_n$CH(OH)R (R = H, CCl$_3$, alkyl; n = 1, 2)	None 5-Cl, Me	D^b/lactone derivatives	287
2-CH(R)CH$_2$OH (R = alkyl)	None	2-CH(R)CO$_2$Et/LiAlH$_4$	144
2-CH$_2$C(OH)(Ar)(CH$_2$)$_2$NR$_2$ (R = alkyl or NR$_2$ = piperidino, etc.)	None	Amino ketone derivative/ ArMgBr	959
1-CH$_2$R (R = hydroxyalkyl)	2-SMe	Alkylation by RCH$_2$Br	590
2-(CH$_2$)$_3$OH	None	D^b/γ-butyrolactone	285
2-(CH$_2$)$_2$CH(OH)R (R = H, alkyl)	None 5-Cl	D^b/furan-2-one derivatives	286

4-SH	6-NO$_2$	4-NH$_2$-6-NO$_2$/
		  (i) diazotize
		  (ii) K ethyl xanthate
			21b

4-SH	6-NH$_2$	4-SH-6-NO$_2$/SnCl$_2$/HCl	21b
2-CH$_2$SH	None	D^b/carboxylic acid	56
2-CH(Me)SH	None	D^b/carboxylic acid	1116
2-(CH$_2$)$_n$SH (n = 1, 2)	None	D^b/carboxylic acid	1116

^{a} For acetoxy derivatives of the type 1-(CH$_2$)$_2$OCOR, 2-CH$_2$OAc, and 4-CH$_2$OAc, see Refs. 771, 390, and 969, respectively.

^{b} D = appropriate o-arylenediamine derivative.

TABLE 1.64. ALKOXY AND ARYLOXY DERIVATIVES

Substituent	Other substituents	Method of preparation	Ref.
2-OMe	None	2-Cl/OMe⁻	882
2-OR (R = Me, Et, i-Pr)	1-C(Me)=CH₂	2-Cl derivative/RO⁻	889
2-OR (R = Me, Et, i-Pr)	None	1-C(Me)=CH₂-2-OR/KMnO₄/OH⁻	889
4-OMe	2-Alkyl 7-OMe	Dᵃ/carboxylic acid	32
4-OMe	2-(CH₂)₂CO₂H-7-OMe	Dᵃ/succinic anhydride	32
4-OMe	2-CH₂Cl-7-OMe	2-CH₃OH derivative/SOCl₂	32
4-OMe	2-CH₂-1-morpholinyl-7-OMe	2-CH₂Cl derivative/morpholine	32
4-OMe	2-Me-6-NH₂	2-Me-4-OMe-6-NO₂/H₂/Pd	38
4-OMe	2-Me-6-NO₂	Dᵃ/CH₃CO₂H	38
4,7-(OMe)₂	1-Me	4,7-(OMe)₂/Me₂SO₄	1020
4,7-(OMe)₂	None 2-Alkyl	Dᵃ/carboxylic acid	1019
4,7-(OMe)₂	5-NO₂	4,7-(OMe)₂/H₂SO₄/HNO₃	1019
4,7-(OMe)₂	1-Me-2-NMe₂	2-Cl derivative/Me₂NH	1020
5-OMe	6-OMe	Dᵃ/HCO₂H	32
5,6-(OMe)₂	None 2-Me, CH₂OH	Dᵃ/carboxylic acid	1019
5,6-(OMe)₂	2-CH₂Cl	5,6-(OMe)₂-2-CH₂OH/SOCl₂	1019
2-OEt	1-Me	1-Me 2-NO₂/OEt⁻	868
2-OEt	5-Me	Dᵃ/C(OEt)₄	1177
2-O(CH₂)ₙNMe₂ (n = 2, 3)	1-CH₂Ar 5-OMe 6-Cl, etc.	2-Cl derivative/NaH/alcohol derivative	890
4-OCH₂CH(OH)CH₂NHR (R = t-Bu, i-Pr)	None 2-Me	Dᵃ/carboxylic acid	34
2-OCH₂CH=CHR (R = H, Me)	1-Me	1-Me-2-SO₃H/NaOCH₂CH=CHR	1102
2-OC₆H₄R (R = H, p-Me)	None	2-Cl/ArO⁻	891
2-OPh	None	5-OPh-1-Ph-1-H-tetrazole/photolysis	428, 429
1-CH₂OR (R = Me, Et, n-Bu)	None	Benzimidazole/ROCH₂Cl	1125
1-CH₂OR (R = alkyl)	None	1-CH₂Cl/RONa	897
1-CH₂OR (R = alkyl)	None	1-CH₂Cl/RONa	898
2-CH₂O(CH₂)₂Cl	1-Me	1-Me-2-CH₂O(CH₂)₂OH/SOCl₂	914
2-CH₂O(CH₂)₂OH	1-Me	1-Me-2-CH₂Cl/HO(CH₂)₂OH	914
2-CH₂O(CH₂)₂NR₂ (R = Me, Et or NR₂ = morpholino)	1-Me	1-Me-2-CH₂Cl/R₂N(CH₂)₂OH/NaNH₂	914
2-CH₂O(CH₂)₂NEt₂	1-Me	2-CH₂Cl/HO(CH₂)₂NEt₂/NaNH₂	141
2-CH₂O(CH₂)₂OH	1-Me	2-CH₂Cl/(HOCH₂)₂/Na	141
1-CH₂OPh	None	1-CH₂Cl/PhONa	1125

TABLE 1.64   (*Continued*)

Substituent	Other substituents	Method of preparation	Ref.
2-CH$_2$OPh	None	Homolytic alkylation by R· (from RCO$_2$H/peroxy-disulfate/Ag$^+$)	986
2-CH$_2$OAr	5-Cl, alkyl, etc.	D^a/carboxylic acid	70, 71
2-CH$_2$O—C$_6$H$_4$CONHCH$_2$-NHCONHAr	None	2-CH$_2$Cl/salicylamide derivative	916
2-CH$_2$-O-(8-quinolinyl) (also substituted deriatives)	None	2-CH$_2$Cl derivative/Na 8-quinolinate	917
2-CH(OEt)$_2$	None	1-CF$_2$CHFCl/KOH/EtOH	939
2-CH(OR)$_2$ (R = Et, Bu)	None	D^a/ester	43
2-C(OMe)$_3$	None	C$_6$H$_5$NHC($=$NCl)CCl$_3$/NaOH	932
2-C(OR)$_3$ (R = Me, Et, aryl)	None	2-CCl$_3$/ROH or ArOH	933
1-(CH$_2$)$_2$OCHPh$_2$	None	1-(CH$_2$)$_2$OH/Ph$_2$CHCl	897, 898
2-CMe$_2$CH$_2$OMe	None	D^c/carboxylic acid/high pressure	40

^{a} D = appropriate o-phenylenediamine derivative.

TABLE 1.65.   ALKYL-, ALKENYL-, AND ARYLTHIO DERIVATIVES

Substituent	Other substituents	Method of preparation	Ref.
2-SMe	1-CH$_2$R (R = hydroxyalkyl, CO-aryl, CO$_2$H)	Alkylation by RCH$_2$Br	590
2-SR (R = H, Me, i-Pr)	1-⟨N⟩—(CH$_2$)$_3$-COC$_6$H$_4$F-p	2-SH derivative/alkylation	1176
2-SR (R = H, alkyl, aryl)	1-⟨N⟩—(CH$_2$)$_3$Z-C$_6$H$_4$F-p (Z = CO, protected CO, CHC$_6$H$_4$F-p)	From appropriate pyridinium salts	1175
6-SR (R = alkyl, Ph, HO(CH$_2$)$_2$, AcOCH$_2$CH(OAc)CH$_2$)	5-Cl	D^a/anhydride	162
1-SCCl$_3$	4-SH	4-SH/Cl$_3$CSCl	635
1-SCCl$_3$	2-Alkyl, aryl, hetaryl 5,6-Me$_2$ 2-Me-5-Cl	1-Unsubstituted derivative/ Cl$_3$SCl	1091
1-SCCl$_3$	2,5-(hetaryl)$_2$ hetaryl = e.g., 4-thiazolyl, 2-furyl	1-Sulfenylation by ClSCCl$_3$	634

TABLE 1.65   (*Continued*)

Substituent	Other substituents	Method of preparation	Ref.
2-SCH=CHCH₃	None	2-SCH₂CH=CH₂/KOH/ aq. MeOH	984
2-SC(CN)=CHAr	1-Me	1-Me-2-SCH₂CN/ArCHO	814
2-SCH₂CH=CHR (R = H, Me)	1-Me	2-SH/BrCH₂CH=CHR	1102
2-SPh	None 5-Me	2-Cl derivative/PhSH or PhSNa	894
2-SC₆H₄R-*p* (R = alkyl, alkoxy, hal, etc.)	1-(CH₂)₂NR₂ (R = alkyl or NR₂ = 4-morpholino)	(i) 2-Cl derivative/ArS⁻ (ii) Alkylation	893
2-SCH₂-2-(5-nitrofuryl)	None	2-SNa (or K)/ICH₂-2- (5-nitrofuryl)	1169
2-S(CH₂)₂CO₂H	None	2-SH/Cl(CH₂)₂CO₂H	1112
2-S(CH₂)ₙCO₂H (n = 1, 2)	5-Cl	2-SH-5-Cl/X(CH₂)ₙCO₂H [X = Cl (for n = 1), Br (for n = 2)]	1174
2-SCHRCO₂H (R = H, Me, Et)	1-H, Ph, Me 5-H, NO₂	2-SH/XCHRCO₂H (X = Cl, Br)	1117
2-SCH₂CONH₂	None	2-SH/ClCH₂CONH₂	1112
2-SCH₂CONHC₆H₄R-*p*	1-CO₂Et	(i) 2-SH derivative/α-chloro- acetanilide derivative (ii) EtO₂CCl	1167
2-SCH₂CONHHPh	None	Thiazolo[3,2-*a*]benzimidazol 3(2*H*)one/PhNHNH₂	1112
2-SCH₂CONHN=CHR (R = aryl, styryl)	None	2-SCH₂CONHNH₂/RCHO	807
2-SCH₂CONHNHCSNHR (R = aryl, allyl)	None	2-SCH₂CONHNH₂/RNCS	808
2-SCH₂Ph	1-Me	2-SH/BrCH₂Ph	1102
2-SCH₂COMe	1-Ac	Ring opening of a thiazolo benzimidazole derivative	1115
2-SCH₂COPh	None	2-SH/BrCH₂COPh	1115
2-SCH₂COAr	None	2-SH/BrCH₂COAr	1114
2-S(CH₂)₂CN	None	2-SH/CH₂=CHCN/HOAc/HCl	1173
2-SC(R¹R²)CH₂COR³ (R¹, R² = H, Me, Ph; R³ = H, Me)	None	2-SH/R¹R²C=CHCOR³/ methanolic HCl	1170
2-S(CH₂)ₙNR¹R² (n = 2, 3; R, R¹ = H, Me)	1-CH₂Ar 5-MeO 6-Cl, etc.	2-Cl derivative/NaH/thiol derivative	890
2-S(CH₂)₃NHCXNHR (X = O, S; R = aryl, cyclohexyl)	1-CONHC₆H₁₁, Ac 5-NO₂	(i) 2-SH derivative/ Br(CH₂)₃NH₂ (ii) RNCX	1166, 1171
2-S-A-NR¹R² (A = alkylene; R¹, R² = H, alkyl, aryl, alkenyl)	1-Alkyl, hydroxyalkyl, etc. Various benzimi- dazole aryl substituents	2-SH derivative/aminoalkyl chloride	1168

TABLE 1.65 (*Continued*)

Substituent	Other substituents	Method of preparation	Ref.
2-SR (R = CO alkyl, CO-aryl, CH$_2$CO$_2$Et, CONHBu)	Various benzimidazole aryl substituents	2-SH derivative/RCl	1172
2-CH(SC$_6$H$_4$Cl-p)$_2$	None	2-C(SC$_6$H$_4$Cl-p)$_3$/NaOEt	933
2-C(SR)$_3$ (R = Me, aryl)	None	2-CCl$_3$/RSH or ArSH	933
2-CH$_2$SR (R = alkyl, aryl)	None	2-CH$_2$SH/RX or (activated) ArX (X = hal)	56
2-CH$_2$SCH$_2$CO$_2$H	None	2-CH$_2$SH/ClCH$_2$CO$_2$H	1112
2-CH$_2$SCH$_2$CONR^{1}R^2 (R, R^1 = H, alkyl, aryl, etc.)	None	2-CH$_2$SK/ClCH$_2$CONR^{1}R^2	827
2-CH$_2$S(CH$_2$)$_n$COCH$_3$ (n = 1, 2)	None	2-CH$_2$SH/Cl(CH$_2$)$_n$COMe/ NaOH	1148
2-CH$_2$SCH$_2$Ar	None	2-CH$_2$SH/ArCH$_2$X	830
2-CH$_2$SCH$_2$COAr	None	2-CH$_2$SH/KOH/XCH$_2$COAr (X = Br, Cl)	829
2-(CH$_2$)$_n$SC(CH$_2$)$_4$CO (n = 1, 2)	None	2-(CH$_2$)$_n$SH/NaOH/2- chlorocyclohexanone	1154
2-CH$_2$SCOAr	None	2-CH$_2$SH/ArCOCl	831
2-CH$_2$SAr	None	2-CH$_2$Cl/ArSH	921
2-CH$_2$-S—C(=NH)NH$_2$·HCl	None	2-CH$_2$Cl/thiourea	1112
1-(CH$_2$)$_2$SEt	2-Me 5-Cl, NO$_2$ 6-Cl, NO$_2$	1-Alkylation by NaH/ Cl(CH$_2$)$_2$SEt	597
2-(CH$_2$)$_n$SR [R = CH$_2$CO$_2$H (n = 2); R = CH$_2$Ph, aryl (n = 1)]	None	D^a + carboxylic acid or 2-(CH$_2$)$_n$SH/ClCH$_2$CO$_2$H	1116

^{a} D = appropriate o-arylenediamine derivative.

bstituent	Other substituents	Method of preparation	Ref.
CHO	1-Me	1-Me/HCONH$_2$/Na (AcOH work-up)	846
CHO	1-Me	(a) 1-Me-2-CH$_2$OH/AgNO$_3$/H$_2$SO$_4$/ K$_2$S$_2$O$_8$ or (b) 1,2-Me$_2$/SeO$_2$	1025
CHO	None 2-Me 6-Me	5-Me derivative/ceric ammonium nitrate	1026
COCH$_3$	1-COAr, SO$_2$Ar	2-COCH$_3$/aroyl or sulfonyl chloride	820
COCH$_2$Br	1-Me, CH$_2$Ph	2-COCH$_3$ derivative/Br$_2$/CCl$_4$	820
COCHBr$_2$	None 1-Me, CH$_2$Ph	2-COCH$_3$ derivative/Br$_2$/AcOH	820
COCHXR (X = halogeno, R = alkyl)	None 1-Me, CH$_2$Ph. Various aryl ring substituents	(i) D^a/carboxylic acid (or derivative) or (ii) Side-chain oxidation	146
COCH(Me)C$_6$H$_4$CH$_2$CHMe$_2$-p	None	Acylation by acyl chloride/Et$_3$N	681
COCHPh$_2$	None	Benzimidazole/Ph$_2$C=C=O	673
COCHPh$_2$	1-Me, CH$_2$Ph	2-C(=CPh$_2$)OCOCHPh$_2$ derivative/ HOCH$_2$NH$_2$	674
COCHPh$_2$	None	(a) 2-CH$_2$CHPh$_2$/SeO$_2$ or (b) 2-CH(OH)CHPh$_2$/DDQ	1030
COCH$_2$CH(Ph)Ar (Ar = Ph, p-MeOC$_6$H$_4$)	None	2-COCH=CHAr/PhMgBr	817
CO(CH$_2$)$_n$Ar (n = 1, 2)	1-(CH$_2$)$_2$NEt$_2$-5-MeO 1-(CH$_2$)$_2$NEt$_2$- 5,6-(MeO)$_2$	2-CH(OH)(CH$_2$)$_n$Ar derivatives/ CrO$_3$ or other oxidants	1029
COCH=CHAr	None	2-COCH$_3$/ArCHO/NaAc	817
COCH=CHPh	2-CH$_2$CO$_2$Et Me, MeO, Cl in benzimidazole aryl ring	Acylation by acyl chloride	682
COCH=CHR (R = aryl, hetaryl)	1-(CH$_2$)$_2$OH	1-(CH$_2$)$_2$OH-2-COCH$_3$/RCHO	819
COCH=CHR (R = 2-thienyl, 2-furyl, aryl)	None	2-COCH$_3$/RCHO/pyridine	961
COCH=CHR (R = hetaryl)	None	2-COCH$_3$/RCHO/EtONa	818
COPh	2-Ph	Thermolysis of 1aH-oxazirino- [2,3-a]quinoxaline	430
COAr	2-CH=CHAr various benzimi- dazole aryl substituents	Aroylation by ArCOX (X = hal)/Et$_3$N	683
COC$_6$H$_4$NO$_2$-p	5,6-Cl$_2$	5,6-Cl$_2$/ArCOCl	706
COPh	None 1-CH$_2$CO$_2$Me	(i) Oxidation of 2-CH(OH)Ph (ii) Alkylation	142
COPh	5-NO$_2$	Photolysis of 1-(2,4-dinitro- phenyl)-4-phenylimidazole	416

TABLE 1.66 (*Continued*)

Substituent	Other substituents	Method of preparation	R
2-COAr	None 1-(CH$_2$)$_2$NEt$_2$	2-CH(OH)Ar derivative/CrO$_3$	10
2-COC$_6$H$_4$CO$_2$H	None	2-CH$_2$C$_6$H$_4$CO$_2$H-*o*/KMnO$_4$	75
2-COC$_6$H$_3$(R)COX (R = H, hal, MeO, etc.; X = NH, alkyl, morpholinyl)	Cl, CF$_3$, NO$_2$ in benzimidazole aryl ring	Ring-opening of benzimidazo- [1,2-b]isoquinolin-6,11-diones	76
5-COC$_6$H$_4$CO$_2$H-*o*	2-NMe$_2$	2-NMe$_2$/phthalic anhydride/AlCl$_3$	57
2-CO-2-pyridyl	None	2-CHO/2-pyridyl MgBr/oxidation in work-up	95
2-COCHO	None	2-COCH$_3$/SeO$_2$	96
2-COCH=NR (R = aryl, 2-thiazolyl)	None	2-COCHO/RNH$_2$	96
1-CH$_2$COR (R = alkyl, aryl)	2-Cl	1-Alkylation by α-haloketone/NaOMe	61
2-CH$_2$COMe (via ketal)	None	D^a + EtO$_2$CCH$_2$COMe (ketal)	11
2-CH$_2$CO[2,4-(OH)$_2$C$_6$H$_4$]	None 1-Me	2-CH$_2$CN/HCl/BF$_3$/resorcinol	11
2-CH$_2$COCHO	None	2-CH$_2$COCH$_3$/SeO$_2$	96
2-CH$_2$COCO$_2$Et	1-Me	1,2-Me$_2$/NaH/(CO$_2$Et)$_2$	75
2-CH$_2$COCH=NR (R = aryl, 2-thiazolyl)	None	2-CH$_2$COCHO/RNH$_2$	96
2-(CH$_2$)$_2$COMe	None	2-CH$_2$CH=CClCH$_3$/H$_2$SO$_4$	84
2-CMe$_2$CO*i*-Pr	None 1-Me, Ph 5-Cl, Me, MeO, NO$_2$, etc.	D^a/dimethylketen dimer	40
1-(CH$_2$)$_2$COAr	2-Me, CH$_2$OH, (CH$_2$)$_3$OH	Alkylation by ArCO(CH$_2$)$_2$-1- piperidinyl	65

^{a} D = appropriate *o*-arylenediamine derivative.

TABLE 1.67. OXIMES

Substituent	Other substituents	Method of preparation	Ref.
2-CH=NOH	None	2-CHCl$_2$/H$_2$NOH·NCl (pH 6)	43
2-CH=NOH	None	1-CF$_2$CHFCl/H$_2$NOH	939
2-CH=NOH	1-Me	1-Me-2-CHO/H$_2$NOH	952
2-CH=NOH	1-(CH$_2$)$_5$Cl-5-R (R = H, Cl, NO$_2$)	*o*-aminoarylpiperidines/ Cl$_3$CCHO/H$_2$NOH·HCl	306c
2-C=NOH | Cl	1-H, Me 5-H, Cl, NO$_2$	2-CH=NOH/Cl$_2$/AcOH	1076
2-C=NOH | NO$_2$	1-H, Me 5-H, Cl, NO$_2$	2-CH=NOH/HNO$_3$	1076
2-CH=NOCON[(CH$_2$)$_2$Cl]$_2$	None	2-CH=NOH/ClCON- [(CH$_2$)$_2$Cl]$_2$/NaH (1:1:1 mole ratio)	833

TABLE 1.68. CARBOXYLIC ACIDS, ESTERS, AND AMIDES

Substituent	Other substituents	Method of preparation	Ref.
2-CO$_2$H	None	2-CCl$_3$/H$_2$O	933
2-CO$_2$H	1-Me	1-Me/HCONH$_2$/Na (EtOH work up)	846
2-CO$_2$H	1-Me	(i) 1-Me/BuLi; (ii) CO$_2$/$-60°$	842
2-CO$_2$H	1-Me	1-Me-2-CN/hydrolysis	1104
2-CO$_2$H	1-Me, Et	(i) 1-Me/PhNa; (ii) CO$_2$	843
2-CO$_2$H	1-Ph	(i) 1-Ph/PhLi; (ii) CO$_2$	845
2-CO$_2$H	5-C$_6$H$_4$NH$_2$-p	2-CN-5-Ar/75% H$_2$SO$_4$/150°	1142
5-CO$_2$H	None 2-Me, Ph	5-Me derivatives/H$_2$Cr$_2$O$_7$	109
5-CO$_2$H	1-Me 1-Me-2-Cl	5-Me derivative/KMnO$_4$/100°	858
5-CO$_2$H	1-Me 1-Me-2-Cl	1,5-Me$_2$ derivative/KMnO$_4$	1024
1-CH$_2$CO$_2$H	None 2-pyridyl	Benzimidazole/ClCH$_2$CO$_2$H/aq. NaOH or BrCH$_2$CO$_2$Me/K$_2$CO$_3$	140
2-CH$_2$CO$_2$H	None	2-CH$_2$CN/OH$^-$	141
2-(CF$_2$)$_n$CO$_2$H (n = 1, 2)	None	D^a/dicarboxylic acid	117
2-CH(R)CO$_2$H (R = alkyl)	None	2-CH(R)CN/alkaline hydrolysis	1095
5-CH(R^1)COR^2 (R^1 = H, Me; R^2 = OH, OEt, NH$_2$	1-Bu 2-alkyl, aralkyl, etc.	D^a/imidate	177
1-(CH$_2$)$_2$CO$_2$H	2-(CH$_2$)$_2$CO$_2$H	1-(CH$_2$)$_2$CN derivative/OH$^-$	659
1-(CH$_2$)$_2$R (R = CONH$_2$, CO$_2$H)	None 2-Me NO$_2$ in aryl ring	(i) Alkylation by CH$_2$=CHCONH$_2$ (ii) hydrolysis	663
2-(CH$_2$)$_2$CO$_2$H	None 5-OMe, OEt	D^a/dicarboxylic acid	67, 111, 112, 1099
2-(CH$_2$)$_2$CO$_2$H	1-CH$_2$Ph	2-(CH$_2$)$_2$CO$_2$H/PhCH$_2$Cl	592
2-(CF$_2$)$_n$CO$_2$H (n = 1, 2)	None	D^a/dicarboxylic acid	342
2-(CF$_2$)$_n$COX (X = NH$_2$, OH, Cl; n = 2, 3)	None	o-H$_2$NC$_6$H$_4$NHCO(CF$_2$)$_n$CONH$_2$/ 170–190° (for X = NH$_2$)	330
1-CH(Me)CH$_2$CO$_2$H	None 2-alkyl, aralkyl. Various substituents in benzimidazole aryl ring	Alkylation by MeCH=CHCO$_2$H	661
5-CH(Me)CH$_2$CO$_2$H	2-NH$_2$	2-NHCO$_2$Me/γ-butyrolactone/AlCl$_3$	573
1-CH$_2$CH(R)CO$_2$H (R = H, Me)	None 2-Me	Alkylation by CH$_2$=C(R)CO$_2$H	662
2-CH(OH)CH$_2$CO$_2$H	None 5-OMe, OEt	D^a/carboxylic acid	1099
2-[CH(OH)]$_n$CO$_2$H (n = 2, 4)	None 5-OMe, OEt	D^a/carboxylic acid	1099

205

TABLE 1.68 (*Continued*)

Substituent	Other substituents	Method of preparation	Ref.
1-CH(CO$_2$H)CH$_2$CO$_2$H	None 2-alkyl, alkoxy, NO$_2$. Various aryl ring substituents	Benzimidazole derivative/maleic acid (mono Na salt)	665
5-CH(Me)(CH$_2$)$_2$CO$_2$H	2-NH$_2$	2-NHCO$_2$Me/γ-valerolactone/AlCl$_3$	574
2-(CH$_2$)$_4$CO$_2$H	None	D^a/dicarboxylic acid	113
2-(CH$_2$)$_8$CO$_2$H	5-NO$_2$, CO$_2$Me	D^a/imidate	186
1-CH$_2$XCO$_2$H (X = hydroxy alkylene)	4-NH$_2$	4-NO$_2$/(i) (X = protected hydroxyalkylene) (ii) hydrolysis, (iii) reduction	645
1-CO$_2$R (R = alkyl, aryl)	2-CF$_3$-5,6,7-Cl$_3$ 2,5-(CF$_3$)$_2$, etc.	1-Alkylation by ClCO$_2$R/NaOEt	622
1-CO$_2$R (R = alkyl, aralkyl, allyl, etc.)	2-CF$_3$, Cl, SMe. Various benzimi- dazole aryl sub- stituents	1-Alkylation by ClCO$_2$R/NaOEt	621
1-CO$_2$R (R = alkyl, aryl, alkenyl, etc.)	2-R^1 R^1 = hal, NO$_2$, CN etc. Hal or poly- haloalkyl in benzi- midazole aryl ring	2-R^1/ClCO$_2$R	690
2-CO$_2$R (R = alkyl)	None	2-CO$_2$H/ROH/HCl or SOCl$_2$/NaOR	945
2-CO$_2$(CH$_2$)$_2$NR$_2$ (R = Me, Et)	None 1-Me 5-NO$_2$, Br, alkyl	2-CO$_2$Na/Cl(CH$_2$)$_2$NR$_2$	832
1-CO$_2$CH$_2$CH=CH$_2$	2-CF$_3$-4,5,6-Cl$_3$ 5-NO$_2$	Alkoxycarbonylation by ClCO$_2$CH$_2$CH=CH$_2$/K$_2$CO$_3$	623 623
1-CO$_2$Ph	2-CF$_3$-4,5,6-Cl$_3$ 5-NO$_2$	Alkoxycarbonylation by ClCO$_2$Ph/K$_2$CO$_3$	623
1-CS$_2$Et	2-CF$_3$-4,5,6-Cl$_3$	2,4,5,6-substituted derivative/ ClCS$_2$Et/Et$_3$N	623
1-CH$_2$CO$_2$Et	None	Alkylation by XCH$_2$CO$_2$Et (X = hal)/ K$_2$CO$_3$	616
2-CH$_2$CO$_2$Me	1-COPh	2-CH$_2$CO$_2$Me(Ag salt)/PhCOCl	142
2-CH$_2$CO$_2$Et	None	D^a/imidate	182
2-CH$_2$CO$_2$Et	None 1-Me	2-CH$_2$CO$_2$H/esterification	1165
2-CH$_2$CO$_2$(CH$_2$)$_2$NR$_2$ (R = Me, Et)	None 1-Me 5-NO$_2$, Br, alkyl	2-CH$_2$CO$_2$Na/Cl(CH$_2$)$_2$NR$_2$	832
1-(CH$_2$)$_2$CO$_2$Me	2-(CH$_2$)$_2$CO$_2$Me	1-(CH$_2$)$_2$CN-2-(CH$_2$)$_2$CO$_2$H/ MeOH/HCl	659
2-(CH$_2$)$_2$CO$_2$Et	None	2-(CH$_2$)$_2$CO$_2$H/EtOH/HCl	946
2-(CH$_2$)$_n$CO$_2$Et (n = 4, 5)	None	D^a + (CH$_2$)$_n$ ⟨structure⟩ CO$_2$Et n = 3,4	1107

TABLE 1.68 (*Continued*)

Substituent	Other substituents	Method of preparation	Ref.
2-CONH$_2$	None	Homolytic amidation by ·CONH$_2$ from HCONH$_2$/t-BuO$_2$H/Fe^{2+}	986
2-CSNH$_2$	None	2-CN/MeCSNH$_2$	942
2-CONH(CH$_2$)$_3$OH	None	2-(2-oxazolinyl)/hydrolysis	934
2-CONH(CH$_2$)$_n$Ar (n = 0–2)	RO, alkyl, Cl, etc., in benzimidazole aryl ring	D^a/(AcO)$_2$CHCONH(CH$_2$)$_n$Ar	302
1-CONHR (R = Me, aryl)	2-NHAc, (5-nitro-furfurylidene)-amino, 5-Me	Acylation by RNCO	741
1-CONHPh	None	R$_2$CO/PhNH$_2$ (R = 1-benzimidazolyl)	1093
2-CONHC$_6$H$_4$NH$_2$-o	None	D^a/Cl$_3$CCO$_2$H	932
1-CONHAr (Ar = p-ClC$_6$H$_4$, 3,4-Cl$_2$C$_6$H$_3$)	None	Benzimidazole/ArNCO	725
2-CONR^{1}R^2 (R^1 = H, Me, allyl R^2 = Me, allyl, (CH$_2$)$_2$Cl, aryl, 5-chloro-2-pyridyl)	None	2-CCl$_3$/amine	926
1-CON[(CH$_2$)$_2$Cl]$_2$	2-CN	2-CH=NOH/ClCON[(CH$_2$)$_2$Cl]$_2$/NaH (1:2:3 mole ratio)	833
1-CON(CH$_2$CH$_2$Cl)$_2$	None 2-Me, MeO, SMe, etc.	Acylation by (ClCH$_2$CH$_2$)$_2$NCOCl	700

^{a} D = appropriate o-arylenediamine derivative.

TABLE 1.69. AMINO ACIDS

Susbtituent	Other substituents	Method of preparation	Ref.
2-CH$_2$CH(NH$_2$)CO$_2$H	None	D^a/carboxylic acid	62
2-CH$_2$CH(NH$_2$)CO$_2$H	1-Me	1-Me-2-[structure: imidazolidinedione ring, CH$_2$—NH, O=, N, =O, H]/NaOH 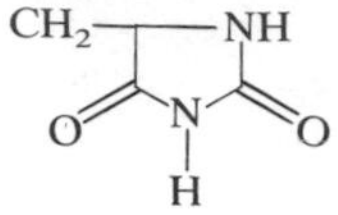	953
2-(CH$_2$)$_n$CH(NH$_2$)CO$_2$H (n = 1–4)	Various aryl ring substituents	D^a/carboxylic acid	63

2-(CH$_2$)$_n$CHCO$_2$H
   |
   NHSO$_2$Ph
 (n = 1, 2)
	None	D^a/dicarboxylic acid	68

2-(CH$_2$)$_2$C(NH$_2$)(R)CO$_2$H (R = H, Me)	None	D^a/carboxylic acid	62
4-NHCH$_2$CONHCH(CH$_3$)CO$_2$H	2-Me-5,7-(NO$_2$)$_2$ 5,7-(NO$_2$)$_2$	4-F derivative/alanyl glycine	565

2-CHCH$_2$CO$_2$H
  |
  NH$_2$
	None	D^a/carboxylic acid	62

2-NHCOXCO$_2$H (X = C(R) = CR (R = H, Me) (CH$_2$)$_2$, phenylene, etc.)	Cl, Me, OMe in benzimidazole aryl ring	2-NH$_2$ derivative/[structure: anhydride ring O=, O, X, =O] 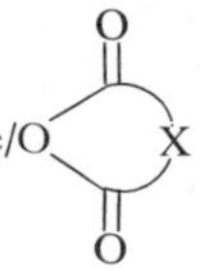	795

^{a} D = appropriate o-arylenediamine derivative.

TABLE 1.70. SULFONIC ACIDS, SULFONYL CHLORIDES, SULFONAMIDES, AND SULFONES

Substituent	Other substituents	Method of preparation	Ref.
2-SO$_3$H	None	2-SH/H$_2$O$_2$	1100
2-SO$_3$H	1-Me	1-Me-3-oxide/NaHSO$_3$	1104
5-SO$_3$H	None 1-Me	Benzimidazole. H$_2$SO$_4$/heat	559
5-SO$_3$H	None 2-Me	(i) D^a/carboxylic acid or (ii) Heat the sulfate	17
5-SO$_3$H	None 2-Me, Ph	ClSO$_3$H/benzimidazole derivative	561
5-SO$_2$R (R = NH$_2$, OH)	1-H, alkyl, aralkyl, Ac, 2-Ac, alkyl, etc.	D^a/carboxylic acid (or derivative)	147

208

TABLE 1.70 (*Continued*)

Substituent	Other substituents	Method of preparation	Ref.
5-SO$_3$K	None 2-alkyl	Sulfonation by H$_2$SO$_4$	560
2-(CH$_2$)$_n$SO$_3$H (n = 1, 2)	None	2-(CH$_2$)$_n$SH/H$_2$O$_2$	1116
5-SO$_2$Cl	2-C$_6$H$_4$R-p R = H, NO$_2$	2-Ar/ClSO$_3$H	561
1-SO$_2$NMe$_2$	2-NH$_2$-5-(and 6-)CHO	Oxidation of 5- and 6-CH$_2$OH derivatives by Jones reagent	1027
1-SO$_2$NR^{1}R^2 (R^1, R^2 = Me, Et or NR^{1}R^2 = morpholino, etc.)	2-NH$_2$	2-NH$_2$/R^{1}R^{2}NSO$_2$Cl	708
1-SO$_2$R (R = alkylamino, dialkylamino, pyrrolidino, etc.)	2-NHR^1 (R^1 = H, Ac, Me, 5-CO$_2$ alkyl, CONH$_2$, etc.)	Sulfonylation by RSO$_2$Cl	709
2-SO$_2$NR$_2$ (R = H, alkyl)	None	2-SO$_2$Cl (*in situ*)/R$_2$NH	1100
2-SO$_2$NH(CH$_2$)$_n$NHO$_2$SR (n = 2–10, 12; R = 2-benzimidazolyl)	None	2-SO$_2$Cl/H$_2$N(CH$_2$)$_n$NH$_2$	929
6-SO$_2$NR$_2$ (R = Et, (CH$_2$)$_2$Cl)	1-(CH$_2$)$_2$Cl-2-Me	D^a/carboxylic acid	39
1-SO$_2$Me	None	Benzimidazole/MeSO$_2$Cl/ pyridine	674
1-SO$_2$R (R = alkyl)	2-CF$_3$	2-CF$_3$/RSO$_2$Cl	707
1-SO$_2$R (R = Me, Ph)	2-CF$_3$-4,5,6-Cl$_3$	Sulfonation by RSO$_2$Cl	702, 703
1-SO$_2$R (R = alkyl, aryl, 2-thienyl)	2-COR^1 (R^1 = Me, chloroalkyl)	Sulfonation by RSO$_2$Cl/NaHCO$_3$	701
1-SO$_2$C$_6$H$_4$R-p (R = H, NO$_2$)	None	Benzimidazole/ArSO$_2$Cl	861
1-SO$_2$Ar	None 2-alkyl, Ph	Sulfonation by ArSO$_2$Cl	704
1-SO$_2$Ar	2-(4-thiazolyl)	Sulfonation by ArSO$_2$Cl	705
1-SO$_2$C$_6$H$_4$NO$_2$-p	5,6-Cl$_2$	Sulfonation by ArSO$_2$Cl	706
2-CH$_2$SO$_2$R (R = Me, Et)	None 5-Cl, NO$_2$	2-CH$_2$R^1/ (i) RONa or RSNa (ii) m-ClC$_6$H$_4$CO$_3$H (R^1 = SH or Cl)	915
2-CH$_2$SO$_2$CHPh$_2$	None	2-CH$_2$SCHPh$_2$/H$_2$O$_2$	1116
2-CH$_2$SO$_2$Ph	5,6-R$_2$ (R = H, Cl)	D^a/Cl$_2$C=CHSO$_2$Ph	290
1-(CH$_2$)$_2$SO$_2$Et	2-Me 5-Cl, NO$_2$ 6-Cl, NO$_2$	1-(CH$_2$)$_2$SEt derivative/ m-ClC$_6$H$_4$CO$_3$H	597

^{a} D = appropriate o-arylenediamine derivative.

209

TABLE 1.71. ORGANOPHOSPHORUS, -SILICON, AND -IRON DERIVATIVES

Substituent (or compound)	Other substituents	Method of preparation	Ref.
2-CH$_2$$\overset{+}{\text{P}}Ph_3$(Cl$^-$)	None	2-CH$_2$Cl/Ph$_3$P	1130
1-P($=$X)R^{1}R^2 (R $=$ NMe$_2$, O-alkyl, OAr, OC$_6$H$_{11}$; R^1 $=$ alkyl, ONa; X $=$ O, S)	2-(4-thiazolyl)	Phosphinylation by R^{1}R^{2}P($=$X)Cl	631
(R)$_2$PR^1 [R $=$ 1-benzimidazolyl R^1 $=$ OH, O(OH), S(OH)]	5,6- and 5′,6′-Me$_2$	5,6-Me$_2$/ (i) PCl$_3$ or POCl$_3$ or PSCl$_3$ (ii) H$_2$O	633
(R)$_3$X (X $=$ P, PO, PS; R $=$ 1-benzimidazolyl)	None	Benzimidazole/PCl$_3$ or POCl$_3$ or PSCl$_3$	632
2-Z-OPO(OR)Me [Z $=$ (CH$_2$)$_2$, CMe$_2$; R $=$ Et, Pr]	None	D^a/imidate	198
2-(CHR^1)$_n$PO(OR)$_2$ (R $=$ alkyl; R^1 $=$ H, alkyl; n $=$ 1, 2)	None	D^a/imidate	197
1-SiMe$_3$	None	Benzimidazole/(Me$_3$Si)$_2$NH	1090
2-(CH$_2$)$_n$-ferrocenyl (n $=$ 1–5)	1-Me, Ph Me, Cl, MeO in benzimidazole aryl ring	D^a/carboxylic acid	73

^{a} D $=$ appropriate o-arylenediamine derivative.

TABLE 1.72. BISBENZIMIDAZOLES[a] (CONTAINING A BENZIMIDAZOL-2-YL GROUP)

Compound	Other substituents	Method of preparation	Ref.
R–R	None	D^c/oxamide	1156
R–R	None	D^c/1-CF_2CHFCl	939
R–R	None	Pyrolysis of aryl derivatives of α-amino acids	385
R–R	1- and 1'-Me	1-Me-3-oxide/hydrolysis	1104
R–R	1- and 1'-Me	1-Me/BuLi/room temp.	842
R–R	1- and 1'-Me, Et, i-Pr	1-Alkyl/K(or Na)/C_6H_6	847
R–R	5-Cl } in one ring, 5,6-Cl_2, 5,6-Me_2 } only	D^c/$Cl_3CC(=NH)OMe$	172
R–R	5- and 5'-Cl	D^c/$Cl_3CC(=NH)OR^1$(R^1 = alkyl)	932
R–R	6- and 6'-Cl, Me	1-$CH_2C_6H_4OMe$-p/Chichibabin	865
$R-(CH_2)_n-R$ ($n = 0, 1$)	1- and 1'-$CH_2C_6H_4OMe$-p	D^c/oxamide	158
$R(CH_2)_n R$ ($n = 0, 1, 4, 8$)	5- and 5'-H, Me	D^c/diamide	158
[structure]	5 (and 5')-Me	1-Me-3-MeO benzimidazolium iodide/reflux in MeCN/Et_3N	971
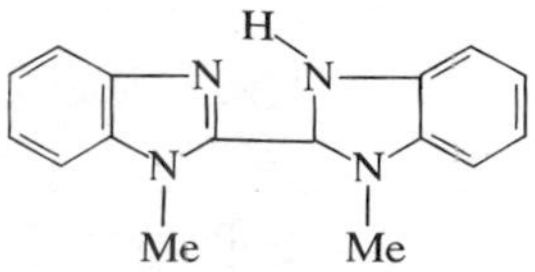 [structure]	None	1-Me/BuLi/room temp.	842
$R-R$ ($2BF_4^-$)	1- and 1'-Me, 3- and 3'-Me	Dialkylation of the bisbenzimidazole by $Me_3O^+BF_4^-$	1156

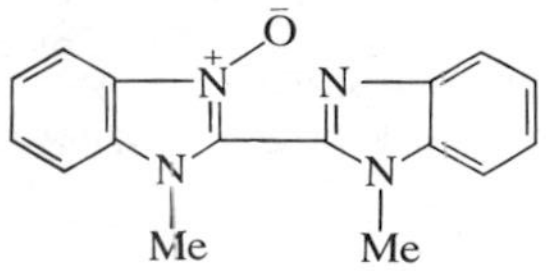

TABLE 1.72 (Continued)

Compound	Other substituents	Method of preparation	Ref.
R—R (2ClO$_4$)$^-$	1- and 1'-Ph 3- and 3'-Ph	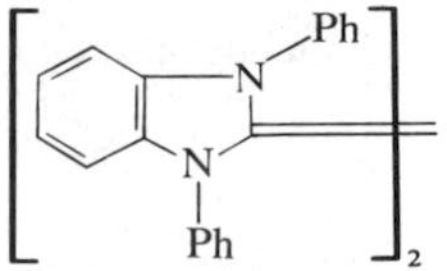 [bis-benzimidazolylidene structure with N—Ph groups] $\Big]_2$ (i) AgNO$_3$ (ii) HCl (iii) HClO$_4$	1106
R—R^1 (R^1 = 1'-benzimidazolyl)	None	2-SO$_3$H derivative/benzimidazole	872
R—R^1 (R^1 = 4'-benzimidazolyl)	1-Me	Photolysis of benzimidazole in EtOH	1004
R—R^1 (R^1 = 5'-benzimidazolyl)	None	Photolysis of benzimidazole in EtOH	1004
R—R^1 (R^1 = 5'-benzimidazolyl)	None	D^c/benzimidazole-5-CO$_2$H	109
R—R^1 (R^1 = 5'-benzimidazolyl)	2-Me, Ph 1-Et 1'-Ph-2'-Me	D^c/benzimidazole-5'-CO$_2$H	110
R—R^1 (R^1 = 5'-benzimidazolyl)	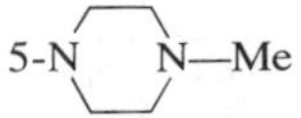 5-N$\langle$piperazine$\rangle$N—Me 2'-C$_6$H$_4$R-p (R=OH, O-alkyl)	D^c/imidate	192
R—CHR^1—R (R^1 = H, alkyl)	1- and 1'-H, alkyl 5'-NH$_2$, NHAc	D^c/imidate	181
R—CHR^1—R (R^1 = H, Me, Bu)	5,6-(and 5', 6')-Me$_2$	D^c/imidate, amide or ester	185
R—CF$_2$—R	5-CF$_3$	D^c/dicarboxylic acid	117
RCH(R^1)R (R^1 = alkyl)	None	D^c/2-CH(R^1)CO$_2$Et	144
R—CHR^1(R) (R^1 = H, alkyl)	5,6-Me$_2$ 1-H, Et 5'-NH$_2$, NHAc	D^c/ester	1086

212

Structure	Substituent	Method	Ref.
$R-(CH_2)_n-R$ ($n = 1, 4, 8$)	None	D^c/diamide	158
$R-(CH_2)_n-R$ ($n = 1-8, 10$)	None	D^c/dicarboxylic acid	124
$R-(CH_2)_n-R$ ($n = 1-20$)	1-R^1 (R^1 = C$_{1-20}$ alkyl)	Dialkylation of the bis benzimidazole/alkyl halide	593
$R-(CH_2)_2-R$	None	D^c/dicarboxylic acid or ciester	111, 126
$R-(CH_2)_n CH(R)NHSO_2Ph$ ($n = 1, 2$)	None	D^c/dicarboxylic acid	68
$R-[CH(OH)]_2-R$	4,7-(OEt)$_2$ 5-OEt	D^c/dicarboxylic acid	115, 116
$R-X-R$ (X = (CH$_2$)$_n$, $n = 1-3$; (CHMe)$_2$; CH$_2$CMe$_2$; (CMe$_2$)$_2$; CH—CH; [CH(OH)]$_n$, $n = 2, 4$; (CHOCOCH$_3$)$_2$)	None	D^c/carboxylic acid	1099
$R-(CH_2)_n-R$ ($n = 2-8$)	None	D^c/carboxylic acid (or derivative)	1096
$R-(CH_2)_n-R \cdot 5H_2O$ ($n = 2-8$)	1- and 1'-NO	R-(CH$_2$)$_n$-R/NaNO$_2$/HCl	1096
$R-(CF_2)_n-R$ ($n = 2-4$)	None 1- and 1'-Me	(i) D/carboxylic acid (ii) Alkylation of bisbenzimidazoles by CH$_2$N$_2$	1088
$R-(CH_2)_n-R$ ($n = 4, 8$)	None	D^c/di-imidate	186
$R-[CH(OH)]_4-R$	None	D^c/dicarboxylic acid	111
$(RCH_2CH_2)_2C(NO_2)_2$	None	D^c/bis imidate	199
$R-(CH_2)_8-R$	None	D^c/dicarboxylic acid	116
RCH=CHR	1-Methyl 1'-alkyl	1-Me-2-CHO/2-Me benziridazole derivative	950
RCH=CHR	1- and 1'-Me 3-Me (I$^-$)	1-Me-2-CHO/1,3-dimethylbenzimidazolium iodide	1025
RCH=CHR (cis and trans)	1- (CH$_2$)$_2$OH, (CH$_2$)$_2$Cl	(i) D^c/dicarboxylic acid	114
R_2Z (2X$^-$) (Z = (CH=CH)$_n$, $n = 1-3$; X$^-$ = e.g., BF$_4^-$)	5,6- and 5',6'-H, Me 1- and 1'-Me 3- and 3-Me	(ii) Alkylation o-(MeNH)$_2$C$_6$H$_4$/R^{1}CO(CH=CH)$_n$COR^1 (R^1 = Cl or i-PrCH$_2$OCO$_2$)	1150

TABLE 1.72 (Continued)

Compound	Other substituents	Method of preparation	Ref.
(R—C≡C)$_2$	1- and 1'-Me	1-Me-2-C≡CH/O$_2$/Cu$_2$Cl$_2$/pyridine	1031
RC≡C—CH=C(OMe)R	1- and 1'-Me	1-Me-2-C≡CH/O$_2$/Cu$_2$Cl$_2$/MeOH/pyridine	1031
R—C$_6$H$_4$—R—p	None	D^c/dicarboxylic acid	126
R—C$_6$H$_4$—R—p	None	Oxidative cyclization of N,N'-terephthalidene bis(o-aminoaniline)	400
R—C$_6$H$_4$—R—p	None	Thermolysis of bis anil derived from o-N$_3$C$_6$H$_4$NH$_2$/p-(OHC)$_2$C$_6$H$_4$	393
R—C$_6$H$_4$—R—p	2,5-(OH)$_2$ in p-phenylene ring	D^c/dialdehyde	211
1,4—R$_2$—2,5—R$_2^1$C$_6$H$_2$ (R^1= OH, OMe)	None	D^c/terephthalaldehyde derivatives	225
R—Q—R (Q= p-C$_6$H$_4$, m-C$_6$H$_4$, 4,4'—C$_6$H$_4$—C$_6$H$_4$, 2,6—C$_{10}$H$_6$, 4,4'—C$_6$H$_4$OC$_6$H$_4$, 4,4'—C$_6$H$_4$SO$_2$C$_6$H$_4$	1- and 1'-Ph	Cyclization of (o-PhNHC$_6$H$_4$NHCO)$_2$Q	347
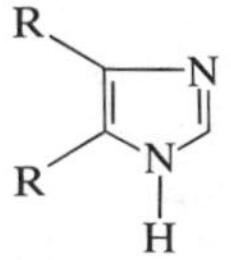 (imidazole structure, N—N—H with R substituents)	None	D^c/4,5-di(trichloromethyl)imidazole	300
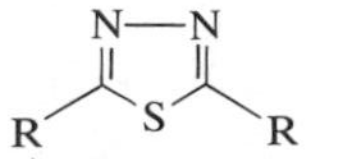 (1,3,4-thiadiazole structure, N—N, S, R substituents)	5-Cl	D^c/2,5-bis[(MeO)$_3$C]-1,3,4-thiadiazole/H$_2$SO$_4$	301
(RCH$_2$)$_2$Z (Z = 1,4-piperazinyl)	1- and 1'-Me	1-Me-2-CH$_2$Cl/piperazine	654
R—S—R	None	(i) 2-SCN/CS$_2$/Et$_3$N (ii) hydrolysis of the tetracyclic product from (i)	918
R—S—R	None	Hydrolysis of a thiadiazino-bis benzimidazole	769
R—S—R	1- and 1'-H, Me Me, NO$_2$ in carbocyclic ring	2-Cl derivative/2-SH derivative	894

214

Reagent / product	Substituent	Method	Ref.
R—SCH(R)COS—R (R = H, Et)	1- and 1'-P$\blacksquare$	Acid hydrolysis of mesoionic thiazolo benzimidazoles	1117
(p—RC$_6$H$_4$)$_2$Z (Z = O, S, SO$_2$)	5- and 5'-NH$_2$	Reductive cyclization of [2,4-(O$_2$N)$_2$C$_6$H$_3$NHCOC$_6$H$_4$]$_2$Z	348
(RCH$_2$S)$_2$Z (Z = alkylene, alkenyl, xylylene)	None	RCH$_2$SM/ZX$_2$ M = alkali metal X = Cl, Br	828
(RCH$_2$S)$_2$CS	1- and 1'-CO$_2$R (R = Me, Et)	2-CH$_2$Cl derivative/(KS)$_2$CS	920
(R—S)$_2$	None	2-SH/I$_2$	1100
(R—S)$_2$	None	2-SH/H$_2$O$_2$	1101
(RCH$_2$S)$_2$	None	2-CH$_2$Cl/thiourea	1112
[R(CH$_2$)$_2$]$_2$SO$_2$	None	R(CH$_2$)$_2$SO$_3$H/SOCl$_2$/NH$_3$ or [R(CH$_2$)$_2$]$_2$S/H$_2$O$_2$	1116
[R(CH$_2$)$_n$C(R^1)(R^2)(CH$_2$)$_m$]SO$_2$ (R^1,R^2 = H, alkyl; n = 0, 1; m = 0-4)	1-alkyl, H, alkyl etc. in aryl ring	D^c/dicarboxylic acid	125

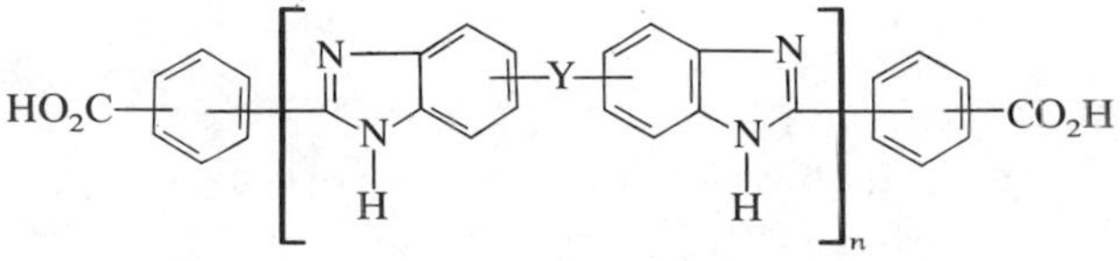

(Y = O, S, SO$_2$, CO, CH$_2$; $n \geq 1$)

Reagent / product	Substituent	Method	Ref.
[benzimidazole structure shown above]	None	Bis diamine/bis dicarboxlic acid	127 (cf, also 128, 129)
(RSO$_2$NH)$_2$(CH$_2$)$_2$	None	2-SO$_2$Cl/H$_2$N(CH$_2$)$_2$NH$_2$	929
(R)$_2$PR^1 [R^1 = OH, O(OH), S(OH)]	5,6- and 5',6'-Me$_2$	5,6-Me$_2$/(i) PCl$_3$ or POCl$_3$ or PSCl$_3$ (ii) H$_2$O	633
RCOR^1 (R^1 = 1-benzimidazolyl)	None	2-CCl$_3$/NaHCO$_3$/H$_2$O	933
R$_2$NH	1- and 1'-Me	(a) 1-Me-2-NHAc/thermolysis or (b) 1-Me-2-NH$_2$ + 1-Me-2-NH$_2$·HCl/250°	996
RN=NR	None	1-CH$_2$Ph-2-NH$_2$/Na/NH$_3$ or 2-NH$_2$/oxidation or	1038
RN=NR	1-Me	(for 1-Me derivative) 2-NH$_3$/Na/oxidation	1040

TABLE 1.72 (Continued)

Compound	Other substituents	Method of preparation	Ref.
RN=NR	None	1-CH₂Ph-2-NH₂ derivative/(i) Na/NH₃, (ii) air	1039
RN=NR	5- and 5'-MeO, Me. 5,6- and 5',6'-Me₂	2-NH₂ derivatives/NaOCl	1035
	1- and 1'-Me, Et, PhCH₂, Ph		
RN=NR¹ [R¹ = 5- + 6 - (1-methyl-2-aminobenzimidazolyl)]	1-Me	1-Me-2-NHNO/conc. H₂SO₄	796
RN=N— 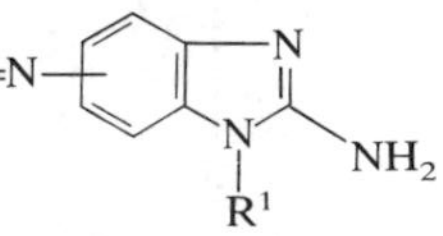 (benzimidazol-2-yl azo structure, 2-NH₂, R¹) [R¹ = Me, Et (2,5'+2,6'-derivatives)]	1-Me, Et	Diazotization of 1-Me-2-NH₂	1119
RNHN=NR	1- and 1'-alkyl, benzyl	1-alkyl-2-NHNO/H₂SO₄ or 1-Me-2-N₂⁺ (BF₄⁻)/1-Me-2-NH₂	801
RNHN=C(R¹)N=NR (R¹ = Me, Ph)	1- and 1'-CH₂Ph	2-NHN=CHR/1-CH₂Ph-2-NHNH₂/[O] in pyridine	1037
RNHN=C(R¹)N=N—2—(1-benzylbenzimidazole) (R¹ = CH=CH₂, CH=CHMe)	None	2-NHN=CHR¹/1-CH₂Ph/2-N₂⁺	826

[a] Note. in this Table, R denotes the benzimidazol-2-yl group.

[b] See also Table 1.2 for other examples of bisbenzimidazoles.

[c] D = appropriate o-arylenediamine derivative.

216

TABLE 1.73. MISCELLANEOUS BIS-, TRIS-, AND TETRAKISBENZIMIDAZOLES

Benzimidazole type	Other substituents	Method of preparation	Ref.
R_2CH_2 (R = 1—benzimidazolyl)	None	Benzimidazole/CH_2Br_2/KOH	596, 1125
R_2Z [R = 1-benzimidazolyl, Z = $(CH_2)_n$ (n = 3–5)]	2- and 2'-NH_2	Chichibabin reaction	866
R_2Z [R = 1-benzimidazolyl, Z = $(CH_2)_n$ (n = 1–5) CH_2OCH_2, p-$CH_2C_6H_4CH_2$]	None	Benzimidazole/ZX_2 (X = hal)	866
R_2CO (R = 1-benzimidazolyl)	None	Benzimidazole/$COCl_2$	1093
[o-$RCOC_6H_4]_2NH$ (R = 1-benzimidazolyl)	None	Benzimidazole derivative/ 2,2'-iminobenzoyl chloride	762
[$RCONH(CH_2)_n]_2$ (n > 2; R = 1-benzimidazolyl)	5,6- and 5',6'-Me_2 2- (and 2'-)-$NHCO_2Me$-5-(and 5'-)-SPh	Acylation by $OCN(CH_2)_n$ NCO or $XCONH(CH_2)_n$ NHCOX (X = hal)	736
$(RCONH)_2Z$ (Z = $(CH_2)_6$, -arylene-CH_2-arylene, R = 1-benzimidazolyl)	2- and 2'-$NHCO_2Me$	Acylation of 2-$NHCO_2Me$ derivatives with $Z(NCO)_2$	738
[$RCONH]_2Z$ (R = 1-benzimidazolyl; Z 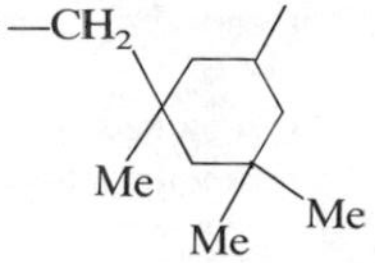 = —CH_2—)	2- and 2'-$NHCO_2Me$	2-$NHCO_2Me$/$(ONC)_2Z$	739
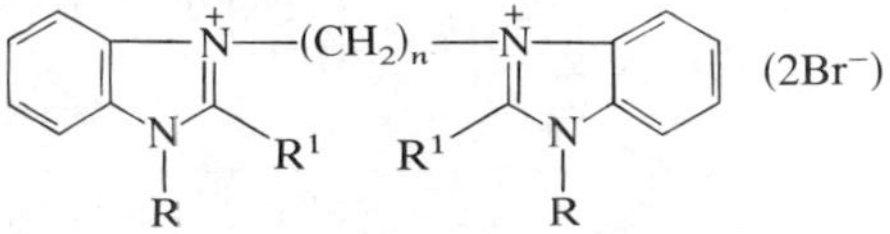 $CH_2CMe_2CH_2CHMe(CH_2)_2$, $(CH_2)_6$, $CH_2(C_6H_4$-p$)_2$, arylene) (R = n-octyl, n-decyl; R^1 = H,NH_2; n = 4,6)	None	Benzimidazole derivative/$(CH_2)_nBr_2$	1161

217

TABLE 1.73   (Continued)

Benzimidazole type	Other substituents	Method of preparation	Ref.
RN=NR (2X⁻) (R = 1-benzimidazolyl; X = Cl, Br, ClO₄)	2- and 2'-Me, aryl	$1\text{-Me-2-R}^1\text{-3-NH}_2(\text{Br}^-)/\text{Br}_2/\text{H}_2\text{O}$ (R^1 = H, Me, aryl)	368
5,5'-Bisbenzimidazole	3- and 3'-Me	(bis)Dᵃ/carboxylic acid	1088
5,5'-Bisbenzimidazole	2- and 2'-R R = C_2F_5, C_3F_7 1 (and 1')-Me, Ph 2 (and 2')-$(CF_2)_2CF_3$	Bis Dᵃ/ester	145
5,5'-Bisbenzimidazole	2- and 2'-Ph	Bis Dᵃ/$PhCO_2Ph$	148
5,5'-Bisbenzimidazole	2- and 2'-Ph	Oxidative cyclization of N,N'-dibenzyl-idene(3,3'-diaminobenzidene)	400
5,5'-Bisbenzimidazole	2- and 2'-Ph, $(CF_2)_2CF_3$	Tetraminobiphenyl/RCO_2Ph (R = Ph, $(CF_2)_2CF_3$)	145, 148
R_3X (X = P, PO, PS; R = 1-benzimidazolyl)	None	Benzimidazole/PCl_3 or $POCl_3$ or $PSCl_3$	632
R_2CHCHR_2 (R = 2-benzimidazolyl)	Various substituents in aryl rings, e.g., alkyl, MeO, etc.	Dᵃ/tetraester	156
$R_2C=CR_2$ (R = 2-benzimidazolyl)	Various substituents in aryl rings, e.g., alkyl, MeO, etc.	$R_2CHCHR_2/K_2CO_3/PhNO_2$	156
1,2,4,5-$R_4C_6H_2$ (R = 2-benzimidazolyl)	None	Dᵃ/pyromellitic dianhydride	164, 165
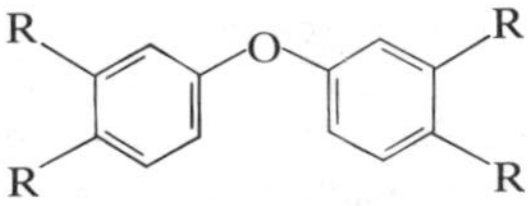 [structure: diphenyl ether with R substituents, R = 2-benzimidazolyl]	None	Dᵃ/diphenyl ether 3,3'-4',4'-tetra-carboxylic acid	165
o-R—C_6H_4—R—R^1—R—C_6H_4—R-o (R^1 = bond, $CH_2(5,5'$-link); R = 2-benzimidazolyl)	None	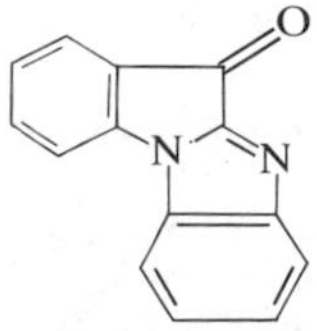 [structure]/appropriate tetramine	165

ᵃ D = appropriate o-arylenediamine derivative.

218

TABLE 1.74. BENZIMIDAZOLIUM COMPOUNDS

Imidazole ring substituents

1	3	Anion	Other substituents	Ref.
Me	CH$_2$OR (R = alkyl)	Cl$^-$	2-NH$_2$	1135
Me	[2,4-(O$_2$N)$_2$C$_6$H$_3$]	Cl$^-$	None 2-Me 5-OMe	625
Me	NH$_2$	$\bar{O}$SO$_2$C$_6$H$_2$Me$_3$-s	None 2-Me	763
Me	H	Tos$^-$	2-CH=CH$_2$ Me	761
Me	H	2ClO$_4^-$	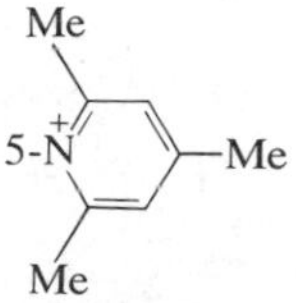	1145
Me	Me	I$^-$	2-CH(OH)Ph	982
Me	Me	I$^-$	2-CH=CHFc (Fc = ferrocenyl)	1136
Me	Me	I$^-$	2-CH(R)OPO(OEt)$_2$ (R = Me, Ph)	982
Me	(CH$_2$)$_2$OH	Cl$^-$	2-Me-5-SO$_2$NEt$_2$-6- (and 7-)NH$_2$	1133
Me	CH=CH$_2$	I$^-$	None 2-alkyl	1159
R (R = Me, Et)	CH$_2$C≡CH	Br$^-$	2-CH$_2$R^1 (R^1 = H, Me, Ph)	981
Me, Et	CH$_2$COR (R = OMe, Ar)	Br$^-$	None	1073
Me	COPh	Cl$^-$	2-NH$_2$	1131
Me, CH$_2$Ph	CH$_2$COR (R = Ph, O-alkyl)	Br$^-$	None	1157
Me, Ph	NH$_2$	$\bar{O}$SO$_2$C$_6$H$_2$Me$_3$-s	None	765
Me, Et	NH$_2$	$\bar{O}$SO$_2$C$_6$H$_2$Me$_3$-s	None 2-Me	1003, 1075
Me, Et	N=CHPh	$\bar{O}$SO$_2$H$_2$Me$_3$-s	None 2-Me	1075
Et	Me	BF$_4^-$	2-R (R = NHPh, OPh, O—Np^a, SPh)	970
Et	CF=CClF CF$_2$CHClF CF=CFCF$_3$ CF$_2$CHFCF$_3$	ClO$_4^-$	2-CH=CHC$_6$H$_4$NMe$_2$-p	499
CH$_2$OAr	CH$_2$OAr	Cr$^-$	2-NH$_2$	1134
CH$_2$Ph	CH$_2$C$_6$H$_4$NO$_2$-m	Cl$^-$	5-Br	1132
1-CH$_2$Ar	CH$_2$CO$_2$Et	Cl$^-$, Br$^-$	2-Me-4-R (R = H, Cl, Me, MeO)	1160

TABLE 1.74 (*Continued*)

Imidazole ring substituents

1	3	Anion	Other substituents	Ref.
aralkyl	$CH_2C_6H_4NO_2$-*m*	Cl^-	None	1158
$CH_2C\!\!=\!\!CH$	$CH_2C\!\!=\!\!CH$	Br^-	2-NH$_2$	603
Ph	Ph	I^-	2-(1-piperidyl), $N\!\!=\!\!C(NMe_2)_2$	1106
Ph	Ph	Cl^-	2-CH$_2$COMe	1106
Tos	Tos	BF_4^-	None	1097
COPh	COPh	BF_4^-	None	1097

a Np = naphthyl

TABLE 1.75. BENZIMIDAZOLIUM BETAINES*a*

1- and 3-substituents

1	3	Other substituents	Method of preparation	R
Me	$\bar{C}HCOR$	None	1-Me-3-CH$_2$COR(Br$^-$)/ K$_2$CO$_3$/DMF	8
Me, Et	$\bar{N}COR$ (R = Me, Ph, OEt)	None 2-Me	1-Alkyl-3-NH$_2$ benzimidazolium salts/acylating agent	
Me	$C(CSNHR^1)\!\!=\!\!C(O^-)R$ (R = aryl: R^1 = alkyl, aryl, etc.)	None	1-Me-3-$\bar{C}$HCOR/R^{1}NCS	8
Et	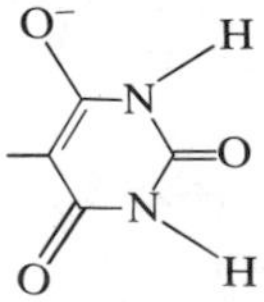	2-Me 2-Me-5-Br	*o*-H$_2$NC$_6$H$_4$NEt$_2$/alloxan	
Me	Me	2-CH=CH—C$_6$H$_4$O$^-$ NO$_2$ substituent(s) in both Ar rings	1,2,3-Trimethylbenzimidazolium derivative/salicylaldehyde derivative/pyridine	9
R (R = alkyl, aryl)	R (R = alkyl, aryl)	2-CHROPO(OMe)(O$^-$) (R = Me, Ph)	1,3-R$_2$(I$^-$)/(MeO)$_2$POCOR	9

a For examples of other betaine derivatives, see structure **1.320** and Refs. 886 and 887.

TABLE 1.76. ISOTOPICALLY-LABELED DERIVATIVES

ompound abeled	Position and type of label	% enrichment or specific activity	Labeled starting material	Ref.
enzimidazole	^{14}C-2	35 μCi/mmole	$H^{14}CO_2H$	36
-NHCO₂Me[a]	^{14}C-2	53.4 mCi/g	$[BaN(CO_2Me)^{14}CN]Cl$	254
-NHCO₂Me[b]	$^{14}CH_3$	6.6 mCi/g	$[BaN(CO_2{}^{14}CH_3)CN]Cl$	254
-NHCO₂Me-5-CO- cyclopropyl	^{14}C-2	21.8 μCi/mg	Not quoted	1082
-(4-Thiazolyl)-5-NHCO₂-i-Pr	Thiazole-^{13}C-4	57%	$H_2N^{13}CH_2CO_2H$	1081
-(4-Thiazolyl)-5-NHCO₂-i-Pr	Thiazole-^{15}N-3	33%	$H_2{}^{15}NCH_2CO_2H$	1081
-(4-Thiazolyl)-5-NHCO₂-i-Pr	Benzimidazole-^{14}C-2	2.81 mCi/mmole	$H_2NCH_2{}^{14}CO_2H$	1081
-(4-Thiazolyl)-5-NHCO₂-i-Pr	All C atoms of carbo- cyclic ring	9.00 mCi/mmole 114.16 mCi/mmole	$H^{14}C{\equiv}{}^{14}CH$	1081
-(4-Thiazolyl)-5-NHCO₂-i-Pr	Carbamoyl-^{14}CO	1.93 mCi/mmole	$^{14}COCl_2$	1081

Treatment of this compound with BuNCO gives the 1-butylcarbamoyl derivative (sp. act. 35 mCi/g).
Treatment of this compound with BuNCO gives the 1-butylcarbamoyl derivative (sp. act. 4.27 mCi/g).

Abbreviations

Am	ammonium salt
At	acetate
Au	chloroaurate
Cl	chloride
CPt	chloroplatinate
F	flavianate
HBr	hydrobromide
HCl	hydrochloride
HI	hydroiodide
HOx	hydrogen oxalate
I	iodide
It	iodate
Ni	nitrate
Ox	oxalate
PCl	perchlorate
Pi	picrate
PI	periodate
PS	picrylsulfonate
S	styphnate
Su	sulfate

Appendix B. Benzimidazole Derivatives Cited During the Period 1919–1950
(Compiled by K. Hofmann and Reprinted from Reference 1)

TABLE 1.77. ALKYL AND ARYL BENZIMIDAZOLES

Compound	M.p.(°C.)	Ref.
Benzimidazole	170–172	1178–1181
1-(p-Acetamidophenylsulfonyl) deriv	197–200	1182
1-Acetyl deriv	113–114,	1183, 1184
	Pi 158, It 205	
1-(p-Aminophenylsulfonyl) deriv	285–288	1182
1-Benzoyl deriv	93, Pi 215	1183
1-Butyryl deriv	45, Pi 145	1183, 1185
1-(1-Naphthylcarbamyl) deriv	141.5–142	1186
1-(Phenylcarbamyl) deriv	153–154	1186
1-Propionyl deriv	125, Pi 228	1183–1185
4,5,6,7-Tetrahydro deriv	149–150,	1187
	Pi 189–190	
1-Benzoyl deriv	131–132	1187
2-Amylbenzimidazole	163–163.5,	1188–1191
	Pi 282	1192–1193
2-Benzhydryl-	218–219,	1194
	HCl 233–238	
1-Benzyl-	115–115.5,	1195,1196
	Pi 161–163	
2-Benzyl-	189, Pi 214–215,	1197–1199, 1194,
	HCl 92–94,	1200, 1201
	phthalate 177	
1-Benzyl-2-methyl-		
4,5,6,7-Tetrahydro deriv	76, Pi 143	1202
2-Benzyl-1-methyl-	Pi 249	1198
2-Benzyl-5(or 6)-methyl-	150–151,	1203
	HCl 61–63	
1-Benzyl-2-phenyl-	134	1204, 1205
2-Butyl-	155–155.5	1189–1192
2-Cyclohexyl-	280	1206
4,5,6,7-Tetrahydro deriv	267	1206
2-Decyl-	114–114.5	1191
1,2-Diethyl-	Pi 207, HCl	1207
	168.5–169.5	
1,2-Dimethyl-	112, Pi 243,	1208–1210, 1207,
	hydrate 66–69	1198, 1211, 1212
4,5,6,7-Tetrahydro deriv	Pi 192	1206
1,5-Dimethyl-	96, Pi 248–250,	1213, 1208, 1211
	tartrate	
	185.2–185.9	
1,6-Dimethyl-	74–75	1208, 1211
1,7-Dimethyl-	68–70.5	1211
2,4(or 2,7)-Dimethyl-	168–169	1211
2,5(or 2,6)-Dimethyl-	203–204	1211, 1208, 1214,
		1215
4,5,6,7-Tetrahydro deriv	184	1187

TABLE 1.77   (*Continued*)

Compound	M.p.(°C)	Ref.
4,5(or 6,7)-Dimethyl-	196–197	1211
5,6-Dimethyl-	204–205	1211, 1216
2-(2,6-Dimethyl-1,5-heptadienyl)-	102	1189, 1190
2,4(or 2,7)-Dimethyl-7(or 4)-isopropyl-	179.5–179.9	1217
2-(1,2-Diphenylvinyl)-	269	1218
2-Dodecyl-	109–109.5	1191
1-Dodecyl-2-methyl-	—	1219
1-Ethyl-	160–162, *Pi* 218–219	1195, 1207
2-Ethyl-	177, *Pi* 137, *phthalate* 197	1220, 1212, 1193, 1201, 1181, 1189, 1191, 1190, 1192
4,5,6,7-Tetrahydro deriv	202, *Pi* 145–146	1206, 1187
1-Ethyl-2-methyl-	*Pi* 236–237	1207
2-Ethyl-1-methyl-	54.5–55.5, *Pi* 235–236	1207
2-Ethyl-5(or 6)-methyl		
4,5,6,7-Tetrahydro deriv	204–205	1187
1-Ethyl-2-phenyl-	88–88.5	1207
2-Ethyl-1-propyl-	*Pi* 212–212.5, *HI* 158–159	1207
5(or 6)-Ethyl-2,4,6,7(or 2,4,5,7)-tetramethyl benzimidazole	205.5	1221
2-Hendecyl-	107.5	1188, 1191
2-Heptadecyl-	93–94.5	1222, 1191, 1188
2-Heptyl-	144.5–145	1191, 1188
2-Hexadecyl-	93.5–94.5	1191
1,2,4,5,6,7-Hexamethyl-	165, *Su* 258–261, *HI* >350	1223
2-Hexyl-	136–138	1189, 1191, 1190
4,5,6,7-Tetrahydro deriv	157–158, *Pi* 142–144	1189, 1191, 1190
	157–158, *Pi* 142–144	1187
2-Isobutyl-	186–187	1190, 1189
4,5,6,7-Tetrahydro deriv	206	1187
2-Isopropyl-	228, *Pi* 136	1190, 1189, 1188, 1193
4,5,6,7-Tetrahydro deriv	240–241, *Pi* 90–93	1187
2-Isopropyl-1-methyl-	*Pi* 225–226	1207
1-Methyl-	66, *Pi* 246–247, *HCl* 226–227	1209, 1212, 1207, 1211, 1224, 1208, 1225, 1226
2-Methyl-	178.5–179, *Pi* 214, *phthalate* 190, *HCl* >300	1227–1229, 1180, 1181, 1191, 1178, 1189, 1190, 1225, 1230, 1210, 1205, 1231, 1192, 1232, 1200, 1194, 1233, 1198, 1201, 1212, 1193, 1208, 1211, 1196
1-(Phenylcarbamyl) deriv	128–129	1186
4,5,6,7-Tetrahydro deriv	224, *Pi* 185–186	1206, 1202, 1187

TABLE 1.77 (Continued)

Compound	M.p.(°C)	Ref.
4(or 7)-Methyl-	145, *HCl* >300	1234, 1211
5(or 6)-Methyl-	114	1214, 1208, 1196, 1211
4,5,6,7-tetrahydro deriv	117–118	1187
1-Methyl-2-phenyl-	98	1207
2-Methyl-1-phenyl-	70	1212
5(or 6)-Methyl-2-phenyl-	249–250, *HCl* 274–275	1235, 1215, 1203
5(or 6)-Methyl-2-(2-phenylethyl)-	134–136, *HCl* 82–83.5	1203
2-Methyl-1-propyl-	*Pi* 218–219	1207
5(or 6)-Methyl-2-propyl-		
4,5,6,7-Tetrahydro deriv	183–184	1187
2-Methyl-4,5,6,7-tetraethyl-	241–242	1236
2-(2-Methylbutyl)-	158–159	1188
2-(*p*-Methylphenyl)-	266–269	1192, 1237
2-(1-Naphthyl)-	271	1194, 1237
2-Nonyl-	127–127.5	1191, 1188
2-Octyl-	139.5–140.5	1191, 1238
2-Pentadecyl-	96.5-97	1191, 1188
2,4,5,6,7-Pentamethyl-	264	1223
1-Phenyl-	98	1212
2-Phenyl-	290, *HCl* 306	1181, 1239, 1204, 1189, 1190, 1237, 1205, 1192, 1194, 1240, 1235, 1196
4,5,6,7-Tetrahydro deriv	290–291, *Pi* 258, *HCl* 249–251	1187
2-(2-Phenylethyl)-	189–190, *HCl* 268–270	1199, 1203
2-(9-Phenylheptadecyl)benzimidazole	—	1241
2-(2-Phenylvinyl)-	201–202	1190, 1189
2-(2-Phenylvinyl)-4,5,6,7-tetramethyl-	278–279	1223
1-*n*-Propyl-	*Pi* 204–206	1195, 1207
2-*n*-Propyl-	157–159, *Pi* 124	1188–1193
4,5,6,7-Tetrahydro deriv	185–186, *Pi* 115–116	1187
2-Tetradecyl-	98.5–99.5	1191
1,2,4,5–Tetramethyl-	144–145	1211
1,2,5,6-Tetramethyl-	164	1211
2,4,5,7(or 2,4,6,7)-Tetramethyl-	233	1242
2-Tridecyl-	105–105.5	1191
1,2,5-Trimethyl-	142	1215, 1208, 1211
1,2,6-Trimethyl-	122–123	1215, 1208, 1211
1,2,7-trimethyl-	146–147	1211
1,4,5-Trimethyl-	95–96	1211
1,5,6-Trimethyl-	140–143	1211
2,4,5(or 2,6,7)-Trimethyl-	188–190	1211
2,5,6-Trimethyl-	233–234	1211
1-vinyl-	*Pi* 194–195, *CPt* 240–245	1243
2-Vinyl-(polymer)	290	1244, 1245

TABLE 1.78.   ALKYL AND ARYL BENZIMIDAZOLIUM SALTS

Compound	M.p. (°C)	Ref.
1-Benzyl-3-ethylbenzimidazolium, iodide	173.5–174.5	1195
1-Benzyl-3-methyl-, iodide	158	1195
1,3-Diethyl-, iodide	225–227,	1195
	Pi 254–257	
1,3-Diethyl-2-methyl-, iodide	200–202	1246
1,3-Dimethyl-, sulfate	128	1212
1,3-Dimethyl-2-ethyl-, iodide	173–173.5	1220
1,3-Dipropyl-, iodide	202–203	1195
1-Ethyl-3-methyl-, iodide	192–193	1195
1-Ethyl-3-propyl-, iodide	171.5–172	1195
1,2,3-Trimethyl-, chloride	226–231	1247
1,2,3-Trimethyl-, iodide	256–259	1247

TABLE 1.79.   MONO AND POLYHYDROXYALKYL AND HYDROXYARYL
BENZIMIDAZOLES, THEIR ETHERS, AND HALOGEN
DERIVATIVES

Compound	M.p. (°C)	Ref.

(a) Hydroxy-, Hydroxyalkyl-, and Hydroxyarylbenzimidazoles, Their Ethers and Sulfur Analogs

Compound	M.p. (°C)	Ref.
4,7-Dihydroxy-2-methyl-1,3-diphenylbenz- imidazolium acetate		
O,O′-Diacetyl deriv	135–136,	1248
	Pi 207,	
	HCl 259	
5(or 6)-Ethoxy-2-benzylbenzimidazole	162–163, HCl 189	1203
6-Ethoxy-5-methyl-1-(p-methylphenyl)-	102.5, Pi 228	1249
1-(4-Ethoxy-3-methylphenyl)-6-methyl-	Pi 186–187	1249
5(or 6)-Ethoxy-2-phenyl-	51–53, HCl 249–250	1203
5(or 6)-Ethoxy-2-(2-phenylethyl)-	151–152,	1203
	HCl 216–217	
2-(2-Ethoxyethyl)-	156.5–157	1245
2-(1-Ethoxyisopropyl)-	203.7–204.4	1250
2-(Ethoxymethyl)-	154.5–155	1196, 1250
1-(4-Hydroxy-3-methoxybenzyl)-2-(4-hydroxy- 3-methoxyphenyl)-	224	1204
2-(4-Hydroxy-3-methoxyphenyl)-	221–222	1189, 1190
1-Hydroxy-2-methyl-	231	1251
5(or 6)-Hydroxy-2-methylbenzimidazole	187.5–188.5	1252
5-Hydroxy-2-methyl-1-phenyl-	243	1253
Acetyl deriv	141	1253
5-Hydroxyl-1-(p-methylphenyl)-	239	1253
Acetyl deriv	133	1253
1-(4-Hydroxy-3-methylphenyl)-6-methyl-	196–197	1249
2-(1-Hydroxy-2-naphthyl)-	>265	1254
5-Hydroxy-1-phenyl-	244	1253
Acetyl deriv	88	1253

TABLE 1.79 (*Continued*)

Compound	M.p.(°C)	Ref.
2-(α-Hydroxybenzyl)	202–203, *Pi* 209	1181, 1193
1-(2-Hydroxyethyl)-	107–108, *Pi* 204, *HCl* 183–184, *CPt* 189–190	1243
DL-2-(1-Hydroxyethyl)-	179–181	1181, 1255, 1250, 1256
Acetyl deriv	152–153	1245
L-2-(1-Hydroxyethyl)-	175–177, *Pi* 131, *HCl* 213–215	1255, 1193
D-2-(1-Hydroxyethyl)-	175–177, *HCl* 213–215	1255
2-(2-Hydroxyethyl)	153.5–154.5	1245
2-(1-Hydroxyethyl)-5-ethoxy-	170–171	1181
2-(1-Hydroxyethyl)-1-methyl-	80	1212, 1250
2-(1-Hydroxyisopropyl)-	227.5–228	1245, 1250
1-(Hydroxymethyl)-	141–143	1257
2-(Hydroxymethyl)-	171–172, *Pi* 214	1193, 1181, 1258, 1250
2-(Mercaptomethyl)-	158	1199
2-(Hydroxymethyl)-1-methyl-	105	1199
2-[*o*-Hydroxymethyl)phenyl]-	237–239	1259
2-(*o*-Hydroxyphenyl)-	233	1260
2-(1-Hydroxypropyl)-	220–221	1250
5(or 6)-Methoxy-	123, *Pi* 191	1261
5(or 6)-Methoxy-2-benzyl-	46, *HCl* 176–178	1203
5(or 6)-Methoxy-2-methyl-	141.5–142.5, *Pi* 191.5–192.5	1262
5-Methoxy-2-methyl-1-phenyl-	103–105	1263
5(or 6)-Methoxy-6(or 5)-phenoxy-2-methyl-	149	1264
5(or 6)-Methoxy-2-phenyl-	142, *HCl* 255–257	1203
5(or 6)-Methoxy-2-(2-phenylethyl)-	128–130, *HCl* 239–241	1203, 1203
2-(Methoxymethyl)-	137	1196, 1199
2-(*p*-Methoxyphenyl)-	228–230	1189, 1190
4,5,6,7-Tetrahydro deriv	236–238, *Pi* 211–212	1187
2-(*p*-Methoxyphenyl)-1-ethyl-	106–106.5	1207
2-(*p*-Methoxyphenyl)-1-methyl-	118	1207
2-(*p*-Methoxyphenyl)-1-propyl-	67.5–68	1207
1-(3,4-Methylenedioxybenzyl)-2-(3,4-methylenedioxyphenyl)-	175	1204
2-(3,4-Methylenedioxyphenyl)-	249	1189, 1190
2-(Phenoxymethyl)-	162	1196, 1199

(*b*) *Hydroxy-, Hydroxyalkyl-, and Hydroxyarylbenzimidazoles Containing Halogen*

4,5(or 6,7)-Dihydroxy-6,7(or 4,5)-dichloro-benzimidazole	196	1253
Diacetyl deriv	213	1253

TABLE 1.79 (Continued)

Compound	M.p.(°C)	Ref.
5,6-Dihydroxy-4,7-dichloro-1,2-dimethyl-	>300, HCl >300	1253
Diacetyl deriv	200	1253
4,5(or 6,7)-Dihydroxy-6,7(or 4,5)- dichloro-2-methyl-	HCl >250	1253
Diacetyl deriv	148	1253
5,6-Dihydroxy-4,7-dichloro-2-methyl-	>300, HCl >300	1253
Diacetyl deriv	267	1253
4,5-Dihydroxy-6,7-dichloro-2-methyl-1- phenyl-	>160	1265
5-Hydroxy-4-bromo-2-methyl-1-phenyl-	266	1253
6-Hydroxy-1-(p-bromophenyl)-	295	1249
Ethyl ether	—	1249
5-Hydroxy-4-chloro-1-(o-chlorophenyl)-2 2-methyl-	250	1253
5-Hydroxy-4-chloro-1-(p-chlorophenyl)- 2-methyl-	270	1253
Acetyl deriv	186	1253
5-Hydroxy-4-chloro-2-methyl-1-phenyl-	287	1253
Acetyl deriv	161	1253
5-Hydroxy-4-chloro-1-(p-methylphenyl)-	217	1253
5-Hydroxy-4-chloro-1-phenyl-	220	1253
5-Hydroxy-1-(o-chlorophenyl)-2-methyl-	245	1253
5-Hydroxy-1-(p-chlorophenyl)-2-methyl-	228	1253
5(or 6)-Hydroxy-4,6(or 5,7)-dibromo- 2-methyl-	—	1266
5-Hydroxy-4,6-dibromo-2-methyl-1-phenyl-	190	1253
5(or 6)-Hydroxy-4,6(or 5,7)-dichloro-	>250	1253
Acetyl deriv	212	1253
5(or 6)-Hydroxy-4,6(or 5,7)-dichloro- 2-methyl-	HCl 250	1265
5-Hydroxy-4,6-dichloro-2-methyl-1-phenyl-	216, HCl 283	1253
Acetyl deriv	183	1253
5(or 6)-Hydroxy-4,6(or 5,7)-dichloro-2-phenyl-	240	1265
2-(Hydroxymethyl)-5(or 6)-chloro-	206–207	1267

(c) Polyhydroxyalkylbenzimidazoles

2-(D-erythro-Trihydroxypropyl)benzimidazole	177–178	1268
2-(L-erythro-Trihydroxypropyl)-	177–178	1268
1-L-Arabityl-2,6-dimethyl-	235–236	1269
2-(L-arabino-Tetrahydroxybutyl)-	235, Pi 158, HCl 230	1270, 1271
Tetraacetyl deriv	141–142	1271
2-(1,4-Anhydro-D-arabino-tetrahydroxybutyl)-	208	1272
Isopropylidene deriv	195–196	1272
2-(D-lyxo-Tetrahydroxybutyl)-	189, Pi 95–99, HCl 191	1273, 1270
2-(L-lyxo-Tetrahydroxybutyl)-	189	1273

TABLE 1.79 (Continued)

Compound	M.p.(°C)	Ref.
2-(1,4-Anhydro-D-*lyxo*-tetrahydroxybutyl)-	200–204, Pi 132–138	1272
2-(D-*ribo*-Tetrahydroxybutyl)-	191, Pi 184–186, HCl 201–203	1274–1277
2-(1,4-Anhydro-D-*ribo*-tetrahydroxybutyl)-	82–83, Pi 120–125	1272
1-D-Xylityl-2,6-dimethylbenzimidazole	179–180	1269
2-(D-*xylo*-Tetrahydroxybutyl)-	141–143, HCl 181–182	1272
2-(1,4-Anhydro-D-*xylo*-tetrahydroxybutyl)-	221–223	1272
2-(1,4-Anhydro-D-*xylo*-tetrahydroxybutyl)- 1-benzyl-	215–217	1272
2-(D-*altro*-Pentahydroxypentyl)-	198	1277
2-(D-*galacio*-Pentahydroxypentyl)-	246, Pi 217, HCl 202–204	1277–1279, 1270
Hexaacetyl deriv	179	1278
2-(L-*galacto*-Pentahydroxypentyl)	250	1274
2-(DL-*galacto*-Pentahydroxypentyl)-	233	1280
1-D-Sorbityl-2,5-dimethyl-	226	1269
2-(D-*gluco*-Pentahydroxypentyl)-	215, Pi 203, HCl 180	1277, 1270, 1279
2-(L-*gluco*-Pentahydroxypentyl)-	215	1268
2-(D-*gluco*-Pentahydroxypentyl)-1-benzyl-	188	1279
2-(D-*gulo*-Pentahydroxypentyl)-	201	1277
2-(D-*ido*-pentahydroxypentyl)-	154–156	1277
2-(D-*manno*-Pentahydroxypentyl)-	227, Pi 205	1270, 1277
2-(D-*talo*-Pentahydroxypentyl)-	190–191	1277
2-(D-*digitoxo*-Trihydroxypentyl)-	207–209, Pi 124–127	1275
2-(L-*galactomethylo*-Tetrahydroxypentyl)-	Pi 248–249, Pi 189–191, HCl 224–225	1275
2-(D-*glucomethylo*-Tetrahydroxypentyl)-	190	1274
2-(L-*mannomethylo*-Tetrahydroxypentyl)-	210, Pi 168, HCl 173–175	1270, 1277
2-(D-*gala*-L-*manno*-Hexahydroxyhexyl)-	218	1277
2-(D-*gluco*-D-*gulo*-Hexahydroxyhexyl)-	215	1277
2-(D-*gluco*-D-*ido*-Hexahydroxyhexyl)-	192	1277
2-(D-*manno*-D-*gala*-Hexahydroxyhexyl)-	241	1277
2-(D-*gala*-L-*galo*-Heptahydroxyheptyl)-	234–235	1281
2-(D-*gluco*-L-*gala*-Heptahydroxyheptyl)-	246–247	1274
2-(D-*gluco*-L-*talo*-Heptahydroxyheptyl)-	191–192	1281
2,2'-(1,2,3,4-Teteahydroxytetramethylene)- dibenzimidazole		
D-(*galactomuco* Dibenzimidazole	298, Pi 250, HCl 318	1282
D-*mannomuco* Saccharo-	250, Pi 241, HCl 256–257	1282
D-*saccharo*-	238, Pi 211, HCl 257–258	1282

TABLE 1.80   BENZIMIDAZOLECARBOXALDEHYDES, KETONES, AND
QUINONES

Compound	M.p. (°C)	Ref.
2-Benzimidazolecarboxaldehyde	235	1272
2,4-Dinitrophenylhydrazone	309–311	1272
Oxime	213–215	1272
3-(2-Benzimidazolyl)-2,4-pentanedione	138–139,	1227
	HCl 243–244	
3-(2-Benzimidazolyl)-4-(phenylimino)-2-pentanone	315–316	1283
2-Benzimidazolylmethyl methyl ketone	148	1284
5-(2-Methylbenzimidazolyl)methyl ketone	190–191	1285
2,4-Dinitrophenylhydrazone	336	1285
4,7-Dioxo-6(or 5)-chloro-5(or 6)-hydroxybenzimidazole	250	1286
4,7-Dioxo-6(or 5)-chloro-5(or 6)-hydroxy-2-methyl-	250	1286
4,5(or 6,7)-Dioxo-6,7(or 4,5)-dichloro-	>280, Ni 162	1286
4,5-Dioxo-6,7-dichloro-1,2-dimethyl-	>300	1253
4,5(or 6,7)-Dioxo-6,7(or 4,5)-dichloro-2-methyl-	280	1286
4,5-Dioxo-6,7-dichloro-2-methyl-1-phenyl-	229	1286
4,5(or 6,7)-Dioxo-6,7(or 4,5)-dichloro-2-phenyl-	305	1286
5,6-Dioxo-4,7-dihydro-1,2-dimethyl-4,4,7,7-tetrachloro-	HCl 200	1253
5,6-Dioxo-4,7-dihydro-2-methyl-4,4,7,7-tetrachloro-	HCl 300	1253
4(or 7)-Oxo-6(or 5)-chloro-5(or 6)-hydroxy-2-methyl-		
7(or 4)-(phenylimino)-	300	1286
4(or 7)-Oxo-6(or 5)-chloro-5(or 6)-hydroxy-7(or 4)-		
(phenylimino)-	280	1286
5(or 6)-Oxo-4,4,6,7(or 4,4,5,7)-tetrachloro-2-phenyl-	HCl 199	1286

TABLE 1.81.   HALOGENO, HALOGENOALKYL, AND HALOGENOARYL
BENZIMIDAZOLES

Compound	M.P. (°C)	Ref.
5(or 6)-Bromobenzimidazole	137	1208
5-Bromo-1,2-dimethyl-	141	1208
6-Bromo-1,2-dimethyl-	180	1208
5(or 6)-Bromo-2-methyl-	215	1208
2-(2-Bromoethyl)-	227–229	1245
5(or 6)-Chloro-	124–126, *Pi* 215–216	1267, 1179
5(or 6)-Chloro-2-(chloromethyl)-	140, *Pi* 195–196, *HCl* 213–214	1287, 1179
5-Chloro-1,6-dimethyl-	154	1288
5(or 6)-Chloro-6(or 5)-methyl-	191	1288, 1196
5-Chloro-6-methyl-1-phenyl-	100	1288
5(or 6)-Chloro-2-phenyl-	210, *HCl* 290–291	1289, 1203
5(or 6)-Chloro-2-propenyl-	184–187	1290
4(or 7)-Chloro-2,5,6,7(or 2,4,5,6)-tetramethyl-	288.5	1242
5(or 6)-Chloro-2,4,6,7(or 2,4,5,7)-tetramethyl-	250–251	1242
2-(1-Chloroethyl)-	134.7–135.4	1250, 1210, 1245
2-(2-Chloroethyl)-	88.5–89, *HCl* 180	1287, 1245
2-(1-Chloroethyl)-1-methyl-	64–65	1250
2-(1-Chloroisopropyl)-	135.5–136.6	1250, 1245
2-(Chloromethyl)-	165	1199, 1291, 1250, 1287
2-(Chloromethyl)-1-methyl-	94.5–95.5	1199, 1250
2-(o-Chlorophenyl)-5(or 6)-methyl-	160, *Pi* 189	1235
2-(1-Chloropropyl)-	144.5–145.5	1250
1-(2-Iodoethyl)-	*Pi* 196–197, *HI* 174–175	1243
2-(Iodomethyl)-	137–139	1250

230

TABLE 1.82. NITRO, NITROALKYL, AND NITROARYL BENZIMIDAZOLES
INCLUDING THOSE CONTAINING ADDITIONAL FUNCTIONAL
GROUPS

Compound	M.p. (°C)	Ref.
4,5-Dinitro-1,2-dimethylbenzimidazole	170	1253
5,6-Dinitro-1,2-dimethyl-	232	1253
2-[2-(2,4-Dinitrophenyl)vinyl]-	215	1292
4(or 7)-Nitro-	238–239, Pi 184–185	1293
5(or 6)-Nitrobenzimidazole	203, Pi 215	1286, 1293
4(or 7)-Nitro-2,6(or 2,5)-dimethyl-	248	1294
5-Nitro-1,2-dimethyl-	226	1253, 1178, 1208
6-Nitro-1,2-dimethyl-	242	1208
5-Nitro-1,2-diphenyl-	181	1295
5(or 6)-Nitro-2-methyl-	221	1178, 1229, 1208
5-Nitro-1-methyl-2-phenyl-	189, HCl 262	1295
5-Nitro-2-methyl-1-phenyl-	—	1212
5-nitro-1-(p-methylphenyl)-2-phenyl-	177–178, HCl 235	1295
5(or 6)-Nitro-2-(p-nitrophenyl)-	358	1296
5(or 6)-Nitro-1-phenyl-	—	1212
5(or 6)-Nitro-2-phenyl-	206	1286
2-(o-Nitrobenzyl)-	217	1199
2-(p-Nitrobenzyl)-	215	1199
2-(o-Nitrophenyl)-	261, HCl 291	1189, 1190, 1260
2-(m-Nitrophenyl)-	207–208	1189, 1190, 1296
2-(p-Nitrophenyl)-	329–330, HCl 310	1296, 1189, 1190
2-(m-Nitrophenyl)-1-ethyl-	117.5–11.8	1207
2-(p-Nitrophenyl)-1-methyl-	213–214	1207
5(or 6)-(2,4-Dinitroanilino)-2-methyl-	HCl 325–326	1297
α-Form	156	
β-Form	241	
5-(2,4-Dinitroanilino)-1-phenyl-	221	1297
5-Nitro-6-amino-1,2-dimethyl-	309	1253
5(or 6)-Nitro-6(or 5)-amino-2-methyl-	292	1178
Acetyl deriv	235	1178
5-Nitro-1-(3-nitro-4-anilinophenyl)-	199	1253
2-methyl-		
2-(o-Nitrophenyl)-1-(p-methylanilino)-		
5-methyl-	230, Pi 182	1235
Nitroso deriv	130	1235
2-(m-Nitrophenyl)-1-(p-methylanilino)-	224–225, Pi 174	1235
Nitroso deriv	120	1235
2-(p-Nitrophenyl)-1-(p-methylanilino)-	—	1235
5-methyl-		
5-Nitro-2-chloro-1-methyl-	186	1298,[a] 1299[a]
6-Nitro-2-chloro-1-methyl-	202–203	1298,[a] 1299[a]
5-Nitro-1-(o-chlorophenyl)-2-methyl-	140	1253
5-Nitro-1-(p-chlorophenyl)-2-methyl-	210	1253
5-Nitro-1-(3-nitro-4-chlorophenyl)-		
2-methyl-	226	1253
6-Nitro-5-hydroxy-4-chloro-2-methyl-		
1-phenyl-	221	1253
4(or 7)-Nitro-6(or 5)-methoxy-	>250, HCl 100	1300
5(or 6)-Nitro-6(or 5)-methoxy-	Ni 204	1261

[a] See Ref. 390 for a reinvestigation of the structure of these products.

231

TABLE 1.83.   AMINOBENZIMIDAZOLES

Compound	M.p. (°C)	Ref.

(1) *Amino, aminoalkyl-, and aminoarylbenzimidazoles*

2-Aminobenzimidazole	—	1301–1303
N-(*m*-Aminophenylsulfonyl) deriv	325	1303
N-(*p*-Aminophenylsulfonyl) deriv	321–322	1303
1-(*p*-Aminophenylsulfonyl) deriv	211–212	1303, 1304
N-(*m*-Nitrophenylsulfonyl) deriv	299–301	1303
1-(*m*-Nitrophenylsulfonyl) deriv	233–234	1303
N-(Phenylsulfonyl) deriv	354–356	1303
1-Carbamyl deriv	*Pi* 250–260	1301
4(or 7)-Aminobenzimidazole	120–121, *Pi* >250	1293
N-Formyl deriv	209	1293
5(or 6)-Amino-	166.5–167, *Pi* 205, *HCl* >210, *Di HCl* >300	1293, 1305
5-Amino-1-anilino-2-phenyl-	228, *Pi* 204, *Po* 186	1235
Diacetyl deriv	105	1235
4(or 7)-Amino-2,6(or 2,5)-dimethyl-	100	1294
Acetyl deriv	160, *Ni* 125, *HCl* 235	1294
N,1-Diacetyl deriv	169	1294
4(or 7)-(2-Amino-1-naphthylazo)-2,6(or 2,5)-dimethyl-	*HCl* 281	1294
4(or 7)-(2-Hydroxy-1-naphthylazo)-2,6(or 2,5)-dimethyl-	*Su* 210	1294
5-Amino-1,2-dimethyl-	167	1253
Acetyl deriv	239	1253
2-Amino-1-methyl-	—	1306
4(or 7)-Amino-2-methyl-	*HCl* >290	1307
N-acetyl deriv	97–98	1307
5(or 6)-Amino-2-methyl-	—	1229
N-Acetyl deriv	250, *HCl* 325	1229
5-Amino-2-methyl-1-phenyl-	153	1253, 1212
Acetyl deriv	233	1253, 1212
5-Amino-1-(*p*-methylphenyl)-	128	1253
Acetyl deriv	210	1253
5-(*p*-Nitrophenylazimino) deriv	214	1253
1-(2-Amino-4-methylphenyl)-2,5-dimethyl-		
N-Acetyl deriv	217	1308
5-Amino-1-phenyl-	131	1212
2-[*o*-(*o*-Aminoanilinomethyl)-phenyl]-	>300	1259
2-(*p*-Aminobenzyl)-	213, *HCl* 310–312	1194, 1199
1-(2-Aminoethyl)-		
N-Phthaloyl deriv	214–215	1182
2-(2-Aminoethyl)-	*Pi* 193, *Di HCl* 270–272	1309
2-(3-Aminoguanidino)-	195–197	1310
2-(Aminomethyl)-	53, *Di HCl* 263	1199
N-Acetyl deriv	200	1199
N-Benzoyl deriv	231	1199

TABLE 1.83 (*Continued*)

Compound	M.p.(°C)	Ref.
1-(*o*-Aminophenyl)-2-methyl-	220	1308
N-Acetyl derivative		
2-(*o*-Aminophenyl)-5(or 6)-methyl-	190	1235
2-(*m*-Aminophenyl)-5(or 6)-methyl-	236	1235
2-(*p*-Aminophenyl)-5(or 6)-methyl-	124	1235
2-(*p*-Aminophenyl)-1-phenyl-	197	1311
2-[(Amylamino)methyl]-	*Di HCl* 190–191	1291
2-Anilino-	188, *HCl* 151–152	1312
1-Anilino-5-(benzylideneamino)-2-phenyl-	242	1235
Nitroso deriv	110	1235
1-Anilino-2-phenylbenzimidazole	211, *Pi* 199, *Po* 220	1235, 1205
Nitroso deriv	137	1235
2-(Anilinomethyl)-	162	1199
2-(Anilinomethyl)-1-methyl-	118	1199
2-[1-(Benzylamino)ethyl]-	155.5–156,	1210
	HCl 218–220	
2-[(Benzylamino)methyl]-	*Di HCl* 211–213	1291
5-(Benzylideneamino)-1-(*p*-methylphenyl)-	142	1253
2-[1-(Butylamino)ethyl]-	120.3 121.7,	1210
	HCl 171.8–172.7	
2-[(Butylamino)methyl]-	*Di HCl* 203–204	1291
2-[(Cyclohexylamino)methyl]-	*Di HCl* 213–214	1291
5,6-Diamino-1,2-dimethyl-	279	1253
4,5(or 6)-Diamino-2-methyl-		
Diacetyl deriv	176	1313
Triacetyl deriv	260	1313
5,6-Diamino-2-methyl-	—	1178
Acetyl deriv	>300	1178
Diacetyl deriv	>300	1178
2-[1-(Dibenzylamino)ethyl]-	222.3–223.2	1210
2-[(Dibenzylamino)methyl]-	169	1291
2-[1-(Dibutylamino)ethyl]-	139.1–139.3	1210
2-[(Dibutylamino)methyl]-	132	1291
2-[1-(Diethylamino)ethyl]-	177.5–178	1210
2-[(2-Diethylaminoethyl)amino]-	126–128	1314
1-[(Diethylamino)methyl]-	—	1257
2-[(Diethylamino)methyl]-	170	1179, 1291, 1244
2-[(Diisopropylamino)methyl]-	178–178.5	1179
1-[2-Dimethylamino)ethyl]-	*Di HCl* 234–236	1315
2-[1-(Dimethylamino)ethyl]-	208–210	1210
1-[2-(Dimethylamino)ethyl]2-isopropyl-	*Di Pi* 235–236	1315
1-[2-(Dimethylamino)ethyl]-2-methyl-	*Di HCl,* 238–289.5	1315
1-[2-(Dimethylamino)ethyl]-2-phenyl-	72.5–74, *Di HCl* 234	1315
2-[(Dimethylamino)methyl]-	132.5–133	1179
2-(*p*-Dimethylaminophenyl)-1-(*p*-methyl-anilino)-5-methyl-	249, *Pi* 177	1235
2-[2-(*p*-Dimethylaminophenyl)ethyl]-	187, *Pi* 198	1316
2-[2-(*p*-Dimethylaminophenyl)vinyl]-	256, *PCl* 232	1316
2-[2-(*p*-Dimethylaminophenyl)vinyl]-1-methyl-	*HI* 234	1316

233

TABLE 1.83   (*Continued*)

Compound	M.p.(°C)	Ref.
2-[(Diphenylamino)methyl]-	215	1199
2-[(Dipropylamino)methyl]	180.5–181	1179
2-[1-(Ethylamino)ethyl]-	149–149.3, *HCl* 225.7–226	1210
2-[(Ethylamino)methyl]-	*Di HCl* 223–225	1291
2-Guanidino-	254, *Pi* 269–270, *Ni* 228, *Di HCl.* H$_2$O 237	1310, 1317, 1318
2-Guanidino-5,6-dimethyl-	191, *Pi* 258–259, *HCl* 265	1317
2-Guanidino-5(or 6)-methyl-	*Pi* 264, *Di HCl* 228–229	1317
2-(3-Isopropylguanidino)benzimidazole	168, *Pi* 263–264, *Di HCl* 230–232	1317
2-(3-Isopropylguanidino)-5,6-dimethyl-	*Pi* 245, *HCl* 138–141	1317
2-(3-Isopropylguanidino)-5(or 6)-methyl-	*Pi* 212, *HCl* 214–217	1317
2-(3-Methyl-3-phenylguanidino)-	163	1319
2-[(Methylamino)methyl]-	*Di HCl* 207–209	1291
2-(*o*-Methylanilino)-	182, *HCl* 89–90	1312
2-(*p*-methylanilino)-	207, *HCl* 174	1312
1-(*p*-Methylanilino)-5-methyl-2-phenyl-	231	1235, 1320
Nitroso deriv	129	1320
2-(*N*-Methylanilinomethyl)-	202	1199
2-(*N*-Methylanilinomethyl)-1-methyl-	145	1199
2-[(2-Phenylethyl)aminomethyl]-	*Di HCl* 238–239	1291
2-(3-Phenylguanidino)-	178, *Ni* 173–174	1310
2-(3-Phenylureido)-	250, *HCl* 193	1310
2-Ureido-	—	1310, 1321
2-(2-Anilinovinyl)-1,3-dimethylbenzimi- dazolium iodide	273	1322
5-(Dimethylamino)-1,2,3-trimethyl-, sulfate	255	1212
2-[2-(*p*-Dimethylaminophenyl)vinyl]- 1,3-diethyl-, iodide	245–247	1246
2[2-(*f*-dimethylaminophenyl)vinyl]- 1,3-dimethyl-, iodide	>310	1323

(2) *Amino-, aminoalkyl-, and aminoarylbenzimidazoles containing additional functional groups*

2[2-(*p*-Dimethylaminophenyl)vinyl]- 1,3-dimethyl-, iodide	>310	1323
5(or 6)-Amino-4(or 7)-bromo-2-phenylbenz- imidazole	238–239	1253
Acetyl deriv	160	1253
2-Amino-5(or 6)-chloro-	167–168, *Ni* 168, *At* 226–227	1302, 1179
1-(*m*-Nitrophenylsulfonyl) deriv	216–218	1302
5(or 6)-Amino-6(or 5)-chloroformyl deriv	205	1253
5-Amino-4-chloro-2-methyl-1-phenyl-	257	1253
Acetyl deriv	228	1253

TABLE 1.83   (*Continued*)

Compound	M.p.(°C)	Ref.
5-Amino-6-chloro-2-methyl-1-phenyl-	208	1253
N-Acetyl deriv	199	1253
5-Amino-4-chloro-1-(*p*-methylphenyl)-	128	1253
Acetyl deriv	209	1253
Formyl deriv	178	1253
5-Amino-6-chloro-1-(*p*-methylphenyl)-	195	1253
Formyl deriv		
5-Amino-4-chloro-1-phenyl-	143	1253
Acetyl deriv	205	1253
5-Amino-1-(*o*-chlorophenyl)-2-methyl-	130	1253
5-Amino-1-(*p*-chlorophenyl)-2-methyl-	170	1253
5-Phenylazimino deriv	194	1253
5(or 6)-Amino-4,6(or 5,7)-dichloro-	225	1253
Formyl deriv		
2-(Anilinomethyl)-5(or 6)-chlorobenzimidazole	Pi 214–215, Di HCl 250–251	1287
2-[(Diethylamino)methyl]-5(or 6)-chloro-	150–151	1290
1-[3-(Diethylamino)propyl]-5-chloro-2-methyl-	53–54, Pi 165, Di Pi 239, Di Po 233–236	1198
1-[3-(Diethylamino)propyl]-6-chloro-2-methyl-	Di Pi 217, Di Po 236, Di HCl 100–110	1198
1-[3-(Diethylamino)propyl]-5,6-dichloro-2-methyl-	Di Pi 226–228	1198
2-Guanidino-5(or 6)-chloro-	207, Pi 260–261, Di HCl 211	1317
2-Guanidino-5,6-dichloro-	244, Pi 319, HCl 287–290	1317
2-(3-Isopropylguanidino)-5(or 6)-chloro-	Pi 248, Di HCl 215–217	1317
2-(3-Isopropylguanidino)-5,6-dichloro-	204, Pi 296, Di HCl 224–225	1317
1-(*p*-Methylanilino)-2-(*o*-chlorophenyl)-5-methyl-	195, Pi 178–179	1235
Nitroso deriv	124	1235
5-Amino-1-anilino-2-(*o*-hydroxyphenyl)-	177	1235
4(or 7)-Amino-5(or 6)-ethoxy-2-methyl-	147	1324
4(or 7)-Amino-6(or 5)-methoxy-	Pi 240	1300
N-(*p*-Methylphenylsulfonyl) deriv	248	1300
5(or 6)-Amino-6(or 5)-methoxy-	—	1261
N-Acetyl deriv	210	1261
4(or 7)-[(2-Diethylaminoethyl)amino]-6(or 5)-methoxy-		
N-(*p*-Methylphenylsulfonyl) deriv	179–181	1300
1-(2-Amino-4-hydroxyphenyl)-2,5-dimethyl), or 1-(2-Amino-4-methylphenyl)-5-hydroxy-2-methyl-	248	1308
N,O-Diacetyl deriv	244	1308

TABLE 1.83   (*Continued*)

Compound	M.p.(°C)	Ref.
1-(2-Amino-4-hydroxyphenyl)-5-methoxy-2-methyl-, or		
1-(2-Amino-4-methoxyphenyl)-5-hydroxy-2-methyl-	278	1308
N,O-Diacetyl deriv	244	1308
1-(2-Amino-4-methoxyphenyl)-2,5-dimethyl-, or	202	1308
1-(2-Amino-4-methylphenyl)–5–methoxy-2-methyl-*N*-Acetyl deriv		
1-(2-Amino-4-methoxyphenyl)-5-methoxy-2-methyl-	148	1308
N-Acetyl deriv	236	1308
2-(2-Aminoethyl)-5(or 6)-ethoxy-	*Di HCl* 251–252	1309
1-Anilino-5-(*o*-hydroxybenzylideneamino)-2-(*o*-hydroxyphenyl)-	242	1235
Nitroso deriv	125	1235
2-(3-Butylguanidino)-5,6-dimethoxy-	115–120. *Pi* 256, *Di HCl* 232	1317
1-(4-Diethylamino-1-methylbutyl)-5-methoxy-	*Pi* 161	1325
1-(4-Diethylamino-1-methylbutyl)-6-methoxy-	*Pi* 135, *Po* 197	1325
1-(4-Diethylamino-1-methylbutyl)-5-methoxy-2-methyl	*Di Pi* 198, *Po* 229	1325
1-(4-Diethylamino-1-methylbutyl)-6-methoxy-2-methylbenzimidazole	*Pi* 192, *Po* 230	1325
1[2-(Diethylamino)ethyl]-5-methoxy-4(or 7)-[2-Diethylaminoethyl(amino)-6(or 5)-methoxy-1-(2-diethylaminoethyl)]-	*HCl* 203	1300
N-(*p*-Methylphenylsulfonyl) deriv	120	1300
1-[3-(Diethylamino)propyl]-5-methoxy-2-methyl-	40, *Di Pi* 238–239	1262, 1198
1-[3-(Diethylamino)propyl]-6-methoxy-2-methyl-	*Pi* 218–219	1262
1-[3-(Diethylamino)propyl]-6-methoxy-2-(2-phenylvinyl)-	*Di HCl* 234–236	1262
1-[(3-Diethylaminopropyl)amino]-5(or 6)-methoxy-	—	1261
5(or 6)-[(3-Diethylaminopropyl)amino]-6(or 5)-methoxy-	*Pi* 206	1261
1-(*p*-Dimethylaminophenyl)-6-ethoxy-	141–143	1249
1-(*p*-Dimethylaminophenyl)-6-ethoxy-2-(*o*-hydroxyphenyl)-	182–183	1249
2-Guanidino-5,6-dimethoxy-	163, *HCl* H_2O 285	1317
2-Guanidino-5(or 6)-methoxy-	203, *Pi* 258, *Di HCl* 219–220	1317
2-(3-Isopropylguanidino)-5,6-dimethoxy-	215, *Pi* 278, *Di HCl* 244–246	1317
2-(3-Isopropylguanidino)-5(or 6)-methoxy-	117–122, *Pi* 224–225, *Di HCl* 207	1317
1-(*p*-Methylanilino)-2-(*o*-hydroxyphenyl)-5-methyl-	197–198	1320
Diacetyl deriv	—	1320

TABLE 1.83   (*Continued*)

Compound	M.p.(°C)	Ref.
1-(2-Amino-4-hydroxyphenyl)-chloro-2,5- dimethyl-, or 1-(2-Amino-4-methylphenyl)-chloro-5- hydroxy-2-methyl-	280	1308
1-(2-Amino-4-hydroxyphenyl)-chloro- 5-methoxy-2-methyl-, or 1-(2-Amino-4-methoxyphenyl)-chloro- 5-hydroxy-2-methyl-	270	1308
1-(4-Diethylamino-1-methylbutyl)- 5-chloro-2-(*p*-methoxyphenyl)-	—	1326
1-(4-Diethylamino-1-methylbutyl)- 2-(*p*-chlorophenyl)-5-methoxy-	—	1327

TABLE 1.84.   CYANOBENZIMIDAZOLES

Compound	M.p. (°C)	Ref.
1-Cyano-2-(cyanoamino)benzimidazole	—	1318
2-(Cyanoamino)-	—	1318
2-(Cyanomethyl)-	209.7–210.7	1258
5(or 6)-(Cyanomethyl)-	158–159	1181
5(or 6)-(Cyanomethyl)-2-methyl-	206	1181
2-(*p*-Cyanophenyl)-	260	1296
1-(*m*-Cyanophenyl)-2-methyl-5-nitro-	260	1328

TABLE 1.85.   BENZIMIDAZOLE CARBOXYLIC ACIDS

1. *Monocarboxylic Acids*

(*a*) *Alkyl- and Arylbenzimidazole Carboxylic Acids including those containing additional functional groups*

Compound	M.p. (°C)	Ref.
1-Benzimidazolecarboxylic acid		
Ethyl ester	107, *Pi* 178, *PCl* 145	1183
2-Benzimidazolecarboxylic acid	174	1258, 1272
Ethyl ester	212.7–213.7	1258
Methyl ester	187.3	1258
Amide	>300	1258
Benzylamide	172.4	1258
n-Butylamide	180.5–181.5	1258
Cylohexylamide	269.5	1258
Dibutylamide	101.2	1258
Diethylamide	124.5	1258
Dimethylamide	223–224	1258
Ethylamide	210–211	1258

TABLE 1.85   (Continued)

Compound	M.p.(°C)	Ref.
(2-Hydroxyethyl)amide	219–220	1258
(2-Methoxyethyl)amide	138	1258
Methylamide	246.5	1258
(4-Morpholinyl)amide	181.2	1258
1-Methyl-	98.99	1329
2-Methyl-4(or 7)-	>300	1330
2-Ethyl-5(or 6)-		
Ethyl ester	151	1189, 1190
2-Hexyl-		
Ethyl ester	*HCl* 238–240	1189, 1190
7(or 4)-Amino-2-methyl-5(or 6)-	310	1294
Acetyl deriv	>375	1294
5(or 6)-Chloro-2-	159	1267
[3-(1-Piperidyl)propyl] amide	173–174	1267
6(or 5)-Bromo-2-methyl-5(or 6)-	323	1331
6(or 5)-Chloro-2-methyl-	324	1332
6(or 5)-Hydroxy-2-methyl-4(or 7)-	dec. 300–350	1333
6(or 5)-Methoxy-2-methyl-	300–305	1333
6(or 5)-(Hydroxymethyl)-2-methyl-5(or 6)-	—	1334
lactone		
7-Nitro-2-Methyl-1-phenyl-5-	289	1335
7(or 4)-Nitro-2-methyl-5(or 6)-	305	1294

(b) *Carboxyalkyl-, Carboxyalkanyl-, and Carboxyarylbenzimidazoles*

Compound	M.p.(°C)	Ref.
2-Benzimidazoleacetic acid	116	1258
Ethyl ester	128.5–129.5	1258
Amide	244–247	1258
(2-Hydroxyethyl)amide	185–190	1336
α-Acetyl-	172, *HCl* 290–291	1283
Ethyl ester	128–129	1283
Anilide	255	1283
1-Methyl-		
Amide	239–240	1250
5(or 6)-Benzimidazoleacetic acid	*HCl* 240–242	1336
Ethyl ester	65–66	1336
(2-Hydroxyethyl)amide	160–162	1336
2-Methyl-	218–219	1214
α,β-Diphenyl-2-benzimidazoleacrylic acid	186	1337
Ethyl ester	—	1337
Anilide	278	1337
β-(2-Benzimidazolylimino)butyric acid	>300	1338
2-Benzimidazolecarbamic acid		
Ethyl ester	320	1339
2-[(o-Carboxyanilino)methyl]benzimidazole		
Methyl ester	216	1199
2-[(p-Carboxyanilino)methyl]-	281, *Pi* 216, *HCl* 268	1287, 1340
Azide	150	1287
Ethyl ester	248, *Pi* 210, *HCl* 255	1287
Benzylidene hydrazide	276, *Pi* 224, *HCl* 283	1287

TABLE 1.85   (Continued)

Compound	M.p.(°C)	Ref.
2-[(p-Carboxyanilino)methyl]-5(or 6)-chloro-	237, Pi 235, HCl 276	1287
2-(2-Carboxycyclohexyl)-	245–247	1341
Ethyl ester	163–164	1341
2-[2-2′-Carboxy)diphenyl]-	206–207	1337
Amide	227	1337
Anilide	248	1337
Ethyl ester	143	1337
Phenylhydrazide	157	1337
2-(8-Carboxy-1-naphthyl)-	265–269	1342
2-(8-Carboxy-1-naphthyl)-5(or 6)-methyl-	273–275	1342
2-(o-Carboxyphenyl)-	270	1343, 1268
Amide	264	1344
Anilide	327	1344
Hydrazide	293	1344
Phenylhydrazide	262	1344
2-(o-Carboxyphenyl)-5(or 6)-chloro-	285	1345, 1346
2-(o-Carboxyphenyl)-5(or 6)-methyl-	260–262	1342
2-(2-Carboxy-3,4,5,6-tetrachlorophenyl)-	236	1344
Phenylhydrazide	295	1344
2-(3-Carboxy-1,2,2-trimethylcyclopentyl)-	233	1347
2-(3-Carboxy-2,2,3-trimethylcyclopentyl)-	203	1347
2-(3-Carboxy-1,2,2-trimethylcyclopentyl)- 5(or 6)-methyl-	239–240	1347
2-(3-Carboxy-2,2,3-trimethylcyclopentyl)- 5(or 6)-methylbenzimidazole	250–252	1347
β-(2-Benzimidazolylamino)crotonic acid	>300	1338
N-(2-Benzimidazolyl)-α,α-diethylmalonamic acid	214	1348
Methyl ester	116	1348
5-(2-Benzimidazolyl)-3,4-dihydroxy-2,5- bis(p-methoxyphenyl)-2,4-pentadienoic acid		
γ-Lactone	334	1197
5-(2-Benzimidazolyl)-3,4-dihydroxy- 2,5-diphenyl-γ-Lactone	312	1197
5-(2-Benzimidazolyl)-4-hydroxy-3- methoxy-2,5-diphenyl-		
γ-Lactone	325	1197
2-Benzimidazolepropionic acid	228, HCl 236–237	1181, 1309, 1341
Ethyl ester	137	1309
Methyl ester	144–145	1309
Amide	259–260	1309
Hydrazide	268	1309
5(or 6)-Ethoxy-	181, HCl 221	1309
Methyl ester	103	1309
Amide	189	1309
1,5,7-Trimethyl-	265–267	1349
1-Methyl-2-benzimidazolepyruvic acid		
Ethyl ester	154–156	1350
1-Phenyl-		
Ethyl ester	151–152, Pi 185–186	1350

TABLE 1.85 (*Continued*)

Compound	M.p.(°C)	Ref.
N-{p-[2-Benzimidazolylmethyl)amino] benzoyl}-DL-methionine	195	1287

2. *Dicarboxylic Acids*

Compound	M.p.(°C)	Ref.
2-Benzimidazolemalonic acid		
Diethyl ester	218	1227
5,7(or 4,6)-Dihydroxy-4,6(or 5,7)-benzimi-dazole-dicarboxylic acid		
Diethyl ester	131–132	1351
6(or 5)-Methoxy-2,4(or 2,7)-	290–295	1333
N-{p-[(2-Benzimidazolylmethyl)amino benzoyl}-L-glutamic acid (benzimidazole analog of folic acid)	167, *HCl* 201	1340, 1287
Diethyl ester	125, *Pi* 179, *HCl* 232	1340, 1287
N-{p-[5(or 6)-Chloro-2-benzimidazolyl methyl)amino]-benzoyl}-L-glutamic acid	*HCl* 199–202	1287
Diethyl ester	123, *Pi* 195, *HCl* 219–220	1287
N-[N-(2-Benzimidazolylmethyl)sulfanilyl]-l-glutamic acid	—	1352

TABLE 1.86. BENZIMIDAZOLESULFONIC ACIDS, SULFOALKYL- AND
SULFOARYLBENZIMIDAZOLES

Compound	M.p.(°C)	Ref.
2-(p-Aminophenyl)-1-phenyl-benzimidazolesulfonic acid	—	1311
1-Benzimidazolesulfonic acid	221–222	1353
2-Benzimidazolesulfonic acid	365	1354
2-(Methylsulfonyl)benzimidazole	202	1314
5(or 6)-Benzimidazolesulfonic acid amide	213–214	1355
2-Methyl-		
Amide	221	1355
d-α-Methyl-2-benzimidazoleëthanesulfonic acid	—	1356
l-α-Methyl-2-benzimidazoleëthanesulfonic acid	—	1356
DL-α-Methyl-2-benzimidazoleëthanesulfonic acid	—	1357
d-α-Ethyl-2-benzimidazolemethanesulfonic acid	—	1358
d-α-Propyl-	—	1359
DL-α-Ethyl-2-benzimidazolemethanesulfonic acid	—	1358, 1360
DL-α-Ethyl-5(or 6)-methyl-	—	1358, 1360
DL-α-Methyl-	>300	1361
DL-α-Phenyl-	—	1362
DL-α-Propyl-	—	1359
2-[(p-Methylsulfonyl)phenyl]benzimidazole	292, *Pi* 235	1363

TABLE 1.87. BENZIMIDAZOLE ARSENICALS

Compound	M.p.(°C)	Ref
5(or 6)-[(3-Amino-4-hydroxyphenyl)arseno]benz-imidazole	*HCl* 206	1364
4(or 7)-Arsono-	277	1365
5(or 6)-Arsono-	297	1365–1368
5(or 6)-Arsono-2-amino-	*Su* 210	1369
5(or 6)-Arsono-6(or 5)-amino-2-methyl-	—	1178
Acetyl deriv	—	1178
5-Arsono-1-(carbamylmethyl), or 6-Arsono-1-(carbamylmethyl)-	—	1370
5(or 6)-Arsono-2-(carbamylmethylmercapto)-	—	1354
5-Arsono-1-(carboxymethyl), or 6-Arsono-1-(carboxymethyl)-	—	1370
5(or 6)-Arsono-2-(carboxymethylmercapto)-	—	1354, 1368
5-Arsono-2,3-dihydro-2-hydroxy-1,2,3-trimethyl-	—	1208
6-Arsono-1,2-dimethyl-	>300	1208
5(or 6)-Arsono-2,7(or 2,4)-dimethyl-	—	1366
5(or 6)-Arsono-2-ethyl-	—	1365
4(or 7)-Arsono-2-(1-hydroxyethyl)-	—	1365
5(or 6)-Arsono-2-(1-hydroxyethyl)-	—	1365
5(or 6)-Arsono-2-(1-hydroxyethyl)-1-phenyl-	—	1212
5(or 6)-Arsono-2(3H)-benzimidazolethione		1371, 1367, 1368
Thiolacetamide deriv	245	1354
2,2'-Dithiobis[5(or 6)-arsonobenzimidazole]	—	1354
4(or 7)-Arsono-2-methyl-	280–282	1365
5(or 6)-Arsono-2-methyl-	275	1366, 1365, 1212
5(or 6)-Arsono-7(or 4)-methyl-	300	1366
5(or 6)-Arsono-2-methyl-6(or 5)-nitro-	>300	1178
5-Arsono-2-methyl-1-phenyl-	—	1212, 1372
5-Arsono-1-phenylbenzimidazole	—	1212
5(or 6)-Arsono-2-sulfo-	—	1354
5(or 6)-Thioarso-2(3H)-benzimidazolethione	—	1367, 1371, 1368
4,4'(or 7,7')-Arsenobisbenzimidazole	—	1365
5,5'(or 6,6')-Arsenobis-	—	1366, 1365
5,5'(or 6,6')-Arsenobis[1-(carbamylmethyl)-	—	1370
5,5'(or 6,6')-Arsenobis[2-(carbamylmethylmercapto)-	—	1354
5,5'(or 6,6')-Arsenobis[1-carboxymethyl)-	—	1370
5,5'(or 6,6')-Arsenobis[2-(carboxymethylmercapto)-	—	1354
5,5'(or 6,6')-Arsenobis[2,7(or 2,4)-dimethyl-	—	1366
5,5'(or 6,6')-Arsenobis[2(3H)-benzimidazolethione]	—	1367
4,4'(or 7,7')-Arsenobis(2-methyl-	—	1365
5,5'(or 6,6')-Arsenobis(2-methyl-	—	1365, 1366
5,5'(or 6,6')-Arsenobis[7(or 4)-methyl-	—	1366
5,5'(or 6,6')-Arsenobis(2-sulfo-	—	1354
5,5'-Arsenobis[2(3H)-benzimidazolone]	—	1367
2(3H)-Benzimidazolone-5-arsinic acid	—	1367
2(3H)-Benzimidazolone-5-arsonous acid	—	1373

TABLE 1.88.   HETERORING-SUBSTITUTED BENZIMIDAZOLES

Compound	M.p.(°C)	Ref.

1. Furan Derivatives

2-(2-Furyl)benzimidazole	286–290	1272, 1189, 1190
4,5,6,7-Tetrahydro deriv	290–300, *Pi* 220–225	1187
2-(2-Furyl)-1-methyl-	56	1207
2-(2-Furyl)-5-methyl-1-(*p*-methylanilino)-	227	1320

2.   Thiophene Derivatives

2-(2-Thienyl)benzimidazole	>280	1374
2,5-Di-2-benzimidazolyl-3,4-dibromothiophene	385	1375

3.   Pyrrole Derivative

2-[2-(2,5-Dimethyl-1-phenyl-3-pyrryl)vinyl]-1,3-diethylbenzimidazolium iodide	236–238	1246

4.   Pyrazole Derivative

4-(2-Benzimidazolyl)-3-methyl-1-phenyl-pyrazolone	172–173, *HCl* 283–284	1283

5.   Pyridine Derivatives

2-(2-Carboxy-3-pyridyl)benzimidazole	246–247	1344
Amide	266–268	1344
Anilide	315–317	1344
Methyl ester	195–196	1344
2-(2-Pyridyl)-	—	1376, 1377
2-(3-Pyridyl)-	310, *HCl* 296–298	1376, 1377, 1344, 1194

6.   Piperidine Derivatives

2-[1-Piperidyl)ethyl]benzimidazole	167–167.2	1210
1-(1-Piperidylmethyl)-	91.5–92.5	1257
2-(1-Piperidylmethyl)-	201, *Di HCl* 204–205	1199, 1244, 1291
2-(1-Piperidylmethyl)-5(or 6)-chloro-	163–164, *Di HCl* >250, *HCl* 249–251	1179, 1290

7.   Morpholine Derivatives

2-[1-(4-Morpholinyl)ethyl]benzimidazole	196.8–197	1210
1-(4-Morpholinylmethyl)-	110.5–111.5	1257
2-(4-Morpholinylmethyl)-	211, *Di HCl* 194–195	1199, 1291

TABLE 1.88 (*Continued*)

Compound	M.p.(°C)	Ref.

8. Xanthyl Derivatives

1-(9-Xanthyl)-2(3*H*)-benzimidazolone	268–270	1378
1,3-Di-9-xanthyl-	283–285	1378
1-(9-Xanthyl-2(3*H*)-benzimidazolethione	252–254	1378
1,3-Di-9-xanthyl-	260–262	1378

9. Phthalide Derivative

3-(1-Phenyl-2-benzimidazolylmethylene)- phthalide	280–281	1350

10. Indane and Indole Derivatives

2-[2(3*H*)-Benzimidazolylidene]- 1,3-indanedione	>350, *Ni* 184	1201
3-(α-2-Benzimidazolylbenzylidene)oxindole	264	1201
3-(2-Benzimidazolylmethylene)oxindole	>350	1201

11. Quinoline Derivatives

2-(2,4-Dimethyl-3-quinolyl)benzimidazole	328–330, *Pi* 250, *HCl* >300	1283
2-(2-Hydroxy-4-methyl-sulfo-3-quinolyl)-	293–295, *HCl* >310	1283

12. Benzodiazepine Derivative

3-(2-Benzimidazolyl)-2,4-dimethyl-1,5- benzodiazepine	>310, *HCl* >300	1283

13. Acenaphthene Derivative

2-(2-Benzimidazolylmethylene)-1- acenaphthenone	295	1201

TABLE 1.89. BISBENZIMIDAZOLES

Compound	M.p.(°C)	Ref.
2,2′-Bisbenzimidazole	—	1379
1,4-Di-2-benzimidazolyl-2,3-butanedione	>300	1380
s-Di-2-benzimidazolylthiourea	208	1348
1,1′-Benzylidenebis(2-benzylbenzimidazole)	171	1218
2,2′-Ethylenebisbenzimidazole	325–330	1381
	Di HCl 312–315	
2,2′-Heptamethylene-	273–275	1381
	Di HCl 269–272	
2,2′-Hexamethylene-	263–266	1381
	Di HCl 296–299	
2,2′-(Iminodiethylidene)-	206.8–210.2,	1210
	Di HCl 236–270	
5,5′-Isopropylidenebis(2-methylbenzimidazole)	225	1382
2,2′-(Methyliminodiethylidene)bisbenzimidazole	205.1–205.9,	1210
	Di HCl 234–237	
2,2′-Octamethylene-	277–279,	1381
	Di HCl 263–265	
2,2′-Pentamethylene-	225–226,	1381
	Di HCl 270–272	
2,2′-o-Phenylenedibenzimidazole	412	1383
2,2′-p-Phenylene-	>300	1375
2,2′-Tetramethylenebisbenzimidazole	259–260,	1381
	Di HCl 305–309	
2,2′-Trimethylene-	258–259,	1381, 1384
	Di HCl 270–273,	
	Di methiodide, 228	
2-[3-(1,3-Diethyl-2(3H)-benzimidazolylidene)pro-penyl)-1,3-diethylbenzimidazolium iodide	278–280	1246
2-[3-(1,3-Dimethyl-2(3H)-benzimidazolylidene)-(1-methyl-2-benzimidazolyl)isobutenyl]-1,3-dimethyl-3-benzimidazolium iodide	230	1385
2-[3-(1,3-Dimethyl-2(3H)-benzimidazolylidene)pro-penyl]-1,3-dimethylbenzimidazolium iodide	303	1385, 1386

REFERENCES

1. K. Hofmann, "Imidazole and Its Derivatives, Part I," Wiley-Interscience New York, 1953.
2. J. H. Lister, "Fused Pyrimidines, Part II: Purines," Wiley-Interscience New York, 1971.
3. Cf. R. Bonnett, Chem. Rev., **63,** 573 (1963).
4. S. Nakamura, Chem. Pharm. Bull., **3,** 379 (1955).
5. See for example, C. Cosar, C. Crisan, R. Horclois, R. M. Jacob, J. Robert, S. Tchelitscheff, and R. Vampre, Arzneim.-Forsch, **16,** 23 (1966).

6. J. B. Wright, *Chem. Rev.*, **48**, 397 (1951).
7. P. N. Preston, *Chem. Rev.*, **74**, 279 (1974).
8. E. S. Schipper and A. R. Day, "Heterocyclic Compounds," Vol 5, R. C. Elderfield, Ed., Wiley-Interscience, New York, 1953, p 267.
9. A. F. Pozharskii, A. D. Garnovskii, and A. M. Simonov, *Russ. Chem. Rev.*, **35**, 122 (1966).
10. M. R. Grimmett, *Int. Rev. Sci. Org. Chem.* Ser. 1, **4**, (1973).
11. E. Ochiai, "Aromatic Amine Oxides," Elsevier, New York, 1967.
12. A. R. Katritzky and J. M. Lagowski, "Chemistry of the Heterocyclic N-Oxides," Academic, New York, 1971.
13. H. Lettau, *Z. Chem.*, **10**, 211 (1970).
14. P. N. Preston and G. Tennant, *Chem. Rev.*, **72**, 627 (1972).
15. O. Meth-Cohn and H. Suschitzky, *Adv. Heterocycl. Chem.*, **14**, 211 (1972).
16. L. B. Townsend and G. R. Revankar, *Chem. Rev.*, **70**, 389 (1970).
17. L. S. Efros, *Zh. Obshch. Khim.*, **23**, 842 (1953); *Chem. Abstr.*, **48**, 4524c.
18. L. S. Efros, *Zh. Obshch. Khim.*, **23**, 957 (1953).
19. A. Sykes and J. C. Tatlow, *J. Chem. Soc.*, **1952**, 4078.
20. (a) F. Montarini and R. Passerini, *Boll. Sci. Fac. Chim. Ind. Bologna*, **11**, 42 (1953); *Chem Abstr.*, **48**, 6436h; (b) Ibid., **11**, 46 (1953); *Chem. Abstr.*, **48**, 6437a.
21. (a) J. R. E. Hoover and A. R. Day, *J. Am. Chem. Soc.*, **77**, 4324 (1955); (b) Ibid., **77**, 5652 (1955).
22. L. S. Efros, *Zh. Obshch. Khim.*, **22**, 1008 (1952); *Chem. Abstr.*, **47**, 12366a.
23. C. A. Haley and P. Maitland, *J. Chem. Soc.*, **1951**, 3155.
24 H. B. Gillespie, M. Engelman, and S. Graff, *J. Am. Chem. Soc.*, **76**, 3531 (1954).
25. L. S. Efros, *Zh. Obshch. Khim.*, **23**, 957 (1953); *Chem. Abstr.*, **48**, 8223a.
26. J. Klosa, *Arch. Pharm.* Weinheim, *Ger.* **287**, 62 (1954); *Chem. Abstr.*, **51**, 14741d.
27. W. Knobloch and H. Schaefer, *J. Prakt. Chem.*, **17**, 187 (1962).
28. Y. Kanaoka, O. Yonemitsu, and Y. Bau, *Chem. Pharm. Bull.* (Tokyo), **12**, 773 (1964).
29. L. S. Efros, V. P. Kumarev, and E. R. Zakhs, *Khim. Geterotsikl. Soedin.*, **1967**, 336; *Chem. Abstr.*, **67**, 116847u.
30. A. M. Simonov, Yu. M. Yutilov, and V. A. Anisimova, *Khim. Geterotsikl. Soedin. Akad. Nauk. Latv.*, *SSR*, **1965**, 913; *Chem. Abstr.*, **64**, 12661a.
31. S. I. Vurmistrov and V. V. Boboshko, *Khim. Tekhnol.* (Kharkov), **1971**, 34; *Chem. Abstr.*, **76**, 14431x.
32. L. Weinberger and A. R. Day, *J. Org. Chem.*, **24**, 1451 (1959).
33. W. T. Smith and E. C. Steinle, *J. Am. Chem. Soc.*, **75**, 1292 (1953).
34. E. Fauland, W. Kampe, M. Thiel, W. Bartsch, and W. Schaumann, German Patent 2,432,269 (1976); *Chem. Abstr.*, **84**, 13566a.
35. L. B. Piotrovskii, N. I. Kudryachova, and N. V. Khromove-Borisov, *Khim.-Farm. Zh.*, **9**, 3 (1975).
36. Luu Duc Cuong and C. Agius-Delord, *Bull. Soc. Chim. Fr.*, **1973**, 3317.
37. H. Depoorter, G. G. Van Mierlo, M. J. Libeer, and J. M. Nys, Belgian Patent 595,327 (1961); *Chem. Abstr.*, **58**, 9085a.
38. H. B. Gillespie, M. Engelman, and S. Graff, *J. Am. Chem. Soc.* **78**, 2445 (1956).
39. A. Dikciuviene, V. Bieksa, and J. Degutis, *Liet TSR Mokslu Akad. Darb. Ser. B.*, **1973**, 93; *Chem. Abstr.*, **79**, 115496q.
40. G. Holan, J. J. Evans, and M. Linton, *J. Chem. Soc. Perkin I*, **1977**, 1200.
41. O. F. Ginsberg, B. A. Porai-Koshits, M. I. Krylova, and S. M. Lotareichik, *Zh. Obshch. Khim.*, **27**, 411 (1957); *Chem. Abstr.*, **51**, 15500d.
42. W. Knobloch, *Chem. Ber.*, **91**, 2557 (1958).
43. H. R. Hensel, *Chem. Ber.*, **98**, 1325 (1965).
44. K. H. Beuchel, *Z. Naturforsch.*, *(B)*, **25**, 945 (1970).
45. H. Lawniczak, *Pr. Inst. Przem. Org.*, **1971**, 21; *Chem. Abstr.*, **79**, 66248z.

46. Fisons Pest Control Ltd., Netherlands Patent 6,603,719 (1966); *Chem. Abstr.*, **66,** 85788y.

47. Shell Internationale Research Maatschappy NV, Netherlands Patent 6,705,527 (1967); *Chem. Abstr.*, **69,** 10437m.

48. Fisons Pest Control Ltd., French Patent 1,522,661 (1968); *Chem. Abstr.*, **71,** 61388v.

49. Q. F. Soper, U.S. Patent 3,443,015 (1969); *Chem. Abstr.*, **71,** 30472p.

50. K. L. Kirk and L. A. Cohen, *J. Org. Chem.*, **34,** 384 (1969).

51. M. H. Fisher, U.S. Patent 3,749,789 (1973); *Chem. Abstr.*, **79,** 115580n.

52. G. Hoerlein, A. Studeneer, P. Langelueddeke, and F. Schwerdtle, German Patent 2,326,624 (1974); *Chem. Abstr.*, **83,** 164175a.

53. J. L. Miesel, U.S. Patent 3,890,343 (1975); *Chem. Abstr.*, **83,** 164176b.

54. Fisons Pest Control Ltd., Belgian Patent 659,384 (1965); *Chem. Abstr.*, **63,** 18101h.

55. U.S. Borax and Chemical Corporation, British Patent (Amended) 1,015,937 (1971); *Chem. Abstr.*, **75,** 151788n.

56. B. P. Fedorov and R. M. Mamedov, *Izv. Akad. Nauk. SSSR, Otd. Khim. Nauk.*, **1962,** 1626; *Chem. Abstr.*, **58,** 9048g.

57. M. D. Nair, *J. Indian Chem.*, **7,** 304 (1969); *Chem. Abstr.*, **71,** 3320f.

58. W. R. Siegart and A. R. Day, *J. Am. Chem. Soc.*, **79,** 4391 (1957).

59. S. Akihama, M. Okude, K. Sato, and S. Iwabuchi, *Yakugaku Zasshi,* **88,** 684 (1968); *Chem. Abstr.*, **69,** 96580n.

60. R. Crawford and J. T. Edward, *J. Chem. Soc.*, **1956,** 673.

61. H. Lettre, W. Fritsch, and J. Porath, *Chem. Ber.*, **84,** 719 (1951).

62. L. A. Cescon and A. R. Day, *J. Org. Chem.*, **27,** 581 (1962).

63. R. Geiger and W. Seidel, German Patent 1,131,688 (1962); *Chem. Abstr.*, **57,** 16627a.

64. Y. Kanaoka, K. Tanizawa, and O. Yonemitsu, *Chem. Pharm. Bull. (Tokyo)*, **17,** 2381 (1969); *Chem. Abstr.*, **72,** 55332b.

65. R. Crawford and J. T. Edward, *J. Chem. Soc.*, **1956,** 673.

66. C. L. Moyle and D. M. Chern, U.S. Patent 3,152,142 (1964); *Chem. Abstr.*, **62,** 566a.

67. Laboratorio Farmaceutico Quimico-Lafarquim SA., Spanish Patent 407,882 (1975); *Chem. Abstr.*, **85,** 5639e.

68. N. N. Ghosh and M. M. Naudi, *J. Indian Chem. Soc.*, **53,** 274 (1976); ibid, **53,** 331 (1976).

69. I. Sekikawa, *Bull. Chem. Soc. Jap.*, **31,** 252 (1958).

70. C. L. Moyle and D. M. Chern, U.S. Patent 3,182,070 (1965); *Chem. Abstr.*, **63,** 4304d.

71. C. L. Moyle and D. M. Chern, U.S. Patent 3,147,274 (1964); *Chem. Abstr.*, **61,** 13319c.

72. S. S. Tiwari and S. B. Misra, *J. Indian Chem. Soc.*, **53,** 310 (1976); *Chem. Abstr.*, **85,** 46510z.

73. D. Heydenhauss and H. Schubert, *Z. Chem.*, **4,** 459 (1964); *Chem. Abstr.*, **62,** 9172a.

74. K. Kondal Reddy, N. V. Subba Rao, and Y. C. Ratnam, *Indian J. Chem.*, **1,** 96 (1963); *Chem. Abstr.*, **59,** 5148h.

75. G. M. Kharkharova, *Zh. Obshch. Khim.*, **26,** 1713 (1956); *Chem. Abstr.*, **51,** 1944a.

76. B. A. Porai-Koshits and G. M. Kharkharova, *Zh. Obshch. Khim.*, **24,** 1651 (1954); *Chem. Abstr.*, **49,** 13224h.

77. (a) M. Covello, M. R. Mazza, N. Sacco, and F. De. Simone, *Rend. Accad. Sci. Fis. Nat. Naples,* **37,** 147 (1970); *Chem. Abstr.*, **76,** 34171r. (b) C. L. Moyle and D. M. Chern, U.S. Patent 3,075,991 (1963); *Chem. Abstr.*, **59,** 1645h.

78. G. Carraz, Luu Buc Cuong, and P. Demence, French Patent 2,229,706 (1976); *Chem. Abstr.*, **68,** 189934x. ibid., 2,292,473 (1976); *Chem. Abstr.*, **86,** 155654r.

79. F. Ackermann and J. Meyer, U.S. Patent 2,515,173 (1950); *Chem. Abstr.*, **45,** 666i.

80. G. Paglietti, V. Boido, and F. Sparatore, *Farmaco Ed. Sci.*, **30,** 505 (1975); *Chem. Abstr.*, **83,** 114300s.

81. Z. V. Esayan and G. T. Tatevosyan, *Arm. Khim. Zh.*, **25,** 969 (1972); *Chem. Abstr.*, **78,** 136174c.

82. N. A. Aliev, Ch. Sh. Kadyrov, and Zh. Eshimbetov, USSR Patent 429,059 (1974); *Chem. Abstr.*, **81**, 105505u.

83. T. Sasaki, S. Eguchi, and T. Ogawa, *Heterocycles*, **3**, 193 (1975).

84. Z. V. Esayan, L. A. Manucharova, and G. T. Tatevosyan, *Arm. Khim. Zh.*, **25**, 345 (1972); *Chem. Abstr.*, **77**, 139889e.

85. K. S. Karagezyan, R. T. Grigoryan, and G. T. Tatevosyan, USSR Patent 455,104 (1974); *Chem. Abstr.*, **82**, 170945x.

86. K. S. Karagesyan, G. T. Tatevosyan, A. E. Agayan, and S. G. Chshmarityan, *Sin. Get. Soedin,* **1972**, 52; *Chem. Abstr.*, **79**, 126400e.

87. G. T. Tatevosyan and Z. F. Esayan, USSR Patent 406,845 (1973); *Chem. Abstr.*, **80**, 120938k.

88. G. Sandera, R. W. Isensee, and L. Joseph, *J. Am. Chem. Soc.*, **76**, 5173 (1954).

89. M. Itaya, Y. Takai, and T. Kaiya, *Yakugaku Zasshi,* **86**, 600 (1966); *Chem. Abstr.*, **65**, 15364c.

90. J. Preston, W. DeWinter, and W. L. Hofferbert, *J. Heterocycl. Chem.*, **6**, 119 (1969).

91. C. M. Orlando, J. G. Wirth, and D. R. Heath, *J. Org. Chem.*, **35**, 3147 (1970).

92. L. N. Pushkina, S. A. Mazalov, and I. Ya Postovskii, *Zh. Obshch. Khim.*, **32**, 2624 (1962); *Chem. Abstr.*, **58**, 9049h.

93. C. H. Shunk, Belgian Patent 621,597 (1963); *Chem. Abstr.*, **59**, 11504c.

94. D. W. Hein, R. J. Alheim, and J. J. Leavitt, *J. Am. Chem. Soc.*, **79**, 427 (1957).

95. N. Hasebe, *Yamagata Daigaku Kiyo*; *Shizen Kagaku,* **7**, 401 (1971); *Chem. Abstr.*, **78**, 136170y.

96. A. B. Sungatova, A. I. Pavlov, V. V. Korshak, G. M. Tseitlin, D. D. Berezin, and V. N. Kulagin, USSR Patent 467,072 (1975); *Chem. Abstr.*, **83**, 114403c.

97. S. S. Vogulyubova, L. I. Rudaya, I. Ya. Kvitko, and A. V. El'tsov, USSR Patent 490,800 (1975); *Chem. Abstr.*, **84**, 74266x.

98. I. Ya. Kvitko, L. I. Rudaya, G. B. Salmina, and A. V. El'tsov, USSR Patent 498,298; *Chem. Abstr.*, **84**, 121836v.

99. J. Sluka, J. Novak, and Z. Budesinsky, *Collect. Czech. Chem. Commun.*, **41**, 3628 (1976).

100. L. Sh. Afanasiadi, B. M. Bolotin, and N. F. Levchenko, *Khim. Geterotsikl Soedin,* **1976**, 673; *Chem. Abstr.*, **85**, 94279c.

101. D. Hoff and H. Peterson, French Patent 1,510,330 (1968); *Chem. Abstr.*, **70**, 77967y.

102. L. H. Sarett and H. D. Brown, U.S. Patent 3,017,415 (1962); *Chem. Abstr.*, **56**, 15517c.

103. H. D. Brown and L. H. Sarett, U.S. Patent 3,055,907 (1962); *Chem. Abstr.*, **58**, 2456c.

104. C. G. Helsley, *French Patent* 2,103,639 (1972); *Chem. Abstr.*, **77**, 164705z.

105. V. I. Isagulyants and N. M. Anufrieva, *Zh. Prikl. Khim.*, **46**, 1389 (1973); *Chem. Abstr.*, **79**, 66244v.

106. M. Ichikawa and T. Hisano, *Chem. Pharm. Bull* (*Tokyo*), **25**, 358 (1977).

107. A. Novelli, *Bol. Soc. Quim. Peru,* **19**, 77 (1953); *Chem. Abstr.*, **49**, 1021f.

108. P. M. Dzadzic, B. L. Bastic, L. Borivoje, and M. V. Piletic, *Glas. Hem. Drus. Beograd,* **36**, 137 (1971); *Chem. Abstr.*, **78**, 16096g.

109. B. A. Porai-Koshits, L. S. Efros, and E. S. Boichibova, *Zh. Obshch. Khim.*, **23**, 835 (1953); *Chem. Abstr.*, **48**, 4523e.

110. V. M. Zubarovskii and Uy. P. Makovetskii, *Ukr. Khim. Zh.*, **34**, 1151 (1968); *Chem. Abstr.*, **70**, 68251h.

111. J. Stanek and V. Wollrab, *Monatsh. Chem.*, **91**, 1064 (1960).

112. H. A. Dumesnil, French Patent 1,179,933 (1959); *Chem. Abstr.*, **55**, 19953g.

113. K. Kakimoto and I. Sekikawa, *Nippon Kagaku Zasshi*, **77**, 480 (1956); *Chem. Abstr.*, **52**, 9084b.

114. K. C. Tsou, D. J. Rabiger, and B. Sobel, *J. Med. Chem.*, **12**, 818 (1969).

115. (a) W. R. Roderick, German Patent 2,063,856 (1971); *Chem. Abstr.*, **75**, 76790b. (b) T. Shen and T. A. Maag, German Patent 2,013,910 (1970); *Chem. Abstr.*, **73**, 120626n.

116. C. Rai, W. E. Kramer, and R. C. Kimble, U.S. Patent 3,222,285 (1965); *Chem. Abstr.*, **64,** 9735b.
117. B. C. Bishop, A. S. Jones, and J. C. Tatlow, *J. Chem. Soc.*, **1964,** 3076.
118. L. Li-Yen Wang and M. M. Jouillé, *J. Am. Chem. Soc.*, **79,** 5706 (1957).
119. A. E. Siegrist and M. Duennenberger, U.S. Patent 2,901,408 (1959); *Chem. Abstr.*, **55,** 1658c.
120. M. Duennenberger and A. E. Siegrist, German Patent 1,086,237 (1960); *Chem. Abstr.*, **56,** 4774g.
121. Ciba Ltd., British Patent 861,431 (1961); *Chem. Abstr.*, **55,** 22342i.
122. K. H. Taffs, L. V. Prosser, F. B. Wigton, and M. M. Jouillé, *J. Org. Chem.*, **26,** 462 (1961).
123. D. G. O'Sullivan and A. K. Wallis, *Nature*, **198,** 1270 (1963).
124. H. Schubert, S. Hoffmann, G. Lehmann, I. Barthold, H. Meichsner, and H. E. Polster, *Z. Chem.*, **15,** 481 (1975); *Chem. Abstr.*, **84,** 150564k.
125. R. D. Haugwitz and V. L. Narayanan, German Patent 2,161,524 (1972); *Chem. Abstr.*, **77,** 101604v.
126. S. Chatterjee and J. Wolski, *J. Indian Chem. Soc.*, **43,** 660 (1966).
127. M. Matsumura and Y. Kitsuda, Japanese Patent 75 154,249 (1975); *Chem. Abstr.*, **85,** 5635a.
128. M. Matsumura and Y. Kitsuda, Japanese Patent 75 154,250 (1975); *Chem. Abstr.*, **85,** 5634z.
129. M. Matsura and Y. Kitsuda, Japanese Patent 75 140,445 (1975); *Chem. Abstr.*, **85,** 33006q.
130. M. A. Phillips, *J. Chem. Soc.*, **1928,** 2393.
131. B. A. Porai-Koshits, O. F. Ginsburg, and L. S. Efros, *Zh. Obshch. Khim.*, **17,** 1768 (1947); *Chem. Abstr.*, **42,** 5903c. See also B. A. Porai-Koshits and G. M. Kharkharova, *Zh. Obshch. Khim.*, **25,** 2138 (1955).
132. B. A. Porai-Koshits, O. F. Ginsburg, and L. F. Efros, *Zh. Obshch. Khim.*, **19,** 1545 (1949); *Chem. Abstr.*, **44,** 1100b.
133. T. Sasaki, S. Eguchi, and T. Toru, *Bull. Chem. Soc. Japan*, **42,** 1617 (1969).
134. G. Holan, E. Samuel, B. C. Ennis, and R. W. Hinde, *J. Chem. Soc.*, **1967,** 20.
135. M. T. Davis, P. Mamalis, V. Petrow, and B. Strugen, *J. Pharm. Pharmacol.*, **3,** 420 (1951).
136. Cf. S. D. Hamann, *Annu. Rev. Phys. Chem.*, **15,** 349 (1964).
137. K. J. Morgan and A. M. Turner, *Tetrahedron*, **25,** 915 (1969).
138. C. G. Overberger and C. J. Podsiadly, *Bioorg. Chem.*, **3,** 16, 35 (1974).
139. L. J. Mathias and C. G. Overberger, *Synth. Commun.*, **5,** 461 (1975).
140. H. Irving and O. Weber, *J. Chem. Soc.*, **1959,** 2296.
141. J. Büchi, H. Zwichy, and A. Aebi, *Arch. Pharm.* Weinheim, *Ger.*, **293,** 758 (1960); *Chem. Abstr.*, **55,** 518f.
142. T. Shen, A. R. Matsuk, and H. Scham, French Patent 1,580,823 (1969); *Chem. Abstr.*, **73,** 25468d.
143. J. Sawlewicz and B. Milczarska, *Pol. J. Pharmacol. Pharm.*, **26,** 639 (1974); *Chem. Abstr.*, **82,** 140010w.
144. N. Vinot, *Bull. Soc. Chim. Fr.*, **1966,** 3989; *Chem. Abstr.*, **66,** 85728d.
145. G. I. Braz, I. E. Kardash, V. V. Kopylov, A. F. Oleinik, G. G. Rozantsev, A. N. Pravednikov, and A. Ya. Yakubovich, *Khim. Geterotsikl. Soedin.*, **1968,** 339; *Chem. Abstr.*, **69,** 96574p.
146. Chimetron, S.a.r.l., French Patent 1,439,113 (1966); *Chem. Abstr.*, **65,** 18595b.
147. Chimetron S.a.r.l., French Patent 1,450,505 (1966); *Chem. Abstr.*, **67,** 3092a.
148. D. N. Gray, *J. Heterocycl. Chem.*, **7,** 947 (1970).
149. M. Hammad and K. El-Bayouki, *Acta Chim. Acad. Sci. Hung.*, **90,** 193 (1976); *Chem. Abstr.*, **86,** 89694a.
150. Ciba Ltd., *British Patent* 864,131 (1961).

151. J. Liebscher and H. Hartmann, East German Patent 94,998 (1973); *Chem. Abstr.*, **79**, 1813a.

152. J. M. Singh, *Indian J. Appl. Chem.*, **32**, 133 (1969); *Chem. Abstr.*, **75**, 20293z.

153. A. M. Sarpeshkar and S. Rajagopal, *Indian J. Chem.*, **13**, 1368 (1975).

154. N. Suzuki, T. Yamabayashi, and Y. Izawa, *Bull. Chem. Soc. Japan*, **49**, 353 (1976).

155. K. Oyamada, T. Matsui, J. Tobitsuka, and M. Nagano, *Nippon Nogei Kagaku Kaishi*, **50**, 29 (1976); *Chem. Abstr.*, **85**, 32921x.

156. R. G. Arnold, U.S. Patent 2,697,711 (1954); *Chem. Abstr.*, **49**, 14036b.

157. H. E. Johnson, U.S. Patent 3,255,202 (1966); *Chem. Abstr.*, **65**, 10595h.

158. E. S. Lane, *J. Chem. Soc.*, **1953**, 2238.

159. J. Dvorak and T. Kudrnova, Czech. Patent 152,868 (1974), *Chem. Abstr.*, **81**, 120627d.

160. S. H. Dandegaonker and G. R. Revankar, *J. Karnatek Univ.*, **6**, 25 (1961); *Chem. Abstr.*, **59**, 10023c.

161. H. B. Gillespie, F. Spano, and S. Graaf, *J. Org. Chem.*, **25**, 942 (1960).

162. O. Riester and M. Hase, German Patent 2,014,324 (1972); *Chem. Abstr.*, **76**, 140828v.

163. R. E. Lyle and J. L. La Mattina, *J. Org. Chem.*, **40**, 438 (1975).

164. B. K. Manukian, *Helv. Chim. Acta*, **47**, 2211 (1964).

165. V. V. Korshak, E. S. Kronganz, A. P. Travnikova, and A. L. Rusanov, *Khim. Geterosikl. Soedin.*, **1972**, 247; *Chem. Abstr.*, **76**, 140644g.

166. Farbwerke Hoechst A. G., French Patent 2,051,911 (1971); *Chem. Abstr.*, **76**, 99667b.

167. Ilford Ltd., French Patent 1,486,322 (1967); *Chem. Abstr.*, **69**, 27424c.

168. V. M. Zubarovskii, R. N. Moskaleva, and M. P. Bachurina, *Zh. Obshch. Khim.*, **32**, 1581 (1962); *Chem. Abstr.*, **58**, 6952b.

169. J. Grimshaw and J. Trocha-Grimshaw, *Tetrahedron Lett.*, **1975**, 2601.

170. A. M. Kuznetsov, S. A. Petrova, and B. M. Krasovitskii, *Stsinstill. Org. Lyuminofory*, **1972**, 74; *Chem. Abstr.*, **79**, 115490h.

171. Cf. V. V. Korshak and M. M. Teplyakov, *J. Macromol. Sci. Rev. Macromol. Chem.*, **5**, 409 (1971).

172. G. Holan, E. L. Samuel, B. C. Ennis, and R. W. Hinde, *J. Chem. Soc. C*, **1967**, 20.

173. Monsanto Chemicals (Aust.) Ltd., Netherlands Patent 6,414,890 (1965); *Chem. Abstr.*, **63**, 16357f.

174. D. Floyd, U.S. Patent 3,501,492 (1970); *Chem. Abstr.*, **72**, 111469r.

175. G. Holan and E. L. Samuel, U.S. Patent 3,655,688 (1972) [Division of US 3,560,195 (1966)]; *Chem. Abstr.*, **77**, 34520z.

176. D. L. Hunter, R. A. Smith, and W. S. Belles, U.S. Patent 3,954,438 (1976); *Chem. Abstr.*, **85**, 21374v.

177. P. Dastert, M. Turin, P. Guerret, T. Imbert, G. Raynaud, and J. Thomas, French Patent 2,291,749 (1976); *Chem. Abstr.*, **86**, 171447f.

178. H. Roechling, G. Hoerlein, and L. Emmel, German Patent 2,022,504 (1971); *Chem. Abstr.*, **76**, 59629j.

179. (a) N. S. Nametkin, G. A. Shvekhgeimer, V. P. Dukhovskoi, and V. D. Tyurin, *Khim. Geterotsikl. Soedin.*, **1969**, 1073; *Chem. Abstr.*, **72**, 13261v. (b) V. P. Dukhovskoi, G. A. Shvekhgeimer, and V. D. Tyurin, *Tr. Mosk. Inst. Neftekhim Gazov.*, Prom. No. 3, 7 (1969); *Chem. Abstr.*, **75**, 140761x.

180. G. A. Shvekhgeimer and G. A. Mikheichev, *Izv. Vyssh. Ucheb. Zaved., Khim. Khim. Tecknol.* **19**, 221 (1976); *Chem. Abstr.*, **85**, 21217w.

181. J. J. Ursprung, U.S. Patent 3,105,837 (1963); *Chem. Abstr.*, **60**, 1763g; *cf.* also British Patent 935,776.

182. P. Loew, H. Schwander, and H. Kristinsson, German Patent 2,632,402 (1977); *Chem. Abstr.*, **86**, 189939c.

183. R. C. DeSelms, *J. Org. Chem.*, **27**, 2165 (1962).

184. M. Mousseron, J. M. Kamenka, and A. Steiger, *Chim. Ther.*, **2**, 95 (1967); *Chem. Abstr.*, **68**, 21883j.

185. Heidenheimer Chemisches Laboratum, British Patent 910,146 (1962); *Chem. Abstr.*, **58,** 9087d.

186. P. R. Thomas and G. J. Tyler, *J. Chem. Soc.*, **1957,** 2197.

187. (i) A. Hunger, J. Kebrle, A. Rossi, and K. Hoffman, *Helv. Chiv. Acta*, **43,** 800 (1960); (ii) ibid., **43,** 1032 (1960); (iii) ibid., **43,** 1727 (1960); (iv) German Patent 1,078,579 (1960); *Chem. Abstr.*, **55,** 18776b; (v) German Patent 1,077,666 (1960); *Chem. Abstr.*, **55,** 19953h; (vi) German Patent 1,075,622 (1960); *Chem. Abstr.*, **55,** 19954e; (vii) German Patent 1,079,059 (1960); *Chem. Abstr.*, **55,** 25988i; (viii) German Patent 1,079,646 (1960); *Chem. Abstr.*, **55,** 25989g; (ix) German Patent 1,081,019 (1960); *Chem. Abstr.*, **55,** 25990c; (x) British Patent 870,385 (1961); *Chem. Abstr.*, **55,** 25988g; (xi) Swiss Patent 361,286 (1962); *Chem. Absrt.*, **58,** 6835c; (xii) F. Sparatore, U.S. Patent 3,394,141 (1968); *Chem. Absrt.*, **69,** 86995n; (xiii) Ciba Ltd., British Patent 871,808 (1961); *Chem. Absrt.*, **57,** 4673; (xiv) F. Sparatori, V. Boido, and F. Fanelli, *Farmaco Ed. Sci.*, **23,** 344 (1968); *Chem. Abstr.*, **69,** 59157j.

188. A. V. El'tsov, E. R. Zakhs, and T. I. Frolova, *Zh. Org. Khim.*, **12,** 1088 (1976); *Chem. Abstr.*, **85,** 78050r.

189. G. Paglietti and F. Sparatori, *Farmaco Ed. Sci.*, **27,** 333 (1972); *Chem. Abstr.*, **77,** 48338h.

190. A. Hunger, J. Kebrle, A. Rossi, and K. Hoffman, U.S. Patent 3,004,982 (1959); *Chem. Abstr.*, **56,** 4772h.

191. M. Mengelberg, *Chem. Ber.*, **92,** 977 (1959).

192. Farberke Hoechst A. G. French M. 6681 (1969); *Chem. Abstr.*, **75,** 5898g.

193. Z. F. Solomko and G. A. Polinovskii, *Khim. Geterotsikl. Soedin.*, **1969,** 874; *Chem. Abstr.*, **72,** 100598f.

194. M. S. Malinovskii, Z. F. Solomko, G. A. Poinovskii, and V. I. Shvets, USSR Patent 237,901 (1969); *Chem. Abstr.*, **71,** 61390q.

195. R. C. DeSelms, *J. Org. Chem.* **27,** 2163 (1962).

196. T-Y. Shen, V. J. Grenda, and R. F. Czaja, German Patent 2,219,408 (1972); *Chem. Abstr.*, **78,** 29771y.

197. S. M. Elfimova, V. E. Shishkin, and B. I. No, USSR Patent 494,386 (1975); *Chem. Abstr.*, **84,** 105595k.

198. V. E. Shishkin, Yu M. Yukhno, and B. I. No, *Zh. Obshch. Khim.*, **47,** 1915 (1977).

199. G. A. Shvekhgeimer and G. A. Mikheichev, *Khim. Geterotsikl, Soedin.*, **1974,** 820; *Chem. Abstr.*, **81,** 105393f.

200. Cf. R. Roger and D. G. Nielson, *Chem. Rev.*, **61,** 179 (1961).

201. F. E. King and R. M. Acheson, *J. Chem. Soc.*, **1949,** 1396; see also E. C. Wagner, *J. Org. Chem.*, **5,** 133 (1940).

202. G. Holan, E. L. Samuel, B. C. Ennis, and R. W. Hinde, *J. Chem. Soc. C,* **1967,** 20.

203. D. Jerchel, H. Fischer, and M. Kracht, *Justus Liebigs Ann. Chem.*, **575,** 162 (1952).

204. M. Rombi and P. R. Dick, French Patent 2,178,385 (1972); *Chem. Abstr.*, **80,** 108526s.

205. G. P. Ellis and R. T. Jones, *J. Chem. Soc., Perkin I,* **1974,** 903.

206. N. V. Subba Rao and C. V. Ratnam, *J. Indian Chem. Soc.*, **38,** 631 (1961).

207. O. Sues, U.S. Patent 3,050,389 (1962); *Chem. Abstr.*, **59,** 11505e.

208. J. R. Bottu, French Patent 1,569,337 (1968); *Chem. Abstr.*, **72,** 100703m.

209. G. Hasegawa and H. Maruyama, Japanese Patent 71 09,581 (1971); *Chem. Abstr.*, **75,** 76797j.

210. J. G. Smith and I. Ho, *Tetrahedron Lett.*, **1971,** 3541.

211. C. M. Orlando, J. G. Wirth, and D. R. Heath, German Patent 2,064,683 (1971); *Chem. Abstr.*, **75,** 110313k.

212. E. Yu Belyaev, V. P. Kumarev, L. I. Kondrat'eva, and E. I. Shakhova, *Khim. Geterotsikl. Soedin.*, **7,** 1293 (1971); *Chem. Abstr.*, **76,** 34169w; Cf. ibid., 1688 (1970).

213. I. Ya Kvitko, L. I. Rudaya, and E. S. Kharlamova, USSR Patent 438,650 (1974); *Chem. Abstr.*, **81,** 120624a.

214. C. V. C. Rao, V. Veeranagaiah, and N. V. Subba Rao, *Curr. Sci.*, **43**, 280a (1974); *Chem. Abstr.*, **81**, 49621q.

215. V. Ts. Bukhaeva, *Mater. Nauchn. Konf. Aspir.*, *Rostov-na-Donu Gos Univ.*, **1968**, 233 V. A. Tishchenko, Ed.; *Chem. Abstr.*, **71**, 13060k.

216. A. F. Pozharskii, V. Ts. Bukhaeva, and A. M. Simonov, *Khim. Geterotsikl. Soedin.*, **1967**, 910; *Chem. Abstr.*, **68**, 105094r.

217. R. J. Alaimi and R. J. Storrin, U.S. Patent 3,899,503 (1975); *Chem. Abstr.*, **84**, 4952u.

218. (a) A. Jurasek, M. Breza, and R. Kada, *Collect. Czech. Chem. Commun.*, **37**, 2246 (1972).

    (b) K. Kada, J. Kovac, A. Jurasek, and L. David, *Collect. Czech. Chem. Commun.*, **38**, 1700 (1973).

219. (a) S. Yoshina, A. Tanaka, C-H. Wu, and H-S. Kuo, *Yakugaku Zasshi*, **95**, 1418 (1975); *Chem. Abstr.*, **84**, 90074p.

    (b) H. Harnisch, German Patent 2,346,316 (1975); *Chem. Abstr.*, **83**, 79240y.

    (c) J. Kokosinsky and B. Hancyk, *Pr. Inst. Przem. Org.*, **5**, 1 (1973); *Chem. Abstr.*, **83**, 192172c; ibid., Polish Patent 83,945 (1976); *Chem. Abstr.*, **86**, 72639p.

220. A. F. Pozharskii, V. Ts. Bukhaeva, A. M. Simonov, L. Ya Bakhmet, and P. M. Aleksan'yan, *Khim. Geterotsikl. Soei.*, **1969**, 325; *Chem. Abstr.*, **71**, 22066u.

221. Merck and Co. Inc., British Patent 966,796 (1964); *Chem. Abstr.*, **62**, 2779h.

222. J. M. McManus and R. M. Herbst, *J. Org. Chem.*, **24**, 1042 (1959).

223. A. L. Gershuns and A. N. Brizitskaya, *Khim. Geterotsikl Soedin.*, **1970**, 835; *Chem. Abstr.*, **73**, 109738e.

224. N. Vinot, *Compt. Rend.*, **256**, 699 (1963).

225. C. M. Orlando, J. G. Wirth, and D. R. Heath, U.S. Patent 3,673,202 (1972); *Chem. Abstr.*, **77**, 101619d.

226. R. Weidenhagen, *Chem. Ber.*, **69**, 2263 (1936).

227. H. F. Ridley, R. G. W. Spickett, and G. M. Timmis, *J. Heterocycl. Chem.*, **2**, 453 (1965).

228. O. Fischer and H. Wreszinski, *Chem. Ber.*, **25**, 2711 (1892).

229. G. T. Morgan and W. A. P. Challenor, *J. Chem. Soc.*, **119**, 1537 (1921).

230. N. A. Kalashnikova, G. N. Kul'bitskii, and B. V. Passet, *Khim.-Farm. Zh.*, **8**, 29 (1974); *Chem. Abstr.*, **81**, 169482s.

231. W. Krueger and G. Kreuger, East German Patent 99,787 (1973); *Chem. Abstr.*, **80**, 95953b.

232. N. P. Jensen, T-Y. Shen, and T. B. Windholz, U.S. Patent 3,655,901 (1972); *Chem. Abstr.*, **77**, 48463v.

233. E. Bellasio, German Patent 2,353,163 (1974); *Chem. Abstr.*, **81**, 77920e.

234. L. L. Shaletzky, U.S. Patent 3,928,596 (1975); *Chem. Abstr.*, **84**, 74270u.

235. S. B. Kadin, U.S. Patent 4,002,623 (1977); *Chem. Abstr.*, **86**, 171452d.

236. R. I.-Fu Ho and A. R. Day, *J. Org. Chem.*, **38**, 3084 (1973).

237. A. V. Zeiger and M. M. Joullié, *J. Org. Chem.*, **42**, 452 (1977); see also *idem*, *Synthetic Comm.*, **6**, 457 (1976).

238. B. Adcock, A. Lawson, and D. Miles, *J. Chem. Soc.*, **1961**, 5120.

239. S. Weiss, H. Michand, H. Prietzel, and H. Krommer, *Angew. Chem. Int. Ed.*, **12**, 841 (1973).

240. K. Sawatari, T. Mukai, K. Suenobu, S. Kamenosono, and T. Ike, Japanese Patent 76 16,669 (1976); *Chem. Abstr.*, **85**, 63069e.

241. German Patent, 2,214,600 (1972).

242. K. Harsanyi, G. Toth, A. Simay, C. Gonczi, K. Tabacs, and I. K. Ajzert, Hungarian Patent 5,800 (1973); *Chem. Abstr.*, **79**, 78801n; *idem*, British Patent 1,348,460; *Chem. Abstr.*, **81**, 3936t.

243. M. Kovacs, M. Nadesy, and G. Pfeifer, Hungarian Patent 2,820 (1971); *Chem. Abstr.*, **76**, 46204s.

244. R. Aries, French Patent 2, 063,815 (1971); *Chem. Abstr.*, **76**, 126984m.
245. E. I. du Pont de Nemours and Co., French Patent (Addition) 2,070,266; *Chem. Abstr.*, **76**, 126977m.
246. P. P. Actor and J. F. Pagano, U.S. Patent (Reissue) 28,403 (1975); *Chem. Abstr.*, **84**, 4950s.
247. R. Schlatter and D. A. Adams, U.S. Patent 3,997,553; *Chem. Abstr.*, **86**, 121335x.
248. R. J. Gyurik and V. J. Theodorides, U.S. Patent 3,915,986 (1975); *Chem. Abstr.*, **84**, 31074r.
249. G. Toth, K. Harsanyi, T. Montay, G. Szabo, E. Pasztor, K. Forsang, and I. Toth, Hungarian Patent 9,700 (1975); *Chem. Abstr.*, **84**, 44059q.
250. J. Moyne, German Patent 2,359,259 (1974); *Chem. Abstr.*, **81**, 91527p.
251. N. M. Burmakin, S. S. Kukalenko, A. M. Savinova, S. D. Voldkovich, N. Yu Berman, G. Ya Amasov, and L. S. Fishman, *Khim. Prom Moscow*, **1974**, 13; *Chem. Abstr.*, **80**, 82806f.
252. M. Akborova, S. A. Khasanov, Ch. Sh. Kadyrov, and S. Yu Yunusov, *Dokl. Akad. Nauk. Vzb. SSR*, **31**, 33 (1974); *Chem. Abstr.*, **84**, 105482w.
253. S. Iwaki, Japanese Patent 75 30,874 (1975); *Chem. Abstr.*, **83**, 193309c.
254. P. S. Kholkor, G. D. Sokolova, N. M. Burmakin, and S. G. Zhemchuzhin, *Khim. Geterotsikl. Soedin.*, **1974**, 1547; *Chem. Abstr.*, **82**, 72881d.
255. P. I. King, R. M. Acheson, and P. C. Spensley, *J. Chem. Soc.*, **1948**, 1366.
256. K. M. Acheson, G. A. Taylor, and M. I. Tomlinson, *J. Chem. Soc.*, **1958**, 3750.
257. F. A. Watts, German Patent 2,257,312 (1973); *Chem. Abstr.*, **79**, 53321t.
258. H. M. Loux, U.S. Patent 3,010,968 (1961); *Chem. Abstr.*, **58**, 1466g.
259. Z. Budesinsky, J. Sluka, J. Norvak, and J. Danek, *Collect. Czech. Chem. Commun.*, **40**, 1089 (1975).
260. H. Loewe, J. Urbanietz, D. Duewel, and R. Kirsch, German Patent 2,441,202 (1976); *Chem. Abstr.*, **85**, 21369x. *idem*, German Patent 2,441,201 (1976); *Chem. Abstr.*, **84**, 180222q.
261. H. Loewe, J. Urbanietz, D. Duewel, and R. Kirsch, German Patent 2,348,120 (1975); *Chem. Abstr.*, **83**, 43327t.
262. H. Loewe, J. Urbanietz, D. Duewel, and R. Kirsch, German Patent 2,348,104 (1975); *Chem. Abstr.*, **83**, 43326s.
263. H. Loewe, J. Urbanietz, D. Duewel, and R. Kirsch, German Patent 2,443,297 (1976); *Chem. Abstr.*, **85**, 78128x.
264. R. F. Lauer and A. Walser, German Patent 2,635,326 (1977); *Chem. Abstr.*, **87**, 23276e.
264a. For alternative procedures, see D. Takiguchi, Japanese Patent 75 12,087 (1975); *Chem. Abstr.* **83**, 43322n; *idem*, Japanese Patent 75 32,175 (1975); *Chem. Abstr.*, **83**, 206261h.
265. H. K. Klopping, U.S. Patent 3,896,230 (1975); *Chem. Abstr.*, **84**, 4948x.
266. R. Aries, French Patent 2,044,262 (1971); *Chem. Abstr.*, **76**, 14544m.
267. R. Aries, French Patent 2,052,901 (1971); *Chem. Abstr.*, **76**, 46203r.
268. R. Aries, French Patent 2,052,899 (1971); *Chem. Abstr.*, **76**, 126980g.
269. G. L. Dunn, U.S. Patent 3,694,455 (1972); *Chem. Abstr.*, **78**, 16183h.
270. R. T. Chang, Japanese Patent 77 19,665 (1977); *Chem. Abstr.*, **87**, 117855v.
271. A. H. M. Raeymaekers and H. L. J. Van Gelder, U.S. Patent 4,032,536 (1977); *Chem. Abstr.*, **87**, 168033h.
272. R. J. Gyurik and W. D. Kingsbury, U.S. Patent 3,969,526 (1976); *Chem. Abstr.*, **85**, 192728t.
273. C. C. Beard, J. A. Edwards, and J. H. Fried, German Patent 2,454,632 (1975); *Chem. Abstr.*, **84**, 44051f.
274. C. C. Beard, J. A. Edwards, and J. H. Fried, German Patent 2,411,295 (1974); *Chem. Abstr.*, **82**, 31324f. *idem*, U.S. Patent 4,005,202 (1977); *Chem. Abstr.*, **86**, 189937a.

275. C. C. Beard, J. A. Edwards, and J. H. Fried, German Patent 2,363,351 (1972); *Chem. Abstr.*, **81**, 105504t; *idem*, U.S. Patent 4,002,640 (1977); *Chem. Abstr.*, **86**, 171446e.

276. C. C. Beard, J. A. Edwards, and J. H. Fried, German Patent 2,363,348 (1974); *Chem. Abstr.*, **81**, 105514w.

277. C. C. Beard, J. A. Edwards, and J. H. Fried, U.S. Patent 3,993,768 (1976); *Chem. Abstr.*, **86**, 106583h. *idem*, U.S. Patent 3,929,824 (1975); *Chem. Abstr.*, **84**, 135653v.

278. C. C. Beard, U.S. Patent 4,031,228 (1977); *Chem. Abstr.*, **87**, 102333z. *idem*, U.S. Patent 3,962,437 (1976); *Chem. Abstr.*, **85**, 192724p.

279. K. Oyamada and J. Tobizuka, Japanese Patent 74 117,469 (1974); *Chem. Abstr.*, **82**, 125404j.

280. J. Solar Montolui, Spanish Patent 434,377 (1976); *Chem. Abstr.*, **86**, 189942y.

281. H. Osieka, K. H. Koenig, G. Bolz, and A. Amann, German Patent 2,012,219 (1971); *Chem. Abstr.*, **76**, 3858r.

282. E. H. Pommer, H. Osieka, K. H. Koenig, and G. Boltz, German Patent 2,012,589 (1971); *Chem. Abstr.*, **76**, 25294f.

283. D. K. Chae and N. T. Woo, *Soul Taehakkyo Yahhak Nonmunjip*, **1**, 1 (1976); *Chem. Abstr.*, **87**, 167940q.

284. F. S. Babichev and L. G. Rudchenko, *Ukr. Khim. Zh.*, **34**, 1269 (1968); *Chem. Abstr.*, **70**, 115062j.

285. V. I. Shvedov, L. B. Altukhova, L. A. Cheruyshova, and A. N. Grinev, *Zh. Org. Khim.*, **5**, 2221 (1969); *Chem. Abstr.*, **72**, 66865d.

286. Kh. A. Suerbaev and Ch. Sh. Kadyrov, *Khim. Geterotsikl. Soedin.*, **1974**, 1137; *Chem. Abstr.*, **81**, 152107a.

287. A. A. Shazhenov, Ch. Sh. Kadyrov, and P. Kurbanov, *Khim. Geterotsikl. Soedin.*, **1972**, 641; *Chem. Abstr.*, **77**, 139892a.

288. A. Botta, German Patent 2,110,227 (1972); *Chem. Abstr.*, **78**, 73476n.

289. D. N. Dhar, R. C. Munjal, and K. Bose, *Ann. Chim. Rome*, **63**, 757 (1973); *Chem. Abstr.*, **82**, 72875e.

290. W. Schroth and F. Raabe, East German Patent 93,559 (1972); *Chem. Abstr.*, **78**, 136288t.

291. V. V. Rudchenko, V. A. Buevich, and V. V. Perekalin, *Khim. Geterotsikl. Soedin.*, **1975**, 1576; *Chem. Abstr.*, **84**, 43940h. V. A. Buevich, U. S. Guneva, V. V. Rudchenko, and V. V. Perekalin, USSR Patent 495,312; *Chem. Abstr.*, **84**, 121834t.

292. V. A. Buevich, V. S. Grineva, V. V. Rudchenko, and V. V. Perekalin, *Zh. Org. Khim.*, **11**, 2620 (1975).

293. H. Schaefer and K. Gewald, *Z. Chem.*, **16**, 272 (1976); *Chem. Abstr.*, **86**, 55339y.

294. M. Hammad and A. Emran, *Z. Naturforsch., B*, **32**, 304 (1977).

295. A. Widdig and E. Kuehle, German Patent 1,932,297 (1971); *Chem. Abstr.*, **74**, 76426s.

296. M. M. Fawzi, U.S. Patent 3,562,290 (1971); *Chem. Abstr.*, **75**, 5896e; *idem*, U.S. Patent 3,839,416 (1974); *Chem. Abstr.*, **82**, 16834d.

297. (i) A. Hunger, J. Kebrle, A. Rossi, and K. Hoffmann, *Helv. Chim. Acta*, **44**, 1273 (1961); (ii) *idem*, U.S. Patent 3,000,898 (1959); *Chem. Abstr.*, **56**, 2456a.

298. Y. Miyazaki, D. Takiguchi, and K. Kato, Japanese Patent 74 28,512 (1974); *Chem. Abstr.*, **82**, 140138u.

299. A. K. Sarojini, N. Sriramulu, and C. V. Ratnam, *Curr. Sci.*, **41**, 776 (1972).

300. K. Takahashi, T. Suzuki, Y. Suzuki, H. Takeda, T. Zaima, and K. Mitsuhashi, *Nippon Kagaku Kaishi*, **1974**, 1595; *Chem. Abstr.*, **81**, 169478v.

301. H. Fleig and H. Hagen, German Patent 2,254,423 (1974); *Chem. Abstr.*, **81**, 25681h.

302. R. J. McCaully and C. Gochamn, U.S. Patent 3,740,413 (1973); *Chem. Abstr.*, **79**, 42504d.

303. O. Meth-Cohn, H. Suschitzky, and M. E. Sutton, *J. Chem. Soc. C*, **1968**, 1722.

304. M. D. Nair and R. Adams, *J. Am. Chem. Soc.*, **83**, 3518 (1961).

305. J. Martin, O. Meth-Cohn, and H. Suschitzky, *Tetrahedron Lett.*, **1973**, 4495.

306. (a) I. N. Somin and A. S. Petrov, *J. Gen. Chem. USSR*, **34**, 3177 (1964); (b) I. N. Somin, A. S. Petrov, and S. G. Kuznetzov, *J. Org. Chem. USSR*, **1**, 1454 (1965); (c) R. Garner and H. Suschitzky, *Chem. Commun.*, **1967**, 129.

307. J. S. Walia, P. S. Walia, L. A. Heindl, and P. Zyblot, *Chem. Commun.*, **1972**, 108.

308. F. Kröhnke and H. Leister, *Chem. Ber.*, **91**, 1479 (1958).

309. N. Ishikawa and T. Muramatsu, *Nippon Kagaku Kaishi*, **1973**, 563; *Chem. Abstr.*, **78**, 147873f.

310. B. George and E. P. Papadopoulos, *J. Org. Chem.*, **42**, 441 (1977).

311. Yu. M. Yutilov and L. I. Kovaleva, USSR Patent 541,846 (1977); *Chem. Abstr.*, **86**, 171448g.

312. (a) A. Rossi, A. Hunger, J. Kebrle, and K. Hoffmann, *Helv. Chim. Acta*, **43**, 1298 (1960); (b) *ibid.*, 1046 (1960).

313. Z. F. Solomko, V. S. Tkachenko, A. N. Kost, V. A. Budylin, and V. L. Pikalov, *Khim. Geterotsikl. Soedin.*, **1975**, 533; *Chem. Abstr.*, **83**, 79150u.

314. R. C. DeSelms, *Tetrahedron Lett.*, **1970**, 3001.

315. A. Allen and M. Habada, Czechoslovakian Patent 84,894 (1955); *Chem. Abstr.*, **50**, 9444h.

316. D. E. Burton, A. J. Lambie, D. W. J. Lane, G. T. Newbold, and A. Percival, *J. Chem. Soc. C*, **1968**, 1268.

317. B. George and E. P. Papadopoulos, *J. Org. Chem.*, **41**, 3233 (1976).

318. E. P. Papadopoulos, *J. Org. Chem.*, **41**, 962 (1976).

319. T. Hisano and H. Koga, *Yakugaku Zasshi*, **91**, 180 (1971); *Chem. Abstr.*, **74**, 125559y.

320. (a) J. H. Markgraf, W. P. Homan, R. J. Kaff, and W. L. Scott, *J. Heterocycl. Chem.*, **6**, 135 (1969). (b) C. W. Koch and J. H. Markgraf, *ibid.*, **7**, 235 (1970).

321. Ciba Ltd., British Patent 887,337 (1962); *Chem. Abstr.*, **57**, 15119g.

322. I. Ganea and R. Taranu, *Stud. Univ. Babes-Bolyai. Ser. Chem.*, 95 (1966); *Chem. Abstr.*, **67**, 32648s.

323. L. B. Piotrovskii, N. I. Kudryashova, N. V. Khromov-Borisov, and V. M. Adanin, *Zh. Org. Khim.*, **11**, 793 (1975); *Chem. Abstr.*, **83**, 43242m.

324. A. Dikciuviene, V. Bieksa, and J. Degutis, *Liet. TSR Mokslu Akad. Darb., Ser. B.* **1974**, 81; *Chem. Abstr.*, **83**, 43243n.

325. A. Dikciuviene, V. Bieksa, and J. Degutis, *Liet. TSR Mokslu Akad. Darb., Ser. B*, **1973**, 105; *Chem. Abstr.*, **79**, 115497r.

326. E. H. Gold, U.S. Patent 3,986,182 (1976); *Chem. Abstr.*, **86**, 72638n.

327. A. A. Shazhenov and Ch. Sh. Kadyrov, *Dokl. Akad. Nauk. Uzb SSR*, **31**, 35 (1974); *Chem. Abstr.*, **84**, 159565m.

328. M. T. LeBris, *Bull. Soc. Chim. Fr.*, **1967**, 3411.

329. D. J. Drain and J. G. B. Howes, British Patent 989,191 (1965); *Chem. Abstr.*, **63**, 609e.

330. N. A. Malichenko, A. P. Krasnoshchek, T. P. Medvedeva, and L. M. Yagupol'skii, *Khim. Geterotsikl. Soedin.*, **1976**, 1262; *Chem. Abstr.*, **86**, 55345x. cf. N. A. Malichenko, USSR Patent 505,639 (1976); *Chem. Abstr.*, **85**, 108638h.

331. V. K. Shchel'tsyn, A. Ya. Kaminski, T. P. Shapirovskaya, I. L. Vaisman, V. F. Andrianov, and S. S. Gitis, *Khim. Geterotsikl. Soedin.*, **1973**, 115; *Chem. Abstr.*, **78**, 97559k.

332. A. Baklien, Australian Patent 257,959 (1965); *Chem. Abstr.*, **68**, 12971j.

333. P. Rohrbach and I. Karadavidoff, German Patent 2,049,377 (1971); *Chem. Abstr.*, **75**, 20398n.

334. L. A. Smolenkova, L. I. Rudaya, I. Ya. Kvitko, and A. V. El'tsov, USSR Patent 486,081 (1975); *Chem. Abstr.*, **84**, 31072p.

335. A. Arsac and P. Frank, German Patent 2,601,041 (1976); *Chem. Abstr.*, **85**, 177423w.

336. M. Finotto, German Patent 2,062,265 (1972); *Chem. Abstr.*, **77**, 61997s.

337. L. Guczoghy, M. Puklics, G. Toth, G. Szabo, and D. Palfi, German Patent 2,241,035 (1973); *Chem. Abstr.*, **78**, 159600c.

338. H. Schlaepfer, Swiss Patent, 577,998 (1976); *Chem. Abstr.*, **85**, 177424x.
339. Y. M. Abou-Zeid, A. A. Abou-Ouf and H. Ragab, *Bull. Fac. Pharm. Cairo Univ.*, **2**, 27 (1963); *Chem. Abstr.*, **62**, 5269b.
340. M. Itaya, *Yakugaku Zasshi*, **82**, 1 (1962); *Chem. Abstr.*, **57**, 9840a.
341. M. H. Fisher, D. R. Hoff, and R. J. Bochis, U.S. Patent 3,705,174 (1972); *Chem. Abstr.*, **78**, 58419v.
342. B. C. Bishop, A. S. Jones, and J. C. Tatlow, *J. Chem. Soc.*, **1964**, 3076.
343. B. A. Porai-Koshits and Ch. Frankovskii, *Zh. Obshch. Khim.*, **28**, 928 (1958); *Chem. Abstr.*, **52**, 17240f.
344. H. D. Brown, Belgian Patent 613,916 (1962); *Chem. Abstr.*, **59**, 10065a.
345. R. L. Ellsworth, D. F. Hinkley, and E. F. Schoenewaldt, French Patent 2,014,421 (1970); *Chem. Abstr.*, **74**, 76425r.
346. R. Belcher, A. Sykes, and J. C. Tatlow, *J. Chem. Soc.*, **1954**, 4159.
347. V. V. Korshak, A. L. Rusanov, D. S. Tugushi, and S. N. Leont'eva, *Khim. Geterotsikl. Soedin.*, **1973**, 252; *Chem. Abstr.*, **78**, 136162x.
348. V. V. Lisunova, L. I. Rudaya, I. Ya Kvitko, A. V. El'tsov, and S. S. Gitis, USSR Patent 462,823 (1975); *Chem. Abstr.*, **83**, 43323p.
349. A. Botta, *Justus Liebigs Ann. Chem.*, **1976**, 336.
350. O. Meth-Cohn and H. Suschitzky, *J. Chem. Soc.*, **1964**, 2609.
351. W. Lehmann and H. Rinke, German Patent 842,065 (1952); *Chem. Abstr.*, **52**, 10207.
352. A. Mohsen and M. E. Omar, *Pharmazie*, **27**, 798 (1972).
353. A. Mohsen, M. E. Omar, M. S. Ragab, A. M. Farghaly, and A. M. Barghash, *Pharmazie*, **31**, 348 (1976).
354. A. Mohsen and M. E. Omar, *Synthesis*, **1973**, 41.
355. A. Mohsen, M. E. Omar, A. M. Farghaly, and Sh. A. Shams El-Dine, *Pharmazie*, **30**, 83 (1975).
356. P. Lugosi, B. Agai, and G. Hornyak, *Period Polytech. Chem. Eng.*, **19**, 307 (1975); *Chem. Abstr.*, **84**, 135542h.
357. A. Mohsen, M. E. Omar, A. M. Farghaly, and Sh. A. Shaims El-Dine, *J. Drug. Res.*, **6**, 253 (1974).
358. A. Mohsen, M. E. Omar, Sh. A. Chams El-Dine, and A. A. B. Hazzaa, *Pharmazie*, **30**, 85 (1975).
359. T. Matsui, M. Nagano, J. Tobitsuka, and K. Oyamada, *Yakugaku Zasshi*, **93**, 977 (1973); *Chem. Abstr.*, **79**, 105139m.
360. K. Oyamada, T. Matsui, J. Hizuka, M. Saito, and M. Nagano, Japanese Patent 74 32,354 (1974); *Chem. Abstr.*, **82**, 140142r.
361. S. Kano and M. Kaeriyama, Japanese Patent 75 01,040 (1975); *Chem. Abstr.*, **83**, 10085j.
362. C de Witt Adams and J. B. Wommack, German Patent 2,133,658 (1972); *Chem. Abstr.*, **76**, 99665z.
363. K. Koyamada, T. Matsui, and J. Tobitsuka, Japanese Patent 71 29,853 (1971); *Chem. Abstr.*, **75**, 140859k.
364. Y. Miyazaki, H. Takiguchi, and K. Kato, Japanese Patent 75 01,039 (1975); *Chem. Abstr.*, **83**, 10084h.
365. G. Depost, R. Salle, and B. Sillion, *C.R. Acad. Sci. Ser. C*, **275**, 697 (1972); *Chem. Abstr.*, **77**, 164599t.
366. R. A. Abramovitch and K. Schofield, *J. Chem. Soc.*, **1955**, 2326.
367. M. N. Sheng and A. R. Day, *J. Org. Chem.*, **28**, 736 (1963).
368. E. E. Glover, K. T. Rowbottom, and D. C. Bishop, *J. Chem. Soc. Perkin I*, **1973**, 842.
369. D. W. S. Latham, O. Meth-Cohn, and H. Suschitzky, *Chem. Commun.*, **1973**, 41.
370. G. V. Garner and H. Suschitzky, *Tetrahedron Lett.*, **1971**, 169.
371. J. Gyurko and J. Kaloczy, Hungarian Patent 149,980 (1963); *Chem. Abstr.*, **59**, 14000e.

372. E. Yu. Belyaev, V. P. Kumarev, L. E. Kondrat'eva, and E. I. Shakhova, *Khim. Geterotsikl. Soedin.*, **6**, 1688 (1970); *Chem. Abstr.*, **74**, 53650w; *ibid.* **7**, 1293 (1971); *Chem. Abstr.*, **76**, 34169w.

373. L. Horner and V. Schwenk, *Justus Liebigs Ann. Chem.*, **579**, 204 (1953).

374. (a) Tanabe Drug Manufacturing Co., Japanese Patent 4728 (1952); *Chem. Abstr.*, **47**, 11257i. (b) J. Kreidland and P. Turisanyi, Hungarian Patent 3304, *Chem. Abstr.*, **76**, 113216n.

375. R. L. Ellsworth, D. F. Hinkley, and E. F. Schoenewaldt, French Patent 2,014,420 (1970); *Chem. Abstr.*, **74**, 53799b.

376. N. S. Koslov and M. N. Stepanova, *Dokl. Akad. Nauk. Beloruss SSR*, **13**, 541 (1969); *Chem. Abstr.*, **71**, 124337a.

377. N. S. Koslov and M. N. Tovshtein, *Vestsi. Akad. Navuk Belaruss SSR, Ser. Khim. Navuk* **1967**, 89; *Chem. Abstr.*, **68**, 49507p.

378. N. S. Koslov and M. N. Tovshtein, *ibid*, **1968**, 72; *Chem. Abstr.*, **70**, 115063k.

379. Shell Internationale Research Maatshappij N. V., Netherlands Patent (App) 6,707,085 (1967); *Chem. Abstr.*, **69**, 27419e.

380. A. M. Simonov, V. M. Mar'yanovskii, and A. F. Pozharskii, *Zh. Org. Khim.*, **4**, 534 (1968); *Chem. Abstr.*, **68**, 105088s.

381. J. C. Watts, British Patent 1,351,883 (1974); *Chem. Abstr.*, **81**, 105512u; cf. also German Patent 2,204,479 (1973) and French Patent 2,170,981 (1973).

382. R. H. Smith and H. Suschitzky, *Tetrahedron*, **16**, 80 (1961).

383. H. Suschitzky and M. E. Sutton, *Tetrahedron Lett.*, **1967**, 3933.

384. C. V. C. Rao, K. K. Reddy, and N. V. Subba Rao, *Curr. Sci.*, **43**, 43 (1974).

385. R. S. Goudie and P. N. Preston, *J. Chem. Soc., C*, **1971**, 1139.

386. R. Kuhn and W. Blau, *Justus Liebigs Ann. Chem.*, **615**, 99 (1958).

387. S. Takahashi and H. Kano, *Chem. Pharm. Bull. (Tokyo)*, **12**, 783 (1964).

388. R. Fielden, O. Meth-Cohn, and H. Suschitzky, *Tetrahedron Lett.*, **1970**, 1229.

389. (a) P. van Romburgh and H. W. Huyser, *Versl. Gewone Vergad. Afd. Natuurk. Kon. Ned. Akad. Wetensch*, **35**, 665 (1926); *Chem. Abstr.*, **21**, 382 (1927); (b) *Recl. Trav. Chim. Pays-Bas*, **49**, 165 (1930); (c) P. van Romburgh and W. B. Deys, *Proc. Acad. Sci. Amsterdam*, **34**, 1004 (1931); *Chem. Abstr.*, **26**, 989 (1932).

390. R. K. Grantham and O. Meth-Cohn, *J. Chem. Soc. C*, **1969**, 70.

391. J. I. G. Cadogan, *Q. Rev.*, **22**, 222 (1968); *Synthesis*, **1**, 11 (1969).

392. J. I. G. Cadogan, R. Marshall, D. M. Smith, and M. J. Todd, *J. Chem. Soc. C*, **1970**, 2441.

393. L. Krbeckek and H. Takimoto, *J. Org. Chem.*, **29**, 3630 (1964).

394. J. H. Hall and D. R. Kamm, *J. Org. Chem.*, **30**, 2092 (1965).

395. R. Weidenhagen, *Chem. Ber.*, **69**, 2263 (1936).

396. F. F. Stephens and J. D. Bower, *J. Chem. Soc.*, **1949**, 2971; *ibid*, **1950**, 1722.

397. I. Bhatnagar and M. V. George, *Tetrahedron*, **24**, 1293 (1968).

398. K. S. Balachandran and M. V. George, *Indian J. Chem.*, **11**, 1267 (1973).

399. M. S. A. E. Meligy and S. A. Mohamed, *J. Prakt. Chem.*, **316**, 154 (1974).

400. N. J. Coville and E. W. Neuse, *J. Org. Chem.*, **42**, 3485 (1977).

401. K. H. Grellman and E. Tauer, *J. Am. Chem. Soc.*, **95**, 3104 (1973).

402. S. Linke and C. Wünsche, *J. Heterocycl. Chem.*, **10**, 333 (1973).

403. M. W. Partridge and H. A. Turner, *J. Chem. Soc.*, **1958**, 2086.

404. V. J. Grenda, R. E. Jones, G. Gal, and M. Sletzinger, *J. Org. Chem.*, **30**, 259 (1965).

405. T. L. Gilchrist, C. J. Moody, and C. W. Rees, *J. Chem. Soc. Perkin I*, **1975**, 1964.

406. M. Osone, S. Tanimoto, and R. Oda, *Yuki Gosei Kagaku Kyokai Shi*, **24**, 562 (1966); *Chem. Abstr.*, **65**, 10577g; S. Tanimoto and T. Ishibashi, *Yuki Gosei Kagaku Kyokai Shi*, **28**, 1073 (1970); *Chem. Abstr.*, **74**, 125563v.

407. (a) Merck and Co. Inc., British Patent 988,784 (1965); *Chem. Abstr.*, **63**, 16357c; Netherlands Patent (App) 6,513,320 (1967); *Chem. Abstr.*, **68**, 59577a. (b) R. Aries, French Patent, 1,604,908 (1972); *Chem. Abstr.*, **81**, 13516v.

408. R. L. Ellsworth, D. F. Hinkley, and E. F. Schoenewaldt, French Patent, 2,014,324 (1970); *Chem. Abstr.*, **74**, 87975w.
409. P. Held, M. Cross, and H. Schubert, *Z. Chem.*, **13**, 292 (1973).
410. I. Baxter and D. W. Cameron, *Chem. Ind. (London)*, **1967**, 1403; *J. Chem. Soc. C*, **1968**, 1747.
411. D. W. Cameron and R. G. F. Giles, *Chem. Commun.*, **1965**, 573; *J. Chem. Soc. C* **1968**, 1461.
412. M. Sprecher and D. Levy, *Tetrahedron Lett.*, **1969**, 4957.
413. Cf. E. V. Dehmlow, *Angew. Chem.*, **86**, 187 (1974); J. Dockx, *Synthesis*, **1973**, 441.
414. H. J. J. Loozen and E. F. Godefroi, *J. Org. Chem.*, **38**, 3495 (1973).
415. E. F. Godefroi, A. Corvers, and A. de Groot, *Tetrahedron Lett.*, **1972**, 2173.
416. P. Bouchet, C. Coquelet, J. Elguero, and R. Jacquier, *Bull. Soc. Chim. France*, **1976**, 192.
417. H. Tiefenthaler, W. Dörscheln, H. Göth, and H. Schmidt, *Helv. Chim. Acta*, **50**, 2244 (1967); *Tetrahedron Lett.*, **1964**, 2999.
418. J. P. Ferris and F. R. Antonucci, *J. Am. Chem. Soc.*, **96**, 2014 (1974); *Chem. Commun.*, **1972**, 126.
419. (a) T. Bucchetti and A. Alemagna, *Atti. Accad. Naz. Lincei, Cl. Sci, Fis., Mat. Nat.*, **22**, 637 (1957); *Chem. Abstr.*, **52**, 15511g. (b) *Atti. Accad. Naz. Lincei, Cl. Sci. Fis., Mat. Nat.*, **28**, 824 (1960); *Chem. Abstr.*, **56**, 7304c. (c) *Rend. Ist. Lomb. Sci. Lett.*, A, **94**, 242 (1960); *Chem. Abstr.*, **55**, 16527a.
420. R. L. Ellsworth, D. F. Hinkley, and E. F. Schoenewaldt, French Patent 2,041,402 (1970); *Chem. Abstr.*, **74**, 76422n.
421. J. H. Boyer and P. J. A. Frints, *J. Heterocycl. Chem.*, **7**, 59 (1970).
422. H. Kato, T. Shiba, E. Kitajima, T. Kiyosawa, F. Yamada, and T. Nichiyama, *J. Chem. Soc. Perkin I*, **1976**, 863; H. Kato, T. Shiba, and Y. Miki, *Chem. Commun.*, **1972**, 498.
423. P. A. S. Smith and E. Leon, *J. Am. Chem. Soc.*, **80**, 4647 (1958).
424. P. A. S. Smith and J. Vaughan, *J. Org. Chem.*, **23**, 1909 (1958).
425. A. W. Chapman and F. A. Fidler, *J. Chem. Soc.*, **1936**, 448.
425 (a) T. L. Gilchrist, C. J. Moody, and C. W. Rees, *Chem. Commun.*, **1976**, 414.
426. L Giammanco, F. P. Invidiata, *Atti Accad. Sci. Lett. Arti Palermo, Part 1*, **33**, 243 (1973); *Chem. Abstr.*, **83**, 97144z.
427. R. M. Moriarty and J. M. Kliegman, *J. Am. Chem. Soc.*, **89**, 5959 (1967).
428. F. L. Bach, J. Karliner, and G. E. Van Lear, *Chem. Commun.*, **1969**, 1110.
429. P. D. Hobbs and P. D. Magnus, *J. Chem. Soc., Perkin I*, **1973**, 469.
430. C. Kaneko, I. Yokoe, S. Yamada, and M. Ishikawa, *Chem. Pharm. Bull. Japan*, **14**, 1316 (1966).
431. R. A. Burrell, J. M. Cox, and E. G. Savins, *J. Chem. Soc., Perkin I*, **1973**, 2707.
432. R. J. Hemingway, unpublished results quoted in ref. 431.
433. C. Grundmann and A. Kreutzuerger, *J. Am. Chem. Soc.*, **77**, 6559 (1955).
434. R. L. Ellsworth, D. F. Hinckley, and E. F. Schoenewaldt, French Patent 2,014,422 (1970); *Chem. Abstr.*, **74**, 76423p.
435. A. C. Barker and R. G. Foster, German Patent 2,246,605 (1973); *Chem. Abstr.*, **79**, 5341c; British Patent 1,350,277 (1974); *Chem. Abstr.*, **81**, 25669k.
436. H. Loewe, J. Urbanietz, R. Kirsch, and D. Duewel, German Patent 2,332,486 (1975); *Chem. Abstr.*, **83**, 43324q.
437. H. Loewe, J. Urbanietz, D. Duewel, and R. Kirsch, German Patent 2,541,751 (1977); *Chem. Abstr.*, **87**, 5976d.
438. B. Bobranski, E. Wagner, and W. Roman, Polish Patent 81,666; *Chem. Abstr.*, **86**, 55447g.
439. C. J. Dik-Edixhoven, H. Schenk, and H. Van der Meer, *Crystallogr. Struct. Commun.*, **2**, 23 (1973), cf. also A. Escande and J. L. Galigné, *Acta Crystallogr.*, **B30**, 1647 (1974).

440. C. Hsu, M. H. Litt, and H. W. Chen, *U.S. NTIS, AD Rep.* (1975), AD-A017644; *Chem. Abstr.*, **84**, 129197h.
441. B. L. Trus and R. E. Marsh, *Acta Crystallogr.*, **B29**, 2298 (1973).
442. D. Chasseau, J. Gaultier, and C. Hauw, *C.R. Acad. Sci. Paris, Ser. C*, **274**, 1434 (1972); *Chem. Abstr.*, **77**, 10758g.
443. A. Quick and D. J. Williams, *Can. J. Chem.*, **54**, 2482 (1976).
444. A. N. Nesmeyanov, Yu. A. Belousov, V. N. Babin, G. G. Aleksandrov, Yu. T. Struchov, and N. S. Kochetkova, *Inorg. Chim. Acta*, **23**, 155 (1977).
445. G. Will, *Z. Kristallogr.*, **129**, 211 (1969).
446. O. A. Osipov, A. M. Simonov, V. I. Minkin, and A. D. Garnovskii, *Tr. Soveshch. Fiz Metodam Issled Organ Soedin. Khim Protsessov Akad Nauk Kirg SSR. Inst. Organ. Khim. Frunze*, **1962**, 61; *Chem. Abstr.*, **62**, 3494g.
447. V. S. Bolornikov, T. V. Lifintseva, S. B. Bulgarevich, V. N. Sheinker, O. A. Osipov, and A. D. Garnovskii, *Zh. Org. Khim.*, **12**, 416 (1976); *Chem. Abstr.*, **84**, 134963j.
448. A. F. Pozharskii and A. M. Simonov, *Zh. Obshch. Khim.*, **34**, 224 (1964); *Chem. Abstr.*, **60**, 10517b.
449. Yu V. Kolodyazhnyi, A. D. Garnovskii, S. V. Serbina, O. A. Osipov, B. S. Tanaseichuk, L. T. Rezepova, and S. V. Yartseva, *Khim. Geterotsikl. Soedin.*, **1970**, 819; *Chem. Abstr.*, **73**, 119967t.
450. C. W. N. Cumper and G. D. Pickering, *J. Chem. Soc. Perkin II*, **1972**, 2045.
451. C. W. N. Cumper and J. W. M. Wood, *J. Chem. Soc. B*, **1971**, 1096.
452. S. B. Bularevich, O. A. Osipov, V. S. Bolotnikov, T. V. Lifintseva, V. N. Sheinker, and A. D. Garnovskii, *Dokl. Akad. Nauk SSR*, **224**, 847 (1975); *Chem. Abstr.*, **84**, 43053w.
453. V. I. Minkin and B. Ya Simkin, *Khim. Geterotsikl. Soedin.*, **7**, 678 (1971); *Chem. Abstr.*, **76**, 78654e.
454. Yu L. Frolov, V. B. Mantsivoda, V. B. Modonov, S. N. Elovskii, E. S. Domnina, and G. G. Skvortsova, *Terr. Eksp. Khim.*, **9**, 238 (1973); *Chem. Abstr.*, **79**, 41757b.
455. K. J. Morgan, *J. Chem. Soc.*, **1961**, 2343.
456. D. J. Rabiger and M. M. Joullié, *J. Chem. Soc.*, **1964**, 915.
457. D. J. Rabiger and M. M. Joullié, *J. Org. Chem.*, **29**, 476 (1964).
458. N. A. Prokoshina and N. N. Khovratovich, *Zh. Prikl. Spektrosk.*, **24**, 174 (1976); *Chem. Abstr.*, **84**, 171634y.
459. G. P. Moiseeva, Ch. Sh. Kadyrov, and M. R. Yagudaev, *Khim. Zh.*, **14**, 41 (1970); *Chem. Abstr.*, **73**, 24461.
460. N. Mariaggi, A. Cornu, and R. Rinaldi; *Ann. Phys. Biol. Med.*, **7**, 51 (1973); *Chem. Abstr.*, **80**, 94777k.
461. N. N. Khovratovich and N. A. Borisevich, *Opt. Spektrosk. Akad. Nauk SSR, Otd. Fiz-Mat. Nauk. Sb. Statei*, **3**, 123 (1967); *Chem. Abstr.*, **68**, 109689m.
462. I. I. Mudretsova and S. L. Mertsalov, *Khim. Geterosikl. Soedin.*, **1975**, 1666; *Chem. Abstr.*, **84**, 134922v.
463. A. F. Pozharskii and A. A. Konstantinchenko, *Khim. Geterotsikl. Soedin.*, **1973**, 363; *Chem. Abstr.*, **78**, 146920g.
464. G. L. Carlson and W. G. Fateley, *Appl. Spectrosc.*, **23**, 374 (1969).
465. G. Leandri, A. Mangini, F. Montanari, and R. Posserini, *Gazz. Chim. Ital.*, **85**, 769 (1955).
466. A. F. Pozharskii, *Zh. Obshch. Khim.*, **34**, 630 (1964); *Chem. Abstr.*, **60**, 13118d.
467. G. H. Beaven, E. R. Holiday, and E. A. Johnson, *Spectrochim. Acta*, **4**, 338 (1951).
468. J. L. Rabinowitz and E. C. Wagner, *J. Am. Chem. Soc.*, **73**, 3030 (1951).
469. J. Durmis, M. Karvas, and Z. Manasek, *Collect. Czech. Chem. Commun.*, **38**, 224 (1973).
470. P. Grammaticakis and H. Texier, *C.R. Acad. Sci., Paris, Ser. C*, **274**, 878 (1972); *Chem. Abstr.*, **77**, 18799s.

471. N. N. Chipanina, N. A. Kazakova, Yu. L. Frulov, T. V. Kashik, S. M. Ponomoreva, E. S. Domnina, G. G. Skvortsova, and M. G. Voronkov, *Khim. Geterotsikl. Soedin.*, **1976**, 823; *Chem. Abstr.*, **85**, 123160q.
472. A. Jurášek and R. Kada, *Collect. Czech Chem. Commun.*, **35**, 3808 (1970).
473. R. D. Gordon and R. F. Yang, *Can. J. Chem.*, **48**, 1722 (1970).
474. T. K. Adler, *Anal. Chem.*, **34**, 685 (1962).
475. H. V. Schütt and H. Zimmermann, *Ber. Bunsenges Phys. Chem.*, **67**, 54 (1963).
476. H. C. Börresen, *Acta Chem. Scand.*, **17**, 921 (1963).
477. M. Kondo and H. Kuwano, *Bull. Chem. Soc. Japan*, **42**, 1433 (1969).
478. K. S. Rogers and C. C. Clayton, *Anal. Biochem.*, **48**, 199 (1972).
479. T. M. Kelly and D. F. O'Brien, *Photogr. Sci. Eng.*, **18**, 68 (1974).
480. L. B. Sanders, J. J. Ceterelli, and J. D. Winefordner, *Talanta*, **16**, 407 (1969).
481. P. J. Black and M. L. Heffernan, *Aust. J. Chem.*, **15**, 862 (1962).
482. J. Elguero, A. Fruchier, and S. Mignonac-Moudon, *Bull. Soc. Chim. Fr.*, **1972**, 2916.
483. Z. Mehesfalvi-Vajna, A. Neszmélyi, E. Baitz, and P. Sohár, *Acta Chim. Acad. Sci. Hung.*, **86**, 159 (1975).
484. P. Dembech, G. Seconi, P. Vivarelli, L. Schenetti, and F. Taddei, *J. Chem. Soc. B*, **1971**, 1670.
485. R. Benassi, P. Lazzeretti, L. Schenetti, and F. Taddei, *Tetrahedron Lett.*, **1971**, 3299.
486. J. Elguero, G. Llouquet, and C. Marzin, *Tetrahedron Lett.*, **1975**, 4085.
487. L. Pappalardo, J. Elguero, and C. Marzin, *C.R. Acad. Sci. Paris, Ser. C*, **277**, 1163 (1973); *Chem. Abstr.*, **81**, 70186r.
488. L. Pappalardo, J. Elguero, and A. Fruchier, *An Quim.*, **71**, 598 (1975).
489. Cf. A. N. Nesmayanov, E. B. Zavelovich, V. N. Babin, N. S. Kochetkova, and E. I. Fedin, *Tetrahedron*, **1975**, 1461.
490. H. A. Staab and A. Mannschreck, *Tetrahedron Lett.*, **1962**, 913.
491. J. Elguero, A. Fruchier, and M. del C. Pardo, *Org. Mag. Resonance*, **6**, 272 (1974).
492. K. F. Turchin, *Khim. Geterotsikl. Soedin.*, **1974**, 828; *Chem. Abstr.*, **81**, 119384r.
493. S. Bradamante, G. Pagani, and A. Marchesini, *J. Chem. Soc., Perkin II*, **1973**, 568.
494. P. J. Black, R. D. Brown, and M. L. Heffernan, *Aust. J. Chem.*, **20**, 1305 (1967).
495. K. Tori, T. Fujii, Y. Yoshimura, and M. Ueyama, *Chem. Lett.*, **1973**, 11.
496. I. Morishima, K. Ishihara, K. Tomishima, and T. Yonezawa, *J. Am. Chem. Soc.* **97**, 2749 (1975).
497. R. J. Pugmire and D. M. Grant, *J. Am. Chem. Soc.*, **93**, 1880 (1971).
498. E. Kleinpeter and R. Borsdorf, *J. Prakt. Chem.*, **315**, 765 (1973).
499. V. I. Troitskaya, V. I. Rudyk, E. V. Konovalov, and L. M. Yagupol'skii, *Zh. Org. Khim.*, **10**, 1524 (1974); *Chem. Abstr.*, **82**, 18601z.
500. M. Witanowski, L. Stefaniak, H. Jannszewski, Z. Grabowski, and G. A. Webb, *Tetrahedron*, **28**, 637 (1972).
501. H. Saito, Y. Tanaka, and S. Nagata, *J. Am. Chem. Soc.*, **95**, 324 (1973).
502. J. M. A. Al-Rawi, J. P. Bloxsidge, C. O'Brien, D. E. Caddy, J. A. Elvidge, J. R. Jones, and E. A. Evans, *J. Chem. Soc. Perkin II*, **1974**, 1635.
503. R. O. Metevosyan, L. A. Petrov, and V. D. Galyaminckikh, *Zh. Org. Khim.*, **2**, 896 (1966); *Chem. Abstr.*, **66**, 65422w.
504. A. T. Balaban, P. J. Halls, and A. R. Katritzky, *Chem. Ind.* (London), **1968**, 651.
505. H. Blum, A. R. McGhie, A. Kawada, and M. M. Labes, *J. Chem. Phys.*, **53**, 4097 (1970).
506. B. Lamotte and P. Servoz-Gavin, *Proc. Tihani Symp. Radiat. Chem.*, *2nd*, **1967**, 233; *Chem. Abstr.*, **67**, 59529.
507. G. F. Pedulli, A. Spisni, P. Vivarelli, P. Dembech, and G. Seconi, *J. Mag. Res.*, **12**, 331 (1973).
508. M. K. V. Nair, K. S. V. Santhanam, and B. Venkataraman, *J. Mag. Res.*, **9**, 229 (1973).
509. D. T. Edmonds and P. A. Speight, *J. Mag. Res.*, **6**, 265 (1972).

510. J. W. Clark-Lewis, J. A. Edgar, J. S. Shannon, and M. J. Thomson, *Aust. J. Chem.* **17,** 877 (1964).

511. R. A. Khmel'nitskii, A. N. Kost, K. K. Reddi, and V. I. Vysotskii, *Zh. Org. Khim.*, **5,** 1153 (1969); *Chem. Abstr.*, **71,** 101107m.

512. A. Kamal, A. A. Qureshi, I. H. Quereshi, and M. Aujum, *Pak. J. Sci. Ind. Res.*, **13,** 341 (1970); *Chem. Abstr.*, **75,** 20286z.

513. J. L. Cooper, S. R. Lipsky, and W. J. McMurray, *Org. Mass Spectrom.*, **3,** 1355 (1970).

514. F. L. Bach, J. Karliner, and G. E. Van Lear, *Chem. Commun.*, **1969,** 1100.

515. R. L. Ellsworth, H. E. Mertel, and W. J. A. Van den Huevel, *J. Agric. Food Chem.*, **24,** 544 (1976).

516. A. Tatematsu, H. Yashizumi, E. Hayashi, and H. Nakata, *Tetrahedron Lett.*, **1967,** 2985.

517. S.-O. Lawesson, G. Schroll, J. H. Bowie, and R. G. Cooks, *Tetrahedron*, **24,** 1875 (1968).

518. J. H. Bowie, R. G. Cooks, S.-O. Lawesson, and G. Schroll, *Aust. J. Chem.*, **20,** 1613 (1967).

519. A. Macquestiau, Y. Van Haverbeke, R. Flammang, M. C. Pardo, and J. Elguero, *Org. Mass Spectrom.*, **9,** 1188 (1974).

520. T. Nishiwaki, *J. Chem. Soc. C,* **1968,** 428.

521. A. Maquestiau, Y. Van Haverbeke, R. Flammang, M. C. Pardo, and J. Elguero, *Org. Mass. Spec.*, **10,** 313 (1975).

522. J. H. Bowie, R. G. Cooks, S.-O. Lawesson, and C. Nolde, *J. Chem. Soc. B,* **1967,** 616.

523. P. Natalis and J. L. Franklin, *J. Phys. Chem.*, **69,** 2943 (1965).

524. H. Walba and R. W. Isensee, *J. Am. Chem. Soc.*, **77,** 5488 (1955).

525. D. O. Jordan and H. F. W. Taylor, *J. Chem. Soc.*, **1946,** 994.

526. H. Walba and R. Ruiz-Velasco, *J. Org. Chem.*, **34,** 3315 (1969).

527. V. I. Minkin, V. A. Bren, A. D. Garnovskii, and R. I. Nikitina, *Khim. Geterotsikl. Soedin.*, **1972,** 552; *Chem. Abstr.*, **77,** 61151m.

528. M. G. Voronkov, T. V. Kashik, S. M. Ponomareva, and N. D. Abramova, *Dokl. Akad. Nauk SSSR*, **222,** 350 (1975); *Chem. Abstr.*, **83,** 57893m.

529. J. L. Cohen and K. A. Connors, *J. Pharm. Sci.*, **59,** 1271 (1970).

530. T. Hibma, P. Dupuis, and J. Kommandeur, *Chem. Phys. Lett.*, **15,** 17 (1972).

531. V. N. Sheinker, A. D. Garnovskii, O. A. Osipov, and L. S. Utkina, *Dokl. Akad. Nauk. SSSR*, **223,** 619 (1975); *Chem. Abstr.*, **83,** 177627z. L. S. Utkina, V. N. Sheinker, A. D. Garnovskii, and O. A. Osipov, *Khim. Geterotsikl. Soedin.*, **1974,** 1292; *Chem. Abstr.*, **82,** 30523h.

532. N. N. Chipanina, N. I. Sherfina, Yu. N. Ivlev, E. S. Domnina, D. D. Taryashinova, L. M. Sinegovskaya, G. G. Skvortsova, and M. G. Voronkov, *Khim. Geterotsikl. Soedin.*, **1973,** 1676; *Chem. Abstr.*, **80,** 81580r.

533. V. A. Chernyshev, V. N. Sheinker, A. D. Garnovskii, O. A. Osipov, and S. B. Bulgarevich, *Khim. Geterotsikl. Soedin.*, **1975,** 1696; *Chem. Abstr.*, **84,** 120889c.

534. L. G. Tishchenko, V. N. Sheinker, A. D. Garnovskii, O. A. Osipov, and S. B. Bulgarevich, *Zh. Obshch. Khim.*, **46,** 666 (1976); *Chem. Abstr.*, **84,** 179430z.

535. K. Szabo and A. H. Freiberg, U.S. Patent 3,592,805 (1971); *Chem. Abstr.*, **75,** 109415p.

536. N. P. Sweeney and K. F. Thom, U.S. Patent, 3,943,146 (1976); *Chem. Abstr.*, **85,** 5637c.

537. G. Favini and A. Gamba, *J. Chim. Phys.*, **64,** 1443 (1967).

538. M. Kamiya, *Bull. Chem. Soc. Japan*, **43,** 3344 (1970).

539. M. J. S. Dewar and N. Trinajstic, *J. Am. Chem. Soc.*, **91,** 6321 (1969).

540. Cf. F. Trischler, K. Kjoller, and H. Levine, *J. Appl. Poly. Sci.*, **11,** 1325 (1967).

541. (a) A. Escande, J. Lapasset, R. Raure, E. J. Vincent, and J. Elguero, *Tetrahedron*, **30,** 2903 (1974). (b) M. Gelus, P. M. Vay, and G. Berthier, *Theor. Chim. Acta*, **9,** 182 (1967).

542. V. I. Minkin, I. I. Zakharov, and L. L. Popova, *Khim. Geterotsikl. Soedin.*, **1971**, 1552; *Chem. Abstr.*, **77**, 11548a.

543. A. D. Garnovskii, A. M. Simonov, and V. I. Minkin, *Khim. Geterotsikl. Soedin.*, **1973**, 99; *Chem. Abstr.*, **80**, 2941h.

544. M. Ciureanu and V. Em. Sahini, *Rev. Roum. Chim.*, **20**, 1037 (1975); *Chem. Abstr.*, **84**, 4270v.

545. H. J. Petrowitz, *Chimia*, **19**, 426 (1965); *Chem. Abstr.*, **63**, 10652d.

546. A. A. Chemerisskaya, I. S. Tubina, and T. N. Egorova, *Khim.-Farm. Zh.*, **3**, 29 (1969); *Chem. Abstr.*, **71**, 6582d.

547. F. G. Von Stryk, *J. Chromatogr.*, **72**, 410 (1972).

548. E. R. Cole, G. Crank, and A. Salem Sheikh, *J. Chromatogr.*, **78**, 323 (1973).

549. H. Sato and M. Shimamine, *Eisei Shikensho Hokoku* No. 82, 88 (1964); *Chem. Abstr.*, **65**, 8690a.

550. J. M. McCall, *J. Med. Chem.*, **18**, 549 (1975).

551. D. J. Austin, K. A. Lord, and I. H. Williams, *Pestic Sci.*, **7**, 211 (1976).

552. H. Kambe, I. Mita, and R. Yokota, *Thermal Anal. Proc. Int. Conf. 3rd*, **3**, 387 (1972); *Chem. Abstr.*, **79**, 5824n.

553. R. Kada, A. Jurášek, J. Kovác, and P. Králik, *Chem. Zvesti*, **28**, 391 (1974); P. Králik, R. Kada, and A. Jurášek, *Thermal Anal.*, **3**, 31 (1971).

554. O. Fischer and H. Hess, *Chem. Ber.*, **36**, 3967 (1903).

555. E. Ochiai and M. Kataga, *J. Pharm. Soc. Jap.*, **60**, 543 (1940), *Chem. Abstr.*, **35**, 1785.

556. K. Fries, *Justus Liebigs Ann. Chem.*, **454**, 121 (1927).

557. G. E. Ficker and D. J. Fry, *J. Chem. Soc.*, **1963**, 736,

558. R. D. Brown and M. L. Heffernan, *J. Chem. Soc.*, **1956**, 3683, 4288.

559. L. S. Efros, *Zh. Obshch. Khim.*, **23**, 842 (1953); *Chem. Abstr.*, **48**, 4524c.

560. K. K. Preobrazhenskii, S. N. Khar'kov, V. I. Shlyakhov, and E. N. Smyalkovskaya, *Zh. Prikl. Khim.* (Leningrad), **48**, 2241 (1975); *Chem. Abstr.*, **84**, 30963t.

561. V. G. Sayapin, A. M. Simonov, and V. V. Kuz'menko, *Khim. Geterotsikl. Soedin.*, **1970**, 681; *Chem. Abstr.*, **73**, 45409p.

562. L. S. Efros, *Zh. Obshch. Khim.*, **23**, 951 (1953); *Chem. Abstr.*, **48**, 8222b.

563. A. F. Casey and J. Wright, *J. Chem. Soc.*, **1966**, 1511.

564. G. E. Ficker and D. J. Fry, *J. Chem. Soc.*, **1963**, 736.

565. K. L. Kirk and L. A. Cohen, *J. Org. Chem.*, **34**, 384 (1969).

566. E. L. Samuel, *Aust. J. Chem.*, **25**, 2725 (1972).

567. A. M. Simonov, Yu. M. Yutinov, and V. A. Anisimova, *Khim. Geterotsikl. Soedin, Akad. Nauk. Latv. SSR.*, 913 (1965); *Chem. Abstr.*, **64**, 12661a.

568. H. R. Hensel, French Patent 1, 403, 128 (1965); *Chem. Abstr.*, **64**, 2094f.

569. M. K. A. Khan, A. Mohammady, and F. Y. Ahmed, *Pak. J. Sci. Ind. Res.*, **15**, 11 (1972); *Chem. Abstr.*, **78**, 58306f.

570. W. E. Frick and T. Wenger, U.S. Patent, 3,686,411 (1972); *Chem. Abstr.*, **77**, 140079x.

571. V. A. Buevich and V. V. Rudchenko, *Khim. Geterotsikl. Soedin.*, **1977**, 848; *Chem. Abstr.*, **87**, 184426u.

572. L. S. Efros, B. A. Porai-Koshits, and S. G. Farbenshtein, *Zh. Obshch. Khim.*, **23**, 1691 (1953); *Chem. Abstr.*, **48**, 13686e.

573. M. Akborova and Ch. Sh. Kadyrov, *Uzb. Khim. Zh.*, **19**, 42 (1975); *Chem. Abstr.*, **83**, 164071p.

574. M. Akbarova and Ch. Sh. Kadyrov, *Dokl. Akad. Nauk. Uzb SSR*, **32**, 30 (1975); *Chem. Abstr.*, **84**, 105483x.

575. M. Ewert and H. Oelschläger, *Arch. Pharm.* (Weinheim), **308**, 550 (1975).

576. O. Süs, *Justus Liebigs Ann. Chem.*, **579**, 133 (1953).

577. S. N. Kolodyazhnaya, A. M. Simonov, and I. G. Uryukina, *Khim. Geterotsikl. Soedin.*, **1972**, 1690; *Chem. Abstr.*, **78**, 72004n.

578. P. B. de la Mare and J. H. Ridd, "Aromatic Substitution", Butterworths, London, 1959, p 203.

579. H. Pauly and K. Gundermann, *Chem. Ber.*, **41,** 3999 (1908).
580. D. Harrison, J. T. Ralph, and A. C. B. Smith, *J. Chem. Soc.*, **1963,** 2930.
581. P. Linda, *Tetrahedron*, **25,** 3297 (1969).
582. A. Grimson, J. H. Ridd, and B. V. Smith, *J. Chem. Soc.*, **1960,** 1352, 1357.
583. J. H. Ridd and B. V. Smith, *J. Chem. Soc.*, **1960,** 1363.
584. A. F. Pozharskii, Candidate's Thesis, Rostov-on-Don University, Rostov-on-Don USSR (1963) quoted in ref. 9.
585. M. A. Phillips, *J. Chem. Soc.*, **1931,** 1143.
586. R. Foster, H. R. Ing, and E. F. Rogers, *J. Chem. Soc.*, **1957,** 1671.
587. (a) P. Vivarelli and F. Taddei, *Gazz. Chim. Ital.*, **105,** 801 (1975); (b) see also K. K. Reddy and N. V. Subba Rao, *Proc. Ind. Acad. Sci., Sect. A*, **71,** 141 (1970); *Chem. Abstr.*, **73,** 56024k.
588. P. M. Kochergin, M. V. Povstyanoi, B. A. Priimenko, E. V. Logachev, and V. S. Ponomar, USSR Patent 477,158 (1975); *Chem. Abstr.*, **83,** 164181z.
589. E. V. Logachev, M. V. Povstyanoi, and P. M. Kochergin, *Ukr. Khim. Zh.*, **42,** 401 (1976); *Chem. Abstr.*, **85,** 46502y.
590. E. V. Logachev, M. V. Povstyanoi, P. M. Kochergin, and Yu I. Beilis, *Izv. Vyssh. Ucheb. Zaved., Khim. Khim. Technol,* **19,** 1039 (1976); *Chem. Abstr.*, **85,** 159984a.
591. W. Ost, K. Thomas, and D. Jerchel, U.S. Patent 3,694,454 (1972); *Chem. Abstr.*, **78,** 16181f.
592. Laboratorio Farmaceutico Quimico-Laforquim SA., Spanish Patent 411,071 (1975); *Chem. Abstr.*, **85,** 46685k.
593. R. M. Bystrova and O. N. Karpov, USSR Patent 372,224 (1973); *Chem. Abstr.*, **79,** 42506f.
594. J. A. Maynard, J. D. Rae, D. Rash, and J. M. Swan, *Aust. J. Chem.*, **24,** 1873 (1971).
595. W. Gauss, F. Grewe, and H. Schweinpflug, German Patent 2,130,030 (1972); *Chem. Abstr.*, **78,** 72149p.
596. A. M. Simonov and A. F. Pozharskii, *Zh. Obshch. Khim.*, **31,** 3970 (1961); *Chem. Abstr.*, **57,** 8559h.
597. R. D. Haugwitz and V. L. Narayanan, German Patent 2,210,113 (1972); *Chem. Abstr.*, **77,** 164695w.
598. D. G. O'Sullivan and A. K. Wallis, *J. Med. Chem.*, **15,** 103 (1972).
599. Chimetron, French Patent 1,450,541 (1966); *Chem. Abstr.* **66,** 85792v.
600. Farbenifabriken Bayer A.-G. British Patent 849,793 (1960); *Chem. Abstr.*, **59,** 7535d.
601. E. L. Ringwald and A. B. Craig, U.S. Patent 2,623,879; *Chem. Abstr.*, **47,** 9367c.
602. W. J. Houlihan, U.S. Patent 3,953,600 (1974); *Chem. Abstr.*, **85,** 78228e.
603. I. I. Popov, P. V. Tkachenko, and A. M. Simonov, *Khim. Geterotsikl. Soedin.*, **1973,** 551; *Chem. Abstr.*, **79,** 31984j.
604. J. Sawlewicz, L. Bukowski, and M. Rogaczewska, *Dissert. Pharm.*, **14,** 297 (1962); *Chem. Abstr.*, **59,** 5149d.
605. K. Hoffman and A. Hunger, Swiss Patent 363,657 (1962); *Chem. Abstr.*, **59,** 11504a.
606. K. Hoffman, A. Hunger, J. Kebrle, and A. Rossi, Swiss Patent 361,286 (1962); *Chem. Abstr.*, **58,** 6835c.
607. A. Hunger, J. Kebrle, A. Rossi, and K. Hoffmann, *Helv. Chim. Acta*, **43,** 800 (1960).
608. K. Hoffman, A. Hunger, J. Kebrle, and A. Rossi, Swiss Patent 362,081 (1962); *Chem. Abstr.*, **59,** 8758f.
609. Y. M. Abou-Zeid, A. A. Abou-Ouf, and B. Abdel-Fattah, *U.A.R.J. Pharm. Sci.*, **11,** 29 (1970); *Chem. Abstr.*, **75,** 129723r.
610. A. F. Casy and J. Wright, *J. Chem. Soc. C,* **1966,** 1167.
611. T. Seki, *Yakugaku Zasshi*, **87,** 301 (1967); *Chem. Abstr.*, **67,** 82159d.
612. K. Hideg, H. O. Hankovsky, G. Mehes, L. Decsi, and M. Varszegi, Hungarian Patent, 152,439 (1965); *Chem. Abstr.*, **64,** 8195d.
613. K. Hideg and O. H. Hankovsky, *Acta Chim. Acad. Sci. Hung.*, **49,** 303 (1966); *Chem. Abstr.*, **66,** 65425z.

614. M. Inoue, K. Arai, S. Kidokoro, T. Saito, and K. Okuzawa, Japanese Patent 74 70,970 (1974); *Chem. Abstr.*, **81,** 136143d.

615. A. N. Krasovskii, P. M. Kochergin, and L. V. Samoilenko, *Khim. Geterotsikl. Soedin.*, **1970,** 827; *Chem. Abstr.*, **73,** 109740z.

616. R. M. Palei, P. M. Kochergin, and S. A. Chernyak, USSR Patent 545,641 (1977); *Chem. Abstr.*, **86,** 189949f.

617. S. Herrling, H. Keller, and H. Mückter, German Patent 1,000,384 (1957); *Chem. Abstr.*, **54,** 1550b.

618. H. L. Yale and J. A. Bristol, US Patent 4,004,016 (1977); *Chem. Abstr.*, **86,** 171450b.

619. O. Ehrmann, F. Zimmerman, and L. Friedrich, German Patent 2,300,018 (1974); *Chem. Abstr.*, **81,** 120632b.

620. C. Fauran, J. Eberle, C. Raynaud, and Y. Bailly, German Patent 2,158,801 (1972); *Chem. Abstr.*, **77,** 126628u.

621. H. Roechling and K. H. Buechel, *Z. Naturforsch. B*, **25,** 1103 (1970); *Chem. Abstr.*, **74,** 13062z.

622. Fisons Pest Control, French Patent 1,459,782 (1966); *Chem. Abstr.*, **67,** 54129a; cf. French Addn. 93,257; *Chem. Abstr.*, **72,** 21691c.

623. Fisons Pest Control, Netherlands Appl. 6,609,819 (1967); *Chem. Abstr.*, **67,** 73609y.

624. H. L. Klopping, French Patent 1,523,597 (1968); *Chem. Abstr.*, **72,** 21692d.

625. A. M. Simonov and N. D. Vitkevich, *Zh. Obshch. Khim.*, **29,** 2404 (1959); *Chem. Abstr.*, **54,** 9896b.

626. D. G. O'Sullivan and A. K. Wallis, *Nature* **198,** 1270 (1963).

627. Cf. A Hunger, J. Kebrle, A. Rossi, and K Hoffmann, *Experientia*, **13,** 400; F. Gross and H Turrian, ibid., p. 401.

628. Cf. H. D. Brown, A. R. Matzuk, I. R. Ilves, L. H. Petersen, S. A. Harris, L. H. Sarett, J. R. Egerton, J. J. Yakstis, W. C. Campbell, and A. C. Cuckler, *J. Am. Chem. Soc.*, **83,** 1764 (1961).

629. M. Th. Le Bris and H. Wahl, *Rev. Textile-Tiba*, **57,** 164 (1958); *Chem. Abstr.*, **53,** 2208e.

630. B. D. Murphy and J. Musco, *J. Org. Chem.*, **36,** 3469 (1971).

631. J. Hannah, E. F. Rogers, and D. W. Graham, German Patent 2,307,519 (1973); *Chem. Abstr.*, **80,** 82982k.

632. P. M. Zavlin and G. L. Matevosyan, *Zh. Obshch. Khim.*, **47,** 483 (1977); *Chem. Abstr.*, **87,** 5862p.

633. P. M. Zavlin, G. L. Matevosyan, and R. M. Matyushicheva, *Zh. Obshch. Khim.*, **47,** 484 (1977); *Chem. Abstr.*, **87,** 23154p.

634. R. Aries, French Patent 2,061,874 (1971); *Chem. Abstr.*, **76,** 126985n; cf. idem, 1,565,347 (1969); *Chem. Abstr.*, **73,** 25466b.

635. I. I. Chizhevskaya, R. V. Skupskaya, and A. N. Kharitonovich, *Sin. Org. Soedin.*, **1970,** 49; *Chem. Abstr.*, **76,** 25173r.

636. (a) J. Sawlewicz, L. Bukowski, J. Jasinska, D. Polaczek, J. Purzycka, Z. Sznigir, and D. Wesolowska, *Acta Pol. Pharm.*, **17,** 85 (1960); *Chem. Abstr.*, **54** 17381g; (b) J. Sawlewicz and Z. Sznigir, *Acta Pol. Pharm.*, **18,** 1 (1961); *Chem. Abstr.*, **55,** 2727i; (c) J. Sawliewicz and M. Wasowska, *Acta Pol. Pharm.*, **17,** 113 (1960); *Chem. Abstr.*, **54,** 21056i.

637. A. Dikciuviene, V. Bieksa, and J. Degutis, *Liet TSR Nokslu Akad. Darb.*, Ser. B, 1973, 49; *Chem. Abstr.*, **80,** 59893y.

638. J. Sawlewicz, D. Kajoto, J. Purzycka, and D. Wyzinska, *Gdanskie Towarzyst. Nauk. Wydzial Nauk. Mat. Przyrodniczych, Rozprawy Wydziak III*, No. 1, 175 (1964); *Chem. Abstr.*, **64,** 15869e.

639. I. I. Chizhevskaya and V. I. Pansevich-Kolyada, *Zh. Obshch. Khim.*, **27,** 1495 (1957); *Chem. Abstr.*, **52,** 3720g.

640. (a) J. Sawlewicz, M. Grzybowski, B. Milczorska, J. Kulikowski, and I. Skwierawska, *Rozpr. Wydz. 3: Nauk. Mat.-Przyr., Gdansk Tow. Nauk.*, **9,** 141 (1973); *Chem. Abstr.*,

81, 152099z; (b) Idem, *Rozpr. Gdansk Tow. Nauk. Wydz.,* **1972,** 141; *Chem. Abstr.,* **80,** 37046x.

641. Z. B. Efendiev, *Azerb. Khim. Zh.,* **1975,** 104; *Chem. Abstr.,* **83,** 206158e.

642. V. S. Ponomer, P. M. Kochergin, O. S. Anisimova, Yu N. Sheinker, B. A. Priimenko, and M. V. Povstyanoi, *Khim. Geterotsikl Soedin.,* **1975,** 1284; *Chem. Abstr.,* **84,** 59300k.

643. J. Rokach, Y. Girard, and J. G. Atkinson, *Canad. J. Chem.,* **51,** 3765 (1973).

644. W. J. Irwin and D. L. Wheeler, *Tetrahedron,* **28,** 1113 (1972).

645. T. Kamiya, Y. Saito, M. Hashimoto, and H. Seki, Japanese Patent 75 77,371 (1975); *Chem. Abstr.,* **84,** 31056m.

646. G. B. Bachman and L. V. Heisey, *J. Am. Chem. Soc.,* **68,** 2496 (1946).

647. K. Hideg and H. O. Hankovsky, *Acta Chim. (Budapest),* **53,** 271 (1967); *Chem. Abstr.,* **69,** 51956h.

648. T. Okuda, *Yakugaku Zasshi,* **80,** 205 (1960); *Chem. Abstr.,* **54,** 13141g.

649. G. R. Revankar and S. Siddappi, *Monatsh. Chem.,* **98,** 169 (1967).

650. A. Novelli, *Bol. Soc. Quim. Peru,* **19,** 99 (1953); *Chem. Abstr.,* **49,** 1021f.

651. S. S. Parmar, R. S. Misra, A. Chaudhari, and T. K. Gupta, *J. Pharm. Sci.,* **1972,** 1322.

652. H. Roechling, K. Haertel, and H. Goebel, German Patent 2,230,182 (1972); *Chem. Abstr.,* **80,** 120951j.

653. S. Bahadur, A. K. Goel, and R. S. Varma, *J. Indian Chem. Soc.,* **53,** 1163 (1976).

654. E. A. Steck and R. P. Brundage, *Org. Prep. Proced. Int.,* **7,** 6 (1975); *Chem. Abstr.,* **83,** 131520f.

655. J. C. Craig, M. Moyle, and L. F. Johnson, *J. Org. Chem.,* **29,** 410 (1964).

656. F. Andreani, R. Andrisano, C. D. Casa, and M. Tramontini, *Tetrahedron Lett.,* **1968,** 1059, See also C. Fauran, J. Eberle, A. Y. Le Cloarec, G. Raynaud, and M. Sergant, French Patent 2,186,251 (1974); *Chem. Abstr.,* **81,** 3938v.

657. R. Zelnic and F. Strehlau, *Experientia,* **21,** 617 (1965). *Idem, Mem. Inst.* Butantan Sao Paulo, **35,** 147 (1971); *Chem. Abstr.,* **77,** 126499c.

658. V. N. Syutkin, A. M. Efros, and S. N. Danilov, *Zh. Prikl Khim.,* **43,** 1367 (1970); *Chem. Abstr.,* **73,** 87843f.

659. L. Bukowski, *Rocz. Chem.,* **46,** 1131 (1972); *Chem. Abstr.,* **77,** 164595p.

660. V. L. Narayanan and R. D. Haugwitz, U.S. Patent 3,929,814 (1975); *Chem. Abstr.* **84,** 105605p.

661. G. Paglietti and F. Sparatori, *Farm. Ed. Sci.,* **27,** 471 (1972); *Chem. Abstr.,* **77,** 126493w. *Idem, Studi Sassar Sez,* **48** (3–4), 125 (1970); *Chem. Abstr.,* **76,** 85749t.

662. N. K. Khakimova, A. A. Shazhenov, and Ch. Sh. Kadyrov, *Dokl. Akad. Nauk. Uzb SSR,* **32,** 35 (1975); *Chem. Abstr.,* **84,** 105485y.

663. A. M. Efros and O. N. Usaevich, *Zap. Leningrad Sel'sk. Khoz. Inst.* **180,** 49 (1972); *Chem. Abstr.,* **78,** 111212a.

664. V. P. Dukhovskoi, V. D. Tyurin, and G. A. Shvekhgeimer, *Tr. Mosk. Inst., Neftekhim Gazov. Prom.,* No. 3, 12 (1969): *Chem. Abstr.,* **75,** 151728t.

665. G. Hasegawa and A. Kotani, Japanese Patent 74 20,173 (1974); *Chem. Abstr.,* **81,** 105503s.

666. G. Hasegawa and A. Kotani, Japanese Patent 74 05,967 (1974); *Chem. Abstr.,* **80,** 120941f.

667. G. Hasegawa and A. Kotani, Japanese Patent 73 103,576 (1973); *Chem. Abstr.,* **81,** 37552e.

668. G. Ahmed, T. Wagner-Jauregg, E. Pretsch, and J. Seibl, *Helv. Chim. Acta,* **56,** 1646 (1973). Q. Ahmed and T. Wagner-Jauregg, Swiss Patent, 583,226 (1976); *Chem. Abstr.,* **87,** 5973a.

669. Yu A. Mansurov, V. V. An, G. G. Skvortsov, and V. K. Voronov, USSR Patent, 380,652 (1971); *Chem. Abstr.,* **79,** 53323v.

670. T. Hisano and M. Ichikawa, *Yakugaku Zasshi,* **91,** 1136 (1971); *Chem. Abstr.,* **76,** 14432y.

671. E. Profft and W. Georgi, *Justus Liebigs Ann. Chem.*, **643**, 136 (1961).
672. A. P. Gray, H. Kraus, and D. E. Heitmeier, *J. Org. Chem.*, **25**, 1939 (1960).
673. R. D. Kimbrough, *J. Org. Chem.*, **29**, 1242 (1964).
674. G. A. Taylor, *J. Chem. Soc.*, **1975**, 1001.
675. H. D. Brown and L. H. Sarett, Belgian Patent 621,596 (1963); *Chem. Abstr.*, **59**, 11504h; U.S. Patent 3,055,907 (1962); *Chem. Abstr.*, **58**, 2456c.
676. C. Chiyomaru, S. Kawada, and K. Takita, Japanese Patent 73 99,171 (1973); *Chem. Abstr.*, **80**, 95952a.
677. D. R. Baker, J. J. Menn, and A. H. Freiberg, U.S. Patent 3,658,812 (1972); *Chem. Abstr.*, **77**, 48473y.
678. G. Dransch, H. Mildenberger, D. Duewel, and R. Kirsch, German Patent, 2,425,704 (1975); *Chem. Abstr.*, **84**, 105593h.
679. W. T. Ashton and E. F. Rogers, German Patent, 2,626,346 (1976); *Chem. Abstr.*, **86**. 121338a.
680. H. Roechling, G. Hoerlein, and K. Haertel, German Patent, 2,231,883 (1974); *Chem. Abstr.*, **80**, 95959h.
681. R. Aries, French Patent 2,177,540 (1973); *Chem. Abstr.*, **80**, 146160a.
682. M. Nakao, S. Katayama, and H. Yamamoto, Japanese Patent 73 34,747 (1973).
683. M. Inoue, T. Saito, K. Arai, S. Kidokoro, and H. Okuzawa, Japanese Patent, 74 76,870 (1974); *Chem. Abstr.*, **82**, 16840c.
684. K. H. Mayer, D. Lauerer, and H. Heitzer, *Synthesis*, **1975**, 673.
685. R. Aries, French Patent, 2,061,876 (1971); *Chem. Abstr.*, **76**, 126983k.
686. H. Roechling and K. H. Buechel, *Z. Naturf. B*, **25**, 1103 (1970).
687. R. Aries, French Patent 2,104,634 (1972); *Chem. Abstr.*, **77**, 140077v.
688. G. T. Newbold, A. Percival, and M. B. Purdew, German Patent 2,157,051 (1972); *Chem. Abstr.*, **77**, 88505s.
689. Fisons Pest Control, French Patent 1,459,782; *Chem. Abstr.*, **67**, 54129a.
690. H. F. W. Röchling, K. H. Büchel, and F. Korte, U.S. Patent, 3,988,465 (1976); *Chem. Abstr.*, **86**, 72641h.
691. E. F. Rogers, W. T. Ashton, and R. A. Dybas, U.S. Patent, 4,017,505 (1977); *Chem. Abstr.*, **87**, 53302c. *Cf. idem*, German Patent, 2,519,987 (1975); *Chem. Abstr.*, **84**, 150627h.
692. E. F. Rogers and D. B. R. Johnstone, German Patent 2,519,979 (1975); *Chem. Abstr.*, **84**, 15062g.
693. Fisons Pest Control, Netherlands Appl. 6,609,19 (1967); *Chem. Abstr.*, **67**, 73609y.
694. W. Daum, H. Scheinpflug, and P. E. Frohberger, German Patent 2,140,863 (1973); *Chem. Abstr.*, **78**, 124589d.
695. H. Grabinger, R. Sehring, and C. Drandorevski, German Patent 2,204,416 (1973); *Chem. Abstr.*, **79**, 115588w.
696. E. I. Du Pont de Nemours and Co., French Patent, 2,155,138 (1973); *Chem. Abstr.*, **79**, 105255w.
697. W. Daum, P. Frohberger, and B. Hamburger, German Patent 2,227,920 (1973); *Chem. Abstr.*, **80**, 82977n.
698. R. Aries, French Patent 2,045,672 (1971); *Chem. Abstr.*, **76**, 3860k.
699. W. T. Ashton and E. F. Rogers, U.S. Patent 4,017,504 (1977); *Chem. Abstr.*, **87**, 23286h.
700. W. Schulze, G. Letsch, W. Jungstand, W. Gutsche, and K. Wohlrabe, East German Patent 96,947 (1973); *Chem. Abstr.*, **79**, 137154d.
701. Chimetron, French Patent 1,439,128 (1966); *Chem. Abstr.*, **65**, 20135d.
702. G. T. Newbold and A. Percival, U.S. Patent 3,430,259 (1969); *Chem. Abstr.*, **70**, 96797j.
703. Fisons Pest Control, Netherlands Appl. 6,610,554 (1967); *Chem. Abstr.*, **67**, 73610s.
704. J. Sawlewicz and J. Jasinska, *Rocz. Chem.*, **38**, 1073 (1964); *Chem. Abstr.*, **61**, 16062e.
705. Chimetron, French Patent, 1,439,129 (1966); *Chem. Abstr.*, **65**, 18594c.

706. P. E. Wittreich, K. A. Folkers, and F. M. Robinson, U.S. Patent 3,056,777; *Chem. Abstr.*, **58**, 9087g.

707. K. H. Büchel and W. Stendel, German Patent 2,242,785 (1974); *Chem. Abstr.*, **80**, 146159g.

708. A. Widdig, E. Kuehle, F. Grewe, and H. Kaspers, German Patent 2,206,010 (1973); *Chem. Abstr.*, **79**, 115589x.

709. C. J. Paget and J. H. Wilkcl, German Patent 2,528,846 (1976); *Chem. Abstr.*, **84**, 135664z. *idem*, U.S. Patent 4,018,790 (1977); *Chem. Abstr.*, **87**, 117857x.

710. Yu S. Simanenko, L. M. Litvinenko, and V. A. Dadali, *Zh. Org. Khim.*, **10**, 1308 (1974); *Chem. Abstr.*, **81**, 77300c.

711. L. M. Litvinenko, V. A. Dadali, S. A. Lapshin, Yu. S. Simanenko, and V. I. Rybachenko, *Zh. Org. Khim.*, **11**, 249 (1975); *Chem. Abstr.*, **82**, 124369q.

712. R. Aries, French Patent 2,048,150 (1971); *Chem. Abstr.*, **76**, 3861m.

713. C. C. Beard, J. A. Edwards, and J. H. Fried, Belgian Patent 817,164 (1975); *Chem. Abstr.*, **83**, 97298c. *idem*, U.S. Patent 4,031,234 (1977); *Chem. Abstr.*, **87**, 135333c.

714. P. S. Khokhlov, S. S. Kukalenko, G. D. Sokolova, E. A. Dvoichenkova, V. D. Tkachev, A. I. Mochalkin, and G. M. Strongin, USSR Patent 487,886; *Chem. Abstr.*, **84**, 17358e.

715. G. Hoerlein, H. Mildenberger, A. Kröniger, and K. Haertel, German Patent 2,125,815 (1972); *Chem. Abstr.*, **78**, 58417t. *Idem*, German Patent 2,202,774 (1973); *Chem. Abstr.*, **79**, 115586u.

716. W. Daum, H. Scheinpflug, P. E. Frohberger, and F. Grewe, British Patent 1,228,108 (1971); *Chem. Abstr.*, **75**, 36053g. *Idem*, U.S. Patent 3,673,210 (1972); *Chem. Abstr.*, **77**, 101614y.

717. K. Szabo, German Patent 2,641,947 (1977); *Chem. Abstr.*, **87**, 135337g.

718. K. Szabo, German Patent 2,641,759 (1977); *Chem. Abstr.*, **87**, 68360u.

719. N. M. Golyshin, British Patent 1,385,123 (1975); *Chem. Abstr.*, **83**, 58830a.

720. A. Widdig, K. Sasse, F. Grewe, H. Scheinpflug, P. E. Frohberger, and H. Kaspers, German Patent 1,936,130 (1971); *Chem. Abstr.*, **74**, 87980u.

721. E. H. Pommer, H. Osieka, K. H. Koenig, and G. Bolz, German Patent 2,012,589; *Chem. Abstr.*, **76**, 25294f.

722. J. E. Moore, German Patent 2,128,013 (1971); *Chem. Abstr.*, **76**, 72517r.

723. J. E. Moore, U.S. Patent 3,732,241 (1973); *Chem. Abstr.*, **79**, 53335a.

724. R. Aries, French Patent 2,043,470 (1971); *Chem. Abstr.*, **76**, 3857q.

725. V. Andriska, K. Gorog, G. Bruckner, Z. Nemessanyi, M. Havasi, B. Baskay, E. Grega, T. Szigethy, and J. Dudas, U.S. Patent 4,046,773 (1977); *Chem. Abstr.*, **87**, 184509y. P. Gribovszki, Z. Pinter, G. Bors, S. Morosvolgyi, G. Szilagy, J. Dudas, T. Szigeti, E. Grega, and B. Raskay, French Patent, 2,233,938 (1975); *Chem. Abstr.*, **83**, 97295z.

726. U. Gebert and H. Mildenberger, German Patent 2,217,169 (1973); *Chem. Abstr.*, **80**, 3521h.

727. W. Daum, H. Scheinpflug, and F. Grieve, German Patent 2,145,512 (1973); *Chem. Abstr.*, **78**, 147967q.

728. E. Hubele, Swiss Patent 574, 936 (1976); *Chem. Abstr.*, **85**, 108642e.

729. G. Dransch, H. Bresenberg, H. Mildenberger, K. Haertel, and R. Kirsch, German Patent 2,303,999 (1974); *Chem. Abstr.*, **81**, 13649k.

730. G. Hoerlein, H. Mildenberger, A. Kröniger, and K. Haertel, German Patent 2,126,838 (1972); *Chem. Abstr.*, **78**, 72146k.

731. G. Hörlein, H. Mildenberger, A. Kröniger, K. Haertel, D. Düwel, and R. Kirsch, German Patent 2,257,184 (1974); *Chem. Abstr.*, **81**, 63627c.

732. G. Hörlein, H. Mildenberger, A. Kröniger, and K. Haertel, German Patent 2,044,205 (1972); *Chem. Abstr.*, **77**, 34522b.

733. J. Perronnet and P. Girault, German Patent 2,347,650 (1974); *Chem. Abstr.*, **81**, 13517w.

734. K. Roeder, R. Puttner, and E. A. Pieroh, German Patent 2,347,386 (1975); *Chem. Abstr.*, **83**, 43325r.

735. All-Union Scientific Research Institute of Chemicals for Plant Protection, Japanese Patent 76 43,760 (1976); *Chem. Abstr.*, **86**, 72637m.

736. G. Dransch, H. Mildenberger, D. Düwel, R. Kirsch, and U. Gebert, German Patent 2,425,705 (1975); *Chem. Abstr.*, **84**, 121826s.

737. B. I. Dittmar, S. African Patent 74 04,351 (1973); *Chem. Abstr.*, **80**, 82980h.

738. A. A. Atakaziev, Ch. Sh. Kadyrov, and S. A. Khasanov, *Usb. Khim. Zh.*, **18**, 49 (1974); *Chem. Abstr.*, **81**, 3834h.

739. B. I. Dittmar, British Patent 1,398,325 (1975); *Chem. Abstr.*, **83**, 131606p. *Idem*, U.S. Patent 3,970,668 (1976); *Chem. Abstr.*, **86**, 16670e. E. I. du Pont de Nemours and Co., French Patent 2,191,555 (1974); *Chem. Abstr.*, **81**, 37551d.

740. R. Aries, French Patent 2,077,443 (1971); *Chem. Abstr.*, **77**, 62000y.

741. G. Toth and I. Toth, German Patent 2,641,416 (1977); *Chem. Abstr.*, **87**, 23284f.

742. B. I. Dittmar, German Patent 2,226,301 (1973); *Chem. Abstr.*, **80**, 82967j.

743. R. D. Haugwitz and V. L. Narayanan, German Patent 2,310,076 (1973); *Chem. Abstr.*, **79**, 126496r.

744. R. D. Haugwitz and V. L. Narayanan, German Patent 2,446,259 (1975); *Chem. Abstr.*, **83**, 79248g.

745. C. J. Paget, J. W. Chamberlin, and J. H. Wikel, German Patent 2,638,553 (1977); *Chem. Abstr.*, **87**, 68362w.

746. J. H. Wikel and C. J. Paget, U.S. Patent 4,008,243 (1977); *Chem. Abstr.*, **86**, 189940w.

747. R. Aries, French Patent 2,067,436 (1971); *Chem. Abstr.*, **77**, 5473f.

748. Chinoin Gyogyszer es Vegyeszeti Termekek Gyara Rt, *Belg.* 841,538 (1976); *Chem. Abstr.*, **87**, 135325b.

749. G. Toth, I. Toth, and T. Montay, German Patent 2,618,853 (1976); *Chem. Abstr.*, **86**, 72651m.

750. R. Aries, French Patent 2,054,799 (1971); *Chem. Abstr.*, **76**, 25292d.

751. B. I. Dittmar, U.S. Patent 3,641, 048 (1972); *Chem. Abstr.*, **76**, 140827u.

752. M. F. Shostakovskii, G. G. Skortsova, N. P. Glaskova, and E. S. Domnina, *Khim. Geterotsikl. Soedin.*, **1969**, 1070; *Chem. Abstr.*, **72**, 132612x.

753. I. E. P. Murray, Ph.D. Thesis, Heriot-Watt University, Edinburgh, 1980.

754. R. M. Acheson, M. W. Foxton, P. J. Abbot, and K. R. Mills, *J. Chem. Soc. C*, **1967**, 882.

755. R. M. Acheson and W. R. Tully, *J. Chem. Soc. C*, **1968**, 1623.

756. R. M. Acheson and M. S. Verlander, *J. Chem. Soc., Perkin I*, **1973**, 2348.

757. Yu P. Andreichikov, G. E. Trukhan, S. N. Lyubchenko, and G. N. Dorofeenko, *Khim. Geterotsikl. Soedin.*, **1976**, 238; *Chem. Abstr.*, **84**, 164682j.

758. V. N. Bubnovskaya, V. A. Ol'shevskaya, and F. S. Babichev, *Ukr. Khim. Zh.*, **43**, 43 (1977); *Chem. Abstr.*, **86**, 171331p.

759. A. R. Katritzky and F. Yates, *J. Chem. Soc., Perkin I*, **1976**, 309.

760. N. Grier, German Patent 1,945,452 (1970); *Chem. Abstr.*, **72**, 132730j.

761. J. Elguero, A. R. Katritzky, B. S. El-Osta, R. L. Harlow, and S. H. Simonsen, *J. Chem. Soc. Perkin I*, **1976**, 312.

762. A. Banerji, J. C. Cass, and A. R. Katritzky, *J. Chem. Soc. Perkin I*, **1977**, 1162.

763. Y. Tamura, H. Hayashi, J. Minamikawa, and M. Ikeda, *Chem. Ind.* (London), **1973**, 952.

764. Y. Tamura, J. Minamikawa, Y. Miki, S. Matsugashita, and M. Ikeda, *Tetrahedron Lett.*, **1972**, 4133.

765. E. E. Glover and K. T. Rowbottom, *J. Chem. Soc. Perkin I*, **1976**, 367.

766. E. Regel and K-H Büchel, *Justus Liebigs Ann. Chem.*, **1977**, 145.

767. J. Hannah, D. W. Graham, and E. F. Rogers, U.S. Patent 3,749,734 (1973); *Chem. Abstr.*, **79**, 78802p. *Idem*, French Patent 2,145,633 (1973); *Chem. Abstr.* **79**, 53322u.

768. L. Bukowski, *Acta Pol. Pharm.*, **31**, 725 (1974); *Chem. Abstr.*, **83**, 28152p.
769. R. D. Haugwitz and V. L. Narayanan, *J. Org. Chem.*, **37**, 2776 (1972).
770. V. M. Pechenina, N. A. Mukhina, K. A. Abaturova, L. P. Grebenschchikova, T. V. Mikhailova, V. M. Kurilenko, and A. P. Gilev., *Khim.-Farm. Zh.* **5**, 13 (1971); *Chem. Abstr.*, **76**, 14430w.
771. V. M. Pechenina, N. A. Mukhina, K. A. Abaturova, V. K. Gorshkova, and A. P. Gilev, *Khim.-Farm. Zh.*, **3**, 18 (1969); *Chem. Abstr.*, **71**, 13062n.
772. V. I. Sokolov, A. F. Pozharskii, and B. I. Ardashev, *Khim. Geterotsikl. Soedin.*, **1973**, 967; *Chem. Abstr.*, **79**, 115493m.
773. N. Grier, U.S. Patent 3,907,700 (1975); *Chem. Abstr.*, **84**, 4959b.
774. E. Bellasio, A. Campi, A. Trani, E. Baldoli, A. M. Caravaggi, and G. Nathansohn, *Farm. Ed. Sci.*, **28**, 164 (1973); *Chem. Abstr.*, **78**, 111214c.
775. L. L. Skaletsky, German Patent 2,127,960 (1971); *Chem. Abstr.*, **76**, 59628h.
776. H. Osicka, E. H. Pommer, B. Zeeh, and B. Girgensohn, German Patent 2,211,903 (1973); *Chem. Abstr.*, **79**, 137151a.
777. G. Toth and I. Toth, German Patent 2,518,188 (1975); *Chem. Abstr.*, **84**, 44064n.
778. J. Koo, S. Avakian, and G. J. Martin, *J. Am. Chem. Soc.*, **77**, 5373 (1955).
779. R. J. Stedman, U.S. Patent, 3,480,642 (1969); *Chem. Abstr.*, **75**, 118314g.
780. K. H. Mayer, F. Grewe, and P. E. Frohberger, German Patent 2,253,324 (1974); *Chem. Abstr.*, **81**, 25673g.
781. R. Aries, French Patent 2,058,611 (1971); *Chem. Abstr.*, **76**, 99661v.
782. Dyanchim S.a.r.l., French Patent 2,046,114 (1971); *Chem. Abstr.*, **75**, 151800k.
783. K. H. Mayer and F. Grewe, German Patent 2,331,183 (1975); *Chem. Abstr.*, **82**, 170935u.
784. R. Aries, French Patent 2,061,875 (1971); *Chem. Abstr.*, **76**, 126979p.
785. H. Loewe, J. Urbanietz, D. Duewel, and R. Kirsch, German Patent 2,332,487 (1975); *Chem. Abstr.*, **82**, 156301c.
786. H. Oewe, J. Urbanietz, D. Duewel, and R. Kirsch, German Patent 2,541,741 (1977); *Chem. Abstr.*, **87**, 39485h.
787. G. Dransch and G. Hörlein, German Patent 2,250,469 (1974); *Chem. Abstr.*, **81**, 13512r.
788. A. M. Simonov and V. N. Komissarov, *Khim. Geterotsikl. Soedin.*, **1976**, 783; *Chem. Abstr.*, **85**, 123818k.
789. H. O. Hankovsky and K. Hideg, *Acta Chim. Acad. Sci. Hung.*, **53**, 405 (1967); *Chem. Abstr.*, **68**, 29642a.
790. A. M. Simonov and N. D. Vitkevich, *Zh. Obshch. Khim.*, **30**, 590 (1960); *Chem. Abstr.*, **54**, 24677i.
791. M. Augustin and K. R. Kuppe, *Tetrahedron*, **30**, 3533 (1974).
792. A. M. Simonov and A. F. Pozharskii, *Khim. Geterotsikl. Soedin. Akad. Nauk. Latv. SSR*, **1965**, 203; *Chem. Abstr.*, **63**, 8343g; *Zh. Obshch. Khim.*, **31**, 3970 (1961).
793. R. D. Haugwitz and V. L. Narayanan, U.S. Patent 3,926,967 (1975); *Chem. Abstr.*, **84**, 90155r.
794. J. R. E. Hoover and R. J. Stedman, U.S. Patent 3,399,212; *Chem. Abstr.*, **70**, 11697d.
795. M. Augustin and K. R. Kuppe, *Z. Chem.*, **14**, 306 (1974).
796. A. M. Simonov, S. N. Kolodyazhnaya, and L. N. Podladchikova, *Khim. Geterotsikl. Soedin*, **1974**, 689; *Chem. Abstr.*, **81**, 77841e.
797. S. N. Kolodyazhnaya, A. M. Simonov, and L. N. Podladchikova, USSR Patent 467,902 (1975); *Chem. Abstr.*, **83**, 114408h.
798. A. M. Simonov, S. N. Kolodyazhnaya, and L. N. Podladchikova, USSR Patent 469,701 (1975); *Chem. Abstr.*, **83**, 114408h.
799. V. S. Misra and N. S. Agarwal, *J. Prakt. Chem.*, **311**, 697 (1969).
800. B. G. Khadse, M. H. Shah, C. V. Deliwala, M. B. Bhide, S. S. Majajani, and M. V. Bhat, *Bull. Haff. Inst.*, **5**, 9 (1977).

801. S. N. Kolodyazhnaya, A. M. Simonov, and L. N. Podladchikova, *Khim. Geterotsikl. Soedin.*, **1975**, 829; *Chem. Abstr.*, **83**, 178929m.

802. R. Aries, French Patent 2,058,435 (1971); *Chem. Abstr.*, **76**, 99666a.

803. A. Arens, P. Zarins, and E. S. Lavrinovich, USSR Patent 382,627; *Chem. Abstr.*, **79**, 78798s.

804. M. V. Provstyanoi, E. V. Logachev, P. M. Kochergin, and Yu. I. Beilis, *Izv. Vyssh. Uchebn. Zaved. Khim. Khim. Technol.*, **19**, 708 (1976); *Chem. Abstr.*, **85**, 78051s.

805. P. E. Frohberger and W. Gauss, German Patent 2,435,210 (1976); *Chem. Abstr.*, **85**, 105398u.

806. L. Bukowski, *Acta Pol. Pharm.*, **32**, 651 (1975); *Chem. Abstr.*, **86**, 72516w.

807. C. H. Budeanu, G. Rusu, Z. Cojocaru, and C. Nistor, *Rev. Med-Chir.*, **80**, 605 (1976); *Chem, Abstr.*, **87**, 102230p.

808. C. H. Budeanu, G. Rusu, Z. Cojocaru, and C. Nistor, *Bull. Inst. Politeh Iasi, Sect. 2*, **22**, 63 (1976); *Chem. Abstr.*, **86**, 171334s.

809. W. R. Sullivan, *J. Med. Chem.*, **13**, 784 (1970).

810. J. Sawlewicz and W. Kuzmierkiewicz, *Acta Pol. Pharm.*, **33**, 661 (1976); *Chem. Abstr.*, **87**, 23152m.

811. J. Ciernik and V. Vystavel, Czechoslovakian Patent 149, 322 (1973); *Chem. Abstr.*, **80**, 37119y.

812. N. K. Chub, E. B. Tsupak, A. M. Simonov, and L. K. Nezhenets, *Khim. Geterotsikl. Soedin.*, **1972**, 130; *Chem. Abstr.*, **76**, 153675t.

813. A. Morimoto, Y. Nakai, and II. Tagasugy, Japanese Patent 73 78,165 (1973); *Chem. Abstr.*, **80**, 82968k.

814. J. Sawlewicz, B. Milczoiska, and W. Manowska, *Pol. J. Pharmacol. Pharm.*, **27**, 187 (1975); *Chem. Abstr.*, **83**, 206171d,

815. B. Milczoiska, J. Sawliewicz, and W. Manowska, *Pol. J. Pharmacol Pharm.*, **28**, 521 (1976); *Chem. Abstr.*, **87**, 58650.

816. J. Sawliewicz and W. Kuzmierkiewicz, *Rozpr. Wyda. 3: Nauk. Mat.-Przyr., Gdansk, Tow. Mauk.*, **9**, 147 (1973); *Chem. Abstr.*, **81**, 152100t.

817. M. M Ali, Abd-Elsamei M. Abd-Elfattah, and H. A. Hammouda, *Z. Naturforsch. B*, **31**, 254 (1976).

818. C. Fauran, J. Eberle, G. Raynaud, and J. Thomas, German Patent 2,159,674 (1972); *Chem. Abstr.*, **77**, 140078w.

819. C. Fauran, J. Eberle, G. Raynaud, and N. Dorme, French Patent 2,140,347 (1973); *Chem. Abstr.*, **79**, 32057w.

820. B. Serafin and E. Szymanowska, *Rocz. Chem.*, **49**, 791 (1975); *Chem. Abstr.*, **83**, 131522h.

821. Yu. A. Sedov and I. Ya. Postovskii, *Khim. Pharm. Zh*, **2**, 16 (1968); *Chem. Abstr.*, **70**, 3940n.

822. N. P. Bednyagina, N. V. Serebryakova, R. I. Ogloblina, and I. I. Mudretsova, *Khim. Geterotsikl. Soedin.*, **1968**, 541: *Chem. Abstr.*, **70**, 3944s.

823. N. P. Bednyagina and G. N. Lipunova, *Khim. Geterotsikl. Soedin.*, **1968**, 902; *Chem. Abstr.*, **71**, 13063p.

824. N. P. Bednyagina, Yu. A. Sedov, I. Ya. Postovskii, and Yu. A. Rybakova, *Zh. Khim. Abstr.*, No. 8Zh, 364 (1968); *Chem. Abstr.*, **70**, 3956x.

825. G. M. Petrova, N. P. Bednyagina, T. G. Malkina, and V. N. Podchainova, *Khim. Geterotsikl. Soedin.*, **4**, 709 (1968); *Chem. Abstr.*, **70**, 37713f.

826. T. A. Kuznetsova, R. I. Ogloblina, and N. P. Bednyagina, *Tezisy Dokl.-Nauchno-Tekh. Konf. "Khim. Primen Formazanov,"* 2nd 1974 (1975) 45–6; *Chem. Abstr.*, **87**, 23158t.

827. J. Sawlewicz, B. Milczorska, A. Czarnocka-Janowicz, H. Kupiec, and C. Kurowicka, *Rozpr. Wydz 3: Nauk Nat.-Pyzyr Gdansk Tow Nauk*, **9**, 135 (1973); *Chem. Abstr.*, **83**, 9897u.

828. T. Ishii, N. Anzai, and T. Ito, Japanese Patent 75 13,264 (1975); *Chem. Abstr.*, **84**, 31057n.
829. T. Ishii and S. Ito, Japanese Patent 74 28,193 (1974); *Chem. Abstr.*, **84**, 44047j.
830. T. Ishii and S. Ito, Japanese Patent 74 28,192 (1974); *Chem. Abstr.*, **82**, 140141q.
831. T. Ishii and T. Ito, Japanese Patent 74 28,511 (1974); *Chem. Abstr.*, **82**, 140140p.
832. L. Bukowski and J. Sawlewicz, *Rozpr. Wydz 3: Nauk. Mat.-Przyr., Gdansk Tow. Nauk*, **7**, 131 (1973); *Chem. Abstr.*, **81**, 152102v.
833. W. Schulze, *J. Prakt. Chem.*, **315**, 189 (1973).
834. R. D. Haugwitz and V. L. Narayanan, U.S. Patent 3,879,414 (1975); *Chem. Abstr.*, **83**, 131592f.
835. V. S. Misra and I. Husain, *J. Indian Chem. Soc.*, **37**, 710 (1960); *Chem. Abstr.*, **55**, 12390f.
836. P. Brenneisen and A. Margot, U.S. Patent 3,586,670 (1971); *Chem. Abstr.*, **77**, 5469j. *Idem*, 3,822,356 (1974); *Chem. Abstr.*, **81**, 91533n.
837. S. A. Agripat, French M. 8207; *Chem. Abstr.*, **78**, 43478q.
838. J. Renault and J. Berlot, *C.R. Acad. Sci., Paris, Ser. C*, **270**, 358 (1970).
839. Cf. I. V. Grachev, *Zh. Obshch. Khim.*, **17**, 2268 (1947).
840. J. V. Hay, D. E. Portlock, and J. F. Wolfe, *J. Org. Chem.*, **38**, 4379 (1973).
841. B. A. Tertov, A. S. Morkovnik, and Yu G. Bogachev, *Khim. Geterotsikl. Soedin.*, **1976**, 1699; *Chem. Abstr.*, **86**, 155724p.
842. P. W. Alley and D. A. Shirley, *J. Org. Chem.*, **23**, 1791 (1958).
843. B. A. Tertov and S. E. Pauchenko, *Zh. Obshch. Khim.*, **33**, 3671 (1963); *Chem. Abstr.*, **60**, 8020e.
844. A. V. Koblik, *Mater. Nauch. Konf. Aspir.*, Rostov.-na-Donu Gos. Univ., 7th, 1967, 235 (1968); *Chem. Abstr.*, **71**, 13061m (1969).
845. B. A. Tertov, N. A. Ivankova, and A. M. Simonov, *Zh. Obshch. Khim.*, **32**, 2989 (1962); *Chem. Abstr.*, **58**, 9048b.
846. B. A. Tertov and A. V. Koblik, *Khim. Geterotsikl. Soedin.*, **1967**, 1123; *Chem. Abstr.*, **69**, 59158k.
847. B. A. Tertov, A. V. Koblik, and Yu. V. Kolodyazhnyi, *Tetrahedron Lett.*, **1968**, 4445.
848. B. A. Tertov, V. V. Burykin, P. P. Onishchenko, A. S. Morkovnik, and V. V. Bessonov, *Khim. Geterotsikl. Soedin.*, **1973**, 1109; *Chem. Abstr.*, **79**, 126395g.
849. B. A. Tertov and V. V. Bessonov, USSR Patent 443,034 (1974); *Chem. Abstr.*, **82**, 31321c.
850. B. A. Tertov, N. F. Vanieva, A. V. Koblik, and P. P. Onishchenko, *Khim.-Farm. Zh.*, **7**, 27 (1973); *Chem. Abstr.*, **79**, 126402g.
851. J. L. Miesel, German Patent 2,029,753 (1970); *Chem. Abstr.*, **74**, 53792u.
852. J. L. Miesel, U.S. Patent 3,927,020 (1975); *Chem. Abstr.*, **85**, 94361y.
853. J. L. Miesel, U.S. Patent 4,000,295 (1976); *Chem. Abstr.*, **86**, 140050s.
854. J. L. Miesel and D. I. Wickiser, U.S. Patent 3,939,166 (1976); *Chem. Abstr.*, **84**, 150630d.
855. A. M. Simonov and N. D. Vitkevich, *Zh. Obshch. Khim.*, **30**, 590 (1960); *Chem. Abstr.*, **54**, 24677i.
856. A. M. Simonov and P. A. Uglov, *Zh. Obshch. Khim.*, **21**, 884 (1951); *Chem. Abstr.*, **46**, 498c.
857. A. M. Simonov, V. G. Sayapin, and V. I. Siderman, *Zh. Vses. Khim. Obschchest*, **15**, 232 (1970); *Chem. Abstr.*, **73**, 4541m.
858. A. M. Simonov and A. N. Lomakin, *Zh. Obshch. Khim.*, **32**, 2228 (1962); *Chem. Abstr.*, **58**, 7923g.
859. A. N. Lomakin, A. M. Simonov, and V. A. Chigrina, *Zh. Obshch. Khim.*, **33**, 204 (1963); *Chem. Abstr.*, **58**, 13936f.
860. A. M. Simonov and A. F. Pozharskii, *Zh. Obshch. Khim.*, **31**, 3970 (1961); *Chem. Abstr.*, **57**, 8559h.

861. N. D. Vitkevich and A. M. Simonov, *Zh. Obshch. Khim.*, **29,** 2614 (1959); *Chem. Abstr.*, **54,** 11002h.
862. A. M. Simonov and A. F. Pozharskii, *Zh. Obshch. Khim.*, **33,** 2350 (1963); *Chem. Abstr.*, **59,** 13967h.
863. A. N. Lomakin, *Materialy 4-oi Chetvertoi Nauchn Konf. Aspirantov (Rostov-on-Don: Rotovsk Univ.)*, **1962,** 108; *Chem. Abstr.*, **60,** 10670e.
864. A. F. Pozharskii, M. M. Medvedeva, E. A. Zvezdina, and A. M. Simonov, *Khim. Geterotsikl. Soedin.*, **7,** 665 (1971); *Chem. Abstr.*, **76,** 126865y.
865. A. F. Pozharskii, V. V. Kuz'menko, and A. M. Simonov, *Khim. Geterotsikl. Soedin.*, **7,** 1105 (1971); *Chem. Abstr.*, **76,** 153676u.
866. M. M. Medvedeva, A. F. Pozharskii, and A. M. Simonov, *Khim. Geterotsikl. Soedin.*, **1972,** 1418; *Chem. Abstr.*, **78,** 43360v.
867. G. A. Mokrushina, R. V. Kunakova, and N. P. Bednyagina, *Khim. Geterotsikl. Soedin.*, **1970,** 131; *Chem. Abstr.*, **72,** 90369r.
868. A. F. Pozharskii, G. N. Pershin, E. A. Zvezdina, T. N. Zykova, S. N. Milovanova, and N. A. Novitskaya, *Khim. Farm. Zh.*, **4,** 14 (1970); *Chem. Abstr.*, **72,** 132623b.
869. I. I. Popov, P. V. Tkachenko, A. M. Simonov, A. F. Pozharskii, and Yu. G. Bogachev, *Zh. Org. Khim.*, **10,** 1789 (1974); *Chem. Abstr.*, **81,** 136056c.
870. A. V. El'tsov and K. M. Krivozheiko, *Zh. Org. Khim.*, **2,** 189 (1966).
871. G. N. Tyurenkova and N. P. Bednyagina, *Zh. Org. Khim.*, **1,** 136 (1965).
872. A. M. Simonov and V. N. Komissarov, *Khim. Geterotsikl. Soedin.*, **1975,** 826; *Chem. Abstr.*, **83,** 193173d.
873. A. M. Simonov and A. N. Lomakin, *Zh. Vses. Khim. Obshchest im D. I. Mendeleeva,* **8,** 234 (1963); *Chem. Abstr.*, **59,** 5130c.
874. E. I. Engelhardt, U.S. Patent 2,971,005 (1961); *Chem. Abstr.*, **55,** 13449b.
875. N. P. Bednyagina and I. Ya Postovskii, *Zh. Obshch. Khim.*, **30,** 1431 (1960); *Chem. Abstr.*, **55,** 1586h.
876. A. Hunger, J. Kebrie, A. Rossi, and K. Hoffmann, *Helv. Chim. Acta,* **44,** 1273 (1961).
877. R. M. Claramunt, R. Granados, and M. C. Repolles, *An. Quim.*, **71,** 206 (1975); *Chem. Abstr.*, **83,** 131524k.
878. T. Kodama, A. Takai, M. Nakabayashi, I. Watanabe, H. Sadaki, T. Kodama, N. Abe, and A. Kurokawa, Japanese Patent 75 126,682 (1975); *Chem. Abstr.*, **84,** 44060h.
879. Chimetron S.a.r.l., French Patent 4,761 (1967); *Chem. Abstr.*, **69,** 67383v.
880. Y. Shiokawa and S. Ohki, *Chem. Pharm. Bull.*, **21,** 981 (1973).
881. Chimetron S.a.r.l., French Patent 1,439,244 (1966); *Chem. Abstr.*, **65,** 18594h.
882. A. Ricci and P. Vivarelli, *Gazz. Chim. Ital.*, **97,** 741 (1967); *Chem. Abstr.*, **67,** 64300p.
883. A. Ricci, G. Seconi, and P. Vivarelli, *Gazz. Chim. Ital.*, **99,** 542 (1969); *Chem. Abstr.*, **71,** 124326w.
884. A. Ricci and P. Vivarelli, *Gazz. Chim. Ital.*, **97,** 750 (1967); *Chem. Abstr.*, **67,** 108597x.
885. A. Ricci and P. Vivarelli, *Gazz. Chim. Ital.*, **97,** 758 (1967); *Chem. Abstr.*, **67,** 108598y.
886. G. V. Boyd, *Tetrahedron Lett.*, **1966,** 3369.
887. I. Ya. Postovskii, V. P. Mamaev, G. A. Mokrushina, O. A. Zagulyaeva, and M. A. Kosareva, *Khim. Geterotsikl Soedin.*, **1975,** 987; *Chem. Abstr.*, **83,** 164074s.
888. D. Harrison and J. T. Ralph, *J. Chem. Soc.*, **1965,** 236.
889. D. Harrison and J. T. Ralph, *J. Chem. Soc. C* **1969,** 886.
890. H. Hasegawa, N. Tsuda, and M. Hasoya, Japanese Patent, 74 41,198 (1974); *Chem. Abstr.*, **82,** 156308k.
891. T. Nagai, Y. Fukushima, T. Kuroda, H. Shimizu, S. Sekiguchi, and K. Matsui, *Bull. Chem. Soc. Japan*, **46,** 2600 (1973).
892. Chimetron S.a.r.l., French Patent 4,806 (1967); *Chem. Abstr.*, **69,** 67386y.

893. Taisho Pharmaceutical Co. Ltd., French Patent 1,438,607 (1966); *Chem. Abstr.*, **65**, 18596e.

894. D. Harrison and J. T. Ralph, *J. Chem. Soc.*, **1965**, 3132.

895. D. Lloyd and J. S. Sneezum, *Tetrahedron*, **3**, 334 (1958).

896. A. F. Pozharskii, A. M. Simonov, E. A. Zvezdina, and N. K. Chub, *Khim. Geterotsikl Soedin.*, **1967**, 889; *Chem. Abstr.*, **68**, 105096t.

897. V. M. Pechenina, N. A. Mukhina, K. A. Abaturova, L. P. Grebenschchikova, T. V. Mikhailova, V. M. Kurilenko, and A. P. Gilev, USSR Patent 319,597 (1971).

898. V. M. Pechenina, N. A. Mukhina, K. A. Abaturova, L. P. Grebenschchikova, T. V. Mikhailova, V. M. Kurilenko, and A. P. Gilev, *Khim. Farm. Zh.*, **5**, 13 (1971); *Chem. Abstr.*, **76**, 14430w.

899. W. J. Welstead and C. G. Helsley, German Patent 2,017,265 (1970); *Chem. Abstr.*, **74**, 3627y.

900. M. Inone, K. Arai, S. Kodikoro, T. Saito, and H. Okuzawa, Japanese Patent 74 70,969 (1974); *Chem. Abstr.*, **81**, 136141b.

901. S. Tatsuoka and H. Hitomi, Japanese Patent 3780 (1952); *Chem. Abstr.*, **48**, 4004i.

902. Schering A.-G., British Patent 703,272 (1954); *Chem. Abstr.*, **49**, 1816e.

903. Schering A.-G., British Patent 703,723.

904. W. Knobloch, *Chem. Ber.*, **91**, 2562 (1958).

905. H. Richter, German Patent 1,078,132 (1960); *Chem. Abstr.*, **55**, 17654a.

906. C. Fauran, J. Eberle, N. Dorme, and C. Raynaud, German Patent 2,552,151 (1976); *Chem. Abstr.*, **85**, 108672q.

907. W. Knobloch, *Chem. Ber.*, **91**, 2557 (1958).

908. O. F. Ginzburg, B. A. Porai-Koshits, M. I. Krylova, and S. M. Lotareichik, *Zh. Obshch. Khim.*, **27**, 411 (1957); *Chem. Abstr.*, **51**, 15500d.

909. M. Schenck, U.S. Patent, 2,728,776 (1955); *Chem. Abstr.*, **50**, 15593c.

910. J. M. Sprague, U.S. Patent 2,567,912 (1951); *Chem. Abstr.*, **46**, 2583c.

911. P. Cheosakul, R. Parker, and H. E. Skipper, *Thai. Sci. Bull.*, **10**, 14 (1959); *Chem. Abstr.*, **59**, 3906f.

912. W. R. Siegart and A. R. Day, *J. Am. Chem. Soc.*, **79**, 4391 (1957).

913. A. J. Charlson and J. S. Harington, *Carbohydr. Res.*, **43**, 383 (1975).

914. J. Büchi, H. Zwicky, and A. Aebi, *Arch. Pharm.* (Weinheim), **293**, 758 (1960); *Chem. Abstr.*, **55**, 518f.

915. R. D. Haugwitz and V. L. Narayanan, German Patent 2,140,496 (1970); *Chem. Abstr.*, **76**, 140820m.

916. S. S. Tiwari and V. K. Paudey, *J. Indian Chem. Soc.*, **52**, 460 (1975).

917. Chimetron S.a.r.l., French Patent 1,503,697 (1967); *Chem. Abstr.*, **70**, 11701a.

918. (i) R. D. Haugwitz and L. V. Narayanan, German Patent 2,110,440 (1971); *Chem. Abstr.*, **75**, 151786k; (ii) R. D. Haugwitz, B. V. Maurer, and V. L. Narayanan, *Chem. Commun.*, **1971**, 1100.

919. A. L. Misra, *J. Org. Chem.*, **23**, 897 (1958).

920. J. Tulecki and B. Golus, *Pol. J. Pharmacol. Pharm.*, **25**, 259 (1973); *Chem. Abstr.*, **79**, 78685c.

921. R. Aries, French Patent 2,157,052 (1973); *Chem. Abstr.*, **79**, 126498t.

922. H. Baganz, *Angew. Chem.*, **68**, 151 (1956).

923. H. Schubert, H. Lettau, and J. Fischer, *Tetrahedron*, **30**, 1231 (1974).

924. P. McCloskey and R. A. Sizeland, German Patent, 2,014,293 (1970); *Chem. Abstr.*, **74**, 3619x.

925. E. L. Samuel and G. Holan, U.S. Patent 3,576,818 (1971); *Chem. Abstr.*, **75**, 36028c.

926. E. L. Samuel and G. Holan, *J. Chem. Soc. C*, **1967**, 25.

927. Chimetron S.a.r.l., French Patent 88,775 (1967); *Chem. Abstr.*, **67**, 73607w.

928. J. Sawlewicz and K. Wisterowicz, *Acta Pol. Pharm.*, **33**, 429 (1976); *Chem. Abstr.*, **87**, 68242g.

929. S. M. Deshpande and K. C. Datta, *J. Indian Chem. Soc.*, **53**, 320 (1976).

930. A. F. Pozharskii, V. Ts. Bukhaeva, A. M. Simonov, and R. A. Savel'eva, *Khim. Geterotsikl. Soedin.*, **1969**, 183; *Chem. Abstr.*, **71**, 3323j.

931. J. R. Corbett and A. Percival, German Patent 2,137,508 (1972); *Chem. Abstr.*, **76**, 140831r. See also J. R. Corbett British Patent 1,356,245 (1974); *Chem. Abstr.*, **81**, 91521g.

932. G. Holan, E. L. Samuel, B. C. Ennis, and R. W. Hinde, *J. Chem. Soc. C* **1967**, 20.

933. B. C. Ennis, G. Holan, and E. L. Samuel, *J. Chem. Soc. C*, **1967**, 30.

934. B. C. Ennis, G. Holan, and E. L. Samuel, *J. Chem. Soc. C*, **1967**, 33.

935. D. W. Wooley, *J. Biol. Chem.*, **152**, 225 (1944).

936. S. Cohen, E. Thom, and A. Bendich, *J. Org. Chem.*, **27**, 3545 (1962); *Biochemistry*, **2**, 176 (1963); S. Cohen and M. Dinar, *J. Am. Chem. Soc.*, **87**, 3195 (1965).

937. Monsanto Chemicals (Australia), British Patent 1,075,259 (1967); *Chem. Abstr.*, **68**, 68988n.

938. G. A. Shvekhgeimer and V. I. Kelarev, *Khim. Geterotsikl. Soedin.*, **1974**, 122; *Chem. Abstr.*, **80**, 95825m.

939. W. Reid and H. Lohwasser, *Angew Chem. Int. Ed. Engl.*, **5**, 835 (1966); *Justus Liebigs Ann. Chem.*, **699**, 88 (1966).

940. R. K. Howe, *J. Org. Chem.*, **34**, 2983 (1969).

941. G. Hasegawa and A. Kotani, Japanese Patent 74 20,174 (1974); *Chem. Abstr.*, **81**, 105508x.

942. E. W. Berndt, H. A. Fratzke, and B. G. Held, *J. Heterocycl. Chem.*, **9**, 137 (1972).

943. Ch. Sh. Kadyrov, S. A Khasanov, M. Akbarova and S. Yu. Yunusov, USSR Patent 520,361; *Chem. Abstr.*, **85**, 177420t

944. V. L. Narayanan and R. D. Haugwitz, German Patent 2,428,061 (1975); *Chem. Abstr.*, **82**, 170928u.

945. V. Brasiunas, *Mater. Nauchn. Konf. Kainas. Med. Inst.*, 22nd, 1972, 137; *Chem. Abstr.*, **84**, 105478z.

946. Laboratorios Lafarquim S. A., Spanish Patent 412,685 (1976); *Chem. Abstr.*, **86**, 189936z.

947. D. M. Pond, R. H. S. Wang, and G. Irick, German Patent 2,550,876 (1976); *Chem. Abstr.*, **85**, 123927v.

948. W. Daum, German Patent, 2,227,919 (1973); *Chem. Abstr.*, **80**, 82979q.

949. K. Sawatari, T. Mukai, K. Sucnobu, and T. Ike, Japanese Patent 75 142,565 (1975); *Chem. Abstr.*, **84**, 164780q.

950. H. Häusermann, German Patent 1,077,222 (1960); *Chem. Abstr.*, **55**, 18776e.

951. A. F. Wagner, P. E. Wittreich, A. Lusi, and K. Folkers, *J. Org. Chem.*, **27**, 3236 (1962).

952. A. M. Simonov and D. D. Dalgatov, *Zh. Obshch. Khim.*, **34**, 3052 (1964); *Chem. Abstr.*, **62**, 1644f.

953. D. D. Dalgatov and A. M. Simonov, *Khim. Geterotsikl. Soedin.*, **1967**, 908; *Chem. Abstr.*, **68**, 105091n.

954. I. I. Chizhevska, L. S. Marisheva, and N. M. Yatsevich, *Vestsi. Akad. Navuk. Belarus. SSR. Ser. Khim. Navuk.*, **1970**, 78; *Chem. Abstr.*, **74**, 111963p.

955. A. A. Zubenko, I. I. Popov, and A. M. Simonov, *Khim. Geterotsikl. Soedin.*, **1974**, 1544; *Chem. Abstr.*, **82**, 72876f.

956. I. I. Popov, A. M. Simonov, V. I. Mihailov, and N. A. Sil'vanovich, *Khim. Geterotsikl. Soedin.*, **1974**, 408; *Chem. Abstr.*, **81**, 25607p.

957. A. M. Simonov, I. I. Popov, V. I. Mikhailov, N. A. Sil'vanovich, and O. E. Shelepin, *Khim. Geterotsikl. Soedin.*, **1974**, 413; *Chem. Abstr.*, **81**, 25601g.

958. V. S. Misra and M. Varshneya, *Indian J. Pharm.*, **37**, 60 (1975); *Chem. Abstr.*, **84**, 59307t.

959. V. S. Misra and K. M. Varshneya, *Indian J. Pharm.*, **39**, 35 (1977); *Chem. Abstr.*, **87**, 53163h.

960. V. S. Misra and S. Prakhash, *Indian J. Chem.*, **13**, 752 (1975).

961. H. H. Zoorob and E. S. Ismail, *Z. Naturforsch, B*, **31**, 1680 (1976).

962. H. H. Zoorob, H. A. Hammouda, and E. Ismail, *Z. Naturforsch, B*, **32,** 443 (1977).
963. C. Fauran, M. Turin, G. Raynaud, and B. Pourrias, French Patent, 2,259,590 (1975); *Chem. Abstr.*, **84,** 59477y.
964. I. I. Popov, A. M. Simonov, and A. A. Zubenko, *Zh. Org. Khim.*, **11,** 1139 (1975); *Chem. Abstr.*, **83,** 79152w.
965. I. I. Popov, A. M. Simonov, and A. A. Zubenko, USSR Patent 433,148 (1974); *Chem. Abstr.*, **81,** 105511t.
966. V. A. Hudrenko, T. F. Grigorenko, L. F. Avramenko, I. A. Ol'shevskaya, and V. Ya. Pochinok, *Ukr. Khim. Zh.*, **41,** 764 (1975); *Chem. Abstr.*, **83,** 147427s.
967. E. Winterfeldt, in "Chemistry of Acetylenes," H. G. Viehe, Ed., Marcel Dekker, New York, 1969, pp. 271–275.
968. I. I. Popov, A. M. Simonov, and A. A. Zubenko, *Khim. Geterotsikl. Soedin*, **1975,** 140; *Chem. Abstr.*, **82,** 140012y.
969. M. Utaka, M. Takatsu, and A. Takeda, *Bull. Chem. Soc.*, Japan, **50,** 3276 (1977).
970. K. Dimroth and K. Severin, *Justus Liebigs Ann. Chem.*, **1973,** 380.
971. S. Takahashi and H. Kano, *Tetrahedron Lett.*, **1965,** 3789.
972. H. Tani, *Chem. Pharm. Bull.*, **12,** 783 (1964).
973. D. M. Lemal and K. I. Kawano, *J. Am. Chem. Soc.*, **84,** 1761 (1962).
974. R. Breslow, *J. Am. Chem. Soc.*, **76,** 1762 (1957).
975. G. Scherowsky, *Justus Liebigs Ann. Chem.*, **739,** 45 (1970).
976. N. K. Beresneva, E. R. Zakhs, and L. S. Efros, *Khim. Geterotskil. Soedin.*, **1971,** 961; *Chem. Abstr.*, **76,** 140647k.
977. L. S. Efros, E. R. Zakhs, and N. K. Beresneva, *Khim. Geterotsikl. Soedin.*, **1970,** 1004.
978. A. M. Simonov and N. D. Vitkevitch, *Zh. Obshch. Khim.*, **29,** 2404 (1959); *Chem. Abstr.*, **54,** 9896a.
979. A. M. Simonov, N. D. Vitkevitch, and S. Ya Zheltonozhko, *Zh. Obshch. Khim.*, **30,** 2688 (1959); *Chem. Abstr.*, **55,** 15467e.
980. O. Meth-Cohn, *J. Chem. Soc.*, **1964,** 5245.
981. P. V. Tkachenko, I. I. Popov, A. M. Simonov, and Yu. V. Medvedev, *Khim. Geterotsikl Soedin*, **1976,** 972; *Chem. Abstr.*, **85,** 159980w.
982. A. Takamizawa, Y. Hamashima, H. Sato, and Y. Matsumoto, *Chem. Pharm. Bull.*, **18,** 1576 (1970).
983. E. B. Tsupak, N. K. Chub, A. M. Simonov, and V. A. Kruchinin, *Khim. Geterotsikl. Soedin*, **1975,** 526; *Chem. Abstr.*, **83,** 79149a.
984. M. Kocevar, B. Stanovnik, and M. Tisler, *Croat. Chem. Acta*, **45,** 457 (1973); *Chem. Abstr.*, **80,** 37043u.
985. F. Minisci, R. Bernardi, F. Bertini, R. Galli, and M. Perchinunno, *Tetrahedron*, **27,** 3575 (1971).
986. F. Bertini, R. Galli, F. Minisci, and A. Porta, *Chem. Ind.* (Milan), **54,** 223 (1972); *Chem. Abstr.*, **77,** 5403h.
987. J. M. Aderson and J. K. Kochi, *J. Am. Chem. Soc.*, **92,** 1651 (1970).
988. A. F. Pozharskii, T. M. Meleshko, and A. M. Simonov, *Khim. Geterotsikl. Soedin. Akad. Nauk. Latv. SSR*, **1966,** 473; *Chem. Abstr.*, **65,** 8895c.
989. T. L. Gilchrist and C. W. Rees, "Carbenes, Nitrenes and Arynes," Nelson, London, 1969, p. 108.
990. R. L. Ellsworth, D. F. Hinkley, and E. F. Schoenewaldt, French Patent 2,014,308 (1970); *Chem. Abstr.*, **74,** 76424q.
991. *Cf.* G. L'abbe, *Chem. Rev.*, **69,** 345 (1969).
992. M. Gelus and J. M. Bonnier, *J. Chim. Phys. Physicochim. Biol.*, **65,** 253 (1968); *Chem. Abstr.*, **69,** 77164p.
993. W. W. Kilgore and E. R. White, *Bull. Environ. Contam. Toxicol.*, **5,** 67 (1970); *Chem. Abstr.*, **73,** 56025m.
994. D. I. Dodds, U.S. Patent 4,001,423 (1977); *Chem. Abstr.*, **86,** 189932v.
995. H. A. Staab and G. Seel, *Justus Liebigs Ann. Chem.*, **612,** 187 (1958).

996. B. I. Kristich and G. G. Turchuk, *Khim. Geterotsikl. Soedin.*, **1974,** 429; *Chem. Abstr.*, **81,** 13438w.

997. B. I. Dittmar, U.S. Patent 3,647,817 (1972); *Chem. Abstr.*, **76,** 140813m.

998. G. Kempter, W. Ehrlichmann, and R. Thomann, *Z. Chem.*, **17,** 220 (1977).

999. W. Daum and P. E. Frohberger, German Patent 2,227,921 (1973); *Chem. Abstr.*, **80,** 82978p.

1000. J. A. Van Allan and G. A. Reynolds, *J. Heterocycle. Chem.*, **5,** 471 (1968).

1001. E. R. Lavagnino and D. C. Thomson, *J. Heterocycl. Chem.*, **9,** 149 (1972).

1002. E. K. Fields, U.S. Patent 3,891,656 (1975); *Chem. Abstr.*, **83,** 96769v.

1003. Y. Tamura, H. Hayashi, J. Minamikawa, and M. Ikeda, *J. Heterocycl. Chem.*, **11,** 781 (1974).

1004. E. R. Cole, G. Crank and A-Salem Sheikh, *Tetrahedron Lett.*, **1973,** 2987.

1005. J. H. M. Hill, *J. Org. Chem.*, **28,** 1931 (1963).

1006. D. A. M. Watkins, *Chemosphere*, **3,** 239 (1974).

1007. D. A. M. Watkins, *Chemosphere*, **5,** 77 (1976).

1008. T. A. Jacob, J. A. Carlin, R. W. Walker, F. J. Wolf, and W. J. A. Van Huevel, *J. Agric. Food Chem.*, **23,** 704 (1975); *Chem. Abstr.*, **83,** 109773u.

1009. A. J. Hubert and H. Reimlinger, *Chem. Ber.*, **103,** 2828 (1970). For other examples of this approach, see A. J. Hubert, *J. Chem. Soc. C*, **1969,** 1334.

1010. J. de Mendoza and J. Elguero, *Bull. Soc. Chim. Fr.*, **1974,** 2987.

1011. T. Nagai, Y. Fukushima, T. Kuroda, H. Shimizu, S. Sekiguchi, and K. Matsui, *Bull. Chem. Soc. Japan*, **46,** 2602 (1973).

1012. S. Udenfriend, C. T. Clark, J. Axelrod, and B. B. Brodie, *J. Biol. Chem.*, **208,** 731 (1954).

1013. E. R. Cole, G. Crank, and A-Salem Sheikh, *Tetrahedron Lett.*, **1974,** 2925.

1014. E. Bamberger, *Ann.*, **273,** 267 (1893).

1015. L. S. Efros, N. V. Khromov-Borisov, L. R. Davidenkov, and M. M. Nedel, *Zh. Obshch. Khim.*, **26,** 455 (1956); *Chem. Abstr.*, **50,** 13881f.

1016. H. von Euler, H. Hasselquist, and O. Heidenberger, *Ark. Kemi*, **14,** 419 (1958); *Chem. Abstr.*, **54,** 12156.

1017. H. Schubert, G. Lehmann, and G. Otterphol. East German Patent 73,768 (1970); *Chem. Abstr.*, **74,** 76417q.

1018. N. F. Yyupalo, V. A. Yakobi, A. A. Stepanyan, L. F. Budennaya, and A. Z. Kozorezov, *Ukr. Khim. Zh.*, **42,** 394 (1976); *Chem. Abstr.*, **85,** 46577b. N. F. Tyupalo, V. A. Yakobi, A. A. Stepanyan, and R. G. Zaika, *Ukr. Khim. Zh.*, **43,** 53 (1977); *Chem. Abstr.*, **86,** 171345w. For a kinetic study see N. F. Tuupalo, V. A. Yakobi and A. A. Stepanyan, *Katal. Katal*, **13,** 53 (1975); *Chem. Abstr.*, **83,** 163345u.

1019. L. Weinberger and A. R. Day, *J. Org. Chem.*, **24,** 1451 (1959).

1020. E. R. Zakhs and L. S. Efros, *Zh. Org. Khim.*, **2,** 1095 (1966); *Chem. Abstr.*, **65,** 15365f.

1021. E. R. Zakhs and L. S. Efros, *Zh. Obshch. Khim.*, **34,** 1633 (1964); *Chem. Abstr.*, **61,** 5636h.

1022. L. C. March and M. M. Joullié, *J. Heterocycl. Chem.*, **7,** 39 (1970).

1023. A. V. El'tsov and L. S. Efros, *J. Gen. Chem.* (*USSR*), **30,** 3319 (1960); *Chem. Abstr.*, **55,** 18711g.

1024. A. M. Simonov and A. N. Lomakin, *Zh. Obshch. Khim.*, **32,** 2228 (1962); *Chem. Astr.*, **58,** 7923g.

1025. D. D. Dalgatov and A. M. Simonov, *Zh. Obshch. Khim.*, **33,** 1007 (1963); *Chem. Abstr.*, **59,** 10024d.

1026. C. N. Talaty, N. Zenker, and P. S. Callery, *J. Heterocycl. Chem.*, **13,** 1121 (1976).

1027. C. J. Paget, J. W. Chamberlin, and J. H. Wilkel, German Patent 2,638,551 (1977); *Chem. Abstr.*, **87,** 53298f.

1028. M. Sato, M. Arimoto, and K. Ueno, Japanese Patent 76, 131,879 (1976); *Chem. Abstr.*, **87,** 5967b.

1029. Donau-Pharmazie G.m.b.H. Austrian Patent 227,693 (1963); *Chem. Abstr.*, **59**, 11503g.
1030. Donau-Pharmazie G.m.b.H, French M 1627 (1963); *Chem. Abstr.*, **59**, 6415c.
1031. I. I. Popov and A. M. Simonov, *Khim. Geterotsikl. Soedin.*, **1974**, 1696; *Chem. Abstr.*, **82**, 14008b, see also I. I. Popov, A. M. Simonov, and V. I. Mikhailov, USSR Patent 427,011 (1974); *Chem. Abstr.*, **81**, 77922g.
1032. V. Lafon, German Patent 2,246,429 (1973); *Chem. Abstr.*, **79**, 18712z.
1033. Farbwerke-Hoechst A.-G., German Patent 2,334, 651 (1975); *Chem. Abstr.*, **83**, 79241z.
1034. H. Loewe, J. Urbanietz, D. Duewel, and R. Kirsch, German Patent 2,432,631 (1976); *Chem. Abstr.*, **84**, 135667c.
1035. A. F. Pozharskii, E. A. Zvezdina, and A. M. Simonov, *Khim. Geterotsikl. Soedin.*, **1967**, 184; *Chem. Abstr.*, **67**, 64302r.
1036. N. P. Bednyagina, Yu. A. Sedov, G. M. Petrova, and I. Ya Postovskii, *Khim. Geterotsikl. Soedin.*, **1972**, 390; *Chem. Abstr.*, **77**, 61885d.
1037. N. P. Bednyagina, G. N. Lipunova, and G. M. Petrova, *Khim. Geterotsikl. Soedin.*, **1973**, 699; *Chem. Abstr.*, **79**, 42407z.
1038. A. F. Pozharskii, E. A. Zvezdina, and A. M. Simonov, *Tetrahedron Lett.*, **1967**, 2219; USSR Patent 193,525; *Chem. Abstr.*, **69**, 27422a.
1039. A. F. Pozharskii, E. A. Zvezdina, Yu. P. Andreichikov, A. M. Simonov, V. A. Anisimova, and S. P. Popova, *Khim. Geterotsikl. Soedin.*, **1970**, 1267.
1040. T. P. Filipskikh, A. F. Pozharskii, and E. A. Zbezdina, *Khim. Geterotsikl. Soedin.*, **1972**, 238; *Chem. Abstr.*, **76**, 140646j.
1041. R. J. Bochis and M. H. Fisher, South African Patent 68,07,992 (1970); *Chem. Abstr.*, **74**, 87984y.
1042. R. J. Bochis and M. H. Fisher, U.S. Patent 3,928,372 (1975); *Chem. Abstr.*, **84**, 121829v.
1043. E. Profft and W. Georgi, *Justus Liebigs Ann. Chem.*, **612**, 187 (1958).
1044. R. D. Haugwitz and V. L. Narayanan, U.S. Patent 3,864,350 (1974); *Chem. Abstr.*, **82**, 156304f.
1045. E. Hayashi, E. Ishiguro, and N. Enomoto: paper presented at the 13th annual meeting of the Pharmaceutical Society of Japan (1960).
1046. D. J. Kew and P. F. Nelson, *Aust. J. Chem.*, **15**, 792 (1962).
1047. G. W. Stacey, B. V. Ettling, and A. J. Papa, *J. Org. Chem.*, **29**, 1537 (1964).
1048. H. Geiseman and G. Hälschke, *Chem. Ber.*, **92**, 92 (1959); H. Geiseman, H. Lettau, and H. G. Manfeldt, ibid, **93**, 570 (1960).
1049. A. Vilsmeier and A. Haack, *Chem. Ber.*, **60**, 119 (1927).
1050. *Cf.* I. Vlattas, I. T. Harrison, L. Tökes, J. H. Fried, and A. D. Cross, *J. Org. Chem.*, **33**, 4176 (1968).
1051. *Cf.* R. G. R. Bacon and J. R. Doggart, *J. Chem. Soc.*, **1960**, 1332.
1052. L. Bolger, R. T. Brittain, D. Jack, M. R. Jackson, L. E. Martin, J. Mills, D. Poynter, and M. B. Tyers, *Nature*, **238**, 354 (1972).
1053. S. Nakamura, *Chem. Pharm. Bull.*, **3**, 379 (1955).
1054. *Cf.* C. Cosar, C. Crisan, R. Horclois, R. M. Jacob, J. Robert, S. Tchelitscheff, and R. Vaupre, *Arzneim.-Forsch.*, **16**, 23 (1966).
1055. I. Butula, German Patent 1,913,184 (1969); *Chem. Abstr.*, **72**, 111472m.
1056. H. Schubert and H. Fritsche, *J. Prakt. Chem.*, **7**, 207 (1958).
1057. I. Butula, German Patent 1,948,795 (1971); (*Chem. Abstr.*, **75**, 129812u.
1058. E. Lebenstedt and W. Schunack, *Arch. Pharm.* (Weinheim), **310**, 455 (1977).
1059. H. Oelschläger and G. Giebenhain, *Arch. Pharm.* (Weinheim), **306**, 485 (1973).
1060. F. Bohlmann, *Chem. Ber.*, **85**, 390 (1952).
1061. T. Kato and M. Daneshtalab, *Chem. Pharm. Bull.*, **24**, 1640 (1976).
1062. C. Fauran, J. Eberle, A. Y. LeCloarec, G. Raynaud, and N. Dorme, French Patent 2,160,719 (1973); *Chem. Abstr.*, **79**, 137156f.

1063. A. Arens, P. Zarins, S. Germane, and E. S. Lavrinovich, USSR Patent 382,628 (1973); *Chem. Abstr.*, **79**, 146525c.

1064. Cf. J. Grimshaw and J. Trocha-Grimshaw, *Tetrahedron Lett.*, **1974**, 993.

1065. I. M. McRobbie, O. Meth-Cohn, and H. Suschitzky, *Tetrahedron Lett.*, **1976**, 929.

1066. Cf. I. M. McRobbie, O. Meth-Cohn, and H. Suschitzky, *Tetrahedron Lett.*, **1976**, 925.

1067. T. Sasaki and T. Ohishi, *Bull. Chem. Soc. Japan*, **41**, 3012 (1968).

1068. (a) R. Huisgen, *Angew. Chem. Int. Ed. Engl.*, **2**, 565 (1963); (b) R. Huisgen, ibid, **7**, 321 (1968).

1069. G. Wallbillich, Diploma Thesis, University of Munich, Germany (1959); quoted in Ref. 1068a.

1070. H. M. R. Hoffman, *Angew. Chem. Int. Ed. Engl.* **8**, 556 (1969).

1071. N. Finch and C. W. Gemenden, *J. Org. Chem.*, **35**, 3114 (1970); *Tetrahedron Lett.*, **1969**, 1203.

1072. K. K. Balasubramanian and B. Venugopalan, *Tetrahedron Lett.*, **1974**, 2643,2645.

1073. H. Ogura and K. Kichuki, *J. Org. Chem.*, **37**, 2679 (1972).

1074. I. Zugravescu, J. Herdan, and I. Druta, *Rev. Roum. Chim.*, **19**, 659 (1974); *Chem. Abstr.*, **81**, 25602h.

1075. Y. Tamura, H. Hayashi, Y. Nishimura, and M. Ikeda, *J. Heterocycl. Chem.*, **12**, 225 (1975).

1076. E. B. Tsupak, N. K. Chub, A. M. Simonov, and N. M. Miroshnichenko, *Khim. Geterotsikl. Soedin.*, **1972**, 812; *Chem. Abstr.*, **77**, 88399k.

1077. H. Ogura, H. Takayanagi, Y. Yamazaki, S. Yonezawa, H. Takagi, S. Kobayashi, I. Kamioka, and K. Kamoshita, *J. Med. Chem.*, **15**, 923 (1972).

1078. A. V. Zeiger and M. M. Joullié, *J. Org. Chem.*, **42**, 542 (1977).

1079. R. I-Fu Ho and A. R. Day, *J. Org. Chem.*, **38**, 3084 (1973).

1080. C. D. Arnett, J. Wright, and N. Zenker, *J. Med. Chem.*, **21**, 72 (1978).

1081. R. L. Ellsworth, H. E. Mertel, and W. J. A. Vanden Huevel, *J. Agric. Food Chem.*, **24**, 544 (1976); *Chem. Abstr.*, **85**, 5547y.

1082. R. R. Brodie, B. C. Mayo, L. F. Chasseaud, and D. R. Hawkins, *Arzneim.-Forsch. (Drug Res.)*, **27**(I), 593 (1977).

1083. V. V. Korshak, A. L. Rusanov, S. N. Leont'eva, and T. K. Dzhashiashvili, *Izv. Akad. Nauk. Gruz. SSR, Ser. Khim.*, **2**, 376 (1976); *Chem. Abstr.*, **87**, 23167v.

1084. R. Garner, G. V. Garner, and H. Suschitzky, *J. Chem. Soc. C*, **1969**, 1146.

1085. E. S. Lane, *J. Chem. Soc.*, **1955**, 534.

1086. J. J. Ursprung, U.S. Patent 3,105,837 (1963); *Chem. Abstr.*, **60**, 1763g.

1087. Yu. P. Andreichikov and A. M. Simonov, *Khim. Geterotsikl. Soedin.*, **1970**, 679; *Chem. Abstr.*, **73**, 45422n.

1088. G. I. Braz, G. G. Rozantzev, A. Ya Yakubovitch, and V. P. Bazov, *Zh. Obshch. Khim.*, **35**, 305 (1965); *Chem. Abstr.*, **62**, 14657a.

1089. R. Aries, French Patent 2,054,799 (1971); *Chem. Abstr.*, **76**, 25292d.

1090. R. Fessenden and D. F. Crowe, *J. Org. Chem.*, **26**, 4638 (1961); L. Birkhofer, P. Richter, and A. Ritter, *Chem. Ber.*, **93**, 2804 (1960).

1091. C. Hennart, *Bull. Soc. Chim. Fr.*, **1967**, 4286.

1091. C. Hennart, *Bull. Soc. Chim. Fr.*, **1967**, 4286.

1092. K. D. Banerji and K. K. Sen, *J. Indian Chem. Soc.*, **50**, 433 (1973).

1093. H. A. Staab and G. Seel, *Justus Liebigs Ann. Chem.*, **612**, 187 (1958).

1094. O. Süs, *Justus Liebigs Ann. Chem.*, **579**, 133 (1953).

1095. N. Vinot, *Bull. Soc. Chim. Fr.* **1966**, 3989.

1096. W. Ried and J. Patschorke, *Justus Liebigs Ann. Chem.*, **616**, 87 (1958).

1097. G. Scherowski, *Justus Liebigs Ann. Chem.*, **739**, 45 (1970).

1098. L. Bukowski, *Roczn. Chem.*, **47**, 1719 (1973).

1099. W. R. Roderick, C. W. Nordeen, A. M. Von Esch, and R. N. Appell, *J. Med. Chem.*, **15**, 655 (1972).

1100. W. Knobloch and K. Rintelen, *Arch. Pharm.* (Weinheim), **291,** 180 (1958); *Chem. Abstr.,* **53,** 3197d.

1101. K. Mori and H. Kunihoro , Japanese Patent 69 28,499 (1970), *Chem. Abstr.,* **72,** 43675c.

1102. K. M. Krivozheiko, M. B. Kolesova, and A. V. El'tsov, *Biol. Aktiv. Soedin.,* **1968,** 300; *Chem. Abstr.,* **72,** 21660s.

1103. Y. Shiokawa and S. Ohki, *Chem. Pharm. Bull.,* **19,** 401 (1971).

1104. S. Takahashi and H. Kano, *Chem. Pharm. Bull.,* **12,** 783 (1964).

1105. H. Braenniger and E. G. Kleinschmidt, *Pharmazie,* **24,** 24 (1969).

1106. J. Bourson, *Bull. Soc. Chim. Fr.,* **1971,** 3541.

1107. A. Rossi, A. Hunger, J. Kebrle, and K. Hoffmann, *Helv. Chim. Acta,* **43,** 1298 (1960).

1107. A. Rossi, A. Hunger, J. Kebrle, and K. Hoffmann, *Helv. Chim. Acta.,* **43,** 1298 (1960).

1108. F. Montanari, *Boll. Sci. Fac. Chim. Ind. Bologna,* **11,** 73 (1953); *Chem. Abstr.,* **49,** 6263h. F. Montanari, *Gazz. Chim. Ital.,* **85,** 981 (1955); *Chem. Abstr.,* **50,** 16111b.

1109. G. A. Reynolds, J. A. Van Allan, and J. F. Tinker, *J. Org. Chem.,* **24,** 1205 (1959).

1110. G. A. Reynolds and J. A. Van Allan, *J. Org. Chem.,* **24,** 1478 (1959).

1111. R. C. Perera and R. K. Smalley, *Chem. Commun.,* **1970,** 1458.

1112. A. L. Misra, *J. Org. Chem.,* **23,** 897 (1958).

1113. R. C. DeSelms, *J. Org. Chem.,* **27,** 2165 (1962).

1114. H. Alper and A. E. Alper, *J. Org. Chem.,* **35,** 835 (1970).

1115. A. E. Alper and A. Taurins, *Can. J. Chem.,* **45,** 2903 (1967); see also V. K. Chadha, H. S. Chaudhary, and H. K. Pujari, *Indian J. Chem.,* **7,** 769 (1969).

1116. E. S. Milner, S. Synder, and M. M. Joullié, *J. Chem. Soc.,* **1964,** 4151.

1117. G. F. Duffin and J. D. Kendall, *J. Chem. Soc.,* **1956,** 361.

1118. M. Ridi and S. Checchi, *Ann. Chim.* (Rome), **44,** 28 (1954); *Chem. Abstr.,* **49,** 4658f.

1119. S. N. Kolodyazhanaya and A. M. Simonov, *Khim. Geterotsikl. Soedin.,* **1967,** 186; *Chem. Abstr.,* **67,** 82162z.

1120. E. C. Fisher and M. M. Joullié, *J. Org. Chem.,* **23,** 1944 (1958).

1121. L. Holmquist and L. Larsson, *Acta Pharm. Sued,* **9,** 602 (1972); *Chem. Abstr.,* **78,** 111215d.

1122. Z. V. Esayan and G. T. Tatevosyan, *Arm. Khim. Zh.,* **27,** 908 (1974); *Chem. Abstr.,* **83,** 79154y.

1123. C. Fauran, M. Turin, T. Imbert, and G. Raynaud, German Patent 2,557,446 (1976); *Chem. Abstr.* **85,** 192729u.

1124. C. Fauran, J. Eberle, G. Raynaud, and N. Dorme, German Patent 2,431,532 (1975); *Chem. Abstr.,* **82,** 156314j.

1125. A. F. Pozharskii, A. M. Simonov, E. A. Zvezdina, and N. K. Chub, *Khim. Geterotsikl. Soedin.,* **1967,** 889; *Chem. Abstr.,* **68,** 105096t.

1126. G. A. Shvekhgeimer and V. I. Kelarev, *Khim. Geterotsikl. Soedin.,* **1974,** 122; *Chem. Abstr.,* **80,** 95825m.

1127. J. R. Corbett, British Patent 1,356,245 (1974); *Chem. Abstr.,* **81,** 91521g, [cf. British Patent 1,356,246 (1974)].

1128. E. R. White, E. A. Bose, J. M. Ogawa, B. T. Manji, and W. W. Kilgore, *J. Agr. Food Chem.,* **21,** 616 (1973); *Chem. Abstr.,* **79,** 101526n.

1129. A. M. Kuznetzov, S. A. Petrova, and B. M. Krasovitskii, *Stsintill. Org. Lyuminofory,* **1972,** 74; *Chem. Abstr.,* **79,** 115490h.

1130. I. I. Popov, A. M. Simonov, and A. A. Zubenko, *Khim. Geterotsikl. Soedin.,* **1976,** 1145; *Chem. Abstr.,* **85,** 177318r.

1131. B. I. Khristich, A. M. Simonov, and G. M. Survorova, *Khim. Geterotsikl. Soedin.,* **1973,** 1293; *Chem. Abstr.,* **79,** 146456f.

1132. I. Ganea, R. Taranu, and A. Popescu, *Stud. Univ. Babes-Bolyai Ser. Chem.,* **17,** 97 (1972); *Chem. Abstr.,* **77,** 114309c.

1133. A. Dikciuviene, V. Bieksa, and J. Degutis, *Liet. TSR Mokslu Akad. Darb. Ser. B,* **1973,** 59; *Chem. Abstr.,* **80,** 59895a.

1134. H. L. Yale and J. A. Bristol, U.S. Publ. Patent Appl. B 542,158 (1976); *Chem. Abstr.*, **84**, 135658a.

1135. Yu. V. Koshchienko, A. M. Simonov, T. N. Vashchenko, G. M. Suvrova, and V. A. Mukarov, *Khim.-Farm. Zh.*, **11**, 14 (1977); *Chem. Abstr.*, **87**, 68234f.

1136. A. M. Osman, Kh. M. Hassan, Z. H. Khalil, and V. D. Turin, *J. Appl. Chem. Biotechnol.*, **26**, 71 (1976); *Chem. Abstr.*, **85**, 95779c.

1137. E. H. Pommer, G. Bolz, and B. Girgensohn, German Patent 2,222,341 (1973); *Chem. Abstr.*, **80**, 37110p.

1138. R. I. Ogloblina, N. P. Bednyagina, and N. N. Gulemina, *Khim. Geterotsikl. Soedin.*, **1972**, 393; *Chem. Abstr.*, **77**, 88386d.

1139. V. P. Khilya, L. G. Grishko, and T. N. Sokolova, *Khim. Geterotsikl. Soedin.*, **1975**, 1593; *Chem. Abstr.*, **84**, 105486a.

1140. A. R. Katritzky and F. Yates, *J. Chem. Soc. Perkin I*, **1976**, 309.

1141. R. D. Haugwitz and V. L. Narayanan, German Patent 2,328,095 (1973); *Chem. Abstr.*, **80**, 82973h.

1142. L. V. Alam, I. Ya. Kvitko, and A. V. El'tsov, USSR Patent 503,864 (1976); *Chem. Abstr.*, **85**, 21364s.

1143. H. Loewe, J. Urbanietz, D. Duewel, and R. Kirsch, German Patent 2,541,752 (1977); *Chem. Abstr.*, **87**, 23283e.

1144. H. Willitzer, H. Braeuniger, D. Engelmann, D. Krebs, W. Ozegowski, and M. Tonew, East German Patent 2,123,466 (1976); *Chem. Abstr.*, **87**, 135328c.

1145. G. N. Dorofeenko, E. A. Zvezdina, M. P. Zhanova, and I. A. Barchan, *Khim. Geterotsikl. Soedin.*, **1973**, 1682; *Chem. Abstr.*, **80**, 828050.

1146. H. Koelling, H Thomas, A. Widdig, and H. Wollweber, German Patent 2,438,120 (1976); *Chem. Abstr.*, **85**, 21363r.

1147. R. J. Gyurik and W. D. Kingsbury, U.S. Patent 4,025,638 (1977); *Chem. Abstr.*, **87**, 85006a.

1148. H. Singh and S. Singh, *Indian J. Chem.*, **9**, 918 (1971); *Chem. Abstr.*, **76**, 3761d.

1149. L. V. Shkrabova, N. A. Mukhina, V. M. Kurilenko, A. P. Gilev, L. P. Basova, V. G. Motovilova, V. T. Romanova, and V. G. Pashinskii, *Khim.-Farm. Zh.*, **10**, 49 (1976); *Chem. Abstr.*, **85**, 5549a.

1150. S. Hünig, D. Scheutzow, H. Schlaf, and A. Schott, *Justus Liebigs Ann. Chem.*, **1974**, 1423.

1151. R. F. Kovar and F. E. Arnold, U.S. Patent 4,001,268 (1977); *Chem. Abstr.*, **86**, 122000w.

1152. H. Loewe, R. Kirsch, J. Urbanietz, and D. Duewel, German Patent, 2,332,398 (1975); *Chem. Abstr.*, **82**, 156311f.

1153. R. Aries, French Patent, 2,063,777 (1971); *Chem. Abstr.*, **76**, 126981h.

1154. H. Singh and S. Singh, *Indian J. Chem.*, **11**, 311 (1973); *Chem. Abstr.*, **79**, 42411w.

1155. R. D. Haugwitz, B. V. Mauer, and V. L. Narayanan, *J. Org. Chem.*, **39**, 1359 (1974).

1156. S. Hünig, D. Scheutzow, H. Schlaf, and H. Quast, *Justus Liebigs Ann. Chem.*, **765**, 110 (1972).

1157. I. Zugravescu, J. Herdan, and I. Druta, *Rev. Roum. Chim.*, **19**, 649 (1974); *Chem. Abstr.*, **81**, 25603j.

1158. I. Ganea, R. Toranu, and A. Popescu, *Stud. Univ. Babes-Bolyai Ser. Chem.*, **18**, 139 (1973); *Chem. Abstr.*, **79**, 105134f.

1159. S. V. Koroleva and V. V. Kalmykov, *Tr. Voronezh Univ.*, **95**, 36 (1972); *Chem. Abstr.*, **78**, 43371z.

1160. M. Nakao, S. Katayama, and H. Yamamoto, Japanese Patent 72 37,521 (1972); *Chem. Abstr.*, **78**, 4250q.

1161. P. N. Edwards, German Patent 2,250,345 (1973); *Chem. Abstr.*, **79**, 18714b.

1162. Y. Tamura, H. Hayashi, and M. Ikeda, *J. Heterocycl. Chem.*, **12**, 819 (1975).

1163. H. Loewe, J. Urbanietz, R. Kirsch, and D. Duewel, German Patent 2,332,343 (1975); *Chem. Abstr.*, **82**, 156310e.

1164. J. Musco and D. B. Murphy, *J. Org. Chem.*, **36,** 3469 (1971).

1165. R. M. Acheson and M. S. Verlander, *J. Chem. Soc. Perkin I*, **1972,** 1577.

1166. V. L. Narayanan and R. D. Haugwitz, U.S. Patent 3,718,662 (1973); *Chem. Abstr.*, **78,** 136292q.

1167. V. L. Narayanan and R. D. Haugwitz, U.S. Patent 3,720,686 (1973); *Chem. Abstr.*, **78,** 147968r.

1168. G. Hosegawa and M. Hosoya, Japanese Patent 72 55,483 (1972); *Chem. Abstr.*, **80,** 108531q.

1169. R. Kada, J. Kovac, and A. Jurasek, Czechoslovakian Patent 149,084 (1973); *Chem. Abstr.*, **79,** 137153c.

1170. G. F. Galenko, A. K. Bagrii, and P. M. Kochergin, *Ukr. Khim. Zh.* **41,** 759 (1975); *Chem. Abstr.*, **83,** 147426r.

1171. V. L. Narayanan and R. D. Haugwitz, U.S. Patent 3,901,909 (1975); *Chem. Abstr.*, **83,** 206273p.

1172. G. Rovnyak, V. L. Narayanan, R. D. Haugwitz, and C. M. Cimarusti, U.S. Patent 3,927,014 (1975); *Chem. Abstr.*, **84,** 105596m.

1173. G. F. Galenko, A. K. Bagrii, V. A. Grin, and P. M. Kochergin, *Ukr. Khim. Zh.*, **41,** 405 (1975); *Chem. Abstr.*, **83,** 79146x.

1174. J. Mohan, U. K. Chadha, and H. K. Pujari, *Indian J. Chem.*, **11,** 1119 (1973); *Chem. Abstr.*, **80,** 108448t.

1175. M. Sato, M. Arimoto, K. Ueno, H. Kojima, T. Yamazaki, and T. Sakurai, Japanese Patent 76 136,673 (1976); *Chem. Abstr.*, **86,** 189933w.

1176. K. Ueno, M. Sato, M. Arimoto, H. Kojima, T. Yamasaki, and T. Sakurai, Belgian Patent 830,171 (1975); *Chem. Abstr.*, **85,** 160091p.

1177. J. Machin and D. M. Smith, *J. Chem. Soc., Perkin I*, **1979,** 1371.

1178. M. A. Phillips, *J. Chem. Soc.* **1930,** 1409.

1179. E. A. Steck, F. C. Nachod, G. W. Ewing, and N. H. Gorman, *J. Am. Chem. Soc.*, **70,** 3406 (1948).

1180. *Organic Syntheses*, Col. Vol. II, John Wiley, New York, 1943, p. 65.

1181. M. A. Phillips, *J. Chem. Soc.*, **1928,** 2393.

1182. S. I. Lur'e, *J. Gen. Chem.* (USSR) **10,** 1909 (1940); *Chem. Abstr.*, **35,** 4022.

1183. B. Oddo, and F. Ingraffia, *Gazz. Chim. Ital.* **62,** 1092 (1932); *Chem. Abstr.*, **27,** 2686.

1184. B. Oddo and L. Raffa, ibid. **67,** 537 (1937); *Chem. Abstr.*, **32,** 1697.

1185. B. Oddo and L. Raffa, ibid. **68,** 199 (1938); *Chem. Abstr.*, **32,** 7455.

1186. R. A. Henry and W. M. Dehn, *J. Am. Chem. Soc.* **71,** 2297 (1949).

1187. R. Weidenhagen and H. Wegner, *Chem. Ber.* **71,** 2124 (1938).

1188. R. Seka and R. H. Müller, *Monatsh. Chem.* **57,** 97 (1931).

1189. R. Weidenhagen., *Chem. Ber*, **69,** 2263 (1936).

1190. R. Weidenhagen and H. Wegner, *Z. Wirtschaftsgruppe Zuckerind*, **87,** Tech. T1., 755 (1937); *Chem. Abstr.*, **32,** 8416.

1191. W. O. Pool, H. J. Harwood, and A. W. Ralston, *J. Am. Chem. Soc.*, **59,** 178 (1937).

1192. E. L. Holljes and E. C. Wagner, *J. Org. Chem.*, **9,** 31 (1944).

1193. E. L. Brown and N. Campbell, *J. Chem. Soc.*, **1937,** 1699.

1194. B. A. Porai-Koshits, O. F. Ginsburg, and L. S. Eiros, *J. Gen. Chem.* (USSR) **17,** 1978 (1947); *Chem. Abstr.*, **42,** 5903.

1195. K. von Auwers and W. Mauss, *Chem. Ber.*, **61,** 2411 (1928).

1196. F. Feigl and H. Gleich, *Monatsh. Chem.*, **49,** 385 (1928).

1197. A. Schönberg and A. Sina, *J. Chem. Soc.*, **1946,** 601.

1198. F. E. King and R. M. Acheson, *J. Chem. Soc.*, **1949,** 1396.

1199. G. K. Hughes and F. Lions, *J. Proc. Roy. Soc. N.S. Wales*, **71,** 209 (1938); *Chem. Abstr.*, **32,** 5830.

1200. R. C. Elderfield and F. J. Kreysa, *J. Am. Chem. Soc.*, **70,** 44 (1948).

1201. J. van Alphen, *Rec. Trav. Chim.*, **59,** 289 (1940).

1202. F. Lions and E. Ritchie, *J. Proc. Roy. Soc. N.S. Wales*, **74,** 365 (1941); *Chem. Abstr.,* **35,** 2890.

1203. B. A. Porai-Koshits, L. S. Efros, and O. F. Ginzburg, *J. Gen. Chem.* (USSR), **19,** 1545 (1949); *Chem. Abstr.,* **44,** 1100.

1204. S. Weil and H. Marcinkowska, *Rocz. Chem.,* **14,** 1312 (1934); *Chem. Abstr.,* **29,** 6233.

1205. G. B. Crippa and S. Maffei, *Gazz. Chim. Ital.,* **71,** 194 (1941); *Chem. Abstr.,* **36,** 2847.

1206. M. Hartmann and L. Panizzon, *Helv. Chim. Acta,* **21,** 1692 (1938).

1207. R. Weidenhagen, G. Train, H. Wegner and L. Nordström, *Chem. Ber.,* **75,** 1936 (1942).

1208. M. A. Phillips, *J. Chem. Soc.,* **1931,** 1143.

1209. F. Krollpfeiffer, W. Graulich, and A. Rosenberg, *Annalen,* **542,** 1 (1939).

1210. C. H. Roeder and A. R. Day, *J. Org. Chem.,* **6,** 25 (1941).

1211. G. R. Beaven, E. R. Holiday, E. A. Johnson, B. Ellis, P. Mamalis, V. Petrow, and B. Sturgeon, *J. Pharm. Pharmacol.,* **1,** 957 (1949).

1212. M. A. Phillips, *J. Chem. Soc.,* **1929,** 2820.

1213. W. M. Lauer, M. M. Sprung, and C. M. Langkammerer, *J. Am. Chem. Soc.,* **58,** 225 (1936).

1214. J. Bloch, *J. Soc. Chem. Indian* **38,** 118T (1919).

1215. H. Green and A. R. Day, *J. Am. Chem. Soc.,* **64,** 1167 (1942).

1216. N. G. Brink and K. Folkers, *J. Am. Chem. Soc.,* **71,** 2951 (1949).

1217. T. F. Doumani and K. A. Kobe, *J. Am. Chem. Soc.,* **62,** 562 (1940).

1218. S. Skraup and K. Böhm, *Chem. Ber.,* **59,** 1007 (1926).

1219. L. P. Kyrides, F. B. Zienty, G. W. Steahly, and H. L. Morrill, *J. Org. Chem.,* **12,** 577 (1947).

1220. C. W. Smith, R. S. Rasmussen, and S. A. Ballard, *J. Am. Chem. Soc.,* **71,** 1082 (1949).

1221. L. I. Smith and M. K. Kiess, ibid. **61,** 284 (1939).

1222. A. W. Ralston, R. J. van der Wal, and M. R. McCorkle, *J. Org. Chem.,* **4,** 68 (1939).

1223. L. I. Smith and S. A. Harris, *J. Am. Chem. Soc.,* **57,** 1289 (1935).

1224. S. Skraup, *Annalen,* **419,** 1 (1919).

1225. K. W. F. Kohlrausch and R. Seka, *Chem. Ber.,* **71,** 985 (1938).

1226. H. Rudy and K. E. Cramer, *Chem. Ber.,* **72,** 728 (1939).

1227. T. N. Ghosh, *J. Indian Chem. Soc.,* **13,** 86 (1936); *Chem. Abstr.,* **30,** 4859.

1228. P. C. Guha and M. K. De, *Q. J. Indian Chem. Soc.,* **3,** 41 (1926); *Chem. Abstr.* **21,** 2133.

1229. M. A. Phillips, *J. Chem. Soc.,* **1928,** 172.

1230. E. C. Wagner, *J. Org. Chem.,* **5,** 133 (1940).

1231. G. Jacini, *Gazz. Chim. Ital.,* **72,** 42 (1942); *Chem. Abstr.,* **37,** 649.

1232. R. A. Baxter and F. S. Spring, *J. Chem. Soc.,* **1945,** 229.

1233. H. C. Waterman and D. L. Vivian, *J. Org. Chem.,* **14,** 289 (1949).

1234. S. Gabriel and A. Thieme, *Chem. Ber.,* **52,** 1079 (1919).

1235. O. Fischer, E. Thiele, F. Stauber, W. Hild, G. Seufert, H. Hojer, F. von Mann-Tiechler, F. Elflein, and K. Müller, *J. prakt. Chem.,* **107,** 16 (1924).

1236. L. I. Smith and C. O. Guss, *J. Am. Chem. Soc.,* **62,** 2635 (1940).

1237. H. Wuyts and J. van Vaerenbergh, *Bull. Soc. Chim. Belg.,* **48,** 329 (1939).

1238. A. F. McKay and A. R. Bader, *J. Org. Chem.,* **13,** 75 (1948).

1239. P. Galimberti, *Gazz. Chim. Ital.,* **63,** 96 (1933); *Chem. Abstr.,* **27,** 3475.

1240. P. C. Guha and S. K. Ray, *Q. J. Indian Chem. Soc.,* **2,** 83 (1925): *Chem. Abstr.,* **20,** 745.

1241. W. Kimura and H. Taniguti, *J. Soc. Chem. Ind. Japan,* **42,** 234 (1939); *Chem. Abstr.,* **33,** 8591.

1242. L. I. Smith and C. L. Moyle, *J. Am. Chem. Soc.,* **58,** 1 (1936).

1243. J. Meisenheimer and B. Wieger, *J. Prakt. Chem.,* **102,** 45 (1921).

1244. A. Ahmed, K. S. Narang, and J. N. Ray, *J. Indian Chem. Soc.,* **15,** 152 (1938).

1245. G. B. Bachman and L. V. Heisey, *J. Am. Chem. Soc.*, **71,** 1985 (1949).
1246. L. G. S. Brooker, A. L. Sklar, H. W. J. Cressman, G. H. Keyes, L. A. Smith, R. H. Sprague, E. Van Lare, G. Van Zandt, F. L. White, and W. W. Williams, ibid., **67,** 1895 (1945).
1247. R. L. Shriner and P. G. Boermans, ibid., **66,** 1810 (1944).
1248. W. Gündel and R. Pummerer, *Annalen*, **529,** 11 (1937).
1249. P. Jacobson, *Annalen*, **427,** 142 (1922).
1250. H. Skolnik, J. G. Miller, and A. R. Day, *J. Am. Chem. Soc.*, **65,** 1854 (1943).
1251. K. Fries and H. Reitz, *Annalen*, **527,** 38 (1936).
1252. S. D. Gerson and G. L. Webster, *J. Am. Chem. Soc.*, **63,** 2853 (1941).
1253. K. Fries, E. Modrow, B. Racke, and K. Weber, *Annalen*, **454,** 191 (1927).
1254. J. A. van Allen, *J. Am. Chem. Soc.*, **69,** 2913 (1947).
1255. R. J. Dimler and K. P. Link, *J. Biol. Chem.*, **143,** 557 (1942).
1256. S. Moore, R. J. Dimler, and K. P. Link, *Ind. Eng. Chem., Anal. Ed.* **13,** 160 (1941).
1257. G. B. Bachman and L. V. Heisey, *J. Am. Chem. Soc.*, **68,** 2496 (1946).
1258. R. A. B. Copeland and A. R. Day. *J. Am. Chem. Soc.*, **65,** 1072 (1943).
1259. M. V. Betrabet and G. C. Chakravarti, *J. Indian Chem. Soc.*, **7,** 495 (1930); *Chem. Abstr.*, **25,** 701.
1260. L. Hunter and J. A. Marriott, *J. Chem. Soc.*, **1941,** 777.
1261. E. Ochiai and M. Katada, *J. Pharm. Soc.* Japan, **60,** 543 (1940); *Chem. Abstr.* **35,** 1785.
1262. A. M. Simonov, *J. Gen. Chem.* (USSR), **10,** 1588 (1940): *Chem. Abstr.*, **35,** 2870.
1263. F. E. King, T. J. King, and I. H. M. Muir, *J. Chem. Soc.*, **1946,** 5.
1264. F. Lions and A. M. Willison, *J. Proc. Roy. Soc. N. S. Wales*, **71,** 435 (1938).
1265. H. Fromherz and H. Spiegelberg, *Helv. Physiol. Acta*, **6,** 42 (1948); *Chem. Abstr.*, **42,** 8956.
1266. G. Heller, W. Dietrich, T. Hemmer, H. Kätzel, E. Rottsahl, and P. G. Zambalos, *J. Prakt. Chem.*, **129,** 211 (1931).
1267. A. F. Crowther, F. H. S. Curd, D. G. Davey, and G. J. Stacey, *J. Chem. Soc.*, **1949,** 1260.
1268. N. K. Richtymer and C. S. Hudson, *J. Am. Chem. Soc.*, **64,** 1609 (1942).
1269. P. Karrer, K. Schöpp, F. Benz, and K. Pfaehler, *Helv. Chim. Acta*, **18,** 69 (1935).
1270. S. Moore and K. P. Link, *Biol. Chem.*, **133,** 293 (1940).
1271. H. Ohle, *Chem. Ber.*, **67,** 155 (1934).
1272. C. F. Huebner, R. Lomar, R. J. Dimler, S. Moore, and K. P. Link, *J. Biol. Chem.*, **159,** 503 (1945).
1273. J. M. Gulland and G. R. Barker, *J. Chem. Soc.*, **1943,** 625.
1274. N. K. Richtymer and C. S. Hudson, *J. Am. Chem. Soc.*, **64,** 1612 (1942).
1275. R. J. Dimler and K. P. Link, *J. Biol. Chem.*, **150,** 345 (1943).
1276. G. R. Barker, K. R. Cooke, and J. M. Gulland, *J. Chem. Soc.*, **1944,** 339.
1277. W. T. Haskins and C. S. Hudson, *J. Am. Chem. Soc.*, **61,** 1266 (1939).
1278. R. Kuhn and F. Bär, *Chem. Ber.*, **67,** 898 (1934).
1279. S. Moore and K. P. Link, *J. Org. Chem.*, **5,** 637 (1940).
1280. D. J. Bell and E. Baldwin, *J. Chem. Soc.*, **1941,** 125.
1281. R. M. Hann, A. T. Merrill, and C. S. Hudson, *J. Am. Chem. Soc.*, **66,** 1912 (1944).
1282. R. Lohmar, R. J. Dimler, S. Moore, and K. P. Link, *J. Biol. Chem.*, **143,** 551 (1942).
1283. T. N. Ghosh, *J. Indian Chem. Soc.*, **15,** 89 (1938); *Chem. Abstr.*, **32,** 6650.
1284. W. A. Sexton, *J. Chem. Soc.*, **1942,** 303.
1285. W. Borsche and J. Barthenheier, *Annalen*, **553,** 250 (1942).
1286. K. Fries, O. Diekmann and A. Fingerling, ibid., **454,** 225 (1927).
1287. F. E. King, R. M. Acheson, and P. C. Spensley, *J. Chem. Soc.*, **1949,** 1401.
1288. G. T. Morgan and W. A. P. Challenor, ibid., **119,** 1537 (1921).
1289. K. Dziewonski and L. Sternbach, *Bull. Intern. Acad. Polonaise, Classe Sci. Math. Nat.*, **1935A,** 333; *Chem. Abstr.*, **30,** 2971.

1290. D. M. Hall and E. E. Turner, *J. Chem. Soc.*, **1948**, 1909.
1291. A. Bloom and A. R. Day, *J. Org. Chem.*, **4**, 14 (1939).
1292. G. M. Bennett and W. L. C. Pratt, *J. Chem. Soc.*, **1929**, 1465.
1293. G. M. van der Want, *Rec. Trav. Chim.*, **67**, 45 (1948).
1294. H. Lindemann and H. Krause, *J. Prakt. Chem.*, **115**, 256 (1927).
1295. K. Brand and E. Wild, *Chem. Ber.*, **56**, 105 (1923).
1296. F. F. Stephens and J. D. Bower, *J. Chem. Soc.*, **1949**, 2971.
1297. M. Stäuble, *Helv. Chim. Acta*, **30**, 224 (1947).
1298. P. van Romburgh and H. W. Huyser, *Rec. Trav. Chim.*, **49**, 165 (1930).
1299. P. van Romburgh and H. W. Huyser, *Verslag Akad. Wetenschappen Amsterdam*, **35**, 665 (1926).
1300. F. F. King, R. J. S. Beer, and S. G. Waley, *J. Chem. Soc.*, **1946**, 92.
1301. G. Pellizzari, *Gazz. Chim. Ital.*, **49**, 16 (1919); *Chem. Abstr.*, **14**, 169.
1302. N. J. Leonard, D. Y. Curtin, and K. M. Beck, *J. Am. Chem. Soc.*, **69**, 2459 (1947).
1303. C. C. Price and R. H. Reitsema, *J. Org. Chem.*, **12**, 269 (1947).
1304. G. W. Raiziss, L. W. Clemence, and M. Freidfelder, *J. Am. Chem. Soc.*, **63**, 2739 (1941).
1305. M. Stäuble, *Helv. Chim. Acta*, **32**, 135 (1949).
1306. D. Wood, Jr., and F. W. Bergstrom, *J. Am. Chem. Soc.*, **55**, 3314 (1933).
1307. K. Auwers and E. Frese, *Chem. Ber.*, **59**, 539 (1926).
1308. M. L. Tomlinson, *J. Chem. Soc.*, **1939**, 158.
1309. B. Chatterjee, *ibid*, **1929**, 2965.
1310. G. Pellizzari, *Gazz. Chim. Ital.*, **51**, 89 (1921); *Chem Abstr.*, **15**, 3076.
1311. M. Schubert, *Annalen*, **558**, 10 (1947).
1312. J. F. Deck and F. B. Dains, *J. Am. Chem. Soc.*, **55**, 4986 (1933).
1313. P. Ruggli and R. Fischer, *Helv. Chim. Acta*, **28**, 1270 (1945).
1314. E. Hoggarth, *J. Chem. Soc.*, **1949**, 3311.
1315. J. B. Wright, *J. Am. Chem. Soc.*, **71**, 2035 (1949).
1316. H. Rupe, F. Pedrini, and A. Collin, *Helv. Chim. Acta.*, **15**, 1321 (1932).
1317. F. E. King, R. M. Acheson, and P. C. Spensley, *J. Chem. Soc.* **1948**, 1366.
1318. G. Pellizzari, *Gazz. Chim. Ital.*, **51**, 140 (1921); *Chem. Abstr.* **15**, 3078.
1319. G. Pellizzari, *ibid.*, **54**, 177 (1924); *Chem. Abstr.*, **18**, 3173.
1320. O. Fischer, W. Meier, H. Schwappacher, and H. Kracker, *J. Prakt. Chem.* **104**, 102 (1922).
1321. G. Pellizzari, *Gazz. Chim. Ital.*, **52**, 199 (1922); *Chem. Abstr.*, **16**, 2508.
1322. A. B. Lal and V. Petrow, *J. Chem. Soc.*, **1948**, 1895.
1323. J. L. B. Smith, *ibid.* **123**, 2288 (1923).
1324. M. T. Bogert and W. H. Taylor, *J. Am. Chem. Soc.*, **49**, 1578 (1927).
1325. G. R. Clemo and G. A. Swan, *J. Chem. Soc.*, **1944**, 274.
1326. R. L. McKee, M. K. McKee, and R. W. Bost, *J. Am. Chem. Soc.*, **68**, 1904 (1946).
1327. R. L. McKee and R. W. Bost, *ibid.*, **69**, 468 (1947).
1328. J. J. Blanksma and E. M. Petri, *Rec. Trav. Chim.*, **66**, 353 (1947).
1329. E. Usherwood and M. A. Whitely, *J. Chem. Soc.*, **123**, 1069 (1923).
1330. T. S. Moore, M. T. Marrack, and A. K. Proud, *ibid.*, **119**, 1786 (1921).
1331. H. Goldstein and G. Gianola, *Helv. Chim. Acta*, **26**, 173 (1943).
1332. H. Goldstein and A. Studer, *ibid.*, **21**, 51 (1938).
1333. R. C. Elderfield, F. J. Kreysa, J. H. Dunn, and D. D. Humphreys, *J. Am. Chem. Soc.*, **70**, 40 (1948).
1334. M. E. Kurilo and M. M. Shemyakin, *J. Gen. Chem.* (USSR) **15**, 704 (1945); *Chem. Abstr.*, **40**, 5714.
1335. H. Lindemann and W. Wessel, *Chem. Ber.*, **58**, 1221 (1925).
1336. R. G. Jones, Q. F. Soper, O. K. Behrens, and J. W. Corse, *J. Am. Chem. Soc.*, **70**, 2843 (1948).

1337. A. Bistrzycki and K. Fässler, *Helv. Chim. Acta,* **6,** 519 (1923).
1338. G. B. Crippa and G. Perroncito, *Gazz. Chim. Ital.,* **65,** 1067 (1935); *Chem. Abstr.,* **30,** 4864.
1339. J. A. Murray and F. B. Dains, *J. Am. Chem. Soc.,* **56,** 144 (1934).
1340. F. E. King, P. C. Spensley and R. H. Nimmo-Smith, *Nature,* **162,** 153 (1948).
1341. M. V. Betrabet and G. C. Chakravarti, *J. Indian Chem. Soc.,* **7,** 191 (1930); *Chem. Abstr.,* **25,** 701.
1342. G. C. Chakravarti and I. S. Gupta, . *J. Indian Chem. Soc.,* **1,** 19 (1924); *Chem. Abstr.,* **19,** 830.
1343. G. B. Crippa and P. Galimberti, *Gazz. Chim. Ital.,* **59,** 825 (1929); *Chem. Abstr.,* **24,** 2115.
1344. A. Bistrzycki and A. Lecco, *Helv. Chim. Acta,* **4,** 425 (1921).
1345. F. M. Rowe, W. C. Dovey, B. Garforth, E. Levin, J. D. Pask, and A. T. Peters, *J. Chem. Soc.,* **1935,** 1796.
1346. F. M. Rowe, D. A. W. Adams, A. T. Peters, and A. E. Gillam, ibid., **1937,** 90.
1347. G. C. Chakravarti and I. S. Gupta, *Q. J. Indian Chem. Soc.* **1,** 329 (1925); *Chem. Abstr.,* **19,** 2493.
1348. G. B. Crippa, G. Perroncito, and G. Sacchetti, *Gazz. Chim. Ital.,* **65,** 38 (1935); *Chem. Abstr.,* **29,** 4007.
1349. R. Adams and N. K. Sundholm, *J. Am. Chem. Soc.,* **70,** 2667 (1948).
1350. W. Borsche and W. Doeller, *Annalen,* **537,** 53 (1938).
1351. H. Leuchs, ibid., **460,** 1 (1928).
1352. P. C. Edwards, D. Starling, A. M. Mattocks, and H. E. Skipper, *Science,* **107,** 119 (1948).
1353. R. Weidenhagen, R. Herrmann, and H. Wegner, *Chem. Ber.,* **70,** 570 (1937).
1354. J. G. Everett, *J. Chem. Soc.,* **1930,** 2402.
1355. C. F. H. Allen, A. Bell, and C. V. Wilson, *J. Am. Chem. Soc.,* **66,** 835 (1944).
1356. H. J. Backer and A. Boemen, *Rec. Trav. Chim.,* **45,** 110 (1926).
1357. H. J. Backer and A. Bloemen, ibid., **45,** 100 (1926).
1358. H. J. Backer and J. H. de Boer, ibid., **43,** 420 (1924).
1359. H. J. Backer and M. Toxopens, ibid., **45,** 890 (1926).
1360. H. J. Backer and J. H. de Boer, *Proc. Acad. Sci. Amsterdam,* **26,** 79 (1923); *Chem. Abstr.,* **17,** 2867.
1361. H. J. Backer, *Rec. Trav. Chim.,* **40,** 582 (1921).
1362. J. Brust, ibid., **47,** 153 (1928).
1363. Ng. Ph. Buu-Hoï and J. Lecocq, *Bull. Soc. Chim., France,* **1946,** 139.
1364. R. C. Fargher, *J. Chem. Soc.,* **117,** 865 (1920).
1365. M. A. Phillips, ibid., **1928,** 3134.
1366. R. R. Baxter and R. G. Fargher, ibid., **115,** 1372 (1919).
1367. J. G. Everett, ibid., **1929,** 670.
1368. J. G. Everett, ibid., **1931,** 3032.
1369. R. W. E. Stickings, ibid., **1928,** 3131.
1370. J. G. Everett, ibid., **1935,** 155.
1371. J. G. Everett, ibid., **1930,** 1691.
1372. H. J. Barber, ibid., **1929,** 471.
1373. G. O. Doak, H. G. Steinman, and H. Eagle, *J. Am. Chem. Soc.,* **63,** 99 (1941).
1374. Ng. Ph. Buu-Hoï and Ng. Hoan, *Rec. Trav. Chim.* **68,** 5 (1949).
1375. W. Steinkopf, R. Leitsmann, A. H. Müller, and H. Wilhelm, *Annalen,* **541,** 260 (1939).
1376. A. Leko and G. Vlajinats, *Bull. Soc. Chim. Roy Yougoslav.,* **3,** 85 (1932); *Chem. Abstr.,* **27,** 3475.
1377. A. Leko and G. Vlajinats, ibid., **4,** 17 (1933); *Chem. Abstr.,* **28,** 3735.
1378. L. Monti, *Gazz. Chim. Ital.,* **72,** 515 (1942); *Chem. Abstr.,* **38,** 4599; *Chem. Zentralb,* **1943,** I, 2197.
1379. G. B. Crippa and P. Galimberti, ibid., **62,** 937 (1932); *Chem. Abstr.,* **27,** 1342.

1380. T. N. Ghosh, *J. Indian Chem. Soc.*, **14,** 713 (1937); *Chem. Abstr.*, **32,** 4166.
1381. R. L. Shriner and R. W. Upson, *J. Am. Chem. Soc.*, **63,** 2277 (1941).
1382. J. von Braun, E. Anton, W. Haensel, G. Irmisch, R. Michaelis, and W. Teuffert, *Annalen* **507,** 14 (1933).
1383. B. A. Porai-Koshits and M. M. Antoshul'skaya, *J. Gen. Chem.* (USSR), **13,** 339 (1943); *Chem. Abstr.*, **38,** 1234.
1384. A. I. Kiprianov and I. K. Ushenko, ibid., **17,** 1538 (1947); *Chem. Abstr.*, **42,** 2253.
1385. T. Ogata, *Proc. Imp. Acad.* (Tokyo), **9,** 602 (1933); *Chem. Abstr.*, **28,** 2007.
1386. T. Ogata, *J. Chem. Soc. Japan*, **55,** 394 (1934); *Chem. Abstr.*, **28,** 5816.

CHAPTER 2

Benzimidazole-*N*-Oxides

DAVID M. SMITH

2.1. INTRODUCTION

Benzimidazole-*N*-oxides were discovered, apparently independently, by von Niementowski in 1887[1] and by Bankiewicz in the following year.[2] They were obtained as by-products from the reduction of *N*-acyl-*o*-nitroanilines with tin and hydrochloric acid or with ammonium sulfide. The parent compound was first isolated by von Niementowski in 1910,[3] by which time at least seven other representatives of the class were known.[3-6]

(2.1)   (2.2)

Thereafter, unaccountably, this area of chemistry suffered almost total neglect for the next fifty years. Admittedly, the "oxbenzimidazoles" of these early papers had been assigned tricyclic structures (e.g., **2.1**), and it was only in 1951 that they were correctly formulated as *N*-oxides (**2.2**) for the first time;[7] but it is nonetheless surprising that the neglect continued throughout the 1950s, a period of intense activity and interest in other aspects of *N*-oxide chemistry.

2.2. PREPARATION

2.2.1. General

Unlike six-membered heterocyclic systems of comparable basicity (e.g., pyridine and quinoline), benzimidazoles do not undergo *N*-oxidation by treatment with peracids.[8-10] Even peroxytrifluoroacetic acid, a reagent which oxidizes the most feebly basic pyridine derivatives,[11] is apparently without effect on benzimidazoles: for example, when 2-(2- and 4-thiazolyl)benzimidazoles are treated with this reagent, only the thiazole nitrogen is oxidized.[12] Another indirect method of *N*-oxidation, namely, addition of bromine followed by dehydrobromination and hydrolysis (eq. 1), may be discounted as a general method for benzimidazole-*N*-oxide formation, since von Niementowski, who discovered the reaction,[5] was subsequently unable to reproduce it.

Almost all the successful synthetic routes to benzimidazole-*N*-oxides

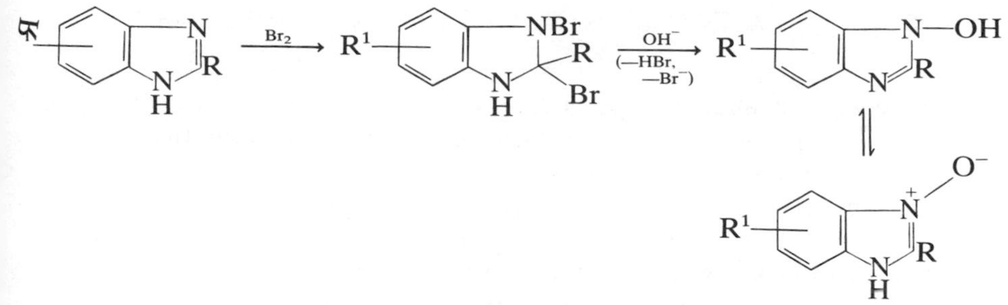

(1)

Type A:

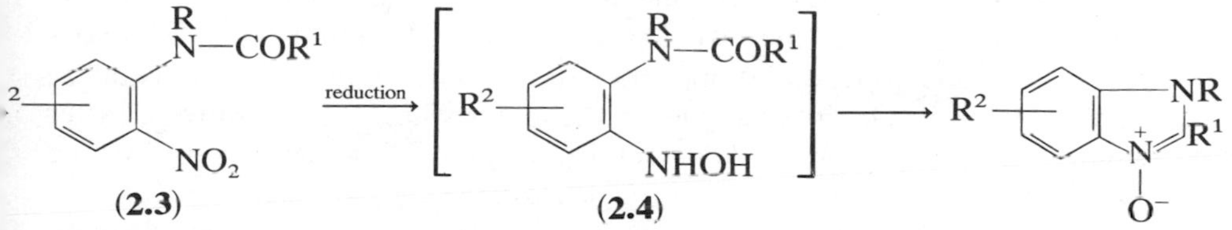

(2.3) (2.4)

(2)

Type B:

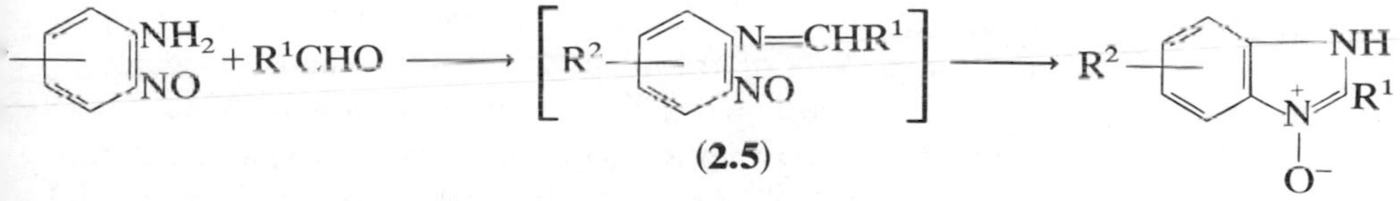

(2.5)

(3)

Type C:

(2.6)

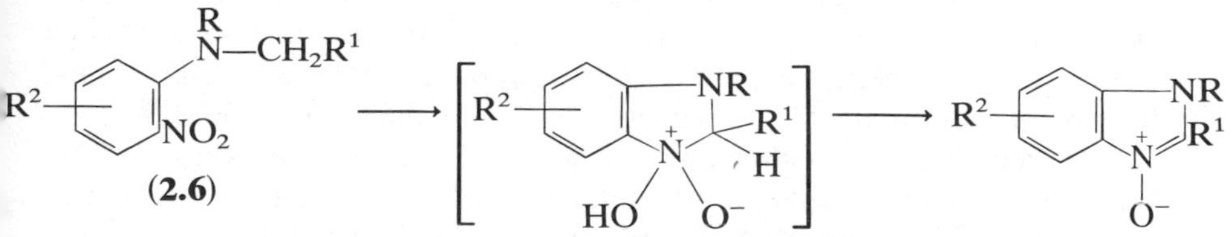

(4)

involve cyclizations in which the C_2–N_3 bond is formed. They fall into three main categories, which are represented by eqs. 2, 3, and 4. Type A cyclizations involve condensation of an amidic carbonyl group with a hydroxylamine; type B cyclizations involve interaction of an azomethine carbon with a nitroso group; and type C cyclizations involve an aldol-like condensation of a reactive methylene center with a nitro group.

2.2.2. Type A Cyclizations. Partial Reduction of *o*-Nitroanilides

Reference has already been made (section 2.1) to the discovery of benzimidazole-*N*-oxides as by-products from the reduction of *o*-nitroanilides. Their formation in such reactions presumably results from the spontaneous cyclization of the *o*-hydroxylaminoanilide (**2.4**) before the latter can be reduced further. By a careful choice of reducing agent and/or reaction conditions, however, cyclization at the hydroxylamine stage may be made to compete successfully with further reduction, and synthetically useful yields of benzimidazole-*N*-oxides may thus be obtained.

Catalytic hydrogenation over platinum or palladium gives moderate to good yields of benzimidazole-*N*-oxides when the reaction is carried out in presence of at least one molar equivalent of acid (usually HCl).[10,13–16] In the absence of added acid, the products are the *o*-aminoanilide and the (unoxidized) benzimidazole, and it appears that the effect of the acid is to catalyze cyclization of the hydroxylamine to the point where ring closure is faster than further reduction.[13] This method is particularly successful for the preparation of 1,2-dialkyl- (or aryl-)benzimidazole-3-oxides, and for the cyclization of anilides with an especially electron-deficient carbonyl group (e.g., a trifluoroacetyl compound).

A useful variant of the catalytic reduction method employs sodium borohydride in presence of palladium, platinum, or Raney nickel.[17,18] This method has been applied to the reduction of a wide range of substituted *o*-nitroanilides, including those of the type (**2.3**, R = H), and the yields of benzimidazole-*N*-oxides so obtained are generally in excess of 50%.[18]

Of the noncatalytic reduction methods, the most widely exploited is that using ammonium sulfide,[2,3,19–21] a process discovered by Bankiewicz and later developed, first by von Niementowski and more recently by Takahashi and Kano. In these reductions, the yields of 1-substituted benzimidazole-3-oxides are generally higher than those of their 1-unsubstituted analogs. A related procedure, with sodium hydrogen sulfide (in presence of a calcium salt) as reducing agent, has been used to prepare a variety of 2-aryl- (and -heteroaryl-)benzimidazole-*N*-oxides, although in unspecified yield.[22]

Other reducing agents that have been used in specific cases are sodium dithionite,[23,24] and zinc and ammonium chloride.[19] The latter reagent, which satisfactorily reduces nitrobenzene to phenylhydroxylamine, has found little application in benzimidazole-*N*-oxide synthesis, presumably because the reaction fails with some simply substituted *o*-nitroanilides.[19]

2.2.3. Type B Cyclizations. Formation and Cyclization of *o*-Nitrosoanils and Related Species

o-Nitrosoanils (**2.5**) are apparently an unstable (i.e., highly reactive) class of azomethine. No representative of the class has yet been isolated: all attempts to prepare such an anil result in the formation of the cyclic isomer, namely, the benzimidazole-N-oxide (**2.2**, R = H).

The simplest form of the reaction, shown in eq. 3, involves the acid-catalyzed condensation of an o-nitrosoaniline with an aldehyde. o-Nitrosoaniline itself reacts in this way with benzaldehyde,[25] and 4-nitro-2-nitrosoaniline reacts similarly with various aromatic aldehydes,[26] the yields of benzimidazole-N-oxides obtained being approx. 50–65%. However, the method has not been widely used, since o-nitrosoanilines are not easily prepared and are somewhat unstable. A more useful form of the reaction involves the photolysis of the N-(2,4-dinitrophenyl) derivatives of α-amino-acids:[20,27] the main products of photolysis are 4-nitro-2-nitrosoaniline and the 5-nitro-benzimidazole-N-oxide (**2.7**), the latter predominating if the photolysis is conducted in an acidic medium (eq. 5). Related photolyses that also yield benzimidazole-N-oxides, apparently via o-nitrosoanils, are those of the o-nitrophenylaziridine derivatives (**2.8** and **2.9**) (eq. 6), in which the yields are almost quantitative[28] (see also "Syntheses from Imidazoles" in section 1.29 in Chapter 1); those of the o-nitrophenylimidazole derivatives (**2.10** and **2.11**) (eq. 7 and 8);[29] and that of the herbicide trifluralin (**2.12**) (eq. 9).[30]

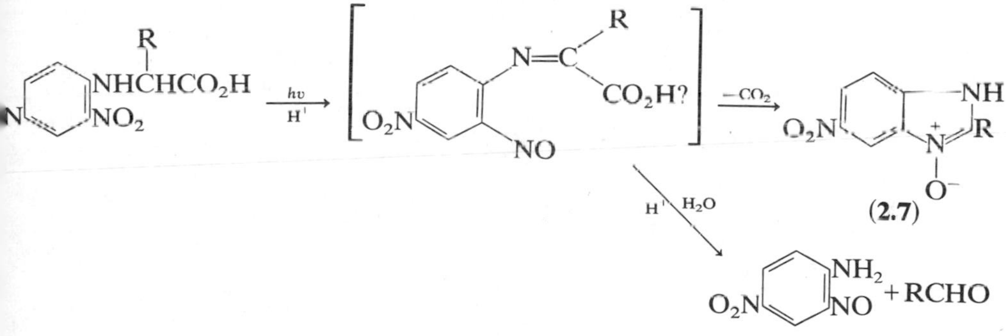

(2.7)

(5)

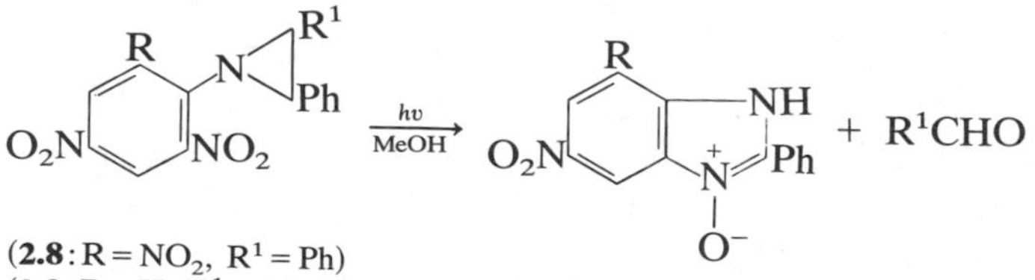

(**2.8**: R = NO$_2$, R^1 = Ph)
(**2.9**: R = H, R^1 = COPh)

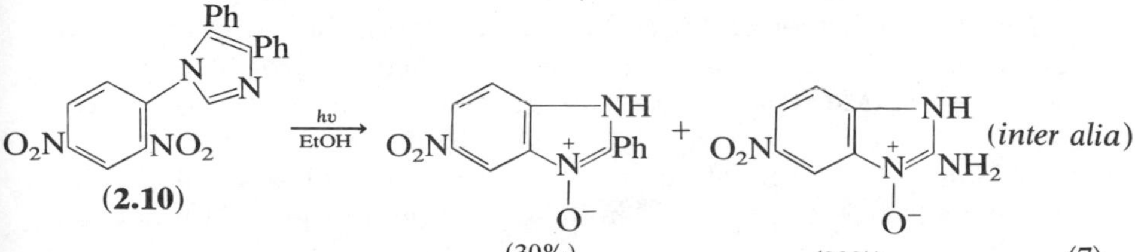

(6)

(**2.10**)

(30%)   (20%)   (7)

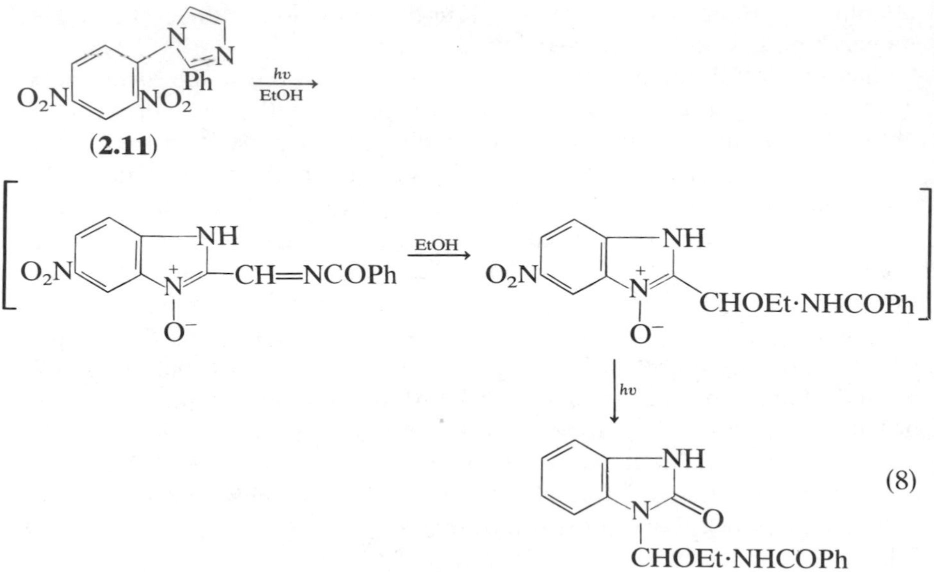

$$(2.11)$$

$$(8)$$

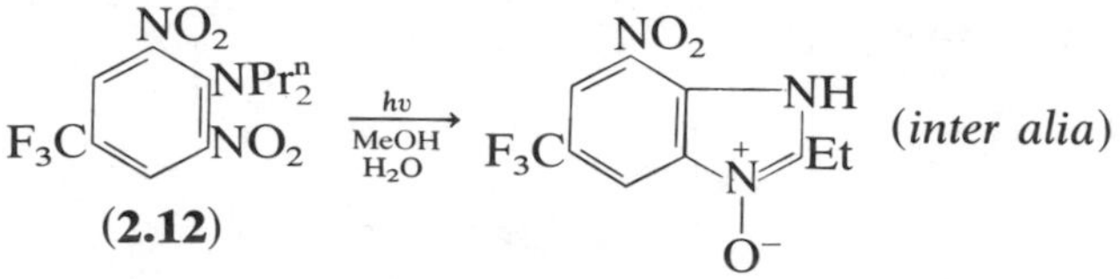

$$(inter\ alia) \quad (9)$$

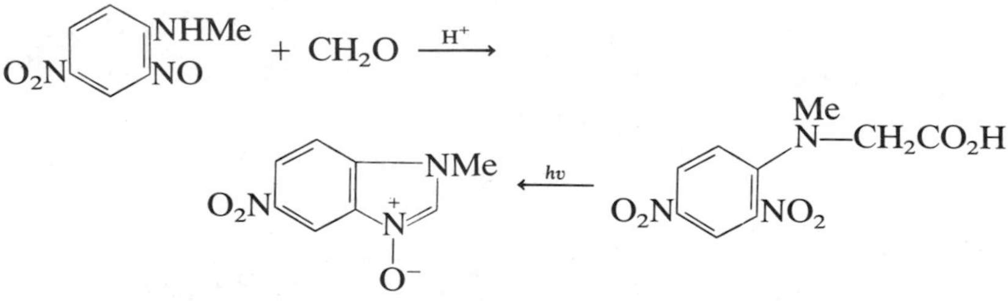

$$(10)$$

With regard to the mechanism for the cyclization of *o*-nitrosoanils, the simplest representation (Scheme 2.1a) is probably incorrect, since a 5-*Endo-Trig* process is a disfavored mode of ring closure.[31] In any case, the species undergoing cyclization in the acid-catalyzed reactions is probably not the anil (**2.5**), but the cation (**2.13**) or the adduct (**2.14**), the cyclization of the latter being a (favored) 5-*Exo-Tet* process (Scheme 2.1b). Similar intermediates, with NMe replacing NH, are presumably involved in the formation of 1-methyl-5-nitrobenzimidazole-3-oxide, from *N*-methyl-4-nitro-2-nitrosoaniline and formaldehyde or from photolysis of *N*-(2,4-dinitrophenyl)sarcosine (eq. 10);[27] in these reactions nitrosoanil formation is

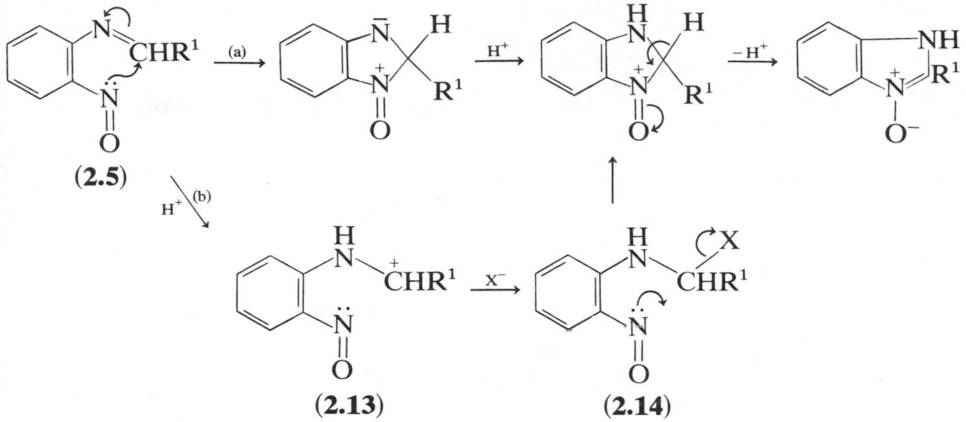

(2.5)

H⁺ (a)

H⁺ (b)

−H⁺

CHR¹

NH

R^1

O⁻

CHR¹

X⁻

CHR¹

(2.13)

(2.14)

Scheme 2.1

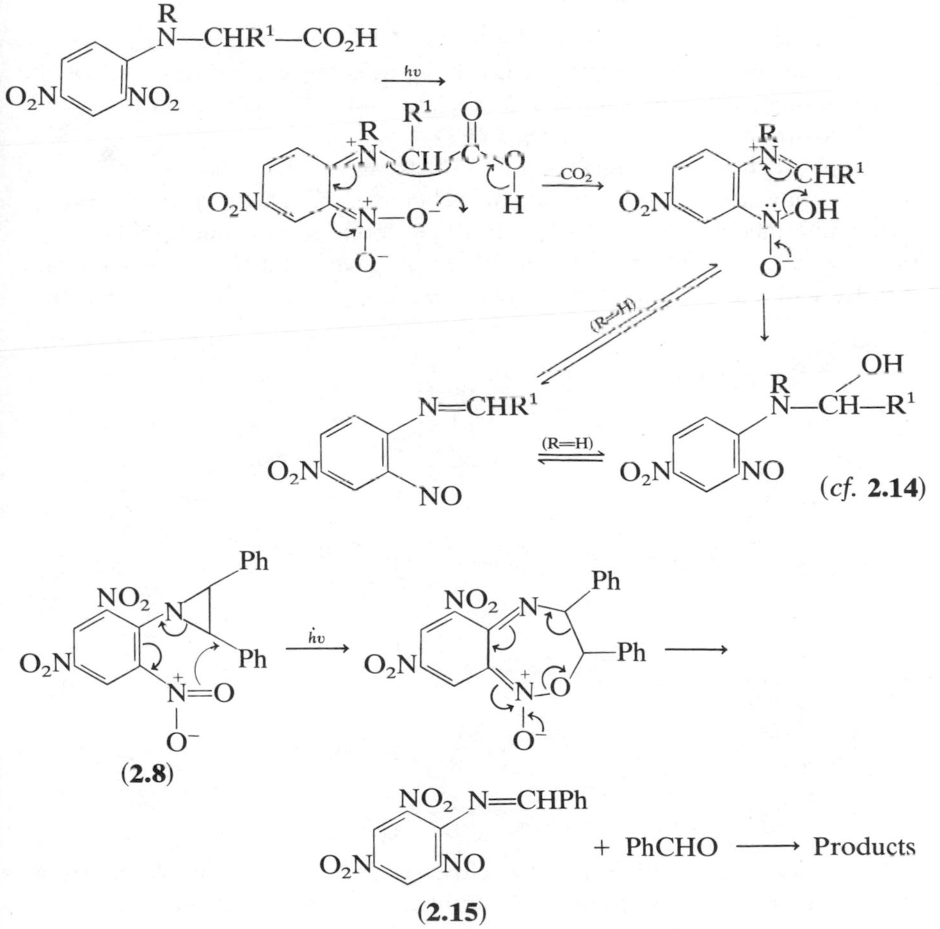

$$R$$
$$N\!-\!CHR^1\!-\!CO_2H$$

O_2N NO_2

$\xrightarrow{h\nu}$

$\xrightarrow{CO_2}$

O_2N

O_2N

CHR¹

OH

(R=H)

$N\!=\!CHR^1$

O_2N NO

(R=H)

O_2N NO

$$N\!-\!CH\!-\!R^1$$

OH

R

(*cf.* **2.14**)

(2.8)

Ph

NO_2

O_2N

Ph

$N\!=\!O$

O⁻

$\xrightarrow{h\nu}$

NO_2

O_2N

Ph

Ph

O⁻

$\longrightarrow$

NO_2 $N\!=\!CHPh$

O_2N NO

(2.15)

$+\ PhCHO \longrightarrow$ Products

Scheme 2.2

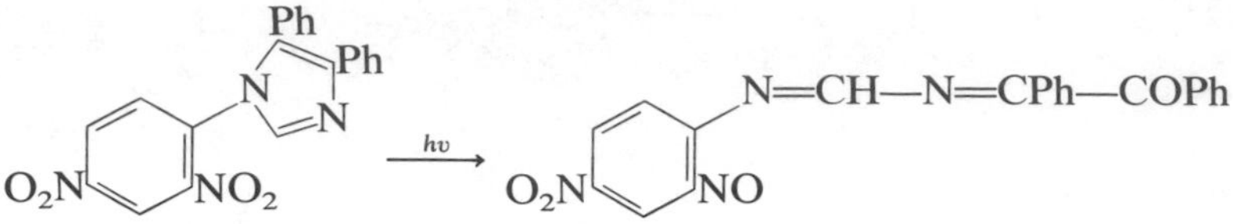

(2.10)   (2.16a)

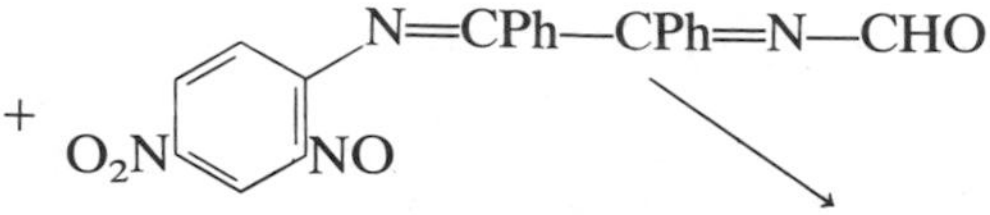

(2.16b)   Products

Scheme 2.2 (continued)

not possible. Nitrosoanil-alcohol adducts (**2.14**, X = OMe or OEt) may be involved in those cyclizations not involving acid but carried out in alcoholic solvents (e.g., those of **2.15** and **2.16**); alternatively, such cyclizations may be electrocyclic processes (cf. eq. 12).

The photochemical generation of *o*-nitrosoanils from *o*-amino-nitro compounds may be rationalized as in Scheme 2.2. There are, of course, alternative mechanisms. Indeed, it has been pointed out that the photolysis of an *N*-(alkylmethylene)-*o*-nitroaniline may give a benzimidazole-*N*-oxide directly without the intermediacy of an *o*-nitrosoanil at all (Scheme 2.3).[32]

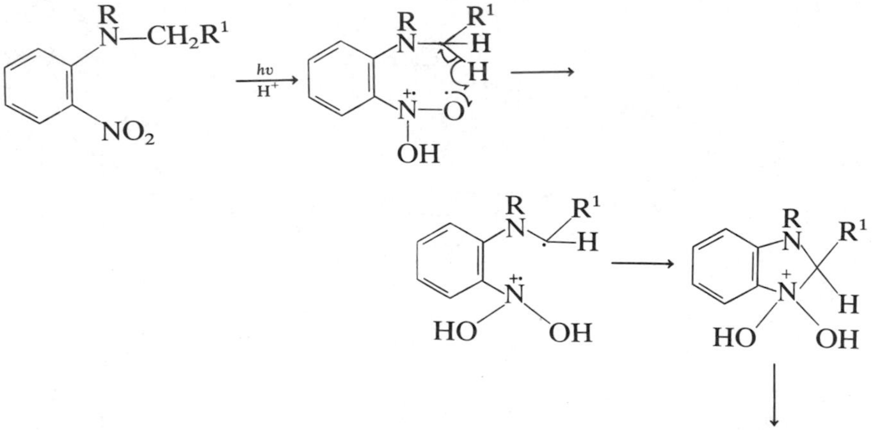

Scheme 2.3   (2.2)

The second group of Type B cyclization uses quinoxaline-*N*-oxides as starting materials. This group has been little exploited to date as a synthetic route to benzimidazole-*N*-oxides, but is of considerable mechanistic interest.

2-Substituted quinoxaline-4-oxides (**2.17**) are oxidized by alkaline hydrogen peroxide to 2-substituted benzimidazole-*N*-oxides and (presumably) formate ion[33,34] (eq. 11). The most plausible mechanism for these reactions involves nucleophilic attack by HO_2^- at the 3-position, followed by ring-opening with the loss of OH^-. The resulting *o*-nitrosoanil (**2.18**) then undergoes recyclization to the benzimidazole-*N*-oxide, with the loss of formate.

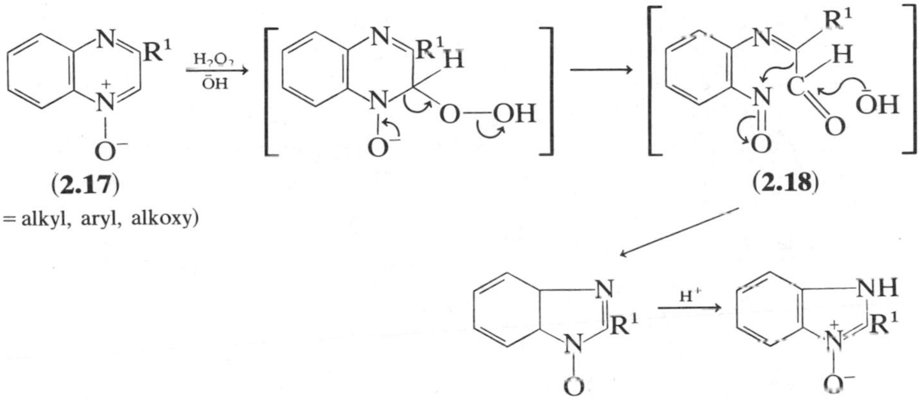

(**2.17**)   (**2.18**)

(R^1 = alkyl, aryl, alkoxy)

(11)

2-Azidoquinoxaline-1-oxide (**2.19**) undergoes thermolysis merely by heating in boiling benzene for 30 min, the products being nitrogen and 2-cyano-benzimidazole-*N*-oxide (**2.20**) (eq. 12) in almost quantitative yield.[35,36] 2-Azidoquinoxaline-1,4-dioxide (**2.21**) similarly gives 2-cyano-1-hydroxy-benzimidazole-3-oxide (**2.22**) in 61% yield,[36] and 2-azido-3-methyl-quinoxaline-1,4-dioxide (**2.23**) gives the rather unstable 2-cyano-2-methyl-2*H*-benzimidazole-1,3-dioxide (**2.24**) along with the oxadiazine derivative

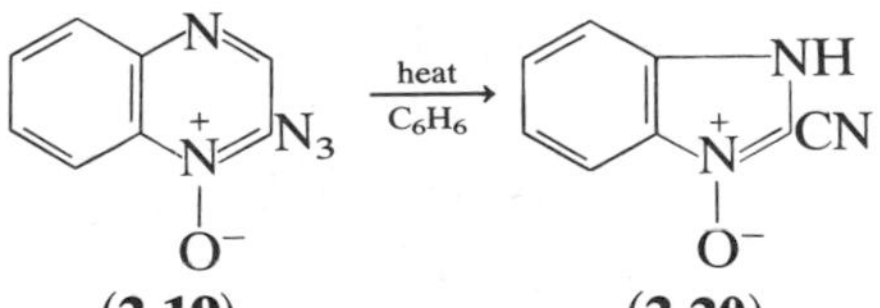

(12)

(**2.19**)   (**2.20**)

(13)

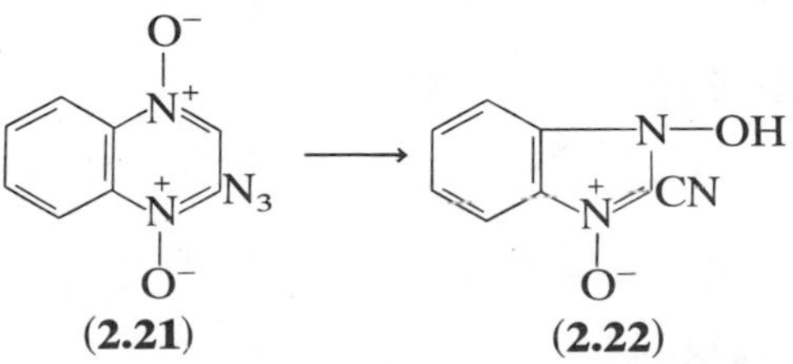

(**2.21**)   (**2.22**)

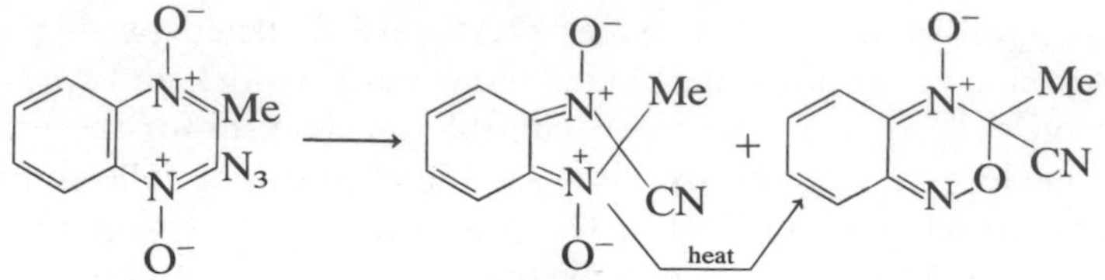

$$(14)$$

(2.23)   **(2.24)**   **(2.25)**

(2.25) (eqs. 13 and 14, respectively).[36] The proposed mechanisms for these reactions are set out in Scheme 2.4, the intermediates again being an *o*-nitrosoanil **(2.26)** or the corresponding nitrone (**2.27** or **2.28**). Ring closure of **(2.26)**, in a hydrocarbon solvent and in absence of acid, is most probably an electrocyclic process, but two other mechanisms are possible for cyclization of the nitrones, namely, eqs. 15 and 16.

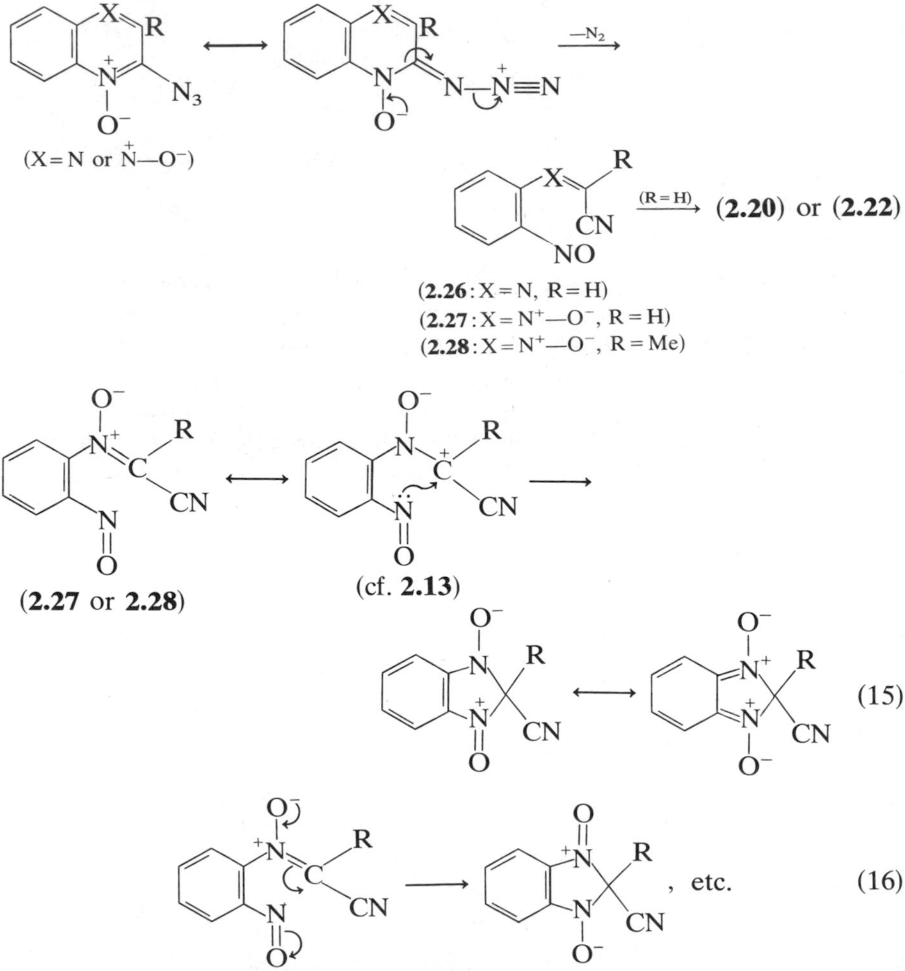

$(X = N$ or $\overset{+}{N}{-}O^-)$

$\xrightarrow{(R=H)}$ **(2.20)** or **(2.22)**

(**2.26**: X = N, R = H)
(**2.27**: X = N⁺—O⁻, R = H)
(**2.28**: X = N⁺—O⁻, R = Me)

(2.27 or 2.28)   (cf. **2.13**)

$$(15)$$

$$(16)$$

, etc.

Scheme 2.4

The final group of type B cyclizations uses benzofuroxans as starting materials. This group of reactions is now beginning to assume considerable significance, since it leads to several types of benzimidazole-N-oxide that are not otherwise accessible.

Benzofuroxan (benzo[c]-1,2,5-oxadiazole-N-oxide, **2.29**) is the cyclic tautomer of o-dinitrosobenzene, and so in principle the reaction of benzofuroxan with a reactive methylene compound should yield an o-nitrosoanil and hence a benzimidazole-N-oxide. In practice, however, such reactions of benzofuroxan are considerably more complicated: the complications arise mainly because nitroso groups act not only as electron acceptors, but as electron donors as well.

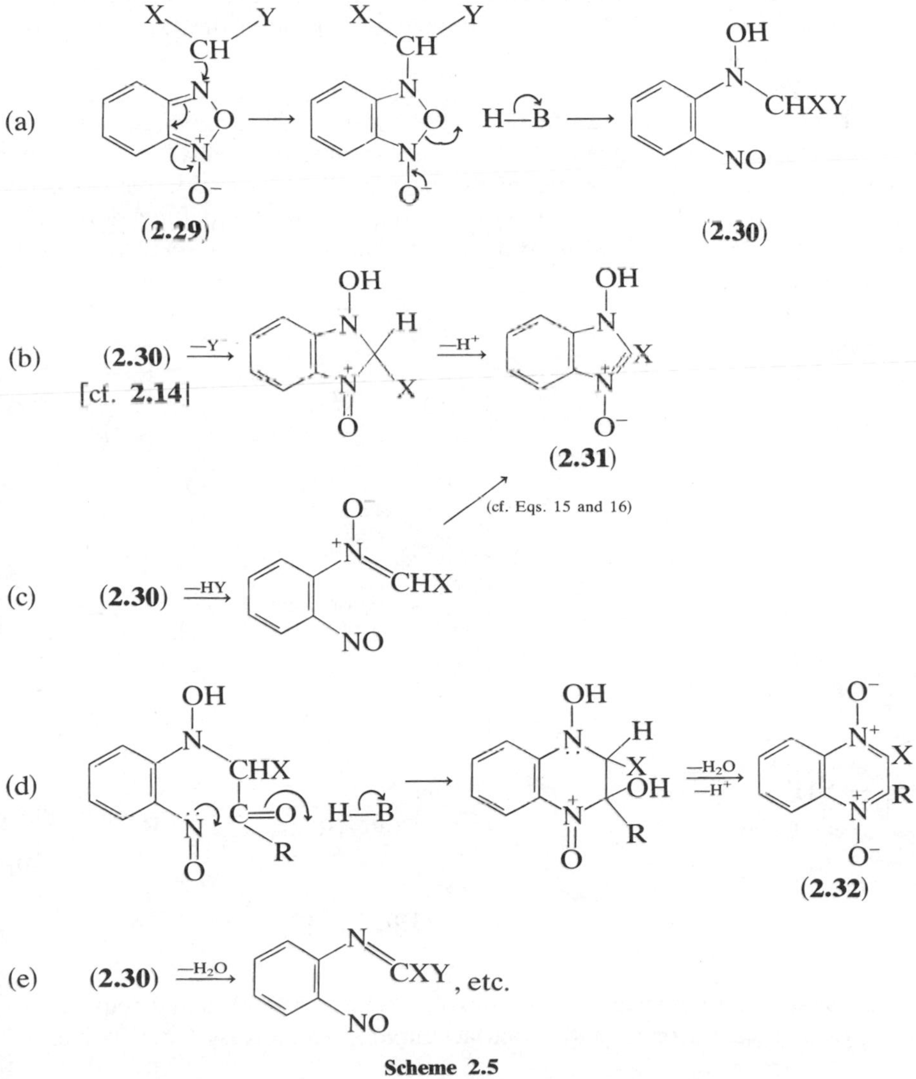

(a)

(2.29)   (2.30)

(b)   (2.30) $\xrightarrow{-Y^-}$
[cf. 2.14]

(2.31)

(cf. Eqs. 15 and 16)

(c)   (2.30) $\xrightarrow{-HY}$

(d)

(2.32)

(e)   (2.30) $\xrightarrow{-H_2O}$, etc.

Scheme 2.5

In general terms, the reaction of benzofuroxan with a reactive methylene compound, XCH_2Y, in presence of base may be represented as in Scheme 2.5a. The resulting intermediate (**2.30**), which is an o-nitrosophenyl-hydroxylamine, may then undergo one or more of the following reactions:

i. If Y is a leaving group (e.g., CN, NO_2, or SO_2R), it may then be lost as Y^-, according to Scheme 2.5b or c and the product obtained is the 1-hydroxy-2-X-benzimidazole-3-oxide (**2.31**).

ii. If Y is a keto-group, the ketonic carbonyl function may interact with the o-nitroso group, and the product is then a quinoxaline-1,4-dioxide (**2.32**) (Scheme 2.5d).

iii. If neither of the above reactions intervenes, the o-nitrosoanil may be formed and may yield the benzimidazole-N-oxide in the expected manner (Scheme 2.5e).

Scheme 2.5 accounts satisfactorily for most of the known reactions of benzofuroxan with reactive methylene compounds. With ethyl acetoacetate, the quinoxaline dioxide is the main product, the benzimidazole-N-oxide being formed in relatively small amount[37] (eq. 17); with a malonyl derivative such as barbituric acid, on the other hand, benzimidazole-2-carboxylic acid 3-oxide is the sole product[38] (eq. 18).

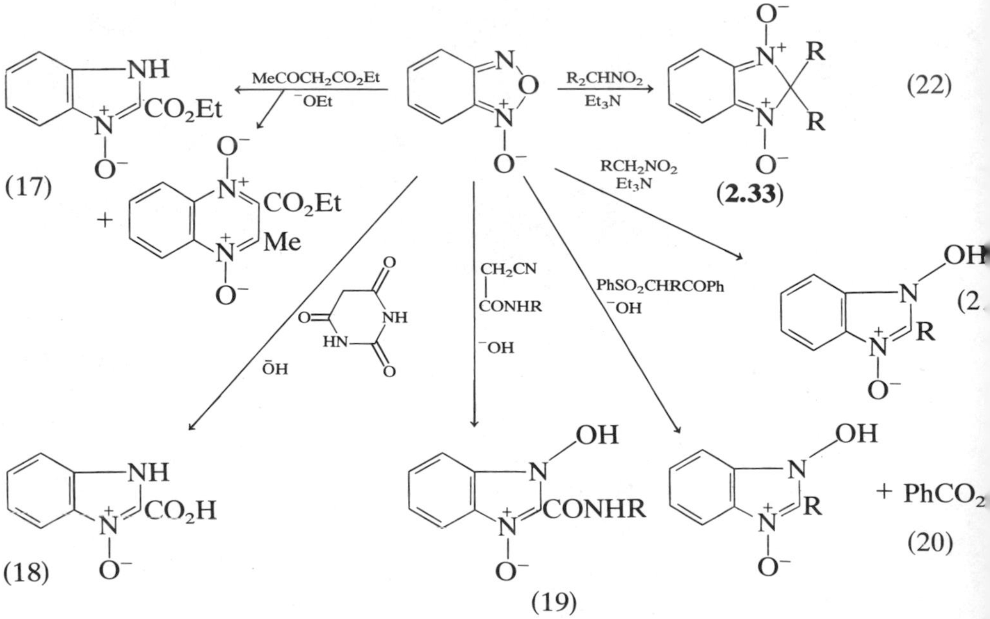

2-Substituted 1-hydroxybenzimidazole-3-oxides are obtained (eqs. 19 to 21) from benzofuroxan and cyanoacetamide derivatives,[39] derivatives of phenylsulfonylacetic acid and phenylsulfonylacetophenone,[40] and nitroalkanes.[41–45] In an extension of this last reaction, benzofuroxan reacts with secondary nitroalkanes (eq. 22) to give 2,2-disubstituted 2H-benzimidazole-1,3-dioxides (**2.33**),[41–45] this being the only general route at present available to these compounds.

The nucleophilic character of the nitroso group (or, alternatively, of the nitrogen at position 3 in benzofuroxan) is utilized in another group of benzimidazole-N-oxide syntheses. Reaction of benzofuroxan with strong alkylating agents (e.g., alkyl trifluoromethylsulfonates) may lead to 1-hydroxybenzimidazole-3-oxides[46] (eq. 23). With formaldehyde (eq. 24), good to excellent yields of 1,3-dihydroxybenzimidazolones (**2.34**: tautomeric with 1,2-dihydroxybenzimidazole-3-oxides) are obtained,[47] and the corresponding reactions with the formaldimine trimers (**2.35**) lead to 1,3-dihydroxybenzimidazolimines (**2.36**) (eq. 25).[48]

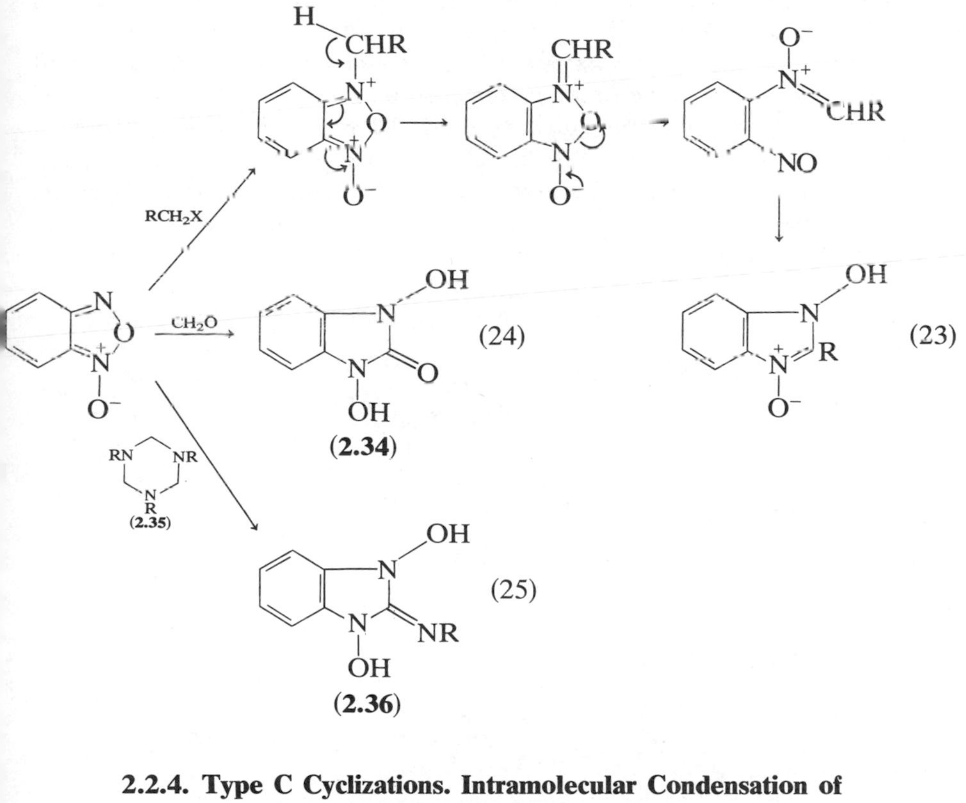

(2.34)

(23)

(25)

(2.36)

2.2.4. Type C Cyclizations. Intramolecular Condensation of N-(Activated Alkyl-)o-nitroanilines

The cyclization of o-substituted nitroarenes, involving a condensation reaction between the nitro group and a nucleophilic center in the *o*-substituent, has come to be recognized as a standard route for the preparation of many different kinds of heterocyclic *N*-oxides.[49] For benzimidazole-*N*-oxide synthesis, the required substrates are *N*-alkyl-*o*-nitroanilines (**2.6**) in which the alkyl substituent is activated by an electron-accepting group, R′; in their most common form the cyclizations are effected by the action of a base.

Base-Induced Condensations

It is convenient to consider these under two headings, namely, those involving monosubstituted *o*-nitroanilines (**2.6**, R = H) and those involving disubstituted *o*-nitroanilines (**2.6**, R ≠ H).

The former group has been much more extensively studied and constitutes a useful and general route to benzimidazole-*N*-oxides with functional groups at position 2. Where the activating group R′ is a strong electron acceptor, e.g., a ketone,[50] a nitrile,[51,52] an ester,[53] or an amide,[54] the cyclizations proceed under extremely mild conditions and often give good yields of the heterocyclic products (eqs. 26 to 29). Even a carboxyl group may serve as the activating substituent if the reaction is conducted in a buffered medium[54] (eq. 30). Weaker electron acceptors such as aryl groups require that a stronger base be used to effect the cyclization, and hydroxide, alkoxide, or hydride ion is normally employed for the cyclization of *N*-benzyl-*o*-nitroaniline and its derivatives[9,10,24,28,55–57] (eq. 31).

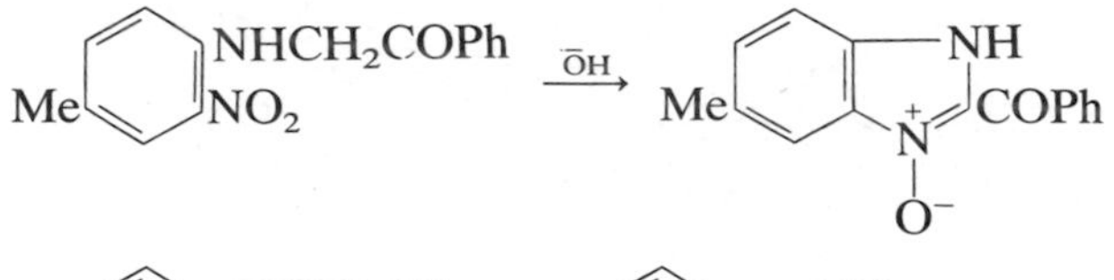

$$\text{(26)}$$

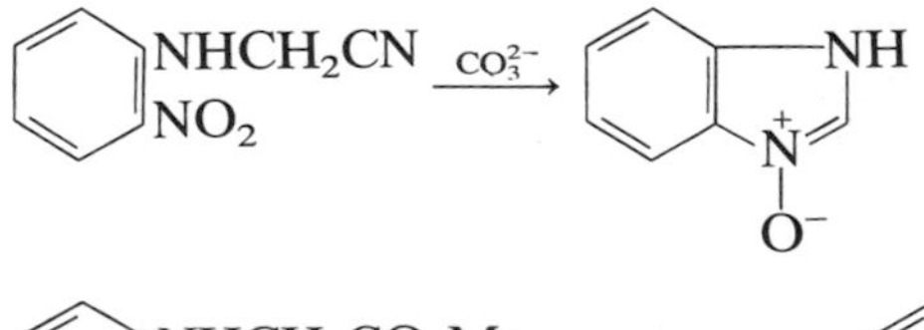

$$\text{(27)}$$

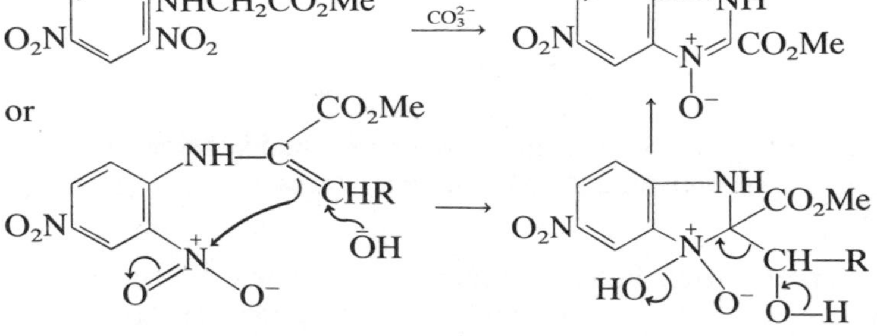

$$\text{(28)}$$

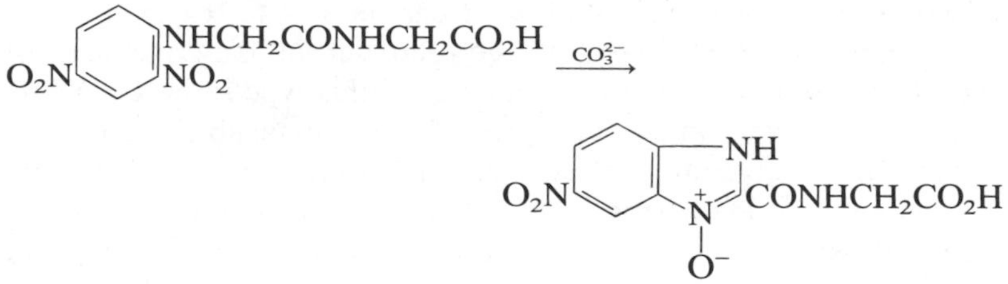

O$_2$N— ⟨ring⟩ NHCH$_2$CONHCH$_2$CO$_2$H / NO$_2$ $\xrightarrow{CO_3^{2-}}$

O$_2$N— ⟨benzimidazole⟩ —NH / N$^+$—CONHCH$_2$CO$_2$H / O$^-$ (29)

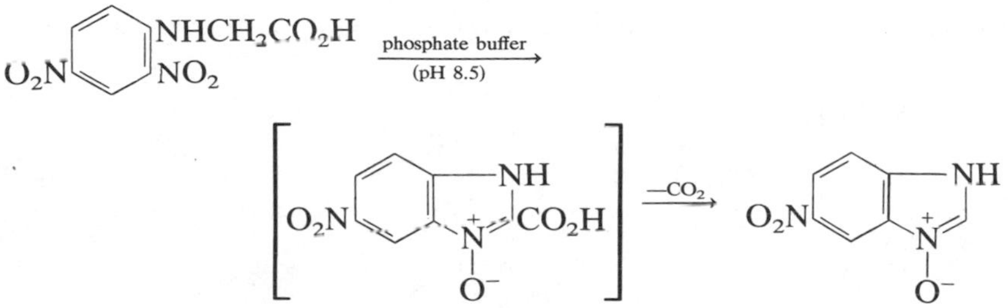

O$_2$N— ⟨ring⟩ NHCH$_2$CO$_2$H / NO$_2$ $\xrightarrow[\text{(pH 8.5)}]{\text{phosphate buffer}}$

$$\left[\text{O}_2\text{N—} \langle\text{benzimidazole}\rangle \begin{array}{c}\text{—NH}\\ \text{N}^+\text{—CO}_2\text{H}\\ \text{O}^-\end{array} \right] \xrightarrow{-CO_2} \text{O}_2\text{N—}\langle\text{benzimidazole}\rangle \begin{array}{c}\text{—NH}\\ \text{N}^+\\ \text{O}^-\end{array}$$ (30)

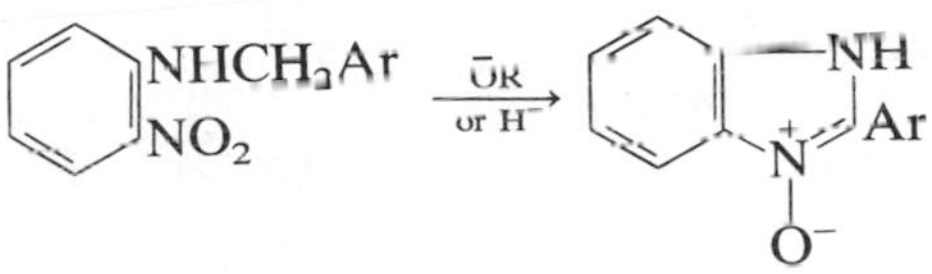

⟨ring⟩ NHCH$_2$Ar / NO$_2$ $\xrightarrow[\text{or H}^-]{\bar{O}R}$ ⟨benzimidazole⟩ —NH / N$^+$—Ar / O$^-$ (31)

If the nucleophilic center is a tertiary carbon, cyclization may still occur, provided that one of the substituents on that carbon may be subsequently removed: thus N-(α-cyanobenzyl)-o-nitroaniline and base (or an N-arylidene-o-nitroaniline and cyanide ion) yield a 2-arylbenzimidazole-N-oxide and, presumably, a cyanate (eq. 32).[51,58,59]

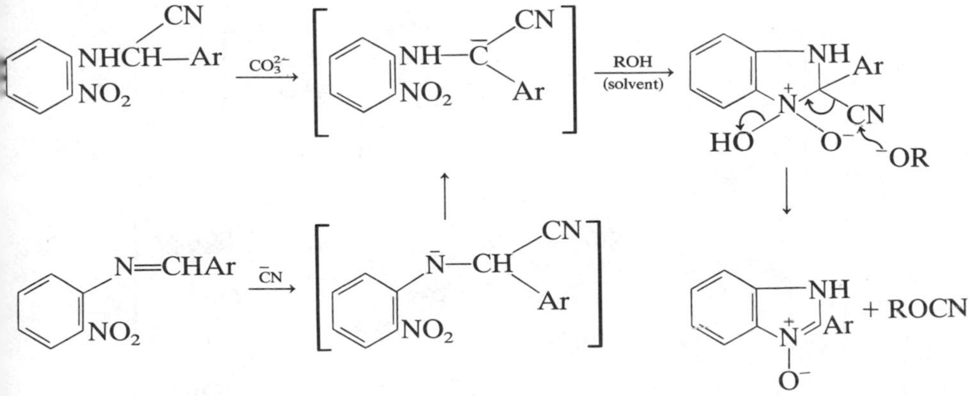

⟨ring⟩ CN / NHCH—Ar / NO$_2$ $\xrightarrow{CO_3^{2-}}$ $\left[\langle\text{ring}\rangle \begin{array}{c}\text{CN}\\ \text{NH—C}\\ \text{NO}_2 \quad \text{Ar}\end{array} \right]$ $\xrightarrow[\text{(solvent)}]{\text{ROH}}$ ⟨benzimidazole⟩ —NH / Ar / N$^+$ CN / HO   O$^-$ OR

↑

⟨ring⟩ N=CHAr / NO$_2$ $\xrightarrow{\bar{C}N}$ $\left[\langle\text{ring}\rangle \begin{array}{c}\text{CN}\\ \bar{\text{N}}\text{—CH}\\ \text{NO}_2 \quad \text{Ar}\end{array} \right]$

↓

⟨benzimidazole⟩ —NH / N$^+$—Ar / O$^-$ + ROCN

(32)

The cyclization of *N,N*-disubstituted *o*-nitroanilines (**2.6**, R ≠ H) has not yet been systematically investigated (the required starting materials are not always easy to obtain), but the known examples of this type show surprising differences from the corresponding reactions of their monosubstituted counterparts. Thus, the *N*-substituted *N*-cyanomethyl-*o*-nitroanilines (**2.37**) are cyclized by sodium carbonate, not to the 2-cyano-, but to the 2-*hydroxy*-benzimidazole-*N*-oxides [⇌ the *N*-hydroxybenzimidazolones (**2.38**)] (eq. 33),[17,51] and *N*-acylmethylene-*N*-arylsulfonyl-*o*-nitroanilines (**2.39**) are cyclized by sodium alkoxides, not to a 2-acyl-1-arylsulfonyl-benzimidazole-3-oxide (**2.40**), but to a 2-alkoxybenzimidazole-*N*-oxide (**2.41**) (eq. 34).[60]

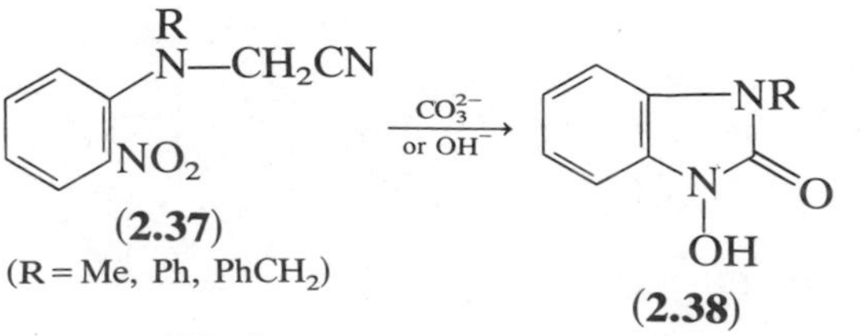

$$(33)$$

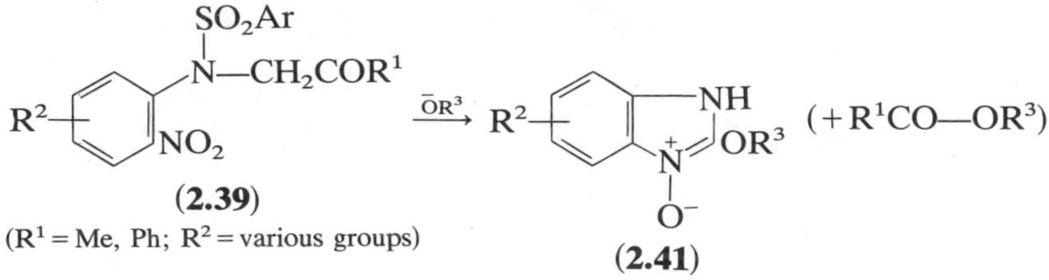

(**2.37**)
(R = Me, Ph, PhCH$_2$)

(**2.38**)

$$(34)$$

(**2.39**)
(R^1 = Me, Ph; R^2 = various groups)

(**2.41**)

In the proposed mechanisms for these reactions (Scheme 2.6), the initial step is assumed to be the "normal" cyclization, yielding the 1-substituted benzimidazole-3-oxide. Whereas a 1-unsubstituted benzimidazole-3-oxide is weakly acidic and is deprotonated (and hence stabilized) in basic solution (cf. section 2.3), a 1-substituted 3-oxide is nonacidic and is susceptible to nucleophilic attack at the 2-position.

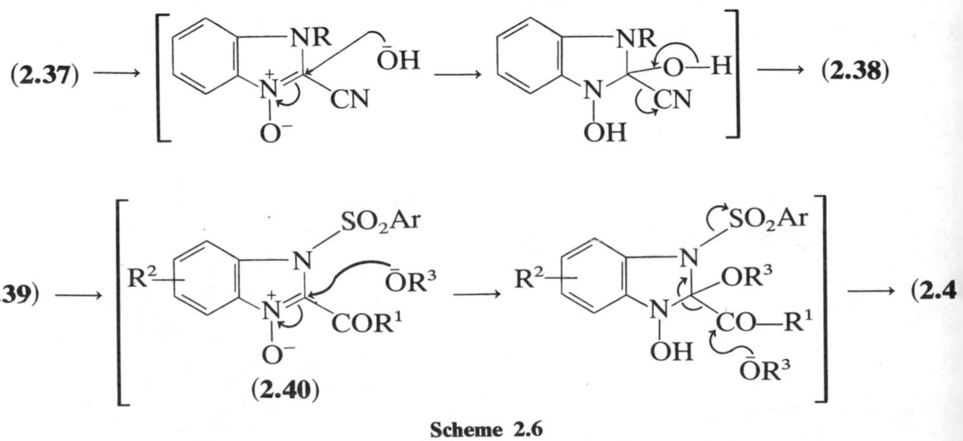

(**2.37**) ⟶ [...] ⟶ (**2.38**)

(**2.39**) ⟶ [... (**2.40**) ...] ⟶ (**2.4**

Scheme 2.6

A further anomaly, of a different kind, is observed in the cyclization of *N*-*p*-nitrobenzyl-*N*-tosyl-*o*-nitroaniline (or its *N*-mesyl analog) with sodium methoxide. In this reaction (eq. 35) the products are the desulfonylated *N*-oxide (**2.42**) and its *O*-methyl derivative (**2.43**). It is assumed that methyl tosylate (or mesylate) is responsible for the methylation of the product.[57,61]

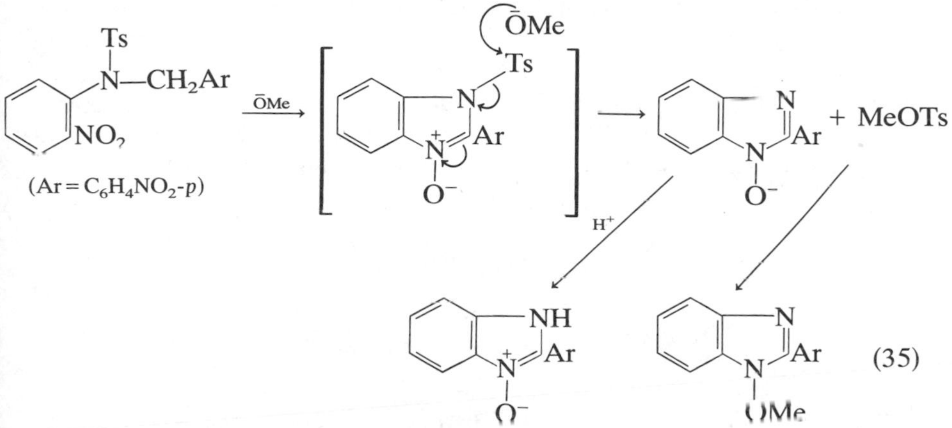

$$(Ar = C_6H_4NO_2\text{-}p) \qquad\qquad\qquad\qquad\qquad\qquad\qquad\qquad (35)$$

Finally, it must be noted that the *N*-methyl, *N*-tosyl, and *N*-mesyl derivatives of *N*-benzyl-*o*-nitroaniline are not cyclized at all by base,[10,57,61] and the corresponding *N*-acyl derivatives are cyclized only because they undergo prior deacylation.[57] These results must cast doubt on the simple condensation mechanism for the cyclization of *N*-benzyl-*o*-nitroaniline itself (eq. 31); it appears that the (removable) hydrogen of the amino group has a part to play in this cyclization, and Scheme 2.7 indicates possible alternative mechanisms in which the mobility of this hydrogen is important.

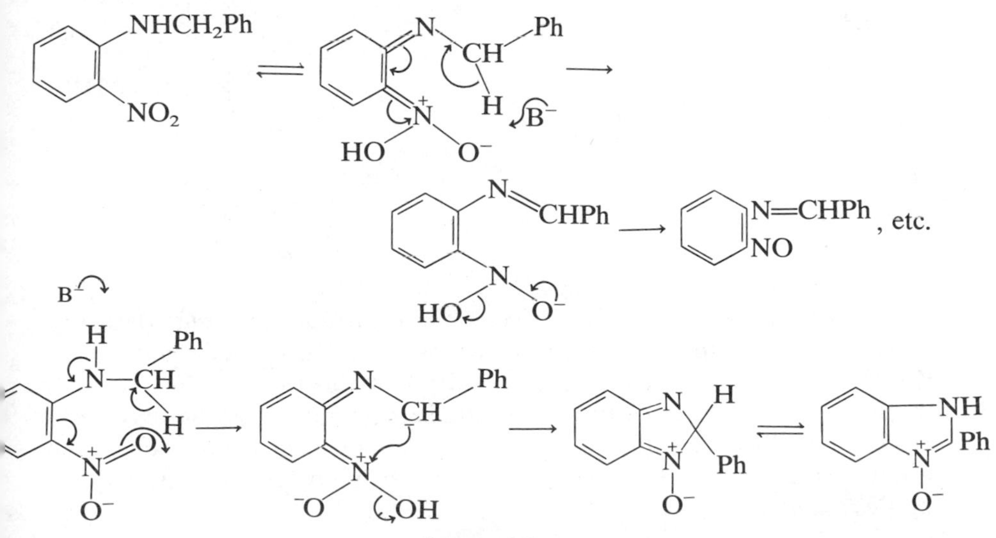

Scheme 2.7

Acid-Induced Condensations

Mention has already been made (section 2.2.3) of the photo-induced conversion of N-(2,4-dinitrophenyl)-α-amino acids into benzimidazole-N-oxides in acidic solution, and of the photocyclization of trifluralin (eqs. 5 and 9). N,N-Dialkyl-o-nitroanilines may also be cyclized to benzimidazole-N-oxides, however, by acid alone.[32] Admittedly, the reaction requires high temperatures (>100°) and prolonged times (usually 12–48 hr), but the yields are good, especially if the two alkyl substituents form part of a ring system (cf. section 8.1.1 in Chapter 8). A possible mechanism (Scheme 2.8) invites comparison with those of Scheme 2.7 and of Schemes 2.2 and 2.3.

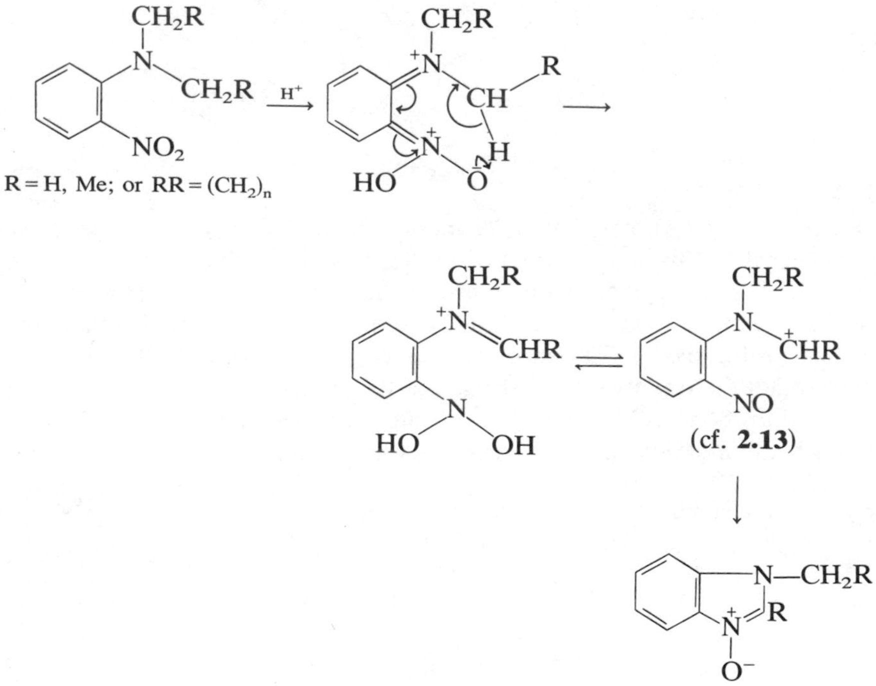

Scheme 2.8

A report in the literature claims that the o-nitrophenylglycine derivative (**2.44**) is cyclized to 5-methylbenzimidazole-N-oxide by hot acetic anhydride.[62] However, this claim appears difficult to justify, since the work-up includes a "hydrolysis" with boiling aqueous ammonia. In the absence of evidence to the contrary, eq. 36 represents the most likely sequence of events, with the cyclization being effected, not by the acetic anhydride, but by the ammonia.

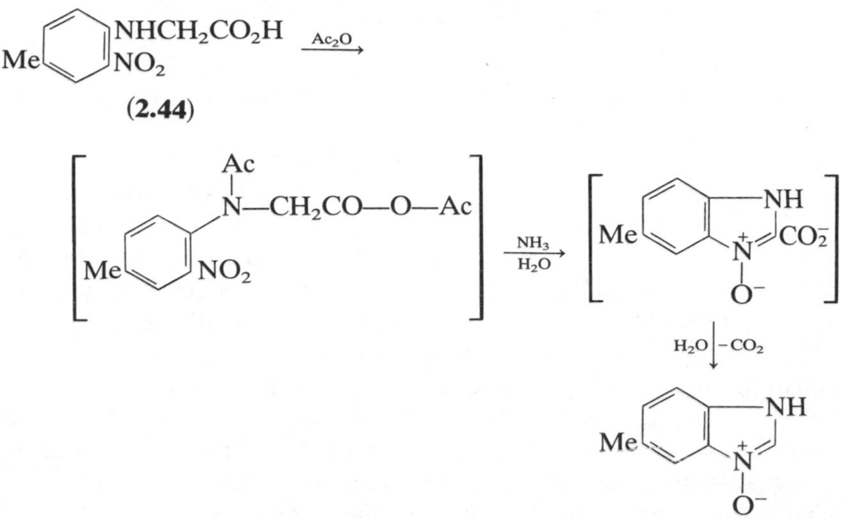

(2.44)

(36)

Uncatalyzed (Thermal) Condensations

2-Phenylbenzimidazole-N-oxide has been obtained in 30% yield from the thermolysis of the amino acid (2.45) in sand at 200° [63] In the thermolyses of related compounds, benzimidazole-N-oxides have been proposed as intermediates, but none has been isolated.[63] The proposed mechanism (eq. 37) again involves an *aci*-nitro intermediate, as in Schemes 2.7 and 2.8.

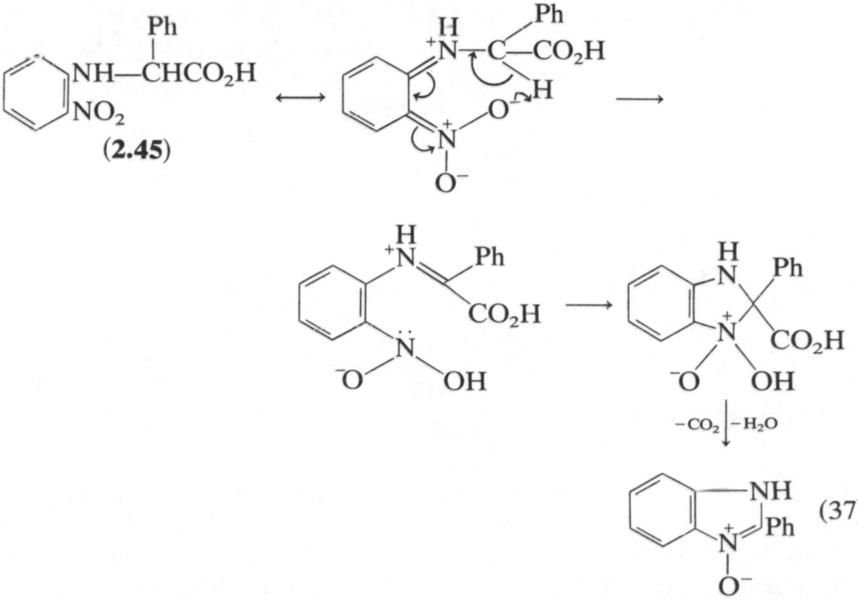

(2.45)

(37)

2.2.5. Other Cyclizations

   Three routes to benzimidazole-*N*-oxides remain to be considered under
this "miscellaneous" heading.

 i. *o*-Nitroaniline and its derivatives react with aromatic aldehydes to give,
    according to the conditions, *o*-nitroanils (**2.46**), arylidenebis-*o*-
    nitroanilines (**2.47**), 2-arylbenzimidazole-*N*-oxides (**2.48**), or 2-
    arylbenzimidazoles (**2.49**), the proportion of each product depending on
    the reaction conditions.[9,64–67] The reactions are normally carried out in
    boiling hydrocarbon solvents, with removal of water by azeotropic
    distillation. For maximum yield of the *N*-oxide (**2.48**) the ratio of
    aldehyde to amine should be slightly greater than 2:1 (the stoichiometric
    proportions), and the correct choice of solvent (i.e., of reaction tempera-
    ture) is absolutely critical. A larger excess of aldehyde, or a higher-
    boiling solvent, favors the formation of the reduced product (**2.49**).
       The mechanism of these cyclizations has not been elucidated, although
    it is obvious that a reduction step is involved at some point. The simplest
    rationalization is provided in eq. 38, whereby the *o*-nitroanil is reduced
    by the aldehyde, and the process would then correspond to a type B
    cyclization. For the present, however, this mechanism is mere conjec-
    ture.

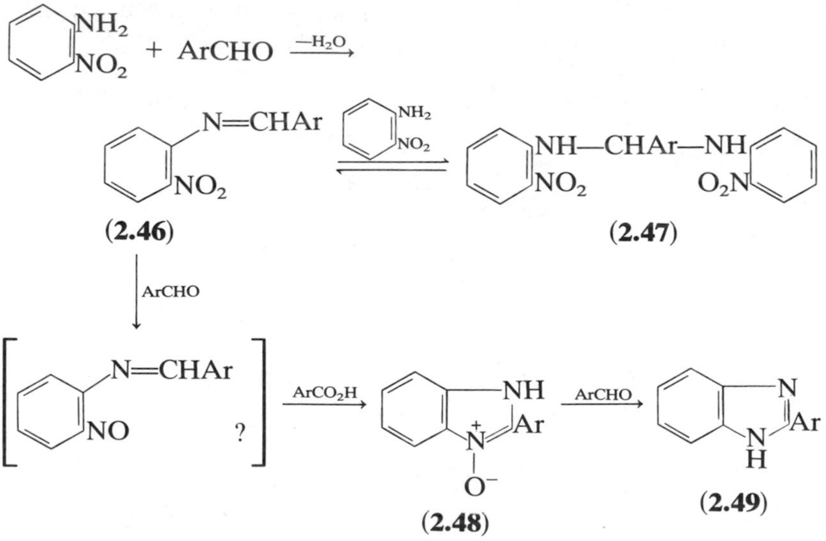

ii. The reaction of *o*-benzoquinonedioxime with acetaldehyde, which gives
    1-hydroxy-2-methylbenzimidazole-3-oxide,[46] may also be rationalized in
    terms of a type B-related process (eq. 39), although other mechanisms
    are possible. The generality of this method has not been explored.

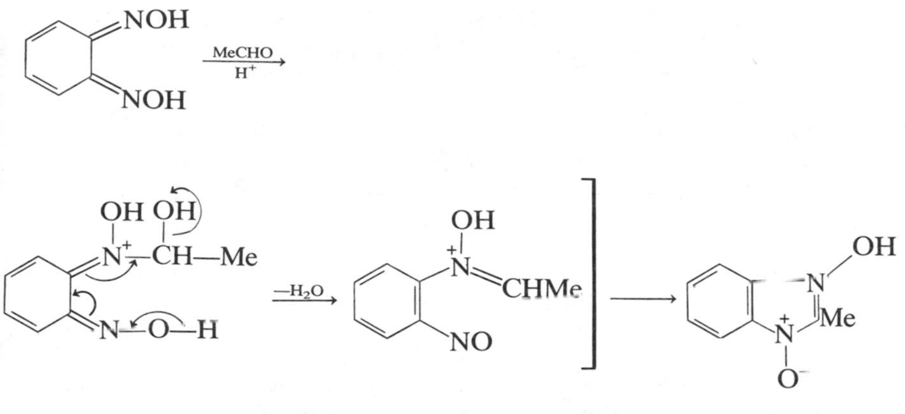

$$\tag{39}$$

iii. The only route to benzimidazole-N-oxides that does not involve formation of the 2,3-bond in the cyclization step is provided by the reaction of
benzonitrile oxides with nitrosobenzenes,[56,68,69] phenylhydroxylamine,[70]
or N-aryl-S,S-dimethylsulfimides.[71] The common intermediates in these
reactions (eqs. 40 to 42) are, apparently, α-nitrosoarylideneanilines or
their nitrone analogs (i.e., **2.50** or **2.51**).

(40) $PhNO + ArC{\equiv}\overset{+}{N}{-}O^{-}$

(41) $PhNHOH + ArC{\equiv}\overset{+}{N}{-}O^{-} \ (+O_2)$

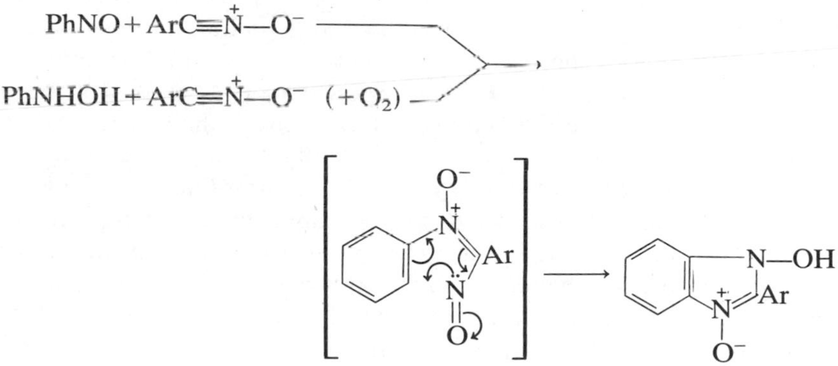

(**2.51**)

(42) $Ph\overset{-}{N}{-}\overset{+}{S}Me_2 + ArC{\equiv}\overset{+}{N}{-}O^{-} \longrightarrow$

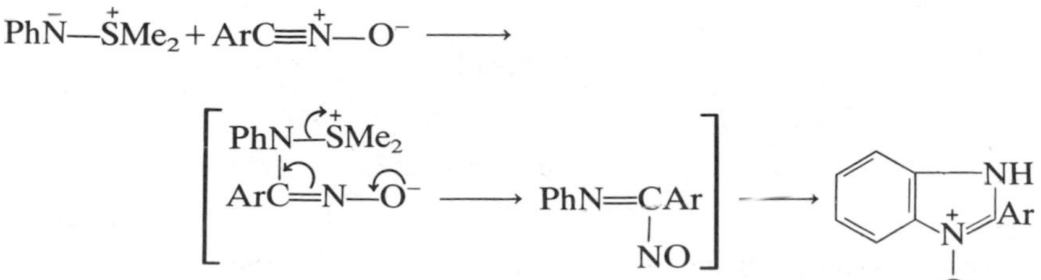

(**2.50**)

2.3. PHYSICOCHEMICAL PROPERTIES INCLUDING SPECTRA

To date there has been relatively little systematic study of the physicochemical properties of benzimidazole-N-oxides. Most of the work that has been undertaken has been incidental to another investigation, for example, that of tautomerism (section 2.4).

2.3.1. Infrared and Ultraviolet Spectra

The infrared spectra of 1-unsubstituted benzimidazole-3-oxides show broad and weak absorption in the region (below $3000\,cm^{-1}$) expected for acidic and strongly hydrogen-bonded OH groups. This absorption is attributable to the –OH group of the N-hydroxybenzimidazole tautomer (see section 2.4). Other weak absorptions above $3000\,cm^{-1}$ have been attributed to N–H and O–H stretching.[10,30] No agreement appears to exist regarding the frequency of N–O stretching.

There is considerably more information available on ultraviolet spectra.[9,10,13,19,20,27,36,42,46,54,57,72] 1-Unsubstituted benzimidazole-3-oxides, which are both acidic and basic in character (cf. section 2.3.4) as well as exhibiting tautomerism (cf. section 2.4) have ultraviolet spectra that vary appreciably with the pH of the solution and the nature of the solvent, since one or more of four species (two neutral tautomers, an anion, and a cation; eq. 43) may contribute to the spectra under a given set of conditions.[10,19,20,27,54,72] The characteristic absorptions of the neutral and cationic species occur at ~280–300 nm (ε ~7000–20,000), and those of the anions at higher wavelength (330–400 nm). The 1-substituted benzimidazole-3-oxides, which are nonacidic and do not exhibit tautomerism, have simpler spectra (only one neutral and one cationic species may contribute), with the principal absorption band at ~280–300 nm. 1-Hydroxybenzimidazole-3-oxides are, of course, acidic and show the shift to higher wavelength in solutions of high pH.[46] 2,2-Disubstituted 2H-benzimidazole-1,3-dioxides, which are quinone-like in nature, are similarly highly colored, with a medium-intensity absorption in the visible (500–540 mm) as well as one in the ultraviolet (240–250 nm) region.[36,42]

2.3.2. NMR Spectra

^1H Spectra

The proton spectrum of benzimidazole-N-oxide itself, which the author has recorded (in d_6-dimethyl sulfoxide), shows a one-proton singlet at δ 8.45

(H-2) and a 4-proton multiplet at δ 7.1–7.8 (H-4, -5, -6, and -7). The 2-proton thus resonates at lower field than in benzimidazole itself (δ 8.29)[73] [see Chapter 1, Table 1.22], thus providing a contrast with pyridine-N-oxide, in which the 2-proton resonates at higher field than in pyridine.[74] Also, the 4-proton in benzimidazole-N-oxide is not significantly deshielded by the N-oxide group, unlike the protons *peri* to the N-oxide function in quinoline- or cinnoline-1-oxides.[75,76]

It can be argued that these differences between the 5- and 6-membered N-oxides arise because the former are tautomeric species and not, therefore, true N-oxides. In the 1-substituted benzimidazole-N-oxides, which cannot tautomerize, the H-2 resonance occurs at even lower field (δ 8.6 to 9.3), and the remaining protons are more deshielded than in their 1-unsubstituted counterparts. The H-4 resonance is occasionally separated from the multiplet,[21,32] having a chemical shift of $\sim\delta$ 8.0.

2,2-Disubstituted 2H-benzimidazole-1-oxides and -1,3-dioxides give spectra in which H-4, -5, -6, and -7 resonate at δ 6.5 to 7.3.[36,42,44]

^{13}C Spectra

Information on these is extremely scanty. They have been used, however, to establish the structures of the 2,2-disubstituted 2H-benzimidazole-1-oxides and -1,3-dioxides:[36,44] typical resonances are shown in formulas (**2.52** and **2.53**).

(**2.52**)   (**2.53**)

2.3.3. Mass Spectra

In common with other heterocyclic N-oxides, those in the benzimidazole series have mass spectra that show prominent (M-16)$^{+\cdot}$ ions,[21,72,77,78] this fragment corresponding to the loss of the N-oxide oxygen. In many cases an (M–17)$^{+}$ ion is also observed, but it is not clear if this arises by loss of hydrogen from the (M-16)$^{+\cdot}$ ion or by loss of $\cdot$OH from the (possibly rearranged) molecular ion. 2,2-Dialkyl-2H-benzimidazole-1,3-dioxides show fragment ions that correspond to the loss of one or both oxygens from the molecular ion.[36]

Another type of fragmentation has been recognized in the case of 2-arylbenzimidazole-N-oxides:[78] this involves a rearrangement in the molecular ion, with the formation of a bond between the N-oxide oxygen and C-2; this leads to the production of a fragment $(ArCO)^+$ (where Ar is the 2-substituent). It is significant to note the $(M\text{-HCO})^+$ is an important fragment in the mass spectrum of benzimidazole-N-oxide itself.[78]

2.3.4. Acidity and Basicity

Benzimidazole-N-oxide is a weak base (pK_a 2.90),[72] that is, weaker than benzimidazole itself by $\sim$2.6 pK units. 1-Methylbenzimidazole-3-oxide is of comparable basicity,[72] and 5-nitrobenzimidazole-3-oxide is weaker still (pK_a 2.2).[54] 1-Unsubstituted benzimidazole-N-oxides are also weakly acidic (cf. eq. 43). For proton loss from the parent compound, the pK_a is 7.86, and in the 5-nitro-analog the corresponding value is 6.1. These N-oxides are thus stronger acids than benzimidazole itself (pK_a 12.3).

2.4. TAUTOMERISM

Reference has already been made in section 2.3 to the possibility of tautomerism between 1-unsubstituted benzimidazole-3-oxides and N-hydroxybenzimidazoles (**2.54** $\rightleftharpoons$ **2.55** in eq. 43). The position of the equilibrium between the tautomers has been studied extensively for the parent compound[19,72] and also for its 2-phenyl[10] and 5-nitro[20] derivatives.

$$\qquad\qquad\qquad (43)$$

$$\text{(2.54)} \qquad\qquad \text{(2.55)}$$

Comparison of spectroscopic and pK_a data for benzimidazole-N-oxide with those for N-methoxybenzimidazole (**2.56**) and 1-methylbenzimidazole-3-oxide (**2.57**) indicates that in aqueous solution the N-oxide tautomer predominates. As the polarity or hydrogen-bonding power of the solvent decreases, the proportion of the N-hydroxy tautomer increases, until in solvents such as acetonitrile, dioxan, or chloroform this tautomer predominates.[72] The same trends are apparent in the 5-nitro derivative, although the investigation has been less extensive. The 2-phenyl compound, however, appears to exist predominantly in the N-hydroxy form even in polar solvents.

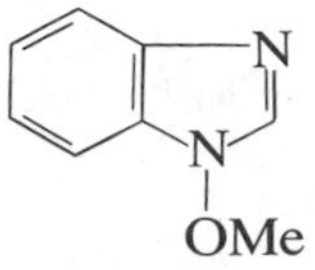

(2.56) 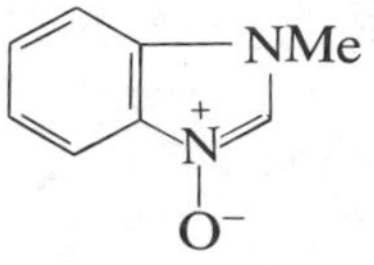   (2.57)

No conclusive evidence is available regarding the position of the equilibrium in the solid state.

2.5. REACTIONS

2.5.1. Protonation, Alkylation, and Acylation

Benzimidazole-N-oxide, its 1-methyl derivative, and N-methoxybenzimidazole have ultraviolet spectra in acid solution that are almost identical,[19] and so there is little doubt that protonation of the N-oxides gives N-hydroxybenzimidazolium cations.

Alkylation is readily achieved using an alkyl halide (or a sulfonate ester) in presence of base,[10,19,24,56,57,61,79–81] and gives an N-alkoxybenzimidazole (or an N-alkoxybenzimidazolium salt). Similarly acylation takes place on the N-oxide oxygen, giving N-acyloxybenzimidazoles,[8,9,60,61] although the acyl derivatives are especially prone to undergo rearrangement (see section 2.5.3).

There is only one report of alkylation occurring on nitrogen:[82] methylation of benzimidazole-N-oxide by diazomethane has been alleged to give both N- and O-methyl derivatives (2.57 and 2.56), but to the author's knowledge the details of this work have never been published. Certainly, diazomethane reacts with 2-phenylbenzimidazole-N-oxide, giving predominantly the O-methylated product.[9]

2.5.2. Oxidation and Reduction

Oxidation

Benzimidazole-N-oxide is oxidized by heating its aqueous solution under pressure for 10 hr at 180° in presence of oxygen.[83] The oxidation product, a deep red-violet compound, is apparently a dehydro-dimer of the starting material: it has been assigned the structure (2.58) and named "NO-indigo." Structure (2.58) has been questioned, however, in view of the compound's color, and an alternative structure (2.59) has been suggested.[84] In the author's opinion, it would be of interest to reinvestigate the claim that

NO-indigo is also formed by peracid oxidation of 2,2'-bibenzimidazolyl (**2.60**), since simple benzimidazoles do not undergo *N*-oxidation under these conditions (cf. section 2.2.1).

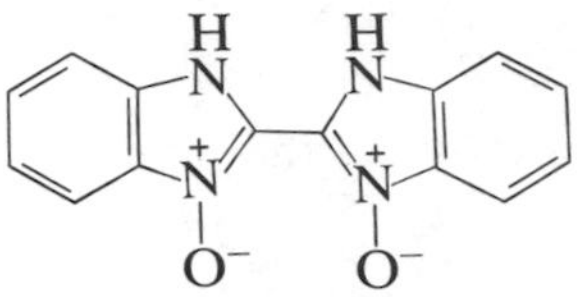

It may well be relevant to the above to note that 1-methylbenzimidazole-3-oxide is converted, simply on heating to 130° (above its m.p.), into 2,2'-bis-(1-methylbenzimidazolyl) and what is alleged to be the mono-*N*-oxide of the latter[85] (eq. 44). It is also of interest that the alleged mono-*N*-oxide is colorless.

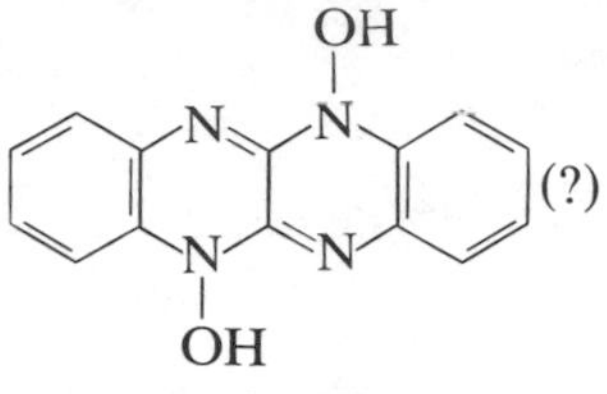

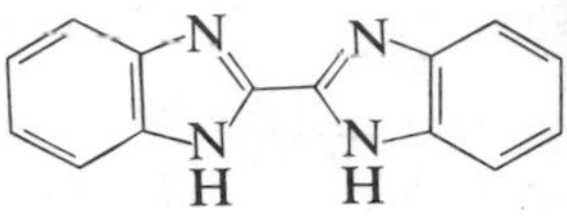

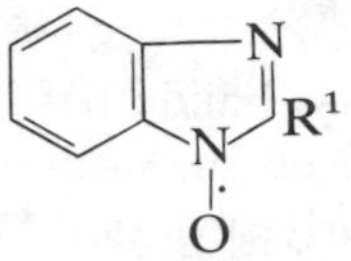

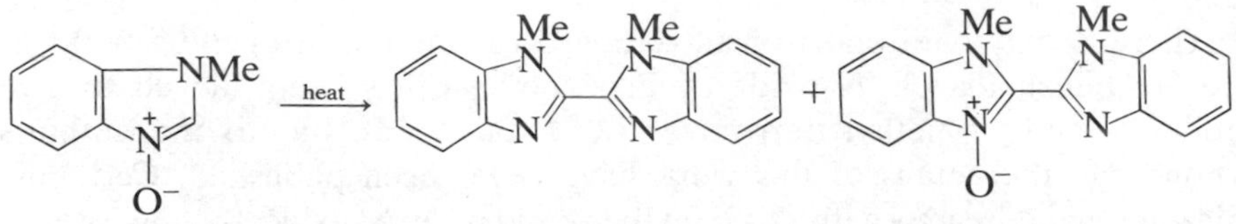

Lead dioxide oxidizes 1-unsubstituted benzimidazole-3-oxides and 1-hydroxybenzimidazole-3-oxides to radicals (**2.61**) and (**2.62**) respectively[69,70] (see also "^{14}N Spectra" in section 1.3.3 in Chapter 1).

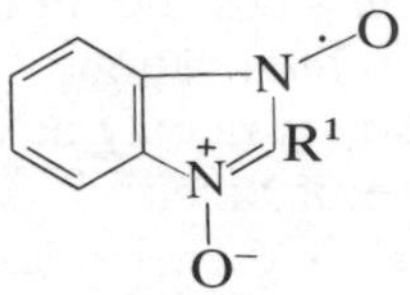

2,2-Dialkyl-2*H*-benzimidazole-*N*-oxides are unusual in that they may be oxidized to the di-*N*-oxides by peracids.[44] These mono-*N*-oxides are also unlike the *N*-oxides of 1*H*-benzimidazoles in that they themselves may be produced by *N*-oxidation of the parent 2*H*-benzimidazoles.[44]

Reduction

In common with other *N*-oxides, those of the benzimidazole series are reduced to the parent heterocycle under a variety of conditions. Phosphorus trichloride,[8,15,37] hydrogen in presence of Raney nickel,[19,86] sulfur dioxide or sodium bisulfite,[15] sodium borohydride,[15] lithium aluminum hydride, trialkyl phosphines and phosphites, sodium (or ammonium) sulfide, iron powder, and ferrous oxalate[86] have all been used. Catalytic hydrogenolysis of the N-alkoxy[79,87] or N-acyloxy derivatives[50,60,61] is a useful alternative. An unusual deoxygenation reaction is that involving carbon disulfide[88,89] (eq. 45): two possible mechanisms have been suggested.

Catalytic hydrogenation of 1-hydroxybenzimidazole-3-oxides gives the benzimidazole directly,[40,68] but the use of stannous chloride[68] or the carbon disulfide method[40] permits the isolation of the intermediate benzimidazole-N-oxide (N-hydroxybenzimidazole). Stepwise reduction of 2H-benzimidazole-1,3-dioxides is also possible using sodium borohydride.[44]

2.5.3. Addition–Elimination Reactions

Very little is known about electrophilic aromatic substitution reactions of benzimidazole-N-oxides and related compounds, except that N-ethoxy-benzimidazole is nitrated at positions 5 and 6, and is chlorinated (at all four positions in the carbocyclic ring, and finally at position 2) by sulfuryl chloride.[79] Simple nucleophilic substitutions of the S_NAr type are unknown among benzimidazole-N-oxides and their derivatives, except for the replacement of a 2-cyano substituent by hydroxyl referred to in section 2.2.4 (eq. 33). Most of the addition–elimination reactions of these compounds are of the so-called "abnormal" or "AE_a" type, in which addition of a nucleophile at one position is followed by elimination of the leaving group from another position. Such reactions are, of course, well known in the chemistry of pyridine-N-oxides.[90]

These reactions may be divided into several groups. In the first, 1-alkoxy-benzimidazoles lacking a 2-substituent react with alkoxides, with hydrazine, and with bisulfite ion (Scheme 2.9) to give 2-alkoxy- and 2-hydrazinobenz-imidazoles, and benzimidazole-2-sulfonic acid, respectively.[79] In the second

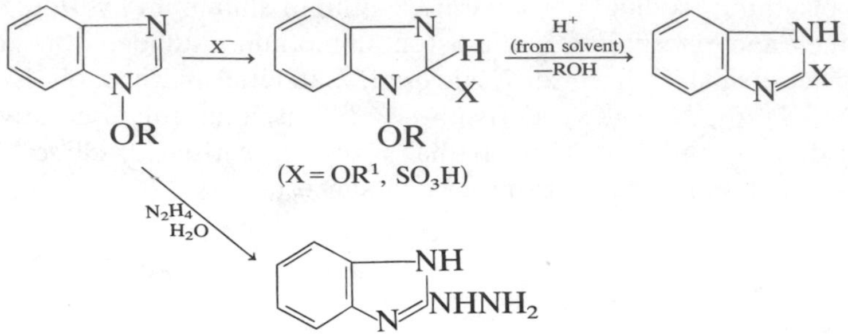

Scheme 2.9

group (Scheme 2.10a), 3-methoxy-1-methylbenzimidazolium salts (prepared from 1-methylbenzimidazole-3-oxide and methyl iodide) react with a wide variety of nucleophiles (OH^-, OR^-, CN^-, BH_4^-, HSO_3^-; also amines, Grignard reagents, and stabilized carbanions) to give the appropriate 2-substituted 1-methylbenzimidazoles.[80] If the 2-position is already substituted, however, the strongly basic nucleophiles (e.g., OH^-, OR^-, RNH_2) attack the methoxy group (Scheme 2.10b) and the less basic cyanide ion may also attack the 6-position (Scheme 2.10c).[80]

(a)

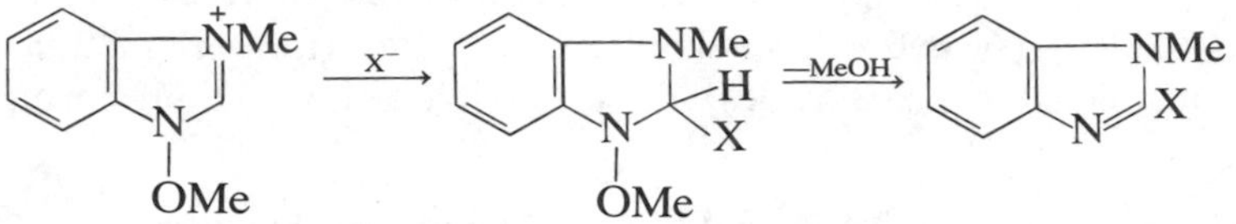

(b)

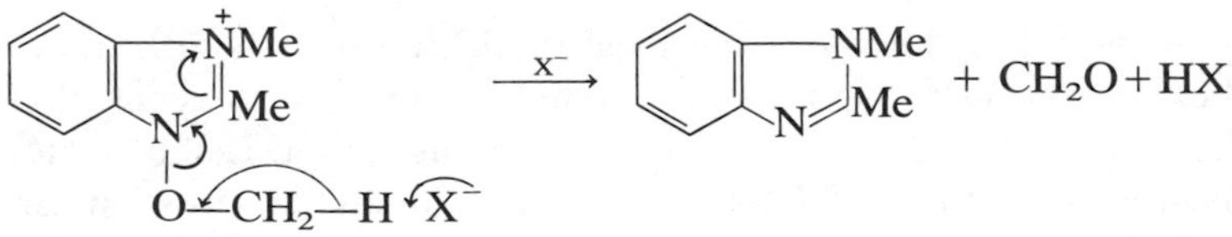

(c)

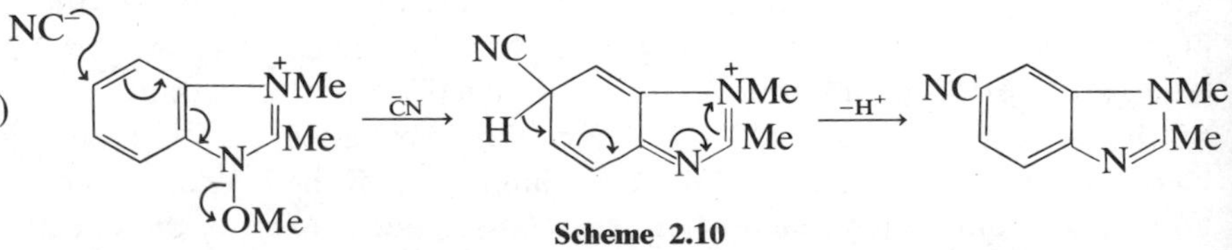

Scheme 2.10

The third group comprises the reactions of 1-substituted benzimidazole-3-oxides with acylating agents. In these, the intermediate *N*-acyloxybenzimidazolium salt (Scheme 2.11) is especially reactive towards nucleophiles, and

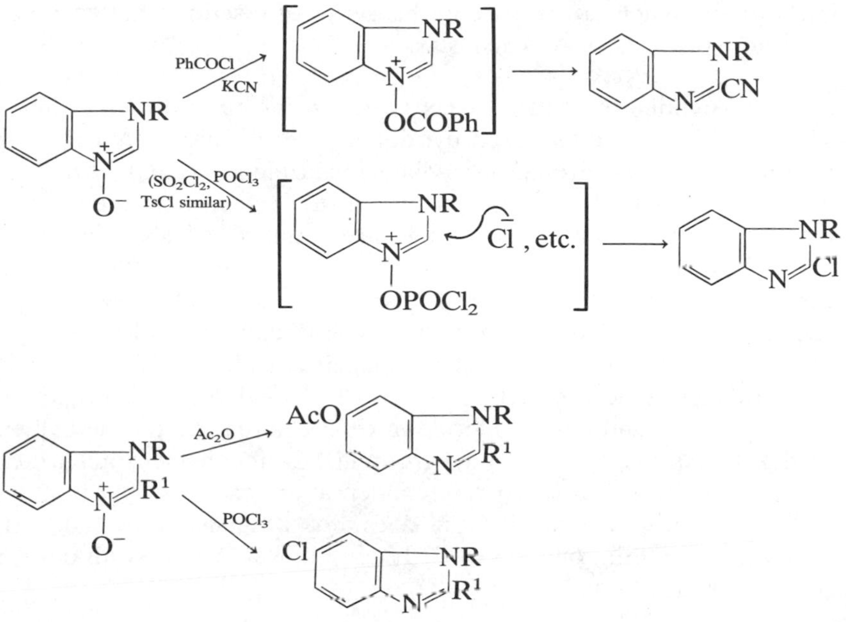

Scheme 2.11

even weakly nucleophilic counterions from the acylating agent may interact with the intermediate. Nucleophilic attack occurs at the 2-position if this is vacant;[85] otherwise it occurs at the 6-position.[15,17,91] This type of reaction is not common for neutral (i.e., uncharged) N-acyloxybenzimidazoles, although notable exceptions are shown in eqs. 46[51] and 47.[92]

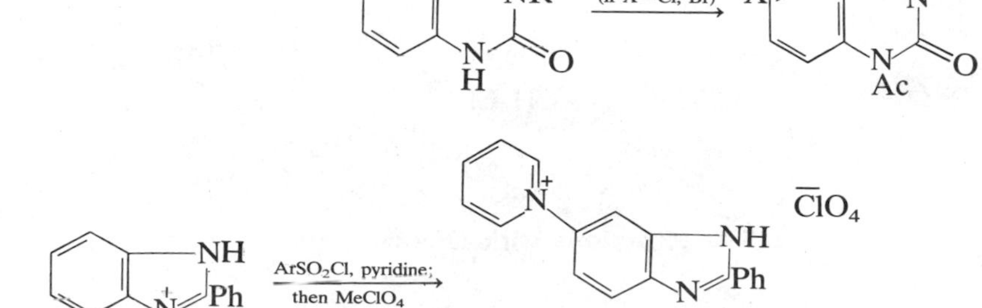

$(X = OAc, Cl, Br)$

(46)

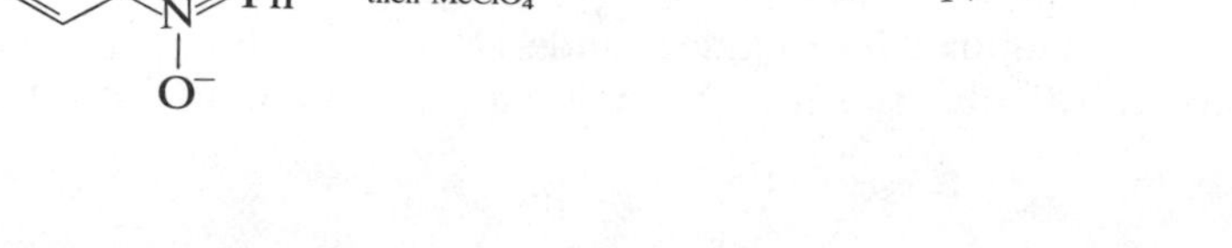

(47)

Two further complications arise in this series of reactions, but each has a parallel in pyridine-N-oxide chemistry.

i. Just as pyridine-N-oxide reacts with acetic anhydride giving 2-acetoxypyridine and thence 2-pyridone,[93] benzimidazole-N-oxide and acetic anhydride give N-acetylbenzimidazolone (**2.63**) or N,N'-diacetylbenzimidazolone (**2.64**) according to the reaction conditions[8,94] (eq. 48). 1-Ethoxybenzimidazole and acetic anhydride also yield (**2.63**) and (**2.64**)[79] (eq. 49).

ii. Just as 2-methylpyridine-N-oxide reacts with acetic anhydride giving 2-acetoxymethylpyridine,[95] and with arenesulfonyl chlorides giving 2-chloromethylpyridine,[96] 1,2-dimethylbenzimidazole-3-oxide and acetic anhydride give 2-acetoxymethyl-1-methylbenzimidazole (eq. 50), and the same N-oxide and tosyl chloride give 2-chloromethyl-1-methylbenzimidazole (eq. 51).[15] (The methyl group in 1,2-dimethylbenzimidazole-3-oxide is, however, less reactive in condensation reactions than that of 2-methylpyridine-N-oxide, since it does not condense with aldehydes, nitrosoarenes, or diazonium salts.[15] It does, however, react with dimethyl oxalate[15] and with alkyl nitrites.[15])

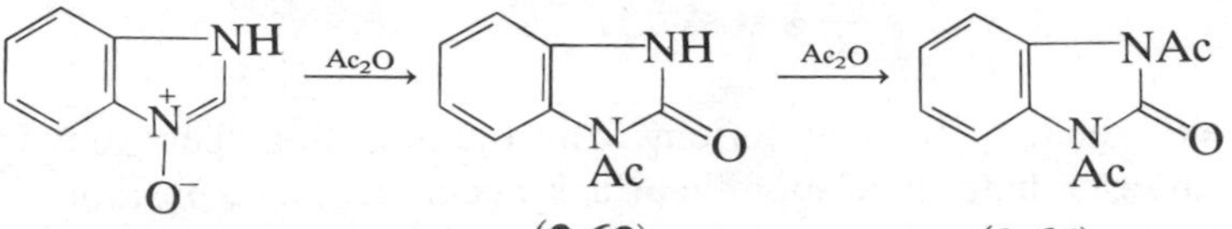

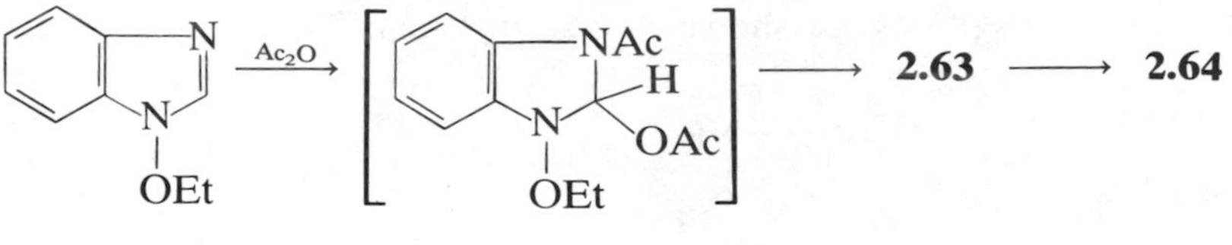

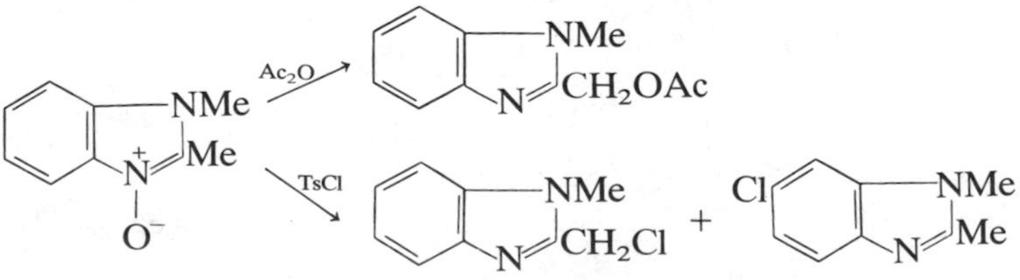

2.5.4. Reactions with Dipolarophiles

1-Substituted benzimidazole-3-oxides undergo cycloaddition reactions characteristic of 1,3-dipolar species, this being a further point of similarity with the 6-membered *N*-oxides. The cycloadducts in both series, however, are unstable and undergo ring opening in such a way that the dipolarophile ultimately forms a 2-substituent.

Typical reactions of 1-alkylbenzimidazole-3-oxides with dipolarophiles[21,89] are shown in Scheme 2.12.

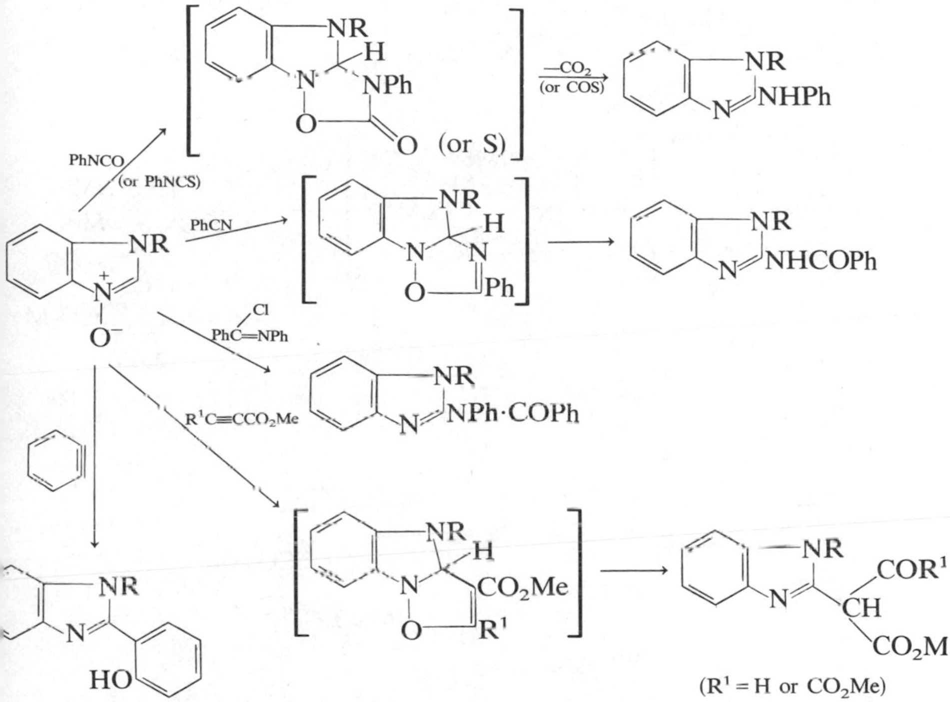

Scheme 2.12

If the 2-position of the *N*-oxide is already substituted, the reactions with dipolarophiles take a different course. 1,2-Dimethylbenzimidazole-3-oxide reacts with phenyl isocyanate to give 6-anilino-1,2-dimethylbenzimidazole[15] (eq. 52), and with acetylenic esters to give the benzimidazolium ylids (**2.65**)[97] (eq. 53).

It might have been anticipated that 2,2-disubstituted 2*H*-benzimidazole-1,3-dioxides, with a quinone-like structure, would undergo Diels–Alder cycloadditions. Such is not, however, the case: there is no reaction between these dioxides and many of the familiar dienophiles (e.g., maleic anhydride), and the reactions with dimethyl acetylenedicarboxylate, with benzyne, and with tetracyanoethylene lead to the truly remarkable products (**2.66**, **2.67**, and **2.68**), respectively[44] (Scheme 2.13).

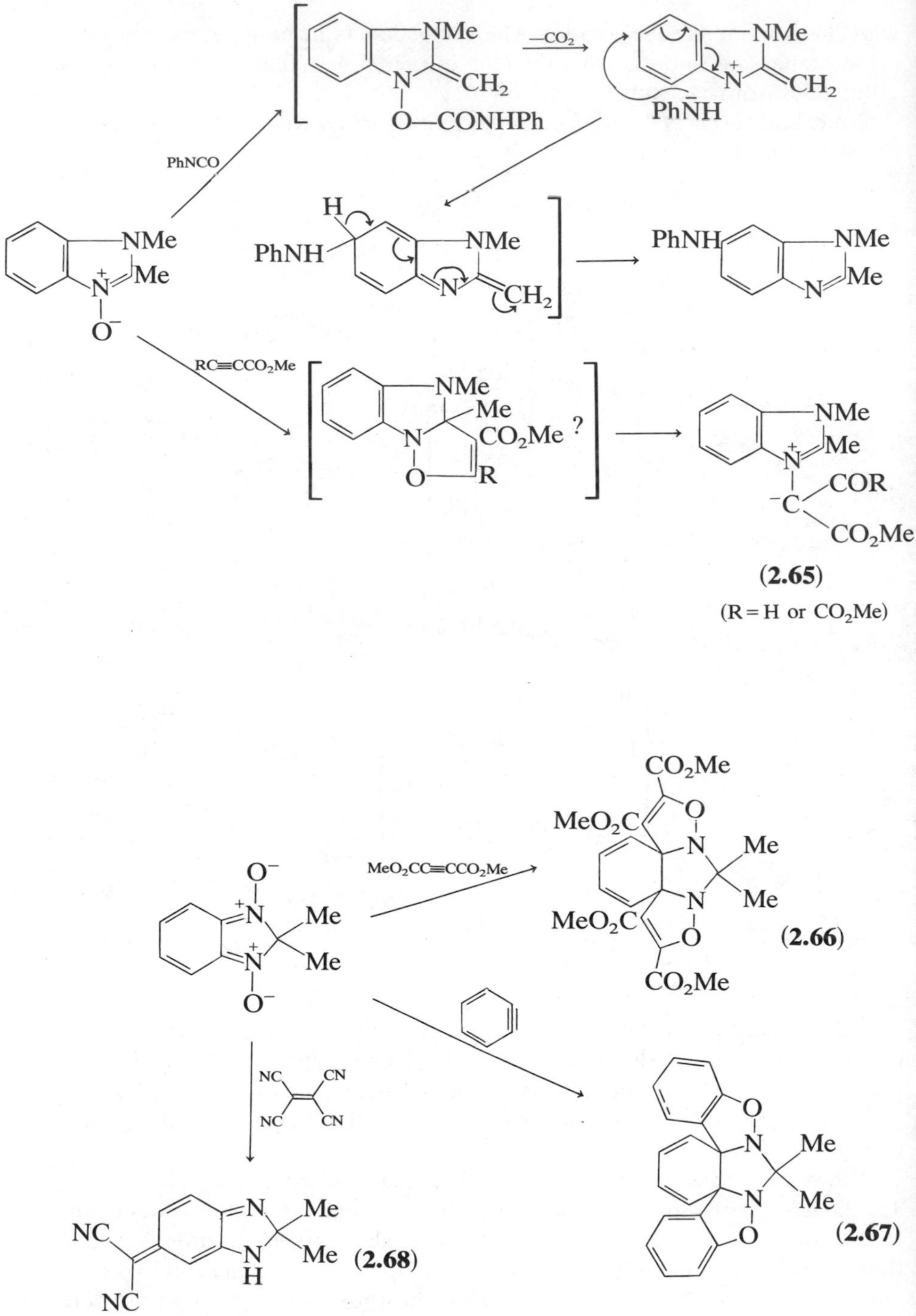
PhNCO
−CO2
RC≡CCO2Me
(2.65)
(R = H or CO2Me)
MeO2CC≡CCO2Me
(2.66)
(2.67)
(2.68)

Scheme 2.13

2.5.5. Rearrangements

There are several recorded instances of rearrangement of benzimidazole-
N-oxides to benzimidazolones, such rearrangements being induced either
thermally or photochemically. 1-Methylbenzimidazole-3-oxide undergoes
rearrangement to 1-methylbenzimidazolone simply by heating it in acetone
or chloroform,[85] a reaction that is most simply rationalized as a water-
catalyzed or a bimolecular AE_a process (eq. 54). By analogy with this, and
with section 2.2.4, the formation of benzimidazolones by thermolysis of the
o-nitrophenylglycine derivatives (**2.69**) (eq. 55) is assumed to involve
benzimidazole-*N*-oxide intermediates.[63]

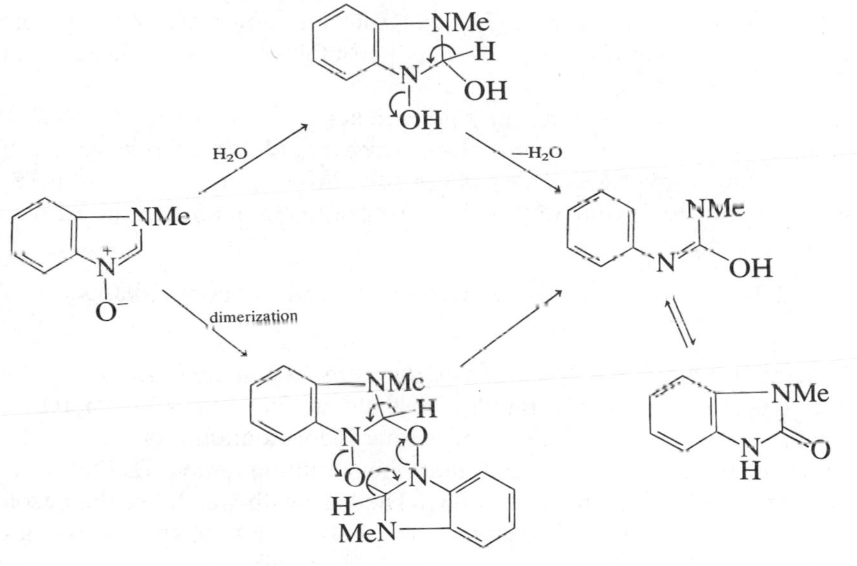

$$(54)$$

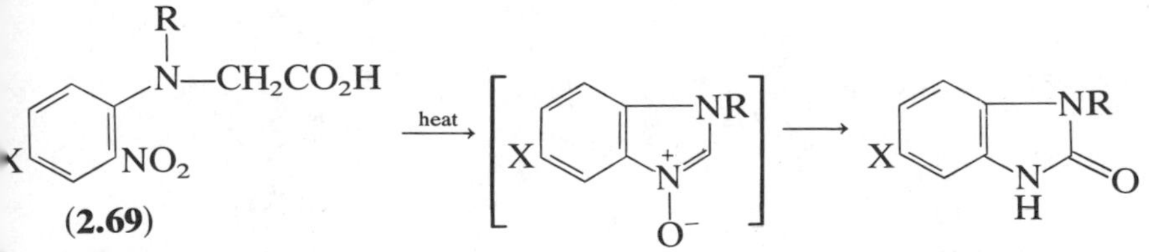

(**2.69**)
X = H: R = H or Me)
X = NO_2; R = H or Ph)

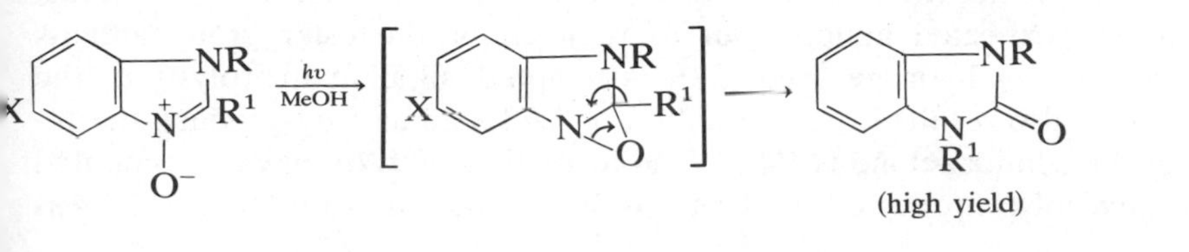

$$(55)$$

(high yield)

$$(56)$$

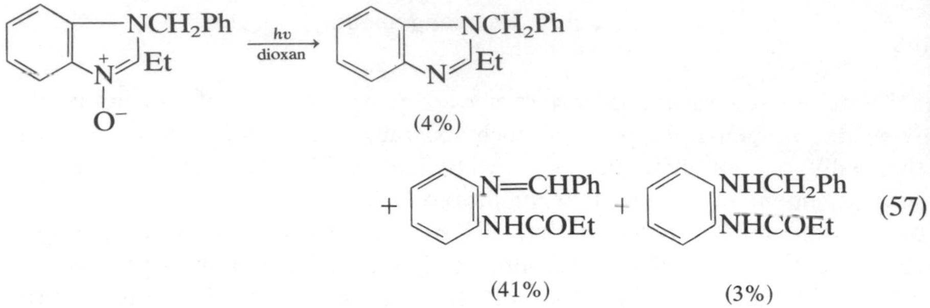

(4%)

(57)

(41%)   (3%)

Photolysis of 1,2-dialkylbenzimidazole-3-oxides in alcoholic solution gives 1,3-dialkylbenzimidazolones,[98,99] the process probably involving an ox-aziridine intermediate (eq. 56). [In other solvents, however, or in presence of sensitisers, other photo-products may be obtained at the expense of the benzimidazolone (eq. 57)].[99]

A rearrangement of a different kind is included in eq. 14 (section 2.2.3): 2-cyano-2-methyl-2*H*-benzimidazole-1,3-dioxide (**2.24**) is rearranged by gentle heating into **2.25**. This presumably involves tautomerism to the ring-opened nitrosonitrone (**2.28**) and recyclization in a different direction.

2.5.6. 2-(*N*-Oxidobenzimidazolyl) Anions as Intermediates

1-Methylbenzimidazole-3-oxide undergoes hydrogen–deuterium exchange at the 2-position relatively rapidly (half-life at 39° = ca. 1 hr) merely by dissolving it in D_2O.[100] This implies that deprotonation of the starting material is a facile process and that the resulting anion (**2.70a**) has a somewhat surprising degree of stability. The ease with which benzimidazole-2-carboxylic acid *N*-oxide is decarboxylated also points to the formation of a relatively stable intermediate (possibly **2.71** in this case).[38]

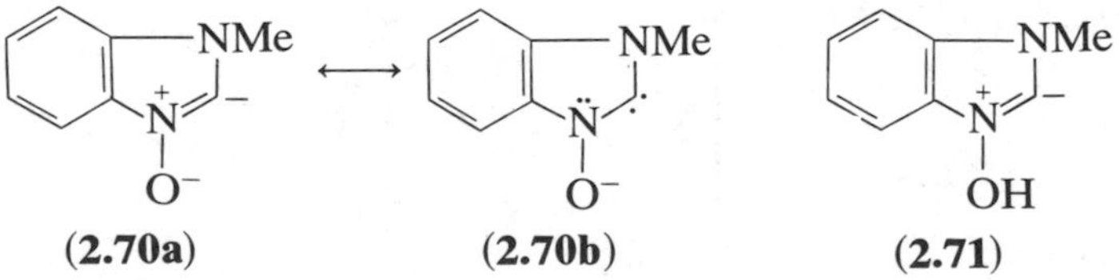

(**2.70a**)   (**2.70b**)   (**2.71**)

Reaction of the ester (**2.72**) with piperidine and *t*-butylamine[17,100] leads, not to the expected amides, but to removal of the ester group and its replacement by hydrogen (eq. 58). Attempted alkaline hydrolysis of the same ester also results in the formation of **2.73**, and as a by-product, the *N*-hydroxybenzimidazolone (**2.74**).[17] The formation of **2.73** may be accounted for by hydrolysis followed by decarboxylation, but the formation of **2.74** is

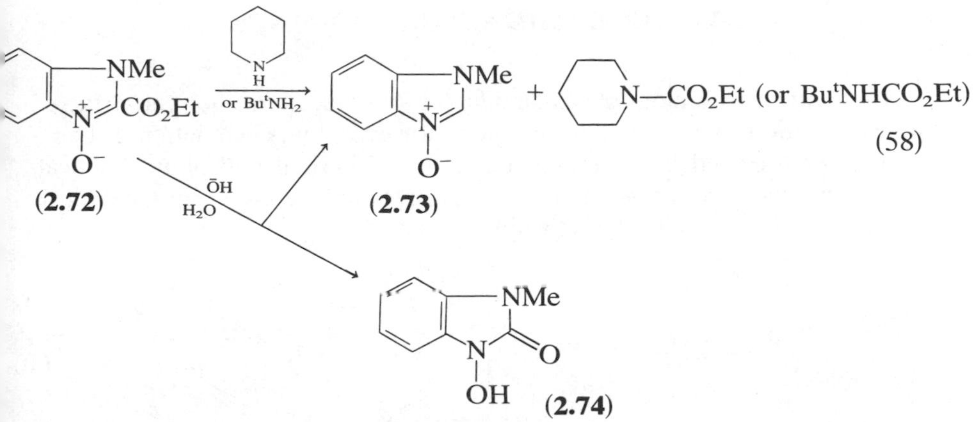

(2.72) (2.73) (2.74)

$$\text{(58)}$$

not so readily explained. It cannot be envisaged as a simple nucleophilic displacement, and the most plausible explanation involves oxidation of intermediate **2.70**, which has a degree of carbene character, as represented by the canonical form (**2.70b**).

Reference has already been made (section 2.5.2, eq. 44) to the conversion of 1-methylbenzimidazole-3-oxide into the mono-*N*-oxide of 2,2-bis(1-methylbenzimidazolyl).[85] This is also envisaged as involving deprotonation of the 2-position followed by an AE$_a$-type of substitution on another molecule of the starting material. The formation of the bibenzimidazolyl derivative (**2.75**) from *N*-ethoxybenzimidazole and sodamide[79] (eq. 59) may be similarly explained.

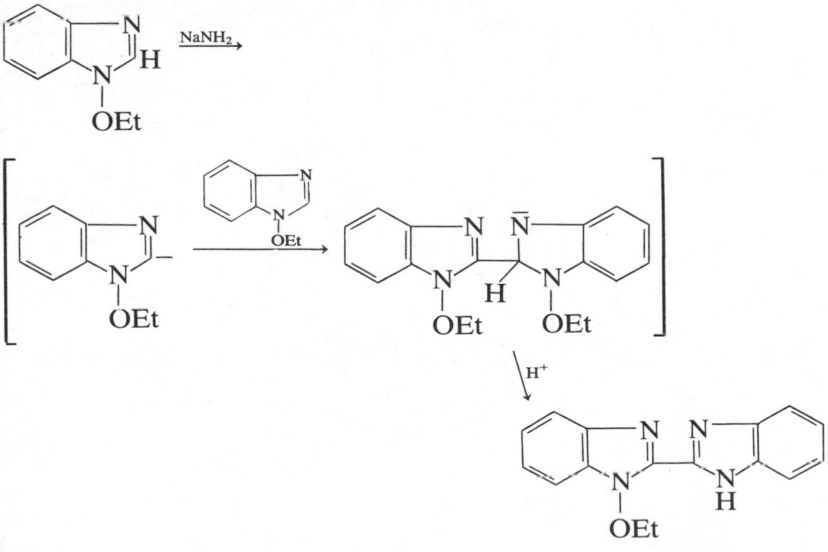

$$\text{(59)}$$

(2.75)

2.6. BENZIMIDAZOLE-*N*-IMINES

These nitrogen analogs of benzimidazole-*N*-oxides have not yet attracted much attention. A few have been made, however, by *N*-amination of benzimidazoles followed by acylation (eq. 60).[101] Thermal and photochemical, and electrocyclic reactions of these derivatives are described in Chapter 1, sections 1.45 and 1.48, respectively).

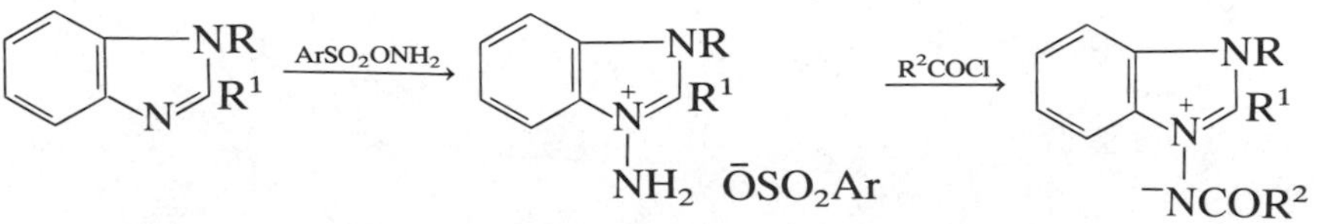

$$\text{(60)}$$

2.7 SYSTEMATIC SURVEY OF BENZIMIDAZOLE-*N*-OXIDE DERIVATIVES

The following tables provide information about selected benzimidazole-*N*-oxides and derivatives. It is emphasized that these lists are not intended to be comprehensive. Only the most important literature references, in which directions for the preparation of these compounds may be found, are given.

TABLE 2.1. 1-UNSUBSTITUTED BENZIMIDAZOLE-3-OXIDES
(⇌ N-HYDROXYBENZIMIDAZOLES)

Substituent(s)	Method of preparation (and % yield)[a]	M.p. (°)	Ref.
—	A1(15); A2(74); A3(60–70); A(Zn/NH$_4$Cl)(43); Decarboxylation of 2-CO$_2$H(85)	210(d), 215(d)	14, 18, 3, 19, 38
2-Me	A2(37), A3(−) A(Na$_2$S$_2$O$_4$)(−), B3(37)	251–252(d) 231, 238	18, 3 23, 34
2-Et	B3(26)	133	34
2-Pr^i	B3(36)	159	34
2-Bu^t	B3(40)	193	34
5-Me	C1(?) (~30)	176–178	62
2,5-Me$_2$	A3(−)	232-234(d)	4
5-NO$_2$	B2(70), C1(73)	279–285(d)	20, 54
2-Me, 5-NO$_2$	A3(25), B2(79)	292–294(d)	20
2-Bu^i, 5-NO$_2$	B2(76)	215–216	20
2-Bu^s, 5-NO$_2$	B2(78)	202	20
2-Ph	A2(−), A(NaSH)(−), B1(52), B3(−), C1(20–80), C3(30), D1(55, 86)	225(d), 220(d), 215–218(d) 222–223(d), 230	18, 22, 25, 33 9, 10, 51, 57, 58, 63, 102
5-Me, 2-Ph	C1(64)	231–233(d)	59
5-Cl, 2-Ph	A2(−), C1(−)	241	18, 24, 56
6-Cl, 2-Ph	A2(−), C1(−)	206	18, 56
5-NO$_2$, 2-Ph	B2(56, 96, 30), C1(−)	273(d), 264–265, 250–252(d)	26, 28, 29, 24, 78
5,7-(NO$_2$)$_2$, 2-Ph	B2(95), C1(−)	281–282	28
2-PhCH$_2$	A2(−)	174–175	18
2-C$_6$H$_4$Me-p	C1(61)	224–226(d)	59
2-(C$_6$H$_2$Me$_3$-2,4,6)-5-NO$_2$	D2(~8)	306	71
2-(C$_6$H$_4$Cl-o)	C1(58)	232–234(d)	59
2-(C$_6$H$_4$Cl-p)	A2(−), B1(64), C1(50, 72)	249–250(d), 216–217 211–213(d) (+HCl+H$_2$O)	18, 26, 24, 56, 59
2-(C$_6$H$_3$Cl$_2$-2,6)-5-NO$_2$	D2(14)	318	71
2-(C$_6$H$_4$OMe-o)	B1(64), C1(72)	255–257(d), 220–222(d)	26, 59
2-(C$_6$H$_4$OMe-m)	B1(62)	267–269(d)	26
2-(C$_6$H$_4$OMe-p)	A2(−), B1(64), C1(70)	273–274(d), 205(d), 189–191	18, 26, 24, 59
2-[C$_6$H$_3$(OMe)$_2$-3,4]	B1(62)	268–270(d)	26
2-(C$_6$H$_4$NO$_2$-p)	C1(30–56)	243–246(d)	57, 59
2-(2-naphthyl)	B1(62)	258–259(d)	26
2-(4-pyridyl)	A2(−), B1(59)	279–280(d), 231–233(d)	18, 26
2-(4-thiazolyl)	A2(−), A(NaSH)(−), C1(−), D1(−)	237–238	18, 22, 55, 64

^{a} Footnotes following Table 2.2

323

TABLE 2.2. 1-SUBSTITUTED BENZIMIDAZOLE-3-OXIDES

1-substituent	Other substituents	Method of preparation (and % yield)[a]	M.p. (°)	Ref.
Me	—	A2(62), A3 (ca. 40), A1(58)	60–62 (dihydrate) 86–88 (hydrate)	18, 19 14
Et	—	A2(−), A3 (ca. 40)	80–82 (hydrate)	18, 19
PhCH$_2$	—	A2(−), A3 (ca. 40)	$\begin{cases}158–159 \text{ (monohydrate)} \\ 47–50 \text{ (trihydrate)}\end{cases}$	18, 19, 21
Me	5-NO$_2$	B1(35), B2(38)	204(d)	27
Me	6-Cl	A2(88)	>250 (dihydrate)	17
Me	2-Me	A2(−), A3(80), A1(62, 74)	65–70 (dihydrate) 162–174(d)	18, 19 14, 15
Me	2,5-Me$_2$	A4(−)	163	1
Me	2-Ph	A1(77, ca. 50)	$\begin{cases}235–240(d) \\ 166–171(d)\end{cases}$	14, 10
Et	2-Me	A2 (ca. 70), A3(47), C2(47)	113–115, 240 80–83 (dihydrate)	18, 19, 32
Ph	2-Me	A1(62, 78), A2(−), A3(69)	$\begin{cases}97–100 \text{ (0.75 H}_2\text{O)} \\ 164–165, 204–209(d), \\ 173\end{cases}$	13, 18, 19, 102
Ph	2-Ph	A1(76)	192–195(d), 203	14, 102
PhCH$_2$	2-Me	A2(−), A3(59)	83–85 (hydrate)	18, 19

[a] A = reduction of the o-nitroanilide: 1 = catalytic hydrogenation, 2 = NaBH$_4$/Pd, 3 = ammonium sulfide, 4 = metal-acid.
B = cyclization of an o-nitrosoanil or similar species: 1 = nitrosoanil + aldehyde, 2 = photolysis of an o-nitroaniline derivative, 3 = 2-X-quinoxaline-4-oxide + H$_2$O$_2$/OH$^-$.
C = cyclization of an N-alkylated o-nitroaniline: 1 = with base, 2 = with acid, 3 = with heat.

D1 = reaction of an o-nitroaniline with an aldehyde. D2 = ArC≡$\overset{+}{N}$—$\bar{O}$ + Ar'$\bar{N}$—$\overset{+}{S}$Me$_2$.

TABLE 2.3. N-ALKOXYBENZIMIDAZOLES

Alkoxy group	Other substituents	M.p. (°)	Ref.
OMe	—	oil (b.p. 98–99/5 mm)	19
OEt	—	oil (b.p. 95/2 mm)	79
OCH$_2$CH=CH$_2$	—	oil (b.p. 108/3 mm)	79
OMe	2-Ph	102–104, 107–108	9, 10
OMe	2-C$_6$H$_4$NO$_2$-p	154–156	61
OEt	2-C$_6$H$_4$NO$_2$-p	107–109	61
OMe	2-(4-thiazolyl)	117–118	81
OMe	Various 2-aryl and heteroaryl	—	81
OCH$_2$CH$_2$NR$_2$	Various 2-aryl and heteroaryl	—	56

TABLE 2.4. BENZIMIDAZOLE-N-OXIDES WITH FUNCTIONAL GROUPS AT POSITION 2

Substituents			Method of preparation (and % yield)	M.p. (°)	Ref.
1-	2-	Other			
—	CO_2H	—	Benzofuroxan + barbituric acid (92)	103–110(d) (hydrate)	38
—	CO_2Et	—	Benzofuroxan + $MeCOCH_2CO_2Et$ (29)	166–167(d)	37
—	CN	—	$o\text{-}O_2NC_6H_4NHCH_2CN$ + base (77, 94?)	232, 240–241	51, 52
—	CN	—	2-N_3-quinoxaline-1-oxide (92)	249–250(d)	36
—	CO_2Me	5-NO_2	2,4-$(O_2N)_2C_6H_3NHCH_2CO_2Me$ or 2,4-$(O_2N)_2C_6H_3NHC(CO_2Me)$=CHR + base (63, 69–79)	205–206(d)	53
—	COPh	5-Me	2-O_2N-4-$MeC_6H_3NHCH_2COPh$ + base (−)	132	50
Me	CO_2H	—	From ethyl ester via hydroxamic acid (−)	70(d)	17
Me	CO_2Me	—	From the nitrile via the imido-ester (−)	160(d)	17
Me	CO_2Et	—	From the nitrile via the imido-ester (−)	148–150	17
Me	CN	—	Dehydration of oxime (−)	206(d)	17
—	OMe	—	2-alkoxyquinoxaline-4-oxide + $H_2O_2/\bar{O}H$ (83)	156	34
—	OEt	—	2-alkoxyquinoxaline-4-oxide + $H_2\bar{O}_2/\bar{O}H$ (70)	166	34
			$o\text{-}O_2NC_6H_4NTs{\cdot}CH_2COR$ + $\bar{O}Et$ (40–80)	162–164	60
—	OR	various	$o\text{-}O_2NC_6H_4NTs{\cdot}CH_2COR$ + OR (15–82)	—	60
Me[a]	OH	—	$o\text{-}O_2NC_6H_4NMeCH_2CN + CO_3^{2-}$	203	51
Ph[a]	OH	—	$o\text{-}O_2NC_6H_4NPhCH_2CN + CO_3^{2-}$	216	51
$PhCH_2$[a]	OH	—	$o\text{-}O_2NC_6H_4N(CH_2Ph)CH_2CN + CO_3^{2-}$	172	51
—[a]	OH	4-Me	6-Me-2-$O_2NC_6H_3NTs{\cdot}CH_2COR$ + $\bar{O}R$ (17–33)	254–257	60
—	NH_2	5-NO_2	1-[2,4-$(NO_2)_2C_6H_3$]-4,5-Ph_2-imidazole photolysis (20)	—	29

[a] Exists as the 1-hydroxybenzimidazolone tautomer.

TABLE 2.5. 1-HYDROXYBENZIMIDAZOLE-3-OXIDES

Substituents	Method of preparation (and % yield)	M.p. (°)	Ref.
—	Benzofuroxan (**2.29**) + nitromethane (62, 40)	223, 203	42, 44
—	**2.29** + PhCOCH$_2$SO$_2$Ph (90)	224(d)	40
—	**2.29** + MeOSO$_2$CF$_3$ (35)	206	46
2-Me	**2.29** + nitroethane (64, 65)	195–196, 204–205	44, 42
2-Me	o-Benzoquinone + MeCHO (41)	199–200 (hemihydrate)	46
2-Me	**2.29** + PhCOCHMe·SO$_2$Ph (35)	200–201	40
2-Et	**2.29** + 1-nitropropane (66)	194–195	42
2-Et	**2.29** + PhCOCHEt·SO$_2$Ph (28)	190–191	40
2-Ph	**2.29** + PhCH$_2$NO$_2$ (52)	215	44
2-Ph	PhC$^-$=N$^+$—Ō + PhNO (−)	216	68
2-CN	2-N$_3$-quinoxaline dioxide (61)	165–166(d)	36
2-CO$_2$H	**2.29** + PhSO$_2$CH$_2$CO$_2$H (40)	>300(d)	40
2-CO$_2$Et	**2.29** + EtO$_2$C·CH$_2$NO$_2$ (35)	156.5	42
2-CONH$_2$	**2.29** + H$_2$NCOCH$_2$SO$_2$Ph (12)	218–220	40
2-CONHMe	**2.29** + MeNHCOCH$_2$CN (59)	168(d)	39
2-OH^a	**2.29** + CH$_2$O (91)	227(d)	47
2-NHR^b	**2.29** + RN⌒NR ring (N-R) (21–91)	—	48
6-NMe$_2$, 2-Ph	PhC≡N$^+$—Ō + p-Me$_2$NC$_6$H$_4$NO (−)	88–91(d)	56

^{a} Exists as 1,3-dihydroxybenzimidazolone.
^{b} Exists as 1,3-dihydroxybenzimidazolimine.

TABLE 2.6. 2,2-DISUBSTITUTED 2H-BENZIMIDAZOLE-N-OXIDES

Substituents	Method of preparation (and % yield)		M.p. (°)	Ref.
1,3-Dioxides				
2,2-Me$_2$	Benzofuroxan + nitroalkane	(86)	136–137	42, 44
2-Me, 2-Et		(52)	127–129	42
2-Me, 2-Ph		(38)	139–140	44
2,2-(CH$_2$)$_4$		(78)	130	44
2,2-(CH$_2$)$_5$		(67)	112–115	42, 44
2,2-Me$_2$, 5-X^a		(80–85)	—	44
2-Me, 2-CN	2-N$_3$-3-Me-quinoxaline dioxide	(46)	111–112(d)	36
1-Oxides				
2,2-Me$_2$	Reduction of dioxide (83)		65	44
2,2-(CH$_2$)$_4$	Reduction of dioxide (77)		72	44
2,2-(CH$_2$)$_5$	Reduction of dioxide (72)		86	44
6-Cl, 2,2-Me$_2$	Reduction of dioxide (79)		126	44

^{a} X = Cl, Br, OMe, CF$_3$

REFERENCES

1. S. von Niementowski, *Berichte*, **20,** 1874 (1887).
2. Z. Bankiewicz, *Berichte*, **21,** 2402 (1888).
3. S. von Niementowski, *Berichte*, **43,** 3012 (1910).
4. Z. Bankiewicz, *Berichte*, **22,** 1396 (1889).
5. S. von Niementowski, *Berichte*, **25,** 860 (1892).
6. S. von Niementowski, *Berichte*, **32,** 1456 (1899).
7. J. B. Wright, *Chem. Rev.*, **48,** 462 (1951).
8. D. J. Kew and P. F. Nelson, *Aust. J Chem.*, **15,** 792 (1962).
9. G. W. Stacy, B. V. Ettling, and A. J. Papa, *J. Org. Chem.*, **29,** 1537 (1964).
10. G. W. Stacy, T. E. Wollner, and T. R. Oakes, *J. Heterocycl. Chem.*, **3,** 51 (1966).
11. Cf. S. M. Roberts and H. Suschitzky, *J. Chem. Soc. C,* **1968,** 1537.
12. R. J. Bochis and M. H. Fisher, U.S. Patent 3,928,372 (1975); *Chem. Abstr.*, **84,** 121829 (1976).
13. J. W. Schulenberg and S. Archer, *J. Org. Chem.*, **30,** 1279 (1965).
14. J. W. Schulenberg and S. Cornrich, *J. Chem. Eng. Data*, **13,** 574 (1968); *Chem. Abstr.*, **69,** 106617 (1968).
15. S. Takahashi and H. Kano, *Chem. Pharm. Bull. (Tokyo)*, **14,** 1219 (1966).
16. G. O. Doherty and K. H. Fuhr, *Ann. N.Y. Acad. Sci.*, **214,** 221 (1973); *Chem. Abstr.*, **79,** 66247 (1973).
17. S. Takahashi and H. Kano, *Chem. Pharm Bull. (Tokyo)*, **16,** 521 (1968).
18. Shionogi & Co. Ltd., British Patent 1,218,397 (1971); French Patent 1,555,336 (1969); *Chem. Abstr.*, **72,** 43679 (1970).
19. S. Takahashi and H. Kano, *Chem. Pharm. Bull. (Tokyo)*, **11,** 1375 (1963).
20. D. J. Neadle and R. J. Pollitt, *J. Chem. Soc. C,* **1967,** 1764.
21. R. A. Abramovitch, R. B. Rogers, and G. M. Singer, *J. Org Chem.*, **40,** 41 (1975)
22. Merck & Co. Inc., British Patent 1,109,784 (1968); Netherlands Patent 6,517,256 (1966); *Chem. Abstr.*, **66,** 2568 (1967).
23. K. Fries and H. Reitz, *Annalen*, **527,** 38 (1937).
24. G. de Stevens, A. B. Brown, D. Rose, H. I. Chernov, and A. J. Plummer, *J. Med. Chem.*, **10,** 211 (1967).
25. M. Z. Nazer, M. J. Haddadin, J. P. Petridou, and C. H. Issidorides, *Heterocycles*, **6,** 541 (1977).
26. D. W. Russell, *J. Med. Chem.*, **10,** 984 (1967).
27. D. J. Neadle and R. J. Pollitt, *J. Chem. Soc. C,* **1969,** 2127.
28. H. W. Heine, G. J. Blosick, and G. B. Lowrie, *Tetrahedron Lett.*, **1968,** 4801.
29. P. Bouchet, C. Coquelet, J. Elguero, and R. Jacquier, *Bull. Soc. Chim. Fr.*, **1976,** 192.
30. E. Leitis and D. G. Crosby, *J. Agric. Food Chem.*, **22,** 842 (1974).
31. J. E. Baldwin, *J. Chem. Soc. Chem. Commun.*, **1976,** 734.
32. R. Fielden, O. Meth-Cohn, and H. Suschitzky, *J. Chem. Soc. Perkin I,* **1973,** 696.
33. E. Hayashi and C. Iijima, *J. Pharm. Soc. Japan*, **82,** 1093 (1962); *Chem. Abstr.*, **58,** 4551 (1963).
34. E. Hayashi and Y. Miura, *J. Pharm. Soc. Japan*, **87,** 648 (1967); *Chem. Abstr.*, **67,** 90775 (1967).
35. R. A. Abramovitch and B. W. Cue, *Heterocycles*, **1,** 227 (1973).
36. J. P. Dirlam, B. W. Cue, and K. J. Gombatz, *J. Org. Chem.*, **43,** 76 (1978).
37. W. Dürckheimer, *Annalen*, **756,** 145 (1972).
38. F. Seng and K. Ley, *Synthesis*, **1975,** 703.
39. F. Seng and K. Ley, *Synthesis*, **1972,** 606.
40. D. P. Claypool, A. R. Sidani, and K. J. Flanagan, *J. Org. Chem.*, **37,** 2372 (1972).
41. C. H. Issidorides and M. J. Haddadin, British Patent 1,215, 815 (1970); *Chem. Abstr.*, **74,** 141873 (1971).

42. M. J. Abu El-Haj, *J. Org. Chem.*, **37**, 2519 (1972).
43. D. W. S. Latham, O. Meth-Cohn, and H. Suschitzky, *J. Chem. Soc. Chem. Commun.*, **1972**, 1040.
44. D. W. S. Latham, O. Meth-Cohn, H. Suschitzky, and J. A. L. Herbert, *J. Chem. Soc. Perkin I*, **1977**, 470.
45. M. J. Haddadin and C. H. Issidorides, British Patent 1,305,138 (1973); *Chem. Abstr.*, **78**, 136339 (1973).
46. A. J. Boulton, A. C. Gripper Gray, and A. R. Katritzky, *J. Chem. Soc. B*, **1967**, 911.
47. F. Seng and K. Ley, *Angew. Chem., Int. Edit.*, **11**, 1009 (1972).
48. F. Seng, K. Ley, and K. Wagner, *Synthesis*, **1975**, 703.
49. P. N. Preston and G. Tennant, *Chem. Rev.*, **72**, 627 (1972).
50. J. D. Loudon and G. Tennant, *J. Chem. Soc.*, **1963**, 4268.
51. D. B. Livingstone and G. Tennant, *J. Chem. Soc. Chem. Commun.*, **1973**, 96.
52. L. Konopski and B. Serafin, *Rocz. Chem.*, **51**, 1783 (1977).
53. A. E. Luetzow and J. R. Vercellotti, *J. Chem. Soc. C*, **1967**, 1750.
54. L. A. Ljublinskaya and V. M. Stepanov, *Tetrahedron Lett.*, **1971**, 4511.
55. Merck & Co. Inc., British Patent 1,133,853 (1968); U.S. Patent 3,265,706 (1966); *Chem. Abstr.*, **65**, 13724 (1966).
56. Ciba Ltd., British Patent 1,101,149–150 (1968); Netherlands Patent 6,515,833 (1966); *Chem. Abstr.*, **65**, 15388 (1966).
57. J. Machin, R. K. Mackie, H. McNab, G. A. Reed, A. J. G. Sagar, and D. M. Smith, *J. Chem. Soc. Perkin I*, **1976**, 394.
58. R. Marshall and D. M. Smith, *J. Chem. Soc. C*, **1971**, 3510.
59. D. Johnston and D. M. Smith, *J. Chem. Soc. Perkin I*, **1976**, 399.
60. J. Machin and D. M. Smith, *J. Chem. Soc. Perkin I*, **1979**, 1371.
61. H. McNab and D. M. Smith, *J. Chem. Soc. Perkin I*, **1973**, 1310.
62. A. F. Aboulezz and M. I. El-Sheikh, *Egypt. J. Chem.*, **17**, 517 (1974); *Chem. Abstr.*, **86**, 170326 (1977).
63. R. S. Goudie and P. N. Preston, *J. Chem. Soc. C*, **1971**, 1139.
64. Merck & Co. Inc., British Patent 1,109,785 (1968); Netherlands Patent 6,517,255 (1966); *Chem. Abstr.*, **66**, 2565 (1967).
65. R. Marshall, D. J. Sears, and D. M. Smith, *J. Chem. Soc. C*, **1970**, 2144.
66. R. Marshall and D. M. Smith, unpublished work.
67. E. Yu. Belyaev, V. P. Kumarev, L. E. Kondrat'eva, and E. I. Shakhova, *Chem. Heterocycl. Comp.*, **6**, 1576 (1970).
68. F. Minisci, R. Galli, and A. Quilico, *Tetrahedron Lett.*, **1963**, 785.
69. H. G. Aurich and W. Weiss, *Chem. Ber.*, **106**, 2408 (1973).
70. H. G. Aurich and K. Stork, *Chem. Ber.*, **108**, 2764 (1975).
71. S. Shiraishi, T. Shigemoto, and S. Ogawa, *Bull. Chem. Soc. Japan*, **51**, 563 (1978).
72. S. O. Chua, M. J. Cook, and A. R. Katritzky, *J. Chem. Soc. B*, **1971**, 2350.
73. J. Elguero, A. Fruchier, and S. Mignonac-Mondon, *Bull. Soc. Chim. France*, **1972**, 2916.
74. T. J. Batterham, "NMR Spectra of Simple Heterocycles," Wiley, 1973, pp. 10 and 60.
75. K. Tori, M. Ogata, and H. Kano, *Chem. Pharm. Bull. (Tokyo)*, **11**, 681 (1963).
76. Ref. 74, p. 326.
77. A. Tatematsu, H. Yoshizumi, E. Hayashi, and H. Nakata, *Tetrahedron Lett.*, **1967**, 2985.
78. D. Johnston, J. Machin, and D. M. Smith, *J. Chem. Res. (S)*, **1978**, 366.
79. S. Takahashi and H. Kano, *Chem. Pharm. Bull. (Tokyo)*, **12**, 282 (1964).
80. S. Takahashi and H. Kano, *Chem. Pharm. Bull. (Tokyo)*, **14**, 375 (1966).
81. Merck & Co. Inc., British Patent 1,109,786 (1968); Netherlands Patent 6,517,267 (1966); *Chem. Abstr.*, **66**, 2564 (1967).
82. E. Hayashi, E. Ishiguro, and M. Enomota, unpublished work (cited in Refs. 10 and 19).
83. R. Kuhn and W. Blau, *Annalen*, **615**, 99 (1958).
84. O. Serafimov, W. Seiffert, R. Krässig, and H. Zimmermann, *Ber. Bunsenges. Phys. Chem.*, **75**, 3 (1971).

85. S. Takahashi and H. Kano, *Chem. Pharm. Bull.* (*Tokyo*), **12**, 783 (1964).
86. Merck & Co. Inc., Netherlands Patent 6,611,581 (1967); *Chem. Abstr.*, **68**, 59586 (1968).
87. Merck & Co. Inc., Netherlands Patent 6,611,580 (1967); *Chem. Abstr.*, **68**, 59584 (1968).
88. S. Takahashi and H. Kano, *Tetrahedron Lett.*, **1963**, 1687.
89. S. Takahashi and H. Kano, *Chem. Pharm. Bull.* (*Tokyo*), **12**, 1290 (1964).
90. D. M. Smith, in "Comprehensive Organic Chemistry," Vol. 4, P. G. Sammes, Ed., Pergamon, Oxford, 1978, p. 28.
91. R. Fielden, O. Meth-Cohn, and H. Suschitzky, *J. Chem. Soc. Perkin I*, **1973**, 705.
92. A. R. Katritzky and J. W. Suwinski, *Tetrahedron*, **31**, 1549 (1975).
93. J. II. Markgraf, H. B. Brown, S. C. Mohr, and R. G. Peterson, *J. Am. Chem. Soc.*, **85**, 958 (1963), and references therein.
94. N. F. Cheetham, W. F. Forbes, D. J. Kew, and P. F. Nelson, *Aust. J. Chem.*, **16**, 729 (1963).
95. T. Cohen and G. L. Deets, *J. Am. Chem. Soc.*, **94**, 932 (1972), and references therein.
96. J. F. Vozza, *J. Org. Chem.*, **27**, 3856 (1962), and references therein.
97. S. Takahashi and H. Kano, *J. Org. Chem.*, **30**, 1118 (1965).
98. R. Fielden, O. Meth-Cohn, and H. Suschitzky, *J. Chem. Soc. Perkin I*, **1973**, 702.
99. M. Ogata, H. Matsumoto, S. Takahashi, and H. Kano, *Chem. Pharm. Bull.* (*Tokyo*), **18**, 964 (1970).
100. S. Takahashi, S. Hashimoto, and H. Kano, *Chem. Pharm. Bull.* (*Tokyo*), **21**, 287 (1973).
101. Y. Tamura, H. Hayashi, J. Minamikawa, and M. Ikeda, *J. Heterocycl. Chem.*, **11**, 781 (1974).
102. C. Berti, M. Colonna, I. Greci, and L. Marchetti, *J. Heterocycl. Chem.*, **16**, 17 (1979).

CHAPTER 3

Dihydrobenzimidazoles, Benzimidazolones, Benzimidazolethiones, and Related Compounds

DAVID M. SMITH

3.1. 2,3-DIHYDROBENZIMIDAZOLES (BENZIMIDAZOLINES)

3.1.1. Preparation

Benzimidazolines may be regarded either as reduced benzimidazoles or as cyclic aminals derived from 1,2-diaminobenzenes and carbonyl compounds, and in practice the reduction of benzimidazoles and the reaction of 1,2-diaminobenzenes with aldehydes and ketones constitute the principal synthetic routes to these compounds.

Reduction of Benzimidazoles

Reduction of benzimidazole itself with lithium aluminium hydride gives an oily product which is presumably impure benzimidazoline[1], but attempts to purify the product have been unsuccessful, and the alleged characterization of benzimidazoline as a picrate[1] is open to question.[2] Reduction of 1,3-dialkylbenzimidazolium salts, however, by lithium aluminum hydride[3,4] or sodium or potassium borohydride[2,5] gives good yields of 1,3-dialkyl-benzimidazolines (eq. 1), and the reduction of 1,3-dimethylbenzimidazolone by lithium aluminum hydride (eq. 2) provides a potentially useful alternative.[3]

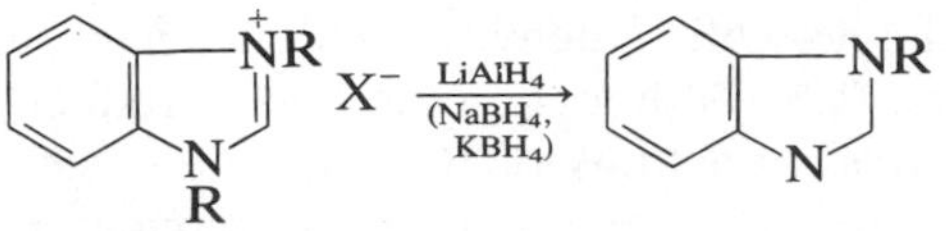

$$(1)$$

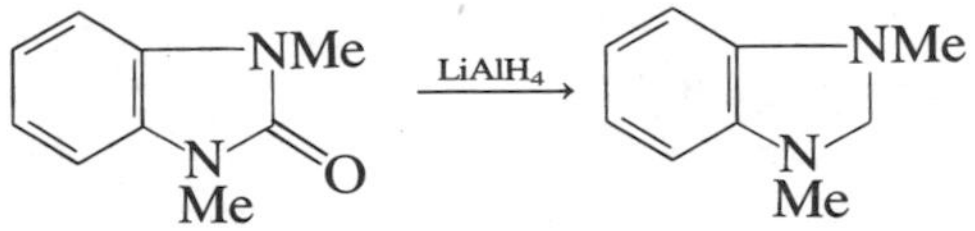

$$(2)$$

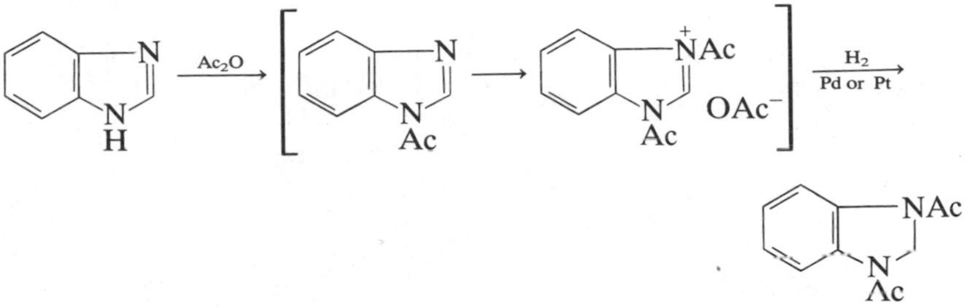

$$(3)$$

Catalytic hydrogenation of benzimidazole in presence of acetic anhydride gives 1,3-diacetylbenzimidazoline in high yield[6,7] (eq. 3). The species undergoing hydrogenation in this reaction is probably 1-acetylbenzimidazole or 1,3-diacetylbenzimidazolium acetate.

Reaction of 1,2-Diaminobenzenes with Carbonyl Compounds

These are conveniently grouped as follows:

a. Those in which both amino-functions are secondary (eq. 4) are simple and straightforward, and give 1,3-disubstituted benzimidazolines in good (more than 60%) yield.[8-15] Most of these proceed simply by warming the reactants in an appropriate solvent, but in some cases acid catalysis is advantageous.[10] 1,3-Diacetylbenzimidazolines cannot be obtained by this method,[16] but a variant of the procedure (eq. 5) permits the preparation of 1,3-di(benzenesulphonyl)-2-phenylbenzimidazoline.[17]

b. Those in which one amino group is secondary and the other primary have been much less studied, but diamines of type (**3.1**) react with aromatic aldehydes, in equimolar proportions, to give Schiff bases (**3.2**) or 1,2-disubstituted benzimidazolines (**3.3**).[8,14] If an excess of aldehyde is present, 1,2,3-trisubstituted benzimidazolines (**3.4**) may also be produced (eq. 6).[14]

c. Those in which both amino groups are primary have been the most extensively studied and are also the most complicated. The reaction of o-phenylenediamine and its analogs with aldehydes seldom leads to simple benzimidazolines of type **3.5**. Although some such products have apparently been isolated from reactions of this kind (see, e.g., Refs. 18–20), the more usual products are the isomeric monoanils (**3.6**) or the benzimidazoles (**3.7**).

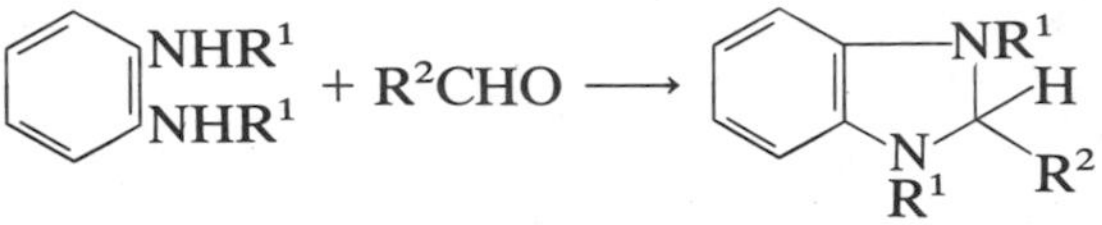

(4)

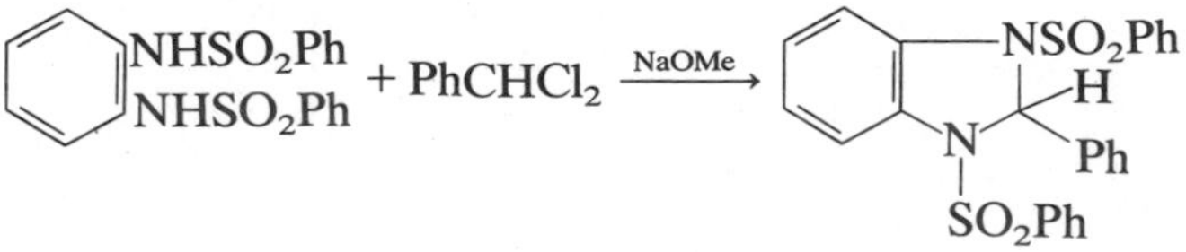

(5)

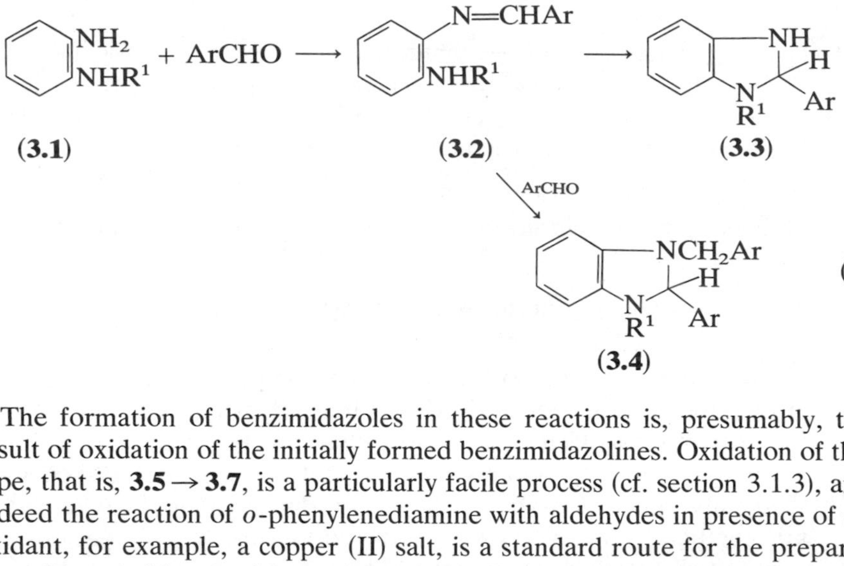

(6)

The formation of benzimidazoles in these reactions is, presumably, the result of oxidation of the initially formed benzimidazolines. Oxidation of this type, that is, **3.5** → **3.7**, is a particularly facile process (cf. section 3.1.3), and indeed the reaction of o-phenylenediamine with aldehydes in presence of an oxidant, for example, a copper (II) salt, is a standard route for the preparation of benzimidazoles (cf. Section 1.2.1 in Chapter 1). In the absence of an added oxidant, the monoanil (3.6) may act as a hydrogen acceptor, and the resulting diamine (cf. **3.1**) may then be transformed into the 1,2-disubstituted benzimidazoline (cf. eq. 6) or, by way of a further sequence involving redox steps, into the 1,2,3-trisubstituted benzimidazoline (cf. **3.4**: eq. 6). The details of the reaction steps[21,22] are given in Scheme 3.1, and the reaction may be used as a preparative route to the trisubstituted benzimidazolines (**3.4**: R′ = aryl),[13,23] although it is obviously not the method of choice in most cases.

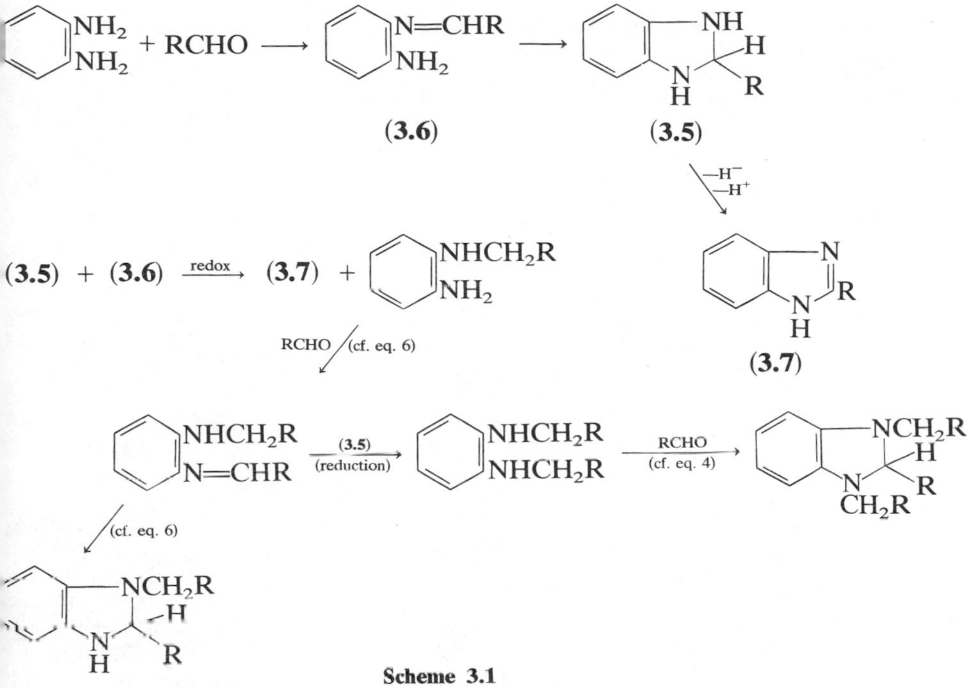

$$\text{(NH}_2, \text{NH}_2) + RCHO \longrightarrow (N=CHR, NH_2) \longrightarrow (NH, H, N, H, R)$$

(3.6) **(3.5)**

$$\xrightarrow[\text{-H}^+]{-\text{H}^-}$$

(3.5) + **(3.6)** $\xrightarrow{\text{redox}}$ **(3.7)** + $(NHCH_2R, NH_2)$

$$\text{RCHO} \diagdown \text{(cf. eq. 6)}$$

$$(N, N, R, H)$$

(3.7)

$$(NHCH_2R, N=CHR) \xrightarrow[\text{(reduction)}]{\textbf{(3.5)}} (NHCH_2R, NHCH_2R) \xrightarrow[\text{(cf. eq. 4)}]{\text{RCHO}} (NCH_2R, H, N, R, CH_2R)$$

$$\diagup \text{(cf. eq. 6)}$$

$$(NCH_2R, H, N, H, R)$$

Scheme 3.1

The reactions of o-phenylenediamines with ketones also present a complicated picture. It is presumed that 2,2-disubstituted benzimidazolines are produced in such reactions (eq. 7), since pyrolysis of the crude reaction products gives 2-substituted benzimidazoles (cf. Section 3.1.3), but with a few notable exceptions, for example, the reactions with cyclohexanone[24,25] and with isatin and its derivatives,[26,27] benzimidazolines have not been isolated from such reactions.[28-30] In the absence of further evidence, it is not clear whether the benzimidazolines are the primary products or merely intermediates produced during the pyrolysis step. Certainly in the reaction with benzophenone, the primary product is the anil (**3.8**),[30] and the corresponding reaction with dibenzyl ketone gives a mixture of isomers: the anil, the benzimidazoline (**3.9**), and the enamine (**3.10**).[30]

The product of reaction of o-phenylenediamine with 2 moles of acetone (or with mesityl oxide), which was once supposed to be the benzimidazoline (**3.11**),[29] has now been reformulated as a benzodiazepine (**3.12**).[2] An analogous compound (**3.13**) is obtained from o-phenylenediamine and cyclopentanone,[31] and the product from the reaction with pentan-2-one is probably also a benzodiazepine.[29]

Two interesting variants of these reactions are worthy of mention. Reaction of o-phenylenediamine with the dithietane (**3.14**) gives 2,2-bis-(trifluoromethyl)benzimidazoline (eq. 8),[32] and 1,2-bis(trimethylsilylamino)-benzene (**3.15**) reacts with cyclohexanone to give the benzimidazoline (**3.16**)

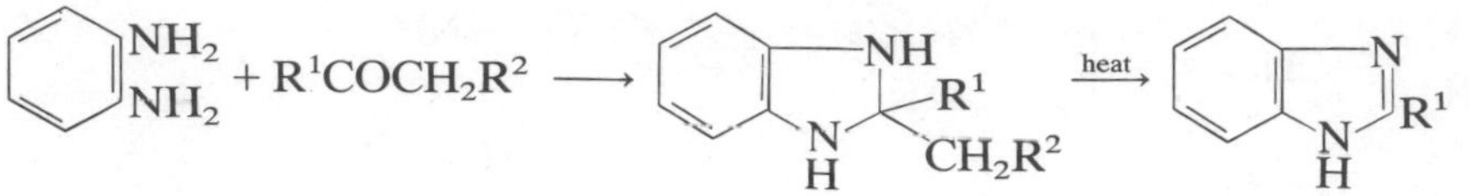

$$(7)$$

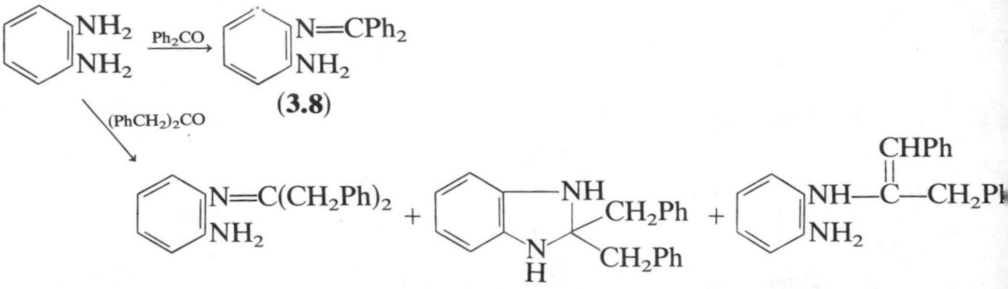

(**3.8**)

(**3.9**)   (**3.10**)

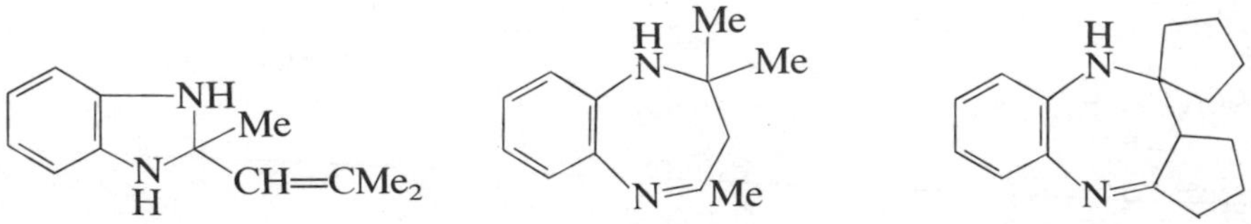

(**3.11**)   (**3.12**)   (**3.13**)

(eq. 9).[33] The corresponding reaction of (**3.15**) with benzaldehyde is alleged to give 2-phenylbenzimidazoline,[33] but this structural assignment is at variance with the quoted ^{1}H NMR spectrum (which shows an azomethine proton at δ 8.40; cf. Section 3.1.2).

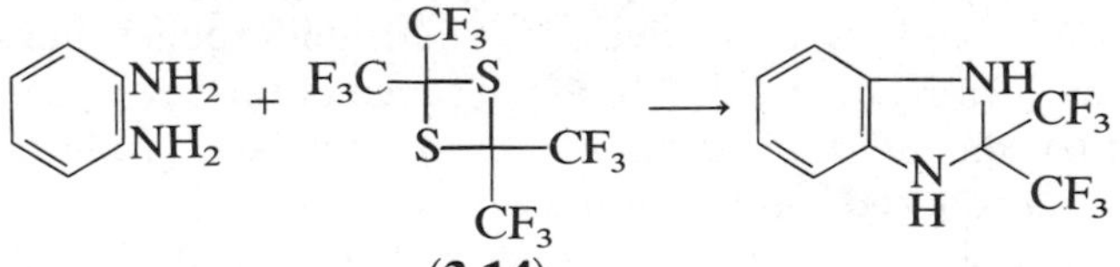

(**3.14**)

$$(8)$$

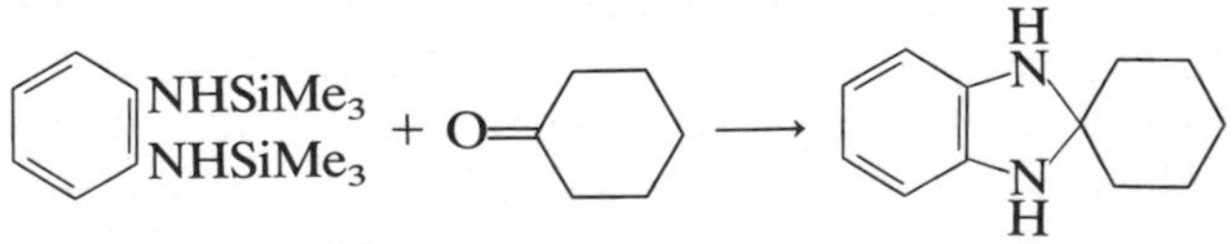

(**3.15**)   (**3.16**)

$$(9)$$

Other Methods

The reduction of benzimidazolium salts with complex metal hydrides described earlier in this section involves nucleophilic attack at the 2-

position in these salts. Other nucleophiles may also react at the 2-position of benzimidazolium salts (and, indeed, of benzimidazoles themselves) to give benzimidazolines. 2-Hydroxybenzimidazolines, which are produced by the action of hydroxide ion, have structures akin to hemiketals and undergo tautomeric ring-opening to amino-amides such as **3.17** (eq. 10).[12,34] 2-Alkoxy-,[35–37] 2-aryl-,[279] and 2-piperidino-[38] 1,3-disubstituted benzimidazolines may, however, be made from the appropriate benzimidazolium salt and nucleophile (eqs. 11–14), and the spiro-analogs (**3.18**: X = O, S, NMe) are prepared similarly (eq. 15).[39] Ylids derived from benzimidazolium salts (**3.19**), or carbanions derived from benzimidazoles (both by deprotonation of the 2-position), may effect nucleophilic additions of this type (eqs. 16 and 17).[36–40] Ylids of type **3.19** may also give benzimidazolines by reaction with ketene acetals and thioacetals (eqs. 18 to 20),[36] and benzimidazole itself reacts with diketene, in presence of acetic acid or acetic anhydride, to give the benzimidazoline **3.20** (eq. 21).[41]

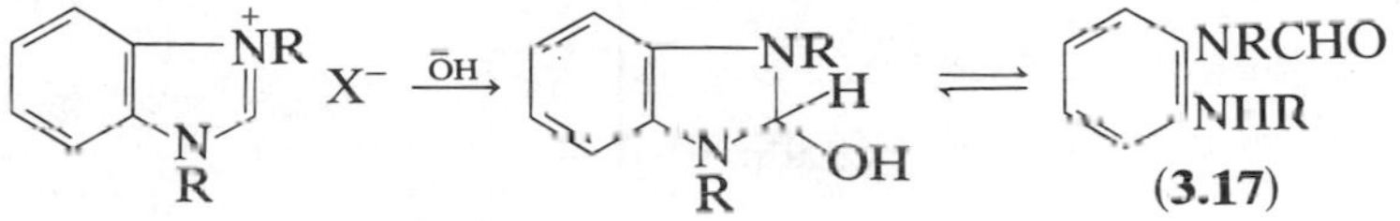

(10)

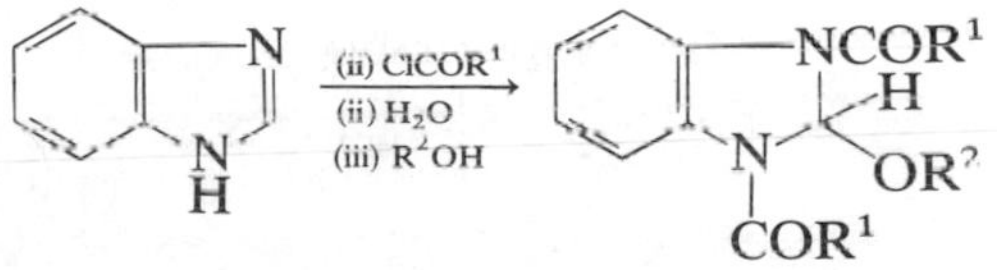

(11)

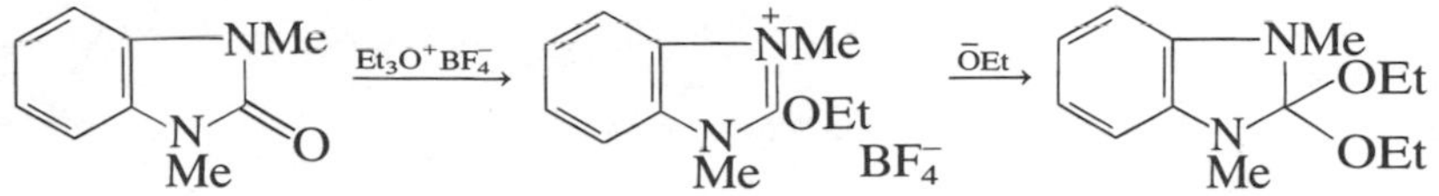

(12)

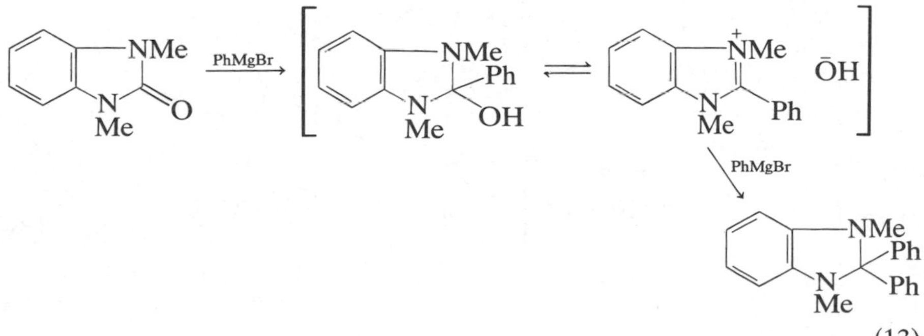

(13)

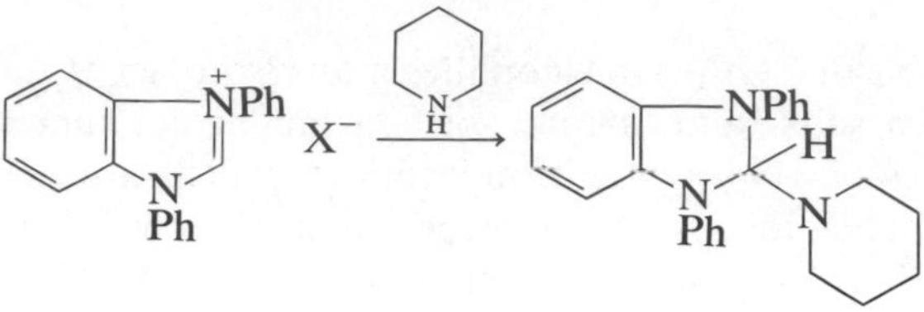

(14)

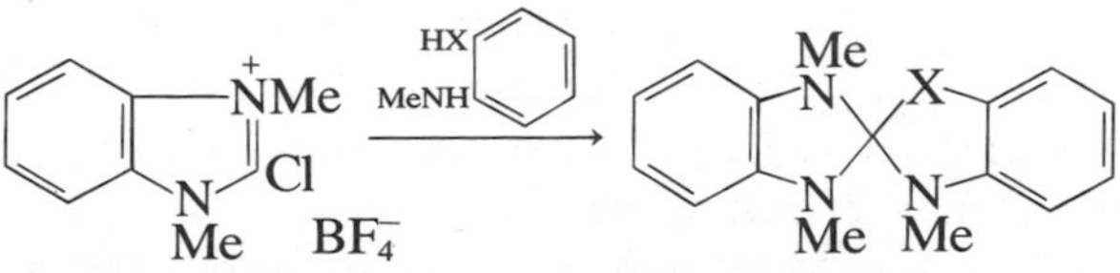

(15)

(**3.18**)

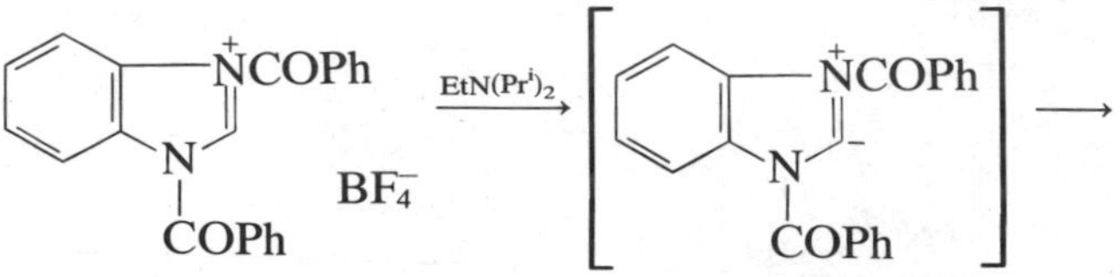

(**3.19**)

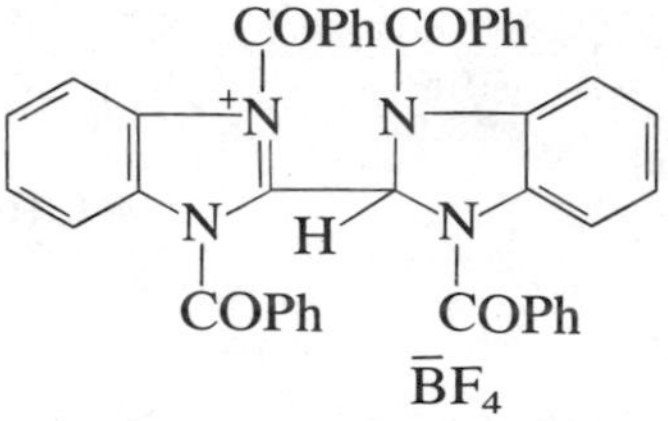

(16)

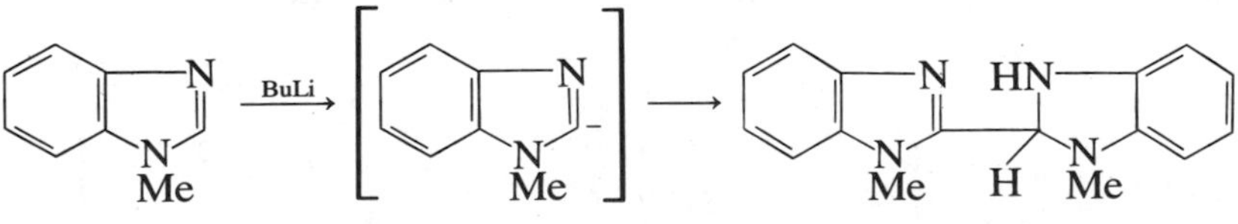

(17)

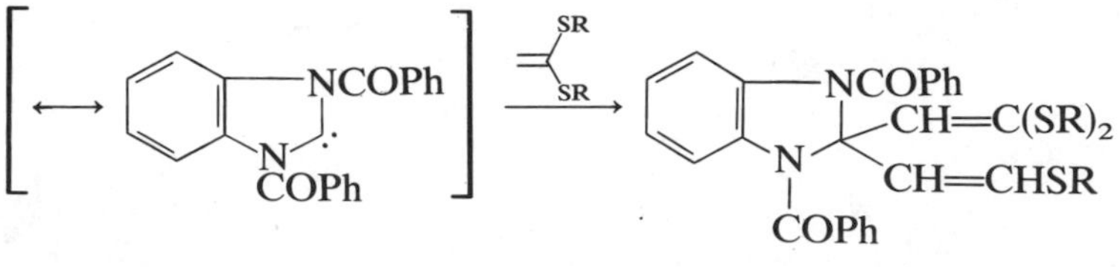

(18)

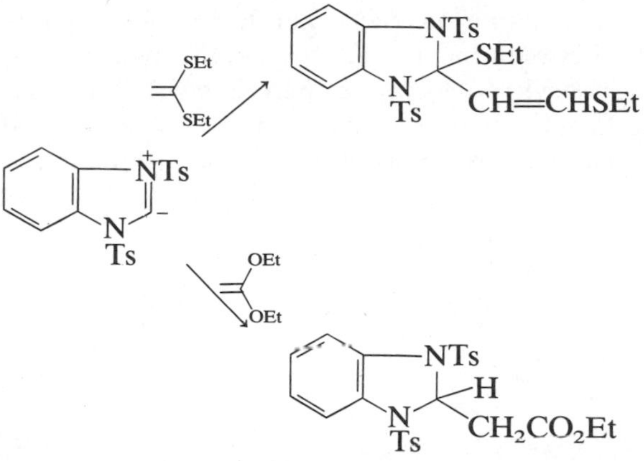

(19)

(20)

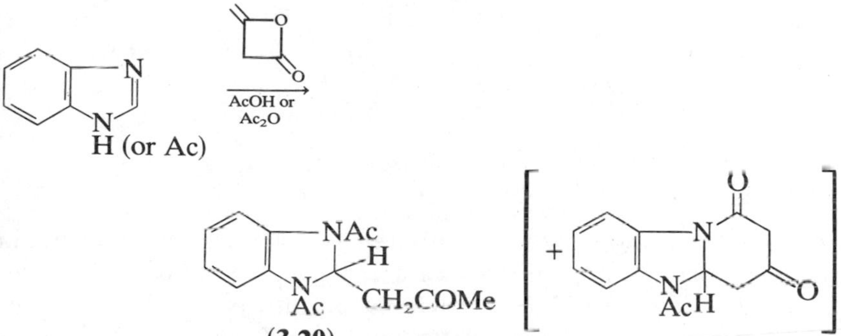

(21)

(3.20)

3.1.2. Physicochemical Studies

Spectra

The infrared and ultraviolet spectra of benzimidazolines are, in general, those expected for 1,2-diaminobenzenes.[4,9,25,32] The ^{1}H NMR spectra show aryl resonances at relatively high field (ca. δ 6.5), as expected; H-2 resonances occur at δ 4.0–4.8 for 2-unsubstituted or 2-monoalkyl compounds, and δ 5.0–6.0 for 2-monoaryl derivatives.[2,4,11,14,25,42]

Conformations

1-Acyl- and 1,3-diacylbenzimidazolines show restricted rotation about the bond(s) between nitrogen and carbonyl carbon atoms. Nuclear magnetic resonance studies[16,43,44] indicate that the *exo,endo* conformer (**3.21b**) is generally preferred over the *exo,exo* (**3.21a**) and *endo,endo* (**3.21c**) conformers in the diacyl series; this in turn indicates that dipole-dipole interactions between the acyl groups are more important factors in determining conformation than steric interactions between the acyl groups and the hydrogens at positions 4 and 7. In the monoacyl series, where there is no such dipole-dipole interaction, the steric factors come into play and so the monoacetyl compound prefers the *endo* and the monobenzoyl the *exo* conformation.

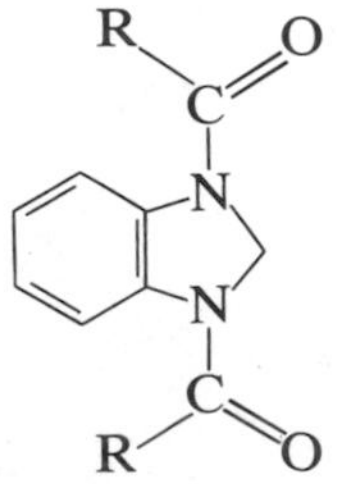

(3.21a) 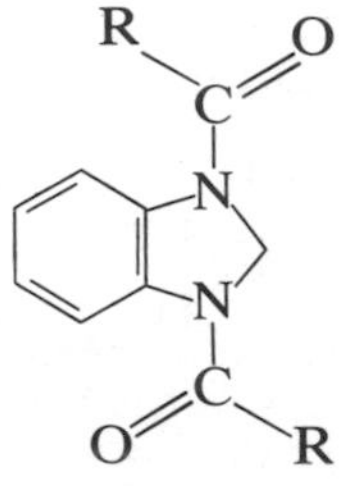   (3.21b) 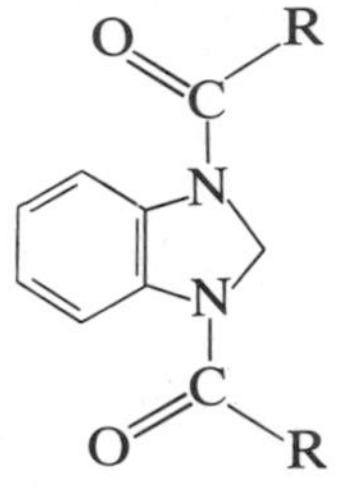   (3.21c)

Tautomerism

Benzimidazolines in which one or both nitrogens are unsubstituted are, in principle, tautomeric with Schiff bases, for example, $3.5 \rightleftharpoons 3.6$. While there is evidence that some Schiff bases of type **3.6** are converted into benzimidazolines on heating or on dissolution in polar solvents,[8,14] the author is unaware of any examples of the reverse reaction, so whether or not this is a genuinely tautomeric system remains an open question. Reference has already been made (section 3.1.1) to the tautomerism of 2-hydroxybenzimidazolines (eq. 10), and in this case it appears that examples of both heterocyclic and ring-opened species have been prepared.[12,34,35,42]

3.1.3. Reactions

Oxidation

Benzimidazolines undergo oxidation of three distinct types, namely;

a. 2-Unsubstituted benzimidazolines are oxidized by oxygen to benzimidazolones (eq. 22), and the corresponding reactions with sulfur, selenium, and tellurium give the thione, selenone, and tellurone derivatives.[5,45] 1,3-Dimethyl-2-phenylbenzimidazoline is oxidized to the amide (**3.22**) (eq. 23).[5] A nitro group in the starting material may serve as an oxidant, as in eq. 24.[46]

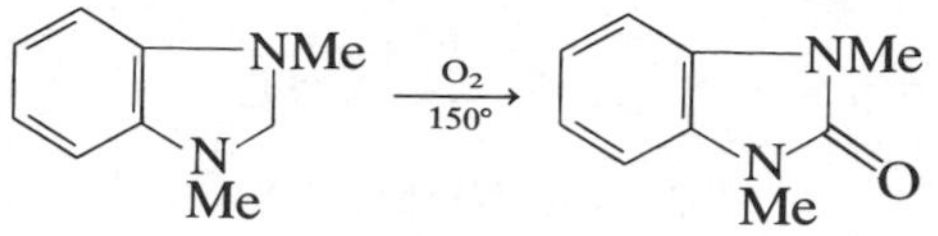

(22)

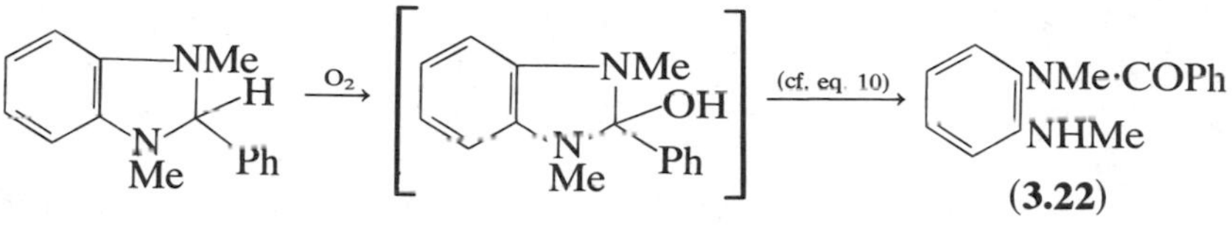

(23)

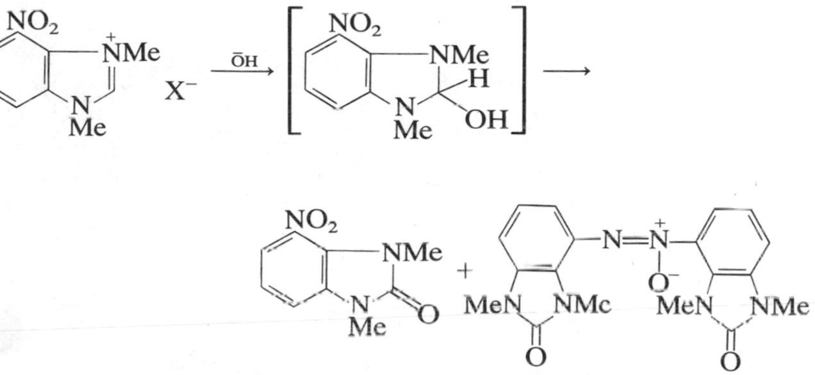

(3.22)

(24)

b. 2,2-Dialkylated benzimidazolines of type **3.16** are oxidized to 2*H*-benzimidazoles (**3.23**) by manganese dioxide.[25,47] [The products may then undergo addition of sulfinic acids giving arylsulfonylbenzimidazolines (**3.24**), and hence by reoxidation and a second addition, 4,5-bis-arylsulfonylbenzimidazolines (**3.25**).[47]] In some cases, oxidation of the type **3.16** → **3.23** may occur spontaneously during the preparation of the former,[48] and a 1:1 mixture of the two components which is highly coloured and reminiscent of a quinhydrone, is produced.

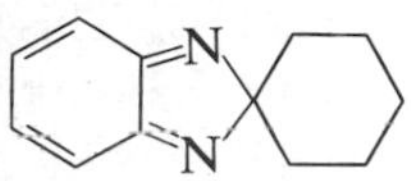

(3.23)

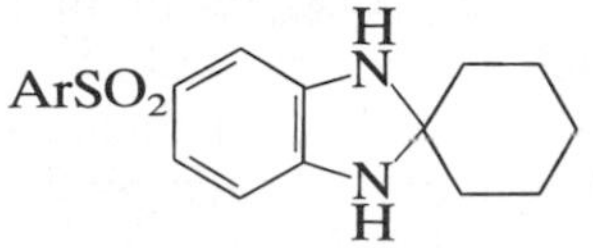

(3.24)

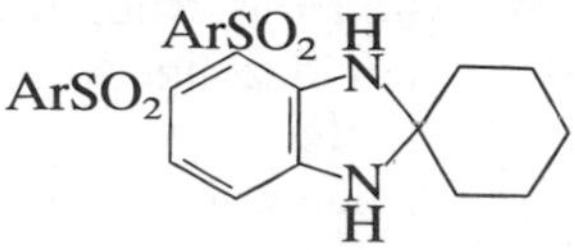

(3.25)

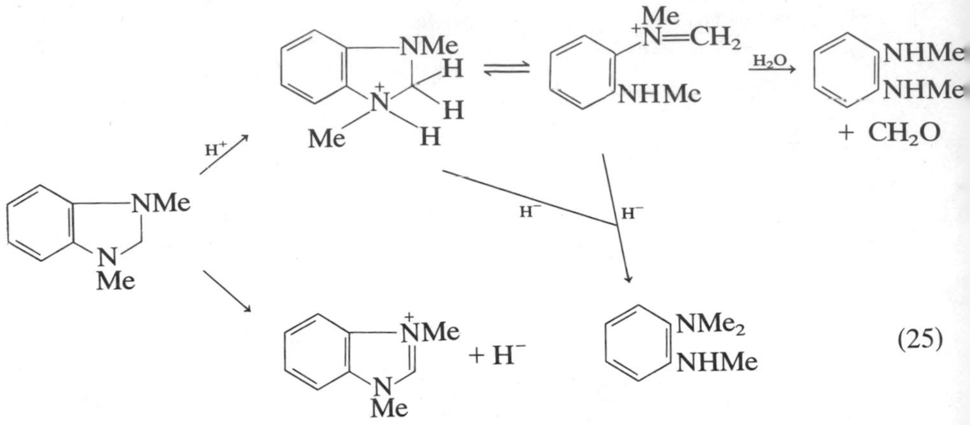

$$\text{(structural schemes for equations 25, 26, 27)}$$

(25)

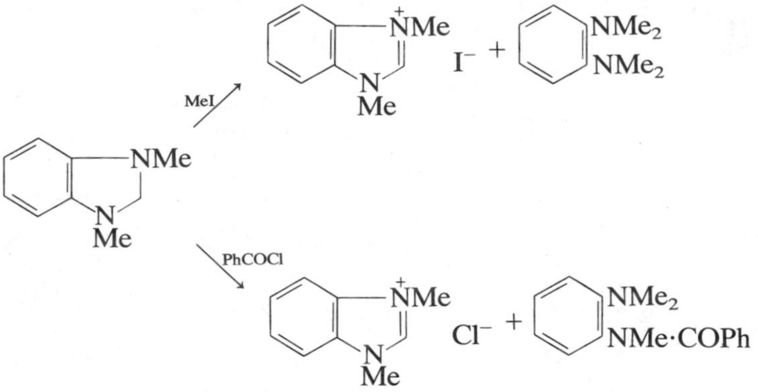

(26)

(27)

c. Benzimidazolines with a hydrogen at the 2-position are, in general, excellent hydride donors, they being converted themselves into benzimidazolium salts.[5,8-10,49-54] Benzimidazolines are evidently the best reducing agents among dihydro-*benzo*-fused heterocycles,[49] and iodine, silver salts, triphenylmethyl chloride, aldehydes, imines, and even carbon tetrachloride may act as hydride acceptors.[8,10,50,53,54] Attempts to protonate or quaternize benzimidazolines may result in an intermolecular hydride transfer (eqs. 25 to 27).[5,50,51,53] Reactions with aqueous acid, of course, may also result in hydrolysis of the benzimidazoline to a 1,2-diaminobenzene and an aldehyde or ketone. This is the only reaction observed when benzimidazolines of type **3.16** are treated with acid,[25] and acylation of such compounds is complicated in some cases only by (tautomeric) ring-opening of the product (eqs. 28 and 29).[25,55]

$$(3.16) \xrightarrow[\text{or (RCO)}_2\text{O}]{\text{RCOCl}} \left[\text{structure} \right] \xrightarrow{\text{H}_2\text{O}} \text{(NHCOR / NHCOR)} + \text{(cyclohexanone)} \quad (28)$$

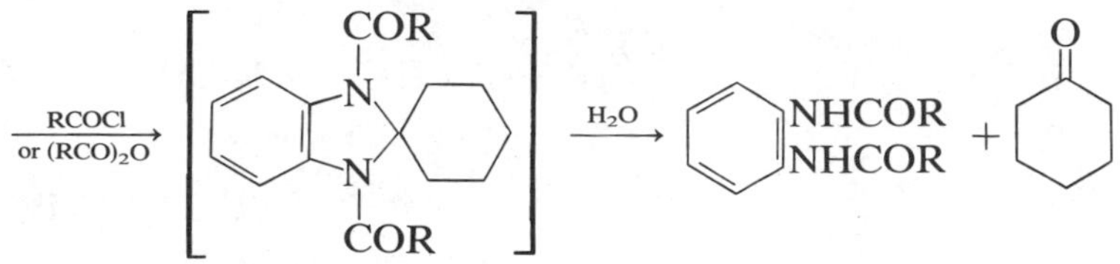

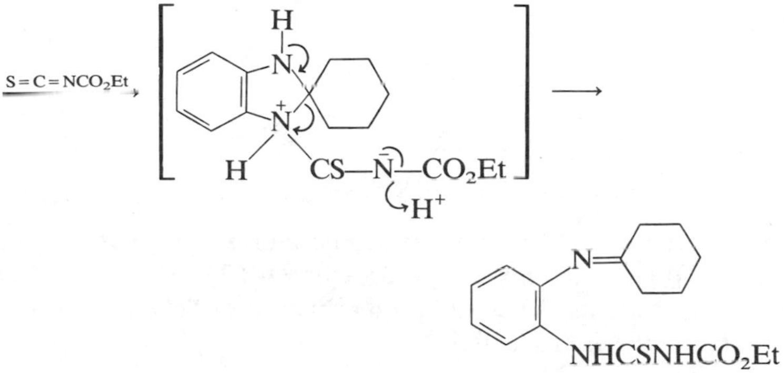

$$(3.16) \xrightarrow{\text{S}=\text{C}=\text{NCO}_2\text{Et}} \left[\text{structure} \right] \longrightarrow$$

$$(29)$$

NHCSNHCO$_2$Et

The hydride-transfer reactions of benzimidazolines are summarized in Scheme 3.2.

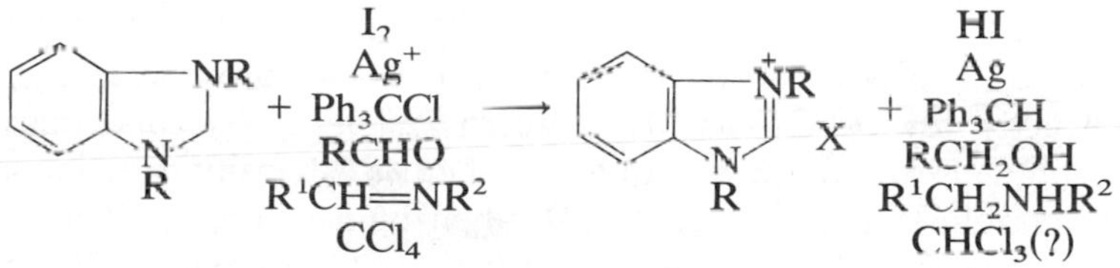

$$\text{structure} \xrightarrow[\substack{\text{Ag}^+ \\ + \text{Ph}_3\text{CCl} \\ \text{RCHO} \\ \text{R}^1\text{CH}=\text{NR}^2 \\ \text{CCl}_4}]{\text{I}_2} \quad \text{structure} \quad \begin{array}{l} \text{HI} \\ \text{Ag} \\ + \text{Ph}_3\text{CH} \\ \text{RCH}_2\text{OH} \\ \text{R}^1\text{CH}_2\text{NHR}^2 \\ \text{CHCl}_3(?) \end{array}$$

Scheme 3.2

Thermolysis

Reference has already been made (section 3.1.1, eq. 7) to the thermolysis of 2,2-disubstituted benzimidazolines, at 200 to 250°, into 2-substituted benzimidazoles and hydrocarbons.[28-30,56,57] The spiro compound (**3.16**) is similarly converted into 2-n-pentylbenzimidazole.[58]

1,2,3-Trisubstituted benzimidazolines of type **3.26** or **3.27** undergo thermolysis with the loss of one of the N-substituents (eq. 30).[59-62] These

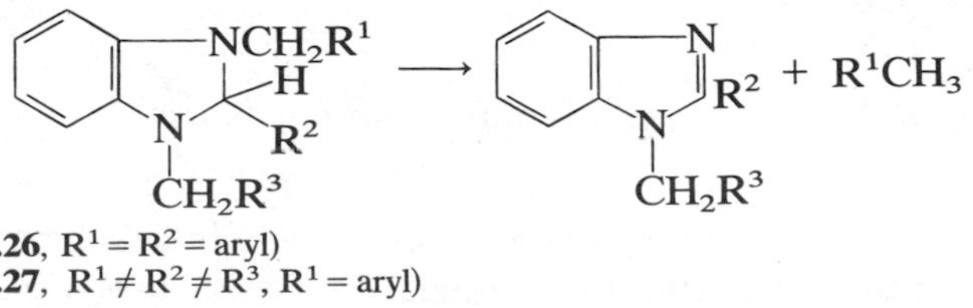

$$\text{structure} \longrightarrow \text{structure} + \text{R}^1\text{CH}_3 \quad (30)$$

(**3.26**, R^1 = R^2 = aryl)
(**3.27**, R$^1 \neq$ R$^2 \neq$ R^3, R^1 = aryl)

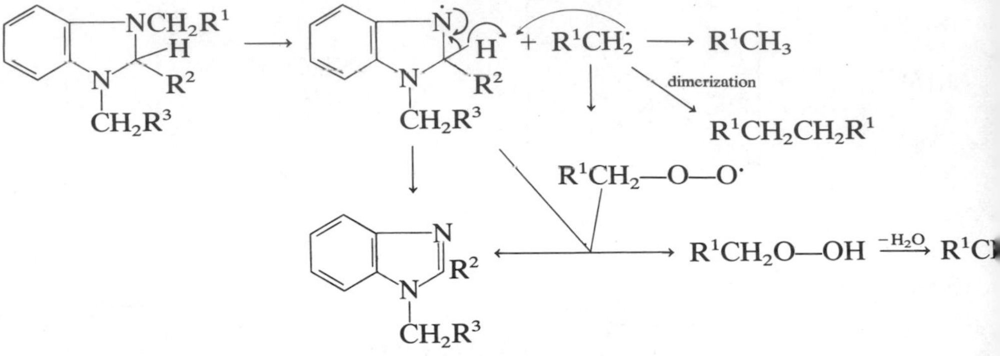

Scheme 3.3

thermolyses proceed at slightly lower temperatures in presence of dibenzoyl peroxide,[59] and give not only the hydrocarbon $R'CH_3$ but also $(R'CH_2)_2$ and $R'CHO$ as by-products.[60] The radical mechanism of Scheme 3.3 satisfactorily accounts for these products.

3.2. BENZIMIDAZOLONES AND BENZIMIDAZOLETHIONES

According to current *Chemical Abstracts* nomenclature, these compounds are correctly described as 1,3-dihydro-2H-benzimidazol-2-ones (and -thiones), but elsewhere they may be described as 2-benzimidazolinones (and 2-benzimidazolinethiones), as 2(3H)-benzimidazolones (and benzimidazolethiones), and as o-phenyleneureas (and thioureas). They are, of course, potentially tautomeric with 2-hydroxy- (and 2-mercapto-) benzimidazoles (cf. section 3.2.3), and the names of the tautomers, including 2-benzimidazolol (and benzimidazolethiol) are also used. For convenience the two series are referred to in this chapter simply as benzimidazolones and benzimidazolethiones.

3.2.1. Preparation of Benzimidazolones

Work in this field prior to 1950 has been reviewed in detail in an earlier volume of this series,[63] and only a brief outline of this early work is incorporated below.

Reaction of 1,2-*Diaminobenzenes with Carbonic Acid Derivatives*

This is the simplest route to benzimidazolones, in practical as well as in mechanistic terms. Of the two original variants, involving the use of phosgene and urea,[63] the latter has been the more popular because it presents fewer handling problems. Carbonate esters,[64,65] diethyl pyrocarbonate (**3.28**),[66] N,N-diethylcarbamyl chloride (**3.29**),[67] substituted ureas including semicarbazide[67] and N,N'-carbonyldiimidazole (**3.30**),[68,69] cyanic acid,[70] and carbon dioxide itself[71,72] may also be used in this reaction (Scheme 3.4) with varying degrees of success. The reaction with phosgene is apparently facilitated by the use of the N,N'-bis-trimethylsilyl derivative (**3.15**) in place of the parent diamine.[73]

Two mechanisms have been proposed for these reactions (Scheme 3.4). Both involve, as the first step, monoacylation of the diamine (it being assumed that the purely thermal reactions involving urea give HNCO as the effective acylating agent). The monoacyl compound (**3.31**)* may then undergo ring closure in one of two ways: (if $R^1 = H$) via o-aminophenyl isocyanate (**3.32**) (pathway a), or (irrespective of the nature of R^1) by the more familiar addition-elimination sequence (pathway b).

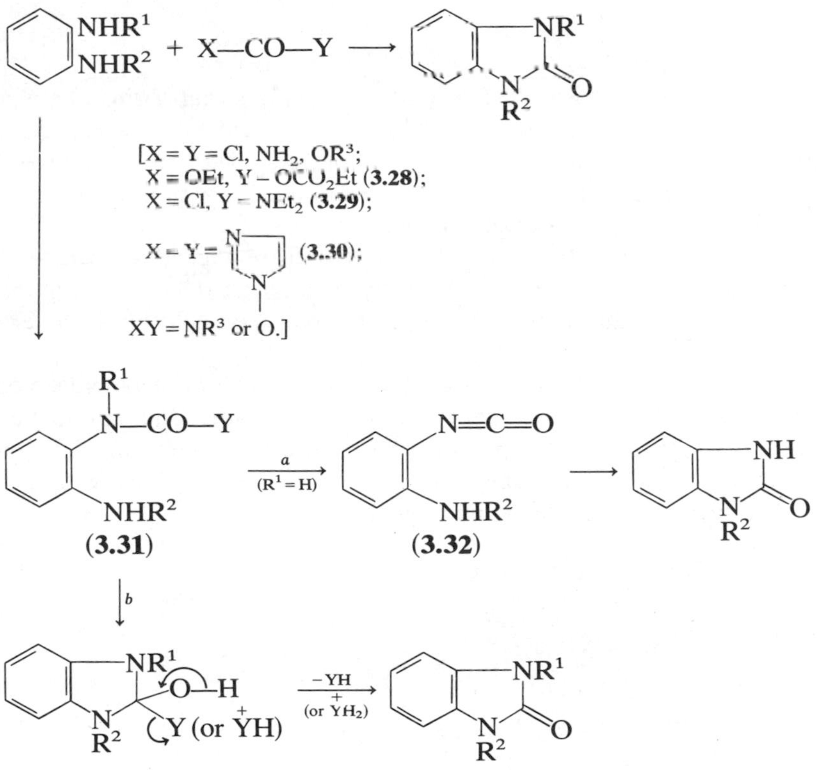

$[X = Y = Cl, NH_2, OR^3;$
$X = OEt, Y - OCO_2Et$ (**3.28**);
$X = Cl, Y = NEt_2$ (**3.29**);

$X = Y =$ ⟨imidazole⟩ (**3.30**);

$XY = NR^3$ or O.]

(**3.31**)   (**3.32**)

Scheme 3.4

* Monoacyl-o-phenylenediamines are, of course, also obtainable in many cases by reduction of the corresponding N-acyl-o-nitroanilines.[63]

Evidence in favor of pathway *a* has been provided in the case of the carbamate (**3.31**, $R^1 = R^2 = H$, $Y = OC_6H_4NO_2$-*p*) by kinetic studies and by trapping of the isocyanate (**3.32**).[74] Thermal cyclization of the urea (**3.31**, $R^1 = R^2 = H$, $Y = NH_2$) is assumed also to follow pathway *a*,[67] and the same may also be true of the carbamyl chloride (**3.31**, $R^1 = R^2 = H$; $Y = Cl$), since prolonged reaction of 3,4-diaminotoluene with phosgene gives 3,4-diisocyanatotoluene (**3.33**).[75] This last result must be interpreted with caution, however, since 5-methylbenzimidazolone itself reacts with phosgene, giving **3.33** (eq. 31). If pathway *a* cannot occur, for example, because $R^1 \neq H$, the

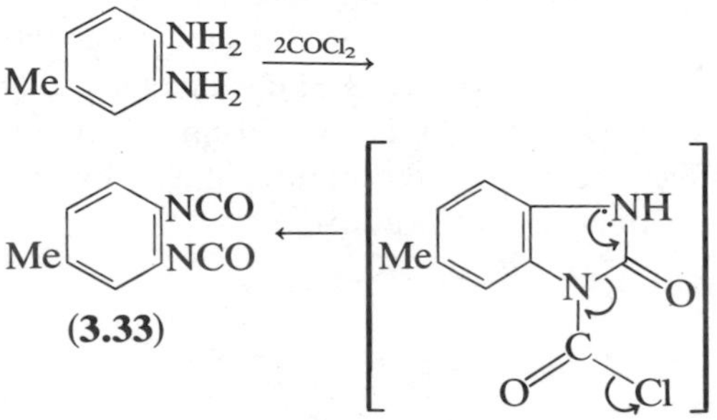

$$\text{(eq. 31)}$$

cyclization follows pathway *b*. This pathway requires that Y must be a good leaving group, or else (as when $Y = NH_2$ or NR_2) acid catalysis is necessary. Thus 1,2-di(methylamino)benzene does not react with urea to give 1,3-dimethylbenzimidazolone, but the corresponding reaction of the amine hydrochloride gives the benzimidazolone almost quantitatively.[67] On the other hand, the formation of 1,3-dimethylbenzimidazolone from 1,2-di(methylamino)benzene and diethylcarbamyl chloride does not require the prior formation of the hydrochloride, since hydrogen chloride is generated in the first step.[67]

Carbon monoxide may be used as the source of C-2 in benzimidazolones: for example, 1,2-diaminobenzene, carbon monoxide, and sulfur heated together in methanol under pressure give benzimidazolone in high (90%) yield,[76] and the cyclopentadienyl cobalt complex (**3.34**) reacts with carbon monoxide giving 1-phenylbenzimidazolone in 60% yield[77] (eq. 32).

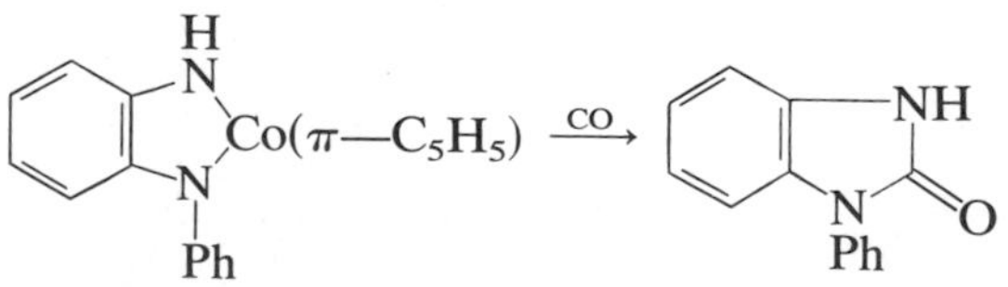

$$\text{(eq. 32)}$$

Curtius and Related Rearrangements

Reference has been made in the last section to the intermediacy of *o*-aminophenyl isocyanate in benzimidazolone syntheses, and so a

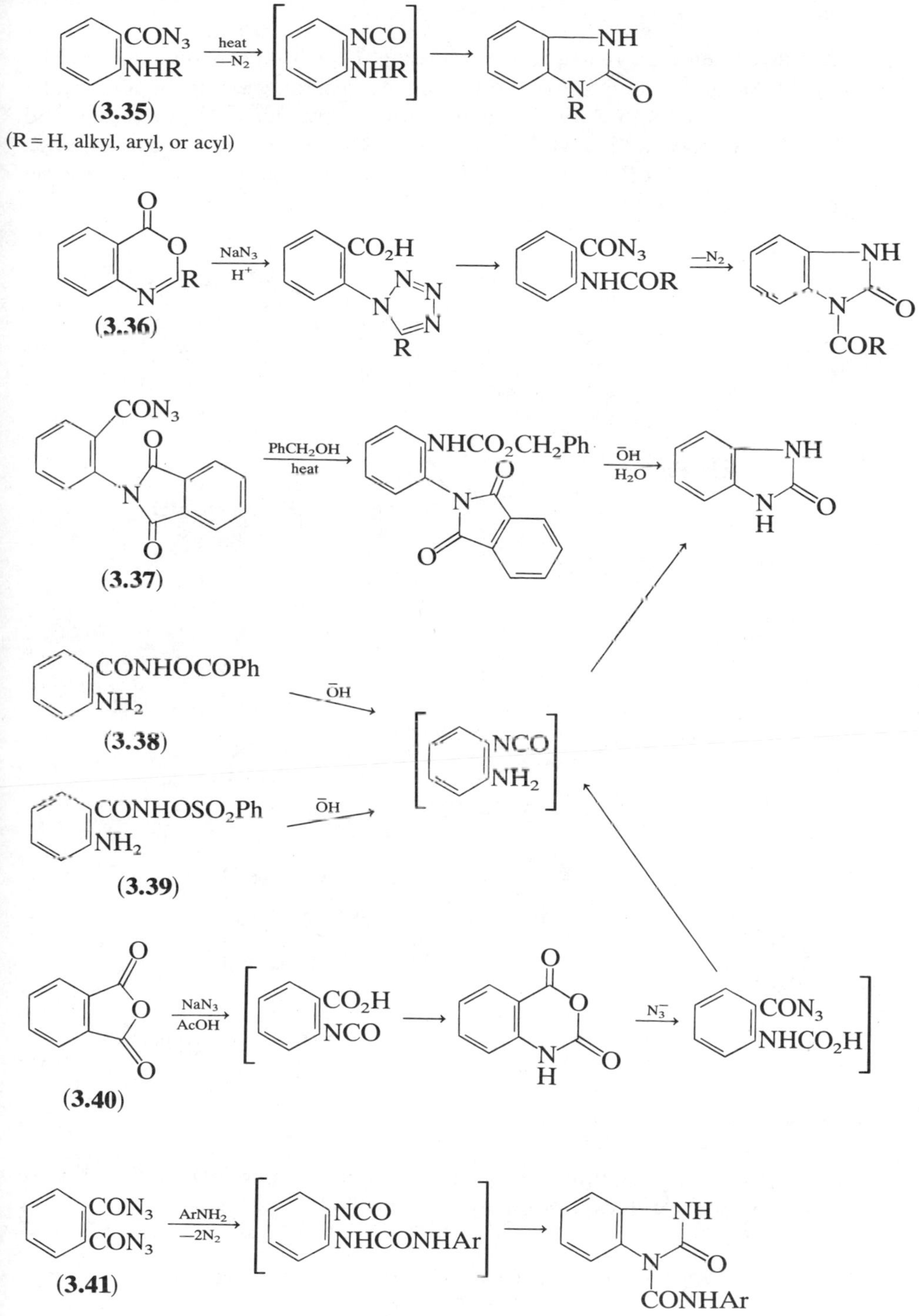

(3.35)

(R = H, alkyl, aryl, or acyl)

(3.36)

(3.37)

(3.38)

(3.39)

(3.40)

(3.41)

Scheme 3.5

Curtius/Lossen-type rearrangement leading to such intermediates should be capable of giving benzimidazolones. Single rearrangements of anthranilic acid derivatives (**3.35–3.39**)[78–83] and double rearrangements of phthalic acid derivatives (**3.40** and **3.41**)[84–86] are both known (Scheme 3.5), although these appear to offer little advantage, in most cases, over the simple reactions based on diamines (Section 3.2.1).

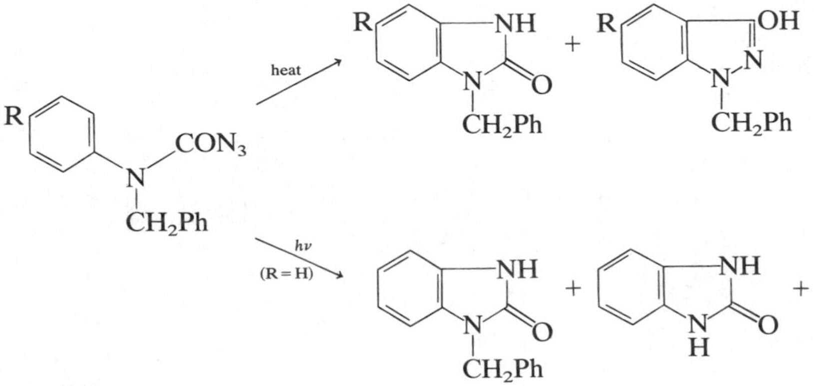

(33)

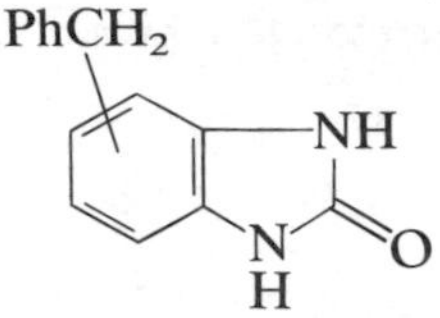

(34)

(4- and 5-substituted)

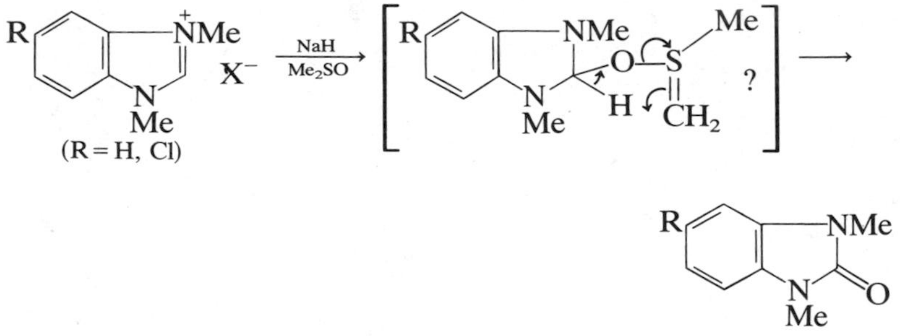

(35)

Other less obvious routes to benzimidazolones which involve decomposition of azides are illustrated in eqs. 33 and 34.[87–89].

Preparation from Other Benzimidazole Derivatives

Direct hydroxylation of benzimidazoles lacking an acidic hydrogen (i.e., those bearing a 1-substituent) may be effected by reaction with potassium hydroxide at high temperatures (200 to 250°).[90-92] 1,2-Disubstituted benzimidazoles also undergo hydroxylation with the loss of the 2-substituent under these conditions.[93,94] The reaction of benzimidazolium salts with hydroxide,[46] or amide ion followed by hydrolysis,[95] gives 2-hydroxybenzimidazolines, which may undergo either ring-opening (cf. eq. 10) or oxidation to the benzimidazolone. Benzimidazolium salts are convertible into benzimidazolones directly by the action of sodium dimsylate (eq. 35).[96]

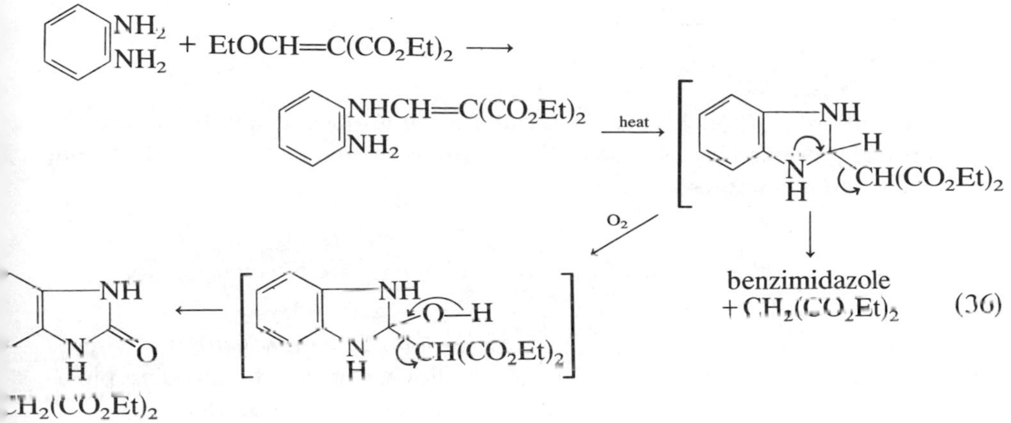

$$(36)$$

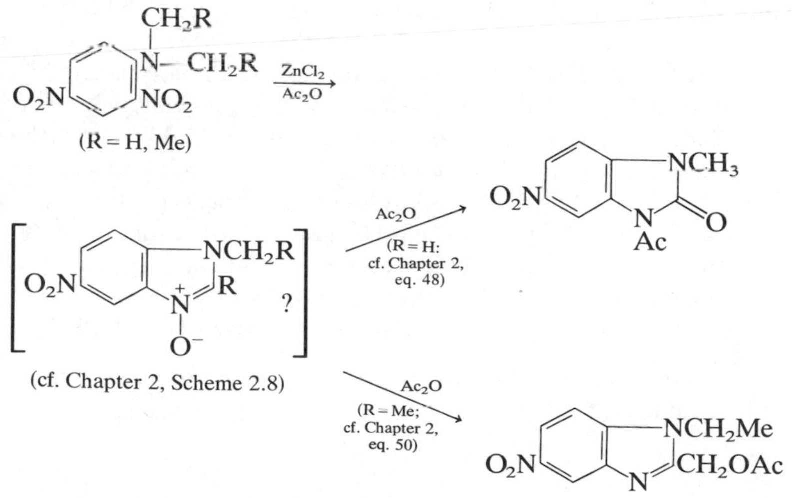

$$(37)$$

Oxidation of benzimidazolines [section 3.1.3; cf. also eq. 36[97]]; rearrangement of benzimidazole-N-oxides [Chapter 2, section 2.5.5; cf. also eq. 37[98]]; the reaction of alkali with benzimidazole-2-sulfonic acids,[99] 1,3-dialkyl-2-(N,N-dimethylamino)benzimidazolium salts,[100] and 2-benzimidazolimines (section 3.3.2) and their salts;[101,102] and acid-induced dealkylation[103] or thermal rearrangement[104,105] of 2-alkoxybenzimidazoles (eq.

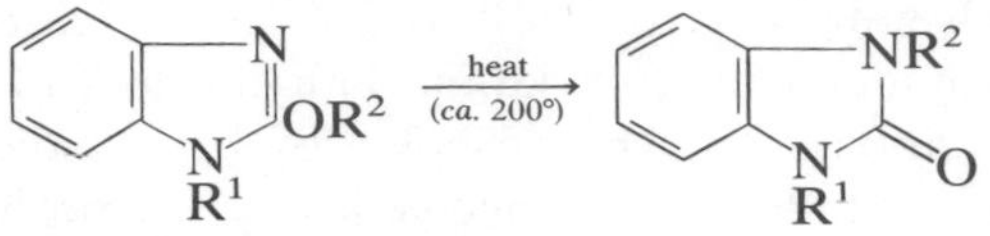

$$(R^1 = H \text{ or alkyl}; R^2 = \text{alkyl})$$

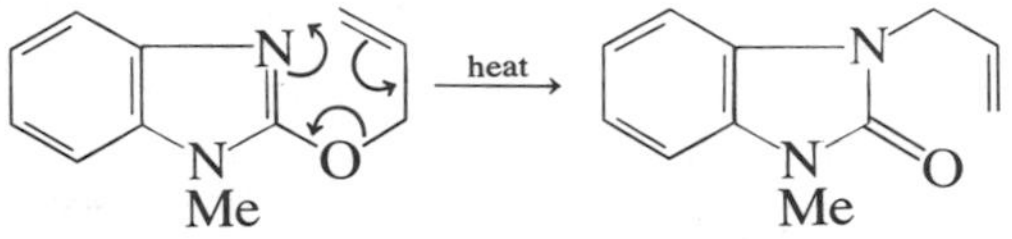

38 and 39), all yield benzimidazolones, but once again these are not normally preferable, as preparative methods, to the simple reactions using 1,2-diaminobenzenes.

Preparation from Other Heterocyclic Systems

FROM BENZODIAZEPINONES. The reaction of 1,2-diaminobenzenes with β-keto-esters leads, depending on the conditions, to a 1,5-benzodiazepin-2-one derivative (3.42) or a 1-alkenylbenzimidazolone (3.43) (eq. 40).[106-114] The benzodiazepinone is, evidently, the primary product in such reactions, since it may be converted into the alkenylbenzimidazolone on heating.[107,110,112,114]

FROM QUINOXALINE-N-OXIDES AND QUINOXALINONES. Photolysis of 2,3-disubstituted quinoxaline-1,4-dioxides (3.44) gives N-acylated benzimidazolones (3.45) (eq. 41),[115-117] the bis-oxaziridines (3.46) being the presumed intermediates. 3-Aryl-1H-quinoxalin-2-one 4-oxides (3.47) are converted into 1-acetyl-3-aroylbenzimidazolones by reaction with acetic anhydride (eq. 42),[118] and 1-hydroxy-1H,4H-quinoxaline-2,3-dione (3.48) is converted into 1,3-dimethylbenzimidazolone by dimethyl sulfate and alkali (eq. 43).[119] Finally, the dihydroquinoxalinone (3.49) undergoes acid-catalyzed oxidation by air (eq. 44) to give the benzimidazolone (3.50).[120]

FROM QUINAZOLINEDIONES AND BENZOTRIAZINONES. These are illustrated by eqs. 45,[121] 46,[122] and 47.[123]

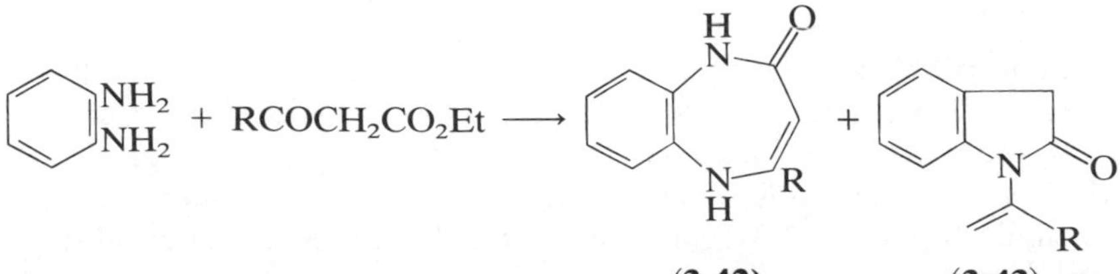

$$(3.42) \qquad (3.43)$$

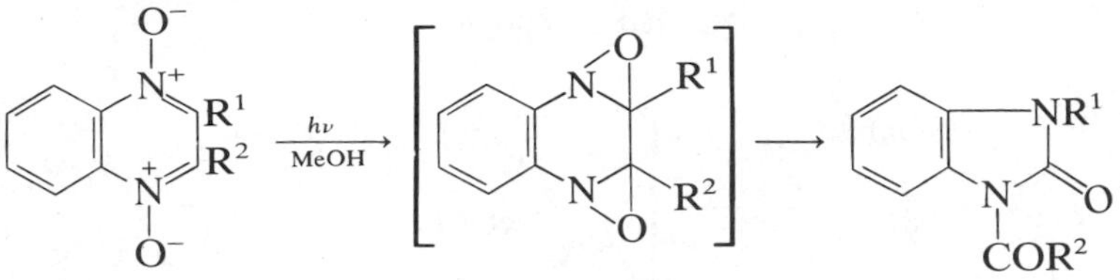

(3.44) (3.46) (3.45) (41)

$(R^1 = PhCH_2, PhCO; R^2 = alkyl \text{ or } aryl)$

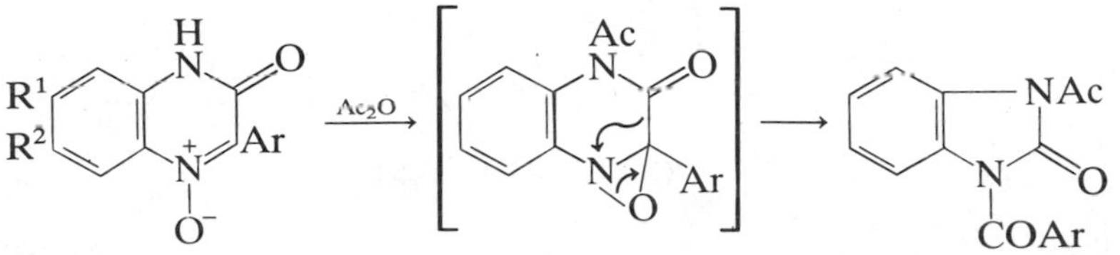

(3.47) (42)

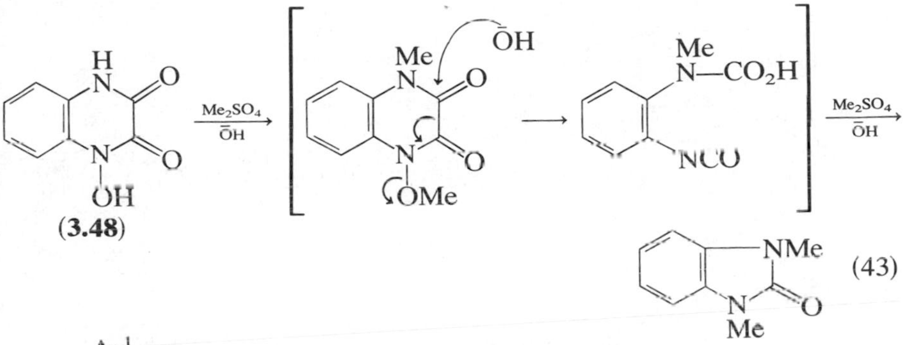

(3.48) (43)

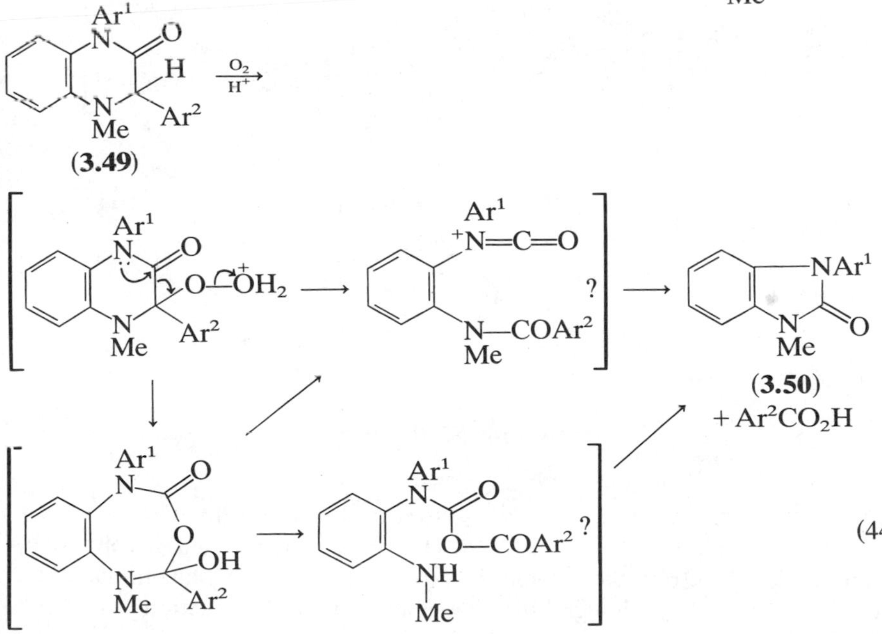

(3.49)

(3.50)

$+ Ar^2CO_2H$ (44)

351

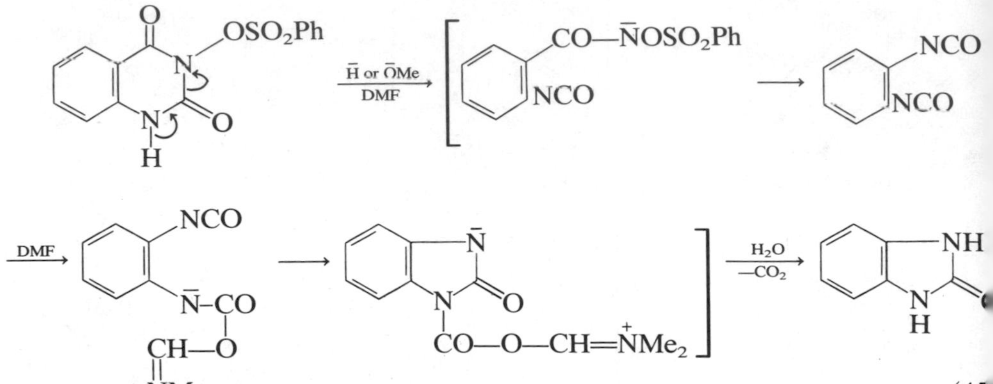

$$(45)$$

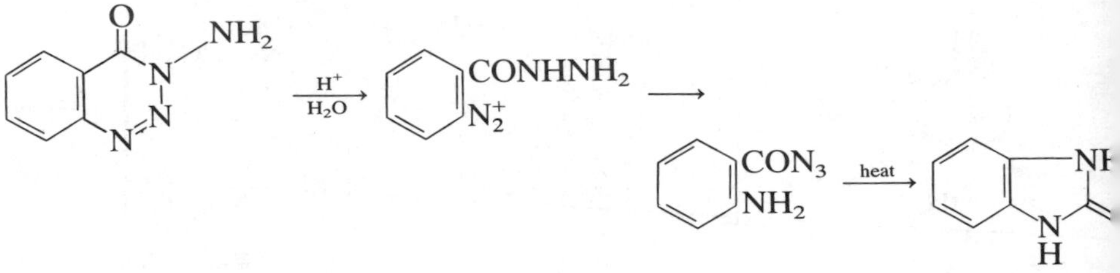

Other Methods

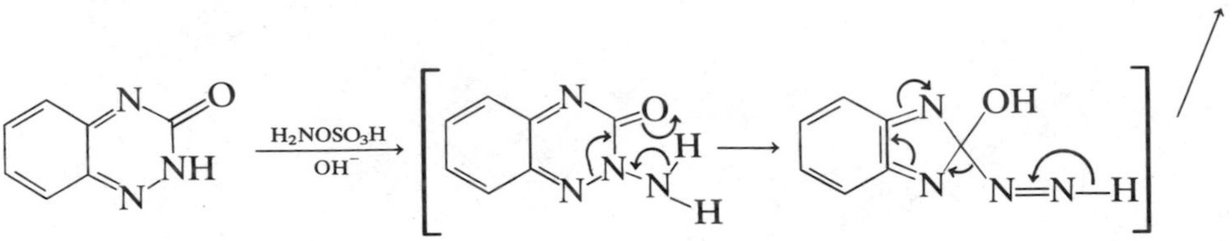

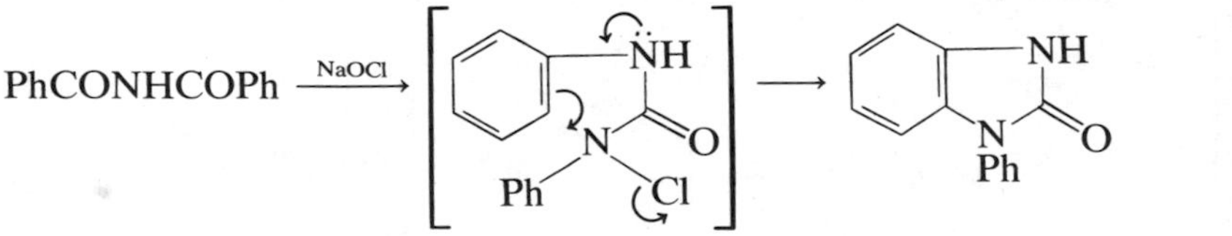

N,N'-Diarylureas react with sodium hypochlorite to give 1-arylbenzimidazolones (eq. 48),[124] the proposed mechanism involving intramolecular electrophilic aromatic substitution. 1-Arylbenzimidazolones are also formed, although in low yield, in the reaction of *N,N'*-diarylthioureas with sodium peroxide in ethanol.[125]

3.2.2. Preparation of Benzimidazolethiones

As in section 3.2.1, the early work in this field has already been reviewed in an earlier volume in this series,[63] and some leading references have also been collected in *Organic Syntheses*.[126]

Predictably, the synthetic routes to benzimidazolethiones are closely related to those that lead to benzimidazolones. The reactions of 1,2-diaminobenzenes with thiophosgene, thiourea, thiocyanates, potassium ethyl xanthate (3.51), and carbon disulfide all give benzimidazolethiones,[63,126] the last of these being the most popular. In the *Organic Syntheses* method, the diamine and carbon disulfide react in presence of potassium hydroxide, but other bases such as pyridine[127] and piperidine[128] have also been used, as has N,N'-dicyclohexylcarbodiimide[129]. Another variant, based on thiophosgene, involves the prior formation of N,N'-thiocarbonyldiimidazole (3.52).[130]

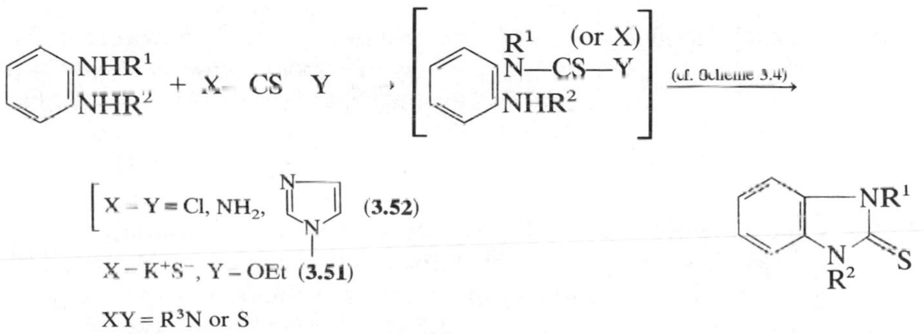

Scheme 3.6

These reactions, which are collected in Scheme 3.6, are assumed to proceed in a similar manner to those involving the oxygen analogs. Evidence in support of this assumption is provided by the thermal decomposition of the o-aminophenylthioureas (3.53) (eq. 49).[131]

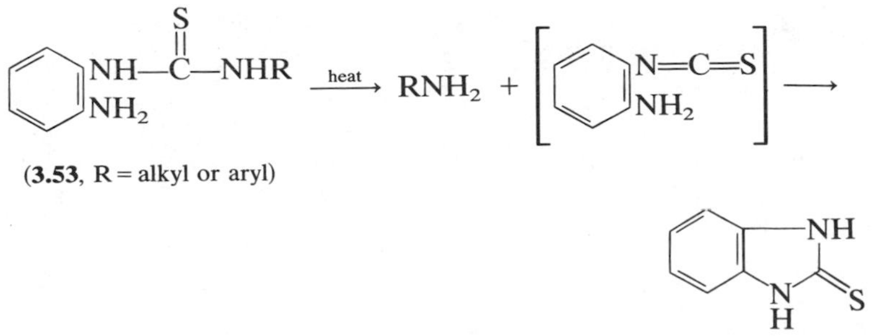

(3.53, R = alkyl or aryl)

(49)

The direct reaction of sulfur with benzimidazoles[132-134] and with ben-zimidazolines[5,45,99] gives benzimidazolethiones in good yield, although the elevated temperatures required for such reactions may limit their general applicability. 2-Chlorobenzimidazoles and their salts are converted into benzimidazolethiones by reaction with sulfur nucleophiles such as hydrogen sulfide ion,[135] thiourea,[135-137] or thiosulfate ion.[137,138]

3.2.3. Physicochemical Studies

As in the case of benzimidazole-N-oxides (Chapter 2), much of the physicochemical investigation of benzimidazolones and benzimidazolethiones has been associated with studies of tautomerism.

Molecular Structure

X-ray crystallographic studies of benzimidazolethione[139] show that in the crystal this molecule is planar and exists as the thione tautomer. The C—S bond length, 1.672 Å, is slightly shorter than that in thiourea (1.71 Å), but very similar to that of 2-pyridinethione (1.68 Å). The molecules are linked in the crystal by N—H · · · S · · · H—N hydrogen bonds, the S—H distance being 2.42 Å.

The value obtained for the C—S bond length has been interpreted as meaning that this bond has approximately 80% "double-bond character",[139] and this in turn may be taken to mean that the thione has some dipolar character as represented by structure **3.54d**. This is also observed in solution, where the dipole moments of benzimidazolethiones are (in some cases, substantially) higher than the vector sum of the individual group moments.[140] For example, the parent compound has a dipole moment of 4.14 D in dioxan, and 3.93 D in a 4:1 benzene-dioxan mixture, compared with the calculated value of 2.49 D for structure **3.54a**, 1.50 D for **3.54b**, and 2.40 D for **3.54c**.

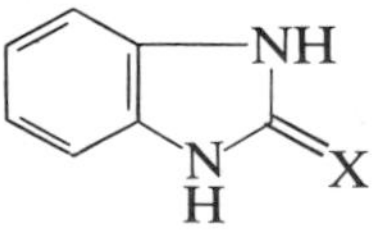

a 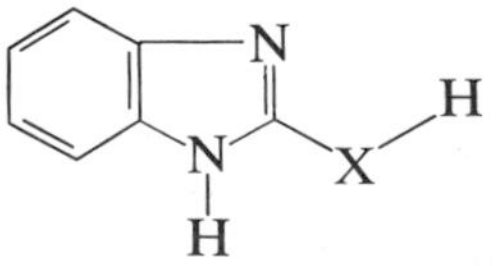   *b* 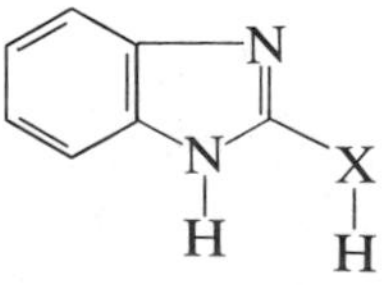   *c*

(**3.54**, X = S)   (**3.55**, X = O)

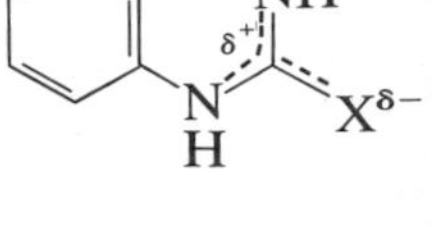

d

In the case of benzimidazolones, crystallographic data are apparently lacking, but the dipole moments in dioxan solution (2.46 D for the parent compound) correspond much more closely with the calculated values (2.15 D for **3.55a**). In the benzene-dioxan mixture, the observed moment (2.0 D) is actually *less* than the calculated value for **3.55a**, and is taken as implying a degree of enolization (μ for **3.55b** = 0.96 D) and/or of molecular association in the less polar medium.

Infrared and Electronic Spectra

The infrared spectra of benzimidazolones have been the subject of considerable interest and the cause of some confusion in the literature. N–H stretching frequencies fall in the region 3100 to 3200 cm^{-1},[79,141] but C=O absorptions ranging from 1675 to 1770 cm^{-1} have been reported.[79,141-147] It appears that 1,3-dialkylbenzimidazolones generally absorb towards the lower end of the range, and 1,3-unsubstituted compounds in the middle. At the upper end are C=O absorptions of 1,3-bis-alkyl- (or aryl-)sulfonyl derivatives. The carbonyl absorptions of 1-acyl- and 1,3-diacylbenzimidazolones have been interpreted in two different ways, and the position of the "2-one" carbonyl absorption is not established beyond doubt.[141,142,146,147] It has been suggested that the "2-one" absorption is shifted to progressively higher frequencies by acylation of one and then the other nitrogen because the nitrogen probably undergoes rehybridization when acylated and the C–O bond acquires more *s* character;[142] alternatively it has been suggested that the two or three absorptions in the carbonyl region are better regarded as associated with the set of carbonyl groups as a whole than with individual groups.[141] [Vibrational interactions of adjacent carbonyl groups (as, e.g. in anhydrides and imides) are, of course, well known.[148]]

Much less is known about the C=S stretching frequencies in benzimidazolethiones; for the parent compound, absorptions at 1180 and at 1351 cm^{-1} have both been attributed to this vibration.[144,149] It has also been claimed that the C=S absorption of *N*-substituted benzimidazolethiones lies in the range 1210 to 1230 cm^{-1}.[149]

The ultraviolet spectra of benzimidazolone and benzimidazolethione strongly resemble those of their *N,N*-dimethyl derivatives[150,151] rather than those of 2-alkoxy- or 2-alkylthiobenzimidazoles. In strongly acidic media, however, benzimidazolone, its 1,3-dimethyl derivative, and 2-ethoxybenzimidazole all have similar spectra, since under these conditions all are present as 2-hydroxy- or 2-ethoxybenzimidazolium salts.[150] The luminescence spectra of benzimidazolone and 1,3-dimethylbenzimidazolone show a fluorescence band at 300 to 350 nm and a phosphorescence band at 370 to 440 nm.

The photoelectron spectrum of 1,3-dimethylbenzimidazolethione reveals seven bands corresponding to ionization potentials of 7.46, 7.98, 8.61, 9.55,

10.49, 11.46, and 12.12 eV.[153] These are associated, respectively, with electrons in the following orbitals: π(C=S), n(S), π(benzene ring: 3 bands), σ(C—C), and σ(C—S).

Nuclear Magnetic Resonance and Mass Spectra

The ^{1}H NMR spectra of benzimidazolones and benzimidazolethiones are unexceptional, with aryl proton resonances in the region of δ 7.0 to 7.5

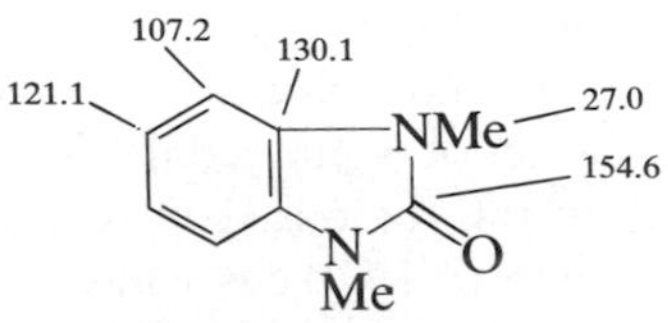

(3.56)

(see, e.g., Refs. 79 and 128). The ^{13}C NMR spectrum of 1,3-dimethyl-benzimidazolone has been interpreted as shown in structure **3.56**.[154]

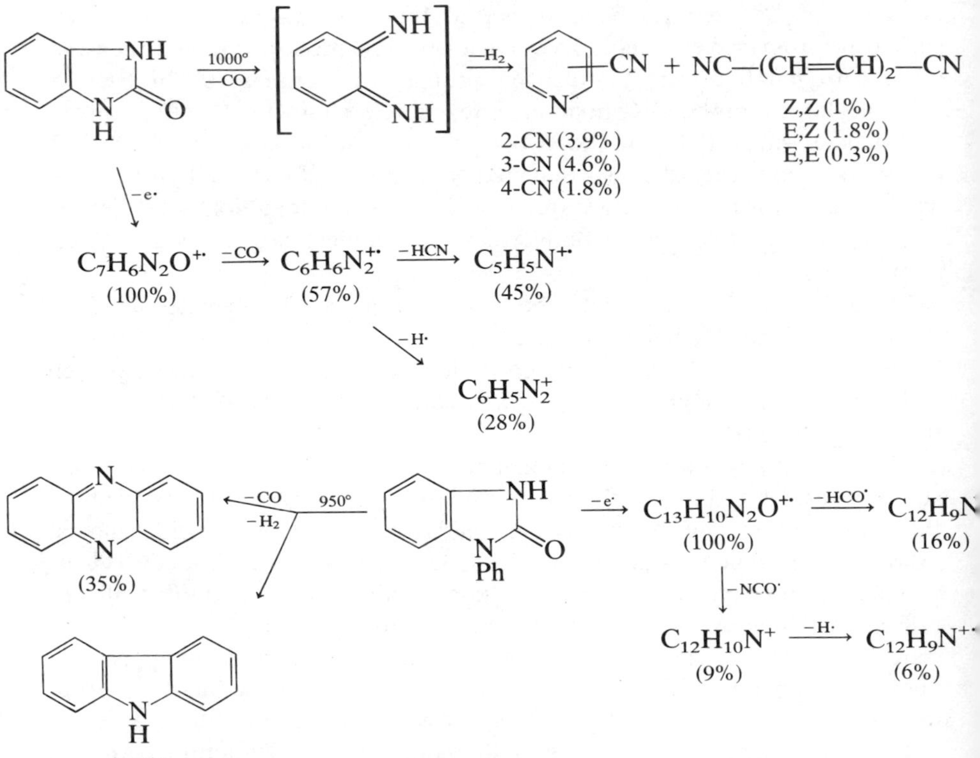

Scheme 3.7(a)

The mass spectra of benzimidazolones and benzimidazolethiones have been studied in detail.[155-159] For the benzimidazolones, the molecular ion is generally intense, and the principal fragmentation pathway, which involves consecutive losses of CO and 2HCN,[156] is easily swamped by fragmentation of substituents.[155] Attempts have been made to correlate the mass spectrometric and thermolytic behavior of benzimidazolones,[155,157] and the results collected in Scheme 3.7a reveal an intriguing similarity between the two degradations. The principal fragmentation of benzimidazolethiones, that is, loss of S from the molecular ion,[159] is also paralleled in the thermolytic degradation, and the same is true of the 1-phenyl analog[157] (Scheme 3.7b).

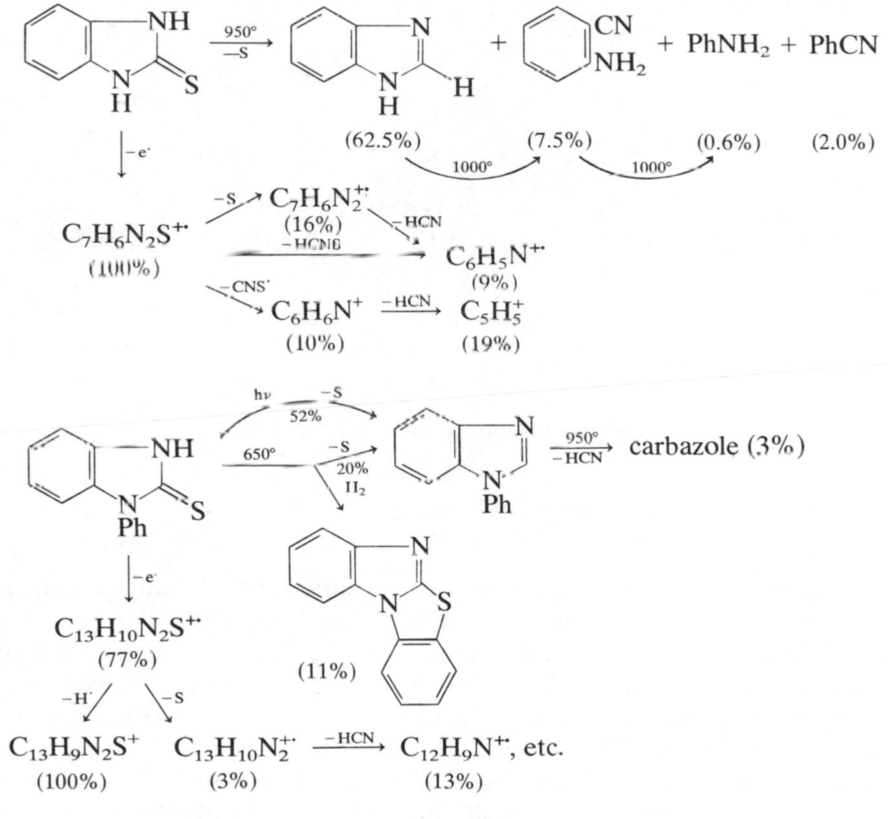

Scheme 3.7(b)

Acidity and Basicity

Benzimidazolones and benzimidazolethiones are very weakly acidic and also very weakly basic. For the protonation of benzimidazolone, pK_a values of 2.4 and -1.7 have been obtained,[150,160,161] and for deprotonation the estimated pK_a values lie between 11.1 and 12.7.[160–162] For ben-zimidazolethione, the corresponding values are 2.60 (for protonation)[163] and 9.8 to 10.4 (for deprotonation).[161,163,164]

Conformations

1,3-Diacylbenzimidazolones exist preferentially in the *endo,endo* confor-mation,[165] and *N*-acyl substituents in benzimidazolethiones have a similar preference[44] in the absence of bulky substituents (R in structure **3.57**). In these *endo* conformers there is a minimal dipole-dipole interaction between the *N*-acyl C=O and the C=O or C=S of the ring system.

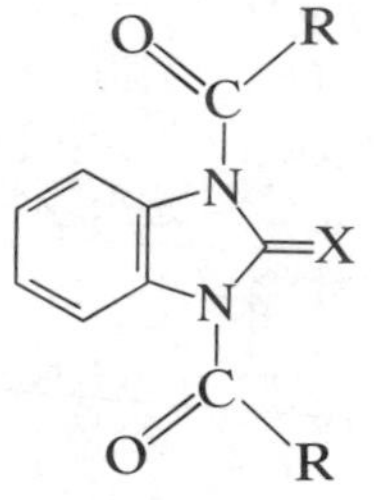

(**3.57**, X = O or S)

3.2.4. Reactions of Benzimidazolones

Alkylation and Acylation

Such reactions occur, almost without exception, at the nitrogen atoms rather than the oxygen atom, and lead to 1-mono- and 1,3-dialkyl (or -acyl) benzimidazolones. In the earlier volume in this series,[63] methylation of benzimidazolone was shown to conform to this pattern, and the same is true for hydroxymethylation (using formaldehyde),[166,167] aminomethylation by the Mannich reaction,[168] hydroxyethylation (using oxirane),[169,170] vinylation (using acetylene),[171–173] glucosidation,[174] trimethylsilylation,[174] and the for-mation of macrocycles like **3.58** and **3.59**.[175] *O*-Alkylation of 1,3-dialkylbenzimidazolones may be achieved only by using strong alkylating agents like triethyloxonium fluoroborate.[37] Alkylation of 1-isopropenyl-benzimidazolones (e.g., **3.43**) occurs at N-3 rather than at the enamine group.[176]

In general, acylation also occurs at the nitrogen atoms, no matter whether the acylating agent is a derivative of a carboxylic (e.g., acetic),[146,147] car-bamic,[177] or sulfonic acid.[146,178] The only exception to the general rule is provided by phosphoryl chloride, which leads eventually to 2-

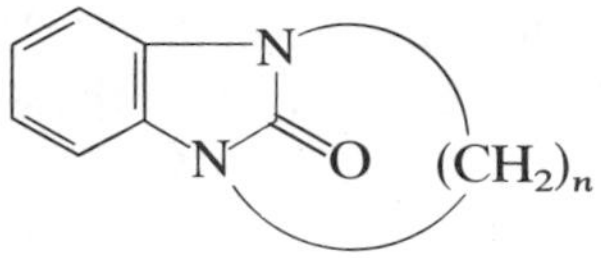

(3.58, n = 10 or 12)

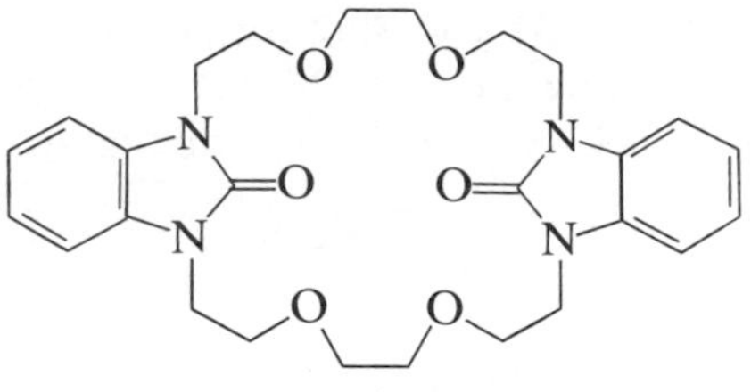

(3.59)

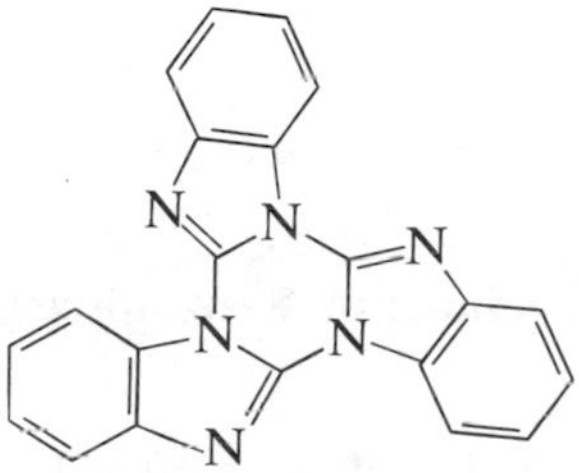

(3.60)

chlorobenzimidazoles,[179] but the products are prone to self-condensation, giving polymers or products such as **3.60**.[179,180]

Electrophilic Substitution in the Carbocyclic Ring

Benzimidazolones undergo the classical electrophilic aromatic substitution reactions with great ease. Nitration gives 5-mono-, 5,6-di-, 4,5,6-tri-, and 4,5,6,7-tetranitro derivatives, according to the severity of the reaction conditions, and each of these may be obtained in good yield if the conditions are carefully controlled.[181–183] Ceric ammonium nitrate is also recorded as effecting the nitration of a benzimidazolone.[184] Chlorination leads to 4,5,6-tri- and 4,5,6,7-tetrachlorobenzimidazolones,[177,185] although N-chlorination is a side reaction. Friedel-Crafts acylation takes place at the 5-position; acetyl[186,187] and aroyl[188] chlorides, and cyclic anhydrides such as maleic,[189] succinic and glutaric,[190] or phthalic,[191] have all beeen used successfully. Carboxyalkylation using γ-lactones[192,193] and Vilsmeier formylation[194,195] also occur at the 5- (or 6-) position (the latter reaction being confined to 1,3-disubstituted benzimidazolones).

4- and 5-Aminobenzimidazolones undergo azo coupling (at the 7- and 6-positions, respectively) by reaction with diazonium salts,[196] and azo coupling of 5-hydroxy-1,3-dimethylbenzimidazolone has also been recorded.[197]

Formation of Quinones

5,6- and 4,7-Dihydroxybenzimidazolones are oxidized by a variety of reagents (e.g., silver oxide and ferric chloride) to the corresponding quinones (**3.61** and **3.62**, respectively,[197–199]) of which the *o*-quinone (**3.61**) is apparently the more stable.[198] These quinones are also obtainable by other oxidative procedures involving, for example, dimethoxy[197,200–205] or amino-hydroxy[197] derivatives. The 4,5-quinone (**3.63**) is, apparently, very unstable.[206]

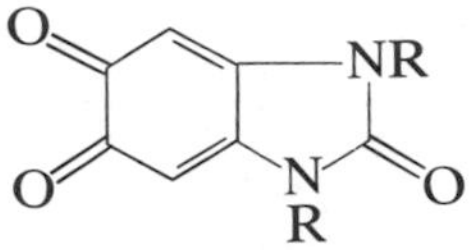

(3.61) 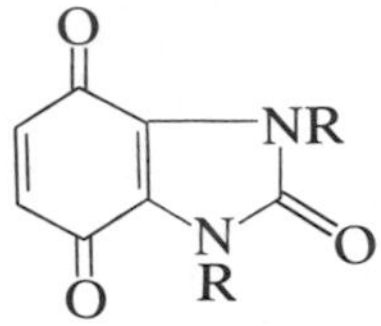   **(3.62)** 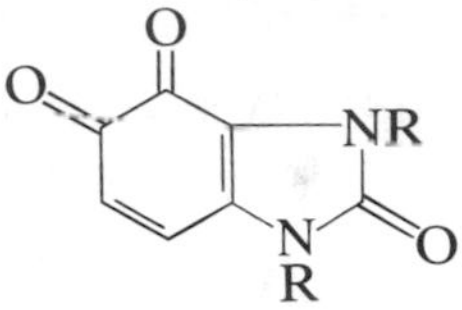   **(3.63)**

3.2.5. Reactions of Benzimidazolethiones

Alkylation and Acylation

Reference has already been made (section 3.2.3) to the fact that benzimidazolethiones have much more dipolar character than benzimidazolones. It is a reflection of this dipolar character that benzimidazolethiones react as genuinely ambident nucleophiles, and (unlike benzimidazolones) may react with electrophiles at both nitrogen and sulfur. The "soft" electrophiles react with the "softer" nucleophilic center, namely, sulfur, and the "harder" electrophiles react at the "harder" nucleophilic center, namely, nitrogen. Thus alkyl halides react with benzimidazolethiones giving 2-*S*-alkylbenzimidazoles: (cf. also Chapter 1, Table 1.65).[63] Dialkylation thus gives *N,S*-dialkyl compounds (e.g., **3.64**) and trialkylation gives *N,N',S*-trialkyl quaternary salts (e.g., **3.65**).[70,207,208] Dealkylation of *S*-alkyl derivatives is, however, a complicating factor in such alkylations, as shown in Scheme 3.8.

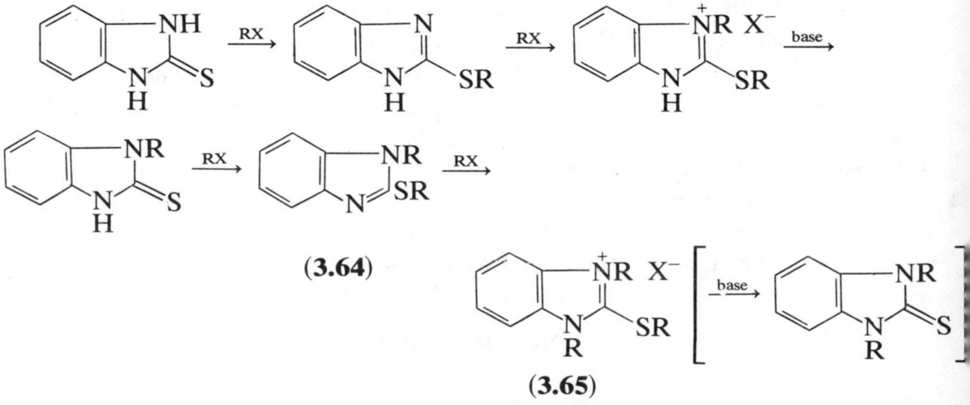

(3.64)

(3.65)

Scheme 3.8

Reactions with reactive aryl halides,[135,209] sulfenyl halides,[210] and *N*-chloroamines[211,212] all occur at the sulfur atom, as do hydroxyethylation[213] and glucosidation.[214,215] Vinylation[216] and trimethylsilylation[136,214] are possibly similar, although in these cases it is the *N,S*-disubstituted derivative which is isolated.

On the other hand, electrophiles, in which the electrophilic center is the carbon of a carbonyl or similar group, react with benzimidazolethiones only at the nitrogen atoms. Acetylation,[217,218] benzoylation,[217] reaction with isocyanates,[217,219] hydroxymethylation,[166,167] the Mannich reaction,[220] and methylthiomethylation (eq. 50)[221] all conform to this pattern.

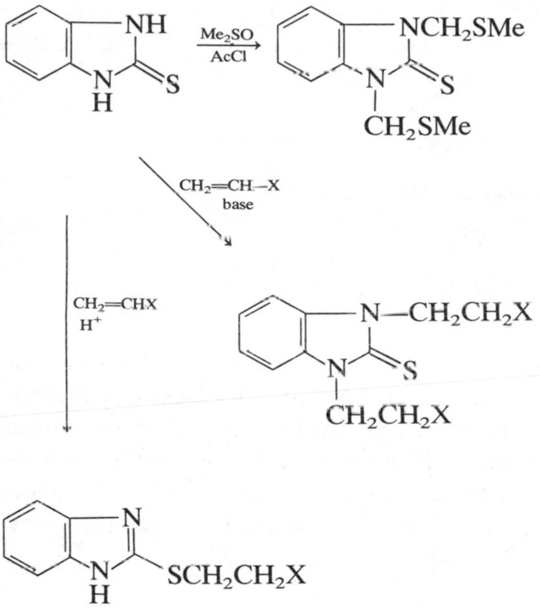

$$(50)$$

$$(51)$$

(X = CN, CHO, COMe)

$$(52)$$

(X = CN, CHO, COMe)

The outcome of reactions between benzimidazolethiones and αβ-unsaturated carbonyl and cyano compounds is much less clearcut. Either *S*- or *N*-substitution may occur in these reactions; although no systematic study has been reported, it appears that basic reaction media favor *N*-substitution[222,223] and acidic media favor *S*-substitution[223,224] (eqs. 51 and 52).

The reactions of benzimidazolethiones with bifunctional electrophiles give rise to a wide variety of tricyclic benzimidazole derivatives. These are discussed in more detail in Chapters 6 and 7; for the purposes of this chapter, the examples in Scheme 3.9 may be taken as representative.

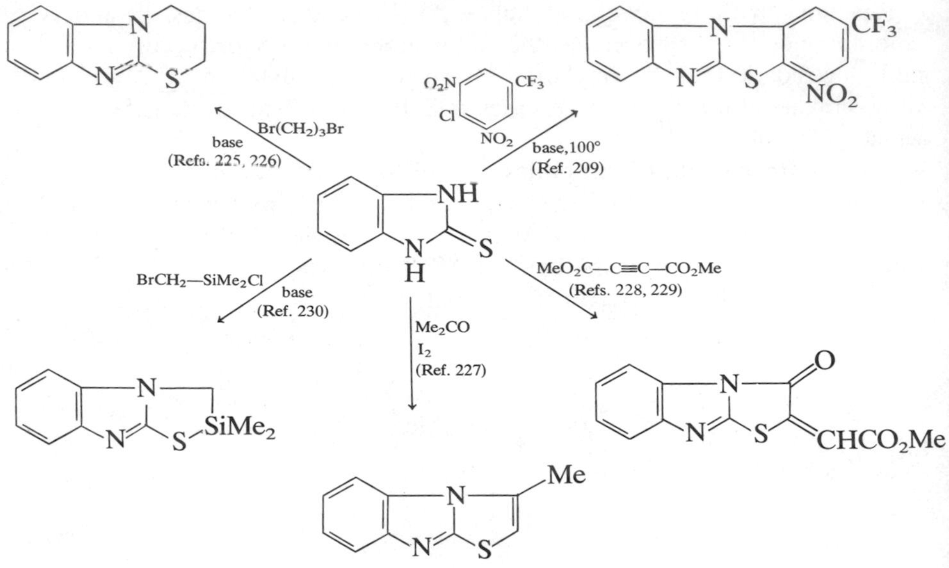

Scheme 3.9

Typical "Thiol" Reactions

OXIDATION. Reaction of iodine with benzimidazolethiones, either direct or via the mercury derivative (**3.66**), leads to di-(2-benzimidazolyl) disulfides.[231–233] Dilute aqueous hydrogen peroxide[232] and electrochemical oxidation[234] produce the same result. More vigorous oxidation with hydrogen peroxide[232] or with potassium permanganate[99,231,235,236] yields benzimidazole-2-sulfonic acids. Benzimidazole-2-sulfonyl chloride is formed by reaction of benzimidazolethione with chlorine.[232]

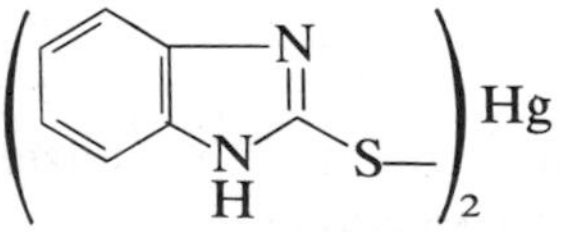

(3.66)

DESULFURIZATION. This may be achieved using Raney nickel or nickel boride,[237] and also, surprisingly, using selenium dioxide.[238] This last reaction is assumed to proceed by oxidation to the sulfonic acid followed by desulfonation. Photochemical desulfurization has also been reported in the case of 1,3-dimethylbenzimidazolethione (eq. 53).[239]

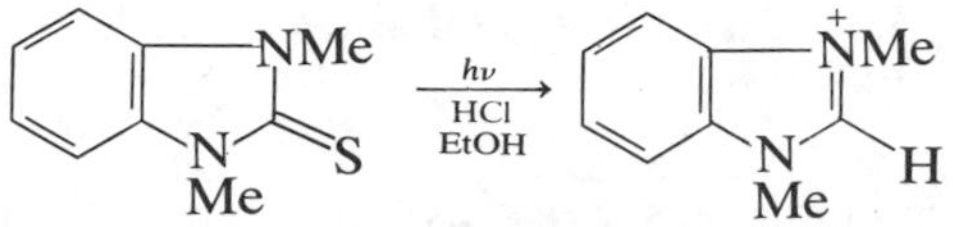

$$\text{(equation 53)} \qquad Cl^- + S \qquad (53)$$

Other Reactions

Nitration of benzimidazolethione takes place at the 5-position.[240] Benzimidazolethione is also a bidentate ligand and has been investigated as a reagent for the quantitative determination of platinum group metals.[241]

3.3. DIHYDROBENZIMIDAZOLES WITH OTHER DOUBLE-BONDED SUBSTITUENTS AT C–2

3.3.1. Benzimidazoleselenones and Benzimidazoletellurones

These are analogs of benzimidazolethiones and are prepared by analogous methods. Members of both series are prepared (the tellurone, albeit, in very low yield) by reaction of 1,3-dimethylbenzimidazolines with elemental selenium or tellurium at high temperatures,[5,45] and the selenones may also be prepared from 1,2-diaminobenzenes and carbon diselenide (eqs. 54 and 55).[242,243] Little is yet known of the chemistry of these compounds, except that the selenones undergo methylation at the selenium atom,[243] and that 1,3-dimethylbenzimidazoleselenone reacts with N,N-dichloroamines to produce only the 2-SeCl derivative (**3.67**) (eq. 56).[212]

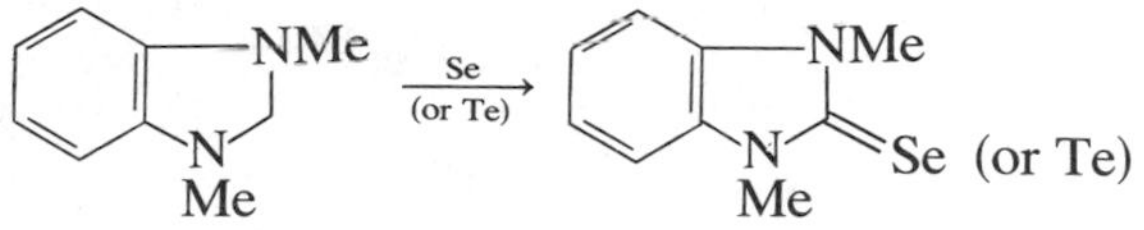

$$\text{(equation 54)} \qquad (54)$$

$$\text{(equation 55)} \qquad (55)$$

(R = H or Me)

$$\text{(equation 56)} \qquad (56)$$

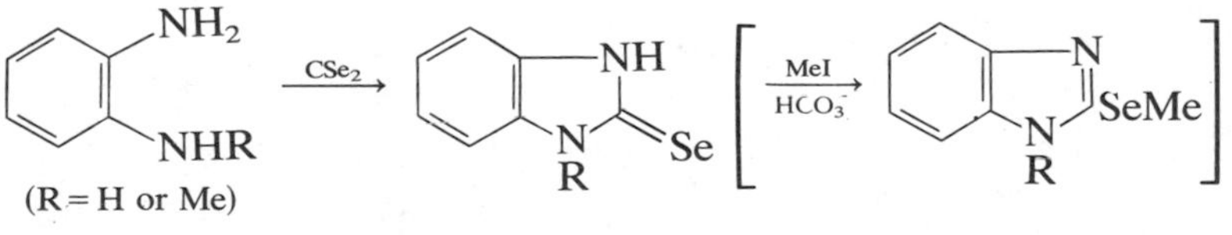

(**3.67**)

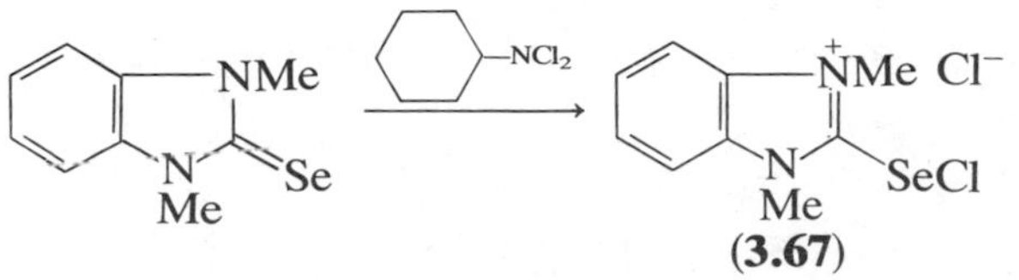

3.3.2. Benzimidazolimines

Although 2-aminobenzimidazoles exist as such and not as the imino tautomers, 1,3-disubstituted benzimidazolimines, for example, **3.68**, are now well known.

Preparation

1-Alkyl-2-aminobenzimidazoles undergo further alkylation on the other hetero-atom in preference to the exocyclic nitrogen,[102,244–252] and basification of the 2-aminobenzimidazolium salts so obtained gives 1,3-dialkylbenzimidazolimines (eq. 57). The use of an excess of alkylating agent leads to 2-alkylamino- and ultimately to 2-dialkylamino-benzimidazolium salts,[100,253,254] and these, on basification, give the 2-alkylimino derivatives (**3.69**) and benzimidazolones, respectively (eq. 58). Acylation of 1-alkyl-2-aminobenzimidazoles also occurs on the other hetero-atom, but the products (**3.70**) are liable to undergo rearrangement on heating, to give 2-acylamino-1-alkylbenzimidazoles (eq. 59).[255] Rearrangement of this kind may be responsible for the formation of products **3.71** and **3.72** in eqs. 60 and 61, respectively.[101,256]

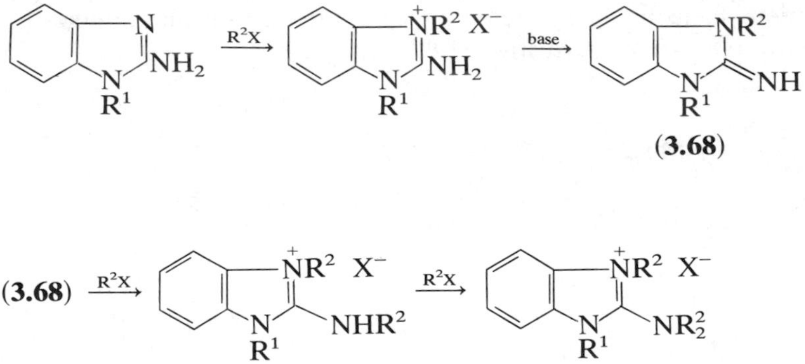

(57)

(**3.68**)

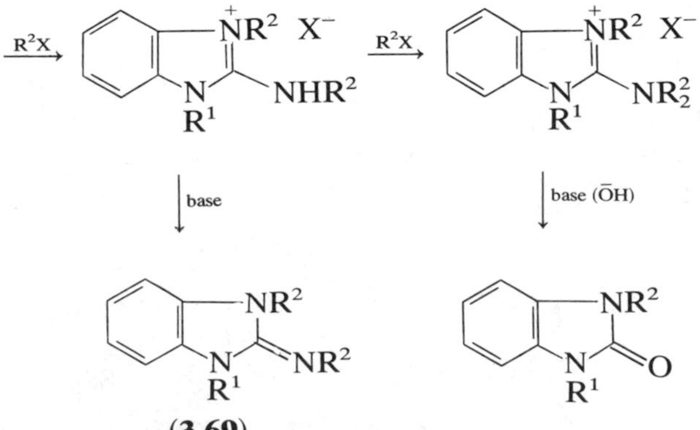

(58)

(**3.69**)

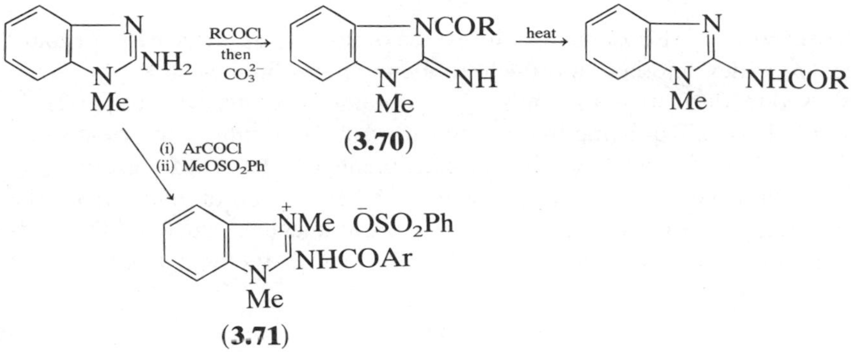

(59)

(3.70)

(i) ArCOCl
(ii) MeOSO$_2$Ph

(60)

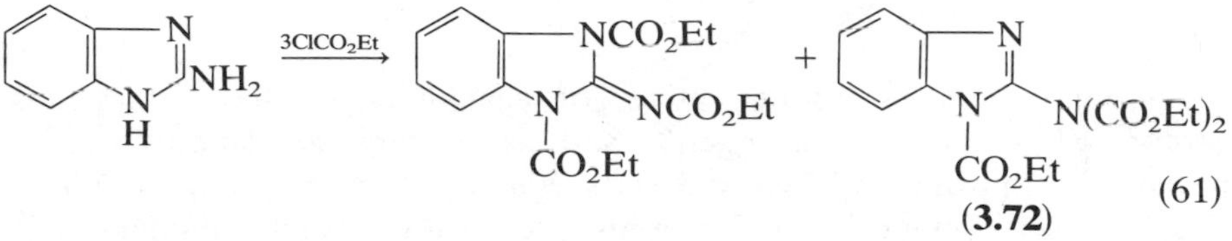

(3.71)

(61)

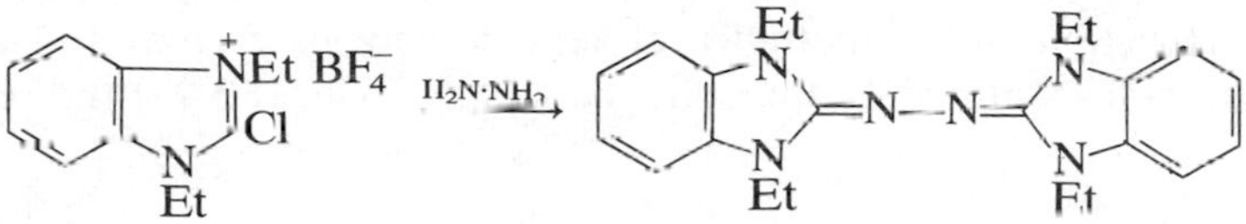

(3.72)

(62)

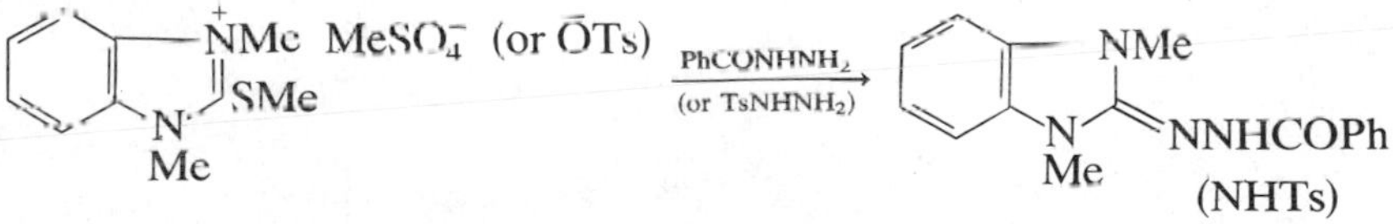

(63)

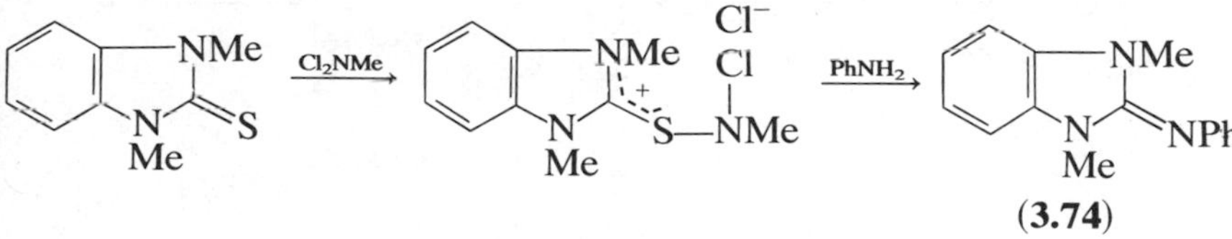

(NHTs)

(64)

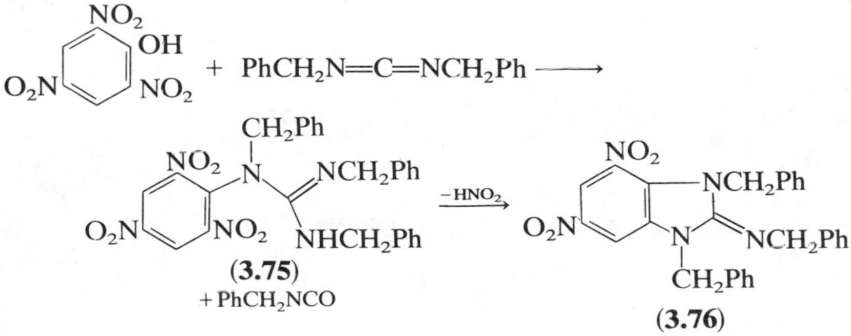

(3.74)

(65)

(3.75)
+ PhCH$_2$NCO

(3.76)

The reaction of 1,3-dialkyl-2-chlorobenzimidazolium salts with primary amines provides another approach to benzimidazolimines, and the azine (**3.73**) is one derivative that has been obtained in this way (eq. 62).[257] Similarly, 1,3-dialkyl-2-(methylthio)benzimidazolium salts may yield benzimidazolimines by nucleophilic displacement of the –SMe group (eq. 63),[258,259] and the phenylimino derivative (**3.74**) may be obtained from the corresponding thione as in eq. 64.[211] Intramolecular nucleophilic substitution in the picrylguanidine derivative (**3.75**) gives the benzimidazolimine (**3.76**) (eq. 65).[260]

Reactions

Benzimidazolimines, as derivatives of guanidine, are basic: they are protonated on the imino nitrogen[261] and also undergo alkylation,[100,253,262] acylation,[253] nitrosation,[253,263] and diazo coupling[264] at this position. Nitration[246] gives 5-mono- and 5,6-di-nitro derivatives, and sulfonation,[265] chlorosulfonation,[265] and electrophilic bromination[100] all occur at the 5-position. Reaction with dimethyl acetylenedicarboxylate gives 1:2 adducts that have been identified as the spiro-benzimidazolines (**3.77**) (eq. 66).[266]

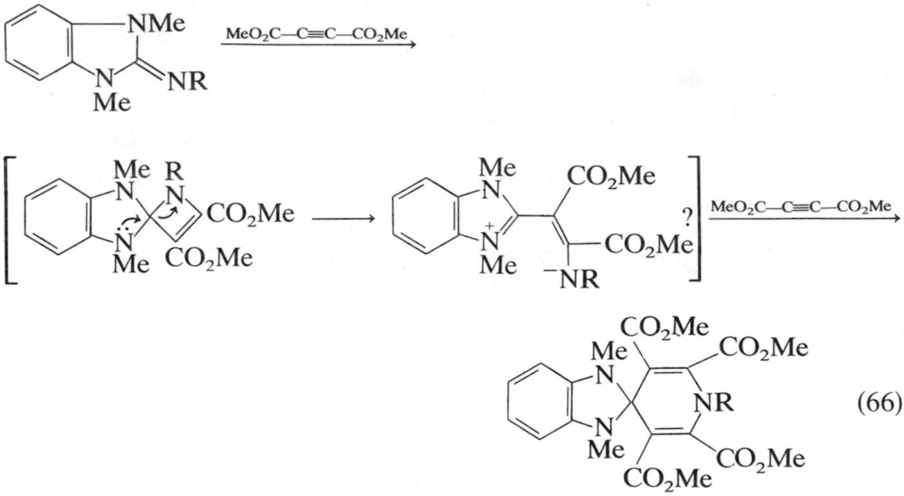

(66)

(**3.77**)

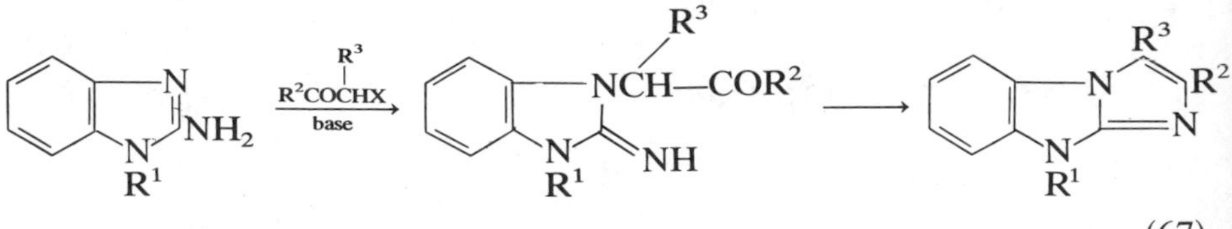

(67)

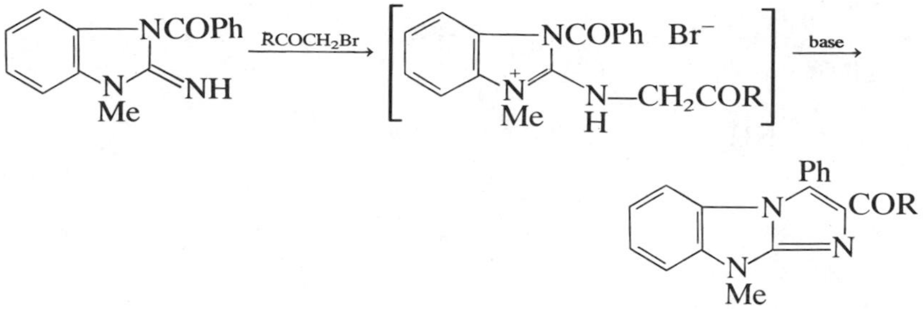

(68)

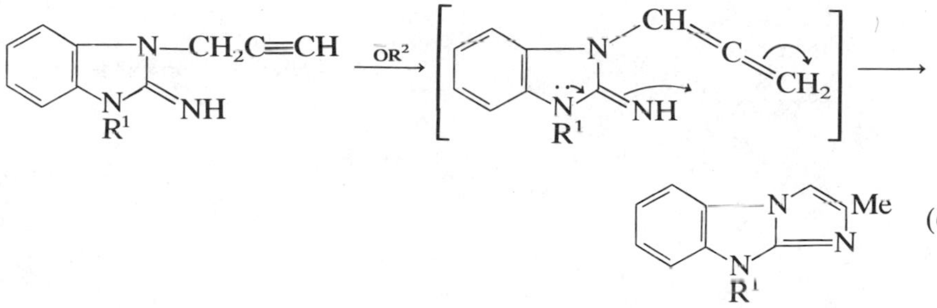

(69)

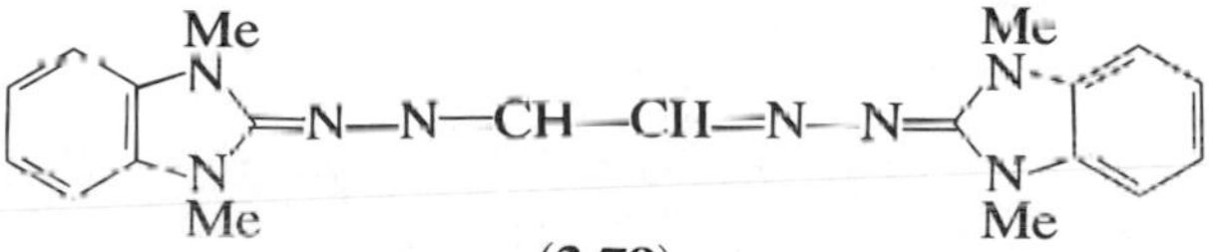

(3.78)

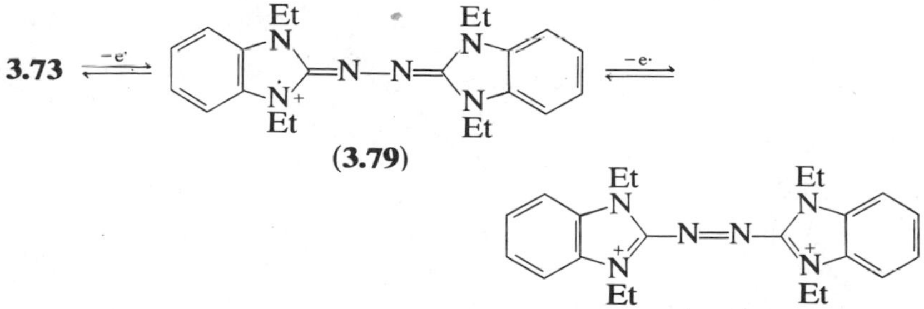

(3.79)

(70)

(3.80)

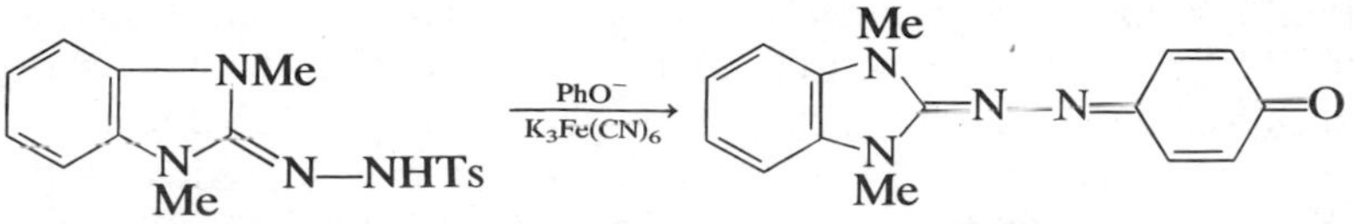

(from equation 63)

(3.81)

(71)

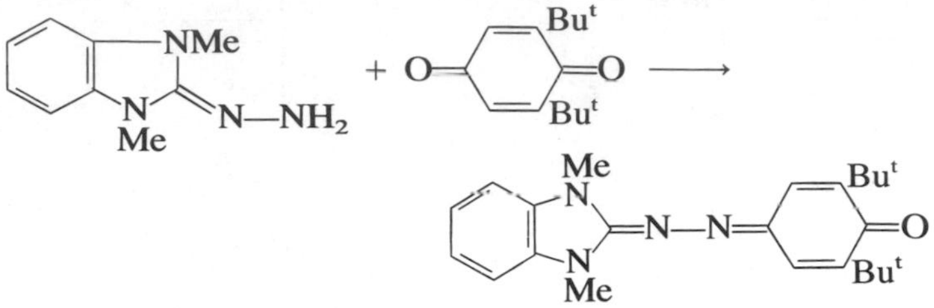

$$(72)$$

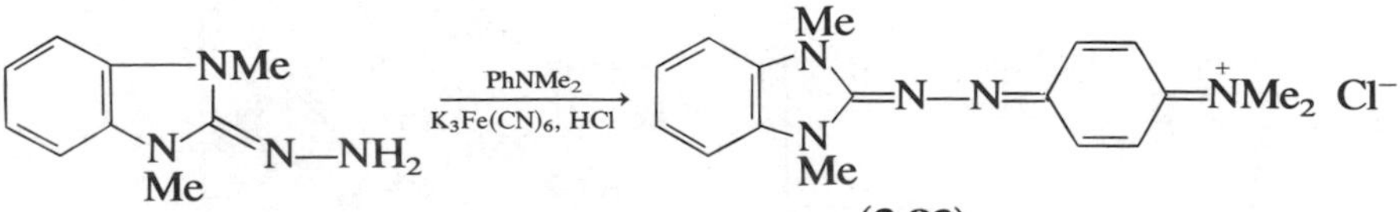

$$(73)$$

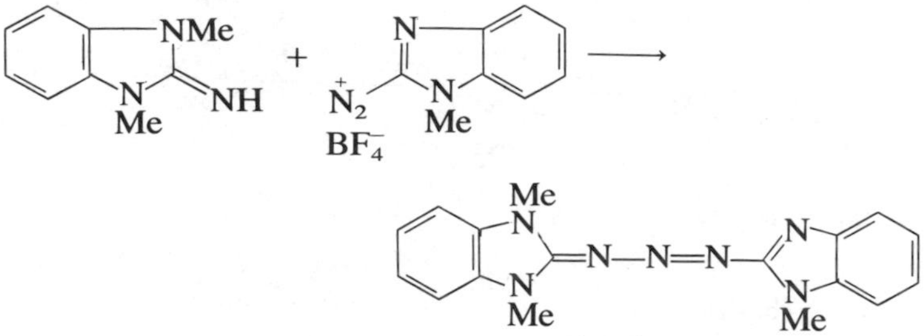

$$(74)$$

$$(75)$$

Cyclization of benzimidazolimines to azolobenzimidazoles is illustrated in eqs. 67–69;[244,248–250,262] these reactions are considered further in Chapter 6.

Azines such as **3.73** and the bis-azine (**3.78**) are the fully reduced forms of two-stage redox systems. [257,267–269] The mono-azines may be oxidized via the radical ions (**3.79**) to the bis-quaternary salts (**3.80**) (eq. 70);[267,268] however, the low solubility of **3.78** precludes a corresponding study of its oxidation.[269] Benzimidazolimine derivatives such as **3.81** to **3.85**, which are obtained by standard methods (eqs. 71–75),[259,264,270–275] are highly colored compounds analogous to cyanine dyes.[276]

3.3.3. 2-Methylenebenzimidazolines

Preparation

The simplest representatives of this group, for example, **3.86**, are obtained by the action of anhydrous base on 1,3-disubstituted 2-methyl (or 2-alkylmethylene)benzimidazolium salts (eq. 76).[38,278,279] Others may be obtained by condensation of 2-methylbenzimidazolium salts with electrophiles such as diazonium salts (eq. 77),[280,281] by the condensation of 2-formylbenzimidazolium salts with hydrazine derivatives (eq. 78),[281,282] and from 1,2-diaminobenzenes by reactions 79 to 82.[283–286] Much of the preparative work in this area has been directed towards the merocyanine dyestuffs; the reader is referred to another volume in this series[277] for details of these and related compounds.

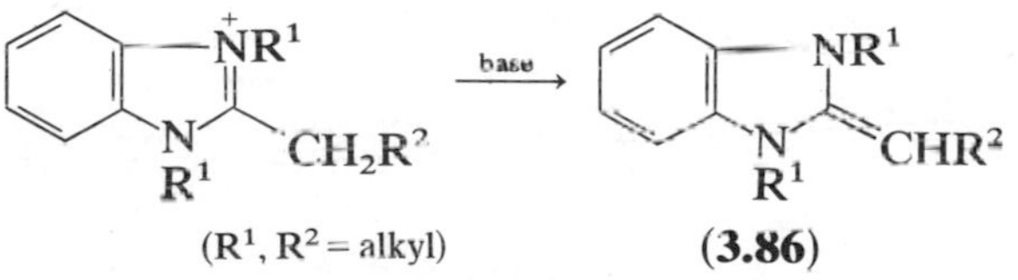

$$
\begin{array}{ccc}
& \xrightarrow{\text{base}} & \\
(\text{R}^1, \text{R}^2 = \text{alkyl}) & & (\textbf{3.86})
\end{array}
\tag{76}
$$

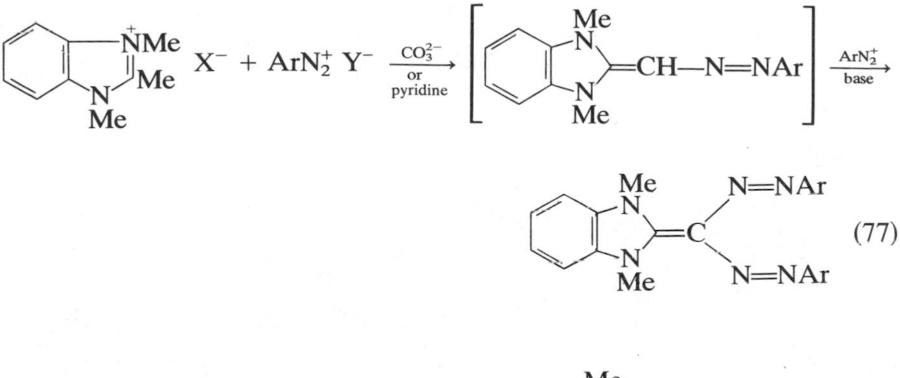

$$
\xrightarrow[\substack{\text{or} \\ \text{pyridine}}]{\text{CO}_3^{2-}} \quad \xrightarrow[\text{base}]{\text{ArN}_2^+}
\tag{77}
$$

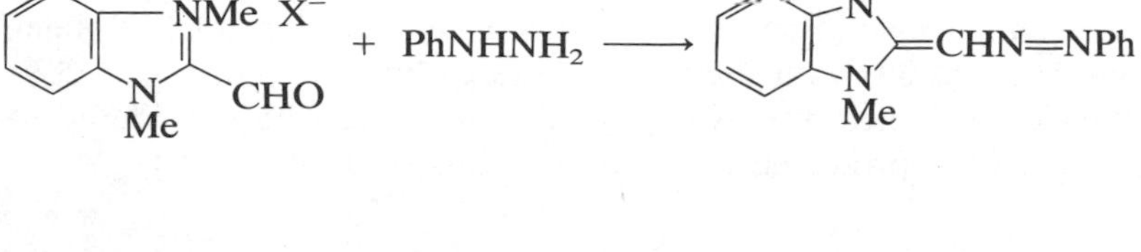

$$
+ \text{PhNHNH}_2 \longrightarrow
\tag{78}
$$

$$(R = CN, CO_2Me) \quad (79)$$

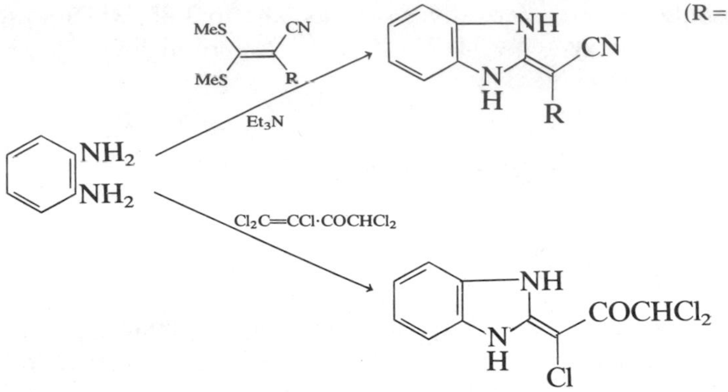

$$(80)$$

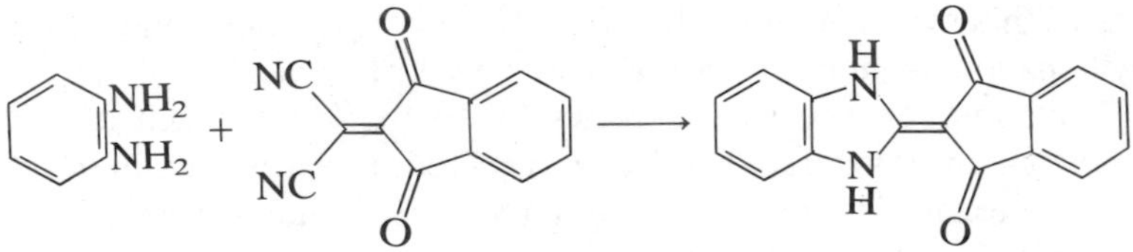

$$(81)$$

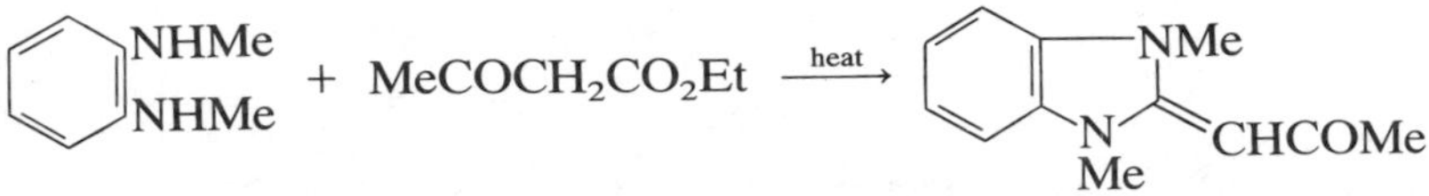

$$(82)$$

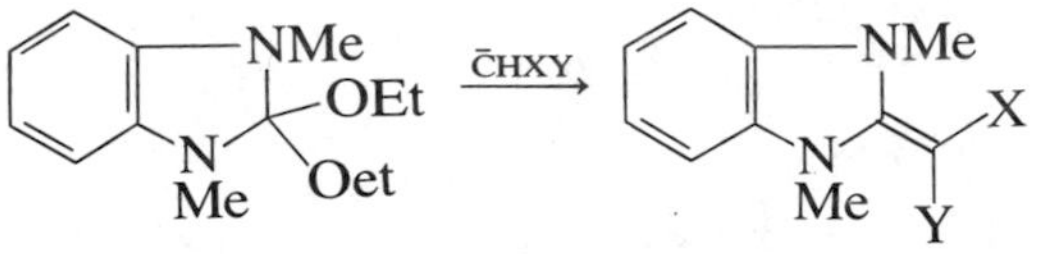

$$XY = (CH{=}CH{-})_2; \\ X = CN,\ Y{=}CO_2Me; \quad (83) \\ X = H,\ Y = NO_2,\ COPh$$

(3.88)   (3.87)

(from eq. 12)

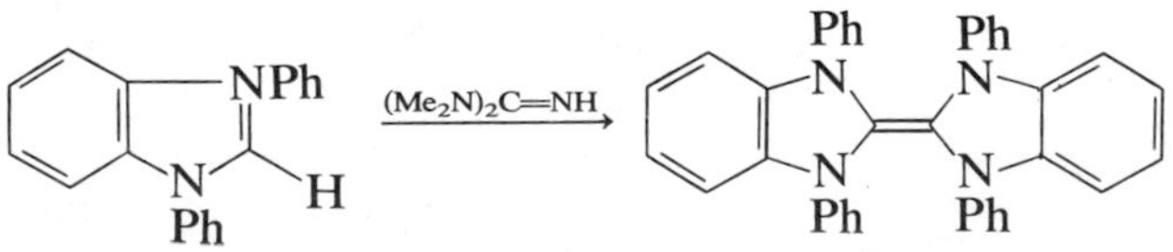

$$(84)$$

(3.89)

Compounds such as **3.87** are formally condensation products of benzimidazolones, and have also been obtained by reaction of the benzimidazolone ketal (**3.88**) with the appropriate carbanions (eq. 83).[37] 2,2′-Bis-benzimidazolinylidenes, for example, **3.89**, must also be considered in this section: these are produced by a carbene dimerization (eq. 84).[38]

Reactions

2-Methylenebenzimidazolines are typical enamines and undergo reactions with electrophiles at the exocyclic carbon atom.[38,279,282,287–289] In the presence of a nucleophilic counterion, the initially formed benzimidazolium cation may then undergo addition at the 2-position (cf. section 3.1.1); the net effect is addition across the exocyclic double bond, and the product is a benzimidazoline. These reactions are exemplified in Scheme 3.10.

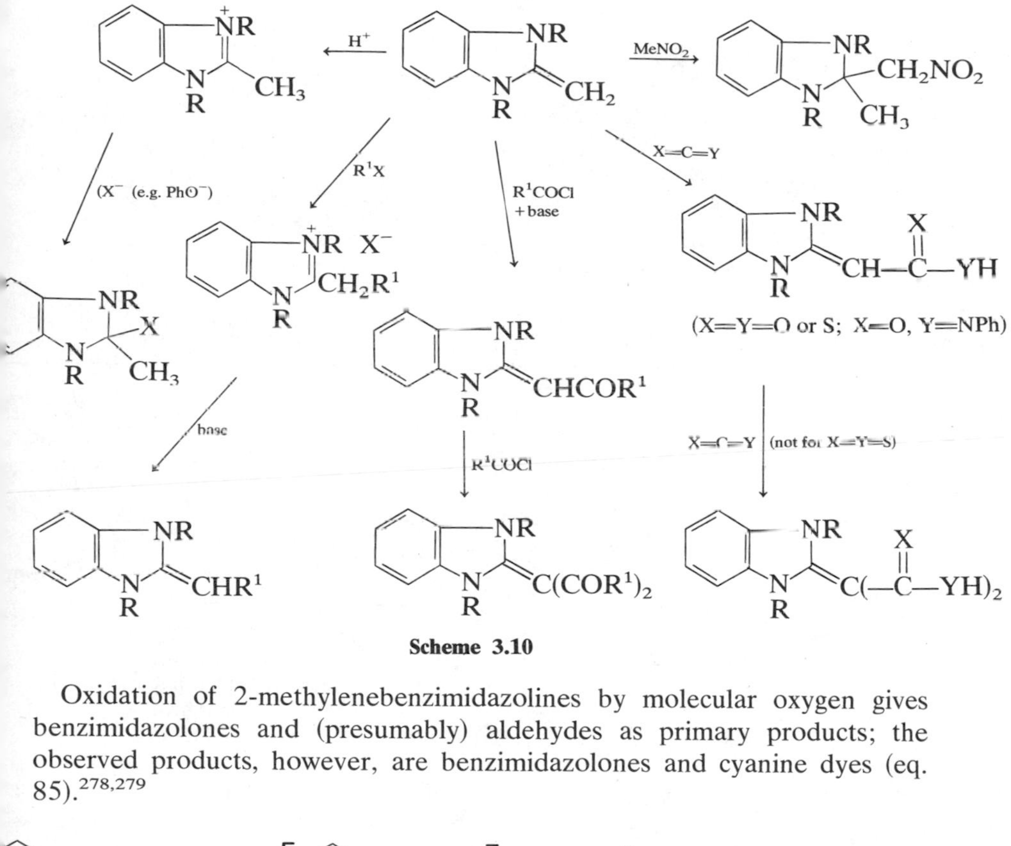

Scheme 3.10

Oxidation of 2-methylenebenzimidazolines by molecular oxygen gives benzimidazolones and (presumably) aldehydes as primary products; the observed products, however, are benzimidazolones and cyanine dyes (eq. 85).[278,279]

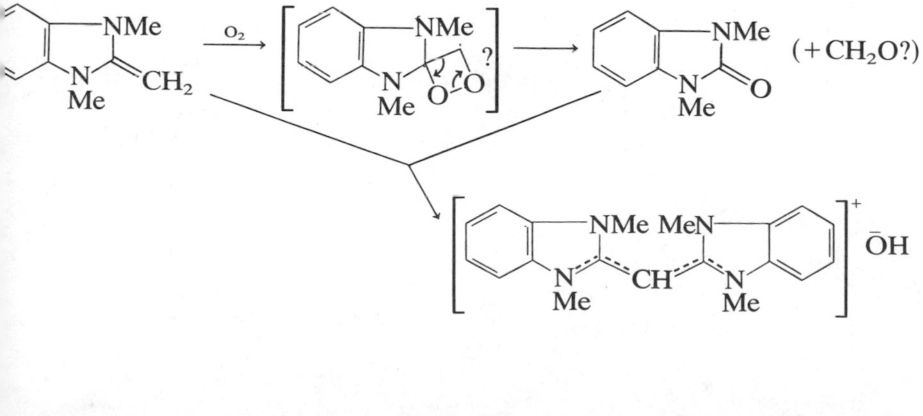

3.4. SYSTEMATIC SURVEY

The following tables, like those in Chapter 2, are not intended to be fully comprehensive, but to refer only to the simpler members of each class of compound. The references cited are generally those that contain preparative information.

Tables 3.1 to 3.3 incorporate material from the corresponding tables (V.K, V.Cl/2) in a previous volume of this series.[63]

TABLE 3.1. 2,3-DIHYDROBENZIMIDAZOLES (BENZIMIDAZOLINES)

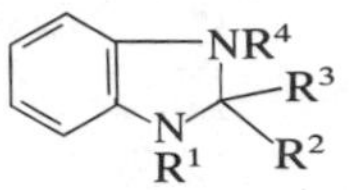

a. Both nitrogens unsubstituted ($R^1 = R^4 = H$) Substituents			Method of preparation[a] (% yield)	M.p. or b.p. (°)	Ref.
R^2	R^3	Other			
H	H	—	A1(−)	liquid (not purified)	1
CF_3	CF_3	—	Eq. 8 (42)	b.p. 113–114/ 28 mm	32
$(CH_2)_5$		—	B(77, 96), eq. 9 (66)	m.p. 130, 131–3, 138	24, 25
$(CH_2)_5$	5-NO_2		B(75)	m.p. 163–4	25
$(CH_2)_6$		—	B(23)	liquid (not purified)	25
$(CH_2)_6$	5-NO_2		B(−)	liquid (not purified)	25
p-MeC_6H_4	H	5-NO_2	B(−)	m.p. 108–10	18
2,6-$Cl_2C_6H_3$	H	—	B(−)	m.p. 109–11	19
5-O_2N-furyl	H	—	B(−)	m.p. 118–9	20
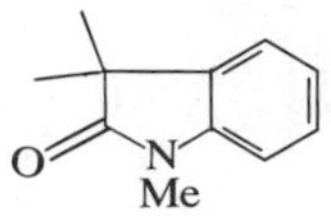			B(70)	m.p. 167–70	27

One nitrogen substituted (R⁴ = H) Substituents				Method of preparation[a] (% yield)	M.p. or b.p. (°)	Ref.
R¹	R²	R³	Other			
Ac	H	H	—	Hydrolysis of 1,3-diacetyl compd. (85)	m.p. 101–3, 105	7, 43
Me	o-O₂NC₆H₄	H	—	B(−)	m.p. 144	8
Me	m-O₂NC₆H₄	H	—	B(−)	m.p. 122	8
PhCH₂	Ph	H	—	B(−)	Not obtained pure	21
Ts	Ph	H	6-OMe	B(80)	m.p. 119–20	290

Both nitrogens substituted Substituents				Method of preparation[a] (% yield)	M.p. or b.p. (°)	Ref.
R²	R³	R⁴	Other			
H	H	Me	—	A1(70–93), A2(81, 90), A4(70)	b.p. 86/1.5 mm, 105/3 mm	2 5
H	H	Ph	—	B(100)	m.p. 178	12
H	H	Ac	—	A3(72, 86)	m.p. 166–8, 168–70	6, 7, 43
H	H	Me	4-NO₂	A2(52), B(75)	m.p. 69–70, 75–6	46, 54
H	H	Me	5-NO₂	B(87)	m.p. 103	11
Me	H	Me	—	A1(35–80), A2(90), B(95)	b.p. 109–10/3–4 mm, m.p. 25–6	2–4, 8
Et	H	Me	—	A1(70–80)	b.p. not quoted	4
Ph	H	Me	—	A2(90), B(87)	m.p. 96, 93–4	5, 8, 291
Aryl	H	Me	—	B(44–95)	—	8, 10
Me	H	Me	5-NO₂	B(87)	m.p. 88–9	11
Me	H	Et	—	A2(−)	b.p. 74.5/ 0.5 mm	2
Ph	H	Ph	—	B(64)	m.p. 115	12
CH₂ Ph	H	Ph	—	B(92)	m.p. 159	13
CH₂ Ar²	H	Ar³CH₂	various	B	—	13, 14, 23, 59–62
SO₂ Ph	H	PhSO₂	—	Eq. 5(91)	m.p. 167–70	17
Aryl	H	b	—	B(>70)	—	9
Ph	Ph	Me	—	Eq. 13(85)	m.p. 231	279
OEt	H	Bz	—	C(10)	m.p. 139	36
OEt	H	Ts	—	B[c](62)	m.p. 133–4	36
OEt	OEt	Me	—	C(76)	m.p. 68	37
piperidino	H	Ph	—	C(52)	m.p. 126	38

[a] = reduction of the corresponding benzimidazole or benzimidazolium salt; 1 = by LiAlH₄, 2 = by NaBH₄ KBH₄, 3 = by catalytic hydrogenation, 4 = reduction of the benzimidazolone by LiAlH₄. B = reaction of appropriate diaminobenzene and aldehyde or ketone. C = nucleophilic attack at the 2-position of the ropriate benzimidazolium salt. [b] R¹R⁴ = (CH₂)ₙ.
thyl orthoformate in place of a carbonyl compound.

TABLE 3.2. BENZIMIDAZOLONES

$$\begin{array}{c}\overset{4}{\underset{6}{5}}\overset{}{\boxed{}}\overset{-NR^2}{\underset{N}{\diagdown}}{=}O\\ R^1\end{array}$$

a. Both nitrogens unsubstituted ($R^1 = R^2 = H$)

Substituents	Method of preparation[a] (% yield)	M.p. (°)	Ref.
—	A1(−), A2(−), A3(57), A4(65–72), A5(15–80), A6(87) B1(85), B2(75)	307–8, 312, 315	63, 73, 292, 2 67, 64–66, € 82, 294, 295
4-Me	A1(>75)	297–300, 302–3	75, 296
5-Me	Al(−)	299–300	75
4,6-Me$_2$	A1(>75)	337	296
5,6-Me$_2$	A2(−), A6(67)	384–6	69, 296
4,5,6,7-Me$_4$	A2(−)	313–4	296
4-Et	A1(>75)	261–2	296
5-Et	A1(>75)	264–5	296
4-Pr^i	C3(−)	232–3	296
5-Pr^i	A1(>75)	270–2	296
Other 5-alkyl	A1 or A2 or reduction of 5-acyl compd.	—	296
4-PhCH$_2$	Eq. 34(4)	206–7	89
5-PhCH$_2$	A2(64)	254–5	89
4-CO$_2$H	A1(−)	>300 Me ester, 260–3)	297
5-CO$_2$H	A1(76)	>300 (Me ester, 312–3)	297
5-Ac	A1(40), C3(63)	294–5	186
5-Bz	C3(47)	303–4	188
Other 5-acyl	C3 (various)	—	186, 188–192
5-F	A1(>75)	303	296
4,5,6,7-F$_4$	A1(87)	302.5–306	298
4-Cl	A1(>75)	335–6	296
5-Cl	A6(89)	324–6	69
4,6-Cl$_2$	A2(−)	>340	296
5,6-Cl$_2$	A2(−), A6(92)	345, >400	69, 296
4,5,6-Cl$_3$	A2(−), C3(88)	342, 333–5	296, 185
4,5,6,7-Cl$_4$	A2(79), C3(50)	314, 326–7, >360	177, 185, 300
5-Br	A1(>75)	336–7	296
5-I	A1(−)	250	299
4,6-I$_2$	A1(−)	230	299
4-NO$_2$	B1(80)	349–50	85
5-NO$_2$	A2(−), B1(82), C3(75), C6(−)	305–6, 308	85, 181–183, 112
5,6-(NO$_2$)$_2$	C3(40–94)	321	181–183
4,5,6-(NO$_2$)$_3$	C3(55–96)	258, 313–4	181–183
4,5,6,7-(NO$_2$)$_4$	C3(65–92)	311(d), 317	182–183
5-NH$_2$	B2(−)	+HCl, >340	296
5-OH	A1(>75)	307–9	296
5,6-(OH)$_2$	Reduction of quinone (−), hydrolysis of 5,6-(OMe)$_2$ (80)	388–93(d)	199, 200

TABLE 3.2 (*Continued*)

a. Both nitrogens unsubstituted ($R^1 = R^2 = H$)

Substituents	Method of preparation[a] (% yield)	M.p. (°)	Ref.
4,7-(OH)$_2$	Hydrolysis of 4,7-(OMe)$_2$ (87)	decomp.	198
5-OMe	A1(>75), A6(75)	256–7	69, 296
4,5-(OMe)$_2$	A2(65)	246	202
5,6-(OMe)$_2$	A1(47), A2(−)	246, 268, 251–4	197, 199, 296
4,7-(OMe)$_2$	A1(56), A2(61)	250–2	198, 199
4,5,6-(OMe)$_3$	A2(70)	226–6.5	203
4,5,7-(OMe)$_3$	A2(57)	183.5–184.5	204
5,6-dioxo	Oxidation of diol (>70)	>300(d)	199, 200
4,7-dioxo	Oxidation of diol (69–85)	>240(d)	198, 199

b. One nitrogen substituted ($R^2 = H$)

Substituents		Method of preparation[a] (% yield)	M.p. (°)	Ref.
R^1	Other			
Me	—	A2(80), A7(52)	188–9	301, 70
		B1(80), C3(>60), C4(90), C6(84)	192, 186	82, 90, 93, 94, 99, 106
Et	—	A1(>75), C6(84)	117–8, 118–20	296, 176
Pr^n	—	C6(74)	104	176
Pr^i	—	C4(23), reduction of (**3.43**)	128–9	91, 108
Other alkyl	—	C6 (various)	—	176
CH$_2$=CMe-	—	Eq. 40(71)	121–2	106 (*cf.* 302)
Other alkenyl	various	Eq. 40(−)	—	107–114
Ph	—	B2(50), C3(20) Eqs. 32(60), 48(−)	204	70, 90, 77, 124
PhCH$_2$	—	A2(96), C3(90)	194–6, 201–2	301, 90
Ac	—	C2(56), B2(60)	205–7	141 (*cf.* 304), 86, 190
Bz	—	B1(75, 84), C2(−)	198–200, 205	141 (*cf.* 304), 82
MeSO$_2$	—	C6(−)	183–4	146
PhSO$_2$		C2(−), C6(−)	234–5, 226–9	146, 178
Ts	—	A1(−), C2(−)	216–8, 211–5	178, 303
Me	5-Me	A1(−)	198–200	296, 305
Me	6-Me	A1(−)	210–2	305
Me	7-Me	A2(−)	233–5(d)	306
Me	5-Cl	A1(−)	225–7	305
Me	6-Cl	A1(−)	224–6	305
Me	7-Cl	A2(−)	209–11	306

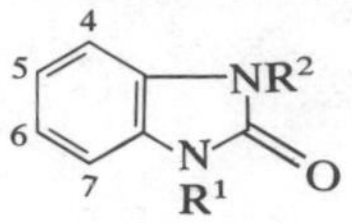

TABLE 3.2 *(Continued)*

b. One nitrogen substituted ($R^2 = H$)

Substituents		Method of preparation[a] (% yield)	M.p. (°)	Ref.
Me	5-NO$_2$	A1(−)	300–1	305
Me	6-NO$_2$	A1(−)	271–2	305
Me	7-NO$_2$	A2(−)	235–6	306
Me	5-OMe	A1(−), C4(85)	199–200, 290–1(?)	305, 90
Me	6-OMe	A1(−)	169–70	305
Me	4,5-(OMe)$_2$	A2(59)	174–5	307
Me	5,6-(OMe)$_2$	A2(75)	194–6	308
Et	various	C6(10–85)	—	176
PhCH$_2$	various (5-sub-stituents)	Eq. 33(<50)	—	87
Ts	5-Me	A1(78)	263–5	303, 309
Ts	5-Cl	A1(60)	252-3	303, 309

c. Both nitrogens substituted

Substituents			Method of preparation[a] (% yield, in parentheses)	M.p. (°)	Ref.
R^1	R^2	Other			
Me	Me	—	A2(95), A4(56), A5(−), C1(88), C4(80), C5(55) Eq. 38(90)	107–9, 113	67, 66, 310, 5, 10
Et	Et	—	C1(62), reduction of divinyl (100)	68–69	296, 17 141
CH$_2$=CH	CH$_2$=CH	—	C1(48)	7–8 (b.p. 140–1/ 1 mm)	171, 17
CH$_2$OH	CH$_2$OH		C1(−)	164–5	166
CH$_2$NMe$_2$	CH$_2$NMe$_2$	—	C1(28)	57	168
CH$_2$NEt$_2$	CH$_2$NEt$_2$	—	C1(46)	42	168
CH$_2$Cl	CH$_2$Cl	—	(CH$_2$OH)$_2$-deriv. + SOCl$_2$ (59)	167(d)	168
CH$_2$CH$_2$OH	CH$_2$CH$_2$OH	—	C1(84)	165–6	169, 17
CH$_2$CH$_2$NR$_2$	CH$_2$CH$_2$NR$_2$	—	C1(−)	—	296
Ph	Ph	—	A1, A2 (ca. 10)	109	12
SiMe$_3$	SiMe$_3$	—	A1(95)	b.p. 110/ 0.3 mm	73
Ac	Me	—	C2(−), B1(78)	120–1	296, 82
Ac	Ph	—	C2(−)	137–8	296
Ac	Ac	—	C2(71–92), B1(75)	149–51, 139	141, 14 190, 304
Bz	Bz	—	C2(71)	212–3, 218	304, 31

376

Both nitrogens substituted

Substituents			Method of preparation[a] (% yield, in parentheses)	M.p. (°)	Ref.
vl (various)	Acyl (various)	—	C2	—	141, 147, 190
SO$_2$	MeSO$_2$	—	C2(−)	247–8	146
SO$_2$	PhSO$_2$	—	C2(23)	201–3, 206–8	146, 178
(CH$_2$)$_n$ (n = 5–12)		—	A1(41–68)	—	312
	Me	5-Me	C1(−)	103–5	296
	Me	5,6-Me$_2$	C1(−)	153–4	296
	Me	5-Cl	C1(−)	163–4	296
	Me	5-CHO	C3(53)	151–1.5	194
	Me	5-CO$_2$H	Oxidation of 5-CHO (−) or 5-Me (−) 5-Ac + NaOCl (~100)	275	187, 194
	Me	5-Ac	C3(61)	147–8	187
	Me	4-NO$_2$	C4(17–40), C5(16)	187	46, 53
	Me	5-NO$_2$	C1(−), C3(60)	208–9, 204–5	296, 182
	Me	5,6-(NO$_2$)$_2$	C3(−)	295	182
	Me	4,5,6-(NO$_2$)$_3$	C3(64)	207	182
	Me	4,5,6,7-(NO$_2$)$_4$	C3(53)	290	182
	Me	5-NH$_2$	Reduction of 5-NO$_2$	+HCl, 310	296
	Me	5-OH	Hydrolysis of 5-OMe (80)	212	197
	Me	4,5-(OH)$_2$	Hydrolysis of 4,5-(OMe)$_2$ (−)	295, 298–301	202, 206
	Me	4,7-(OH)$_2$	Hydrolysis of 4,7-(OMe)$_2$ (71)	decomp.	198
	Me	5,6-(OH)$_2$	Hydrolysis of 5,6-(OMe)$_2$ (low)	266	197
	Me	5-OMe	C1(−)	92–3	296
	Me	4,5-(OMe)$_2$	C1(60)	63.5	202
	Me	4,7-(OMe)$_2$	C1(−)	120–0.5	198
	Me	5,6-(OMe)$_2$	C1(−)	170.5–2	197
	Me	4,5,6-(OMe)$_3$	C1(−)	76	203
	Me	4,5,7-(OMe)$_3$	C1(87)	87	204
	Me	4,5-dioxo	Oxidation of diol	decomp.	206
	Me	4,7-dioxo	Oxidation of diol	195–6(d)	198
	Me	5,6-dioxo	Oxidation of diol or dimethoxy	243–4	197

= reaction of the appropriate diaminobenzene with a carbonic acid derivative; 1 = phosgene, 2 = urea, substituted urea, 4 = dialkylcarbamyl halide, 5 = a carbonate ester, 6 = N,N'-carbonyldiimidazole, 7 = ate ion. B = reaction involving the appropriate o-aminophenyl isocyanate or related species; 1 = Curtius similar rearrangement, 2 = decomposition of an o-aminophenyl-carbamyl compound. C = substitution of other benzimidazole; 1 = N-alkylation, 2 = N-acylation, 3 = C-substitution, 4 = hydroxylation, 5 = lation (of a benzimidazoline), 6 = hydrolysis of 3-isopropenyl derivative.

TABLE 3.3. BENZIMIDAZOLETHIONES

a. Both nitrogens unsubstituted ($R^1 = R^2 = H$)

Substituents	Method of preparation[a] (% yield)	M.p. (°)	Ref.
—	A1(84–87), A2(92), A3(90), A4(68) B1(82), C1(92), C2 (ca. 70)	303–4, 309–10 312–3	126, 129, 130, 133, 135, 138
5-Me	C1(81), A4(20)	285–6(d), >295(d)	135, 314
5,6-Me$_2$	A1(80), B1(92), C2 (ca. 70)	328(d)	313, 134, 138
5-Cl	A1(86), A4(48)	295–7, 290	314, 318
4,5,6,7-Cl$_4$	A1(84), C3(60)	368	300, 177
5-Br	A1(23)	300-1	314
4-I	A1(80)	277–8	315
5-I	A1(29)	361(d)	314
5-NO$_2$	A1(51), nitration of 5-H analog (82), C1 (ca. 100)	280–2	181, 240, 135,3
5,6-(NO$_2$)$_2$	C1(85)	260–1	135
5-NH$_2$	Reduction of 5-NO$_2$-analog	+HCl, 220	181
5-OMe	A1(68–78), A4(25)	252, 261	317, 314
5-OEt	A1(−)A4(22)	244, 251–2	167, 314
5,6-(OCH$_2$O-)	A1(95)	350	180

b. One nitrogen substituted ($R^2 = H$)

R^1	Other	Method of preparation[a] (% yield)	M.p. (°)	Ref.
Me	—	A1(60–83), A4(15) B1(98), C1(91), C2 (ca. 70) Scheme 3.8(−)	195, 199–200 193, 188–90 191	70, 127, 128, 99, 135, 138 207
Et	—	A1(76), Scheme 3.8(−)	163–4, 166	235, 207
Pr^n	—	Scheme 3.8(−)	107–11	207
Pr^i	—	C1(91)	146–7	135
Bu^n	—	A1(66), A4(7)	104, 100–1	127, 314
CH$_2$=CH–	—	N,S-divinyl + EtSH (72)	175	319
CH$_2$=CH—CH$_2$—	—	Scheme 3.8	102–4	207
CH$_2$CH$_2$OH	—	A1(89)	161–1.5	213
Ph	—	A1(70–80)	195–6	70, 127
PhCH$_2$	—	A1(60, 67)	180–1, 185	127, 320
Ac	—	Acylation (62)	205–7	217
Bz	—	Acylation (82)	192–5	217
CO$_2$Et	—	Acylation (68)	162–3	217
Me	5,6-Me$_2$	C2 (ca. 70)	240–2	138
Me	5-NO$_2$	A1(93)	304–5	70
Me	5,6-(NO$_2$)$_2$	C1 (ca. 100)	253–4	135
PhCH$_2$	various (5-)	A1 (various)	—	236
Ts	—	A1(−)	153	303
Ts	5-Me	A1(62)	148–50	303, 309
Ts	5-Cl	A1(60)	143	303, 309

Both nitrogens substituted

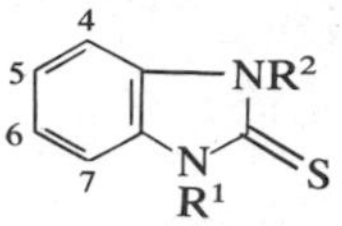

Substituents			Method of preparation[a] (% yield)	M.p. (°)	Ref.
	R²	Other			
	Me	—	Scheme 3.8(63)	153–4, 151–2,	70, 207, 208,
			B2(81)	148–50	5
	Et	—	Scheme 3.8(−), C1 or C2(−)	76	207, 137
	Et	—	Scheme 3.8(−)	39–41 (b.p. 185/7 mm)	207
2OH	CH₂OH	—	Parent thione+CH₂O (−)	160–2	166
2NR₂	CH₂NR₂	—	Mannich reaction (70–85)	—	220
2SMe	CH₂SMe	—	Eq. 50(11)	124–7	221
2CH₂OH	CH₂CH₂OH	—	A1(44)	168–70	213
2CH₂CN	CH₂CH₂CN	—	Eq. 51(−)	—	223
2CH₂COMe	CH₂CH₂COMe	—	A1(−), eq. 51(86)	132–3	222
(CH₂)ₙ (n = 5–12)		—	A1(53–77)	—	312
	Me	5-NO₂	B2(69)	222–3	45
	Ac	—	Acylation (−)	—	44
	Bz	—	Acylation (78)	182–4	217
	Ac	—	Acylation (72–91)	144	218

= reaction of the appropriate diaminobenzene; 1 = with CS_2 or $K^+\bar{S}CSOEt$; 2 = with CS_2 and N,N'-clohexylcarbodi-imide, 3 = with N,N'-thiocarbonyldi-imidazole, 4 = with thiourea. B = from sulfur and 1: parent benzimidazole, 2: the benzimidazoline. C − from the 2-chlorobenzimidazole; 1 = with thiourea, with thiosulfate ion.

TABLE 3.4.   BENZIMIDAZOLESELENONES AND BENZIMID-AZOLETELLURONES

Substituents			Method of preparation (% yield)	M.p. (°)	Ref.
1-	3-	5-			
a. Selenones					
H	H	H	diamine+CSe_2 (85)	234–6 (d)	242
Me	H	H	diamine+CSe_2 (50)	184–5	243
Me	Me	H	benzimidazoline+Se (70)	174.5–5.5	5
Me	Me	NO₂	benzimidazoline+Se (34)	243–4	45
b. Tellurones					
Me	Me	H	benzimidazoline+Te (9)	199–200°	45

379

TABLE 3.5. BENZIMIDAZOLIMINES

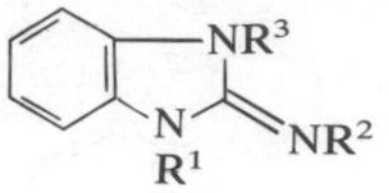

Substituents				Method of preparation (% yield)	M.p. (°)	Ref.
R^1	R^2	R^3	Others			
Me	H	Me	—	Eq. 57	117	246, ?
Me	Me	Me	—	Eq. 58 (87) and analog of eq. 57 (47)	60, 62–4	253, ?
Me	$n\text{-}C_5H_{11}$	Me	—	Eq. 58 (70)	b.p. 200/ 15 mm	253
Me	Ph	Me	—	Analog of eq. 57 (76) Eq. 64 (−)	197–8	254, ?
Me	$2,4\text{-}(NO_2)_2C_6H_3$	Me	—	Eq. 58	232–3	253
Me	Ac	Me	—	Eq. 58 (80)	133	253
Me	Bz	Me	—	Eq. 58 (76)	185	253
Me	$PhSO_2$	Me	—	Eq. 58 (81)	163.5	253
R	H	$CH_2\text{—}C{\equiv}CH$	—	Eq. 57 (−)	—	249, ?
Me	H	CH_2OR	—	Eq. 57 (35–50)	—	252
$PhCH_2$	$PhCH_2$	$PhCH_2$	$4,6\text{-}(NO_2)_2$	Eq. 65 (−)	168–9	260
Me	H	Me	$5\text{-}NO_2$	Nitration of parent imine	⎰239–40	246
Me	H	Me	$5,6\text{-}(NO_2)_2$		⎱258	246
Me	H	Me	5-Br	Bromination of parent imine (78)	122	100
Me	H	Me	$5,6\text{-}Br_2$	Bromination of parent imine (85)	207–208	100
Me	H	Me	$5\text{-}SO_3H$	Sulfonation of parent imine (90)	—	265
Me	H	Me	$5\text{-}SO_2Cl$	Chlorosulfonation of parent imine (85)	—	265
Me	NHBz	Me	—	Eq. 63 (78)	95–7	258
Me	NHTs	Me	—	Eq. 63 (20)	181–3(d)	259
Me	NO	Me	—	2-imine + $NaNO_2$/ H^+ (85)	126(d) (explosive)	253

TABLE 3.6.   2-METHYLENEBENZIMIDAZOLINES

	Substituents			Method of preparation		
R^1	R^2	X	Y	(% yield)	M.p. or b.p. (°)	Ref.
Me	Me	H	H	Eq. 76 (50)	b.p. 108–110/1.0 mm m.p. 57–8	279
Et	Me	H	H	Eq. 76 (83)	b.p. 95–6/0.8 mm m.p. 24	278
Ph	Ph	H	H	Eq. 76 (60–70)	m.p. 60	38
Me	Me	H	2,4-dinitro-phenyl	Scheme 3.10 (80)	m.p. 264	288
Me	Me	H	Ac	Scheme 3.10 (40) Eq. 82 (−)	m.p. 133	279, 286
Me	Me	H	Bz	Scheme 3.10 (−) Analog of eq. 82 (−) Eq. 83 (−)	m.p. 163	37, 286, 288
Me	Me	H	SO₂Me	Scheme 3.10 (−)	(unstable)	288
Me	Me	Ac	Ac	Scheme 3.10 (−)	m.p. >250	279, 288
Me	Me	SO₂Me	SO₂Me	Scheme 3.10 (8)	m.p. >250	288
Ph	Ph	H	Ac	Analog of eq. 76 (87)	m.p. 146	38
Me	Me	H	CO₂H	Scheme 3.10 (−)	m.p. not quoted	287
Me	Me	CO₂H	CO₂H	Scheme 3.10 (−)	m.p. ca. 150	287
Me	Me	H	CS₂H	Scheme 3.10 (−)	m.p. 210	287
Me	Me	H	CONHPh	Scheme 3.10 (−)	m.p. 203	287
Me	Me	CONHPh	CONHPh	Scheme 3.10 (−)	m.p. 250	287
H	H	CN	CN	Eq. 79 (50)	m.p. >320(d)	283
H	H	CN	CO₂Me	Eq. 79 (94)	m.p. 286–8(d)	283
Me	Me	CN	CO₂Me	Eq. 83 (−)	m.p. not quoted	37
Me	Me	H	N=NPh	Eqs. 77, 78 (−)	m.p. 160.5–1.5	282
Me	Me	N=NPh	N=NPh	Eq. 77 (−)	m.p. 240.5	280–282
Me	Me	N=NAr	N=NAr	Eq. 77 (−)		280–282

REFERENCES

1. F. Bohlmann, *Chem. Ber.*, **85,** 390 (1952).
2. J. W. Clark-Lewis, J. A. Edgar, J. S. Shannon, and M. J. Thompson, *Aust. J. Chem.*, **17,** 877 (1964).
3. A. V. El'tsov, *J. Org. Chem. USSR*, **1,** 1121 (1965).
4. J. L. Aubagnac, J. Elguero, R. Jacquier, and R. Robert, *Bull. Soc. Chim. France*, **1971,** 2184.
5. A. V. El'tsov, *J. Org. Chem. USSR*, **3,** 191 (1967).
6. H. Bauer, *J. Org. Chem.*, **26,** 1649 (1961).
7. I. Butula, *Annalen*, **718,** 260 (1968).
8. A. V. El'tsov and Kh. L. Muravich-Aleksandr, *J. Org. Chem. USSR*, **1,** 1321 (1965).
9. A. V. El'tsov and Kh. L. Muravich-Aleksandr, *J. Org. Chem. USSR*, **1,** 1695 (1965).

10. A. V. El'tsov and V. S. Kuznetsov, *J. Org. Chem. USSR*, **2**, 1465 (1966).
11. A. V. El'tsov, Kh. L. Muravich-Aleksandr, and L. M. Roitshtein, *J. Org. Chem. USSR*, **3**, 196 (1967).
12. J. Bourson, *Bull. Soc. Chim. France*, **1970**, 1867.
13. V. Veeranagaiah, C. V. Ratnam, and N. V. Subba Rao, *Indian J. Chem.*, **6**, 279 (1968).
14. P. S. Reddy, V. Veeranagaiah, and C. V. Ratnam, *Proc. Indian Acad. Sci., Sect. A*, **81**, 124 (1975).
15. K. S. Rao and C. V. Ratnam, *Curr. Sci.*, **37**, 611 (1968).
16. H. Suschitzky, G. V. Garner, and O. Meth-Cohn, *J. Chem. Soc. C*, **1971**, 1234.
17. E. Negishi and A. R. Day. *J. Org. Chem.*, **30**, 43 (1965).
18. N. A. Zakharova and B. A. Poraï-Koshits, *Trudy Leningrad Tekhnol. Inst. im Lensoveta*, **40**, 163 (1957), *Chem. Abstr.*, **54**, 19653 (1960).
19. Shell Research Ltd., British Patent 1,013,441 (1965); Belgian Patent 623,714 (1963); *Chem. Abstr.*, **60**, 9299 (1964).
20. T. Ishii and S. Ito, Japanese Patent 70 39,540 (1970); *Chem. Abstr.*, **75**, 5899 (1971).
21. J. G. Smith and I. Ho, *Tetrahedron Letters*, **1971**, 3541.
22. V. Veeranagaiah, C. V. Ratnam, and N. V. Subba Rao, *Indian J. Chem.*, **10**, 133 (1972).
23. V. Veeranagaiah, C. V. Ratnam, and N. V. Subba Rao, *Indian J. Chem.*, **8**, 790 (1970).
24. H. A. Staab and F. Vögtle, *Chem. Ber.*, **98**, 2681 (1965).
25. R. Garner, G. V. Garner, and H. Suschitzky, *J. Chem. Soc. C*, **1970**, 825.
26. F. D. Popp, *J. Heterocycl. Chem.*, **6**, 125 (1969).
27. V. M. Dziomko, A. V. Ivashchenko and R. V. Poponova, *J. Org. Chem. USSR*, **10**, 1330 (1974).
28. R. C. Elderfield and F. J. Kreysa, *J. Am. Chem. Soc.*, **70**, 44 (1948).
29. R. C. Elderfield and J. R. McCarthy, *J. Am. Chem. Soc.*, **73**, 975 (1951).
30. R. C. Elderfield and V. B. Meyer, *J. Am. Chem. Soc.*, **76**, 1887 (1954).
31. P. S. Reddi, C. V. Ratnam, and N. V. Subba Rao, *Indian J. Chem.*, **10**, 982 (1972).
32. T. Kitazume and N. Ishikawa, *Bull. Chem. Soc. Japan*, **47**, 785 (1974).
33. H. Suzuki, M. Ohashi, K. Itoh, I. Matsuda, and Y. Ishii, *Bull. Chem. Soc. Japan*, **48**, 1922 (1975).
34. C. W. Smith, R. S. Rasmussen, and S. A. Ballard, *J. Am. Chem. Soc.*, **71**, 1082 (1949).
35. A. Patchornik, A. Berger, and E. Katchalski, *J. Am. Chem. Soc.*, **79**, 6416 (1957).
36. G. Scherowsky, *Annalen*, **739**, 45 (1970).
37. H. Meerwein, W. Florian, H. Schön, and G. Stopp, *Annalen*, **641**, 1 (1961).
38. J. Bourson, *Bull. Soc. Chim. France*, **1971**, 3541.
39. H. Quast and E. Schmitt, *Chem. Ber.*, **101**, 1137 (1968).
40. P. W. Alley and D. A. Shirley, *J. Org. Chem.*, **23**, 1791 (1958).
41. T. Kato and M. Daneshtalab, *Chem. Pharm. Bull. (Tokyo)*, **24**, 1640 (1976).
42. A. B. Turner and H. C. S. Wood, *J. Chem. Soc.* **1965**, 5271.
43. W. Walter and U. Sewekow, *Annalen*, **1974**, 274.
44. M. R. Ibrahim, A. A. Jarrar, and S. S. Sabri, *J. Heterocycl. Chem.*, **12**, 11 (1975).
45. M. Z. Girshovich and A. V. El'tsov, *J. Gen. Chem. USSR*, **39**, 913 (1969).
46. M. Z. Girshovich and A. V. El'tsov, *J. Org. Chem. USSR*, **10**, 619 (1974).
47. M. V. Gorelik and T. Kh. Gladysheva, *J. Org. Chem. USSR*, **13**, 1817 (1977).
48. J. A. L. Herbert and H. Suschitzky, *Chem. and Ind.*, **1973**, 482.
49. A. V. El'tsov and M. Z. Girshovich, *J. Org. Chem. USSR*, **3**, 1292 (1967).
50. A. V. El'tsov, M. Z. Girshovich, and Kh. L. Muravich-Aleksandr, *J. Org. Chem. USSR*, **5**, 547 (1969).
51. M. Z. Girshovich and A. V. El'tsov, *J. Org. Chem. USSR*, **6**, 638 (1970).
52. V. S. Sorokina and L. A. Pavlova, *J. Org. Chem. USSR*, **7**, 377 (1971).
53. M. Z. Girshovich and A. V. El'tsov, *J. Org. Chem. USSR*, **10**, 385 (1974).
54. S. M. Hecht, B. L. Adams, and J. W. Kozarich, *J. Org. Chem.*, **41**, 2303 (1976).
55. A. Widdig, F. Grewe, H. Scheinpflug, and P. E. Frohberger, German Patent 2,114,882 (1972); *Chem. Abstr.*, **78**, 3985 (1973).

56. R. C. Elderfield and K. L. Burgess, *J. Am. Chem. Soc.*, **82**, 1975 (1960).
57. G. E. Risinger, *Nature*, **209**, 1022 (1966).
58. P. S. Reddy, C. V. Ratnam, and N. V. Subba Rao, *Indian J. Chem.*, **10**, 240 (1972).
59. V. Veeranagaiah, C. V. Ratnam, and N. V. Subba Rao, *Indian J. Chem.*, **7**, 776 (1969).
60. V. Veeranagaiah, C. V. Ratnam, and N. V. Subba Rao, *Indian J. Chem.*, **12**, 346 (1974).
61. V. Veeranagaiah, C. V. Ratnam, and N. V. Subba Rao, *Proc. Indian Acad. Sci. Sect. A*, **79**, 230 (1974).
62. P. S. Reddy, V. Veeranagaiah, and C. V. Ratnam, *Proc. Indian Acad. Sci. Sect. A*, **81**, 132 (1975).
63. K. Hofmann, "Imidazole and Its Derivatives, Part 1," Wiley-Interscience, New York, 1953, pp. 285–291.
64. E. P. Nesynov and P. S. Pel'kis, *J. Org. Chem. USSR*, **3**, 831 (1967).
65. G. Illuminati and U. Romano, German Patent 2,528,368 (1976); *Chem. Abstr.*, **84**, 164777 (1974).
66. Yu. M. Yutilov and I. A. Svertilova, *Chem. Heterocycl. Compounds*, **12**, 1057 (1976).
67. A. V. El'tsov, V. S. Kuznetsov, and M. B. Kolesova, *J. Org. Chem. USSR*, **1**, 1126 (1965).
68. H. A. Staab, *Annalen*, **609**, 75 (1957).
69. W. B. Wright, *J. Heterocycl. Chem.*, **2**, 41 (1965).
70. G. F. Duffin and J. D. Kendall, *J. Chem. Soc.*, **1956**, 361.
71. T. L. Cairns, D. D. Coffman, and W. W. Gilbert, *J. Am. Chem. Soc.* **79**, 4405 (1957).
72. Mobay Chemical Co., French Patent 1,470,892 (1967), *Chem. Abstr*, **68**, 87298 (1968).
73. L. Birkhofer, H. P. Kühlthau, and A. Ritter, *Chem. Ber.*, **93**, 2810 (1960).
74. A. F. Hegarty and L. N. Frost, *J. Chem. Soc. Perkin II*, **1973**, 1719.
75. W. J. Schnabel and E. Kober, *J. Org. Chem.*, **34**, 1162 (1969).
76. F. Applegarth and R. A. Franz, U.S. Patent 2,874,149 (1959), *Chem. Abstr.*, **53**, 12187 (1959).
77. T. Joh, N. Hagihara, and S. Murahashi, *Bull. Chem. Soc. Japan*, **40**, 661 (1967).
78. D. Cattapan, U. Valcavi, and C. G. Alberti, *Gazz. Chim. Ital.*, **88**, 13 (1958).
79. F. M. Hershenson, L. Bauer, and K. F. King, *J. Org. Chem.*, **33**, 2543 (1968).
80. A. W. Scott and B. L. Wood, *J. Org. Chem.*, **7**, 508 (1942).
81. H. Behringer and H. J. Fischer, *Chem. Ber.*, **94**, 1572 (1961).
82. R. K. Smalley and T. E. Bingham, *J. Chem. Soc. C*, **1969**, 2481.
83. A. Vedres and G. Balogh, Hungarian Patent 2209 (1971); *Chem. Abstr.*, **75**, 118316 (1971).
84. H. Lindemann and W. Schultheis, *Annalen*, **464**, 237 (1928).
85. S. Maffei and G. F. Bettinetti, *Ann. Chim. (Rome)*, **49**, 1809 (1959).
86. A. F. M. Fahmy and M. A. Elkomy, *Indian J. Chem.*, **13**, 652 (1975).
87. T. Kametani, K. Sota, and M. Shio, *J. Heterocycl. Chem.*, **7**, 807 (1970).
88. T. Kametani and M. Shio, *J. Heterocycl. Chem.*, **7**, 831 (1970).
89. T. Kametani and M. Shio, *J. Heterocycl. Chem.*, **8**, 545 (1971).
90. I. S. Kashparov and A. F. Pozharskii, *Chem. Heterocycl. Compounds*, **7**, 116 (1971).
91. A. F. Pozharskii, M. M. Medvedeva, E. A. Zvezdina, and A. M. Simonov, *Chem. Heterocycl. Compounds*, **7**, 624 (1971).
92. M. M. Medvedeva, A. F. Pozharskii, and A. M. Simonov, *Chem. Heterocycl. Compounds*, **8**, 1280 (1972).
93. I. I. Popov, P. V. Tkachenko, A. M. Simonov, A. F. Pozharskii, and Yu. G. Bogachev, *J. Org. Chem. USSR*, **10**, 1804 (1974).
94. I. I. Popov, Yu. G. Bogachev, P. V. Tkachenko, A. M. Simonov, and B. A. Tertov, *Chem. Heterocycl. Compounds*, **12**, 437 (1976).
95. A. F. Pozharskii, V. V. Kuz'menko, I. S. Kashparov, Z. I. Sokolov, and M. M. Medvedeva, *Chem. Heterocycl. Compounds*, **12**, 304 (1976).
96. A. Takamizawa, K. Hirai, Y. Hamashima, and H. Sato, *Chem. Pharm. Bull. (Tokyo)*, **17**, 1462 (1969).

97. K. S. Sardesai and S. V. Sunthankar, *Curr. Sci.*, **26,** 250 (1957); *J. Sci. Ind. Res.* (India), **18B,** 158 (1959).

98. R. K. Grantham and O. Meth-Cohn, *J. Chem. Soc. C,* **1969,** 70.

99. A. V. El'tsov and K. M. Krivozheiko, *J. Org. Chem. USSR,* **2,** 183 (1966).

100. S. N. Kolodyazhnaya and A. M. Simonov, *Chem. Heterocycl. Compounds,* **5,** 529 (1969).

101. N. D. Vitkevich and A. M. Simonov, *J. Gen. Chem. USSR,* **30,** 2847 (1960).

102. A. M. Simonov and Yu. M. Yutilov, *J. Gen. Chem. USSR,* **32,** 2629 (1962).

103. T. Sandmeyer, *Berichte,* **19,** 2650 (1886).

104. A. V. El'tsov, K. M. Krivozheiko, and M. B. Kolesova, *J. Org. Chem. USSR,* **3,** 1475 (1967).

105. K. M. Krivozheiko and A. V. El'tsov, *J. Org. Chem. USSR,* **4,** 1074 (1968).

106. J. Davoll, *J. Chem. Soc.,* **1960,** 308.

107. A. Rossi, A. Hunger, J. Kebrle, and K. Hoffmann, *Helv. Chim. Acta,* **43,** 1046 (1960).

108. A. Rossi, A. Hunger, J. Kebrle, and K. Hoffmann, *Helv. Chim. Acta,* **43,** 1298 (1960).

109. R. Barchet and K. W. Merz, *Tetrahedron Lett.,* **1964,** 2239.

110. A. N. Kost, Z. F. Solomko, V. A. Budylin, and T. S. Semenova, *Chem. Heterocycl. Compounds,* **8,** 632 (1972).

111. A. T. Ayupova and Ch. Sh. Kadyrov, *Chem. Heterocycl. Compounds,* **10,** 207 (1974).

112. B. A. Puodzhyunaite and Z. A. Talaikite, *Chem. Heterocycl. Compounds,* **10,** 724 (1974).

113. O. Hromatka, D. Binder, and K. Eichinger, *Monatsh. Chem.,* **106,** 375 (1975).

114. C. Guidon, V. Loppinet, and P. Greiveldinger, *Sci. Pharm. Biol. Lorraine,* **3,** 50 (1975); *Chem. Abstr.,* **85,** 159979 (1976).

115. M. J. Haddadin, G. Agopian, and C. H. Issidorides, *J. Org. Chem.,* **36,** 514 (1971).

116. A. A. Jarrar, S. S. Halawi, and M. J. Haddadin. *Heterocycles,* **4,** 1077 (1976).

117. A. A. Jarrar and Z. A. Fataftah, *Tetrahedron,* **33,** 2127 (1977).

118. Yusuf Ahmad, M. S. Habib, A. Mohammady, B. Bakutiari, and S. A. Shamsi, *J. Org. Chem.,* **33,** 201 (1968).

119. J. K. Landquist, *J. Chem. Soc.,* **1953,** 2830.

120. D. F. Morrow and L. A. Regan, *J. Org. Chem.,* **36,** 27 (1971).

121. Kou-Yi Tserng and L. Bauer, *J. Org. Chem.,* **38,** 3498 (1973).

122. M. S. Gibson and M. Green, *Tetrahedron,* **21,** 2191 (1965).

123. C. W. Rees and A. A. Sale, *J. Chem. Soc. Perkin I,* **1973,** 545.

124. M. L. Oftedahl, R. W. Radue, and M. W. Dietrich, *J. Org. Chem.,* **28,** 578 (1963).

125. J. Shibasaki, T. Koizumi, and S. Matsumura, *J. Pharm. Soc. Japan,* **88,** 491 (1968); *Chem. Abstr.,* **69,** 86908 (1968).

126. J. A. Van Allan and B. D. Deacon, *Org. Synth.,* **30,** 56 (1950); *Collective Volume* **4,** 569 (1963).

127. G. N. Tyurenkova, E. I. Silina, and I. Ya. Postovskii, *J. Appl. Chem. USSR,* **34,** 2203 (1961).

128. J. J. D'Amico, K. Boustany, A. B. Sullivan, and R. H. Campbell, *Int. J. Sulfur Chem. A,* **2,** 37 (1972).

129. J. C. Jochims, *Chem. Ber.,* **101,** 1746 (1968).

130. H. A. Staab and G. Walther, *Annalen,* **657,** 98 (1962).

131. A.-Mohsen M. E. Omar, Sh. A. Shams El-Dine, and A. A. B. Hazzaa, *Pharmazie,* **28,** 682 (1973).

132. W. Treibs, *Naturwissenschaften,* **49,** 13 (1962).

133. A. Giner-Sorolla, E. Thom, and A. Bendich, *J. Org. Chem.,* **29,** 3209 (1964).

134. I. Ya. Postovskii and N. N. Vereshchagina, *Chem. Heterocycl. Compounds,* **1,** 418 (1965).

135. D. Harrison and J. T. Ralph, *J. Chem. Soc.,* **1965,** 3132.

136. G. R. Revankar and L. B. Townsend, *J. Heterocycl. Chem.,* **5,** 477, 615 (1968).

137. I. I. Chizhevskaya and R. V. Skupskaya, USSR Patent 430,099 (1974); *Chem. Abstr.,* **81,** 77919 (1974).

138. I. I. Chizhevskaya, R. V. Skupskaya, and N. M. Yatsevich, *J. Org. Chem. USSR,* **8,** 2490 (1972).

139. G. R. Form, E. S. Raper, and T. C. Downie, *Acta Crystallogr.*, **B32**, 345 (1976).
140. C. W. N. Cumper and G. D. Pickering, *J. Chem. Soc. Perkin II*, **1972**, 2045.
141. D. Harrison and A. C. B. Smith, *J. Chem. Soc.*, **1961**, 4827.
142. W. A. Seth Paul and P. J. A. Demoen, *Bull. Soc. Chim. Belges*, **75**, 524 (1966).
143. Ya. V. Rashkes, *Zh. Prikl. Spektrosk*, **6**, 505 (1967); *Chem. Abstr.*, **67**, 77596 (1967).
144. M. H. Hussain and E. J. Lien, *Spectrosc. Lett.*, **6**, 97 (1973).
145. D. Harrison and A. C. B. Smith, *J. Chem. Soc.*, **1959**, 3157.
146. R. M. Anderson and D. Harrison, *J. Chem. Soc.*, **1964**, 5231.
147. D. J. Kew and P. F. Nelson, *Aust. J. Chem.*, **15**, 792 (1962).
148. Cf. L. J. Bellamy, "Advances in Infrared Group Frequencies," Methuen, London, 1968, pp. 128–130.
149. D. Harrison and J. T. Ralph, *J. Chem. Soc. B*, **1967**, 14.
150. L. S. Efros and A. V. El'tsov, *J. Gen. Chem. USSR*, **27**, 755 (1957).
151. P. Nuhn, G. Wagner, and S. Leistner, *Z. Chem.*, **9**, 152 (1969).
152. R. N. Nurmukhametov, I. L. Belaits, and D. N. Shigorin, *J. Phys. Chem. USSR*, **41**, 1032 (1967).
153. C. Guimon, M. Arbelot, and G. Pfister-Guillouzo, *Spectrochim. Acta A*, **31**, 985 (1975).
154. H.-O. Kalinowski and H. Kessler, *Org. Magn. Res.*, **6**, 305 (1974).
155. T. Kametani, S. Hirata, S. Shibuya, and M. Shio, *Org. Mass Spectrom.*, **4**, 395 (1970).
156. M. L. Thomson and D. C. DeJongh, *Can. J. Chem.*, **51**, 3313 (1973).
157. D. C. K. Lin, M. L. Thomson, and D. C. DeJongh, *Can. J. Chem.*, **52**, 2359 (1974).
158. S. O. Lawesson, G. Schroll, J. H. Bowie, and R. G. Cooks, *Tetrahedron*, **24**, 1875 (1968).
159. D. C. DeJongh and M. L. Thomson, *J. Org. Chem.*, **38**, 1356 (1973).
160. L. S. Efros and B. A. Poraǐ-Koshits, *J. Gen. Chem. USSR*, **23**, 725 (1953).
161. D. J. Brown, *J. Chem. Soc.*, **1958**, 1974.
162. W. Jähnig, *J. Prakt. Chem.*, **314**, 621 (1972).
163. W. O. Foye and J.-R. Lo, *J. Pharm. Sci.*, **61**, 1209 (1972); *Chem. Abstr.*, **77**, 100546 (1972).
164. J. Le Coarer, M. Wone, and A. Broche, *Ann. Fac. Sci. Univ. Dakar*, **6**, 25 (1961); *Chem. Abstr.*, **58**, 6244 (1963).
165. M. J. Haddadin and A. A. Jarrar, *Tetrahedron Lett.*, **1971**, 1651.
166. L. Monti and M. Venturi, *Gazz. Chim. Ital.*, **76**, 365 (1946).
167. L. Spirer, *Rocz. Chem.*, **28**, 455 (1965).
168. H. Zinner and B. Spangenberg, *Chem. Ber.*, **91**, 1432 (1958).
169. J. Sawlewicz and L. Sawlewicz, *Acta Polon. Pharm.*, **19**, 299 (1962); *Chem. Abstr.*, **60**, 4129 (1962).
170. Ciba-Geigy A.-G., German Patent 2,453,326 (1975); *Chem. Abstr.*, **83**, 132568 (1975).
171. B. I. Mikhant'ev and V. V. Kalmykov, *Tr. Lab. Khim. Vysokomolekul. Soedin., Voronezh. Univ.*, **1964**, 38; *Chem. Abstr.*, **65**, 3856 (1966).
172. V. V. Kalmykov, *Sb. Nauch. Rab. Aspir. Voronezh. Gos. Univ.*, No. 2, 80 (1965); *Chem. Abstr.*, **67**, 21869 (1967).
173. B. I. Mikhant'ev and V. V. Kalmykov, *Tr. Probl. Lab. Khim. Vysokomol. Soedin., Voronezh. Gos. Univ.* No. 4, 51 (1966); *Chem. Abstr.*, **69**, 27336 (1968).
174. H. Zinner and K. Peseke, *J. Prakt. Chem.*, **312**, 307 (1969).
175. M. M. Htay and O. Meth-Cohn, *Tetrahedron Lett.*, **1976**, 79.
176. J. Davoll and D. H. Laney, *J. Chem. Soc.*, **1960**, 314.
177. H. Röchling, E. Frasca, and K. H. Büchel, *Z. Naturforsch. B*, **25**, 954 (1970).
178. J. Sawlewicz and J. Jasinska, *Rocz. Chem.*, **38**, 1073 (1964).
179. D. Harrison, J. T. Ralph, and A. C. B. Smith, *J. Chem. Soc.*, **1963**, 2930.
180. E. R. Lavagnino and D. C. Thompson, *J. Heterocycl. Chem.*, **9**, 149 (1972).
181. A. T. James and E. E. Turner, *J. Chem. Soc.*, **1950**, 1515.
182. L. S. Efros and A. V. El'tsov, *J. Gen. Chem. USSR*, **27**, 143 (1957).
183. H. Schindlbauer and W. Kwiecinski, *Monatsh. Chem.*, **107**, 1307 (1976).
184. C. N. Talaty, N. Zenker, and P. S. Callery, *J. Heterocycl. Chem.*, **13**, 1121 (1976).

185. D. F. Kutepov and D. N. Khokhlov, *J. Org. Chem. USSR*, **1**, 186 (1965).
186. J. R. Vaughan and J. Blodinger, *J. Am. Chem. Soc.*, **77**, 5757 (1955).
187. A. V. El'tsov and I. M. Ginzburg, *J. Gen. Chem. USSR*, **34**, 1634 (1964).
188. Yu. A. Rozin, E. P. Darienko, and Z. V. Pushkareva, *Chem. Heterocycl. Compounds*, **4**, 510 (1968).
189. M. N. Kosyakovskaya, Ch. Sh. Kadyrov, A. V. Gordeeva, V. N. Balikhina, and V. V. Filippov, *Dokl. Akad. Nauk Uzb. SSR*, **32**, 34 (1975); *Chem. Abstr.*, **84**, 105500 (1976).
190. M. N. Kosyakovskaya, A. V. Gordeeva, and Ch. Sh. Kadyrov, *Chem. Heterocycl. Compounds*, **8**, 351 (1972).
191. L. S. Efros, B. A. Poraǐ-Koshits, and S. G. Farbenshtein, *J. Gen. Chem. USSR*, **23**, 1779 (1953).
192. Ch. Sh. Kadyrov, M. N. Kosyakovskaya, and M. R. Yagudaev, *Uzb. Khim. Zhur.*, **12**, 34 (1968); *Chem. Abstr.*, **70**, 28870 (1969).
193. M. N. Kosyakovskaya and Ch. Sh. Kadyrov, *Chem. Heterocycl. Compounds*, **6**, 199 (1970).
194. A. V. El'tsov, *J. Gen. Chem. USSR*, **32**, 1511 (1962).
195. A. V. El'tsov, *J. Gen. Chem. USSR*, **33**, 1297 (1963).
196. L. S. Efros and A. V. El'tsov, *J. Gen. Chem. USSR*, **28**, 433 (1958).
197. A. V. El'tsov and L. S. Efros, *J. Gen. Chem. USSR*, **29**, 3655 (1959).
198. A. V. El'tsov, V. S. Kuznetsov, and L. S. Efros, *J. Gen. Chem. USSR*, **33**, 3901 (1963).
199. L. C. March and M. M. Joullié, *J. Heterocycl. Chem.*, **7**, 39 (1970).
200. A. V. El'tsov and L. S. Efros, *J. Gen. Chem. USSR*, **30**, 3287 (1960).
201. A. V. El'tsov and L. S. Efros, *J. Gen. Chem. USSR*, **31**, 1123 (1961).
202. A. V. El'tsov and L. S. Efros, *J. Gen. Chem. USSR*, **31**, 1469 (1961).
203. A. V. El'tsov and L. S. Efros, *J. Gen. Chem. USSR*, **31**, 3726 (1961).
204. A. V. El'tsov and L. S. Efros, *J. Gen. Chem. USSR*, **32**, 191 (1962).
205. A. V. El'tsov, V. S. Kuznetsov, and L. S. Efros, *J. Gen. Chem. USSR*, **34**, 195 (1964).
206. A. V. El'tsov, E. R. Zakhs, and L. S. Efros, *J. Gen. Chem. USSR*, **34**, 3788 (1964).
207. K. Futaki, *J. Pharm. Soc. Japan*, **74**, 1365 (1954); *Chem. Abstr.*, **49**, 15876 (1954).
208. B. H. Klanderman, *J. Org. Chem.*, **30**, 2469 (19/5).
209. J. J. D'Amico, C. C. Tung, and W. E. Dahl, *J. Org. Chem.*, **42**, 600 (1977).
210. H. Oda, H. Kawaoka, and K. Shimanoe, Japanese Patent 73 15,878 (1973); *Chem. Abstr.*, **78**, 147961 (1973).
211. A. V. El'tsov and V. E. Lopatin, *J. Org. Chem. USSR*, **7**, 1324 (1971).
212. A. V. El'tsov and V. E. Lopatin, *J. Org. Chem. USSR*, **9**, 2622 (1973).
213. J. Sawlewicz and W. Rzeszotarski, *Rocz. Chem.*, **36**, 865 (1962).
214. H. Zinner and K. Peseke, *J. Prakt. Chem.*, **311**, 997 (1969).
215. H. Zinner and K. Peseke, *J. Prakt. Chem.*, **312**, 185 (1969).
216. G. G. Skvortsova, N. D. Abramova, and B. V. Trzhtsinskaya, *Chem. Heterocycl. Compounds*, **10**, 1217 (1974).
217. E. Dyer and C. E. Minnier, *J. Heterocycl. Chem.*, **6**, 23 (1969).
218. K. Sasse and F. Grewe, South African Patent 67 05,057 (1968); *Chem. Abstr.*, **70**, 47452 (1969).
219. K. A. Nuridzhanyan and G. V. Kuznetsova, *Chem. Heterocycl. Compounds*, **9**, 639 (1973).
220. H. Zinner, O. Schmitt, W. Schritt, and G. Rembarz, *Chem. Ber.*, **90**, 2852 (1957).
221. K. Anzai and S. Suzuki, *Bull. Chem. Soc. Japan*, **40**, 2854 (1967).
222. H. Irai, S. Shima, and N. Murata, *J. Chem. Soc. Japan* (*Ind. Chem. Sect.*), **62**, 82 (1959); *Chem. Abstr.* **58**, 5659 (1963).
223. G. F. Galenko, A. K. Bagrii, V. A. Grin, and P. M. Kochergin, *Ukr. Khim. Zhur.*, **41**, 405 (1975); *Chem. Abstr.*, **83**, 79146 (1975).
224. G. F. Galenko, A. K. Bagrii, and P. M. Kochergin, *Ukr. Khim. Zh.*, **41**, 759 (1975); *Chem. Abstr.*, **83**, 147426 (1975).
225. K. Hideg, O. Hankovzsky, E. Palosi, G. Hajos, and L. Szporny, German Patent 2,429,290 (1975); *Chem. Abstr.*, **82**, 156307 (1975).

226. A. K. Bagrii, G. F. Galenko, and P. M. Kochergin, *Dopov. Akad. Nauk. Ukr. RSR, Ser. B.*, **1975**, 801; *Chem. Abstr.*, **84**, 43959 (1976).
227. V. K. Chadha, H. S. Chaudhary, and H. K. Pujari, *Indian J. Chem.*, **7**, 769 (1969).
228. E. I. Grinblat and I. Ya. Postovskii, *J. Gen. Chem. USSR*, **31**, 357 (1961).
229. A. McKillop, G. C. A. Bellinger, P. N. Preston, A. Davidson, and T. J. King, *Tetrahedron Lett.*, **1978**, 2621.
230. H. Alper and M. S. Wolin, *J. Org. Chem.*, **40**, 437 (1975).
231. J. G. Everett, *J. Chem. Soc.* **1930**, 2402.
232. W. Knobloch and K. Rintelen, *Arch. Pharm.*, **291**, 180 (1958).
233. Sang-Woo Park, W. Ried, and W. Schuckmann, *Annalen*, **1977**, 106.
234. H. Berge, H. Millat, and B. Strübing, *Z. Chem.*, **15**, 37 (1975).
235. H. Balli and F Kersting, *Annalen*, **647**, 1 (1961).
236. G. A. Mokrushina and N. P. Bednyagina, *Chem. Heterocycl. Compounds*, **6**, 1304 (1970).
237. J. Clark, R. K. Grantham, and J. Lydiate, *J. Chem. Soc. C*, **1968**, 1122.
238. L. Monti and G. Franchi, *Gass. Chim. Ital.*, **81**, 764 (1951).
239. A. V. El'tsov and K. M. Krivozheiko, *J. Org. Chem. USSR*, **6**, 637 (1970).
240. M. Semonský, J. Kuňák, and A. Černý, *Chem. Listy*, **47**, 1633 (1953); *Chem. Abstr.*, **49**, 233 (1955).
241. A. D. Garnovskii, O. A. Osipov, L. I. Kuznetsova, and N. N. Bogdashev, *Russ. Chem. Rev.*, **42**, 89 (1973).
242. J. S. Warner, *J. Org. Chem.*, **28**, 1642 (1963).
243. J. S. Warner and T. F. Page, *J. Org. Chem.*, **31**, 606 (1966).
244. A. M. Simonov and P. M. Kochergin, *Chem. Heterocycl. Compounds*, **1**, 210 (1965).
245. Yu. M. Yutilov, V A Anisimova, and A. M. Simonov, *Chem. Heterocycl. Compounds*, **1**, 278 (1965).
246. A. M. Simonov, Yu. M. Yutilov, and V. A. Anisimova, *Chem. Heterocycl. Compounds*, **1**, 622 (1965).
247. Yu. N. Sheinker, A. M. Simonov, Yu. M. Yutilov, V. N. Sheinker, and E. I. Perel'shtein, *J. Org. Chem. USSR*, **2**, 911 (1966).
248. A. M. Simonov and V. A. Anisimova, *Chem. Heterocycl. Compounds*, **4**, 801 (1968).
249. I. I. Popov, P. V. Tkachenko, and A. M. Simonov, *Chem. Heterocycl. Compounds*, **11**, 347 (1975).
250. I. I. Popov, P. V. Tkachenko, and A. M. Simonov, *Chem. Heterocycl. Compounds*, **11**, 461 (1975).
251. A. M. Simonov, T. A. Kuz'menko, and L. G. Nachinennaya, *Chem. Heterocycl. Compounds*, **11**, 1394 (1975).
252. Yu. M. Koshchienko, A. M. Simonov, T. N. Vashchenko, G. M. Suvorova, and V. A. Makarov, *Khim. Farm. Zh.*, **11**, 14 (1977); *Chem. Abstr.*, **87**, 68234 (1977).
253. V. G. Sayapin and A. M. Simonov, *Chem. Heterocycl. Compounds*, **3**, 868 (1967).
254. J. Musco and D. B. Murphy, *J. Org. Chem.*, **36**, 3469 (1971).
255. B. I. Khristich, G. M. Suvorova, and A. M. Simonov, *Chem. Heterocycl. Compounds*, **10**, 1225 (1974).
256. K. H. Mayer, D. Lauerer, and H. Heitzer, *Synthesis*, **1975**, 673.
257. S. Hünig, H. Balli, H. Conrad, and A. Schott, *Annalen*, **676**, 36 (1964).
258. S. Hünig and H. Balli, *Annalen*, **609**, 160 (1957).
259. S. Hünig, R. D. Rauschenbach, and A. Schütz, *Annalen*, **623**, 191 (1959).
260. J. Bartos, *Bull. Soc. Chim. France*, **1965**, 3694.
261. O. E. Shelepin, V. G. Sayapin, N. K. Chub, and A. M. Simonov, *Chem. Heterocycl. Compounds*, **6**, 624 (1970).
262. Yu. V. Koshchienko, G. M. Suvorova, and A. M. Simonov, *Chem. Heterocycl. Compounds*, **11**, 124 (1975).
263. A. M. Simonov and S. N. Kolodyazhnaya, *Chem. Heterocycl. Compounds*, **6**, 1459 (1970).
264. S. N. Kolodyazhnaya, A. M. Simonov, and L. N. Podladchikova, *Chem. Heterocycl. Compounds*, **11**, 726 (1975).

265. A. M. Simonov and V. G. Sayapin, *Chem. Heterocycl. Compounds*, **5**, 402 (1969).
266. H. Quast and E. Spiegel, *Tetrahedron Lett.*, **1977**, 2705.
267. S. Hünig, G. Kiesslich, F. Linhart, and H. Schlaf, *Annalen*, **752**, 182 (1971).
268. S. Hünig, G. Kiesslich, F. Linhart, and H. Schlaf, *Annalen*, **752**, 196 (1971).
269. S. Hünig, D. Scheutzow, H. Schlaf, and A. Schott, *Annalen*, **1974**, 1423.
270. S. Hünig, H. Geiger, G. Kaupp, and W. Kniese, *Annalen*, **697**, 116 (1966).
271. S. Hünig and G. Kaupp, *Annalen*, **700**, 65 (1966).
272. S. Hünig and H. Balli, *Annalen*, **628**, 56 (1959).
273. S. Hünig and H. Nöther, *Annalen*, **628**, 69 (1959).
274. D. C. Pati and M. K. Rout, *J. Indian Chem. Soc.*, **45**, 425 (1968).
275. A. I. Kiprianov and T. M. Verbovskaya, *J. Org. Chem. USSR*, **2**, 1816 (1966).
276. S. Hünig, F. Brühne, and E. Breither, *Annalen*, **667**, 72 (1963).
277. F. M. Hamer, "The Cyanine Dyes and Related Compounds," Wiley-Interscience, New York, 1964.
278. J. Metzger, H. Larivé, R. Dennilauler, R. Baralle, and C. Gaurat, *Bull. Soc. Chim. France*, **1969**, 3156.
279. J. Bourson, *Bull. Soc. Chim. France*, **1971**, 152.
280. H. Wahl, *Bull. Soc. Chim. France*, **1954**, 251.
281. M. T. Le Bris, *Rev. Textile-tiba*, **57**, 164 (1958); *Chem. Abstr.*, **53**, 2208 (1959).
282. M. T. Le Bris, H. Wahl, and T. Jambu, *Bull. Soc. Chim. France*, **1959**, 343.
283. R. Gompper and W. Toepfl, *Chem. Ber.*, **95**, 2871 (1962).
284. A. Roedig and H. J. Becker, *Annalen*, **597**, 214 (1955).
285. H. Junek, H. Fischer-Colbrie, and H. Sterk, *Chem. Ber.*, **110**, 2276 (1977).
286. V. Denes and G. Ciurdar, Romanian Patent 52,982 (1971); *Chem. Abstr.*, **77**, 61999 (1972).
287. J. Bourson, *Bull. Soc. Chim. France*, **1973**, 2373.
288. J. Bourson, *Bull. Soc. Chim. France*, **1974**, 525.
289. J. Bourson, *Bull. Soc. Chim. France*, **1975**, 644.
290. R. C. Elderfield, W. J. Gensler, T. H. Bembry, T. A. Williamson, and H. Weisl, *J. Am. Chem. Soc.*, **68**, 1589 (1946).
291. F. Krollpfeiffer, W. Graulich, and A. Rosenberg, *Annalen*, **542**, 1 (1939).
292. A. Hartmann, *Berichte*, **23**, 1046 (1890).
293. O. Kym, *J. Prakt. Chem.*, **75**, 323 (1907).
294. C. Rudolph, *Berichte*, **12**, 1295 (1879).
295. A. Darapsky and B. Gaudian, *J. Prakt. Chem.*, **147**, 43 (1937).
296. R. L. Clark and A. A. Pessolano, *J. Am. Chem. Soc.*, **80**, 1657 (1958).
297. J. P. English, R. C. Clapp, Q. P. Cole, I. F. Halverstadt, J. O. Lampen, and R. O. Roblin, *J. Am. Chem. Soc.*, **67**, 295 (1945).
298. G. M. Brooke, J. Burdon, and J. C. Tatlow, *J. Chem. Soc.*, **1961**, 802.
299. B. N. Feitelson, P. Mamalis, R. J. Moualim, V. Petrow, O. Stephenson, and B. Sturgeon, *J. Chem. Soc.*, **1952**, 2389.
300. D. E. Burton, A. J. Lambie, D. W. J. Lane, G. T. Newbold, and A. Percival, *J. Chem. Soc. C*, **1968**, 1268.
301. A. Hunger, J. Kebrle, A. Rossi, and K. Hoffmann, *Helv. Chim. Acta*, **44**, 1273 (1961).
302. W. A. Sexton, *J. Chem. Soc.*, **1942**, 303.
303. Upjohn Co., Netherlands Patent 6,512,029 (1966); *Chem. Abstr.*, **65**, 7186 (1966).
304. G. Heller, A. Buchwaldt, R. Fuchs, W. Kleinicke, and J. Kloss, *J. Prakt. Chem.*, **111**, 1 (1925).
305. A. Ricci and P. Vivarelli, *Gazz. Chim. Ital.*, **97**, 758 (1967).
306. A. Ricci, G. Seconi, and P. Vivarelli, *Gazz. Chim. Ital.*, **99**, 542 (1969).
307. E. R. Zakhs, T. R. Strelets, and L. S. Efros, *Chem. Heterocycl. Compounds*, **6**, 1294 (1970).
308. E. R. Zakhs, V. I. Minkin, and L. S. Efros, *J. Org. Chem. USSR*, **1**, 1486 (1965).
309. J. B. Wright, *J. Med. Chem.*, **8**, 539 (1965).

310.   J. Pinnow and C. Sämann, *Berichte*, **32,** 2181 (1899).
311.   O. Christmann, *Chem. Ber.*, **98,** 1282 (1965).
312.   R. J. Hayward and O. Meth-Cohn, *J. Chem. Soc. Perkin I*, **1975,** 212.
313.   K. Takatori, Y. Yamada, and O. Kawashima, *J. Pharm. Soc. Japan*, **75,** 881 (1955); *Chem. Abstr.*, **50,** 4920 (1956).
314.   W. G. Bywater, D. A. McGinty, and N. D. Jenesel, *J. Pharmacol.*, **85,** 14 (1945).
315.   K. D. Banerji and K. K. Sen, *J. Indian Chem. Soc.*, **50,** 433 (1973).
316.   J. Mohan, V. K. Chadha, K. S. Sharma, and H. K. Pujari, *Indian J. Chem.*, **14B,** 723 (1976).
317.   J. Mohan and H. K. Pujari, *Indian J. Chem.*, **10,** 274 (1972).
318.   J. Mohan, V. K. Chadha, and H. K. Pujari, *Indian J. Chem.*, **11,** 1119 (1973).
319.   N. D. Abramova, B. V. Trzhtsinskaya, and G. G. Skvortsova, *Chem. Heterocycl. Compounds*, **11,** 1413 (1975).
320.   A. P. Terent'ev, I. G. Il'ina, E. G. Rukhadze, and I. G. Vorontsova, *J. Gen. Chem. USSR*, **40,** 1592 (1970).

CHAPTER 4

Condensed Benzimidazoles of Type 5-6-5

G. TENNANT

4.1. TRICYCLIC 5-6-5 FUSED BENZIMIDAZOLES WITH NO ADDITIONAL HETERO ATOMS

Only two frameworks belonging to this category are possible (Scheme 4.1). These are the linear indeno[5,6-d]imidazole (**4.1**) and angular indeno[4,5-d]imidazole (**4.2**) ring systems, the latter also being capable of existing in both 6H and 8H isomeric forms (**4.2a** and **b**). Although isolated reports of both ring systems have appeared from time to time in the literature, the properties of neither have been investigated to any extent.

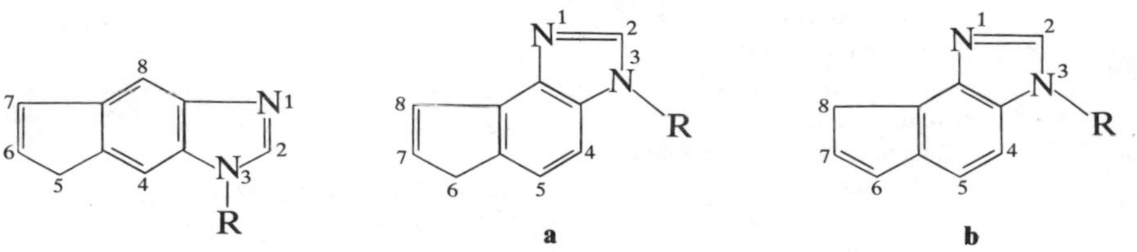

(4.1)   a   **(4.2)**   b

Scheme 4.1

4.1.1. Synthesis

Syntheses leading to derivatives of the indeno[5,6-*d*]imidazole (**4.1**) and indeno[4,5-*d*]imidazole (**4.2**) ring systems are rare. A green product formed by reaction of the benzimidazole di-*N*-oxide (**4.3**) or its malononitrile condensate (**4.4**) with malononitrile (Scheme 4.2) has been tentatively

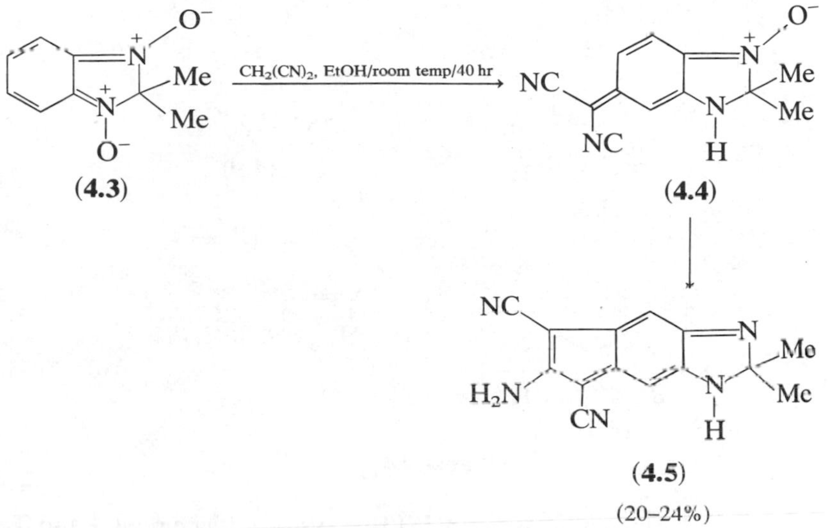

Scheme 4.2

assigned the interesting indeno[5,6-*d*]imidazole structure (**4.5**) on the basis of its spectroscopic properties.[1] Unfortunately, rigorous proof of the structure was precluded by the decomposition of the compound on attempted purification.[1] The intermediate produced by the reaction of benzimidazol-2(1*H*,3*H*)-one with γ-butyrolactone is reported to undergo cyclization to an indano[5,6-*d*]imidazole derivative.[2] 7,8-Dihydro-2-phenyl-6*H*-indeno-[4,5-*d*]imidazole (**4.7**) [m.p. 224 to 225° (decomp.)] has been prepared (Scheme 4.3)[3] in unspecified yield by the reductive cyclization of 5-benzoyl-amino-4-nitroindane (**4.6**).

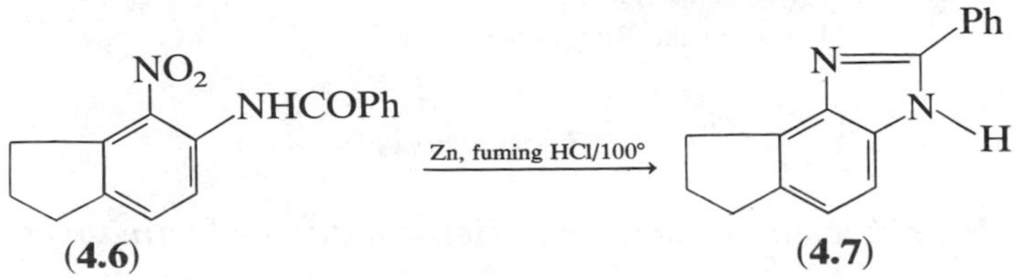

Scheme 4.3

4.2. TRICYCLIC 5-6-5 FUSED BENZIMIDAZOLES WITH ONE ADDITIONAL HETERO ATOM

Of the several possible tricyclic 5-6-5 fused benzimidazole structures containing one additional hetero atom, only the furo[3,4-*f*]benzimidazole (**4.8**), thieno[3,2-*e*]benzimidazole (**4.9**), pyrrolo[2,3-*f*]benzimidazole (**4.10**),

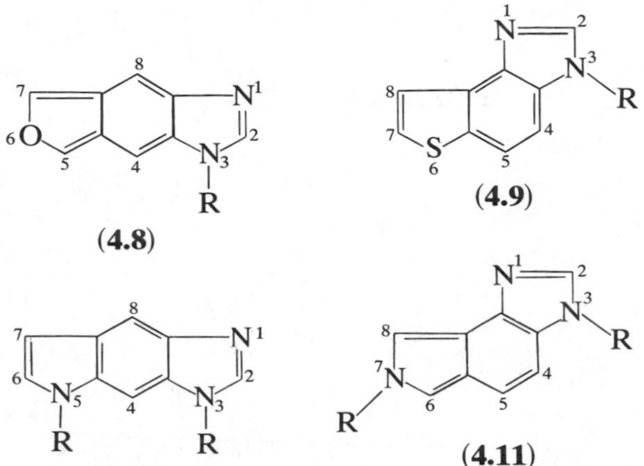

Scheme 4.4

and pyrrolo[3,4-*e*]benzimidazole (**4.11**) ring systems (Scheme 4.4 and Table 4.1) appear to have been investigated to any extent.

TABLE 4.1.   TRICYCLIC 5-6-5 FUSED BENZIMIDAZOLE RING SYSTEMS WITH ONE ADDITIONAL HETERO ATOM

Structure[a]	Name[b]
(**4.8**)	Furo[3,4-*f*]benzimidazole
(**4.9**)	Thieno[3,2-*e*]benzimidazole
(**4.10**)	Pyrrolo[2,3-*f*]benzimidazole
(**4.11**)	Pyrrolo[3,4-*e*]benzimidazole

[a] See Scheme 4.4.
[b] Based on the Ring Index.

4.2.1. Synthesis

Ring-Closure Reactions of Benzimidazole Derivatives

Ring closure of a benzimidazole derivative to a tricyclic 5-6-5 fused benzimidazole structure having one additional hetero atom is exemplified in one instance only: namely, by the conversion of benzimidazole-4,5-dicarboxylic acid (**4.12**) into the phthalimide derivative (**4.13**) in unspecified yield by heating with aniline (Scheme 4.5).[4] Other examples of the ring closure of benzimidazole derivatives to tricyclic 5-6-5 fused systems with one additional hetero atom are rare, if not entirely lacking.

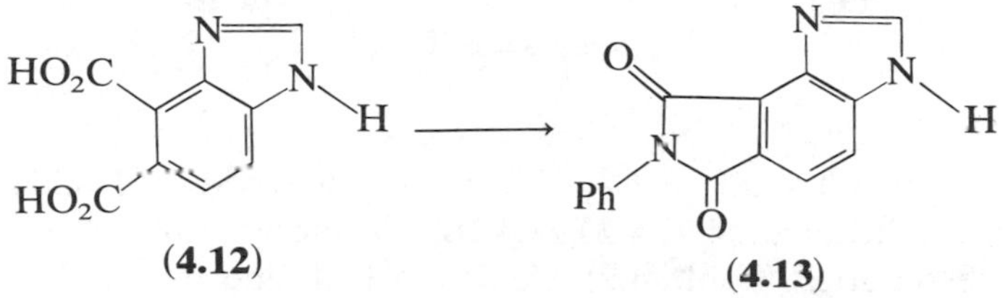

(**4.12**)   (**4.13**)

Scheme 4.5

Ring-Closure Reactions of Other Heterocycles

Ring closure of the 5,6-diaminophthalan derivative (**4.15**) [produced *in situ* by stannous chloride reduction of the corresponding 5-nitro compound (**4.14**)] by heating with acetyl chloride in benzene provides one of the few reported examples of the synthesis of the furo[3,4-*f*]benzimidazole ring system (Scheme 4.6, **4.15 → 4.16**).[5]

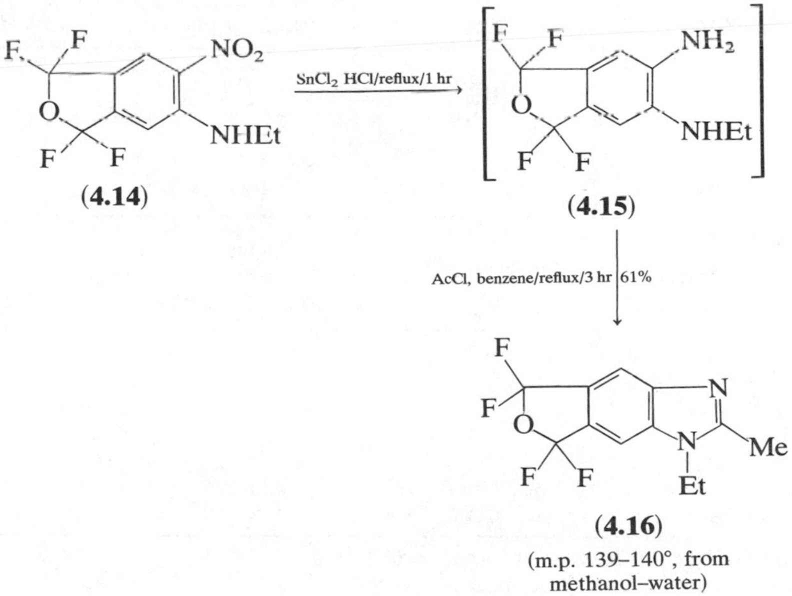

(**4.14**)   (**4.15**)

AcCl, benzene/reflux/3 hr │61%

(**4.16**)

(m.p. 139–140°, from
methanol–water)

Scheme 4.6

4,5-Diaminobenzo[*b*]thiophen derivatives (**4.17**, R^1 = H) undergo smooth ring closure on heating with carboxylic acids such as formic acid to provide a

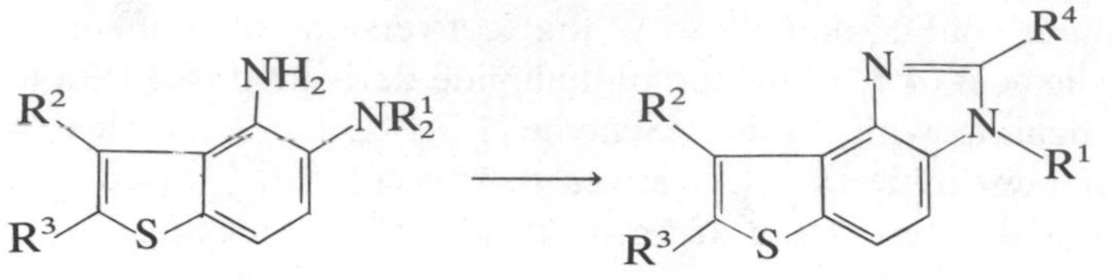

(4.17) **(4.18)**

Scheme 4.7

potentially general and high-yield route[6] to N-1(3)H-thieno[3,2-e]-benzimidazoles (Scheme 4.7, **4.17** → **4.18**). Of the examples of this type of ring closure reported (Table 4.2), yields are high and the products readily isolated as stable crystalline solids. Moreover, carrying out the ring closure in the presence of an oxidant such as performic acid (prepared *in situ* from hydrogen peroxide and formic acid) permits the successful cyclization of 4-amino-5-N,N-dialkylaminobenzo[b]thiophens (**4.17**, R^1 = alkyl) to the corresponding C(2)-alkylthieno[3,2-e]benzimidazoles (**4.18**, R^4 = alkyl) albeit in low yield (Table 4.2).[7] The thermal cyclization (at 250°) of 5-acylamino-4-aminobenzo[b]thiophens to derivatives of the N-1(3)H-thieno[3,2-e]-benzimidazole ring system is the subject of a patent.[8]

TABLE 4.2. RING-CLOSURE REACTIONS OF 4,5-DIAMINOBENZO[b]THIOPHEN DERIVATIVES (**4.17**) TO N-1(3)H-THIENO[3,2-e]BENZIMIDAZOLES (**4.18**)

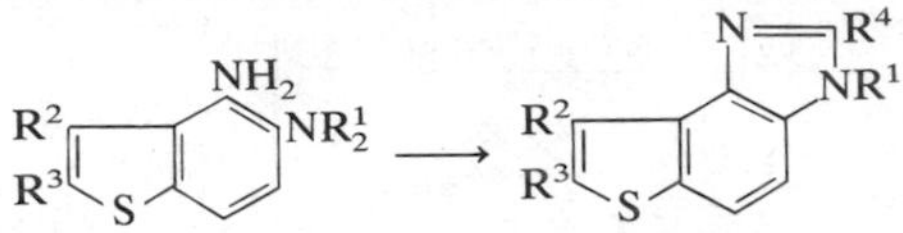

(4.17) **(4.18)**

Starting material (4.17)			Reaction conditions[a]	Product (4.18)				Yield (%)	Melting point (°C)	Solvent of crystalliza-tion	Crystal form	R
R^1	R^2	R^3		R^1	R^2	R^3	R^4					
H	H	H	A	H	H	H	H[b]	86	138–139	benzene-light petroleum (b.p. 40–60°)	—[c]	6
H	Me	H	A	H	Me	H	H[d]	84	220–221	—[e]	—[c]	6
Et	H	CO₂Et	B	Et	H	CO₂Et	Me	31	117–120	—[f]	—[c]	7

[a] A, HCO₂H/reflux/3 hr. B, HCO₂H, 30% H₂O₂/100°/15 min.
[b] Forms a picrate, m.p. 244–245° (decomp.).
[c] Crystal form not specified.
[d] Forms a picrate, m.p. 276–277°.
[e] Solvent not specified.
[f] Purified by distillation, b.p. 160–165°/0.2 mm Hg.

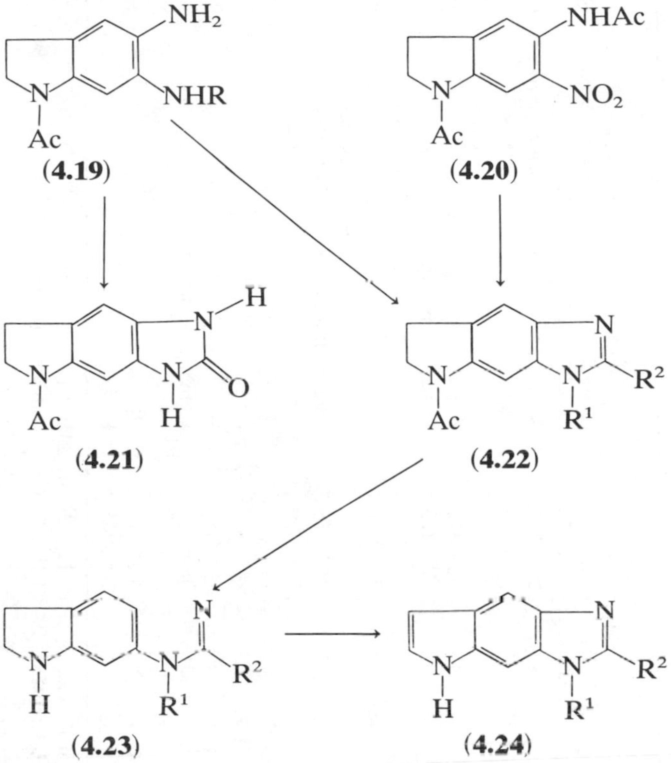

Scheme 4.8

5-Acetyl-6,7-dihydropyrrolo[2,3-f]benzimidazole derivatives (**4.22**) are readily accessible by three procedures (Scheme 4.8) based on (a) ring-closure reactions of 1-acetyl-5,6-diaminoindolines (**4.19** → **4.22**),[9,10] (b) reductive cyclization of 1-acetyl-5-acylamino-6-nitroindolines (**4.20** → **4.22**),[10] and (c) base-catalyzed cyclization of 1-acetyl-6-benzylamino-5-nitroindolines (Scheme 4.9, **4.26** → **4.27**).[9] Moreover, since the acetyl compounds (**4.22**) are readily hydrolyzed[9] to indolines (**4.23**) capable of undergoing dehydrogenation[9] to 5H-pyrrolo[2,3-f]benzimidazoles (**4.24**), all of these procedures serve as synthetic entries to compounds of the latter type.

Ring-closure reactions of the type (**4.19** → **4.22**) may be effected in moderate to excellent yield (ca. 40 to 80%) (Table 4.3) by heating the substrates (**4.19**) under reflux with *ortho* esters (e.g., ethyl orthoformate),[9] or with aliphatic carboxylic acids (e.g., acetic acid).[9,10] However, aromatic carboxylic acids react only reluctantly and in very low yield.[9] Moreover, attempts to circumvent this difficulty by carrying out the ring closure with an aromatic aldehyde in the presence of an oxidizing agent such as Cu(II) acetate[9] or nitrobenzene[9] affords the anticipated 2-arylpyrrolobenzimidazoles (**4.22**, R[2] = aryl) in only moderate yield (ca. 30 to 40%) (Table 4.3). Nor does

TABLE 4.3. SYNTHESES OF PYRROLO[3,4-e]BENZIMIDAZOLES BY RING-CLOSURE REACTIONS OF 5,6-DIAMINOINDOLINE DERIVATIVES

Starting material	R	Reaction conditions	Product	R^1	R^2	Yield (%)	Melting point (°C)	Solvent of crystallization	Crystal form	Ref.
(4.19)	H	A	(4.22)	H	H	89	232–233.5	Methanol	Yellow crystals	9
(4.19)	CH₂Ph	A	(4.22)	CH₂Ph	H	72	203–204	Methanol	Colorless crystals	9
(4.19)	H	B	(4.22)	H	Me	45	329–330	Ethanol	b	10
(4.19)	CH₂Ph	B	(4.22)	CH₂Ph	Me	92	222–223.5	Dimethylformamide	Colorless crystals	9
(4.19)	H	C	(4.22)	H	Ph	35	299.5–301.5	Methanol	Colorless crystals	9
(4.19)	CH₂Ph	C	(4.22)	CH₂Ph	Ph	77	257–258.5	Dimethylformamide	Colorless crystals	9
(4.19)	H	D	(4.22)	H	Ph	43	—	—	—	9
(4.19)	CH₂Ph	D	(4.22)	CH₂Ph	Ph	65	—	—	—	9
(4.19)	H	E	(4.22)	CH₂Ph	Ph	50	—	—	—	9
(4.19)	H	F	(4.21)	—	—	84	380	Dimethylformamide	Light grey crystals	9
(4.32)	Hᶜ	G	(4.33) (R = H)	—	—	85	>270	Ethanol	b	12
(4.32)	Acᶜ	G	(4.33) (R = Ac)	—	—	86	201–205	Ethanol	b	12

[a] A, HC(OEt)₃/100°/4 hr. B, CH₃CO₂H/reflux/3 hr. C, PhCHO, Cu(II) acetate H₂O, EtOH/reflux/3 hr, then room temp./14 hr. D, PhCHO, PhNO₂, MeOH/reflux/4–10 hr. E, PhCHO, PhNO₂, MeOH/reflux/20 min. F, Urea/180–190°/15 hr. G, CS₂, pyridine/100°/3 hr.
[b] Crystal form not specified.
[c] Prepared in situ by reduction of the respective 5,6-dinitroindolines (4.25; R = H or Ac).

398

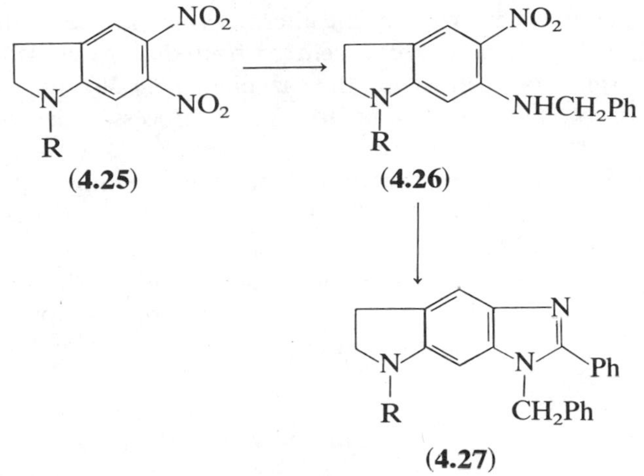

Scheme 4.9

increase in the reaction temperature aid the nitrobenzene-catalyzed process but rather tends to result in complicating arylmethylation of the initial product at $N(3)$ (Table 4.3).[9]

In a variant (Scheme 4.10) of the 4,5-diaminoindoline ring-closure method for the synthesis of 6,7-dihydropyrrolo[2,3-f]benzimidazoles (**4.22**), condensation[11] of the diaminoindoline (**4.28**) with ethyl acetoacetate affords either directly or via the crotonate (**4.29**), the diazepinone (**4.30**), convertible by heating with sulphuric acid into the ring-contracted dihydropyrrolo-[2,3-f]benzimidazole (**4.31**) (Scheme 4.10).[11] However, this compound is

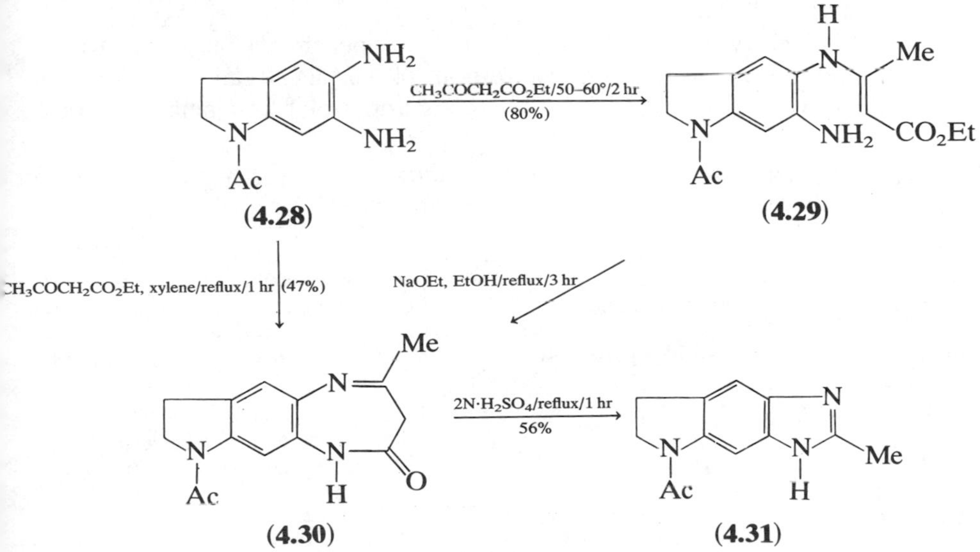

Scheme 4.10

more conveniently synthesized by the thermal ring closure of the diaminoindoline (**4.19**; R = H) with acetic acid or by reductive cyclization (Table
4.3).[10] The ring contraction of the azepinone (**4.30**) to the pyrrolo-
benzimidazole (**4.31**) is analogous to similar processes described in the
benzene series.[11]

Ring closure of 5,6-diaminoindolines (**4.19**) by heating with urea[9] provides a high-yield method (Table 4.3) for the synthesis of pyrrolo[2,3-*f*]-
benzimidazolones (**4.21**) (Scheme 4.8). This type of cyclization has been
extended[12] to the synthesis of the corresponding thioxo-derivatives, which
are obtained in very good yield (Table 4.3) by the reaction of 5,6-diamino-
indolines with carbon disulphide in the presence of pyridine (Scheme 4.11,
4.32 → **4.33**).

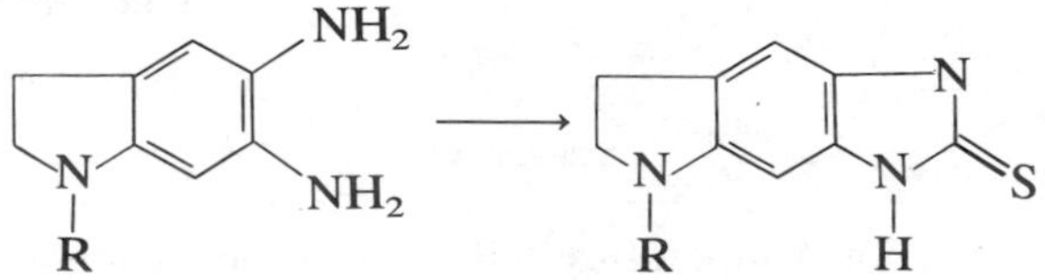

(**4.32**)   (**4.33**)

Scheme 4.11

6,7-Dihydropyrrolo[2,3-*f*]benzimidazoles are also accessible by the direct
reductive cyclization of 5-acylamino-6-nitroindolines. For example, 1-
acetyl-5-acetylamino-6-nitroindoline (**4.20**) is smoothly converted by reduction with stannous chloride in methanolic hydrochloric acid into 1-acetyl-
6,7-dihydro-2-methylpyrrolo[2,3-*f*]benzimidazole (**4.22**; R^1 = H, R^2 = Me)
in good yield (Table 4.4).[10]

Synthetic entry to 6,7-dihydropyrrolo[2,3-*f*] benzimidazoles is also provided by the base-catalyzed cyclization of certain 6-aminated-5-nitro-
indolines. Thus 3-benzyl-2-phenyl-6,7-dihydropyrrolo[2,3-*f*]benzimidazoles

TABLE 4.4. SYNTHESES OF PYRROLO[3,4-*e*]BENZIMIDAZOLES BY CYCLIZATION
REACTIONS OF 5- AND 6-NITROINDOLINES

Starting material	Reaction conditions[a]	Product	Yield (%)	Melting point (°C)	Solvent of crystalliza-tion	Crystal form	Ref.
(**4.20**)	A	(**4.22**) (R^1 = H, R^2 = Me)	55	329–331	Ethanol	Colorless solid	10
(**4.25**) (R = Ac)	B	(**4.27**) (R = Ac)	55	257–258	Dimethyl-formamide	Colorless crystals	9
(**4.25**) (R = H)	C	(**4.27**) (R = H)	95	219–220	Ethanol	Colorless needles	9
(**4.26**) (R = Ac)[b]	—	(**4.27**) (R = Ac)	—	—	—	—	9

[a] A, SnCl$_2$, HCl, ethanol/100°. B, PhCH$_2$NH$_2$/reflux/1–1,5 hr. C, PhCH$_2$NH$_2$/reflux/3 hr.
[b] Details not given.

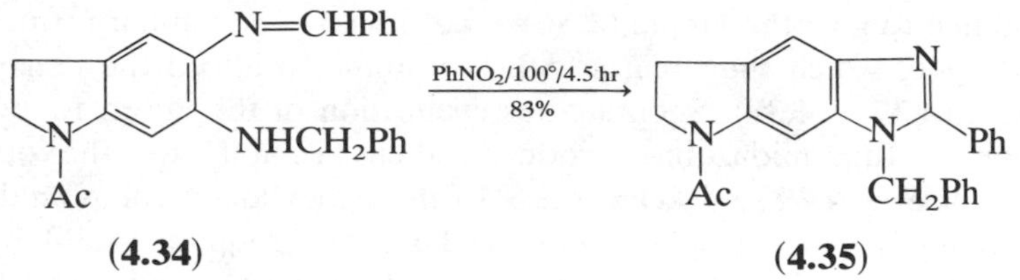

(4.34)   **(4.35)**

Scheme 4.12

are formed in moderate to excellent yield (Table 4.4) by the benzylamine catalyzed cyclization of 6-benzylamino-5-nitroindolines either preformed or prepared *in situ* by the action of benzylamine on 5,6-dinitroindolines (Scheme 4.9, **4.25 → 4.26 → 4.27**).[9] Since the position of the substituted amino group in substrates **4.26** is known with certainty, cyclizations of this type have the practically useful feature of leading to N(3)-substituted pyrrolo[2,3-f]benzimidazoles of unequivocal structure. The mechanistic pathway involved in these interesting cyclizations has not been established, but the demonstration that the benzylideneaminoindoline (**4.34**) is readily and presumably oxidatively cyclized to the pyrrolo[2,3-f]benzimidazole (**4.35**) by heating with nitrobenzene (Scheme 4.12)[9] argues in favor of a mechanism (Scheme 4.13) involving initial reduction by the excess of

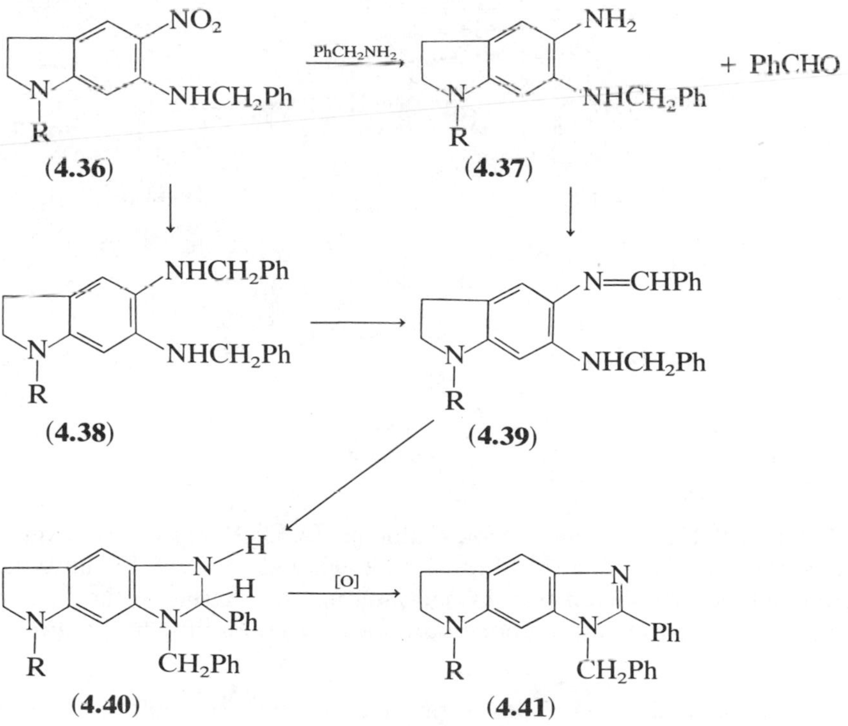

(4.36)   **(4.37)**

(4.38)   **(4.39)**

(4.40)   **(4.41)**

Scheme 4.13

benzylamine to give the amine (**4.36** → **4.37**) with concomitant formation of benzaldehyde, which then reacts with the amine to afford the benzylidene derivative (**4.37** → **4.39**). Spontaneous cyclization of the latter, followed by oxidation of the imidazoline produced then accounts for the observed product (**4.39** → **4.40** → **4.41**). Access to the benzylidene compound (**4.39**) by direct nucleophilic displacement of the nitro-group in (**4.36**) by benzylamine, followed by partial oxidation of the diamino compound formed (**4.36** → **4.38** → **4.39**) is less likely because of the anticipated low reactivity of the nitro group in (**4.36**) toward nucleophilic displacement.

4.2.2. Physicochemical Properties

Spectroscopic Studies

INFRARED SPECTRA. The infrared spectra of simple N-1(3)H-thieno[3,2-e]-benzimidazoles are characterized[6] by the presence of a broad band at 3120 to 2580 cm^{-1}, attributable to the N-1(3)H group. The ester group in the thienobenzimidazole [Scheme 4.14, (**4.42**)][7] absorbs in the infrared region at 1710 cm^{-1}.

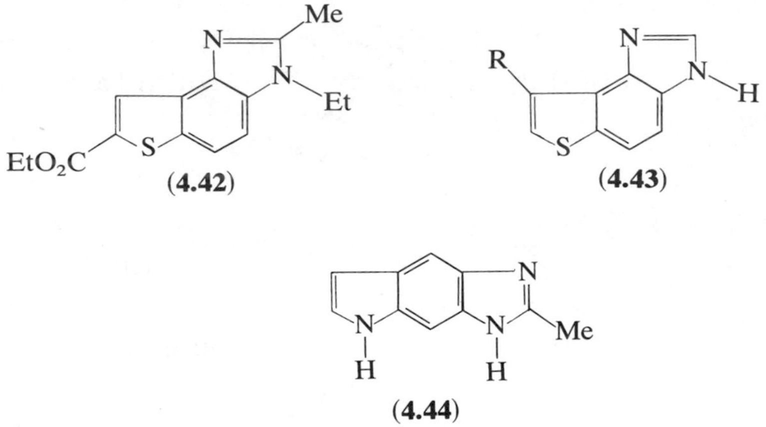

(**4.42**)   (**4.43**)

(**4.44**)

Scheme 4.14

The pyrrolo[2,3-f]benzimidazole-2-thiones (**4.33**; R = H or Ac) exhibit a series of infrared bands at 3240 to 3120, 2490, and 1310 cm^{-1} akin to those of related five-membered thiones and interpreted in terms of the existence of the imidazole ring in the thione (as opposed to the sulfhydryl) tautomeric form.[12]

ULTRAVIOLET SPECTRA. The ultraviolet spectra[6] of simple N-1(3)H-thieno[3,2-e]benzimidazoles (**4.43**) (Table 4.5) are typified by the presence

TABLE 4.5.   ULTRAVIOLET SPECTRA OF N-1(3)H-THIENO[3,2-e]BENZIMIDAZOLES[6]

(4.43)

Compound (4.43) R	λ_{max}, nm (log ε)^a
H	206(4.32), 224(4.39), 266(3.83), 273(3.88), 284(3.78), 294(4.04), 305(4.21)
Me	206(4.45), 227(4.36), 269(4.02), 277(4.05), 288(3.97), 299(4.13), 310(4.20)

^{a} Measured in 0.01 M HCl.

of a series of absorption maxima demonstrating the marked delocalization in these molecules. In contrast, the ultraviolet spectrum (Table 4.6) of 2-methyl-N-1(3)H-pyrrolo[2,3-f]benzimidazole (**4.44**) is relatively simple.[9] However, the increased delocalization in the fully unsaturated molecule (**4.44**) compared with its 6,7-dihydro derivatives is amply demonstrated by the bathochromic shift and increase in intensity of the ultraviolet absorption in the former compared with that in the latter (Table 4.6).[9,10] Indeed the ultraviolet spectra (Table 4.6) of 6,7-dihydropyrrolo[2,3-f]benzimidazoles

TABLE 4.6.   ULTRAVIOLET SPECTRA OF PYRROLO[3,4-e]BENZIMIDAZOLE DERIVATIVES (**4.44** to **4.46**)

Compound	R	R^1	R^2	R^3	X	Solvent^a	λ_{max}, nm (log ε)	Ref.
(4.44)	—	—	—	—	—	A	299(4.14)	9
(4.45)	—	H	H	Me	—	B	237(3.62), 280(3.90), 307(2.35)	9
(4.45)	—	H	H	Ph	—	C	328(3.89)	9
(4.45)	—	Ac	H	H	—	D	272(3.75), 305–314(4.00)	9
(4.45)	—	Ac	H	Me	—	C	232(4.42), 260(3.90), 270(3.87), 302(4.12), 312(4,14)	10
(4.45)	—	Ac	H	Ph	—	D	337(4.41)	9
(4.45)	—	Ac	CH$_2$Ph	H	—	D	276(4.00), 315(4.01)	9
(4.45)	—	Ac	CH$_2$Ph	Me	—	D	276(3.90), 310–315(4.18)	9
(4.45)	—	Ac	CH$_2$Ph	Ph	—	D	279(4.08), 325–329(4.55)	9
(4.46)	Ac	—	—	—	O	D	271(4.05), 318(4.13)	9
(4.46)	H	—	—	—	S	E	325(3.95), 341(3.98)	12
(4.46)	Ac	—	—	—	S	A	223(4.15), 262(3.85), 335(4.28)	12

^{a} A, ethanol. B, 0.01 M HCl. C, methanol. D, dimethylformamide. E, dimethyl sulfoxide.

TABLE 4.7. ^{1}H N.M.R. SPECTRA OF N-1(3)H-THIENO[3,2-e]BENZIMIDAZOLE DERIVATIVES (**4.42**) AND (**4.43**)^{a,b}

(**4.42**)   (**4.43**)

Compound	R	H(2)	H(4)	H(5)	H(7)	H(8)	Me	Others	Ref.
(**4.42**)	—	—	7.59d^{c,d}	7.38d^{c,d}	—	8.72	2.65^e	4.34q^{d,g}	7
	—	—	—	—	—	—	1.40t^{d,f}	4.22q^{d,g}	
	—	—	—	—	—	—	1.38t^{d,f}	—	
(**4.43**)	H	8.24	7.75d^{c,h}	7.65d^{c,h}	7.69d^{c,i}	7.45d^{c,i}	—	9.35^j	6
(**4.43**)	Me	8.11	7.72d^{c,h}	7.58d^{c,h}	7.77^k	—	2.84^l	—	6

^{a} δ values in p.p.m. measured from TMS. in CDCl$_3$ as solvent.
^{b} Signals were sharp singlets unless denoted as d = doublet; t = triplet; q = quartet.
^{c} These assignments may be interchanged.
^{d} J = 7 Hz.
^{e} Me(2).
^{f} Me of ethyl.
^{g} CH$_2$ of ethyl.
^{h} J = 8.5 Hz.
^{i} J = 5.5 Hz.
^{j} NH.
^{k} broad singlet.
^{l} Me(8).

resemble those of analogous benzimidazole derivatives. 6,7-Dihydro-pyrrolo[2,3-f]benzimidazole derivatives exhibit fluorescence in incident ultraviolet light.[9]

Nuclear Magnetic Resonance Spectra. Detailed information on the NMR absorption of the known tricyclic 5-6-5 fused benzimidazole ring systems having one additional hetero atom is almost entirely lacking. For example, despite a number of studies[9–12] relating to pyrrolo[2,3-f]-benzimidazole derivatives, data on the ^{1}H NMR absorption of such molecules does not appear to have been recorded to date. On the other hand ^{1}H NMR spectral data for several N-1(3)H-thieno[3,2-e]-benzimidazole derivatives (Table 4.7) have been reported.[6,7] Noteworthy features are the close correspondence in the chemical shifts of H(4) and H(5) throughout, the enhanced deshielding of H(2), and also the marked difference in the H(4)–H(5) and H(7)–H(8) coupling constants (Table 4.7).

General Studies

Quantitative data on the acidity and basicity of N-1(3)H-thieno[3,2-e]-benzimidazole derivatives is not available. However their ability to form well-defined (though not necessarily salt-like) picrates (see Table 4.2) implies their moderately basic character.

The acidity and basicity of pyrrolo[2,3-*f*]benzimidazole derivatives has likewise not been investigated beyond the qualitative demonstration[11] of the weakly acidic character of the 6,7-dihydro derivatives.

4.2.3. Reactions

The almost total lack of information concerning the chemical reactivity of the known tricyclic 5-6-5 fused benzimidazole ring systems with one additional hetero atom precludes any worthwhile comment under this heading.

4.3. TRICYCLIC 5-6-5 FUSED BENZIMIDAZOLES WITH TWO ADDITIONAL HETERO ATOMS

Because of the various permutations of the common hetero elements (oxygen, sulfur, nitrogen) possible, the fusion of a five-membered heterocyclic ring containing two such hetero atoms across the 4:5, 5:6, and 6:7 positions in benzimidazole results in a large number of ring systems belonging to the title category. A search of the literature reveals that derivatives of *nine* such ring systems have been reported (Scheme 4.15 and Table 4.8). Oxygen- and sulfur-containing ring systems of this type include one (**4.45**) of the two possible [1,3]dioxolobenzimidazoles, two (**4.46** and **4.47**) of the three possible imidazobenzoxazoles, and all three of the possible imidazobenzothiazoles (**4.48**, **4.49**, and **4.50**). Fully nitrogenous tricyclic 5-6-5 fused benzimidazole frameworks containing two additional hetero atoms

TABLE 4.8. TRICYCLIC 5-6-5 FUSED BENZ-
IMIDAZOLE RING SYSTEMS
WITH TWO ADDITIONAL
HETERO ATOMS

Structure[a]	Name[b]
(**4.45**)	[1,3]Dioxolo[4,5-*f*]benzimidazole
(**4.46**)	Imidazo[4,5-*f*]benzoxazole
(**4.47**)	Imidazo[4,5-*g*]benzoxazole
(**4.48**)	Imidazo[4,5-*f*]benzothiazole
(**4.49**)	Imidazo[4,5-*g*]benzothiazole
(**4.50**)	Imidazo[4,5-*e*]benzothiazole
(**4.51**)	Imidazo[4,5-*e*]indazole
(**4.52**)	Imidazo[4,5-*g*]indazole
(**4.53**)	Benzo[1,2-*d*: 4,5-*d'*]diimidazole
(**4.54**)	Benzo[1,2-*d*: 3,4-*d'*]diimidazole

[a] See Scheme 4.15.
[b] Based on the Ring Index.

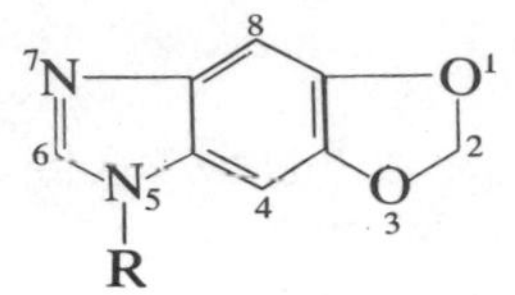

(4.45)

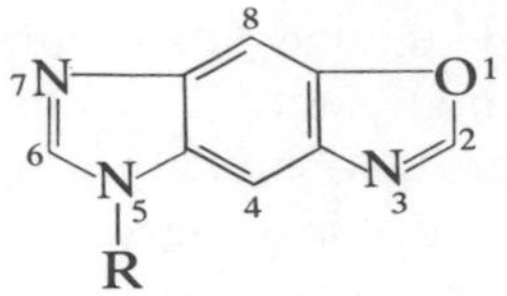

(4.46)

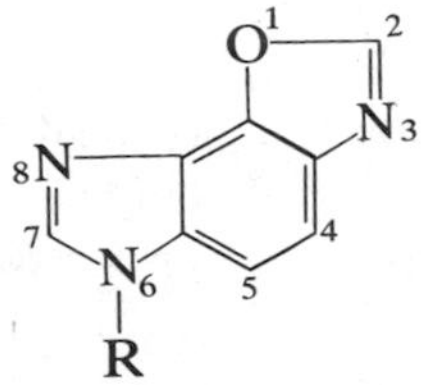

(4.47)

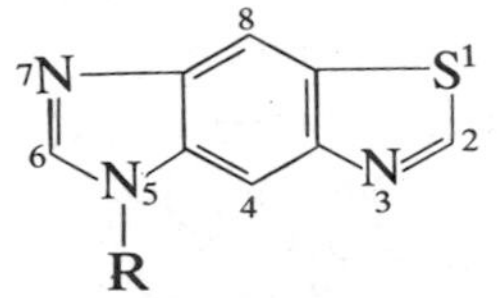

(4.48)

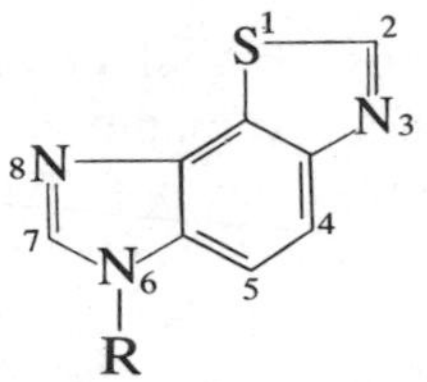

(4.49)

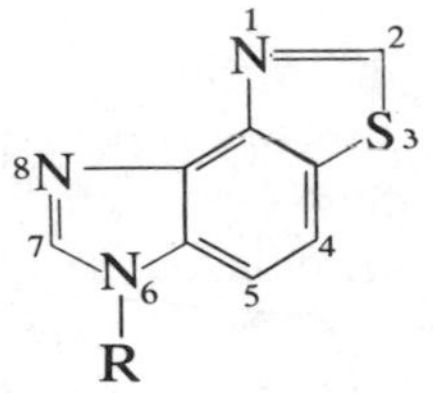

(4.50)

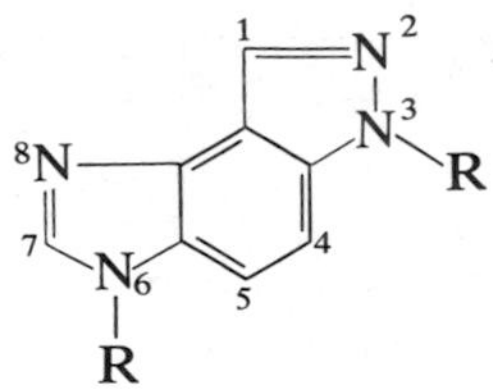

(4.51)

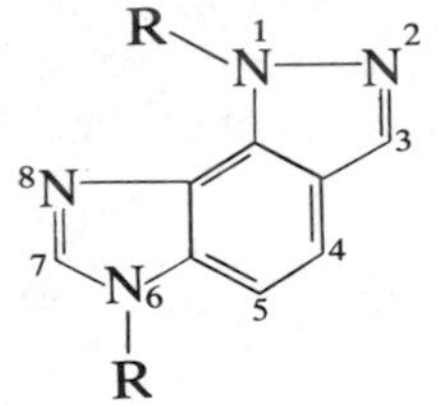

(4.52)

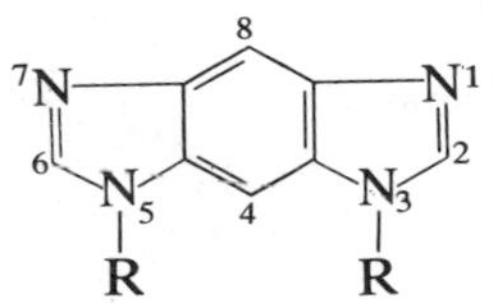

(4.53)

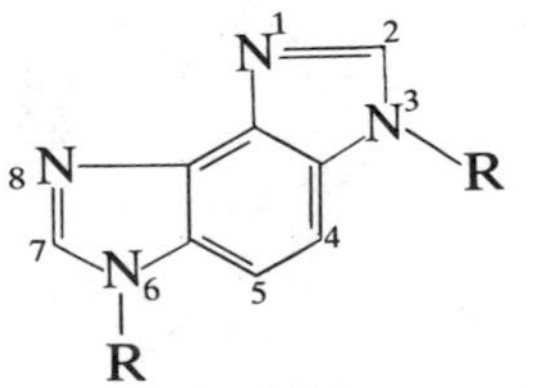

(4.54)

Scheme 4.15

are represented by the imidazo[4,5-e]indazole (**4.51**), imidazo[4,5-g]-indazole (**4.52**), benzo[1,2-d:4,5-d']diimidazole (**4.53**), and benzo[1,2-d:3,4-d']diimidazole (**4.54**) ring systems (Scheme 4.15).

4.3.1. Synthesis

Ring-closure Reactions of Benzimidazole Derivatives

What appears to be the sole example of a derivative of the imidazo[4,5-g]-benzoxazole ring system has been prepared by the application of an aryl azide-based benzoxazole synthesis to 5-azidobenzimidazole.[13] Heating the latter compound with polyphosphoric acid in acetic acid affords 2-methyl-6(8)H-imidazo[4,5-g]benzoxazole in good yield (60%) (Scheme 4.16, **4.55** → **4.56**). The angular (**4.56**) as opposed to linear (**4.57**) structure for this product follows from its hydrolysis to 5-amino-4-hydroxybenzimidazole (Scheme 4.16, **4.56** → **4.58**).[13]

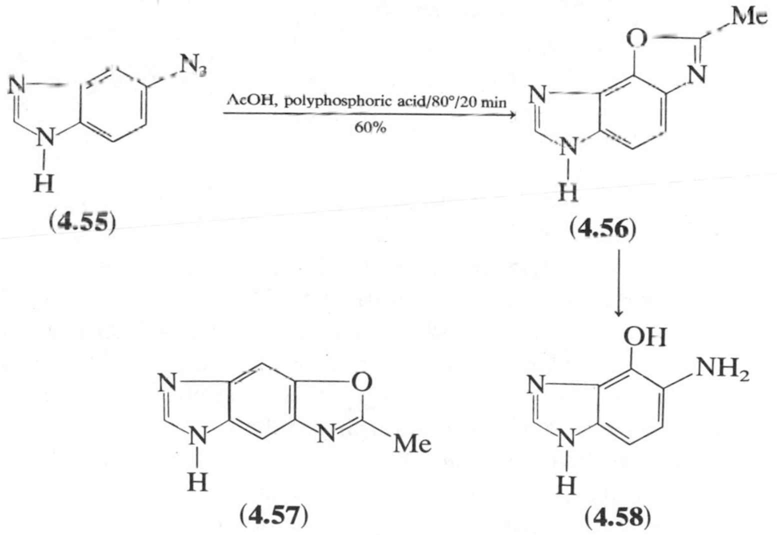

Scheme 4.16

Synthetic entry to the imidazo[4,5-g]benzothiazole ring system is provided by the acylative ring-closure of 5-amino-4-mercaptobenzimidazole alkali metal salts and by the rhodanation of suitable 5-aminobenzimidazole derivatives (Scheme 4.17). Thus reaction of the sodium salt of 5-amino-4-mercapto-1,2-dimethylbenzimidazole (**4.59**; R = Me, M = Na) with benzoyl chloride in alkaline solution affords the corresponding imidazo[4,5-g]-benzothiazole (**4.60**; R^1 = Me, R^2 = Ph) in moderate yield (Table 4.9).[14]

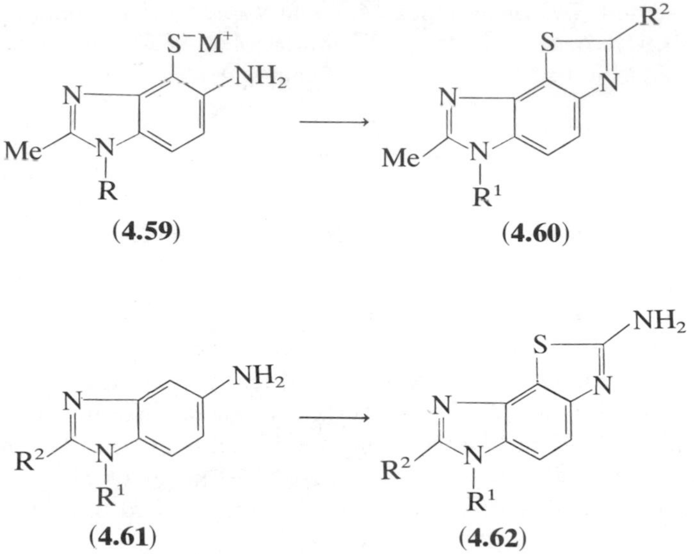

(4.59)   (4.60)

(4.61)   (4.62)

Scheme 4.17

Similarly, heating 5-amino-4-mercapto-2-methyl-1-phenylbenzimidazole potassium salt (**4.59**; R = Ph, M = K) with acetic anhydride in acetic acid results in ring closure, albeit in only moderate yield (Table 4.9), to the imidazo[4,5-g]benzothiazole (**4.60**; R^1 = Ph, R^2 = Me).[16] 2-Aminoimidazo-[4,5-g]benzothiazoles are readily accessible[14-16] in acceptable yield (Table 4.9) by rhodanation of 5-aminobenzimidazoles using ammonium thiocyanate in the presence of bromine (**4.61** → **4.62**).

Syntheses of benzo[1,2-d:4,5-d']diimidazole derivatives are commonly based on acid-catalyzed bis-acylative ring closure reactions with 1,2,4,5-tetraaminobenzene (Scheme 4.18; **4.63**) or its tetrahydrochloride. Since these reactions bear an obvious relationship to the acylative ring closures of 5,6-diaminobenzimidazole derivatives that also afford benzo[1,2-d:4,5-d']-diimidazoles they are discussed later in this section.

1,2,4,5-Tetraaminobenzene reacts in general with aliphatic carboxylic acids in the presence of hydrochloric acid to give usually good yields (Table 4.10) of the corresponding 2,6-disubstituted 1(3)H,5(7)H-benzo[1,2-d:4,5-d']-diimidazoles (**4.63** → **4.64**).[17-19] Ring closure is normally readily achieved by heating the tetraamine (**4.63**) tetrahydrochloride under reflux with the requisite aliphatic carboxylic acid in the presence of 1.5 to 7.0 M hydrochloric acid (Table 4.10).[17-19] A careful study of condensation reactions of this type has shown that yields are only low in the absence of the hydrochloric acid and that effectively no ring closure occurs under buffered conditions (e.g., in the presence of sodium acetate) (Table 4.10).[19] Moreover, optimum yields are best obtained by carrying out the ring closure under pressure in

TABLE 4.9. SYNTHESES OF IMIDAZO[4,5-g]BENZOTHIAZOLES (4.60 AND 4.62) BY RING-CLOSURE REACTIONS OF 5-AMINO- AND 5-AMINO-4-MERCAPTOBENZIMIDAZOLES (4.59 AND 4.61)

Starting material	R	R¹	R²	M	Reaction conditions[a]	Product	R¹	R²	Yield (%)	Melting point (°C)	Solvent of crystallization	Ref.
(4.59)	Me	—	—	Na	A	(4.60)	Me	Ph	40	249	Nitrobenzene	14
(4.59)	Ph	—	—	K	B	(4.60)	Ph	Me	35	246	Ethanol	15
(4.61)	—	Ph	Me	—	C	(4.62)	Ph	Me	57	319	Dioxane	15
(4.61)	—	Me	Me	—	D	(4.62)	Me	Me	23	>300	—[b]	14
(4.61)	—	H	H	—	E	(4.52)	H	H[c]	44	236–237	Water	16
(4.61)	—	Me	H	—	E	(4.52)	Me	H	40	300–301	Water	16
(4.61)	—	CH₂Ph	H	—	E	(4.62)	CH₂Ph	H[d]	34	248–249	Isopropanol	16
(4.61)	—	CH₂CH₂OH	H	—	E	(4.62)	CH₂CH₂OH	H[e]	54	277–278	Water	16

^{a} A, PhCOCl, 5% NaOH. B, Ac$_2$O, AcOH/reflux/2 hr. C, NH$_4$CNS, Br$_2$/$-10°$/3 hr. D, NH$_4$CNS, Br$_2$/-8 to $-10°$/4 hr. E, NH$_4$CNS, Br$_2$/0–5°/2 hr.
^{b} Solvent not specified.
^{c} Forms a dihydrochloride, m.p. 300° (from ethanol–water).
^{d} Forms a dihydrochloride, m.p. 262–264° (from ethanol–water).
^{e} Forms a dihydrochloride, m.p. 258° (decomp.) (from ethanol).

TABLE 4.10. SYNTHESES OF BENZO[1,2-d: 4,5-d']DIIMIDAZOLE DERIVATIVES (**4.64**, **4.68**, **4.69**, AND **4.71**) BY RING-CLOSURE REACTIONS OF 1,2,4,5-TETRAAMINOBENZENE (**4.63**) AND 2,3,5,6-TETRAAMINO-1,4-BENZOQUINONE (**4.70**)

Starting material	Reaction conditions[a]	Product	R	Yield (%)	Melting point (°C)	Solvent of crystallization	Crystal form	Ref.
(**4.63**)	A	(**4.64**)	H	—[b]	>300	—[c]	Colorless amorphous solid	17
(**4.63**)	B	(**4.64**)	H	43	>450	—[c]	Colorless needles	19
(**4.63**)	C	(**4.64**)	Me	24	>450	Ethanol	Colorless crystals	19
(**4.63**)	D	(**4.64**)	Me	49	—	—	—	19
(**4.63**)	E	(**4.64**)	Me	63	>350	—[d]	Colorless needles	21
(**4.63**)[f]	F	(**4.64**)	CF_3	48	>400	Ethanol–water	—[g]	66
(**4.63**)	G	(**4.64**)	CH_2Ph	39	>450	Ethanol	Yellow needles	19
(**4.63**)	H	(**4.64**)	CH=CHPh	—[b]	—[e]	—[h]	—[g]	19
(**4.63**)	I	(**4.64**)	CH_2OH	55[i]	—[e]	—[h]	Red needles	18
(**4.63**)	J	(**4.64**)	CH_2Cl	55	>400	Water	—[g]	18
(**4.63**)	K	(**4.64**)	Ph	38	>550	Ether–acetic acid	—[g]	24
(**4.63**)	L	(**4.64**)	(structure: NH_2–aryl–O–tolyl) 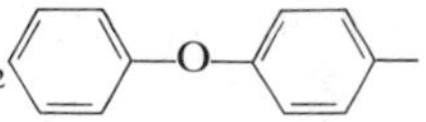	75	423	Pyridine–water	—[g]	20
(**4.63**)	M	(**4.68**)	—	—[b]	—[e]	—[c]	Grey solid	26
(**4.63**)	N	(**4.69**)	—	—[b]	—[e]	—[c]	Colorless powder	26

410

(4.70)		(4.71)						
(4.70)	O	(4.71)	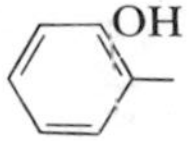 [2-hydroxyphenyl]	82	>300	—[c]	Brown powder	27
(4.70)	P	(4.71)	n-C$_6$H$_{11}$	65	290	—[c]	Light brown powder	27
(4.70)	Q	(4.71)	Bu^{i}O—[furanyl] 	48	>300	—[c]	—[g]	27
(4.70)	Q	(4.71)	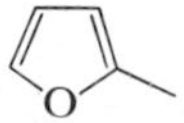 [2-methylfuranyl]	69	>300	—[c]	—[g]	27
(4.70)	Q	(4.71)	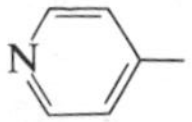 [4-methylpyridinyl]	78	>300	—[c]	—[g]	27

O: [2-hydroxyphenyl]CO$_2$Ph

a A, HCO$_2$H, 5M HCl/reflux/0.5 hr. B, HCO$_2$H, 5% HCl/180° (sealed tube)/1 hr. C, 90% HCO$_2$H/100°/20 hr. D, AcOH, 5% HCl/180° (sealed tube)/1 hr. E, Ac$_2$O/reflux/25 min. F, CF$_3$CO$_2$H, 20% HCl/reflux/6 hr, G, PhCH$_2$CO$_2$H, 15% HCl/180° (sealed tube)/1 hr. H, PhCH=CHCO$_2$H, 5% HCl/180° (sealed tube)/3 hr. I, HOCH$_2$CO$_2$H, 4M HCl/reflux/1 hr. J, ClCH$_2$CO$_2$H, 7M HCl/reflux/1 hr. K, PhCO$_2$Ph/290°/1 hr. L, p-(p'-Aminophenoxy)benzoic acid/170°/17 hr. M, COCl$_2$, H$_2$O. N, CS$_2$, NaOAc, EtOH/heat.

P, CH$_3$(CH$_2$)$_5$CHO/reflux/30 min. Q, RCHO (reaction conditions not specified).

b Yield not recorded.
c Solvent not specified.
d Purified by precipitation from HCl solution using ammonia.
e Melting point not recorded.
f Tetrahydrochloride.
g Crystal form not recorded.
h Not purified.
i Dihydrochloride.

411

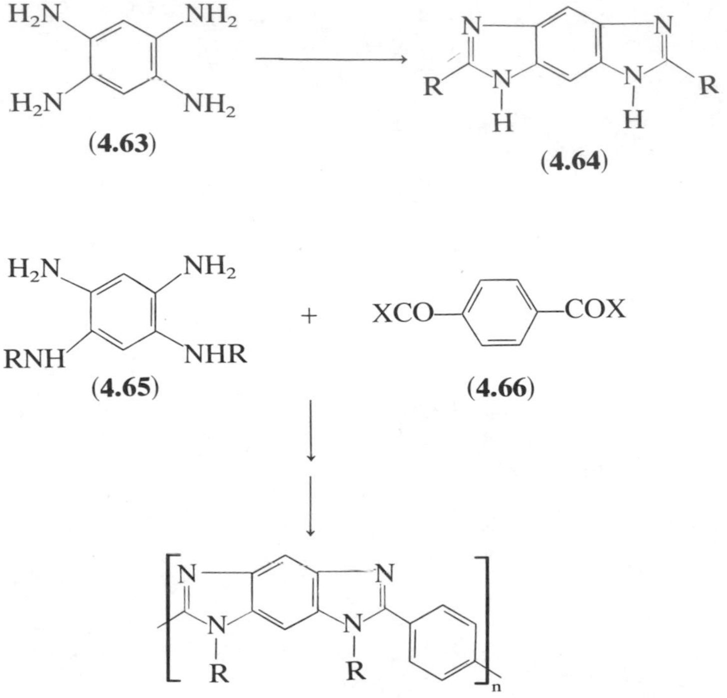

(4.63) → (4.64)

(4.65) + (4.66)

(4.67)

Scheme 4.18

the presence of 1.5 to 2.5 M hydrochloric acid.[19] The yield falls markedly (Table 4.10) as the concentration of the hydrochloric acid is increased,[19] presumably as a result of inhibition of the partial dissociation of the tetrahydrochloride of (4.63) to the free amine, which appears to be a necessary prerequisite of successful condensation. Reactions of type (4.63 → 4.64) are most successful with simple aliphatic carboxylic acids (e.g., formic and acetic acids)[17–19] but can also be accomplished (Table 4.10) using functionalized acids such as glycolic acid and chloroacetic acid,[18] as well as aryl-substituted and α,β-unsaturated acids such as phenylacetic acid and cinnamic acid,[19] though in the latter case the presumed benzodiimidazole derivative could not be purified for identification purposes.[19] Aryl carboxylic acids do not appear to have been investigated to any extent as substrates in ring closures of type (4.63 → 4.64) (Scheme 4.18). However, one report[20] describes the condensation of 1,2,4,5-tetraaminobenzene with p-(p'-aminophenoxy)benzoic acid in hot polyphosphoric acid to give the corresponding benzo[1,2-d:4,5-d']diimidazole derivative in good yield (Table 4.10).

A further variant of the general cyclization process (4.63 → 4.64) involves the reaction of the tetraamine (4.63) with acetic anhydride in the presence of hydrochloric acid, in this case the product being 2,6-dimethylbenzo[1,2-

d:4,5-*d*′]diimidazole (**4.64**; R = Me) (Table 4.10).[21] Simple carboxylic acid chlorides do not appear to have been used for effecting benzo[1,2-*d*:4,5-*d*′]diimidazole syntheses of type (**4.63 → 4.64**). However, reaction of N(1),N(5)-diphenyl-1,2,4,5-tetraaminobenzene (**4.65**, R = Ph) with terephthaloyl chloride (**4.66**; X = Cl) affords polyamides that can be thermally cyclized to polybenzo[1,2-*d*:4,5-*d*′]diimidazoles of type (**4.67**; R = Ph) (Scheme 4.18).[22] Similar polymers (**4.67**; R = toluene-*p*-sulphonyl) having high potential tensile strength and stiffness have been prepared[23] by reaction of N(1),N(5)-ditoluene-*p*-sulphonyl-1,2,4,5-tetraaminobenzene (**4.65**; R = toluene-*p*-sulphonyl) with terephthalic acid (**4.66**; X = OH) in polyphosphoric acid (Scheme 4.18). Analogous polymers are produced by heating 1,2,4,5-tetraaminobenzene (**4.63**) with phthalate esters,[24] while the corresponding thermal condensation of the tetraamine (**4.63**) with phenyl benzoate affords the monomeric 2,6-diphenylbenzo[1,2-*d*:4,5-*d*′]-diimidazole (**4.64**; R = Ph) in moderate yield (Table 4.10). Isophthalaldehyde (as the bis-bisulfite adduct) has also been used as a substrate for the conversion of 1,2,4,5-tetraaminobenzene derivatives into polybenzo[1,2-*d*:4,5-*d*′]diimidazoles.[25]

Condensation of 1,2,4,5-tetraaminobenzene (**4.63**) tetrahydrochloride with phosgene is reported to afford a product in unspecified yield (Table 4.10) formulated as benzo[1,2-*d*:4,5-*d*′]diimidazole-2,6[1(3)H,5(7)H]-dione (Scheme 4.19, **4.68**).[26] The product of the reaction of the tetraamine (**4.63**) with carbon disulfide is analogously assigned the dithione structure (**4.69**) (Table 4.10).[26]

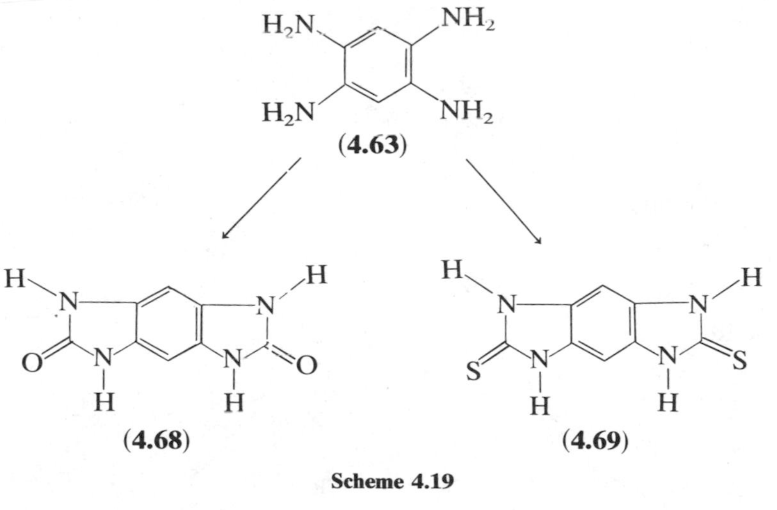

Scheme 4.19

Benzo[1,2-*d*:4,5-*d*′]diimidazole syntheses of type (**4.63 → 4.64**) (Scheme 4.18) have also been achieved with tetraamino substrates other than 1,2,4,5-tetraaminobenzene (**4.63**). Most notably, the reactions of tetraamino-1,4-benzoquinone (TABC) (Scheme 4.20, **4.70**) with aromatic carboxylic esters

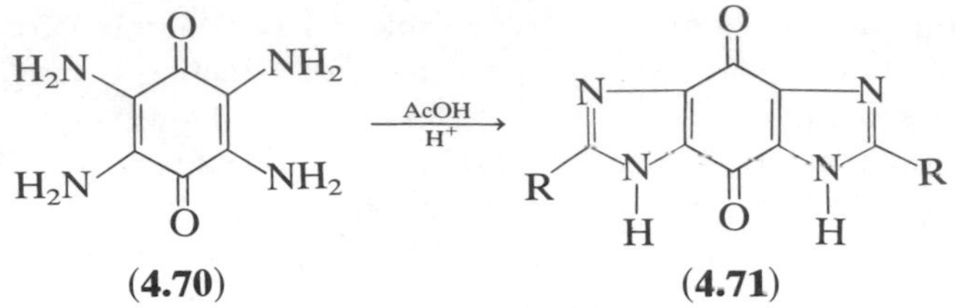

(4.70)   **(4.71)**

Scheme 4.20

in the melt, or with aliphatic and aryl and hetaryl aldehydes under reflux, provide useful routes (Table 4.10) to benzo[1,2-*d*:4,5-*d'*]diimidazole-4,8-quinones (**4.70 → 4.71**).[27]

The foregoing syntheses of benzo[1,2-*d*:4,5-*d'*]diimidazoles based on the acid-catalyzed condensation of carboxylic acids with 1,2,4,5-tetraaminobenzene (**4.63**) are most readily explained in terms of the intermediate formation and subsequent cyclodehydration of bis-acylated derivatives of types (**4.74**) or (**4.75**) (Scheme 4.21). Support for this suggestion is provided by the demonstration[19] that heating 1,2,4,5-tetraacetyl-aminobenzene (**4.72**) with aqueous hydrochloric acid results in a moderate yield (Table 4.11) of 2,6-dimethylbenzo[1,2-*d*:4,5-*d'*]diimidazole (**4.77**; R = H). Formation of the same product by the reductive cyclization (Table

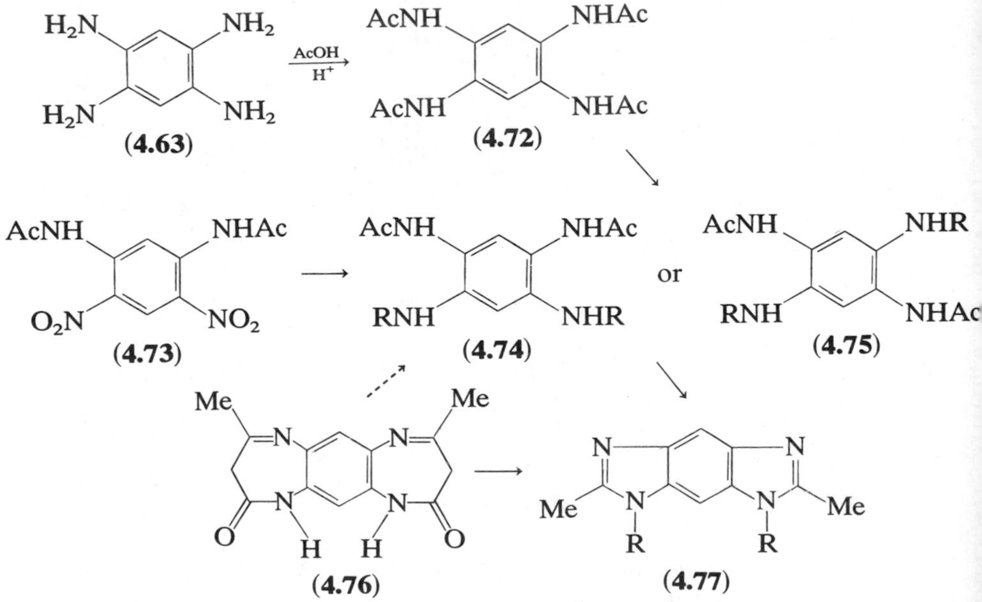

Scheme 4.21

Starting material	R	Reaction conditions[a]	Product	R	Yield(%)	Melting point(°C)	Solvent of crystallization	Crystal form	Ref.
(4.72)	—	A	(4.77)	H	50	>300	—[b]	—[c]	19
(4.73)	—	B	(4.77)	H	—[d]	>300	—[b]	—[c]	28a
(4.76)	—	C	(4.77)	H	—[d]	>300	Methanol–water	Tan plates	29
(4.74)	Ph	D	(4.77)	Ph	—[d]	308	Ethanol–water	—[c]	19
(4.80)	H	E	(4.81)	H[f,g]	72	>300	—[e]	Grey crystals	27
(4.78)	H	F	(4.81)	H[f,g]	Quant.	>300	6 M HCl	—[c]	33
(4.78)	Me	G	(4.81)	Me[f,g]	75	>300	6 M HCl	—[c]	33
(4.78)	n-Bu	H	(4.81)	n-Bu[f]	—[d]	313 (decomp.)	—[b]	—[c]	33
(4.78)	$(CH_2)_{11}$-Me	I	(4.81)	$(CH_2)_{11}Me$[f,g]	89	298–301	Ethanol–chloroform	—[c]	33
(4.78)	$(CH_2)_2NEt_2$	J	(4.81)	$(CH_2)_2NEt_2$[f,i]	82	308–310 (decomp.)	Ethanol–ether	—[c]	33
(4.78)	$(CH_2)_2OH$	K	(4.81)	$(CH_2)_2OH$[h]	—[a]	287–290 (decomp.)	Ethanol–ethyl acetate	Colorless crystals	33
(4.78)	$(CH_2)_2N$⟨morpholine⟩	L	(4.81)	$(CH_2)_2N$⟨morpholine⟩	—[d]	304–306 (decomp.)	—[b]	—[c]	33
(4.78)	$(CH_2)_3NMe_2$	—[j]	(4.81)	$(CF_2)_3NMe_2$[h]	—[d]	291–293 (decomp.)	—[b]	—[c]	33
(4.78)	Ph	M	(4.79)	Ph	—[d]	360	Acetic acid	Yellow crystals	34

[a] A, 2.5 M HCl/reflux/7 hr. B, Sn, SnCl$_2$/HCl. C, 2 N H$_2$SO$_4$/reflux/1 hr. D, concn. HCl/heat. E, concn. HCl/100°/3 hr or HCl gas, HCl, H$_2$O. F, (i) Raney Ni, H$_2$/room temp./atm. pressure; (ii) 6 M HCl/room temp. G, (i) Raney Ni, H$_2$/room temp./atm. pressure; (ii) 2 M HCl/room temp. H, (i) Raney Ni, H$_2$/room temp./atm. pressure; (ii) 2.4 M HCl EtOH. I, (i) Raney Ni, H$_2$/45–50°/atm. pressure; (ii) 2.4 M HCl. J, (i) Raney Ni, H$_2$/room temp./slight excess pressure; (ii) 2.75 M HCl, EtOH. K, (i) Raney Ni, H$_2$/room temp/atmos. pressure; (ii) 6 M HCl, EtOH/reflux/5 min. L, (i) 1 M HCl, EtOH, Raney Ni, H$_2$/room temp./atm. pressure; (ii) 2.4 M HCl, EtOH. M NaOH, EtOH/reflux/3 hr.

[b] Solvent not specified.
[c] Crystal form not specified.
[d] Yield not specified.
[e] Not purified.
[f] Dihydrochloride.
[g] Free base has m.p. >300°.
[h] Tetrahydrochloride.
[i] Free base has m.p. 198–201°.
[j] Reaction conditions not specified.

4.11) of 1,5-diacetylamino-2,4-dinitrobenzene[28a] is likewise plausibly explained in terms of the intermediate formation and cyclodehydration of 1,5-
diacetylamino-2,4-diaminobenzene (**4.73** → **4.74** → **4.77**; R = H). The latter
compound may also be implicated as intermediate in the acid-catalyzed
ring contraction[29] of 2,10-dimethyl-5,7-dihydrobenzo-bis[1,2-*b* : 4,5-*b'*][1,5]-
diazepine-4,8-dione   (**4.76** → **4.74** → **4.77**;   R = H)   (Table   4.11).   The
thermal[30,31]   or   pyridine-acetic   anhydride   mediated[32]   cyclizations   of
monomeric and polymeric amides derived from 1,2,4,5-tetraaminobenzene
derivatives have been used to synthesize monomeric 2,6-diarylbenzo[1,2-
d : 4,5-*d'*]diimidazoles,[32] analogous polymeric systems,[30] and dyestuffs incorporating the benzo[1,2-*d* : 4,5-*d'*]diimidazole nucleus.[31]

Marxer,[33] and independently Winkelmann,[27] have shown that acid-
catalyzed   cyclization   of   2,5-diacetylamino-3,6-diamino-1,4-dihydroxy-
benzenes [preformed[27] or generated *in situ* by catalytic reduction of the
corresponding quinones (**4.78**) (Scheme 4.22)] proceeds smoothly and
affords good yields (Table 4.11) of 4,8-dihydroxybenzo[1,2-*d* : 4,5-*d'*]-
diimidazoles which are readily oxidized to the corresponding quinones
(see section 4.3.3.) (**4.80** → **4.81** → **4.79**). The facility of the cyclizations
(**4.80** → **4.81**) contrasts with the low yields (Table 4.11) of the quinone
(**4.79**; R = Ph) obtained[33,34] by thermal, acid, or base-catalyzed cyclization of
the 1,4-benzoquinone derivative (**4.78**; R = Ph). The reason for the greater
ease of cyclization of the hydroquinones (**4.80**) compared with the parent
quinones (**4.78**) is not clear.[33]

The benzo[1,2-*d* : 4,5-*d'*]diimidazole syntheses described so far share the
common feature that both imidazole rings are constructed simultaneously
from symmetrically substituted, nonheterocyclic 1,2,4,5-tetraaminated benzene precursors. It follows that such synthetic routes lead to symmetrically

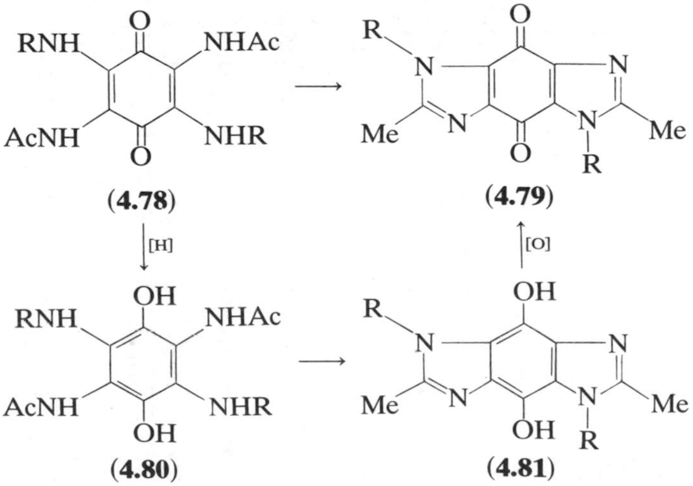

Scheme 4.22

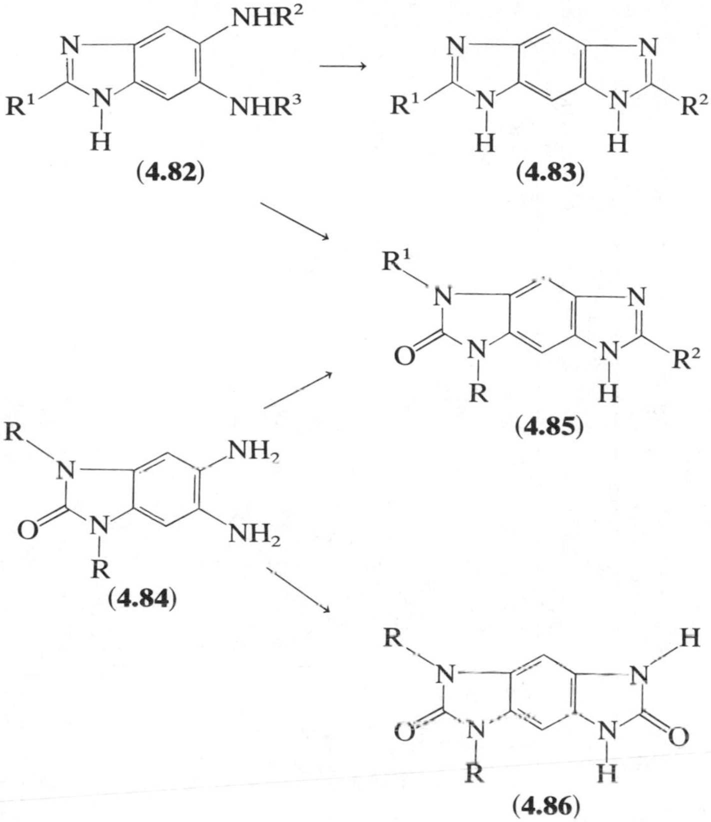

Scheme 4.23

substituted, rather than unsymmetrically substituted benzo[1,2-d:4,5-d']-diimidazole derivatives. However, compounds of the latter type are readily accessible by ring closure reactions of suitable 5,6-diaminobenzimidazole derivatives (Scheme 4.23) (Table 4.12). Thus whereas 5,6-diaminobenzimidazole (**4.82**; $R^1 = R^2 = R^3 = H$) condenses (as the dihydrochloride) with formic acid in the presence of aqueous hydrochloric acid to afford[35] the parent benzo[1,2-d:4,5-d']diimidazole (**4.83**; $R^1 = R^2 = H$), reaction with other aliphatic and aromatic acids under similar conditions leads to unsymmetrically substituted benzodiimidazoles (**4.83**; $R^1 = H$, $R^2 =$ alkyl or aryl) in good yield (Table 4.12).[35] 5,6-Diamino-2-methylbenzimidazole (**4.82**; $R^1 = Me$, $R^2 = R^3 = H$) likewise reacts as the dihydrochloride with acetic acid in the presence of hydrochloric acid[17,35] or sodium acetate[36] to afford the symmetrical 2,6-dimethyl-1(3)H,5(7)H-benzo[1,2-d:4,5-d']diimidazole (**4.83**; $R^1 = R^2 = Me$) in excellent yield (Table 4.12) and with other carboxylic acids (both aliphatic and aromatic) in the presence of hydrochloric acid to give the expected unsymmetrically

TABLE 4.12. SYNTHESES OF BENZO[1,2-d: 4,5-d']DIIMIDAZOLES (**4.83**, **4.85**, **4.86**, AND **4.88**) BY RING-CLOSURE REACTIONS OF 5,6-DIAMINOBENZIMIDAZOLE DERIVATIVES (**4.82**, **4.84**, AND **4.87**)

Starting material[a] R	R¹	R²	R³	Reaction conditions[b]	Product	R	R¹	R²	Yield (%)	Melting point (°C)	Solvent of crystallization	Crystal form	Ref.
(**4.82**) —	H	H	H	A	(**4.83**)	—	H	H	76	360	—[c]	Colorless needles	35
(**4.82**) —	H	H	H	B	(**4.83**)	—	H	Me	70	388–390	—[c]	Colorless solid	35
(**4.82**) —	Me	H	H	A	(**4.83**)	—	H	Me	85	—	—	—	35
(**4.82**) —	Me	H	H	C	(**4.83**)	—	H	Me	—[d]	>300	—[e]	Amorphous solid	17
(**4.82**) —	Me	H	H	D	(**4.83**)	—	Me	Me	—[d]	>300	—[e]	Amorphous solid	17
(**4.82**) —	Me	Ac	H	E	(**4.83**)	—	Me	Me	—[d]	—	—	—	17
(**4.82**) —	Me	Ac	Ac	E	(**4.83**)	—	Me	Me	—[d]	—	—	—	17
(**4.82**) —	Me	H	H	B	(**4.83**)	—	Me	Me	80	438–9	—[c]	—[f]	35
(**4.82**) —	Me	H	H	F	(**4.83**)	—	Me	Ph	60	—[g]	—[h]	—	35
(**4.82**) —	Me	H	H	G	(**4.83**)	—	Me	CH₂Ph	77	—[g]	—[i]	—	35
(**4.82**) —	Me	H	H	H	(**4.83**)	—	Me	(CH₂)₂Ph	73	—[g]	—[i]	—	35
(**4.82**) —	Me	H	H	I	(**4.83**)	—	Me	_2-phenyl-5-methylbenzimidazole structure_	—[d]	—[g]	—[i]	—	35
(**4.82**) —	H	H	H	F	(**4.83**)	—	H	Ph	51	—[g]	—[i]	—	35
(**4.82**) —	H	H	H	G	(**4.83**)	—	H	CH₂Ph	69	—[g]	—[i]	—	35
(**4.82**) —	H	H	H	H	(**4.83**)	—	H	(CH₂)₂Ph	58	—[g]	—[i]	—	35
(**4.82**) —	H	H	H	I	(**4.83**)	—	H	_2-phenyl-5-methylbenzimidazole structure_	49	—[g]	—[j]	—	35

				Method	Product			Yield	m.p. (°C)	Solvent	Crystal form	Ref.
(4.84)	H	—	—	J	(4.85)	H	Me	—[d]	>300	—[e]	—[f]	36
(4.82)	—	Me	—	K	(4.85)	H	Me	—[d]	>300	—[e]	—[f]	36
(4.84)	Me	—	H	L	(4.85)	Me	H	63	276	Acetic acid	Colorless powder	37
(4.84)	Me	—	—	M	(4.85)	—	Me	65	>320	Ethanol	Colorless powder	37
(4.84)	H	—	H	N	(4.86)	H	—	94	>320	—[k]	Grey amorphous powder	37
(4.84)	Me	—	Me	N	(4.86)	Me	—	76	>320	Acetic acid	Colorless powder	37
(4.87)	—	—	—	O	(4.88)	—	—	Quant.	>350	Water	Colorless solid	38

[a] Dihydrochloride.
[b] A, HCO$_2$H, 15% HCl/180° (sealed tube)/1 hr. B, MeCO$_2$H, 15% HCl/80° (sealed tube)/1 hr. C, HCO$_2$H, 5 M HCl/reflux/30 min. D, MeCO$_2$H, 5 M HCl/reflux. E, 5 M HCl/reflux. F, PhCO$_2$H, 15% HCl/180° (sealed tube)/1 hr. G, PhCH$_2$CO$_2$H, 15% HCl/180° (sealed tube)/1 hr. H, Ph(CH$_2$)$_2$CO$_2$H, 15% HCl/180° (sealed tube)/1 hr.

I, [indole structure with N–Ph, NH], 15% HCl/180° (sealed tube)/1 hr.

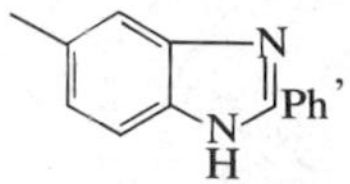

J, MeCO$_2$H, NaOAc/reflux/3 hr. K, urea/180°/few minutes. L, 99% HCO$_2$H/reflux/2 hr. M, Ac$_2$O/reflux/2 hr. N, urea/170°. O, urea/170–190°/30 min.

[c] Purified by precipitation from aqueous acetic acid by ammonia.
[d] Yield not quoted.
[e] Method of purification not specified.
[f] Crystal form not specified.
[g] Melting point not specified.
[h] Purified as the hydrochloride, colorless needles (from water), m.p. unspecified.
[i] Purified as the dihydrochloride, colorless needles (from water), m.p. unspecified.
[j] Purified as the trihydrochloride, colorless needles (from water), m.p. unspecified.
[k] Purified by precipitation from alkaline solution by acid, m.p. unspecified.

substituted benzodiimidazoles (**4.83**; R^1 = Me, R^2 = alkyl or aryl), again in excellent yield (Table 4.12).[17,35]

Benzo[1,2-d:4,5-d']diimidazoles (**4.85**) in which one of the heterocyclic rings is in the imidazolone form are readily prepared (Table 4.12) by reaction of 5,6-diaminobenzimidazol-2-one (**4.84**) dihydrochlorides with formic acid,[37] acetic acid in the presence of sodium acetate,[36] or acetic anhydride[37] (**4.84** → **4.85**). Conversely, compounds of this type can be obtained by the reaction of simple 5,6-diaminobenzimidazoles (e.g., 5,6-diamino-2-methylbenzimidazole) as the dihydrochlorides, with urea in a melt [**4.82** (R^1 = Me, R^2 = R^3 = H) → **4.85** (R^1 = H, R^2 = Me)] (Table 4.12).[36,37] Symmetrical and unsymmetrical benzo[1,2-d:4,5-d']diimidazoles containing two imidazolone rings (**4.86**) are likewise formed (Table 4.12) on fusion of 5,6-diaminobenzimidazolones with urea (cf. Scheme 4.23, **4.84** → **4.86**,[37] and Scheme 4.24, **4.87** → **4.88**).[38] The intermediacy of mono or diacyl derivatives of the starting 5,6-diaminobenzimidazoles in transformations of types (**4.82** → **4.83**) or (**4.84** → **4.85**) is supported by the finding[17] that the mono- and di-N-acetyl derivatives of 5,6-diaminobenzimidazole (**4.82**; R^1 = H, R^2 = Ac, R^3 = H or Ac) cyclize on treatment with hydrochloric acid to afford in both cases 2,6-dimethyl-1(3)H,5(7)H-benzo[1,2-d:4,5-d']-diimidazole (**4.83**; R^1 = R^2 = Me), albeit in unspecified yield (Table 4.12).[17]

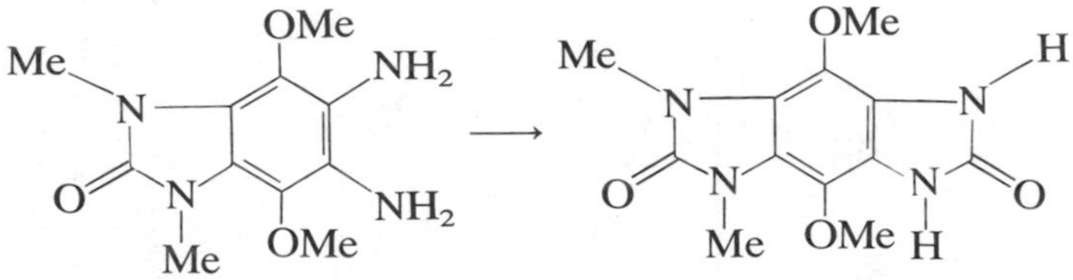

(**4.87**)   (**4.88**)

Scheme 4.24

Synthetic routes to the angular benzo[1,2-d:3,4-d']diimidazole ring system are unexceptional (Scheme 4.25) (Table 4.13) and parallel the methods already discussed at some length for the synthesis of derivatives of the linear benzo[1,2-d:4,5-d']diimidazole ring system. For example, acid-catalyzed cyclization[28b] of 4,5-diacetylamino-2-methylbenzimidazole (**4.89**) leads to the formation of 2,7-dimethyl-1(3)H,6(8)H-benzo[1,2-d:3,4-d']diimidazole (**4.89** → **4.90**) (Table 4.13).

Correspondingly, simple alkyl-substituted benzo[1,2-d:3,4-d']diimidazoles are most commonly synthesized in good-to-excellent yield (Table 4.13) by heating a 4,5-diaminobenzimidazole derivative under reflux with an aliphatic carboxylic acid (**4.91** → **4.92**).[39–42] Variants of this type of ring closure (Scheme 4.25) include fusion of a 4,5-diaminobenzimidazole with urea to yield the corresponding benzo[1,2-d:3,4-d']diimidazol-2(1H,3H)-ones (**4.91** → **4.93**)[41] and the ring closure of 4,5-diaminobenzimidazoles

TABLE 4.13. SYNTHESES OF BENZO[1,2-d:3,4-d']DIIMIDAZOLE DERIVATIVES (**4.90**, **4.92**, **4.93**, AND **4.94**) BY RING-CLOSURE REACTIONS OF 4,5-DIAMINOBENZIMIDAZOLE DERIVATIVES (**4.89** AND **4.91**)

Starting material	Reaction conditions[a]	Product	Yield (%)	Melting point (°C)	Solvent of crystallization	Crystal form	Ref.
(**4.89**)	A	(**4.90**)	—[b]	145	—[a]	Colorless needles	28b
(**4.91**) ($R^1 = R^2 = H$)	B	(**4.92**) ($R^1 = R^2 = R^3 = R^4 = H$)	27	288–296	Water	—[c]	39
(**4.91**) ($R^1 = Me, R^2 = R^3 = H$)	B	(**4.92**) ($R^1 = Me, R^2 = R^3 = R^4 = H$)	94	312–314	Water	Colorless needles	39
(**4.91**) ($R^1 = Me, R^2 = R^3 = H$)	C	(**4.92**) ($R^1 = R^4 = Me, R^2 = R^3 = H$)	81	292–294	Water	Colorless needles	39
(**4.91**) ($R^1 = H, R^2 = R^3 = Me$)	D	(**4.92**) ($R^1 = R^4 = H, R^2 = R^3 = Me$)	80	237–238	Benzene	Colorless needles	40
(**4.91**) ($R^1 = R^3 = H, R^2 = Me$)	D	(**4.92**) ($R^1 = R^3 = R^4 = H, R^2 = Me$)	89	318–319	Water	Colorless needles	40
(**4.91**) ($R^1 = H, R^2 = R^3 = Et$)	D	(**4.92**) ($R^1 = R^4 = H, R^2 = R^3 = Et$)	71	192–193	Benzene	Colorless crystals	40
(**4.91**) ($R^1 = R^3 = H, R^2 = Et$)	D	(**4.92**) ($R^1 = R^3 = R^4 = H, R^2 = Et$)	90	323–324	Benzene	Colorless crystals	40
(**4.91**) ($R^1 = H, R^2 = R^3 = Me$)	E	(**4.92**) ($R^1 = H, R^2 = R^3 = R^4 = Me$)	69	292–293	—[d]	Colorless crystals	41
(**4.91**) ($R^1 = R^2 = R^3 = Me$)	E	(**4.92**) ($R^1 = R^2 = R^3 = R^4 = Me$)	80	359–360	Water	Colorless crystals	41
(**4.91**) ($R^1 = H, R^2 = R^3 = Me$)	F	(**4.93**) ($R^1 = H, R^2 = R^3 = Me$)	63	>360	Ethanol	Colorless crystals	41
(**4.91**) ($R^1 = R^2 = R^3 = Me$)	G	(**4.94**) ($R^1 = R^2 = R^3 = Me$)	88	334 (decomp.)	Water	Rose needles	41
(**4.91**) ($R^1 = H, R^2 = R^3 = Me$)	G	(**4.94**) ($R^1 = H, R^2 = R^3 = Me$)	63	252	Water	—[c]	41

[a] A, H_2SO_4/100°. B, HCO_2H/reflux/2 hr. C, AcOH/180°. D, 85% HCO_2H/reflux/3 hr. E, AcOH/reflux/6 hr. F, urea/150–160°/30 min. G, Br_2, NaCN/0°/1 hr, then 15–20°/1 hr.
[b] Yield not recorded.
[c] Crystal form not specified.
[d] Solvent not specified.

421

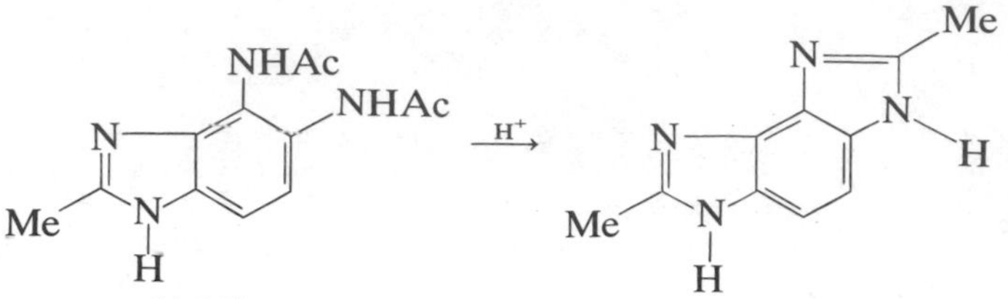

(4.89)   (4.90)

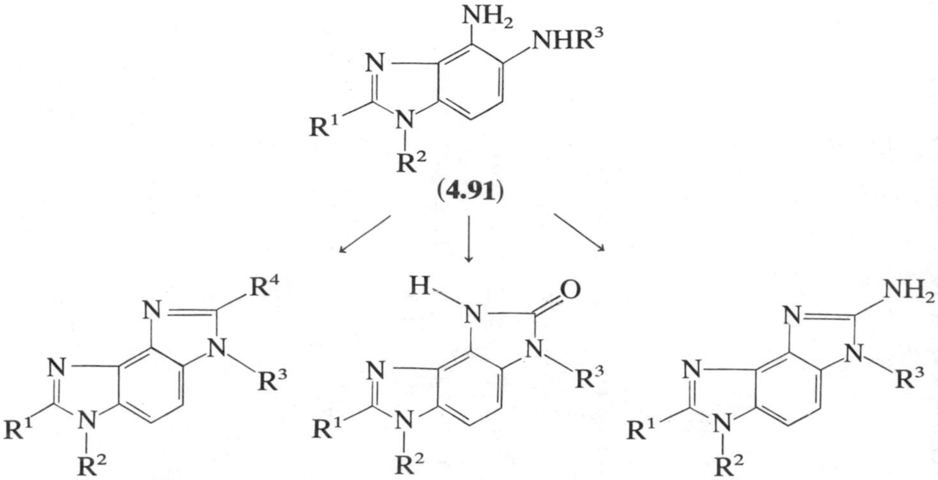

(4.91)

(4.92)   (4.93)   (4.94)

Scheme 4.25

with cyanogen bromide, which provides a useful route to 2-aminobenzo-[1,2-d:3,4-d']diimidazoles (**4.91 → 4.94**).[41]

Ring-Closure Reactions of Other Heterocycles

Methods for the synthesis of the [1,3]dioxolo[4,5-f]benzimidazole ring system are based exclusively on acylative ring-closure reactions of 5,6-diaminobenzo[1,3]dioxoles (Scheme 4.26) (Table 4.14). In most instances, the diamine is not preformed but is more conveniently prepared *in situ* by reduction of the corresponding 5,6-dinitro- or 5-amino-6-nitrobenzo[1,3]dioxole derivatives (Scheme 4.26). For example, tin and hydrochloric acid reduction of 5,6-dinitrobenzo[1,3]dioxole affords the expected 5,6-diamino compound,[43] which is condensed without isolation with formic acid to afford N-5(7)H-[1,3]dioxolo[4,5-f]benzimidazole in very good yield [**4.95 → 4.97** (R^1 = R^2 = H) → **4.98** (R^1 = R^2 = R^3 = H)] (Table 4.14).

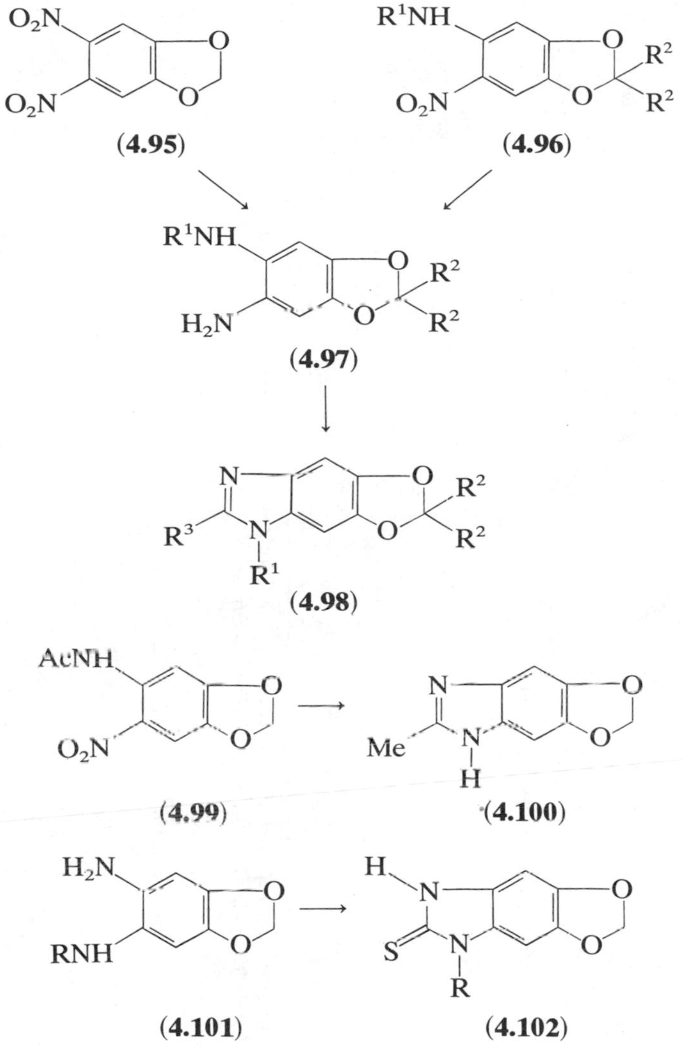

(4.95)   (4.96)

(4.97)

(4.98)

(4.99)   (4.100)

(4.101)   (4.102)

Scheme 4.26

The analogous stannous chloride reduction of 5-amino-[5] or substituted-amino-[44] 6-nitrobenzo[1,3]dioxoles (e.g., **4.96**; R^1 = H or Et, R^2 = F or H) to the corresponding amino compounds (**4.97**; R^1 = H or Et, R^2 = F or H) followed by *in situ* condensation with acetic anhydride leads to 6-methyl[1,3]dioxolo[4,5-*f*]benzimidazoles (**4.98**; R^1 = H or Et, R^2 = F or H, R^3 = Me) though in low overall yield (Table 4.14). Variants of such syntheses (Scheme 4.26) include the reductive cyclization (Table 4.14) of 5-acetylamino-6-nitrobenzo[1,3]dioxoles   to   6-methyl[1,3]dioxolo[4,5-*f*]-benzimidazoles (**4.99 → 4.100**)[45] and the reaction of preformed 5,6-diaminobenzo[1,3]dioxoles with carbon disulfide under basic conditions

TABLE 4.14. SYNTHESES OF [1,3]DIOXOLO[4,5-f]BENZIMIDAZOLE DERIVATIVES (**4.98**, **4.100**, AND **4.102**) BY RING-CLOSURE REACTIONS OF 5,6-DISUBSTITUTED BENZO[1,3]DIOXOLE DERIVATIVES

Starting material	R	R^1	R^2	Reaction conditions[a]	Product	R	R^1	R^2	R^3	Yield (%)	Melting point (°C)	Solvent of crystallization	Crystal form	Ref.
(**4.95**)	—	—	—	A	(**4.98**)	—	H	H	H	77	209–210	Ethanol–ether	Colorless needles	43
(**4.96**)	—	H	F	B	(**4.98**)	—	H	F	Me	16	226–227 (decomp.)	—[b]	—[c]	5
(**4.96**)	—	Et	H	C	(**4.98**)	—	Et	H	Me	50	160	Acetone–ether	—[c]	44
(**4.99**)	—	—	—	D	(**4.98**)[d]	—	—	—	—	—[e]	226–227	Water	Colorless needles	45
(**4.101**)	H	—	—	E	(**4.102**)	H	—	—	—	73	350	Dimethylformamide–water	Yellow crystals	46
(**4.101**)	Et	—	—	F	(**4.102**)	Et	—	—	—	—[e]	273–275	Acetic acid–water	—[c]	44

[a] A, Sn, HCl, HCO_2H/100°. B, (i) $SnCl_2$, concn. HCl, MeOH; (ii) Ac_2O, 20% HCl/reflux/4 hr. C, (i) $SnCl_2$, HCl, MeOH/room temp./1.5 hr; (ii) Ac_2O/140–150°/0.5 hr. D, Sn, $SnCl_2$, AcOH/100°/1 hr. E, CS_2, KOH, EtOH, H_2O/reflux/2 hr. F, CS_2, NaOEt, EtOH/100°/1 hr.
[b] Solvent not specified.
[c] Crystal form not specified.
[d] Forms a picrate, yellow needles, m.p. 230–250 (decomp.) (from methanol).
[e] Yield not quoted.

(Table 4.14) to afford [1,3]dioxolo[4,5-*f*]benzimidazole-6-thiones (**4.101** → **4.102**, R = H or Et).[44,46]

What appears to be the sole example of an imidazo[4,5-*f*]benzoxazole derivative has been prepared by the action of phosgene on 5,6-diaminobenzoxazol-2(3*H*)-one (Scheme 4.27, **4.103** → **4.104**).[47] The properties and yield of this product were not recorded, however.[47]

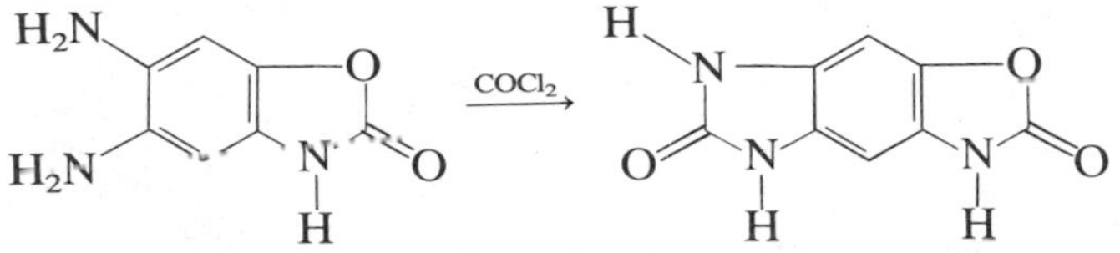

(4.103)   **(4.104)**

Scheme 4.27

In contrast to the paucity of examples of tricyclic 5-6-5 fused benzimidazoles containing an oxazole unit, similar frameworks incorporating a thiazole ring are well documented. Synthetic routes to the three possible 5-6-5 fused thiazolobenzimidazole ring system (cf. Scheme 4.15, **4.48–4.50**) rely heavily on ring-closure reactions (Scheme 4.28) of suitable 4,5-, 5,6-, and 6,7-diaminobenzothiazole derivatives.

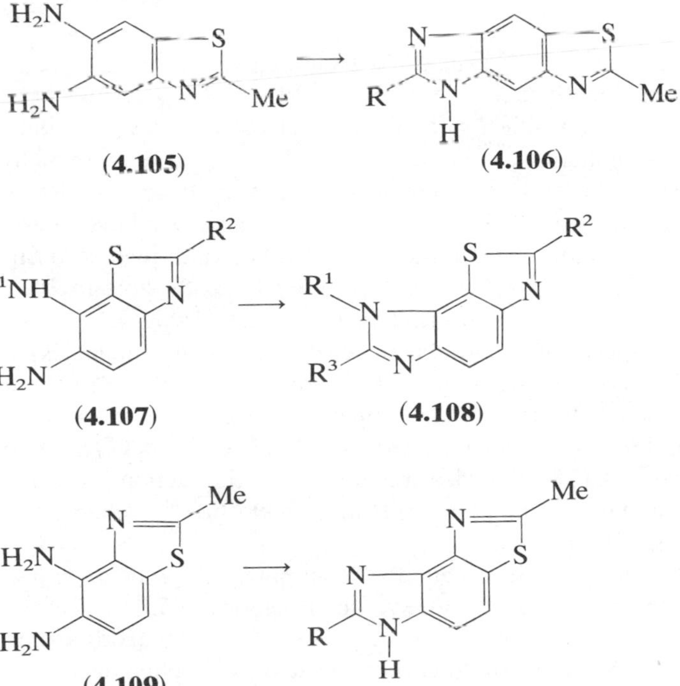

(4.105)   **(4.106)**

(4.107)   **(4.108)**

(4.109)   **(4.110)**

Scheme 4.28

TABLE 4.15. SYNTHESES OF IMIDAZO[4,5-*f*]BENZOTHIAZOLE DERIVATIVES (**4.106**, **4.111**, AND **4.112**) BY RING-CLOSURE REACTIONS[48] OF 5,6-DIAMINO-2-METHYLBENZOTHIAZOLE (**4.105**)

Reaction conditions[a]	Product[b]	R	Yield (%)	Melting point (°C)	Solvent of crystallization
A	(**4.106**)	H[c,d]	61	283	70% Ethanol–water
A	(**4.106**)	Me[e,f]	58	244	70% Ethanol–water
A	(**4.106**)	Et[g]	71	215–216	Methanol
A	(**4.106**)	CH_2OH	65	234–235	Water
B	(**4.106**)	Ph	35	254–255	70% Ethanol–water
C	(**4.106**)	CH_2Ph	26	171	Ethanol
A	(**4.106**)	$(CH_2)_2CO_2H$	35	182 (decomp.)	Water
D	(**4.111**)[h]	—	70	>300	[i]
E	(**4.112**)	—	51	>300	[i]

[a] A, RCO_2H, 15% HCl/reflux/2 hr. B, PhCHO, $PhNO_2$, EtOH/200°/1 min. C, $PhCH_2CO_2H$/125–130°/30 min.

[b] Colorless crystalline solids.

[c] Dihydrate.

[d] Forms a picrate, yellow needles, m.p. 234° (from methanol).

[e] Forms a dihydrochloride, m.p. 298° (decomp.).

[f] Forms a picrate, yellow needles, m.p. 246° (from ethanol).

[g] Forms a picrate, yellow needles, m.p. 238° (decomp.) (from methanol).

[h] Precipitation from 5% Na_2CO_3 with AcOH.

[i] Precipitation from 1% NaOH with AcOH.

Ring closure of 5,6-diamino-2-methylbenzothiazole to 2-methylimidazo[4,5-*f*]benzothiazoles (Scheme 4.28, **4.105** → **4.106**) occurs readily and in moderate-to-high yield (Table 4.15) by heating the diamine under reflux with aliphatic carboxylic acids in the presence of 15% hydrochloric acid,[48] or in the case of phenylacetic acid by fusion of the reactants at elevated temperature.[48] Ring closure with benzoic acid even in the presence of 25% hydrochloric acid under sealed tube conditions is unsatisfactory.[48] However, 6-phenyl-2-methylimidazo[4,5-*f*]benzothiazole (**4.106**; R = Ph) is readily accessible by the alternative procedure (Table 4.15) (also employed for the synthesis of 6,7-dihydropyrrolo[2,3-*f*]benzimidazoles; see section 4.2.1) of treating the diamine (**4.105**) with benzaldehyde in the presence of nitrobenzene.[48] Fusion of the diamine (**4.105**) with urea affords[48] 2-methylimidazo-[4,5-*f*]benzothiazol-6(5*H*,7*H*)-one (**4.111**), whose thione counterpart (**4.112**) likewise results from the action of boiling carbon disulfide on the diamine (**4.105**). Both ring closures[48] (Scheme 4.29) proceed in good yield (Table 4.15).

6,7-Diaminobenzothiazoles also react smoothly with aliphatic carboxylic acids in the presence of aqueous hydrochloric acid to give moderate to excellent yields (Table 4.16) of imidazo[4,5-g]benzothiazoles (**4.107** → **4.108**).[48,49] Aryl-substituted carboxylic acids (e.g., phenylacetic acid) require fusion conditions[48] to react, while benzoic acid derivatives fail to react at all,

TABLE 4.16. SYNTHESES OF IMIDAZO[4,5-g]BENZOTHIAZOLE DERIVATIVES (**4.108**, **4.114**, **4.115**, AND **4.117**) BY RING-CLOSURE REACTIONS OF 6,7-DIAMINOBENZOTHIAZOLE DERIVATIVES (**4.107** AND **4.113**)

Starting material	R^1	R^2	Reaction conditions[a]	Product[b]	R^1	R^2	R^3	Yield (%)	Melting point (°C)	Solvent of crystallization	Ref.
(4.107)	H	H	A	**(4.108)**	H	H	H[c]	84	221	Water	49
(4.107)	H	Me	B	**(4.108)**	H	Me	H[d,e]	74	213	Water	48
(4.107)	H	Me	B	**(4.108)**	H	Me	Me	90	204	Water	48
(4.107)	H	Me	B	**(4.108)**	H	Me	Et	65	210–213	Water	48
(4.107)	H	Me	B	**(4.108)**	H	Me	CH_2OH	74	230	Water	48
(4.107)	H	Me	B	**(4.108)**	H	Me	$CH(OH)Me$	58	140 (decomp.)	70% Ethanol–water	48
(4.107)	H	Me	C	**(4.108)**	H	Me	Ph	35	139 (decomp.)	Ethanol–water	48
(4.107)	H	Me	D	**(4.108)**	H	Me	CH_2Ph	38	263–264	Ethanol	48
(4.107)	H	Me	B	**(4.108)**	H	Me	$(CH_2)_2CO_2H$	16	244–245	Water	48
(4.107)	Ph	Me	E	**(4.108)**	Ph	Me	H	92	142	Ethanol–water	50
(4.107)	Ph	Me	F	**(4.108)**	Ph	Me	Me	75	229	Ethanol	50
(4.113)	—	—	G	**(4.114)**	—	—	—	49	234 (decomp.)	Ethanol	51
(4.116)	—	—	H	**(4.115)**	—	—	—	77	>330	—[f]	48
(4.116)	—	—	I	**(4.117)**	—	—	—	63	>330	—[g]	48

[a] A, HCO_2H, 4 M HCl/reflux/2 hr. B, R^3CO_2H, 15% HCl/reflux/2 hr. C, PhCHO, $PhNO_2$, EtOH/200°/1 min. D, $PhCH_2CO_2H$/120–130°/30 min. E, HCO_2H/reflux/1 hr. F, Ac_2O/reflux/1 hr. G, 85% HCO_2H/reflux/3 hr. H, urea/150–5°/30 min. I, CS_2, EtOH/reflux/3 hr.

[b] Colorless crystalline solids.

[c] Forms a picrate, yellow needles, m.p. 274–275° (decomp.) (from ethanol).

[d] Forms a dihydrochloride, m.p. 346–347° (decomp.).

[e] Forms a picrate, yellow needles, m.p. 246° (from ethanol).

[f] From 5% Na_2CO_3 by precipitation with acetic acid.

[g] From 1% NaOH by precipitation with acetic acid.

427

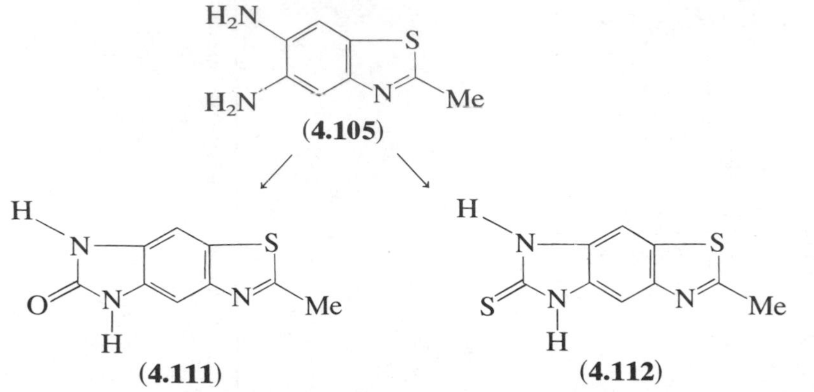

(**4.105**)

(**4.111**)   (**4.112**)

Scheme 4.29

thus necessitating the synthesis of 7-arylimidazo[4,5-g]benzothiazoles (e.g., **4.108**; R^3 = Ph) by reaction of the diamino compound (**4.107**; R^1 = H) with an aromatic aldehyde (e.g., benzaldehyde) in the presence of an oxidizing agent such as nitrobenzene (Table 4.16).[48]

Acylative cyclizations of type (**4.107** → **4.108**) (Scheme 4.28) can also be accomplished in the case of the corresponding N-substituted 6,7-diaminobenzothiazoles. Thus 6-amino-2-methyl-7-N-phenylaminobenzo-thiazole (**4.107**; R^1 = Ph, R^2 = Me) heated with formic acid or acetic anhy-dride affords[50] the 8-phenylimidazo[4,5-g]benzothiazoles (**4.108**; R^1 = Ph, R^2 = Me, R^3 = H or Me) in excellent yield (Table 4.16). The 3-N-ethyl quaternary salt (**4.114**) on the other hand is the end product (Table 4.16) of the reaction (Scheme 4.30) of the N-ethylbenzothiazolium iodide (**4.113**) with hot formic acid.[51]

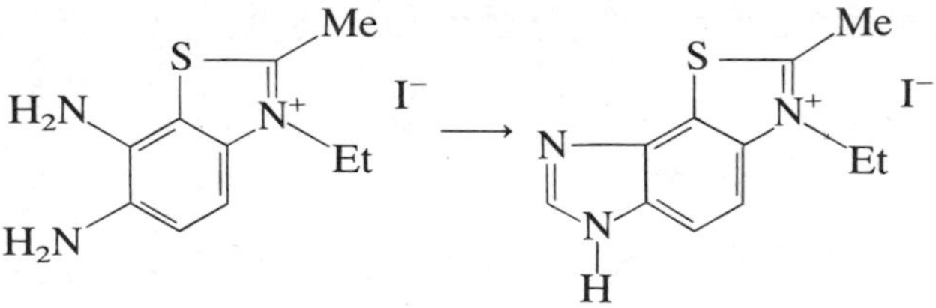

(**4.113**)   (**4.114**)

Scheme 4.30

As in the case of 5,6-diamino-2-methylbenzothiazole discussed earlier in this section the 6,7-diamino isomer (**4.116**) reacts readily with urea under fusion conditions (Scheme 4.31), and with carbon disulfide under reflux to afford the imidazo[4,5-g]benzothiazolone (**4.115**)[48] and its sulfur analog (**4.117**), respectively (Table 4.16).

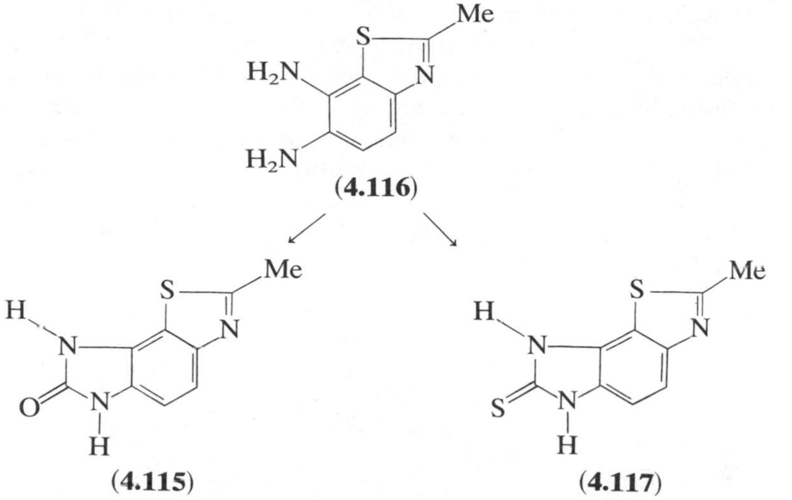

(4.116)

(4.115)   **(4.117)**

Scheme 4.31

Extension of the ring closure reactions described already for 5,6-diaminobenzothiazoles and 6,7-diaminobenzothiazoles, to 4,5-diamino-2-methylbenzothiazole (**4.109**) has been used[48] for the construction of a series of 7-substituted-2-methylimidazo[4,5-*e*]benzothiazoles (**4.109 → 4.110**) (Table 4.17). As in the analogous syntheses of imidazo[4,5-*f*]benzothiazoles and imidazo[4,5-*g*]benzothiazoles, ring closure to imidazo[4,5-*e*]benzothiazoles (**4.109 → 4.110**) succeeds best (Table 4.17) by heating with

TABLE 4.17. SYNTHESES OF IMIDAZO[4,5-*e*]BENZOTHIAZOLE DERIVATIVES (**4.110**, **4.118**, AND **4.119**) BY RING-CLOSURE REACTIONS OF 4,5-DIAMINO-2-METHYLBENZOTHIAZOLE (**4.109**)[48]

Reaction conditions[a]	Product[b]	R	Yield (%)	Melting point (°C)	Solvent of crystallization
A	(4.110)	H[c]	80	297–298	70% Ethanol–water
A	(4.110)	Me	68	242	70% Ethanol–water
A	(4.110)	Et	76	194	70% Ethanol–water
A	(4.110)	CH_2OH	54	245–246	Ethanol–water
A	(4.110)	CH(OH)Me	37	222	Ethanol–water
B	(4.110)	Ph	47	118–119	Ethanol–water
C	(4.110)	CH_2Ph	37	244–245	Ethanol
A	(4.110)	$(CH_2)_2CO_2H$	12	251–252	Water
D	(4.118)	—	53	325	—[d]
E	(4.119)	—	57	325	—[e]

[a] A, HCO_2H, 15% HCl/reflux/2 hr. B, PhCHO, $PhNO_2$, EtOH/200°/1 min. C, $PhCH_2CO_2H$/125–130°/30 min. D, urea/150–155°/30 min. E, CS_2, EtOH/reflux/3 hr.
[b] Colorless crystalline solids.
[c] Forms a picrate, yellow needles, m.p. 233° (decomp.) (from ethanol).
[d] From 5% Na_2CO_3 by precipitation with acetic acid.
[e] From 1% NaOH by precipitation with acetic acid.

aliphatic acids in the presence of hydrochloric acid. Ring closure fails under
similar conditions with aryl-substituted carboxylic acids,[48] and 7-arylmethyl-
or 7-arylimidazo[4,5-*e*]benzothiazoles are best synthesized in the former
case by fusion with the arylacetic acid or in the latter case by reaction with
an aromatic aldehyde in the presence of nitrobenzene as oxidant (Table
4.17).[48] Ring closure of the diamine (**4.109**) also occurs readily on fusion
with urea or on heating with carbon disulfide and leads to the fused
benzimidazolone (**4.118**) and the benzimidazolethione (**4.119**), respectively
(Scheme 4.32) (Table 4.17).[48]

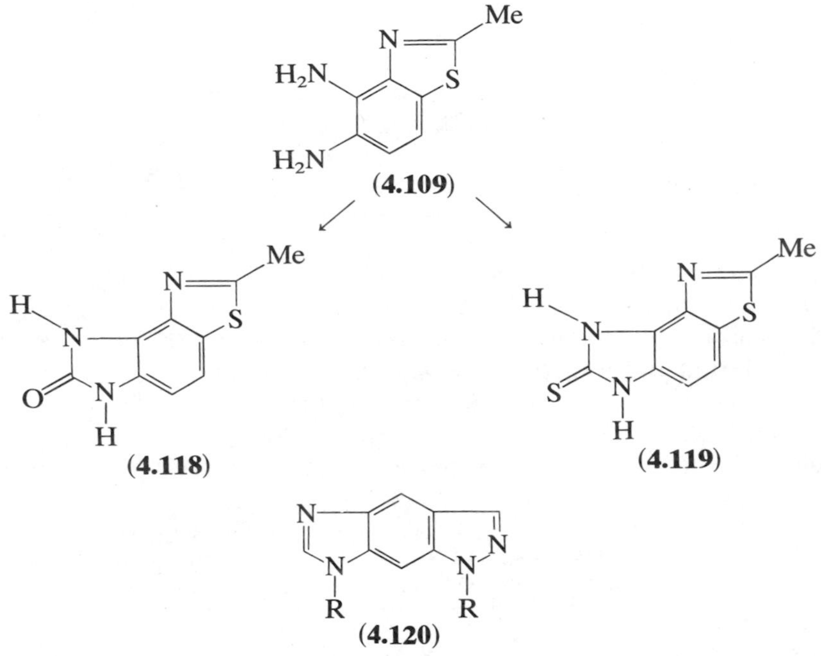

Scheme 4.32

In contrast to the relatively well documented benzo[1,2-*d*:4,5-*d'*]-
diimidazole and benzo[1,2-*d*:3,4-*d'*]diimidazole ring systems, derivatives
of the isosteric, linear imidazo[4,5-*f*]indazole ring system (Scheme 4.32,
4.120) appear to be unknown, and examples of the angular imidazo[4,5-*e*]-
indazole (**4.51**) and imidazo[4,5-g]indazole (**4.52**) ring systems (Scheme
4.15) are rare.[52–54] However, synthetic access to derivatives of the latter ring
system would appear to be provided by orthodox ring closure reactions
(Scheme 4.33) of 6,7-diaminoindazole derivatives (cf. Scheme 4.33, **4.121** →
4.122 and **4.123** → **4.124**).[52,53] Condensation of the cyclic α-diketone
(**4.125**) with aromatic aldehydes in the presence of ammonium acetate
(Scheme 4.33) is reported to afford derivatives (**4.126**) of the imidazo[4,5-*e*]-
indazole ring system in good yield.[54]

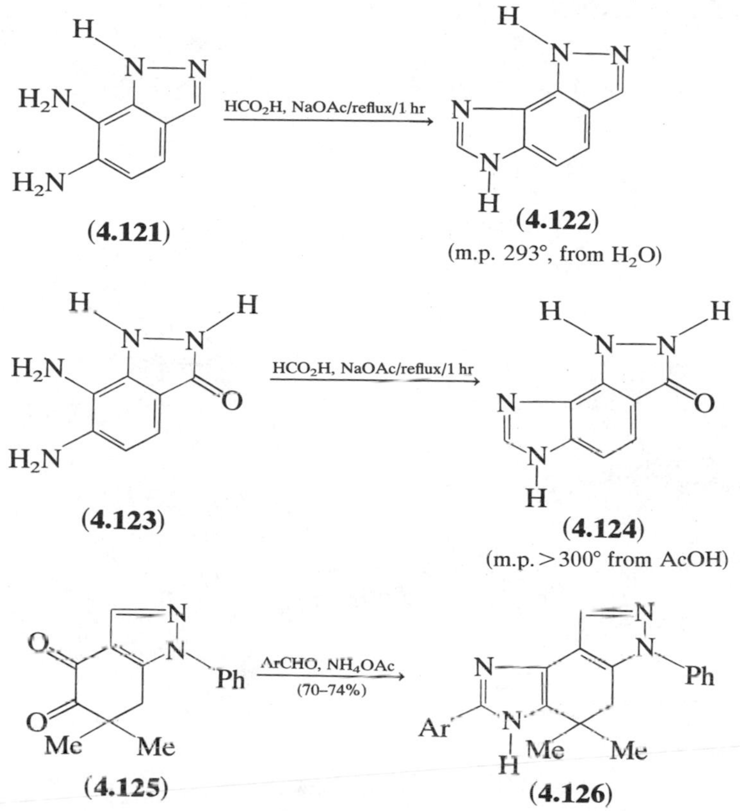

(4.121)   HCO$_2$H, NaOAc/reflux/1 hr   **(4.122)**
(m.p. 293°, from H$_2$O)

(4.123)   HCO$_2$H, NaOAc/reflux/1 hr   **(4.124)**
(m.p. > 300° from AcOH)

(4.125)   ArCHO, NH$_4$OAc (70–74%)   **(4.126)**

Scheme 4.33

4.3.2. Physicochemical Properties

Spectroscopic Studies

INFRARED SPECTRA. Routine identification of functional groups apart, little information is available on specific aspects of the infrared absorption characteristics of tricyclic 5-6-5 fused benzimidazole ring systems with two additional hetero atoms.

N-unsubstituted benzo[1,2-d:4,5-d']diimidazole derivatives show well-defined NH-absorption as illustrated by 2,6-dimethyl-1(3)H,5(7)H-benzo[1,2-d:4,5-d']diimidazole whose infrared spectrum (Nujol suspension)[29] contains a well-defined band at 3150 cm^{-1} due to the NH groups. Comparison of the solid-state infrared spectra of the 2-aminobenzo[1,2-d:3,4-d']diimidazoles [Scheme 4.34, (**4.127**, R = H or Me)] with those of their N-deuterio derivatives demonstrates a marked weakening in the NH deformation vibration consequent on deuteriation, and hence diagnostic for

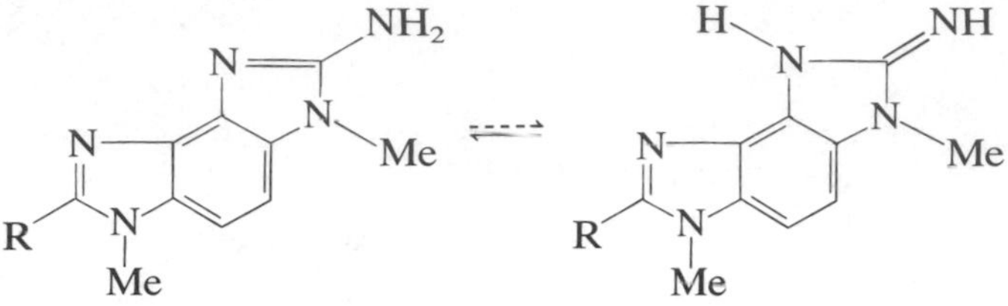

(a)   (b)

(**4.127**)

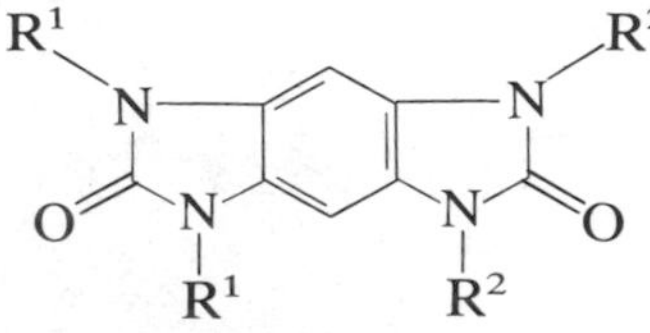

(**4.128**) 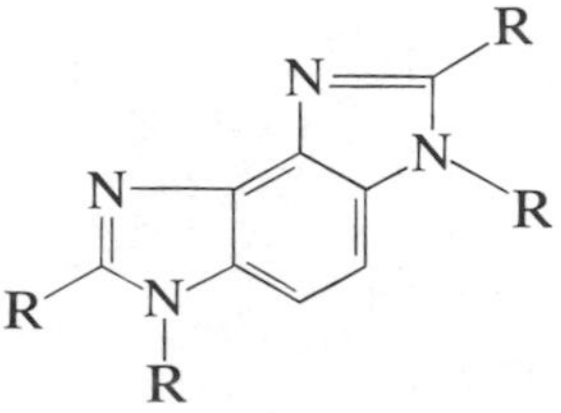   (**4.129**)

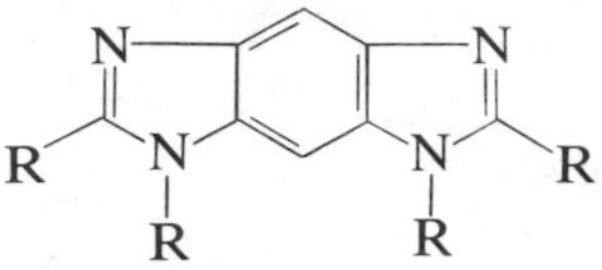

(**4.130**)

Scheme 4.34

the predominance (at least in the solid state) of the amino tautomers (**4.127a**; R = H or Me) as distinct from the imino forms (**4.127b**; R = H or Me).[55] Nuclear amino groups at the C-4(5) position of benzo[1,2-d:3,4-d']-diimidazoles show characteristic infrared absorption at 3334 and 3215 cm^{-1}.[42]

Benzo[1,2-d:4,5-d']diimidazole-2,6-diones of the general structure (**4.128**) (Scheme 4.34) show well-defined carbonyl absorption in the range 1720 to 1680 cm^{-1}, which in N-unsubstituted compounds (e.g., **4.128**; R^1 = H) is accompanied by one or more NH bands in the region 3400 to 3200 cm^{-1}, consistent with the tautomeric nature of such molecules.[38] In accord with their essentially 1,2,3,4-tetrasubstituted benzene character, angular benzo[1,2-d:3,4-d']diimidazole structures (**4.129**) may be differentiated from their linear benzo[1,2-d:4,5-d']diimidazole counterparts (**4.130**) by the presence of strong bands in the infrared at 800 to 790 cm^{-1}.[40] Conversely, strong bands at 900 to 860 cm^{-1} corresponding to a 1,2,4,5-tetrasubstituted benzene pattern, are characteristic of the infrared spectra of derivatives of the linear benzo[1,2-d:4,5-d']diimidazole ring system (**4.130**).[40]

ULTRAVIOLET AND VISIBLE SPECTRA. In view of the highly conjugated frameworks present in fully unsaturated, tricyclic 5-6-5 fused benzimidazole ring systems with one additional hetero atom, it is not surprising that such ring systems are often utilized as integral components of cyanine dyestuffs (see later), and consequently that their ultraviolet and visible absorption characteristics have been extensively investigated.

The ultraviolet spectra (Table 4.18)[48,51] of simple derivatives of the imidazo[4,5-f]benzothiazole, imidazo[4,5-g]benzothiazole, and imidazo[4,5-e]-benzothiazole ring systems are typified by the presence of one or more

TABLE 4.18.   ULTRAVIOLET SPECTRA[a] OF
TYPICAL IMIDAZOBENZOTHIAZOLES
(**4.131**, **4.132**, AND **4.133**)[48,51]

Compound	λ_{max}, nm (log ε)
(**4.131**)	242(4.64)
(**4.132**)	272(3.96), 294(3.78)
(**4.133**)	241(4.33), 288(4.08)

[a] Measured for ethanolic solutions.

intense absorption bands in the 240 to 290 nm region of the near ultraviolet. These bands become more intense and shift into the visible region in cyanine and carbocyanine dyestuffs incorporating imidazobenzothiazole nuclei.[48,50,51,56,57] Typical ultraviolet spectra of such cyanine and carbocyanine dyestuffs are recorded in Tables 4.19 to 4.21.

The extensive conjugation present in benzo[1,2-d:4,5-d']diimidazoles[24,58] and benzo[1,2-d:3,4-d']diimidazoles[40–42,55] gives rise to ultraviolet spectra usually characterized by the presence of two or more intense bands at relatively long wavelengths (Table 4.22). The respective ultraviolet absorption of the linear benzo[1,2-d:4,5-d']diimidazole and angular benzo[1,2-d:3,4-d']diimidazole ring systems is significantly different (Table 4.22) and indeed allows the differentiation of derivatives of both structural types. The absorption bands in the ultraviolet spectrum of 2,6-diphenylbenzo[1,2-d:4,5-d']diimidazole (**4.139**; R = Ph) (Table 4.22) show a marked decrease in intensity in acidic solution, demonstrating the effect on conjugation of protonation of the imidazole ring nitrogen atoms. Comparison of the ultraviolet spectra of 2-aminobenzo[1,2-d:3,4-d']diimidazoles with those of alkylated derivatives having fixed amino or imino structures demonstrates

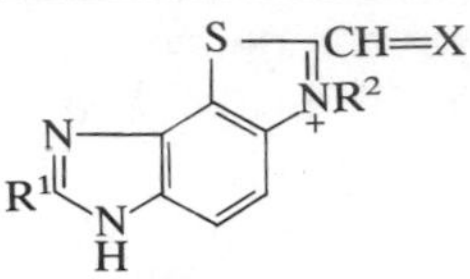

(**4.134**) 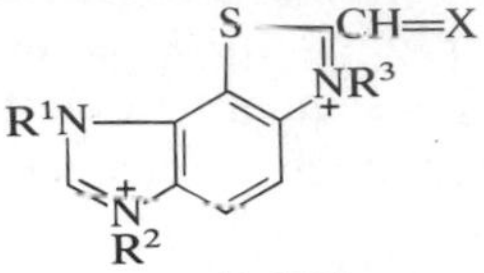   (**4.135**)

Compound	R^1	R^2	R^3	X^a	λ_{max}	Ref.
(**4.134**)	H	Et	—	A	579	51
(**4.134**)	Me	Et	—	B	562	51
(**4.135**)	Ph	Me	Me	C	572	50
(**4.135**)	Ph	Me	Me	D	576	50
(**4.135**)	Et	Et	Et	E	428	56
(**4.135**)	Et	Et	Et	F	576	56
(**4.135**)	Et	Et	Et	B	568	56
(**4.135**)	Et	Et	Et	C	548	56

^{a} $A =$

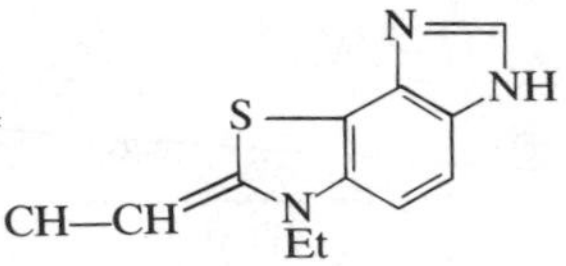

$B =$

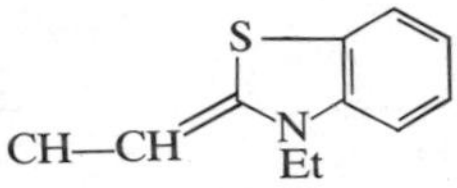

$C = CH$⟨⟩NMe_2.

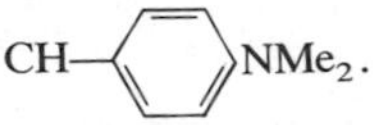

$D =$

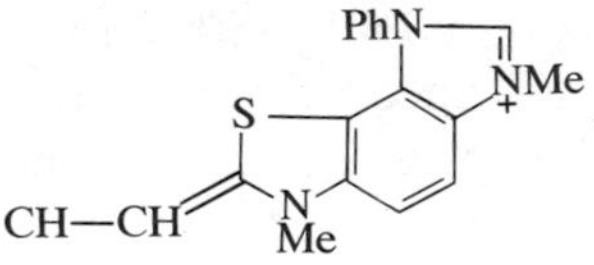

$E =$

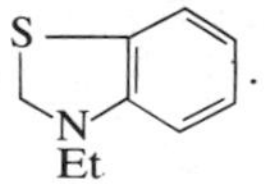

$F =$

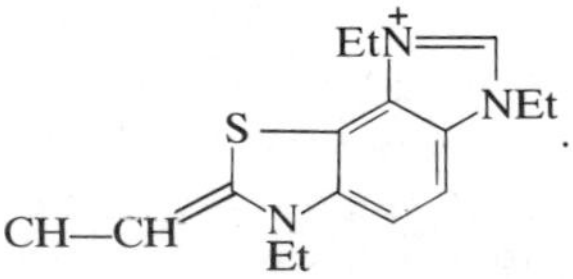

TABLE 4.20.   ULTRAVIOLET ABSORPTION OF IMIDAZO[4,5-*e*]BENZOTHIAZOLE AND IMIDAZO[4,5-*f*]BENZOTHIAZOLE CYANINES (**4.136** AND **4.137**)

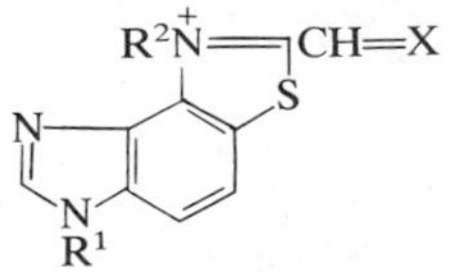

(**4.136**) 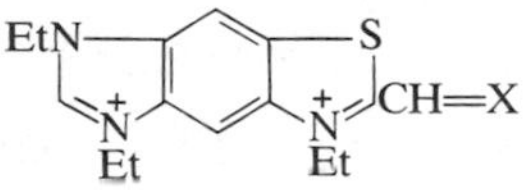   (**4.137**)

Compound	R^1	R^2	X^a	λ_{max}	Ref.
(**4.136**)	Et	Et	A	434	56
(**4.136**)	Et	Et	B	591	56
(**4.136**)	Me	Me	C	564	56
(**4.136**)	Me	Me	D	520	56
(**4.137**)	—	—	D	546	56
(**4.136**)	H	Et	E	590	51
(**4.136**)	Me	Et	C	569	51

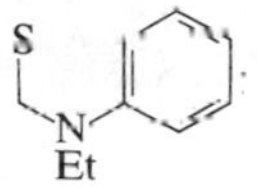

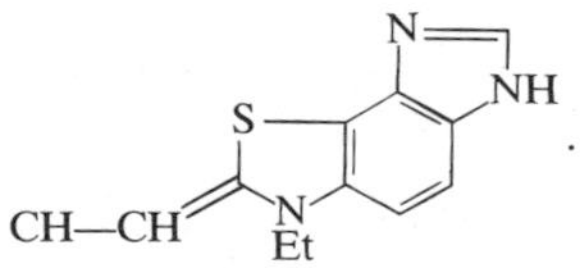

that the parent amines exist in the amino as opposed to the imino tautomeric form.[55]

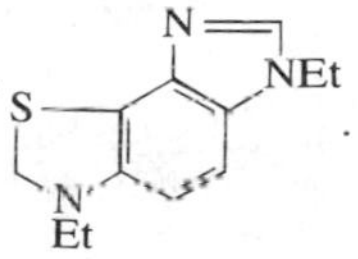

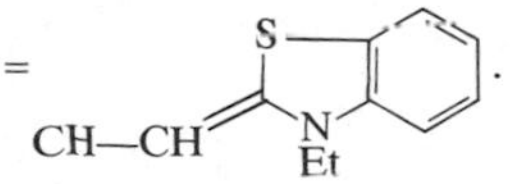

As in the case of imidazobenzothiazoles, incorporation of the benzo[1,2-*d*:4,5-*d'*]diimidazole nucleus within a cyanine or carbocyanine framework produces dyestuffs exhibiting intense absorption in the visible region (Table 4.23).[15]

TABLE 4.21. ULTRAVIOLET SPECTRA OF IMIDAZO[4,5-g]BENZOTHIAZOLE BISCYANINES (**4.138**)

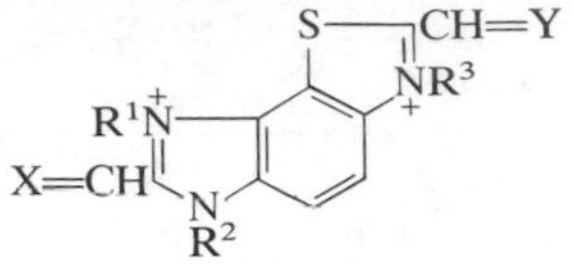

(**4.138**)

R^1	R^2	R^3	X^a	Y^a	λ_{max}	Ref.
Me	Me	Me	H_2	*A*	556	15
Me	Me	Me	*A*	H_2	464	15
Me	Me	Me	H_2	*B*	562	15
Me	Me	Me	*B*	H_2	500	15
Et	Ph	Et	H_2	*B*	564	15
Et	Ph	Et	*B*	H_2	532	15
Me	Me	Me	*C*	*C*	476, 520	15
Me	Me	Me	*D*	*D*	492, 566	15
Et	Ph	Et	*D*	*D*	500, 590	15
Me	Me	Me	H_2	*E*	485	57
Me	Me	Me	*D*	H_2	500	57
Me	Me	Me	*E*	H_2	433	57
Me	Me	Me	H_2	*D*	562	57
Me	Me	Me	*C*	H_2	464	57
Me	Me	Me	H_2	*C*	553	57
Me	Me	Me	*D*	*E*	466, 538	57
Me	Me	Me	*E*	*D*	445, 572	57
Me	Me	Me	*C*	*E*	460, 520	57
Me	Me	Me	*E*	*C*	445, 564	57

A =

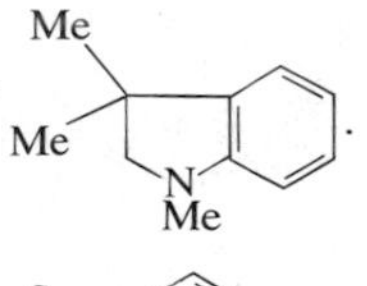

B =

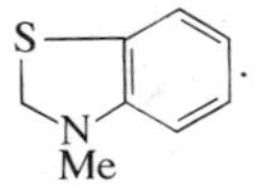

C =

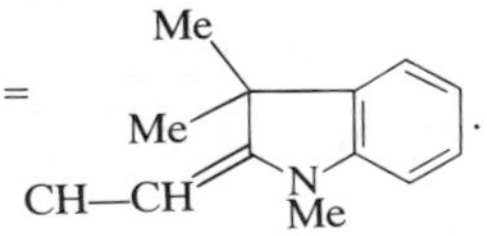

D =

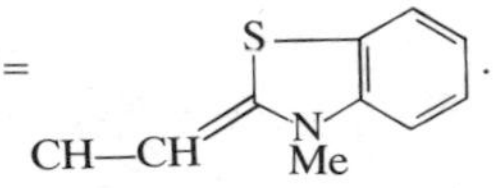

E =

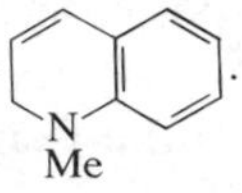

TABLE 4.22. ULTRAVIOLET SPECTRA OF BENZO[1,2-d: 4,5-d']DIIMIDAZOLES
(**4.139**) AND BENZO[1,2-d: 3,4-d']DIIMIDAZOLES (**4.140** AND
4.141)

(**4.139**)   (**4.140**)   (**4.141**)

Compound	R	R^1	R^2	R^3	R^4	Solvent^a	λ_{max}. nm (log ε)	Ref.
(**4.139**)	H	—	—	—	—	A	295^b	24, 59
(**4.139**)	Ph	—	—	—	—	A	224(3.00), 339(3.29)	24, 59
(**4.139**)	Ph	—	—	—	—	B	203(4.75), 225sh, 344(4.68)	58
(**4.140**)	—	Me	H	H	Mc	B	265(4.12), 273(4.26), 284(4.21), 294(3.80)	40
(**4.140**)	—	Me	H	H	H	B	265(4.00), 279(3.93), 289(3.74)	40
(**4.140**)	—	Et	H	H	Et	B	265(4.01), 273(4.18), 284(4.09), 295(3.69)	40
(**4.140**)	—	Et	H	H	H	B	265(3.94), 279(3.86), 289(3.68)	40
(**4.140**)	—	Me	Br	H	Mc	C	270(3.34), 287(3.45), 283(3.39), 289(3.39)	42
(**4.140**)	—	Me	Br	Br	Me	C	273(3.41), 280(3.50), 292(3.46), 296(3.46)	42
(**4.141**)	—	—	—	—	—	C	248(3.89), 257(3.94), 265(3.98), 297(4.02)	41

^{a} A, aq. H$_2$SO$_4$. B, ethanol. C, methanol.
^{b} Intensity not specified.

NUCLEAR MAGNETIC RESONANCE SPECTRA. Apart from sporadic reports
relating to individual compounds, little detailed information on the nuclear
magnetic resonance absorption of tricyclic 5-6-5 fused benzimidazole ring
systems containing two additional hetero atoms is available.

H(4) and H(8) in the [1,3]dioxolo[4,5-f]benzimidazolethione (**4.144**)
(Scheme 4.35) absorb[46] in trifluoroacetic acid as a singlet at δ 6.75 while the
protons of the methylenedioxy group appear at δ 6.02. A singlet at δ 12.33
is assigned[46] to the protons of the NH-groups. The chemical shift (δ 7.75;
CDCl$_3$) of H(6) in [1,3]dioxolo[4,5-f]benzimidazole (**4.145**) has been used
as a measure of the electron density at the C(6) position and hence of the
tendency for nucleophilic attack (i.e., amination) to occur at this site.[43]
Interesting in this context is the apparently enhanced electron deficiency at
C(7) in the imidazobenzoxazole (**4.146**), as evidenced by the marked
deshielding of H(7) that absorbs as a one-proton singlet in trifluoroacetic
acid at δ 9.53.[13] An accompanying two-proton singlet at δ 8.28 is assigned
to superposed H(4) and H(5) protons in this molecule.[13]

TABLE 4.23. ULTRAVIOLET SPECTRA[a] OF BENZO[1,2-*d*: 4,5-*d'*]DIIMIDAZOLE CYANINES (**4.142**) AND CARBOCYANINES (**4.143**)[15]

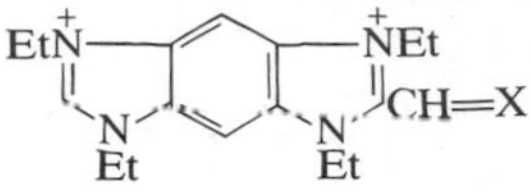

(**4.142**)

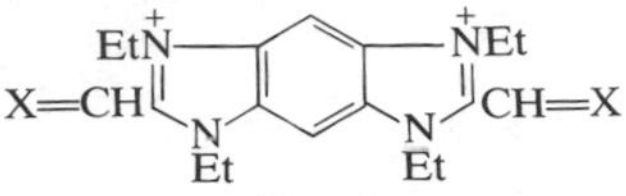

(**4.143**)

Compound	X[b]	λ_{max}, nm
(**4.142**)	*A*	516
(**4.142**)	*B*	533
(**4.143**)	*A*	458, 560
(**4.143**)	*B*	470, 600

[a] Measured for solutions in methanol.

[b] A = 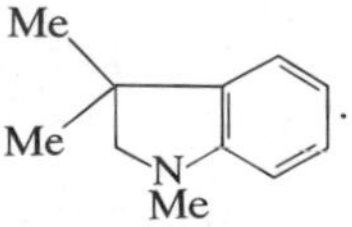

B = 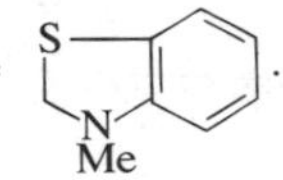

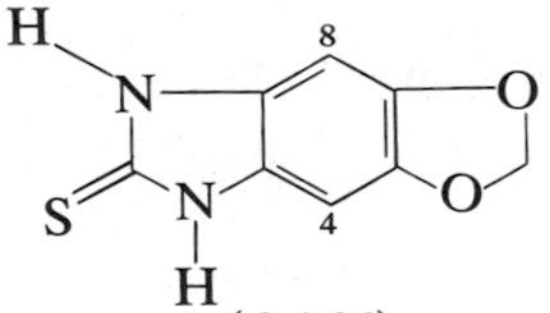

(**4.144**)

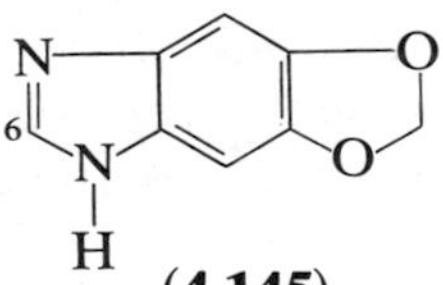

(**4.145**)

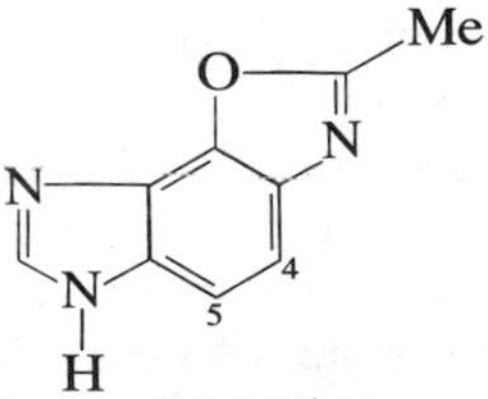

(**4.146**)

Scheme 4.35

MASS SPECTRA. Apart from the routine determination of molecular weights, studies of the mass spectral characteristics, and in particular of the mass spectral fragmentation patterns, of tricyclic 5-6-5 fused benzimidazole ring systems with two additional hetero atoms are lacking. Further comment under this heading therefore, is, not possible.

General Studies

DIPOLE MOMENTS. Derivatives of the benzo[1,2-d:4,5-d']diimidazole and benzo[1,2-d:3,4-d']diimidazole ring systems have been the subject of several theoretical studies,[59-61] all of which emphasize the significant dipolar character of such molecules. Experimental support for the dipolar nature of simple benzo[1,2-d:4,5-d']diimidazoles is provided by the appreciable dipole moments exhibited by 2,6-disubstituted 3,5-dimethylbenzo[1,2-d:4,5-d']diimidazoles (**4.147**) (Scheme 4.36).[62]

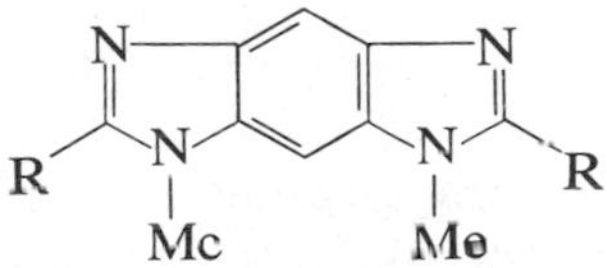

R	μ(D)
H	5.96
Me	6.10
Ph	6.52

(**4.147**)

Scheme 4.36

IONIZATION CONSTANTS. The amphoteric character of derivatives of the [1,3]dioxolo[4,5-f]benzimidazole,[45] imidazobenzothiazole,[40] and benzo[1,2-d:4,5-d']diimidazole[18,19,26,33,36] ring systems is clearly demonstrated by their ability to form well-defined salts (hydrochlorides and picrates) with acids and their solubility (when unsubstituted on the imidazole ring nitrogen atoms) in aqueous alkali.[19,36] However, apart from the demonstration (by potentiometric titration) that 2,6-dimethyl-N-1(3)H,5(7)H-benzo[1,2-d:4,5-d']diimidazole (**4.148**) (Scheme 4.37) is dibasic[19] and the measurement of pK_a data for certain [1,3]dioxolo[4,5-f]benzimidazole derivatives

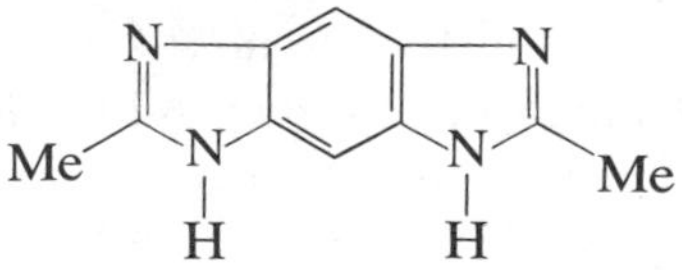

(**4.148**)

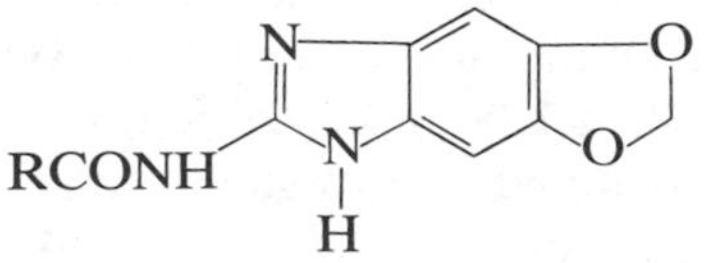

R	ρKa
Me	>11, 4.5
Et	>11, 4.8

(**4.149**)

Scheme 4.37

(**4.149**) (Scheme 4.37),[63] quantitative information on the acidity and basicity of tricyclic 5-6-5 fused benzimidazole ring systems with two additional hetero atoms is lacking.

4.3.3. Reactions

Reactions with Electrophiles

PROTONATION. By virtue of the presence of at least one imidazole ring, all of the known tricyclic 5-6-5 fused benzimidazole ring systems containing two additional hetero atoms are sufficiently basic to undergo protonation with the formation of well-defined salts. For example, [1,3]dioxolo[4,5-f]-benzimidazole derivatives form picrates and hydrochlorides,[45] and imidazobenzothiazoles of all three structural types dissolve readily in dilute hydrochloric acid and form well-defined hydrochlorides and picrates[48] (cf. Tables 4.15–4.17).

Benzo[1,2-d:4,5-d']diimidazoles also are relatively basic substances,[19,36] forming stable sulfates and hydrochlorides,[64] although the basicity is markedly diminished by the presence in the benzene ring of electron-withdrawing (e.g., halogen) substituents. Thus 4,8-dichlorobenzo[1,2-d:4,5-d']-diimidazoles form dihydrochlorides that are readily hydrolyzed to the free bases on warming with water.[64] The particular instability of the sulfates and hydrochlorides of benzo[1,2-d:4,5-d']diimidazole-4,8-quinones is likewise demonstrated by their hydrolysis to the parent bases simply on standing in 96% ethanol.[64] Electron-donating groups at the C(4) and C(8) positions of the benzo[1,2-d:4,5-d']diimidazole ring, on the other hand, have the reverse effect, so that 4,8-dihydroxybenzo[1,2-d:4,5-d']diimidazoles are relatively strong bases.[33]

ALKYLATION. N-Alkylation of imidazobenzothiazoles and benzodiimidazoles occurs readily and is well documented as a method for the synthesis of cyanine dye precursors. In most cases, alkylation is achieved simply and usually in high yield by heating the substrate with an alkyl iodide, sulfate, or tosylate alone or in a suitable inert solvent.

Alkylation of 6(8)H-imidazo[4,5-g]benzothiazoles (**4.150**) with alkyl iodides, sulfates, or tosylates in the absence of base results in preferential attack on the thiazole ring at nitrogen to give the corresponding monoquaternary salts (Scheme 4.38, **4.150 → 4.151**) (Table 4.24).[65] On the other hand, alkylation in the presence of a base such as potassium hydroxide results in preferential replacement of the acidic imidazole NH to give an N(6)-alkyl derivative (Scheme 4.38, **4.150 → 4.152**).[15] Formation of the N(8)-alkyl product is presumably inhibited due to steric hindrance. Subsequent alkylation allows the conversion of either (**4.151**) or (**4.152**) into an N(6)-alkylated monoquaternary salt (**4.153**) (Scheme 4.38).[15] Similar quaternary salts (**4.153**; $R^3 = R^4$) can be formed directly from the parent

TABLE 4.24. N-ALKYLATION OF IMIDAZO[4,5-g]BENZOTHIAZOLES, IMIDAZO[4,5-f]BENZOTHIAZOLES, AND [4,5-e]BENZOTHIAZOLES

Starting material	Reaction conditions[a]	Product	Yield (%)	Melting point (°C)	Solvent of crystallization	Crystal form	Ref.
(4.150) ($R^1 = R^2 = H$)	A	(4.151) ($R^1 = R^2 = H$, $R^3 = Me$, $X = MeSO_4$)	—[b]	253	Ethanol or benzene	—[c]	65
(4.150) ($R^1 = R^2 = Me$)	B	(4.152) ($R^1 = R^2 = R^3 = Me$)	70	173	Benzene	Colorless needles	15
(4.152) ($R^1 = R^2 = R^3 = Me$)	C	(4.153) ($R^1 = R^2 = R^3 = R^4 = Me$, $X = I$)	—[b]	314–315	Water	Colorless needles	15
(4.152) ($R^1 = R^2 = R^3 = Me$)	D	(4.154) ($R^1 = R^2 = R^3 = R^4 = R^5 = Me$, $X = TSO_3$)[d]	—[b]	—[d]	Water	Colorless needles	15
(4.152) ($R^1 = R^2 = Me$, $R^3 = Ph$)	E	(4.153) ($R^1 = R^2 = Me$, $R^3 = Ph$, $R^4 = Et$, $X = I$)	—[b]	267	—[e]	—[c]	15
(4.153) ($R^1 = R^2 = Me$, $R^3 = Ph$, $R^4 = Et$, $X = I$)	F	(4.154) ($R^1 = R^2 = Me$, $R^3 = Ph$, $R^4 = R^5 = Et$, $X = ClO_4$)	—[b]	—[d]	—[e]	—[c]	15
(4.150) ($R^1 = H$, $R^2 = Me$)	G	(4.154) ($R^1 = H$, $R^2 = Me$, $R^3 = R^4 = R^5 = Et$, $X = ClO_4$)	90	—[d]	Ethanol	Colorless needles	56
(4.155)	H	(4.156)[f]	97	110 (decomp.)	—[e]	—[c]	50
(4.157)	I	(4.158)	—[b]	—[d]	—[e]	—[c]	56
(4.159) ($R^1 = H$)	A	(4.160) ($R^1 = H$, $R^2 = Me$, $X = MeSO_4$)	—[b]	219	Ethanol or benzene	—[c]	65
(4.159) ($R^1 = Me$)	J	(4.161) ($R^1 = Me$, $R^2 = Et$, $X = EtSO_4$)	—[b]	—[d]	—[e]	—[c]	56

a A, Me_2SO_4/100°/1 hr. B, Me_2SO_4, KOH, EtOH/reflux/1 hr. C, TSO_3Me/140°/45 min, then treat with KI (T = p-tolyl). D, TSO_3Me/140°/5 hr (T = p-tolyl). E, TSO_3Et/heat, then treat with KI (T = p-tolyl). F, TSO_3Et (reaction conditions not specified), then treat with $HClO_4$ (T = p-tolyl). G, Et_2SO_4 or TSO_3Et (reaction conditions not specified), then treat with $HClO_4$ (T = p-tolyl). H, Me_2SO_4, benzene/reflux/1 hr. I, Reaction conditions not specified. J, Et_2SO_4 or TSO_3Et (reaction conditions not specified) (T = p-tolyl).

b Yield not recorded.

c Crystal form not reported.

d No melting point.

e Solvent not specified.

f Iodide (yield 85%), has m.p. 258° (decomp.), colorless prisms (from water).

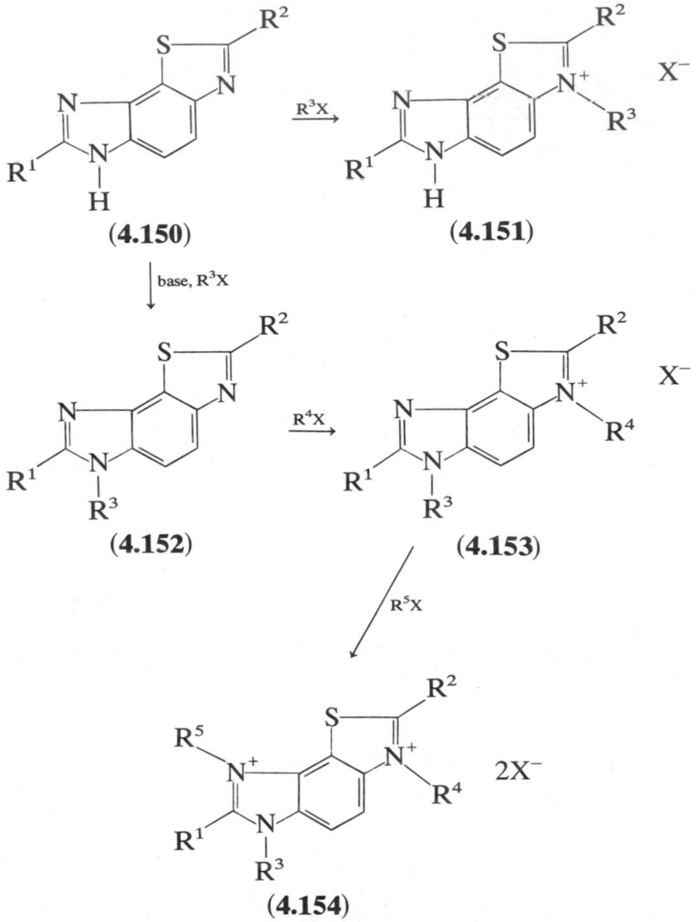

(4.150)   **(4.151)**

base, R³X

(4.152)   **(4.153)**

R⁵X

(4.154)

Scheme 4.38

base (**4.150**)[15,56] and can be reacted further with alkylating agent to afford bis-quaternary salts (Scheme 4.38; **4.153** → **4.154**).[15] Under appropriate conditions (Table 4.24), salts of the latter type may be obtained by direct alkylation of the parent bases (Scheme 4.38; **4.150** or **4.152** → **4.154**, $R^4 = R^5$) (Table 4.24).[15,56] $N(8)$-substituted imidazo[4,5-g]benzothiazoles (**4.155**)[50] and imidazo[4,5-f]benzothiazoles (**4.157**)[56] can likewise be directly alkylated to the corresponding bis-quaternary salts, (**4.156**) and (**4.158**), respectively (Scheme 4.39) (Table 4.24).

In contrast to the behavior of the imidazo[4,5-g]benzothiazole and imidazo[4,5-f]benzothiazole ring systems already discussed, N-6(8)H-imidazo[4,5-e]benzothiazole derivaties undergo stepwise or direct alkylation only as far as a monoquaternary salt (Scheme 4.40, **4.159** → **4.160** → **4.161**),[56,65] presumably as a result of steric hindrance to further alkylation at $N(8)$ in the latter.

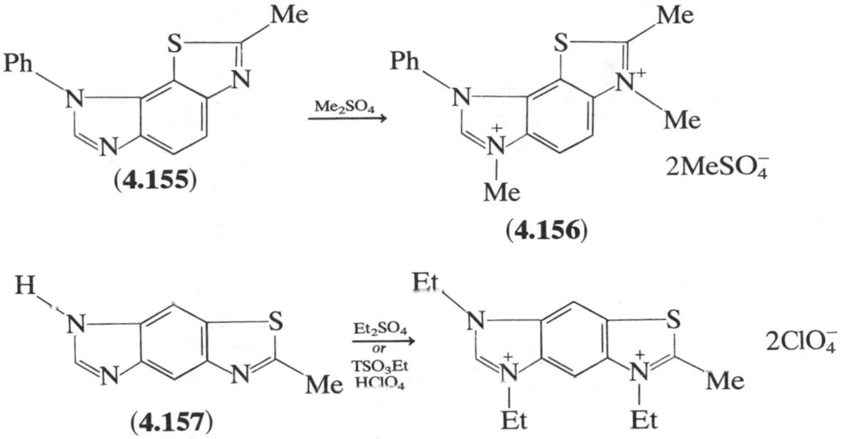

(4.155)   (4.156)

(4.157)   (4.158)

(T = toluene-p-sulfonyl)

Scheme 4.39

The *N*-alkylation of benzo[1,2-*d*:4,5-*d'*]diimidazoles and benzo[1,2-*d*:3,4-*d'*]diimidazoles follows the same pattern as that discussed previously for imidazobenzothiazoles. Thus *N*-1(3)*H*,5(7)*H*-benzo[1,2-*d*:4,5-*d'*]diimidazoles undergo alkylation in the presence of potassium hydroxide to give bis-*N*-alkyl derivatives (Scheme 4.41, **4.162** → **4.163**) (Table 4.25)[15] convertible by reaction with excess of the alkylating agent into *bis*-quaternary salts (Scheme 4.41, **4.163** → **4.164**) (Table 4.25).[15] The latter products are

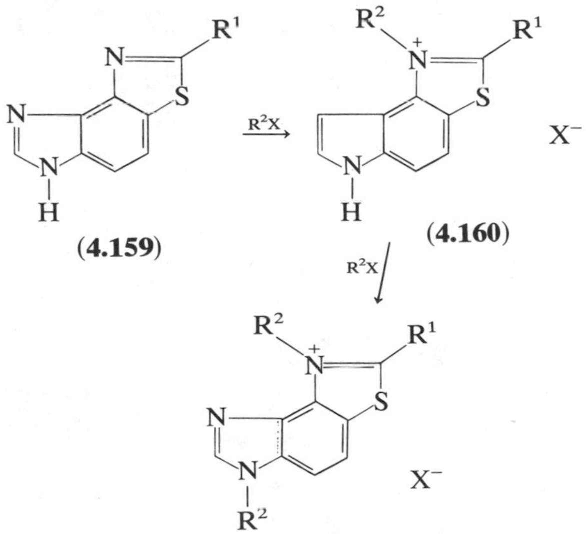

(4.159)   (4.160)

(4.161)

Scheme 4.40

TABLE 4.25. N-ALKYLATION OF BENZO$[1,2\text{-}d{:}4,5\text{-}d']$DIIMIDAZOLES AND BENZO$[1,2\text{-}d{:}3,4\text{-}d']$DIIMIDAZOLES

Starting material	Reaction conditions[a]	Product	Yield (%)	Melting point (°C)	Solvent of crystallization	Crystal form	Ref.
(4.162) ($R^1 = R^2 = Me$)	A	(4.163) ($R^1 = R^2 = Me$, $R^3 = Et$)	45	248	1,2-Dichloroethane	Colorless needles	15
(4.163) ($R^1 = R^2 = Me$, $R^3 = Et$)	B	(4.164) ($R^1 = R^2 = Me$, $R^3 = Et$, $X = ClO_4$)	—[b]	—[c]	—[d]	—[e]	15
(4.162) ($R^1 = H$, $R^2 = Me$)	A	(4.163) ($R^1 = H$, $R^2 = Me$, $R^3 = Et$)	60	227	Nitromethane	Colorless needles	15
(4.163) ($R^1 = H$, $R^2 = Me$, $R^3 = Et$)	B	(4.164) ($R^1 = H$, $R^2 = Me$, $R^3 = Et$)	—[b]	—[c]	—[d]	—[e]	15
(4.162) ($R^1 = R^2 = H$)	C	(4.164) ($R^1 = R^2 = H$, $R^3 = Me$)	65	330–333[f]	Ethanol	Colorless crystals	66
(4.162) ($R^1 = R^2 = CF_3$)	D	(4.164) ($R^1 = R^2 = CF_3$, $R^3 = Me$)	68	276–278	Ethanol–methanol–dioxane	—[e]	66
(4.165) ($R = H$)	E	(4.166) ($R = H$)	24	>320	Acetone–water	Colorless needles	38
(4.165) ($R = OMe$)	F	(4.166) ($R = OMe$)	27	309–310	Xylene	Colorless needles	38
(4.167) ($R^1 = H$)	G	(4.168) ($R^1 = H$, $R^2 = Me$, $X = I$)	—[b]	—[c]	—[d]	—[e]	42
(4.167) ($R^1 = Me$)	G,H	(4.168) ($R^1 = R^2 = Me$, $X = I$ or $PhSO_3$)	—[b]	—[c]	—[d]	—[e]	42

[a] A, Et$_2$SO$_4$, KOH, EtOH/100°/1 hr. B, TSO$_3$Et (no other conditions given), then treat with HClO$_4$ (T = p-tolyl). C, MeBr/170° (autoclave)/6 hr. D, MeBr/155° (autoclave)/6 hr. E, Me$_2$SO$_4$, 5.1% aq. KOH/98°/2 hr. F, Me$_2$SO$_4$, 4.5% aq. KOH/90°/1 hr 40 min. G, MeI. H, PhSO$_3$Me (reaction conditions not specified).
[b] Yield not specified.
[c] No melting point.
[d] Solvent not specified.
[e] Crystal form not specified.
[f] Monohydrate.

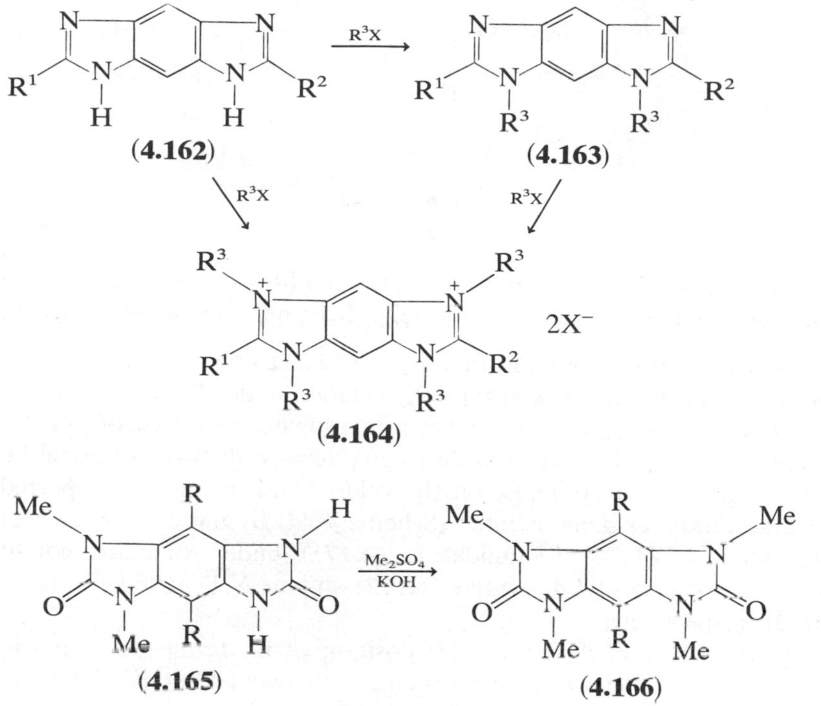

Scheme 4.41

also directly accessible from the parent bases by reaction with an alkyl halide under autoclave conditions (Scheme 4.41, **4.162 → 4.164**) (Table 4.25).[66] Not unexpectedly (in view of the enhanced acidity of the component imidazolone rings), N-unsubstituted benzo[1,2-d:4,5-d']diimidazole-2,6-diones are N-alkylated, in orthodox fashion, under basic conditions (Scheme 4.41, **4.165 → 4.166**) (Table 4.25).[38]

In contrast to their benzo[1,2-d:4,5-d']diimidazole counterparts, benzo[1,2-d:3,4-d']diimidazoles undergo N-alkylation only as far as the mono-quaternary salt stage (Scheme 4.42, **4.167 → 4.168**) (Table 4.25),[42] presumably (as in the case of imidazo[4,5-e]benzothiazoles) due to steric hindrance to further N-alkylation at N(8) in the mono-quaternary salt (**4.168**).

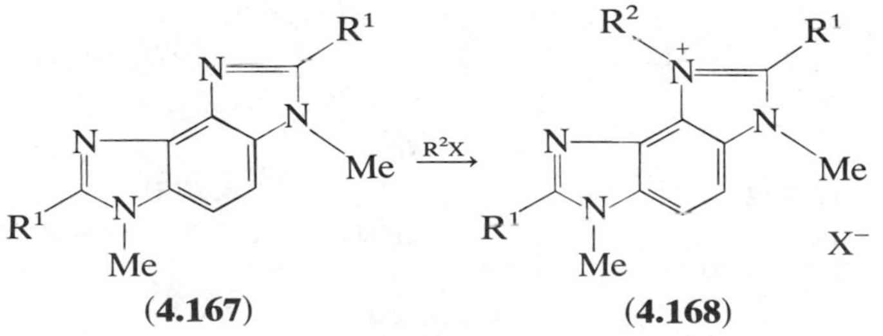

Scheme 4.42

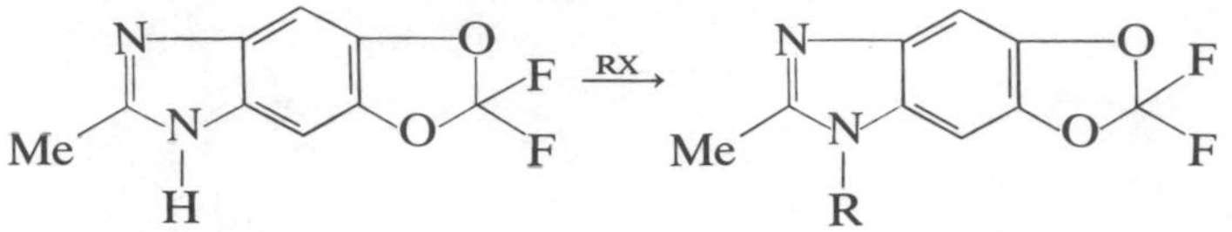

(4.169)   **(4.170)**

Scheme 4.43

Alkylation of N-unsubstituted [1,3]dioxolo[4,5-f]benzimidazoles under basic conditions occurs as expected (e.g., Scheme 4.43, **4.169** → **4.170**).[5]

ACYLATION. In contrast to other types of electrophilic aromatic substitution (i.e., halogenation and nitration), acylation under Friedel–Crafts conditions does not appear to have been investigated for any of the known tricyclic 5-6-5 fused benzimidazole ring systems with two additional hetero atoms. Attached substituents, on the other hand, behave as expected towards acylation, as demonstrated (Scheme 4.44) by the conversion[41] of the aminobenzo[1,2-d:3,4-d']diimidazole (**4.171**) under standard conditions into an N-mono-acetyl derivative (**4.172**) and an N-benzylidene derivative (**4.173**), respectively.

Methyl groups sited at the $C(2)$ position of imidazole rings in tricyclic 5-6-5 fused benzimidazole ring systems with two additional hetero atoms

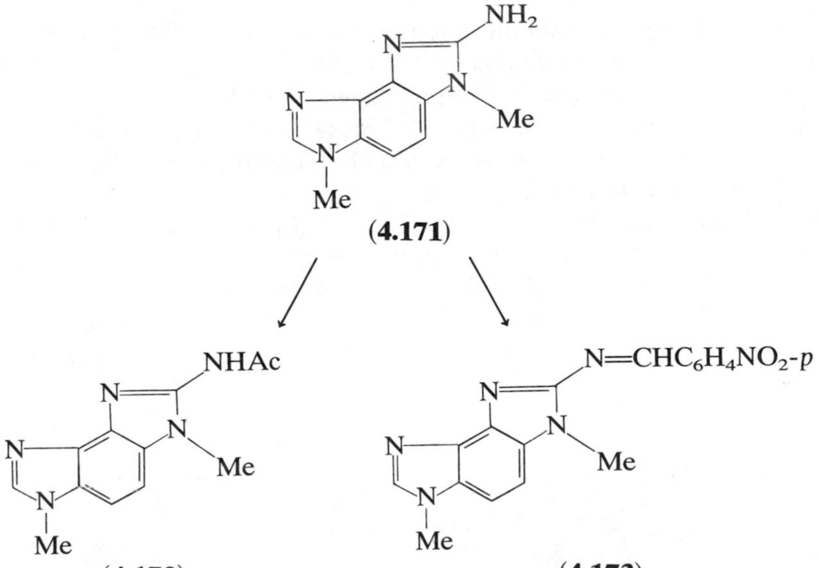

(4.171)

(4.172)   **(4.173)**

(colorless prisms,   (red plates, m.p. 267°, from n-BuOH)
m.p. 291°, from H_2O)

Scheme 4.44

exhibit methylene reactivity, as demonstrated by their ability to participate in both acid and base-catalyzed aldol type condensations. Processes of this type have been employed for the synthesis of 2,6-di-styryl derivatives of the benzo[1,2-d:4,5-d']diimidazole ring system[67] and have been adapted for the synthesis of polymers containing the latter[68] (Scheme 4.45).

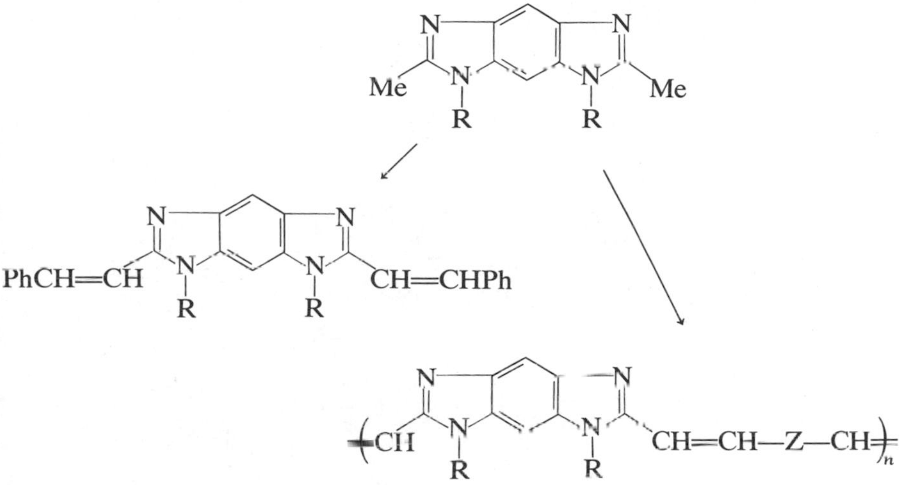

Scheme 4.45

Aldol-type condensation reactions of imidazole C(2)-methyl substituents attached to tricyclic 5-6-5 fused benzimidazole frameworks with two additional hetero atoms are also crucial for the construction of cyanine dyes incorporating such nuclei.[15,50,56,57,65] The efficiency of condensations of this type, which have been used principally for the synthesis of cyanines containing linear or angular imidazobenzothiazole and benzodiimidazole frameworks, is usually dependent on the use of catalysts (e.g., triethylamine, pyridine, and acetic anhydride) and the enhancement of the methyl-group reactivity by prior quaternization at an adjacent nitrogen center. Selected examples[15,50,56,57,65] of such processes are shown in Scheme 4.46.

HALOGENATION. Theoretical studies[59] support the contention[37,64] that the C(4) and C(8) ('meso') positions in benzo[1,2-d:4,5-d']diimidazoles (Scheme 4.47, **4.174**) resemble the C(9) and C(10) positions in anthracene in terms of their reactivity towards electrophilic attack. Thus just as anthracene is readily halogenated at C(9) and C(10), so benzo[1,2-d:4,5-d']-diimidazoles are substituted by chlorine in aqueous hydrochloric acid[64] to give high yields (Table 4.26) of the hydrochlorides of the corresponding 4,8-dichloro derivatives (Scheme 4.47, **4.174** → **4.175**, X = Cl), whose orientation is established unambiguously by oxidation to the corresponding

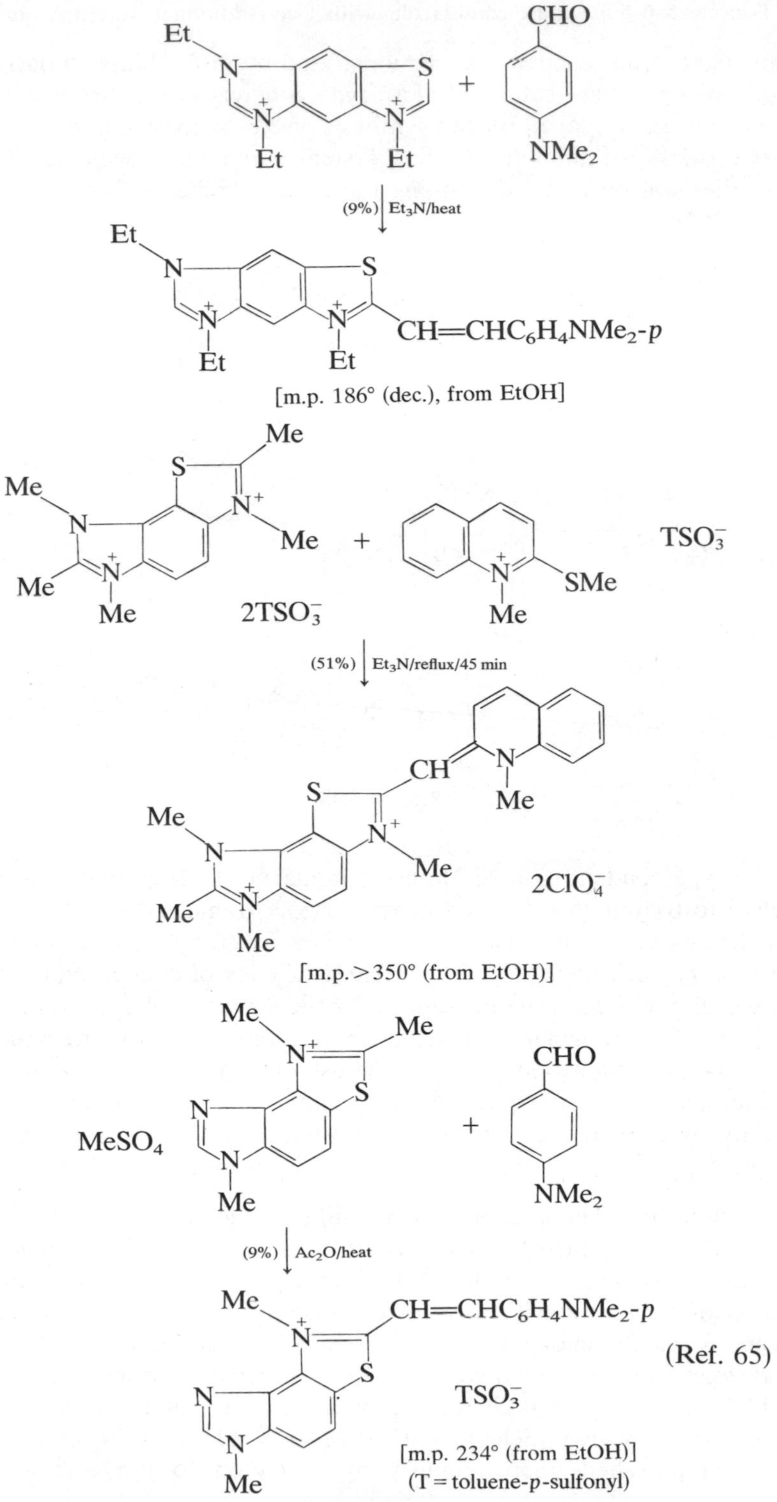

(9%) | Et₃N/heat

[m.p. 186° (dec.), from EtOH]

2TSO₃⁻   +   TSO₃⁻

(51%) | Et₃N/reflux/45 min

2ClO₄⁻

[m.p. >350° (from EtOH)]

MeSO₄   +

(9%) | Ac₂O/heat

(Ref. 65)

CH=CHC₆H₄NMe₂-*p*

TSO₃⁻

[m.p. 234° (from EtOH)]
(T = toluene-*p*-sulfonyl)

Scheme 4.46

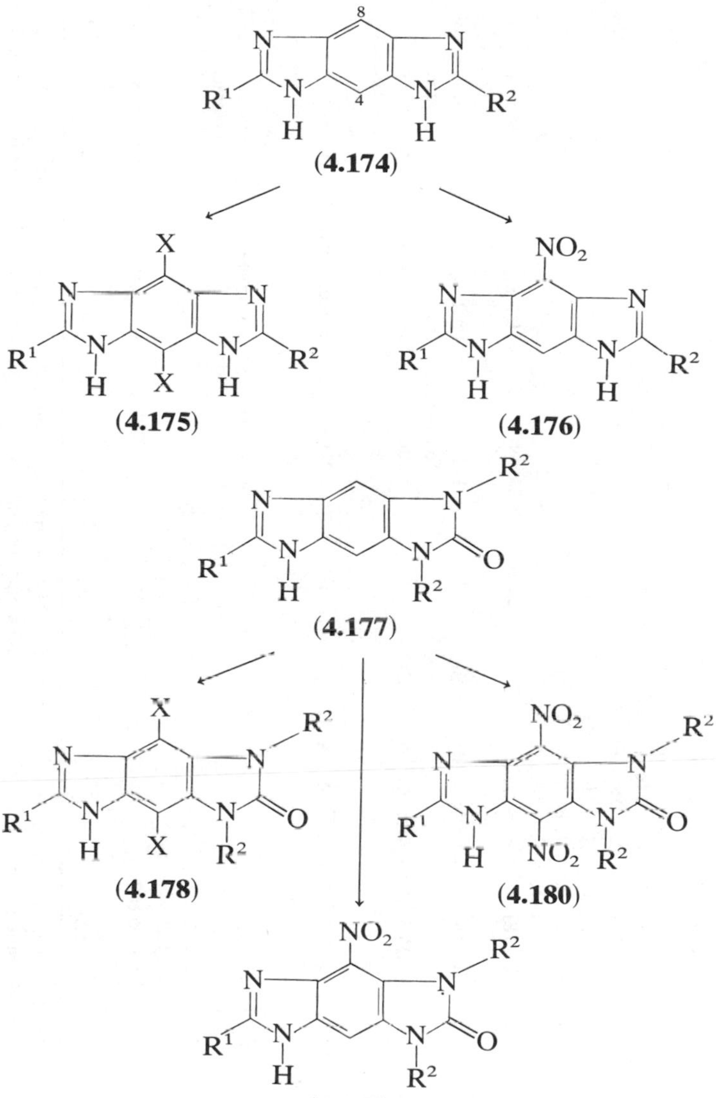

(4.174)

(4.175)   (4.176)

(4.177)

(4.178)   (4.180)

(4.179)

Scheme 4.47

benzo[1,2-*d*:4,5-*d'*]diimidazole-4,8-quinone (see later section, "Oxidation").[64]

Bromination (Scheme 4.47) of benzo[1,2-*d*:4,5-*d'*]diimidazoles (**4.174**) using bromine in 3% aqueous sulfuric acid[64] or of the related benzo[1,2-*d*:4,5-*d'*]diimidazol-2(1*H*,3*H*)-one (**4.177**) dihydrochlorides using bromine in aqueous solution[37] also proceeds smoothly and affords the expected 4,8-dibromo derivatives (**4.175**; X = Br), and (**4.178**; X = Br), albeit in unspecified yield (Table 4.26).[37,64]

TABLE 4.26. HALOGENATION OF BENZO[1,2-d:4,5-d']DIIMIDAZOLES AND BENZO[1,2-d:3,4-d']DIIMIDAZOLES

Starting material	Reaction conditions[a]	Product	Yield (%)	Melting point (°C)	Solvent of crystallization	Crystal form	Ref.
(4.174) (R^1=R^2=H)	A	(4.175) (R^1=R^2=H, X=Cl)[b]	83	—[c]	—[d]	—[e]	64
(4.174) (R^1=H, R^2=Me)	A	(4.175) (R^1=H, R^2=Me, X=Cl)[b]	80	—[c]	—[d]	—[e]	64
(4.174) (R^1=R^2=Me)	A	(4.175) (R^1=R^2=Me, X=Cl)[b]	88	—[c]	—[d]	—[e]	64
(4.174) (R^1=Me, R^2=Ph)	A	(4.175) (R^1=Me, R^2=Ph, X=Cl)[b]	69	—[c]	—[d]	—[e]	64
(4.174) (R^1=R^2=H)	B	(4.175) (R^1=R^2=H, X=Br)[b]	—[f]	—[c]	—[d]	—[e]	64
(4.174) (R^1=H, R^2=Me)	B	(4.175) (R^1=H, R^2=Me, X=Br)[b]	—[f]	—[c]	—[d]	—[e]	64
(4.174) (R^1=R^2=Me)	B	(4.175) (R^1=R^2=Me, X=Br)[b]	—[f]	—[c]	—[d]	—[e]	64
(4.177) (R^1=R^2=H)[b]	C	(4.178) (R^1=R^2=H, X=Br)[g]	—[f]	>320	HCl aq.	Colorless needles	37
(4.177) (R^1=H, R^2=Me)[g]	C	(4.178) (R^1=H, R^2=Me, X=Br)[h]	—[f]	311 (decomp.)	HCl aq.	Colorless needles	37
(4.177) (R^1=Me, R^2=H)	C	(4.178) (R^1=Me, R^2=H, X=Br)	—[f]	>320	Ethanol–water	Colorless amorphous solid	37
(4.177) (R^1=R^2=Me)	C	(4.178) (R^1=R^2=Me, X=Br)	—[f]	>320	Ethanol–water	Colorless amorphous solid	37
(4.181) (R=H)	D	(4.183)[i]	74	198–199	Dioxane–ether	Yellow crystals	42
(4.181) (R=H)	E	(4.183)[i]	66	—	—	—	42
(4.181) (R=H)	F	(4.183)[i] + (4.182)[i]	36 / 26	— / 283–284	Chloroform–ether	Colorless prisms	42
(4.183)	F	(4.182)[j]	69	—	—	—	42

[a] A, Cl$_2$, aq. HCl/room temp. B, Br$_2$, 3% H$_2$SO$_4$/room temp. C, Br$_2$, H$_2$O/warm/few minutes. D, Br$_2$, CHCl$_3$/20–30°/2 hr, then boil the resulting perbromide with water for 2 hr. E, dioxane dibromide, dioxane/20–25°/36 min, then boil the resulting perbromide with H$_2$O for 2 hr. F, KBrO, HBr (d = 1.4)/20°/2 hr, then 60°/2 hr, then reflux the resulting perbromide with water for 2 hr.
[b] Dihydrochloride.
[c] No melting point recorded.
[d] Solvent not specified.
[e] Crystal form not specified.
[g] Monohydrochloride.
[h] Monohydrate.
[i] Hemihydrate.

450

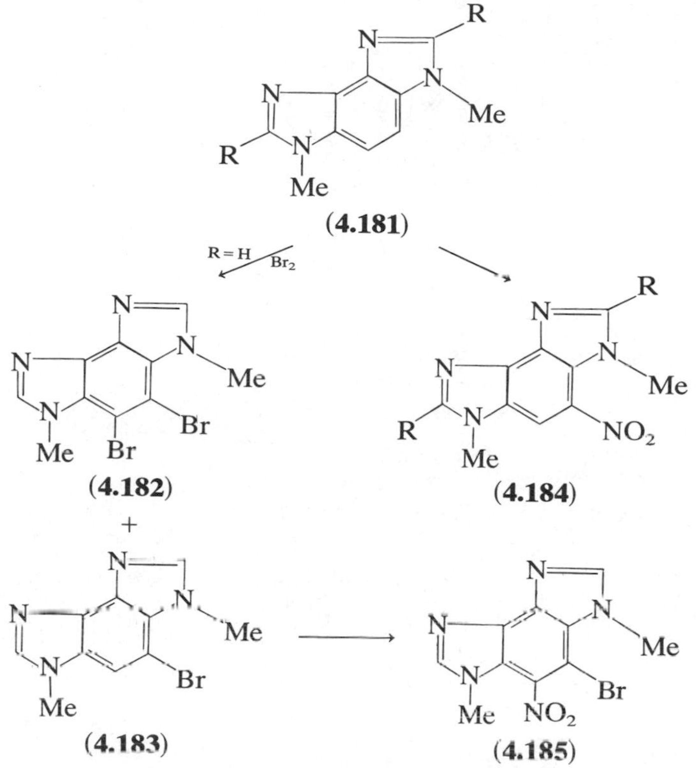

Scheme 4.48

Studies of the bromination of benzo[1,2-d:3,4-d']diimidazoles[42] are in accord with the predictions of molecular orbital calculations,[39] which indicate that such molecules should be prone to electrophilic attack at $C(4)$ and $C(5)$ (cf. Scheme 4.48, **4.181**). Thus bromination[42] of compound (**4.181**; R = H) using bromine in chloroform solution, or with dioxane dibromide in dioxane, results in the formation of a perbromide, which on boiling with water is converted in good yield (Table 4.26) into the 4-bromo derivative (**4.183**) (Scheme 4.48), whose orientation is established unequivocably by ultraviolet and ^{1}H NMR data. Bromination of **4.181**; R = H can also be accomplished using potassium bromate in hydrobromic acid, but under these conditions the mono-bromobenzo[1,2-d:3,4-d]diimidazole (**4.183**) is accompanied by the 4,5-dibrominated product (**4.182**), which is also formed when the former compound is subjected to further bromination (Table 4.26).[42]

NITRATION. With the exception of an isolated report[69] concerning the nitration (KNO₃–H₂SO₄, 100°, 1 hr) of an imidazo[4,5-g]benzoxazole derivative to give an unidentified mono-nitro derivative in 50% yield [yellow

TABLE 4.27. NITRATION OF BENZO[1,2-d:4,5-d']DIIMIDAZOLES AND BENZO[1,2-d:3,4-d']DIIMIDAZOLES

Starting material	Reaction conditions[a]	Product	Yield (%)	Melting point (°C)	Solvent of crystallization	Crystal form	Ref.
(4.174) (R¹ = R² = H)	A	(4.176) (R¹ = R² = H)	—[b]	—[c]	—[d]	—[e]	64
(4.174) (R¹ = H, R² = Me)	A	(4.176) (R¹ = H, R² = Me)	—[b]	—[c]	—[d]	—[e]	64
(4.174) (R¹ = R² = Me)	A	(4.176) (R¹ = R² = Me)	—[b]	—[c]	—[d]	—[e]	64
(4.177) (R¹ = R² = H)	B	(4.179) (R¹ = R² = H)[f]	—[b]	—[g]	—[d]	—[e]	37
(4.177) (R¹ = Me, R² = H)	B	(4.179) (R¹ = Me, R² = H)[h]	—[b]	—[g]	—[d]	Yellow needles	37
(4.177) (R¹ = H, R² = Me)	B	(4.179) (R¹ = H, R² = Me)	—[b]	—[c]	—[d]	Orange needles	37
(4.177) (R¹ = R² = Me)	C	(4.180) (R¹ = H, R² = Me)	—[b]	287	Chlorobenzene	Yellow crystals	37
(4.177) (R¹ = R² = Me)	C	(4.180) (R¹ = R² = Me)	—[b]	284	Pyridine–water	Orange needles	37
(4.177) (R¹ = R² = H)	C	(4.180) (R¹ = R² = H)	—[b]	—[c]	—[d]	Red powder	37
(4.177) (R¹ = Me, R² = H)	C	(4.180) (R¹ = Me, R² = H)	—[b]	—[c]	—[d]	Red powder	37
(4.181) (R = H)	D	(4.184) (R = H)	77	267	Water	Yellow needles	42
(4.181) (R = Me)	D	(4.184) (R = Me)	—[b]	231–232	Acetone	Yellow prisms	42
(4.183)	E	(4.185)	79	276	n-Butanol	Yellow needles	42

[a] A, HNO₃ (d = 1.48), concn. H₂SO₄/15–20°/5 hr. B, HNO₃ (d = 1.50), concn. H₂SO₄/–5°/few min. C, HNO₃ (d = 1.50), concn. H₂SO₄/20°/1 hr. D, HNO₃ (d = 1.52), concn. H₂SO₄/0–5°/30 min, then 20–30°/1.5 hr. E, HNO₃ (d = 1.52), concn. H₂SO₄/0–5°/5 min, then 20–30°/2 hr.

[b] Yield not recorded.

[c] >320.

[d] Solvent not specified.

[e] Yellow solid.

[f] Isolated as the sulfate.

[g] Melting point not recorded.

[h] Hydrochloride.

452

needles, m.p. 300° (decomp.) (from ethanol)], studies of the nitration of tricyclic 5-6-5 fused benzimidazole ring systems with two additional hetero atoms have been confined to benzo[1,2-d:4,5-d']diimidazole and benzo[1,2-d:3,4-d']diimidazole derivatives (Table 4.27).

Mixed acid nitration of benzo[1,2-d:4,5-d']diimidazoles (Scheme 4.47, **4.174**) at 15 to 20° (Table 4.27) leads in unspecified yield to yellow, high-melting solids formulated as 4-mono-nitro derivatives (**4.176**).[64] All attempts to effect the dinitration of benzo[1,2-d:4,5-d']diimidazoles (**4.174**), even under severe conditions, were unsuccessful.[64] In contrast, the greater reactivity of the imidazolones (**4.177**) is demonstrated[37] by their mixed acid nitration (Table 4.27) at −5° to give mono-nitro products (**4.179**) and at 20° to give the corresponding 4,8-dinitro compounds (**4.180**), in both cases in unstated yield.

Treatment of the benzo[1,2-d:3,4-d']diimidazole derivatives (**4.181**; R = H or Me) at low temperature with mixed acid[42] leads to the 4-nitrobenzo[1,2-d:3,4-d']diimidazoles (**4.184**; R = H or Me) (Table 4.27), which resist further nitration even on heating under reflux with mixed nitric and sulfuric acids.[42] In contrast, the mono-bromo compound (**4.183**) is smoothly nitrated by mixed acid to give the bromo-nitro compound (**4.185**) (Table 4.27).[42]

The nitrobenzo[1,2-d:3,4-d']diimidazoles (**4.184**; R = H or Me) are readily reduced (stannous chloride-hydrochloric acid) to give the corresponding amines (cf. Scheme 4.53, **4.201**)[12] This orthodox behavior contrasts with that of the mono-nitro product (hemihydrate, m.p. 276°) of unspecified structure,[28a] formed by the action of fuming nitric acid on the benzo[1,2-d:3,4-d']diimidazole derivative (**4.181**; R = Me, H for Me). The attempted reduction of this compound leads to its reconversion into the parent base (**4.181**; R = Me, H for Me),[28a] suggesting that it may in fact be an N- rather than a C-nitro derivative.

DIAZOTIZATION. 2-Amino-3,6-dimethylbenzo[1,2-d:3,4-d']diimidazole (**4.171**) is readily diazotized, and the resulting diazonium salt couples in the usual fashion with β-naphthol.[42]

Reactions with Nucleophiles

DEPROTONATION. As well as being basic in character (see earlier in this section), benzo[1,2-d:4,5-d']diimidazoles unsubstituted at the N-1(3) and N-5(7) positions behave as weak acids as demonstrated by the solubility of 2,6-dimethyl-N-1(3)H,N-5(7)H-benzo[1,2-d:4,5-d']diimidazole (**4.174**, $R^1 = R^2 = $ Me)] in dilute aqueous alkali.[19,36] This acidic character is markedly enhanced when either the central benzene ring or the flanking imidazole nuclei incorporate electron-withdrawing groups that can stabilize by delocalization the anion produced by deprotonation at N-1(3) or N-5(7).

Thus benzo[1,2-d : 4,5-d']diimidazole-4,8-quinones (see later in this section) are highly acidic, forming violet solutions of isolable alkali metal salts in aqueous alkali.[33,64] The acidity of 4-nitrobenzo[1,2-d : 4,5-d']diimidazoles (**4.176**), as might be expected, is also enhanced, and these compounds are freely soluble in even aqueous ammonia or sodium carbonate solution.[64] In accord with their cyclic lactam structures, benzo[1,2-d : 4,5-d]-diimidazolones of type (**4.177**; $R^2 = H$) and their thione counterparts (**4.177**; $R^2 = H$, S for O) readily dissolve in aqueous alkali due to the formation of water-soluble alkali metal salts from which the parent acids can be regenerated on treatment with mineral acid.[26]

AMINATION. Theoretical studies suggest enhanced electrophilic character for the $C2(6)$ and $C2(7)$ positions, respectively, in benzo[1,2-d : 4,5-d']-diimidazoles and benzo[1,2-d : 3,4-d']diimidazoles.[41,59,60] Paradoxically, the aminative capacity of these positions as predicted from theoretical studies[11,41,60] and favored by pK_a data[41] is not borne out in practice.[41] For example, the benzo[1,2-d : 3,4-d']diimidazole derivatives (**4.181**; R = H or Me) are more or less completely inert to heating with sodamide in xylene, dimethylaniline, or mineral oil at 180 to 195°. This unanticipated inertness is paralleled by the stability of [1,3]dioxolo[4,5-f]benzimidazoles to amination with sodamide,[11] and as in the case of the latter, is attributed to inhibition of attack by the amide anion as a consequence of complex formation between the hetero atoms in the substrate and the sodamide reagent.[41]

Halogen atoms at the $C2(7)$ positions in benzo[1,2-d : 3,4-d']diimidazoles are activated to nucleophilic displacement for example, by amines, affording methods for the synthesis of $C2(7)$ amino derivatives (e.g., Scheme 4.49, **4.190** → **4.191**) (Table 4.28).[55] Side-chain amination in benzo[1,2-d : 4,5-d']-diimidazoles is exemplified by the reactions (Scheme 4.50) of the 2,6-bis-chloromethyl derivative (**4.192**) with amines to afford products of type (**4.193**) (Table 4.28).[18]

HYDROXYLATION. The susceptibility of the $C(2)$ and $C(7)$ positions in benzo[1,2-d : 3,4-d']diimidazoles to nucleophilic attack as implied by the results of molecular orbital calculations[41,59,60] is borne out by the hydroxylation at these positions that occurs (accompanied by hydrogen evolution) when certain benzo[1,2-d : 3,4-d']diimidazole derivatives (Scheme 4.49) are fused with potassium hydroxide (Table 4.28).[41] When either the $C(2)$ or the $C(7)$ position is unsubstituted, the product is a mono imidazolone [**4.186** ($R^1 = Me$, $R^2 = H$) → **4.187**] (Table 4.28),[41] whereas if both the $C(2)$ and $C(7)$ positions are free a separable mixture of a mono and a diimidazolone results [**4.186** ($R^1 = R^2 = H$) → **4.188** + **4.189**] (Table 4.28).[41]

HALOGENATION. Benzo[1,2-d : 3,4-d']diimidazolones of type (**4.187**) (Scheme 4.49) behave as typical cyclic lactams and undergo[55] high-yield (Table 4.28) nucleophilic chlorination on heating with phosphorus oxychloride to give the anticipated chloro derivatives (e.g., **4.190**).[55]

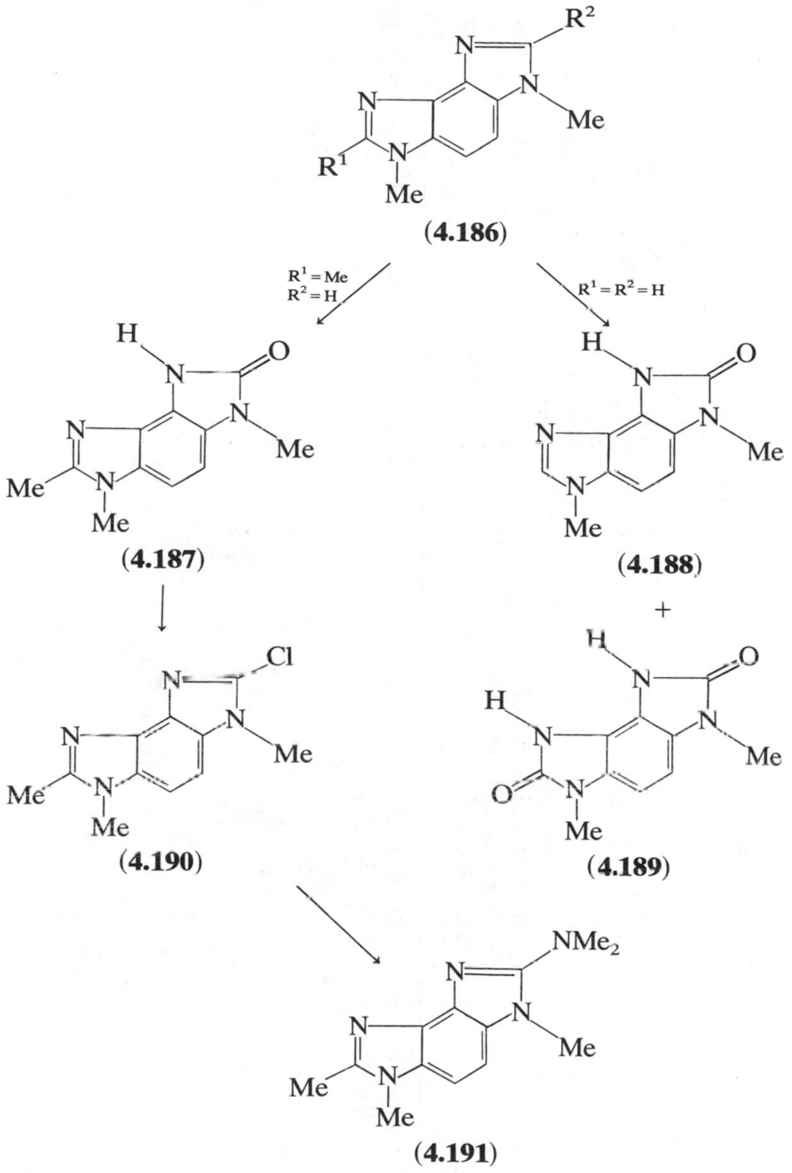

(4.186)

R^1 = Me
R^2 = H

R^1 = R^2 = H

(4.187)

(4.188)

+

(4.190)

(4.189)

(4.191)

Scheme 4.49

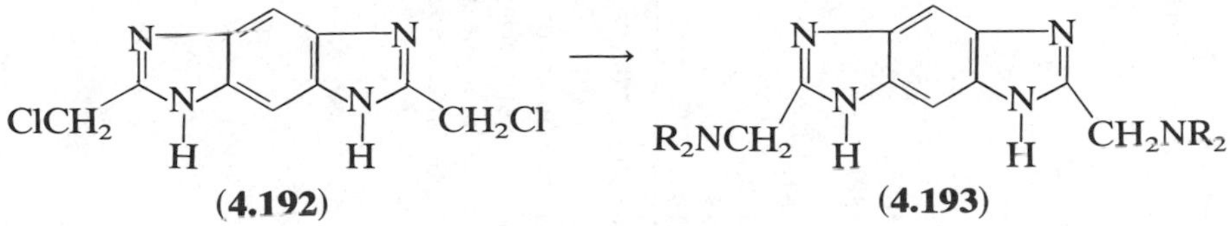

(4.192)

(4.193)

Scheme 4.50

TABLE 4.28. NUCLEOPHILIC SUBSTITUTION REACTIONS OF BENZO[1,2-d:4,5-d']DIIMIDAZOLES AND BENZO[1,2-d:3,4-d']DIIMIDAZOLES

Starting material	Reaction conditions[a]	Product	Yield (%)	Melting point (°C)	Solvent of crystallization	Crystal form	Ref.
(**4.186**) (R^1 = R^2 = H)	A	(**4.188**)	43	>360	Ethanol	Colorless crystals	41
		+(**4.189**)	35	>360	n-Butanol	Colorless crystals	41
(**4.186**) (R^1 = Me, R^2 = H)	B	(**4.187**)	81	332–333	Water	Colorless crystals	55
(**4.187**)	C	(**4.190**)	90	300–301	Ethanol	Colorless crystals	55
(**4.190**)	D	(**4.191**)	75	145–146	Toluene–light petroleum	Yellow crystals	55
(**4.192**)	E	(**4.193**) (R = CH$_2$CH$_2$OH)	77	—[b]	—[c]	—[d]	18
(**4.192**)	E	(**4.193**) (R = —(CH$_2$)$_5$—)[e]	60	—[b]	—[c]	—[d]	18
(**4.192**)	E	(**4.193**) (R = C$_5$H$_5$$\overset{+}{\text{N}}$)	87	—[b]	—[c]	—[d]	18
(**4.194**)	F	(**4.195**)	78	Decomp.	—[c]	—[d]	38

[a] A, KOH/270–278°/30 min, then 290°/1 hr. B, KOH/300–310°/45 min, then 320°/1 hr. C, POCl$_3$/reflux/10 hr. D, Me$_2$$\overset{+}{\text{N}}H_2Cl^-$, NaOAc/140° (autoclave)/10 hr. E, R$_2$NH, EtOH/reflux/2 hr. F, 48% HBr/reflux/1 hr.
[b] Melting point not quoted.
[c] Solvent not specified.
[d] Crystal form not specified.
[e] Tetrahydrochloride.

456

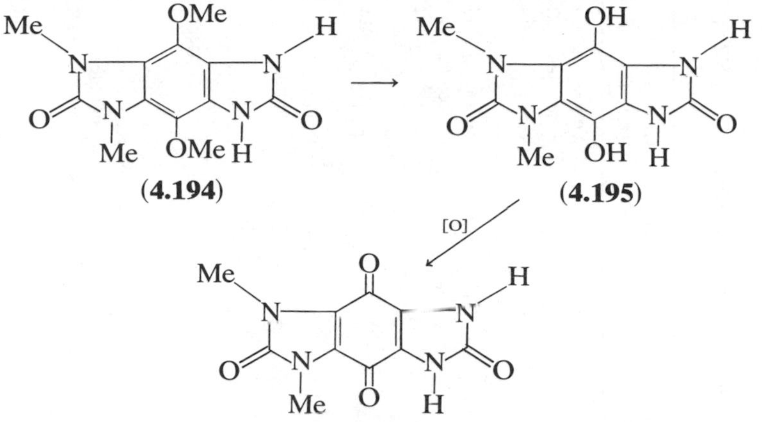

(4.194) → (4.195)

[O]

(4.196)

Scheme 4.51

Nuclear methoxy derivatives of benzo[1,2-d:4,5-d']diimidazoles can be demethylated in good yield (Table 4.28) to the corresponding hydroxy compounds by heating with 48% hydrobromic acid (e.g., Scheme 4 51, **4.194 → 4.195**).[38]

Oxidation

The chemical analogy between the benzo[1,2-d:4,5-d']diimidazole and anthracene ring systems as manifested by the propensity of derivatives of the former to undergo electrophilic attack at C(4) and C(8) (see earlier in this section) is also apparent in the case of oxidation of benzo[1,2-d:4,5-d']-diimidazole derivatives to the corresponding 4,8-quinones (Scheme 4.52, **4.197 → 4.199**).[19,27,64] Oxidation of this type[64] is readily accomplished in good to excellent yield (Table 4.29) by boiling the requisite 4,8-diunsubstituted benzo[1,2-d:4,5-d']diimidazole (**4.197**) with chromic acid in

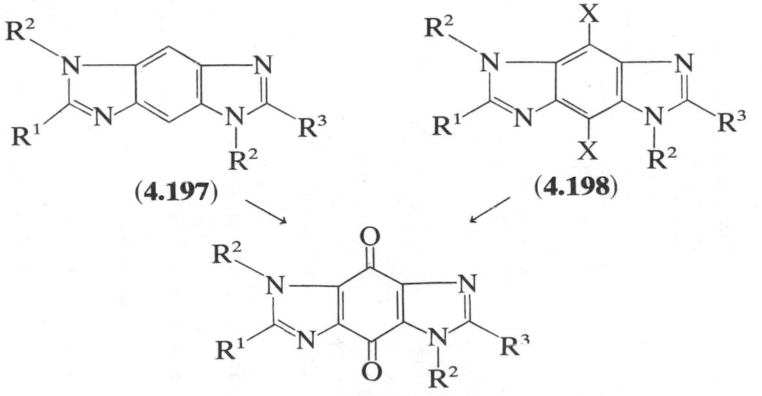

(4.197)

(4.198)

(4.199)

Scheme 4.52

TABLE 4.29. OXIDATION OF BENZO[1,2-d:4,5-d']DIIMIDAZOLE DERIVATIVES (**4.197**, **4.198**, AND **4.195**) TO BENZO[1,2-d:4,5-d']DIIMIDAZOLE-4,8-QUINONES (**4.199**, AND **4.196**)

Starting material	Reaction conditions[a]	Product	Yield (%)	Melting point (°C)	Solvent of crystallization	Crystal form	Ref.
(**4.197**) ($R^1 = R^2 = R^3 = H$)	A	(**4.199**) ($R^1 = R^2 = R^3 = H$)	91	—[b]	—[c]	Yellow needles	64
(**4.198**) ($R^1 = R^2 = R^3 = H$, $X = Cl$)	—[d]	(**4.199**) ($R^1 = R^2 = R^3 = H$)	—[e]	—[b]	—[c]	—	64
(**4.198**) ($R^1 = R^2 = R^3 = H$, $X = Br$)	—[d]	(**4.199**) ($R^1 = R^2 = R^3 = H$)	—[e]	—[b]	—[c]	—	64
(**4.197**) ($R^1 = R^3 = Me$, $R^2 = H$)	A	(**4.199**) ($R^1 = R^3 = Me$, $R^2 = H$)[f]	Quant.	—[b]	40% Sulfuric acid aq.	Yellow needles	64
(**4.198**) ($R^1 = R^3 = Me$, $R^2 = H$, $X = Cl$)	—[d]	(**4.199**) ($R^1 = R^3 = Me$, $R^2 = H$)	—[e]	—[b]	—	—	64
(**4.198**) ($R^1 = R^3 = Me$, $R^2 = H$, $X = Br$)	—[d]	(**4.199**) ($R^1 = R^3 = Me$, $R^2 = H$)	—[e]	—[b]	—	—	64
(**4.197**) ($R^1 = R^2 = H$, $R^3 = Me$)	A	(**4.199**) ($R^1 = R^2 = H$, $R^3 = Me$)[f]	80	—[b]	40% Sulfuric acid aq.-ethanol	Yellow needles	64
(**4.198**) ($R^1 = R^2 = H$, $R^3 = Me$, $X = Cl$)	—[d]	(**4.199**) ($R^1 = R^2 = H$, $R^3 = Me$)	—[e]	—[b]	—	—	64
(**4.198**) ($R^1 = R^2 = H$, $R^3 = Me$, $X = Br$)	—[d]	(**4.199**) ($R^1 = R^2 = H$, $R^3 = Me$)	—[e]	—[b]	—	—	64
(**4.197**) ($R^1 = Me$, $R^2 = H$, $R^3 = Ph$)	B	(**4.199**) ($R^1 = Me$, $R^2 = H$, $R^3 = Ph$)[g]	43	—[b]	—[c]	Red needles	64
(**4.198**) ($R^1 = Me$, $R^2 = H$, $R^3 = Ph$, $X = Cl$)	—[d]	(**4.199**) ($R^1 = Me$, $R^2 = H$, $R^3 = Ph$)	—[e]	—[b]	—[c]	—	64
(**4.198**) ($R^1 = Me$, $R^2 = H$, $R^3 = Ph$, $X = Br$)	—[d]	(**4.199**) ($R^1 = Me$, $R^2 = H$, $R^3 = Ph$)	—[e]	—[b]	—[c]	—	64
(**4.197**) ($R^1 = Me$, $R^2 = H$, $R^3 = CH_2Ph$)	A	(**4.199**) ($R^1 = Me$, $R^2 = H$, $R^3 = COPh$)	—[e]	—[b]	—[c]	—[h]	64

458

Reactant	Method[a]	Product	Yield (%)	m.p. (°C)	Solvent	Form	Ref.
(4.197) (R^1 = Me, R^2 = H, R^3 = 3-pyridyl)	C	(4.199) (R^1 = Me, R^2 = H, R^3 = 3-pyridyl)	35	—[b]	—[c]	Red solid	64
(4.198) (R^1 = R^3 = Me, R^2 = H, X = OH)	D	(4.199) (R^1 = R^3 = Me, R^2 = H)	40	—[b]	—[c]	Brown powder	27
(4.198) (R^1 = R^3 = Me, R^2 = Bu^n, X = OH)	A	(4.199) (R^1 = R^3 = Me, R^2 = Bu^n)	—[e]	181–183	Ether	Orange powder	33
(4.198) (R^1 = R^3 = Me, R^2 = CH_2CH_2OH)	E	(4.199) (R^1 = R^3 = Me, R^2 = CH_2CH_2OH)	80	>300	—[c]	Yellow needles	33
(4.198) (R^1 = R^3 = Me, R^2 = $CH_2CH_2NEt_2$)	E	(4.199) (R^1 = R^3 = Me, R^2 = $CH_2CH_2NEt_2$)[i]	—[e]	208–210	—[c]	Yellow needles	33
(4.198) (R^1 = R^3 = Me, R^2 = $CH_2CH_2NEt_2$) 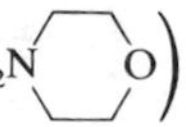	E	 (4.199) (R^1 = R^3 = Me, R^2 = $CH_2CH_2NEt_2$)	—[e]	241–243	—[c]	Yellow needles	33
(4.195) $R^2 = CH_2CH_2N$⟨morpholino⟩	F	(4.196) $R^2 = CH_2CH_2N$⟨morpholino⟩[j]	29	352–353 (decomp.)	Toluene	Green needles	38

[a] A, H_2CrO_4, 40% H_2SO_4/boil. B, H_2CrO_4, 50–60% H_2SO_4/boil. C, H_2CrO_4, concn. H_2SO_4. D, $2M$ NaOH, air/100°. E, H_2O, air/room temp./2 hr. F, $NaNO_2$, concn. H_2SO_4/heat/0.5 hr.

[b] Melting point not quoted.

[c] Solvent not specified.

[d] Reaction conditions not specified.

[e] Yield not recorded.

[f] Sulfate; boiling with water gives the free base, as an orange solid, m.p. >400°.

[g] Hydrochloride; boiling with water gives the free base as an orange solid.

[h] Crystal form not recorded.

[i] Forms a dihydrochloride, m.p. 283–286°.

[j] Forms a dihydrochloride, m.p. 301–303°.

sulfuric acid, conditions which attest to the indestructability of the quinone products (**4.199**)! Benzo[1,2-*d* : 4,5-*d'*]diimidazole-4,8-quinones (**4.199**) are high-melting, highly colored (orange) solids that form violet alkali metal salts with aqueous alkalis.[64] Such quinones are also the stable end-products of the oxidation of other types of benzo[1,2-*d* : 4,5-*d'*]diimidazole derivatives, in particular the 4,8-dihalogeno and 4,8-dihydroxy compounds. [Scheme 4.51, **4.195** → **4.196**, and Scheme 4.52, **4.198** (X = Cl, Br, or OH) → **4.199**] (Table 4.29).[27,33,38,64] Oxidation of the dihydroxy derivatives (hydroquinones) (**4.195**) and (**4.198**; X = OH) takes place under particularly mild conditions (e.g., air oxidation in aqueous acidic or alkaline solution alone[22,33] or in presence of nitrite ion[38]); in view of the ready accessibility[27,33] of the hydroquinone substrates this reaction provides a useful general method (Table 4.29) for the synthesis of benzo[1,2-*d* : 4,5-*d'*]diimidazole-4,8-quinones (**4.199**).

In contrast to the ease of oxidation of 4,8-dihalogenobenzo[1,2-*d* : 4,5-*d'*]-diimidazoles, nitro-substituted derivatives (e.g., **4.176**) are stable to oxidation even under forcing conditions.[64] Conversely, benzo[1,2-*d* : 4,5-*d'*]-diimidazole derivatives incorporating an imidazolone nucleus are destroyed by chromic acid oxidation.[27]

Reduction

The general stability of tricyclic 5-6-5 fused benzimidazole ring systems with two additional hetero atoms to metal-proton donor reducing agents is indicated by the successful synthesis of such ring systems in a number of instances (see earlier in this section) by reductive cyclization using reagents such as tin or stannous chloride in conjunction with hydrochloric acid. Stability to metal-proton donor and similar reducing agents is further substantiated by the fact that ring substituents can be reduced without affecting the ring itself. For example, in Scheme 4.53 the nitrobenzo[1,2-*d* : 3,4-*d'*]diimidazole (**4.200**) is smoothly reduced by stannous chloride in hydrochloric acid to give the expected amine (**4.201**) in high yield.[42]

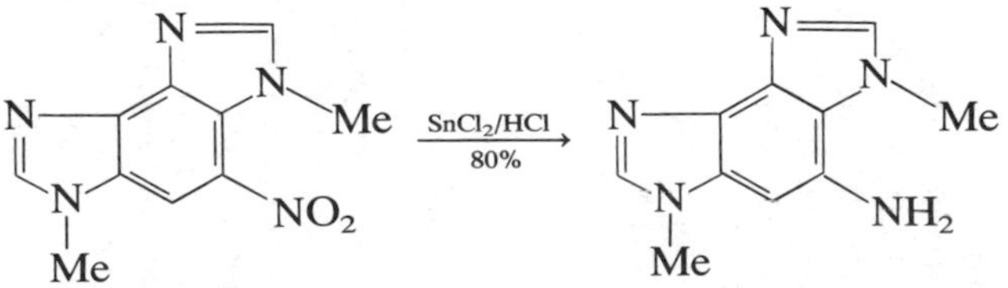

(**4.200**)   (**4.201**)

[colorless needles, m.p. 254°
(from *n*-BuOH)]

Scheme 4.53

Correspondingly, hydrosulfite reduction[64] of benzo[1,2-d:4,5-d']diimidazole-4,8-quinones affords the corresponding hydroquinones [Scheme 4.52, **4.199 → 4.198** (X = OH)].

On the other hand, further support for the chemical analogy between the benzo[1,2-d:4,5-d']diimidazole and anthracene ring systems is provided by the demonstration that benzo[1,2-d:4,5-d']diimidazoles as in the case of anthracene derivatives can be reduced to dihydro products.[70] Thus 2,6-dimethyl-N-1(3)H,5(7)H-benzo[1,2-d:4,5-d']diimidazole is catalytically reduced at normal temperature and pressure to the 4,8-dihydro derivative (Scheme 4.54, **4.202 → 4.203**).[70]

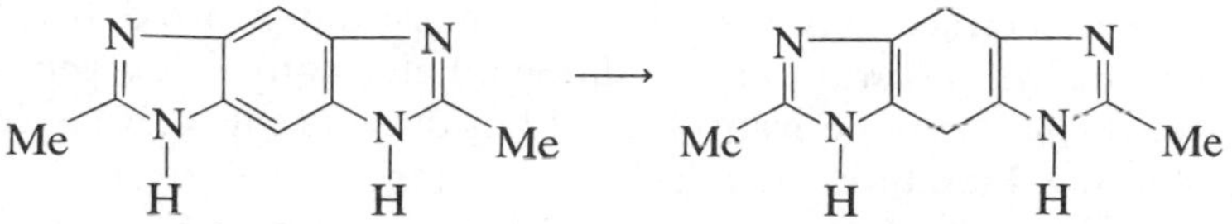

(**4.202**)   (**4.203**)

[m.p. >360°; picrate, m.p. 325–350° (from EtOH)]

Scheme 4.54

4.3.4. Practical Applications

Biological Properties

4,8-Dihydroxybenzo[1,2-d:4,5-d']diimidazoles and the corresponding quinones exhibit a wide spectrum of biological activity, including anticancer and antibacterial properties as well as being effective in the treatment of rheumatoid arthritis.[33]

The biosynthesis of nucleotide cyanocobalamins containing the benzo-[1,2-d:4,5-d']diimidazole nucleus has been described.[71]

Dyestuffs

As discussed previously, the imidazobenzothiazole and benzodiimidazole ring systems have found widespread use as the chromophoric units in many cyanine dyestuffs.[15,50,51,56,57,65] The benzo[1,2-d:4,5-d']diimidazole ring system has also been utilized as a component of anthraquinone dyestuffs[31] and in fluorescors.[67] Additionally, derivatives of the [1,3]dioxolo[4,5-f]-benzimidazole[44] and imidazo[4,5-g]indazole[53] ring systems have been patented as photographic sensitizers and color-coupling agents, respectively.

Polymers

The benzo[1,2-*d*:4,5-*d'*]diimidazole ring system is an important monomer unit in polymers that combine high thermal and chemical stability with high tensile strength.[20,22-25,32,68,72]

4.4 TRICYCLIC 5-6-5 FUSED BENZIMIDAZOLES WITH THREE ADDITIONAL HETERO ATOMS

Of the twelve structurally feasible, fully unsaturated 5-6-5 fused benzimidazole ring systems with three additional hetero atoms (oxygen, sulfur, nitrogen) (Scheme 4.55), derivatives of all but three (Table 4.30) have been reported in the literature. The lack of reference to derivatives of the imidazo[4,5-*f*][1,2,3]benzoxadiazole (**4.205**), imidazo[4,5-*e*][1,2,3]benzoxadiazole (**4.207**), and imidazo[4,5-g][1,2,3]benzoxadiazole (**4.208**) ring systems (Scheme 4.55) is not surprising in view of the well-known[73] instability of 1,2,3-benzoxadiazole itself. Of the known tricyclic 5-6-5 fused benzimidazole ring systems with three additional hetero atoms (Scheme 4.55 and Table 4.30), only the five possible sulfur-containing, imidazobenzothiadiazole structures (**4.209** and **4.211** to **4.214**) and the two possible fully nitrogenous imidazo-1,2,3-benzotriazole frameworks (**4.215** and **4.216**) have received much attention. Additionally, a few derivatives of the imidazo[4,5-*f*][2,1,3]benzoselenadiazole ring system (**4.210**) have been reported (see section 4.4.1).

TABLE 4.30.   TRICYCLIC 5-6-5 FUSED BENZIMIDA-
              ZOLES WITH THREE ADDITIONAL
              HETERO ATOMS

Structure[a]	Name[b]
(**4.204**)	Imidazo[4,5-*f*][2,1,3]benzoxadiazole
(**4.206**)	Imidazo[4,5-*e*][2,1,3]benzoxadiazole
(**4.209**)	Imidazo[4,5-*f*][2,1,3]benzothiadiazole
(**4.210**)	Imidazo[4,5-*f*][2,1,3]benzoselenadiazole
(**4.211**)	Imidazo[4,5-*f*][1,2,3]benzothiadiazole
(**4.212**)	Imidazo[4,5-*e*][2,1,3]benzothiadiazole
(**4.213**)	Imidazo[4,5-*e*][1,2,3]benzothiadiazole
(**4.214**)	Imidazo[4,5-g][1,2,3]benzothiadiazole
(**4.215**)	Imidazo[4,5-*f*][1,2,3]benzotriazole
(**4.216**)	Imidazo[4,5-*e*][1,2,3]benzotriazole

[a] See Scheme 4.55.
[b] Based on the Ring Index.

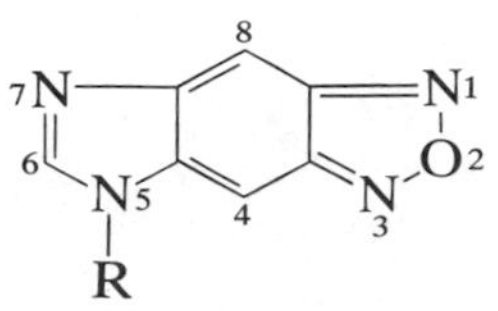

(4.204)

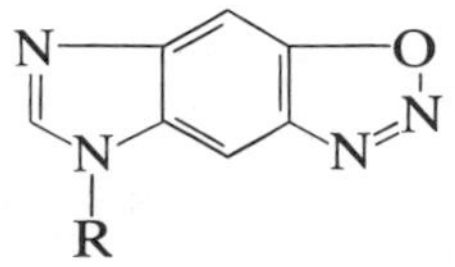

(4.205)

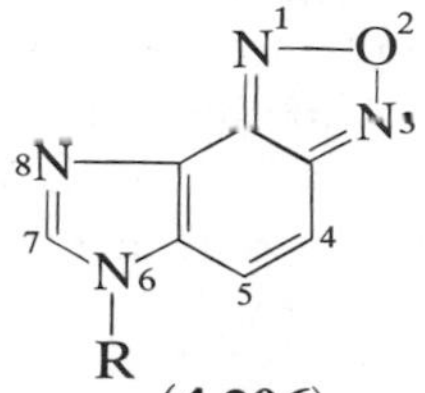

(4.206)

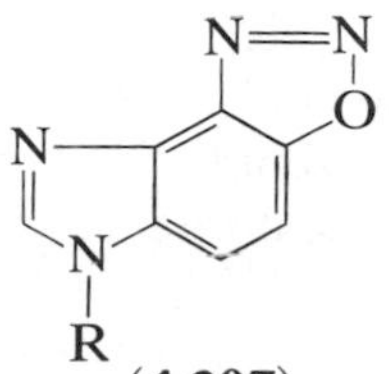

(4.207)

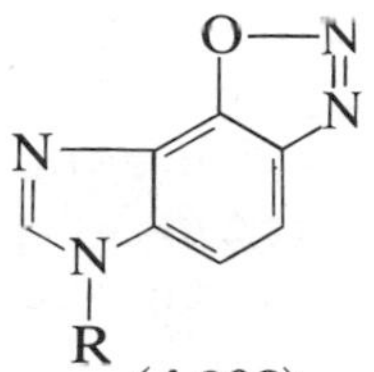

(4.208)

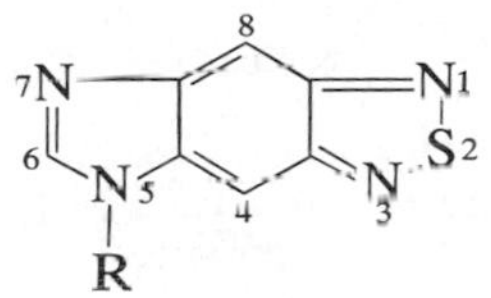

(4.209)

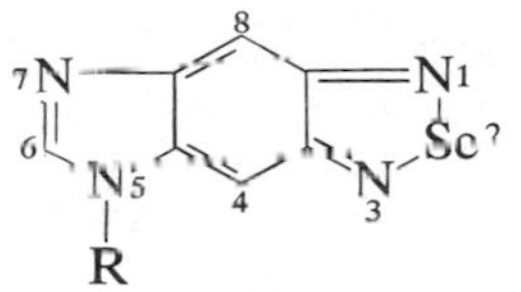

(4.210)

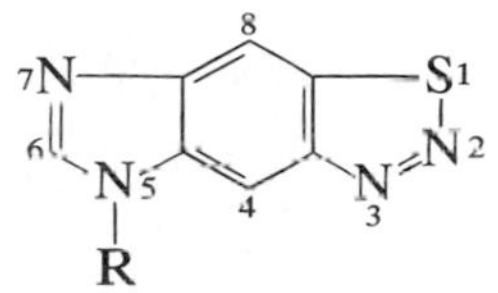

(4.211)

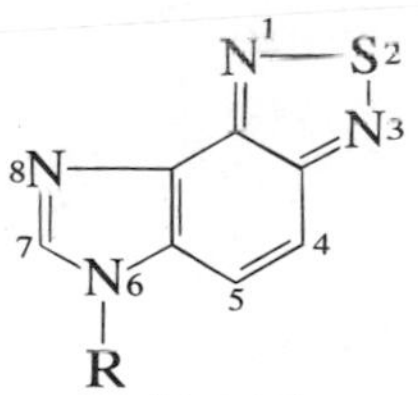

(4.212)

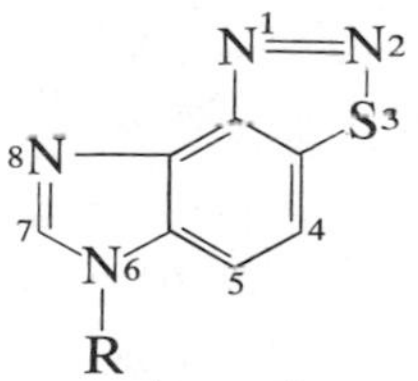

(4.213)

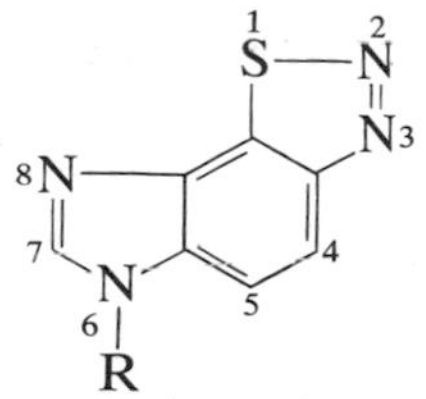

(4.214)

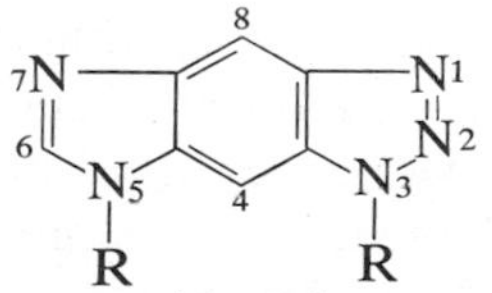

(4.215)

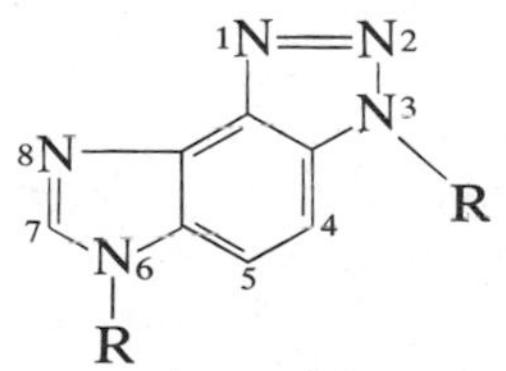

(4.216)

Scheme 4.55

4.4.1. Synthesis

Ring Closure Reactions of Benzimidazole Derivatives

Thermolysis[74] of 5-azido-1-methyl-6-nitrobenzimidazole (**4.217**) proceeds smoothly and efficiently (Scheme 4.56) in acetic acid to afford the tricyclic furazan *N*-oxide (**4.218**), apparently the only extant example of the imidazo[4,5-g][2,1,3]benzoxadiazole ring system. Attempts to effect the analogous ring closure (Scheme 4.56) of 5-amino-1-methyl-6-nitro-benzimidazole (**4.219**) to the imidazo[4,5-f][2,1,3]benzoxadiazole 1-*N*-oxide (**4.218**) using sodium hypochlorite or (diacetoxyiodo)benzene as oxidants were unsuccessful.[74]

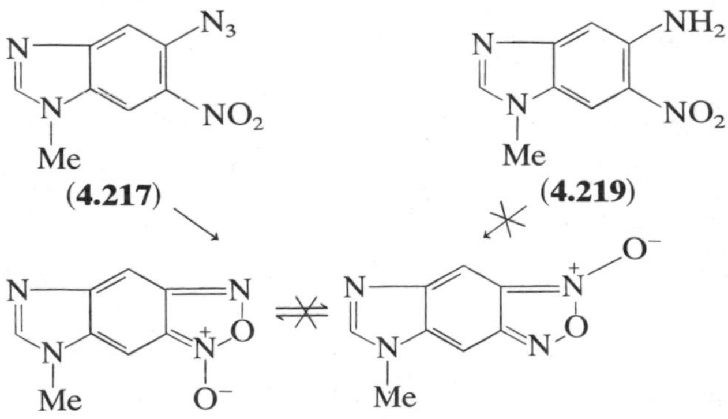

(4.217)   (4.219)

(a)   (b)

(4.218)

[orange needles, m.p. 195–197° (from diethylene glycol dimethyl ether)]

Scheme 4.56

Ring closure of 5,6-diaminobenzimidazole derivatives with thionyl chloride or thionylaniline, or using selenious acid provides useful synthetic routes to derivatives of the imidazo[4,5-f][2,1,3]benzothiadiazole and imidazo[4,5-f][2,1,3]benzoselenadiazole ring systems, respectively (Scheme 4.57). Thus heating the diamine (**4.220**; R = Me) under reflux in the absence of solvent with thionyl chloride or with thionylaniline in the presence of pyridine affords 5,6-dimethylimidazo[4,5-f][2,1,3]benzothiadiazole (**4.221**) in good yield (Table 4.31).[75] The use of a cosolvent (benzene or methanol) in transformations of this type markedly lowers the yield.[75] In analogous processes, 5,6-diaminobenzimidazole derivatives (**4.220**) react with selenious acid at room temperature to give excellent yields (Table 4.31) of the corresponding imidazo[4,5-f][2,1,3]benzoselenadiazole derivatives (**4.222**).[75]

Starting material	Reaction conditions[a]	Product	Yield (%)	Melting point (°C)	Solvent of crystallization	Crystal form	Ref.
(4.220) (R = Me)	A	(4.221)	78	182	Water	Yellow needles	75
(4.220) (R = Me)	B	(4.221)	54	—	—	—	75
(4.220) (R = H)	C	(4.222) (R = H)[b]	61	310 (decomp.)	Water	Yellow plates	75
(4.220) (R = Me)	C	(4.222) (R = Me)	96	243	Water	Yellow needles	75
(4.223)[c]	D	(4.224)	56	246	Xylene	Yellow solid	38
(4.239)	E	(4.240)	—[d]	230 (decomp.)	Water	Colorless needles	81
(4.241) (R^1 = R^2 = H)	F	(4.242) (R^1 = CF$_3$, R^2 = H)	—[d]	264–266	—[e]	—[f]	80
(4.241) (R^1 = R^2 = H)	G	(4.242) (R^1 = CF$_2$CF$_3$, R^2 = H)	—[d]	—[g]	—[e]	—[f]	80
(4.241) (R^1 = R^2 = H)	H	(4.242) (R^1 = Pr-n, R^2 = H)	—[d]	146	—[e]	—[f]	80
(4.241) (R^1 = Ac, R^2 = Me)	I	(4.242) (R^1 = R^2 = Me)	—[d]	189	—[e]	—[f]	82
(4.225)	J	(4.226)	42	97	Water	—[f]	14
(4.243)	K	(4.244) (R = Me)	80	233	Ethanol–water	—[f]	14
(4.243)	F	(4.244) (R = CF$_3$)	—[d]	259–260	—[e]	—[f]	80
(4.243)	G	(4.244) (R = CF$_2$CF$_3$)	—[d]	238–240	—[e]	—[f]	80
(4.243)	H	(4.244) (R = Pr-n)	—[d]	192	—[e]	—[f]	80
(4.245)	L	(4.247) (R = Me)	60	262	Ethanol–water	Colorless needles	14
(4.246) (R = CF$_3$)	M	(4.247) (R = CF$_3$)	—[d]	247–248	—[e]	—[f]	80

[a] A, SOCl$_2$, pyridine, benzene/100°/5 hr. B, thionylaniline, pyridine/130–135°/2 hr. C, selenious acid, H$_2$O/room temp./few minutes. D, SeO$_2$, H$_2$O/100°/few minutes. E, HCO$_2$H/reflux. F, (CF$_3$CO)$_2$O, CF$_3$CO$_2$H/reflux/3 hr. G, CF$_3$CF$_2$CO$_2$H/reflux/6 hr. H, CH$_3$CH$_2$CH$_2$CO$_2$H, 4 M HCl/reflux/7 hr. I, HCl/reflux. J, NaNO$_2$, HCl/0°/1 hr. K, AcOH, dil. HCl/reflux/1 hr. L, AcOH, dil. HCl/reflux/2 hr. M, NaHCO$_3$, dimethylformamide/reflux/3 hr.

[b] Hydrochloride; treatment with ammonia gives the free base, yellow needles, m.p. 275–276° (from water).

[c] Hydrochloride.   [d] Yield not recorded.

[e] Solvent not specified.   [f] Crystal form not specified.

[g] Melting point not recorded.

465

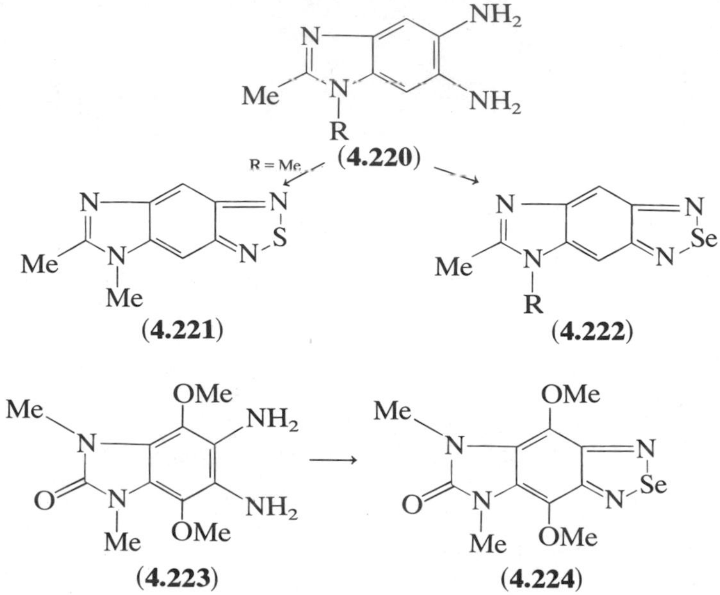

Scheme 4.57

The use of selenium dioxide in aqueous solution at elevated temperature to effect the transformation (**4.223** → **4.224**), has also been reported (Table 4.31).[38]

The construction of the imidazo[4,5-g][1,2,3]benzothiadiazole ring system by formation of the thiadiazole ring via diazotative ring closure in a 5-amino-4-mercaptobenzimidazole derivative is exemplified by transformation (**4.225** → **4.226**) (Scheme 4.58),[14] which proceeds in good yield (Table 4.31). The attractiveness of this route lies in the accessibility of the mercaptobenzimidazole precursors (e.g., **4.225**) by thiocyanation of 5-aminobenzimidazole derivatives followed by reduction.[14]

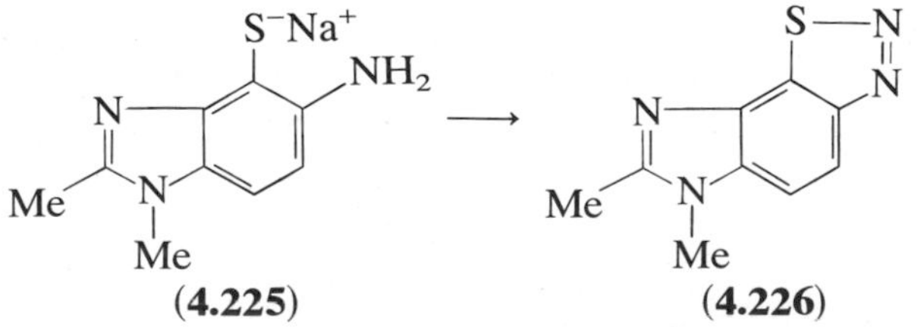

Scheme 4.58

Derivatives of the linear imidazo[4,5-f][1,2,3]benzotriazole ring system have been synthesized almost exclusively by diazotative ring closure in suitable 5,6-diaminobenzimidazole derivatives. The reactions shown in Scheme 4.59 are illustrative (Table 4.32).[76,77] Ring closure of this type in a

TABLE 4.32. SYNTHESES OF IMIDAZO[4,5-f][1,2,3]BENZOTRIAZOLES AND IMIDAZO[4,5-e][1,2,3]BENZOTRIAZOLES

Starting material	Reaction conditions[a]	Product	Yield (%)	Melting point (°C)	Solvent of crystallization	Crystal form	Ref.
(4.227) (R = H)	A	(4.228) (R = H)	—[b]	>300	—[c]	Red-brown needles	76
(4.227) (R = Me)	A	(4.228) (R = Me)	—[b]	311	—[d]	Yellow plates	77
(4.229)	A	(4.230)	—[b]	>300	—[d]	—[e]	76
(4.231)	B	(4.232)	—[b]	—[f]	—[d]	—[e]	78
(4.232)	C	(4.233)	—[b]	—[f]	—[d]	—[e]	78
(4.233)	D	(4.235) +(4.236)	63[g]	—[f]	—[d]	—[e]	78
(4.248)	E	(4.249)	—[b]	256	—[d]	needles	83
(4.250) (R = H)	F	(4.251) (R^1 = R^2 = H)	—[b]	>350	Methanol	—[e]	85
(4.250) (R = Me)	F	(4.251) (R^1 = H, R^2 = Me)	—[b]	276	Dioxane	—[e]	85
(4.250) (R = H)	G	(4.251) (R^1 = Me, R^2 = H)	—[b]	310	Methanol	—[e]	85
(4.250) (R = Me)	G	(4.251) (R^1 = R^2 = Me)	—[b]	305	Toluene	—[e]	85
(4.252) (R = Me)	F	(4.253) (R^1 = H, R^2 = Me)	—[b]	295	Dioxane	—[e]	85
(4.252) (R = Ph)	F	(4.253) (R^1 = H, R^2 = Ph)	—[b]	305	Xylene	—[e]	85
(4.252) (R = Me)	G	(4.253) (R^1 = R^2 = Me)	—[b]	222	Benzene	—[e]	85
(4.252) (R = Ph)	G	(4.253) (R^1 = Me, R^2 = P$_1$)	—[b]	283	Xylene	—[e]	85
(4.254)	F	(4.255) (R = H)	48	282	Dioxane	—[e]	84
(4.254)	G	(4.255) (R = Me)	55	282	Toluene	—[e]	84

[a] A, NaNO$_2$, HCl. B, NaNO$_2$, AcOH. C, concn. HCl. D, Pb(OAc)$_4$. PhN$_3$. E, concn. HCl/reflux. F, HCO$_2$H, concn. HCl/reflux/3 hr. G, Ac$_2$O, concn. HCl/reflux/4 hr.

[b] Yield not recorded.

[c] Solvent not specified.

[d] Purified by precipitation from HCl solution with ammonia.

[e] Crystal form not recorded.

[f] Melting point not recorded.

[g] Product ratio by NMR: (4.235), 38%; (4.236), 62%.

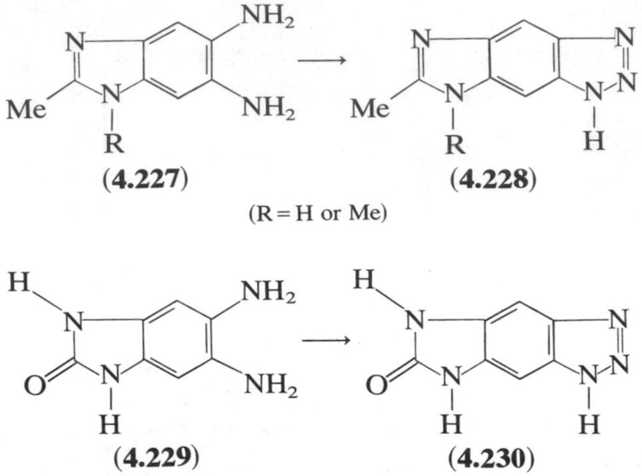

(4.227) → (4.228)

(R = H or Me)

(4.229) → (4.230)

Scheme 4.59

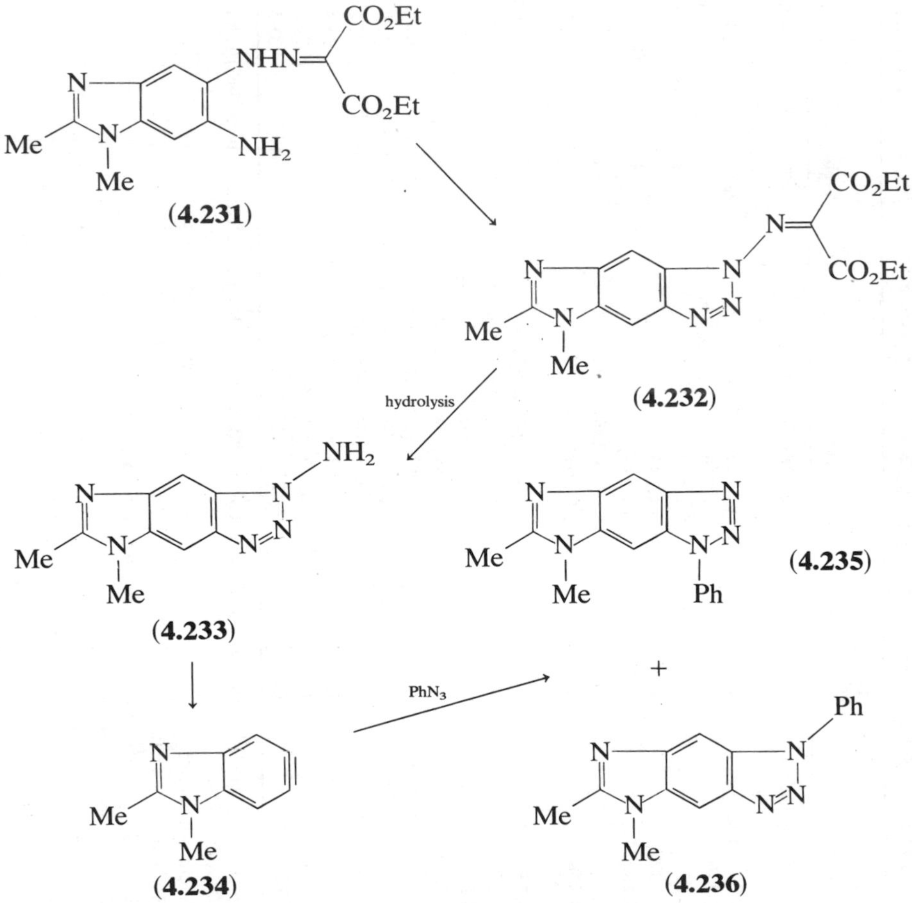

(4.231)

(4.232)

hydrolysis

(4.233)

(4.235)

(4.234)

PhN₃

(4.236)

Scheme 4.60

468

6-amino-5-hydrazonobenzimidazole has been employed as the key step in the synthesis (Table 4.32) of a 1-aminoimidazo[4,5-*f*][1,2,3]benzotriazole derivative of interest as an aryne precursor (Scheme 4.60, **4.231** → **4.232** → **4.233**).[78] In accord with this role, lead tetraacetate oxidation of the *N*-aminotriazole (**4.233**) in the presence of phenyl azide affords a mixture of the isomeric *N*(1)- and *N*(3)-phenylimidazo[4,5-*f*][1,2,3]benzotriazoles (**4.235** and **4.236**).[78] These products were not separated, but their formation implies the intermediacy of the aryne (**4.234**) and its trapping by 1,3-dipolar cycloaddition to phenyl azide.

Ring Closure Reactions of Other Heterocycles

The acid or base-catalyzed cyclization of 4,5-diacylaminobenzofurazan derivatives provides a synthetic entry to the imidazo[4,5-*e*][2,1,3]-benzoxadiazole ring system (Scheme 4.61). Thus hydrochloric-acid-catalyzed cyclization of 4,5-diacetylaminobenzofurazan (**4.237**, R = H)[79] or sodium-hydrogen-carbonate-mediated ring closure of the trifluoroacetyl analog (**4.237**, R = F)[80] affords the corresponding imidazo[4,5-*e*][2,1,3] benzoxadiazole derivatives (Scheme 4.61), albeit in both cases in unspecified yield.

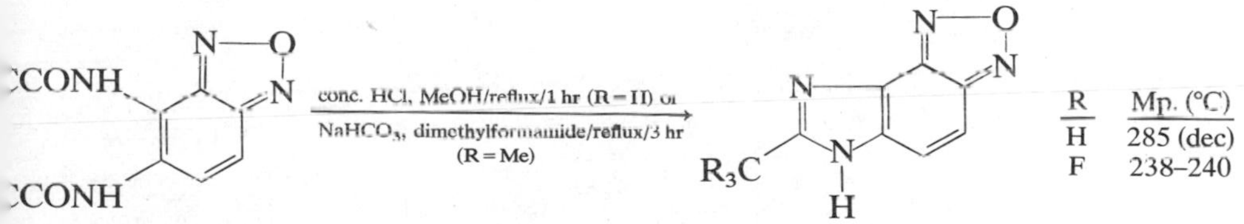

R	Mp. (°C)
H	285 (dec)
F	238–240

Scheme 4.61

Acylative ring closure reactions of *ortho*-diamino derivatives of 1,2,3- and 2,1,3-benzothiadiazoles (Scheme 4.62) are also a fruitful source of deriva-tives (Table 4.31) of the various known imidazobenzothiadiazole ring sys-tems (Table 4.30). Ring closure is variously effected by heating under reflux with a carboxylic acid alone or in the presence of hydrochloric acid or the corresponding anhydride, or by cyclization of the preformed *ortho*-diacylaminobenzothiadiazole. Illustrative examples (Scheme 4.62) (cf. also Table 4.31) include the reaction of 5,6-diamino-1,2,3-benzothiadiazole (**4.239**) with hot formic acid to afford (in unspecified yield) what appears to be the sole example of the imidazo[4,5-*f*][1,2,3]benzothiadiazole ring sys-tem (**4.240**),[81] the analogous ring closures of 4,5-diamino-2,1,3-benzothiadiazole derivatives (**4.241**; R¹ = R² = H)[80] and (**4.241**; R¹ = Ac, R² = Me),[82] 6,7-diamino-1,2,3-benzothiadiazole (**4.243**),[14,80] and 4,5-diamino-1,2,3-benzothiadiazole (**4.245**),[14] to give derivatives of the

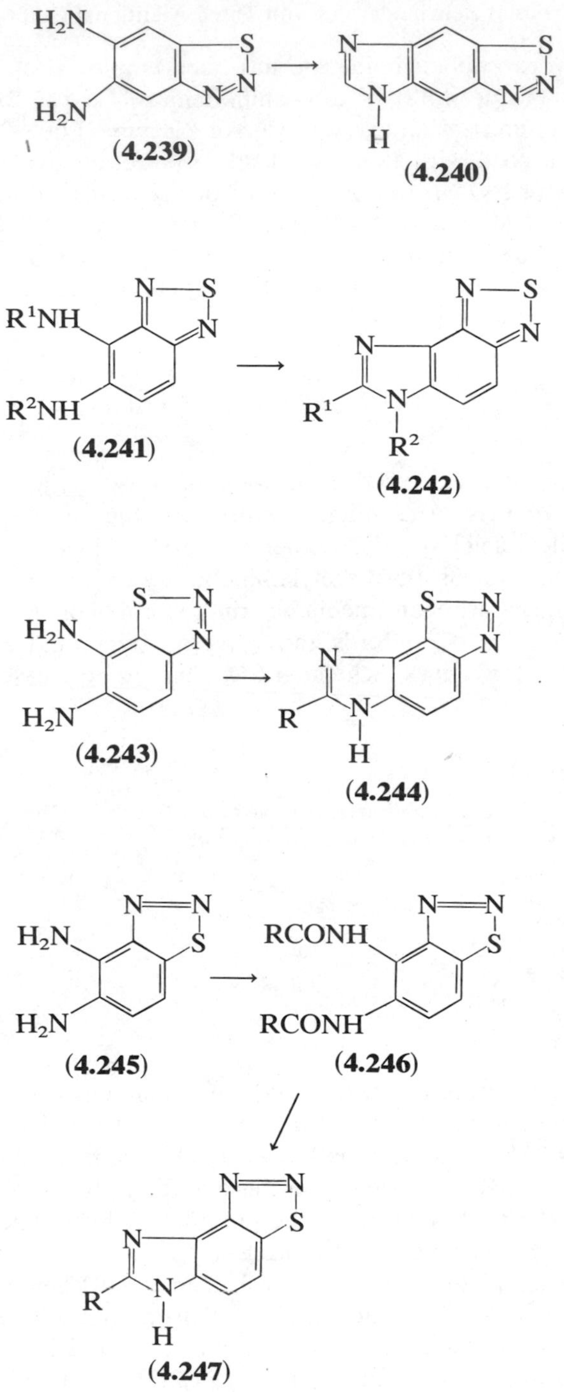

(4.239) → (4.240)

(4.241) → (4.242)

(4.243) → (4.244)

(4.245) → (4.246) → (4.247)

Scheme 4.62

imidazo[4,5-*e*][2,1,3]benzothiadiazole, imidazo[4,5-*g*][1,2,3]benzothiadiazole, and imidazo[4,5-*e*][1,2,3]benzothiadiazole ring systems (**4.242**, **4.244**, and **4.247**), and also the base-catalyzed cyclization of 4,5-diacylamino-1,2,3-benzothiadiazoles (**4.246**),[80] which offers an alternative route to derivatives of the imidazo[4,5-*e*][1,2,3]benzothiadiazole ring system.

Acylative ring closure in a 5,6-diamino-1,2,3-benzotriazole to give an imidazo[4,5-*f*][1,2,3]benzotriazole derivative is exemplified by the cyclization (Scheme 4.63) of 5,6-diacetylaminophenyl-1,2,3-benzotriazole (**4.248**)

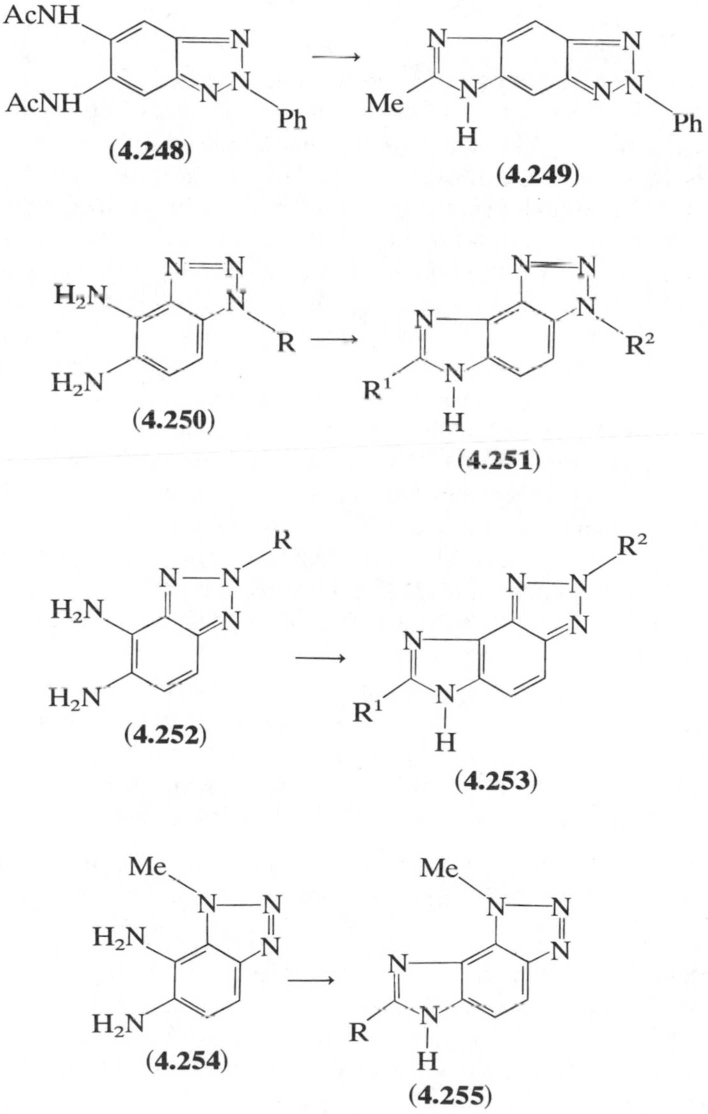

(**4.248**) (**4.249**) (**4.250**) (**4.251**) (**4.252**) (**4.253**) (**4.254**) (**4.255**)

Scheme 4.63

to give the interesting *ortho*-quinonoid structure (**4.249**) (Table 4.32).[83] Related cyclization reactions (Scheme 4.63) of 1-[**4.250**]-, 2-[**4.252**]-, and 3-[**4.254**]-substituted 4,5-diamino-1,2,3-benzotriazoles provide routes (Table 4.32) to various derivatives of the imidazo[4,5-*e*][1,2,3]benzotriazole ring system.[84,85]

4.4.2. Physicochemical Properties

Spectroscopic Studies

Relatively few spectroscopic studies of tricyclic 5-6-5 fused benzimidazoles with two additional hetero atoms have been reported. Accordingly, a cohesive account, under individual headings, of the spectroscopic properties of such molecules is not feasible. In the following account an attempt is made to highlight the points of interest in the available data.

In accord with its benzofurazan *N*-oxide structure, the imidazobenzoxadiazole *N*-oxide (**4.218**) (Scheme 4.56) displays four bands in its infrared spectrum at 1635, 1595, 1550, and 1520 cm^{-1}, and contains a peak (100%) at M$^+$-16 mass units in its mass spectrum corresponding to the loss of the oxygen atom.[74] Correspondingly, its ultraviolet spectrum has features in common with those of benzofurazan *N*-oxide and benzimidazole.[74] However, the absence in the molecule of the tautomerism characteristic of benzofurazan *N*-oxides (see **4.218**, a ⇌ b) is demonstrated by its ^{1}H NMR spectrum, which at room temperature and below shows only a single pair of para-coupled aromatic protons in accord with the presence of only a single tautomeric form (**4.218 a** or **b**).[74] The demonstration of possible benzofurazan *N*-oxide tautomerism in (**4.218**) at elevated temperature is precluded by its insolubility in suitable solvents and its tendency to undergo thermal decomposition.[74]

The presence of several intense maxima at relatively long wavelength in the ultraviolet spectra of imidazobenzothiadiazoles,[75,82] imidazobenzoselenadiazoles,[75] and imidazobenzotriazoles,[75] is indicative of the extensive conjugation present in such molecules. Not unexpectedly, cyanine dyes containing these nuclei absorb strongly in the visible region.[14,75,82]

General Studies

pK_a data for derivatives of tricyclic 5-6-5 fused benzimidazole ring systems with two additional hetero atoms is lacking. However, the amphoteric character of, for example, *N*-unsubstituted imidazo[4,5-*f*][1,2,3]-benzotriazoles is amply demonstrated by their ready solubility in both dilute acids and dilute alkalis.[36,37]

4.4.3. Reactions

Reactions with Electrophiles

With the exception of alkylation and acylation processes, information on the behavior of 5-6-5 fused benzimidazole derivatives with two additional hetero atoms toward electrophilic reagents is sparse.

ALKYLATION. Information on the alkylation reactions of imidazobenzothiadiazoles, imidazobenzoselenadiazoles, and imidazobenzotriazoles has accrued[14,75,82] because of the utility of the resulting quaternary salts as cyanine dye precursors.

In general, both linear and angular N-unsubstituted imidazobenzothiadiazoles and imidazobenzoselenadiazoles are alkylated[14,75] preferentially, under basic conditions at the acidic imidazole NH-group (Table 4.33 and Scheme 4.64, **4.258** → **4.259**;[75] Scheme 4.67, **4.267** → **4.268**;[14] and Scheme 4.68, **4.270** → **4.271**).[14] Bis-N-unsubstituted imidazo[4,5-f][1,2,3]benzotriazoles are likewise alkylated at both the imidazole and triazole rings (Scheme 4.65, **4.261** → **4.262**) (Table 4.33). Conversely, under effectively

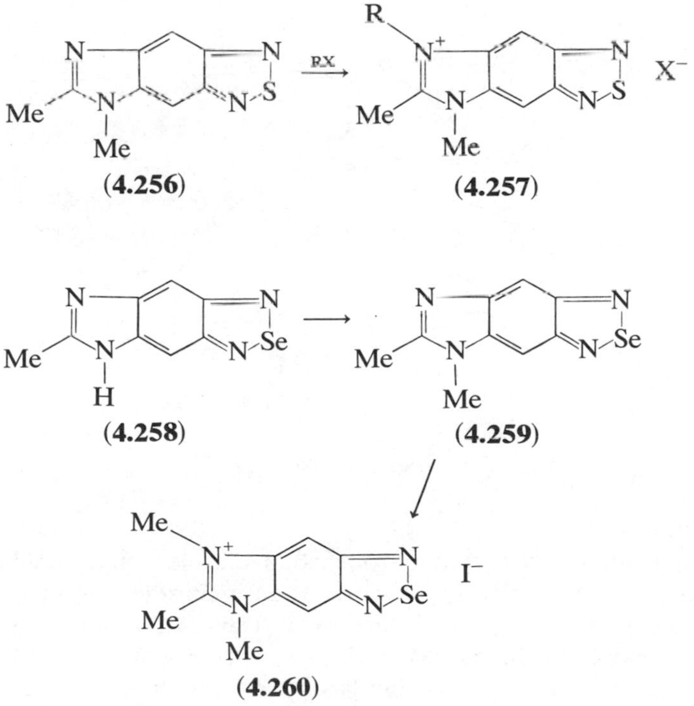

(4.256)   **(4.257)**

(4.258)   **(4.259)**

(4.260)

Scheme 4.64

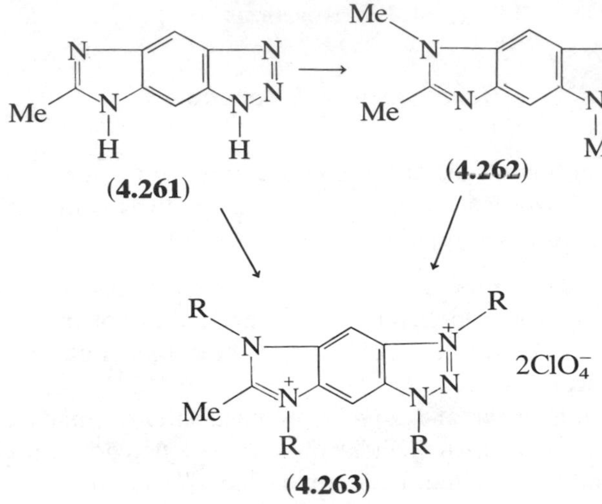

Scheme 4.65

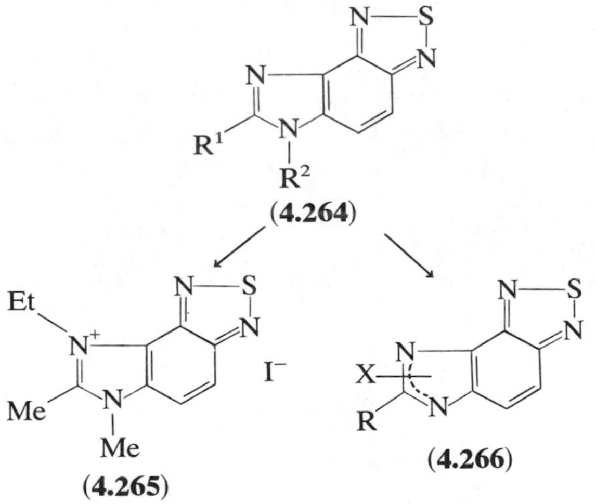

Scheme 4.66

neutral conditions,[14,75,82] imidazobenzothiadiazoles and imidazobenzo-selenadiazoles are selectively alkylated at an imidazole ring nitrogen atom, giving the corresponding quaternary salts (Table 4.33 and Schemes 4.64, 4.66, 4.67, and 4.68). Imidazo[4,5-f][1,2,3]benzotriazole derivatives, on the other hand, react at both the imidazole and the triazole ring, affording *bis*-quaternary salts (**4.261**) or (**4.262** → **4.263**). (Table 4.33).[75]

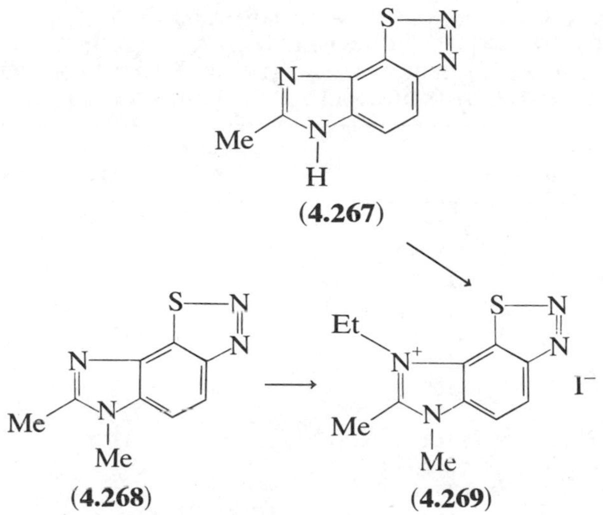

(4.267)

(4.268)   (4.269)

Scheme 4.67

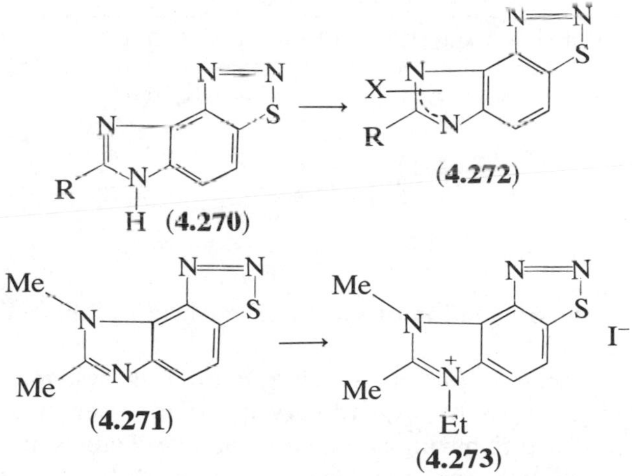

H (4.270)

(4.272)

(4.271)   (4.273)

Scheme 4.68

ACYLATION. Imidazole *NH*-groups in imidazo[4,5-*e*][2,1,3]benzoxa-diazoles, imidazo[4,5-*e*][2,1,3]benzothiadiazoles, and imidazo[4,5-*e*][1,2,3]-benzothiadiazoles are readily acylated[80] by chloroformate esters and by cyanogen bromide giving the respective *N*-acyl or *N*-cyano derivatives [Table 4.34 and Scheme 4.66, 4.264 ($R^2 = H$) → 4.266 (X = CO_2R or CN); Scheme 4.68, 4.270 → 4.272 (X = CO_2R or CN); and Scheme 4.69, 4.274 → 4.275]. Because of the unsymmetrical character of the ring systems in-volved, such reactions lead to isomer mixtures or single products (Table

TABLE 4.33.   ALKYLATION REACTIONS OF IMIDAZO[4,5-*f*][2,1,3]BENZOTHIADIAZOL
IMIDAZO[4,5-*f*][2,1,3]BENZOSELENADIAZOLES, IMIDAZO[4,5-*e*][2,1,3]BEN
THIADIAZOLES, IMIDAZO[4,5-*e*][1,2,3]BENZOTHIADIAZOLES, IMIDAZO[4
g][1,2,3]BENZOTHIADIAZOLES, AND IMIDAZO[4,5-*f*][1,2,3]BENZOTRIAZO:

Starting material	Reaction conditions[a]	Product	Yield (%)	Melting point (°C)	Solvent of crystallization
(4.256)	A	(4.257) (R = Me, X = ClO₄)[b]	—[c]	—[d]	Water
(4.256)	B	(4.257) (R = Et, X = I)	—[c]	282	—[e]
(4.258)	C	(4.259)	80	237–238	Water
(4.259)	A	(4.260)	—[c]	291	Methanol
(4.264) (R¹ = R² = Me)	B	(4.265)	—[c]	278	Methanol or ethanol
(4.267)	D	(4.268)	51	197	Water
(4.268)	E	(4.269)	—[c]	282	Ethanol
(4.270) (R = Me)	F	(4.271)	54	165 (decomp.)	—[e]
(4.271)	E	(4.273)	61	275–276	—[e]
(4.261)	C	(4.262)	51	266–267	Water
(4.262)	G	(4.26) (R = Me)[b]	—[c]	—[d]	Water
(4.261)	H	(4.263) (R = Et)[b]	—[c]	—[d]	Ethanol

[a] A, TSO₃Et/140°/3 hr, then treat with HClO₄ or KI (T = *p*-tolyl). B, EtI/130–140°/2 hr. C, Me₂SO₄
NaOH. D, Me₂SO₄, 5% NaOH. E, TSO₃Et/120–130°/3 hr, then treat with KI (T = *p*-tolyl). F, Me₂SO
NaOH, EtOH/room temp./5 hr, then 95°/30 min. G, TSO₃Me/heat, then treat with NaClO₄ (T = *p*-toly
Et₂SO₄/120°/1 hr, then treat with NaClO₄.
[b] Colorless needles.
[c] Yield not recorded.
[d] Melting point not recorded.
[e] Solvent not specified.

4.34) in which the orientation of the entering acyl or cyano substituent has
not been established.[80]

As already discussed for other 5-6-5 fused ring systems (see section
4.3.3), methyl groups α to a quaternary nitrogen center in a 5-6-5 fused
benzimidazole structure having two additional hetero atoms are activated to
acylative condensation reactions widely exploited for the synthesis of
cyanine dyes.[14,75,82]

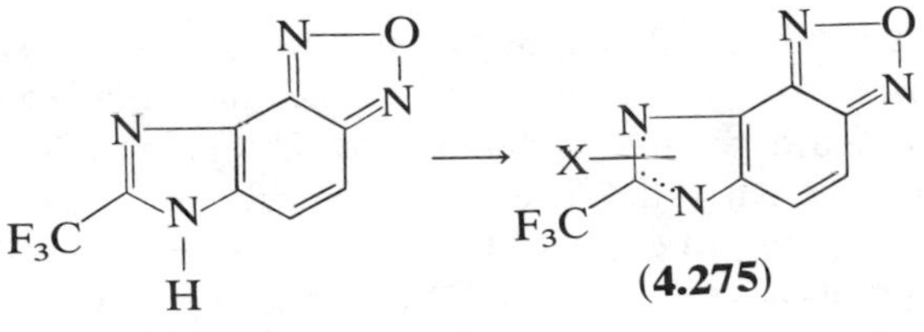

(4.274)

(4.275)

Scheme 4.69

TABLE 4.34. ACYLATION REACTIONS[80] OF IMIDAZO[4,5-*e*]
[2,1,3]BENZOTHIADIAZOLES, IMIDAZO[4,5-*e*]
[1,2,3]BENZOTHIADIAZOLES, AND IMIDAZO[4,5-*e*]
[2,1,3]BENZOXADIAZOLES

Starting material	Reaction conditions[a]	Product	Melting point (°C)
(4.264) ($R^1 = CF_3$, $R^2 = H$)	A	(4.266) ($R = CF_3$, $X = CO_2Et$)	149–150[b]
(4.264) ($R^1 = CF_3$, $R^2 = H$)	A	(4.266) ($R = CF_3$, $X = CO_2Pr$-n)	88–90
(4.264) ($R^1 = CF_3$, $R^2 = H$)	A	(4.266) ($R = CF_3$, $X = CO_2Ph$)	120–121
(4.264) ($R^1 = CF_3$, $R^2 = H$)	A	(4.266) ($R = CF_3$, $X = COPh$)	175–177
(4.264) ($R^1 = CF_3$, $R^2 = H$)	B	(4.266) ($R = CF_3$, $X = CN$)	159–161[c,d]
(4.264) ($R^1 = CF_3$, $R^2 = H$)	B	(4.266) ($R = CF_3$, $X = CN$)	125–126[c,d]
(4.270) ($R = CF_3$)	A	(4.272) ($R = CF_3$, $X = CO_2Pr$-n)	128–129[b,d]
(4.270) ($R = CF_3$)	A	(4.272) ($R = CF_3$, $X = CO_2Pr$-n)	78–79[b,d]
(4.270) ($R = CF_3$)	A	(4.272) ($R = CF_3$, $X = CO_2Ph$)	159–161[d]
(4.270) ($R = CF_3$)	A	(4.272) ($R = CF_3$, $X = CO_2Ph$)	129–130[d]
(4.270) ($R = CF_3$)	A	(4.272) ($R = CF_3$, $X = COPh$)[d]	111–137
(4.270) ($R = CF_2CF_3$)	A	(4.272) ($R = CF_2CF_3$, $X = CO_2Pr$-i)	110–111
(4.270) ($R = CF_2CF_3$)	A	(4.272) ($R = CF_2CF_3$, $X = CO_2Pr$-i)[e]	64–72
(4.270) ($R = CF_3$)	B	(4.272) ($R = CF_3$, $X = CN$)	117–119
(4.270) ($R = CF_2CF_3$)	B	(4.272) ($R = CF_2CF_3$, $X = CN$)	105–106
(4.274)	A	(4.275) ($R = Pr$-n)	60–61
(4.274)	A	(4.275) ($R = Me$)	128–129
(4.274)	A	(4.275) ($R = Ph$)	123–125

[a] A, RCOCl, NaH, MeCN/room temp./18 hr. B, BrCN, NaH, MeCN/room temp./18 hr.
[b] From benzene–hexane.
[c] From benzene–light petroleum.
[d] One of a pair of separable isomers.
[e] Isomer mixture.

Reactions with Nucleophiles

No bona fide examples of the reactions with nucleophilic reagents of
5-6-5 fused benzimidazoles having two additional hetero atoms have been
reported.

Oxidation and Reduction

As is the case with other types of reactivity, information on the behavior
of 5-6-5 fused benzimidazoles with two additional hetero atoms towards
oxidation and reduction is sparse.

The oxidation of an *N*-amino imidazo[4,5-*f*][1,2,3]benzotriazole derivative with lead tetraacetate in the presence of phenyl azide to give products
derivable from an aryne-type intermediate (Scheme 4.60) demonstrates the
stability of the imidazo[4,5-*f*][1,2,3]benzotriazole ring system to this type of

oxidant.[78] Correspondingly, the failure of the imidazo[4,5-f][2,1,3]benz-oxadiazole N-oxide (Scheme 4.56, **4.218**) to undergo deoxygenation on heating with triethyl phosphite indicates the stability of the imidazo[4,5-f][2,1,3]benzoxadiazole ring system to reduction at least under these (admittedly rather special) conditions.[74]

4.4.4. Practical Applications

Biological Properties

Derivatives of the imidazo[4,5-e][2,1,3]benzothiadiazole, imidazo[4,5-e][1,2,3]benzothiadiazole, and imidazo[4,5-g][1,2,3]benzothiadiazole ring systems have been patented as pesticides and herbicides.[80]

Dyestuffs

Derivatives of the various 5-6-5 fused benzimidazole ring systems with two additional hetero atoms have been utilized as chromophoric units in cyanine dyestuffs.[14,75,82]

REFERENCES

1. D. W. S. Latham, O. Meth-Cohn, H. Suschitzky, and J. A. L. Herbert, *J. Chem. Soc. Perkin I*, **1977**, 470.
2. C. S. Kadyrov, M. N. Kosyakovskaya, and M. R. Yagudaev *Uzb. Khim. Zh.* **12**, 34 (1968); *Chem. Abstr.*, **70**, 28870 (1969).
3. W. Borsche and G. John, *Chem. Ber.*, **57**, 656 (1924).
4. O. Fischer, *Chem. Ber.*, **32**, 1312 (1899).
5. L. M. Yagupol'skii, G. I. Klyushnik, and V. I. Troitskaya, *J. Gen. Chem. USSR*, **34**, 304 (1964); *Zh. Obshch. Khim.*, **34**, 307 (1964); *Chem. Abstr.*, **60**, 13352 (1964).
6. N. B. Chapman, K. Clarke, and K. S. Sharma, *J. Chem. Soc. C*, **1971**, 919.
7. B. Iddon, H. Suschitzky, and D. S. Taylor, *J. Chem. Soc. Perkin I*, **1974**, 579.
8. Z. I. Moskalenko, USSR Patent, 230,826; *Chem. Abstr.*, **70**, 87815 (1969).
9. A. P. Terent'ev, E. V. Vinogradova, V. P. Chetverikov, and S. N. Dashkevich, *Chem. Heterocycl. Compounds*, **1970**, 710; *Khim. Geterotsikl. Soedin.*, **1970**, 770; *Chem. Abstr.*, **73**, 109741 (1970); A. P. Terent'ev, E. V. Vinogradova, V. P. Chetverikov, and S. N. Dashkevich, USSR Patent, 230,160; *Chem. Abstr.*, **71**, 13125 (1969).
10. A. P. Terent'ev, E. V. Vinogradova, V. P. Chetverikov, and V. S. Lenenko, *Chem. Heterocycl. Compounds*, **1969**, 196; *Khim. Geterotsikl. Soedin.*, **1969**, 258; *Chem. Abstr.*, **71**, 21963 (1969).
11. A. N. Kost, Z. F. Solomko, N. M. Prikhod'ko, and A. P. Terent'ev, *Chem. Heterocycl. Compounds*, **1971**, 734; *Khim. Geterotsikl. Soedin.*, **1971**, 787; *Chem. Abstr.*, **76**, 25254 (1972).
12. I. G. Il'ina, N. B. Kazennova, V. G. Bakhmut-Skaya, and A. P. Terent'ev, *Chem. Heterocycl. Compounds*, **1973**, 1028; *Khim. Geterotsikl. Soedin.*, **1973**, 1112; *Chem. Abstr.*, **79**, 126396 (1973).

13. R. Garner, E. B. Mullock, and H. Suschitzky, *J. Chem. Soc. C*, **1966**, 1980.
14. S. G. Fridman and L. I. Kotova, *Chem. Heterocycl. Compounds*, **3**, 399 (1967); *Khim. Geterotsikl. Soedin.*, **3**, 497 (1967); *Chem. Abstr.*, **68**, 40976 (1968).
15. S. G. Fridman and A. I. Kiprianov, *J. Org. Chem.* (*USSR*), **4**, 678 (1968); *Zh. Org. Khim.*, **4**, 696 (1968); *Chem. Abstr.*, **69**, 20394 (1968).
16. E. A. Kuznetsova, N. T. Pryanishnikova, L. I. Gaidukova, I. V. Fedina, and S. V. Zhuravlev, *Chem. and Pharm. J.*, **9**, 759 (1975); *Khim. Farm. Zh.*, **9**, 11 (1975); *Chem. Abstr.*, **84**, 130139 (1976).
17. M. A. Phillips, *J. Chem. Soc.*, **1930**, 1409.
18. W. Knobloch and H. Niedrich, *Chem. Ber.*, **91**, 2562 (1958).
19. J. Arient, J. Marhan, and H. Taublova, *Coll. Czech. Chem. Comm.*, **25**, 1602 (1960); *Chem. Abstr.*, **54**, 17381a (1960). Cf. also K. Fries and J. Empson, *Annalen*, **389**, 366 (1912).
20. V. P. Evstavef, G. I. Braz, A. Y. Yakubovich, and G. F. Shalygin, *Vysokomol. Soedin. Ser. A.*, **13**, 2565 (1971); *Chem. Abstr.*, **77**, 20069 (1972).
21. P. Ruggli and R. Fischer, *Helv. Chim. Acta*, **28**, 1270 (1945).
22. V. V. Korshak, A. L. Rusanov, D. S. Tugushi, and G. M. Cherkasova, *Macromolecules*, **5**, 807 (1972); *Chem. Abstr.*, **78**, 72645 (1973); D. S. Tugushi, V. V. Korshak, A. L. Rusanov, V. G. Danilov, G. M. Cherkasova, and G. M. Tseitlin, *Vysokomol. Soedin. Ser. A*, **15**, 969 (1973); *Chem. Abstr.*, **79**, 42884 (1973).
23. R. F. Kovar and F. E. Arnold, *J. Polymer. Sci., Polymer. Chem. Ed.*, **14**, 2807 (1976); *Chem. Abstr.*, **86**, 30110 (1977); *Polymer. Prepr. Am. Chem. Soc. Div. Polymer. Chem.*, **16**, 683 (1975); *Chem. Abstr.*, **86**, 30112 (1977).
24. H. Vogel and C. S. Marvel, *J. Polymer Sci.*, **50**, 511 (1961); *Chem. Abstr.*, **55**, 19307g (1961); *J. Polymer Sci, Part. A-1*, **1963**, 1531; *Chem. Abstr.*, **59**, 6526e (1963).
25. J. Higgins and C. S. Marvel *J. Polymer. Sci., Part A-1*, **8**, 171 (1970); *Chem. Abstr.*, **72**, 55947 (1970).
26. R. Niotzki and E. Muller, *Chem. Ber.*, **22**, 440 (1889).
27. E. Winkelmann, *Tetrahedron*, **25**, 2427 (1969).
28. (a) R. Nietzki and E. Hugenbach, *Chem. Ber.*, **20**, 328 (1887); (b) R. Nietzki and L. Schmidt, *Chem. Ber.*, **22**, 1648 (1889).
29. R. L. Williams, J. Schuller, and D. Lloyd, *J. Heterocycl. Chem.*, **5**, 147 (1968).
30. V. V. Korshak, A. L. Rusanov, I. Batirov, D. S. Tugushi, and I. Y. Kalontarov, *Dokl. Akad. Nauk Tadzh. SSR*, **20**, 26 (1977); *Chem. Abstr.*, **88**, 152492 (1978).
31. T. Holbro and W. Kern, U.S. Patent, 2,807,622; *Chem. Abstr.*, **52**, 1635 (1958).
32. V. P. Evstaf'ev, V. S. Yakubovich, G. F. Shalygin, V. I. Selikhova, Y. A. Zubov, G. I. Braz, and A. Y. Yakubovich, *Vysokomol. Soedin. Ser. A*, **14**, 2174 (1972); *Chem. Abstr.*, **78**, 72740 (1973).
33. A. Marxer, *Helv. Chim. Acta*, **44**, 762 (1961); German Patent 1,109,176; *Chem. Abstr.*, **56**, 10157h (1962).
34. K. Fries and H. Reitz, *Annalen*, **527**, 58 (1937).
35. L. S. Efros, *J. Gen. Chem. USSR*, **22**, 1063 (1952); *Chem. Abstr.*, **48**, 2690b (1954).
36. O. Kym and L. Ratner, *Chem. Ber.*, **45**, 3238 (1912).
37. L. S. Efros and A. V. El'tsov, *J. Gen. Chem. USSR*, **28**, 2214 (1958); *Zh. Obshch. Khim.*, **28**, 2174 (1958); *Chem. Abstr.*, **53**, 2208 (1959); A. T. James and E. E. Turner, *J. Chem. Soc.*, **1950**, 1515.
38. A. V. El'tsov, V. S. Kuznetsov, and L. S. Efros, *J. Gen. Chem. USSR*, **34**, 195 (1964); *Zh. Obshch. Khim.*, **34**, 197 (1964); *Chem. Abstr.*, **60**, 10671 (1964).
39. L. S. Efros, *J. Gen. Chem. USSR*, **23**, 989 (1953); *Zh. Obshch. Khim.*, **23**, 951 (1953); *Chem. Abstr.*, **48**, 8222b (1954).
40. A. M. Simonov, Y. V. Koshchienko, V. G. Poludenko, and V. E. Khorunzhev, *Chem. Heterocycl. Compounds*, **1970**, 1313; *Khim. Geterotsikl. Soedin.*, **1970**, 1406; *Chem. Abstr.*, **74**, 99942 (1971).
41. Y. V. Koshchienko, A. M. Simonov, and A. F. Pozharskii, *Chem. Heterocycl. Compounds*,

     7, 1064 (1971); *Khim. Geterotsikl. Soedin.*, **7**, 1132 (1971); *Chem. Abstr.*, **77**, 19030 (1972).

42. A. M. Simonov, Y. V. Koshchienko, and T. G. Belenko, *Chem. Heterocycl. Compounds*, **1973**, 95; *Khim. Geterotsikl. Soedin.*, **1973**, 107; *Chem. Abstr.*, **78**, 97558 (1973).

43. A. F. Pozharskii, A. M. Simonov, V. M. Mar'yanovskii, and R. P. Zinchenko, *Chem. Heterocycl. Compounds*, **1970**, 987; *Khim. Geterotsikl. Soedin.*, **1970**, 1060; *Chem. Abstr.*, **75**, 48982 (1971).

44. J. Goetze, H. Depoorter, and T. Ghys, German Patent 1,805,548; *Chem. Abstr.*, **74**, 4707 (1971); German Patent, 1,923,992; *Chem. Abstr.*, **75**, 5900 (1971).

45. T. G. H. Jones and R. Robinson, *J. Chem. Soc.*, **111**, 903 (1917).

46. E. R. Lavagnino and D. C. Thompson, *J. Heterocycl. Chem.*, **9**, 149 (1972).

47. R. L. Clark and A. A. Pessolano, *J. Am. Chem. Soc.*, **80**, 1662 (1958).

48. S. G. Fridman, *J. Gen. Chem. USSR*, **30**, 1534 (1960); *Zh. Obshch. Khim.*, **30**, 1520 (1960); *Chem. Abstr.*, **55**, 1629f (1961).

49. J. T. Ralph and C. E. Marks, *Tetrahedron*, **22**, 2487 (1966).

50. I. P. Federova and G. F. Mironova, *J. Gen. Chem. USSR*, **32**, 1871 (1962); *Zh. Obshch. Khim.*, **32**, 1893 (1962); *Chem. Abstr.*, **58**, 4535c (1963).

51. S. G. Fridman, *J. Gen. Chem. USSR*, **33**, 201 (1963); *Zh. Obshch. Khim.*, **33**, 207 (1963); *Chem. Abstr.*, **58**, 14155b (1963).

52. L. F. Fieser, *J. Am. Chem. Soc.*, **48**, 1097 (1926).

53. J. J. Jennen, U.S. Patent 2,673,801; *Chem. Abstr.*, **48**, 9850c (1954).

54. J. Sliede, A. Y. Strakov, and E. Gudriniece, *Latv. PSR Zinat. Akad. Vestis, Khim. Ser.*, **1977**, 111; *Chem. Abstr.*, **87**, 23156 (1977).

55. Y. V. Koshchienko and A. M. Simonov, *Chem. Heterocycl. Compounds*, **1973**, 740; *Khim. Geterotsikl. Soedin.*, **1973**, 807; *Chem. Abstr.*, **79**, 105136 (1973).

56. S. G. Fridman, *J. Gen. Chem. USSR*, **32**, 1448 (1962); *Zh. Obshch. Khim.*, **32**, 1461 (1962); *Chem. Abstr.*, **58**, 1564d (1963).

57. S. G. Fridman and A. I. Kiprianov, *J. Org. Chem. USSR*, **5**, 358 (1969); *Zh. Org. Khim.*, **5**, 373 (1969); *Chem. Abstr.*, **71**, 4498 (1969).

58. M. Gelus and J.-M. Bonnier, *J. Chim. Phys.*, **64**, 1602 (1967); *Chem. Abstr.*, **68**, 100244 (1968).

59. G. I. Kagan, V. A. Kosobutskii, V. K. Belyakov, and O. G. Tarakanov, *Chem. Heterocycl. Compounds*, **1973**, 1265; *Khim. Geterotsikl. Soedin.*, **1973**, 1396; *Chem. Abstr.*, **80**, 26666 (1974).

60. A. D. Garnovskii, A. M. Simonov, and V. I. Minkin, *Chem. Heterocycl. Compounds*, **1973**, 88; *Khim. Geterotsikl. Soedin.*, **1973**, 99; *Chem. Abstr.*, **80**, 2941 (1974).

61. L. S. Efros, V. I. Minkin, and K. V. Veksler, *Zh. Vses. Khim. Obshchest.*, **13**, 710 (1968); *Chem. Abstr.*, **70**, 77169 (1969).

62. C. Pigenet and H. Lumbroso, *Bull. Soc. Chim. France*, **1972**, 3743.

63. E. Bellasio, A. Campi, A. Trani, E. Baldoli, A. M. Caravaggi, and G. Nathansohn, *Farmaco. Ed. Sci.*, **28**, 164 (1973); *Chem. Abstr.*, **78**, 111214 (1973).

64. L. S. Efros, *J. Gen. Chem. USSR*, **22**, 1069 (1952); *Chem. Abstr.*, **48**, 2690b (1954).

65. S. G. Fridman, *J. Gen. Chem. USSR*, **35**, 1370 (1965); *Zh. Obshch. Khim.*, **35**, 1364 (1965); *Chem. Abstr.*, **64**, 3730e (1966).

66. H. Roechling, *Z. Naturforsch., B*, **25**, 931 (1970); *Chem. Abstr.*, **74**, 22760 (1971).

67. C. Granacher and F. Ackermann, U.S. Patent 2,463,264; *Chem. Abstr.*, **43**, 8173; (1949).

68. G. Manecke, L. Brandt, and G. Kossmehl, *Makromol. Chem.*, **178**, 1745 (1977); *Chem. Abstr.*, **87**, 53626 (1977).

69. R. Garner and H. Suschitzky, *J. Chem. Soc. C*, **1967**, 74.

70. H. Schubert and H. Fritsche, *J. Prakt. Chem.*, **7**, 207 (1958).

71. J. Pawelkiewicz, *Congr. Intern. Biochim., Resumés communs, 3ᵉ Congr. Brussels*, **1955**, 100; *Chem. Abstr.*, **51**, 7506c (1957); J. Pawelkiewicz and K. Nowakowska, *Acta Biochim. Polon.*, **2**, 259 (1955); *Chem. Abstr.*, **51**, 13070g (1957).

72. French Patent 1,537,444; *Chem. Abstr.*, **71,** 4077 (1969); G. I. Kudryavtsev, L. F. Balakleitseva, A. M. Shchetinin, and L. V. Chikurina, *Vysokomol. Soedin. Ser. A,* **12,** 2205 (1970); *Chem. Abstr.*, **74,** 13540 (1971); L. I. Rudaya, I. Y. Kvitko, and A. V. El'tsov, USSR Patent 459,465; *Chem. Abstr.*, **82,** 170950 (1975); H. Cherdron, H. J. Leugering, E. Fischer, and K. D. Asmus, German Patent 1,694,078; *Chem. Abstr.*, **84,** 5985 (1976); K. Uno, K. Niiume, and T. Nakayama, *Nippon Kagaku Kaishi,* **1975,** 1584; *Chem. Abstr.*, **84,** 5428 (1976); T. E. Helminiak, US Patent Appl. 638, 211; *Chem. Abstr.*, **85,** 193776 (1976); V. S. Voishchev, O. V. Kolninov, B. V. Kotov, V. I. Berendyaev, N. N. Voznesenskaya, A. N. Pravednikov, and B. I. Sazhin, *Vysokomol. Soedin. Ser. B,* **19,** 203 (1977); *Chem. Abstr.*, **86,** 190550 (1977); V. K. Belyakov and V. A. Kosobutskii, *Vysokomol. Soedin. Ser. A,* **18,** 2452 (1976), *Chem. Abstr.*, **86,** 44175 (1977).

73. E. Hoggarth, "Chemistry of Carbon Compounds," E. H. Rodd, Ed., Elsevier, Amsterdam, 1957, Vol. IVA, Chap. VI, pp. 464–465.

74. R. C. Perera, R. K. Smalley, and L. G. Rogerson, *J. Chem. Soc. C,* **1971,** 1348.

75. S. G. Fridman and L. I. Kotova, *J. Gen. Chem. USSR,* **32,** 2829 (1962); *Zh. Obshch. Khim.,* **32,** 2871 (1962); *Chem. Abstr.*, **58,** 11495g (1963).

76. O. Kym and L. Ratner, *Chem. Ber.,* **45,** 3238 (1912).

77. K. Fries, *Annalen,* **454,** 221 (1927).

78. R. C. Perera and R. K. Smalley, *Chem. Commun.,* **1970,** 1458.

79. W. Borsche and H. Weber, *Annalen,* **489,** 270 (1931).

80. P. Kirby, R. E. Woodall, and B. R. J. Devlin, British Patent 1,294,562; *Chem. Abstr.,* **78,** 58420 (1973).

81. E. R. Ward and D. D. Heard, *J. Chem. Soc.,* **1963,** 4794.

82. D. Dal Monte, E. Sandri, and P. Mazzaracchio, *Boll. Sci. Fac. Chim. Ind. Bologna,* **25,** 3 (1967); *Chem. Abstr.*, **68,** 96789 (1968).

83. K. Fries and E. Roth, *Annalen,* **389,** 337 (1912).

84. M. Kamel, M. I. Ali, and M. M. Kamel, *Annalen,* **733,** 115 (1970).

85. M. M. Kamel, M. M. Abdel Hamid, and M. Kamel, *Annalen,* **746,** 76 (1971).

CHAPTER 5

Condensed Benzimidazoles of Type 6-6-5

P. N. PRESTON AND G. TENNANT

5.1. NAPHTHIMIDAZOLES

An imidazole ring may be fused to a naphthalene ring in an angular (e.g., **5.1**) or in a linear fashion (e.g., **5.2**). The nomenclature indicated (**5.1** and **5.2**) has been used in *Chemical Abstracts* from 1947, but during the period 1937 to 1946, the angular systems (**5.1a** and **b**) were referred to as 1-[1,2]- and 3-[1,2]naphthimidazoles, respectively, and the linear system was denoted as 1-[2,3]naphthimidazole. Prior to 1937, **5.1a** and **b** and **5.2** were designated as αβ- and ββ-naphthimidazoles, respectively.

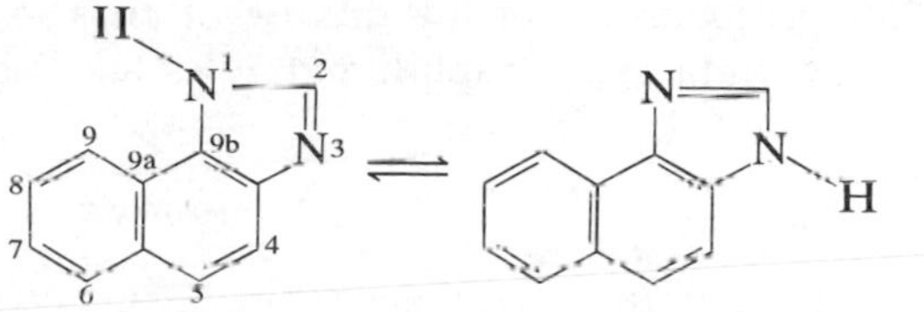

(a)   (b)

(5.1)

1*H*-Naphth[1,2-d]imidazole   3*H*-Naphth[1,2-d]imidazole

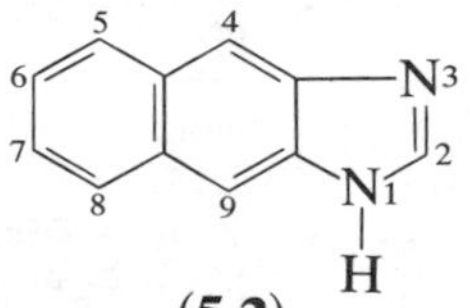

(5.2)

1*H*-Naphth[2,3-d]imidiazole

Much of the chemistry of naphthimidazoles closely resembles that of benzimidazoles, and this section contains numerous cross-references to information in Chapters 1 to 3. It will be noted that the synthesis and reactions of compounds in the [1,2-*d*] series are more extensively documented than those of the [2,3-*d*] series, which reflects the relative ease of preparation of 1,2-naphthylene diamines compared to their 2,3-analogs.

5.1.1. Synthesis of Naphthimidazoles

Synthesis from the Reaction of Naphthylene Diamines and:

CARBOXYLIC ACIDS. Examples of naphth[1,2-*d*][1–16] and naphth[2,3-*d*]-imidazoles[2,9,17–27] which have been prepared from the reaction of naphthylene diamines with carboxylic acids are summarized in Tables 5.1 and 5.2, respectively. The conventional procedures described in section 1.2.1 of Chapter 1 have been applied. The diamine is heated under reflux with the carboxylic acid, preferably in the presence of 4–6 N hydrochloric acid (Phillips procedure). Reasonable yields are obtained by this method (e.g., 71 and 68% yields for 2-$CH_2CH_2NH_2$ derivatives in the [1,2-*d*] and [2,3-*d*] systems, respectively),[9] but cleaner products are obtained by using the improved procedure in which polyphosphoric acid is used as a condensing agent. Thus condensation of 2,3-diaminonaphthalene with terephthalic acid and in the presence of polyphosphoric acid at approximately 150° gives the bis-naphth[2,3-*d*]imidazole derivative in 91% yield.[25] Boric acid has also been used as a condensing agent,[12] but it may be noted that the improved Phillips procedure involving the use of an acid resin has not yet been applied to the synthesis of naphthimidazoles (cf. Ref. 138 in Chapter 1).

TABLE 5.1.   SYNTHESIS OF NAPHTH[1,2-*d*]IMIDAZOLES FROM THE REACTION OF 1,2-NAPHTHYLENE DIAMINES WITH CARBOXYLIC ACIDS

R in RCO$_2$H (naphth[1,2-*d*]imidazole 2-substituent)	Other substituents in the naphth[1,2-*d*]imidazole product	Ref.
H	None	1–4
H	3-Me	5
H	1-Ph-5-OMe	6
Me, Et	None	7
ClCH$_2$	1-CH$_2$Ph	8
H$_2$N(CH$_2$)$_2$	None	9
CH(OH)CH$_3$ } CH(OH)CH$_2$Cl } CH$_2$Cl }	3-Me	10
CH(OH)Me	None	11
p-C$_6$H$_4$CH=CHPh	None	12
3-Pyridyl	None	13
2-Quinolinyl	None	14
2-Phenyl-4-quinolinyl	None	15·
CH$_2$-ferrocenyl	None	16

TABLE 5.2. SYNTHESIS OF NAPHTH[2,3-d]IMIDAZOLES
FROM THE REACTION OF 2,3-NAPHTHY-
LENE DIAMINES WITH CARBOXYLIC ACIDS

R in RCO$_2$H (naphth[2,3-d]imidazole 2-substituent)	Other substituents in the naphth[2,3-d]imidazole product	Ref.
H	None	2
H	6,7-Me$_2$	17
H, Me	None	18, 19
H	5,6,7,8-Tetrahydro	20, 21
Me	5,6,7,8-Tetrahydro	22
CF$_3$	None	23
(CH$_2$)$_2$NH$_2$	None	9
3-Hydroxy-2-naphthyl	None	24
p-C$_6$H$_4$[a]	None	25
5-Phenyl-2-thienyl	None	26
2-Pyridyl	None	27

[a] The starting material is a dicarboxylic acid and the product is a bis-naphth[2,3-d]imidazole.

CARBOXYLIC ACID DERIVATIVES (ANHYDRIDES AND ACID
CHLORIDES). Reactions in which naphthimidazoles are produced from the
reaction of naphthylene diamines and carboxylic acid anhydrides and an acid
chloride are shown in Table 5.3.[4,28-32] Much of the material collected in
Table 5.3 is from the older literature, and it is surprising that improved
variants of these reactions have not been used in naphthimidazole synthesis.
For example, the synthesis from arylene diamines and thioesters has proved
to be valuable in benzimidazole synthesis (see Ref. 154 in section 1.2.1 of
Chapter 1).

TABLE 5.3. SYNTHESIS OF NAPHTHIMIDAZOLES BY THE
REACTION OF NAPHTHYLENE DIAMINES WITH
CARBOXYLIC ACID ANHYDRIDES AND AN ACID
CHLORIDE

Carboxylic acid derivative used	Type of ring system synthesized	Substituents in the product	Ref.
Ac$_2$O	[1,2-d]	2-Me-4-SO$_3$H-7-OH	28
(PhCO)$_2$O	[1,2-d]	2-C$_6$H$_4$CO$_2$H-o	29
(PhCO)$_2$O[a]	[1,2-d]	2-C$_6$H$_4$-(naphth[1,2-d]imidazol-2-yl)-o	30
Hexahydrophthalic anhydride	[1,2-d]	2-(2-Carboxycyclohexyl)	31
(o-HO$_3$SC$_6$H$_4$CO)$_2$O	[2,3-d]	2-C$_6$H$_4$SO$_3$H-o	32
C$_6$H$_4$(COCl)$_2$-o	[1,2-d]	2-C$_6$H$_4$-(naphth[1,2-d]imidazol-2-yl)-o	4
1,5-(COCl)$_2$C$_{10}$H$_6$	[1,2-d]	2-C$_{10}$H$_6$-(naphth[1,2-d]imidazol-2-yl)	4

[a] Two moles of the naphthylene diamine are used.

TABLE 5.4.   SYNTHESIS OF NAPHTHIMIDAZOLES BY THE REACTION OF NAPHTHYLENE DIAMINES WITH ALDEHYDES

Aldehyde used	Type of ring system synthesized	Substituents in the product	Ref
$(CH_3)_2CHCHO$ n-$C_6H_{13}CHO$	[1,2-d]	2-i-Pr, n-C_6H_{13}	33
CH_3CHO PhCHO p-$Me_2NC_6H_4CHO$	[2,3-d]	2-Me, Ph, $C_6H_4NMe_2$-p, 1,3-Me_2-4,9-dioxo^a	34
PhCHO PhCH=CHCHO	[1,2-d]	2-Ph, CH=CHPh	35
2,5-(OH)(R)C_6H_3CHO^b (R = H, OH, OMe, Ph) 2-(OH)$C_{10}H_6$CHO^b	[2,3-d]	2-Aryl 2-(2-Hydroxynaphthyl)	24
$(ClCH_2CH_2)_2NC_6H_4$CHO-p	[2,3-d]	2-$C_6H_4N(CH_2CH_2Cl)_2$	36
(5-X-2-Furyl)CHO X = H, Br, NO_2	[1,2-d]	2-(5-X-Furyl)	37
2,3-Me_2-1-Ph-5-oxo-3-pyrazoline-4-CHO	[1,2-d]	2-(2,3-Me_2-1-Ph-5-oxo-3-pyrazolin-4-yl)	38

^{a} Isolated as naphthimidazolium perchlorates (see text).
^{b} Used in the form of bisulfite addition products (see text).

ALDEHYDE DERIVATIVES. The reaction of arylene diamines with aldehydes described in section 1.2.1 of Chapter 1 has been applied to the synthesis of naphthimidazoles (see Table 5.4).[24,33–38] The Weidenhagen method, in which the diamine is condensed with the aldehyde in the presence of cupric acetate, has been used[33] but the superior route consists of using the bisulfite addition products of the aldehyde (see Ref. 291c and 227 in section 1.2.1 of Chapter 1). Using the latter procedure,[24] the naphthylene diamine is heated with the bisulfite addition adduct in dimethylformamide at 150°, although yields are lower in these reactions than in analogous benzimidazole syntheses (13 to 54% compared to yields in excess of 90%).

The reaction of 2,3-bismethylamino-1,4-naphthoquinone (5.3) with aldehydes is more complex, and the isolated products in this case are 1,2,3-trisubstituted 4,9-dioxo-1H-naphth[2,3-d]imidazolium salts (5.6);[34] naphth[2,3-d]imidazolines (5.4) are probable intermediates in these reactions, in which the quinone moiety is thought to effect aromatization of the imidazoline ring (cf. 5.3 → 5.6 and the use of tetrachlorobenzoquinone for the conversion of analogous benzimidazolines into benzimidazolium compounds).[34]

MISCELLANEOUS COMPOUNDS. A variety of compounds has been condensed with naphthylene diamines, often with the intention of providing specific routes to 2-hydroxy-(naphthimidazolones), 2-mercapto-(naphthimidazol-2-thiones), and 2-amino derivatives. Procedures leading to these classes of

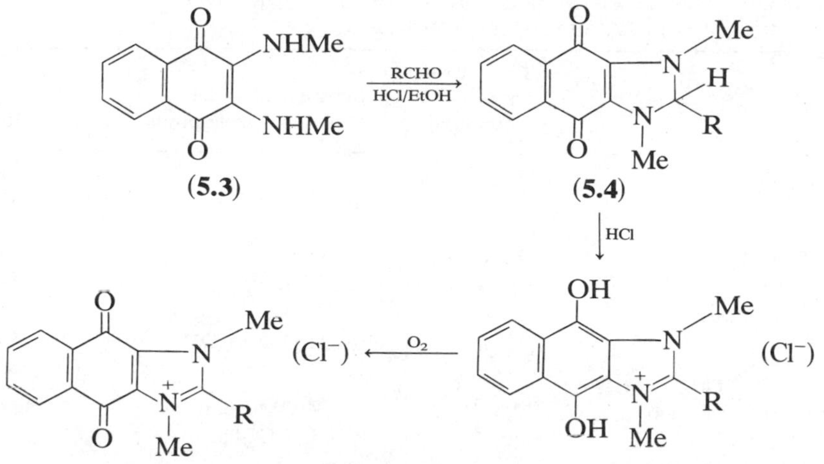

(5.3) → (5.4)

RCHO
HCl/EtOH

(5.6) ← (5.5)

O₂

R = Me, aryl (73 and 90% yields for R = aryl)

compounds are summarized in Table 5.5[39-50] and involve routine methods described in sections 3.2.1, 3.2.2, and 1.2.1 of Chapters 3 and 1.

Other syntheses based on the reactions of diamines and miscellaneous derivatives are shown in Table 5.6.[51-57] Reactions of this type also have precedent in the synthesis of benzimidazoles, and a discussion of such reactions and presentation of a wider variety of methods leading to 2-substituted derivatives is given in section 1.2.1 and Table 1.6 in Chapter 1.

TABLE 5.5.   REACTIONS OF NAPHTHYLENE DIAMINES
LEADING TO NAPHTHIMIDAZOLONES,[a]
NAPHTHIMIDAZOL-2-THIONES,[b] AND 2-
AMINONAPHTHIMIDAZOLES

Reagent	Type of ring system synthesized	Substituents in the naphthimidazole product	Ref.
N,N'-Carbonyl-diimidazole	[2,3-d]	2-OH	39
COCl₂	[1,2-d]	2-OH	40
CO(NH₂)₂	[1,2-d]	2-OH-3-(CH₂)₂OH	41
CO(NH₂)₂	[1,2-d]	2-OH	42
		2-OH-3-CH₂Ph	
CO(NH₂)₂	[2,3-d]	2-OH-4,9-dioxo	43
CO(NH₂)₂	[2,3-d]	2-OH	44
CS₂	[2,3-d]	2-SH	45
CS₂	[1,2-d]	2-SH	46, 47
CS(NH₂)₂	[2,3-d]	2-SH	44
BrCN	[1,2-d]	2-NH₂	48, 49
BrCN	[2,3-d]	2-NH₂	49, 50

[a,b] Designated for convenience as 2-hydroxy and 2-mercapto derivatives, respectively.

TABLE 5.6.   REACTION OF NAPHTHYLENE DIAMINES WITH
              MISCELLANEOUS COMPOUNDS

Reagent	Type of ring system synthesized	Substituents in the naphthimidazole product	Ref.
MeC(OEt)$_3$	[1,2-d]	2-Me	51
(RO)$_4$C	[2,3-d]	2-OR	52
(R = Et, Pr)			
CF$_3$CF=CF$_2$	[1,2-d]	2-CHFCF$_3$	53
γ-Butyrolactone	[2,3-d]	2-(CH$_2$)$_3$OH	54
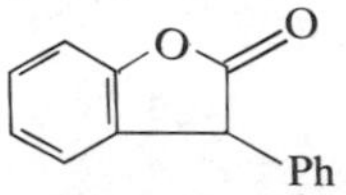	[1,2-d]	2-CH(Ph)C$_6$H$_4$OH-o	55
MeCOCH$_2$CO$_2$Et and PhCOCH$_2$CO$_2$Et	[2,3-d]	2-CH$_2$OMe and 2-CH$_2$COPh	56
PhCH=$\overset{+}{N}$($\bar{O}$)C$_6$H$_4$NMe$_2$-p	[1,2-d]	2-Ph	57

The reaction of naphthylene diamines with isatin derivatives provides one of the few examples in this type of process from which a condensed imidazoline derivative may be isolated[58] (see section 3.1.1 in Chapter 3). Thus 1,2-diaminonaphthalene reacts with isatin (**5.7**) to give a spiro compound (**5.8**), although 2,3-diaminonaphthalene under the same conditions affords a linear adduct (**5.9**). However, N-methylisatin reacts with the 2,3-diamino derivative to provide a spiro compound analogous to **5.8**, but

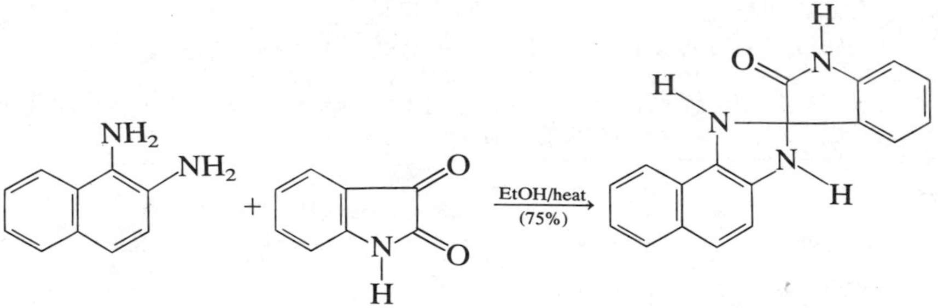

(5.7)   (5.8)

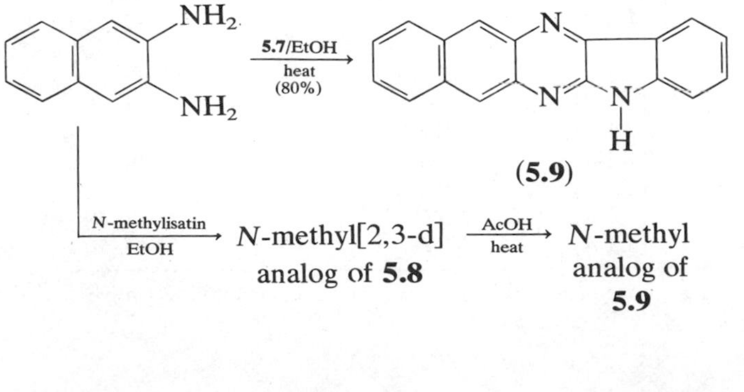

(5.9)

N-methyl[2,3-d] analog of **5.8** → AcOH/heat → N-methyl analog of **5.9** interestingly this is converted into an analog of the linear adduct (**5.9**) by recrystallisation from glacial acetic acid.[58]

Synthesis from o-(N-acylamino and -aroylamino)naphthylamines and -nitronaphthalenes

The detail and scope of this type of synthesis has been described in relation to the chemistry of benzimidazoles in section 1.2.2 of Chapter 1. Analogous naphthimidazole syntheses based on naphthylamines and nitronaphthalenes are collected in Tables 5.7[59–72] and 5.8,[61,73–75] respectively. Cyclization of phenylene diamines to benzimidazoles is usually effected under acidic conditions (see Table 1.7 in Chapter 1), but both acidic and basic media have been employed in naphthimidazole synthesis (see Table 5.7); good yields may be obtained by either procedure. The base-induced cyclization method has been used[63,64] to make a large number of bacteriostatic 1*H*-naphth[2,3-*d*]imidazole-4,9-diones, but the value of an alternative acid-catalyzed approach is well-illustrated by the synthesis of 2-(4′-chlorobutyl)-1*H*-naphth[2,3-*d*]imidazole-4,9-dione (**5.11**):[66] treatment of the *N*-acylnaphthylene diamine (**5.10**) with formic acid gives the desired product in

TABLE 5.7.   SYNTHESIS OF NAPHTHIMIDAZOLES FROM *o*-(*N*-ACYLAMINO AND -AROYLAMINO) NAPHTHYLAMINES

Naphthalene substituents	Reaction conditions	Type of ring system synthesized	Substituents in the naphthimidazole product	Ref.
1-NHAc-2-NHAc	NaOH/EtOH/heat	[1,2-*d*]	2-Me-6-Br	59
1-NHAc-2-NHAc	1 *N* HCl/heat	[1,2-*d*]	2-Me	60
1-NHAc-2-NH₂	AcOH/heat	[1,2-*d*]	2-Me-5-NHAc	61
1-NHAc-2-NH₂	20% HCl/AcOH/heat	[1,2-*d*]	2-Me-5-NH₂	61
1-NHAc-2-NH₂	Ac₂O/HCl/heat	[1,2-*d*]	2-Me-5-OH	62
2-NHCOR-3-NH₂ (R = alkyl, aminoalkyl hydroxyalkyl, etc.)	NaOH/EtOH/heat	[2,3-*d*]	2-R-4,9-dioxo	63,[a] 64[a]
2-NHCOR-3-NHR¹ (R = alkyl, aryl; R¹ = alkyl, aryl, allyl)	NaOH/EtOH/heat	[2,3-*d*]	2-R-3-R¹-4,9-dioxo	65
2-NHCOR-3-NH₂ (R = alkyl, aralkyl, chloroalkyl)	NaOH/heat or HCO₂H/heat	[2,3-*d*]	2-R-4,9-dioxo	66[a]
1-NHR-2-NHCOR¹ (R = alkyl, aralkyl R¹ = alkyl, alkenyl, aryl)	AcOH/heat	[1,2-*d*]	1-R-2-R¹-4,5-dioxo	67[a], 68[a]
1-NH₂-2-NHCOPh	4*N*·HCl/heat	[1,2-*d*]	2-Ph	69
1-NHCOPh-2-NHPh	AcOH/heat	[1,2-*d*]	2,3-Ph₂-5-OCOPh	70
2-NHPh-3-NHAc	NaOH/EtOH	[2,3-*d*]	1-Ph-2-Me-4,9-dioxo	71
1-NHPh-2-NHCOPh	AcOH	[1,2-*d*]	1,2-Ph₂-4,5-dioxo	72

[a] See text

TABLE 5.8. SYNTHESIS OF NAPHTH[1,2-d]IMIDAZOLES FROM o-(N-ACYLAMINO AND -AROYLAMINO) NITRONAPHTHALENES

Nitronaphthylamine	Reaction conditions	Substituents in the naphthimidazole product	Ref.
1-NO$_2$-2-NHCOPh			
1-NHCOPh-2-NO$_2$	Zn/HCl/EtOH	2-Ph	73a–c
1-NO$_2$-2-NHAc	Raney Ni/H$_2$/dioxan/room temp.	2-Me-4-cyclohexyl	74
1-NHAc-2-NO$_2$	Pd/C/H$_2$/AcOH	2-Me-3-OH-5-NHAc^{a,b}	61
1-NO$_2$-2-NHCHO	Sn/HCl	9-NH$_2$	75
1-NO$_2$-2-NHAc	H$_2$/Raney Ni	2-Me-4-Br, 2-Me-9-NH$_2$	75

^{a} Tautomeric with the 2-Me-3-N-oxide derivative (see text).
^{b} For an alternative electrochemical synthesis of naphth[1,2-d]imidazole-3-N-oxide derivatives, see M. Jubault and D. Peltier, *Bull. Soc. Chim. France*, **1972**, 1561.

39% yield, whereas reaction of the former with aqueous ethanolic sodium hydroxide gives rise to a tetracyclic derivative (**5.12**).[66]

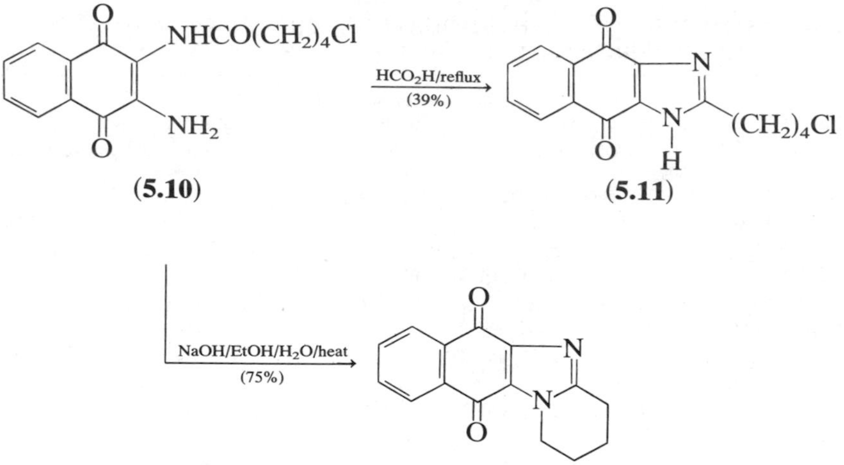

(**5.10**)   (**5.11**)

(**5.12**)

A restriction imposed upon cyclization of 4-alkylamino-3-acylaminonaphtho-1,2-quinones (**5.13**) has been observed.[67] Cyclization proceeds normally under acid-catalyzed conditions for **5.13** [R^1= e.g., n-pentyl (99%) or p-ClC$_6$H$_4$ (100%)], but a cyclized product is not obtained for **5.13** (R^1= t-Bu). Such retardation may occur because of steric repulsion of the 2-methyl and 9-hydrogen substituents by the t-butyl group within the transition state or intermediate dihydro derivatives leading to the dione (**5.14**).

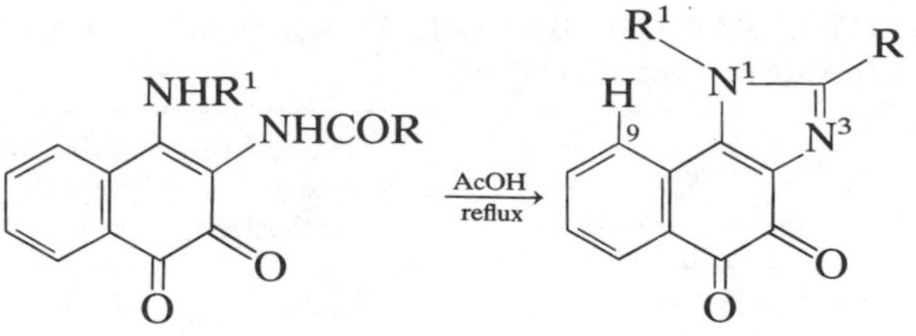

(5.13)   **(5.14)**

Synthetic methods based on cyclization of (N-acylamino)-nitronaphthalenes are restricted to reactions leading to compounds in the [1,2-d] series. As in analogous reactions in the benzimidazole series (see Table 1.8 in Chapter 1), a variety of reductants may be used. By carefully controlling the conditions it is possible to effect partial reduction, and this approach can be used to prepare 3-N-oxide derivatives[61] (see Scheme 5.1 and section 2.2.2 in Chapter 2 for a discussion of the utility of such processes in benzimidazole N-oxide syntheses).

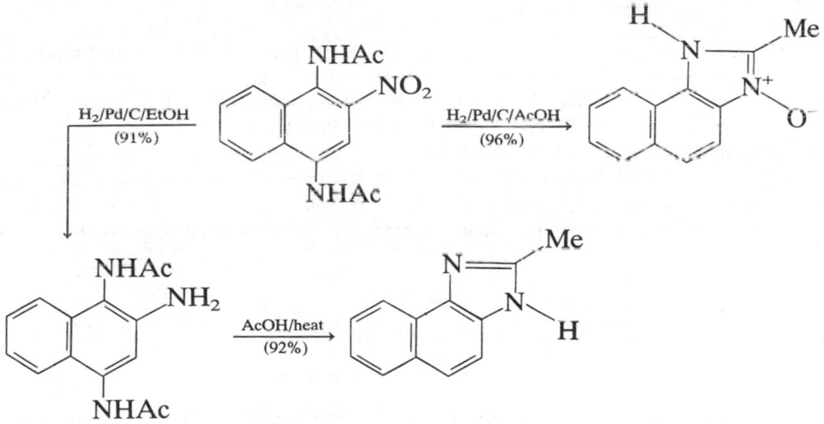

Scheme 5.1

Synthesis from Miscellaneous Naphthylamine Derivatives

Naphthimidazoles have been obtained from the reactions of acyl compounds with a variety of naphthylamine derivatives including azo, nitroso, and carbamate derivatives (see Table 5.9).[56,76–86] Some of these routes are of interest from a mechanistic as well as a synthetic aspect and are discussed in detail below.

Thermal cleavage of 3-aroylaziridines (**5.15**) generates azomethine ylides, and these species are thought to react with 1-nitroso-2-naphthylamine by 1,3-dipolar cycloaddition;[81] cleavage of intermediate oxadiazolidines in

TABLE 5.9. SYNTHESIS OF NAPHTHIMIDAZOLES FROM MISCELLANEOUS NAPHTHYLAMINE DERIVATIVES

Naphthalene substituents	Reaction conditions	Type of ring system synthesized	Substituents in the naphthimidazole product	Ref.
1-$N_2C_6H_4NO_2$-p-2-NHCH$_2$CO$_2$Et	AcOH/heat	[1,2-d]	2-CO$_2$Et	76
1-N$_2$Ph-2-NH$_2$	CH$_3$COCO$_2$H/heat	[1,2-d]	1-NHPh-2-Me	77
1-N$_2$Ph-2-NH$_2$	BrCH$_2$COAr/heat	[1,2-d]	2-COAr	78
1-NO-2-NHMe[a]	ZnCl$_2$/Ac$_2$O	[1,2-d]	None, 7-SO$_3$H	79, 80
1-NO-2-NH$_2$[a]	C_6H_{11}|N/ \\Ar'CH—CHAr^2/C$_6$H$_6$/heat	[1,2-d]	2-C(Ar^2)=NC$_6$H$_{11}$ +2-Ar'	81
1,2-(NHCSNHPh)$_2$	15% KOH/heat	[1,2-d]	2-SH[b]	82
1,2-(NHCO$_2$Et)$_2$	NaOH 100°	[2,3-d]	2-OH[c]	83
2-NHCONHCOPh-3-NHR (R = Bu, Ph)	NaOH/EtOH/heat	[2,3-d]	2-OH-3-R-4,9-dioxo[c]	84
1-NH$_2$-2-N=CHPh[d]	PhCHO/HCl	[1,2-d]	2-Ph-3-CH$_2$Ph	85
2-N=C(Me)CH$_2$COPh-3-NH$_2$[d]	Heat	[2,3-d]	2-Me	56
2-NH$_2$-3-Cl-1,4-dioxo	PhNCO/Et$_3$N	[1,2-d]	2,5-dioxo-3-Ph-4-Cl	86[e]

[a] Nitrosonaphthylamines are probably also intermediates in the conversion of 1-nitroso-2-naphthol in naphth[1,2-d]imidazole 1-oxides by HCHO/MeOH/NH$_4$OH at 45°/6 hr (see Ref. 87).
[b] Tautomeric with naphth[1,2-d]imidazol-2-thiones.
[c] Tautomeric with naphth[2,3-d]imidazol-2-ones.
[d] See text for a further example of synthesis from anil derivatives.
[e] See text, and Ref. 91 in section 5.1.1 for further examples of compounds in this group.

three different modes is suggested as a rationale for the formation of naphth[1,2-d]imidazole and 3-arylbenzo[f]quinoxaline derivatives (see Scheme 5.2). An additional feature of interest in the proposed mechanism is that the steps immediately preceding naphthimidazole formation constitute intramolecular examples of the reaction of arylene diamines with nitrones described in section 1.2.1 of Chapter 1 (see Ref. 308 in Chapter 1). A useful observation from a synthetic viewpoint is that electron-donating substituents in the aryl groups of the aziridine ring (e.g., methoxyl) increase the yield of the naphthimidazole products in relation to the benzo[f]quinoxaline.[81]

4-Chloro-2,5-dioxo-3-phenyl-3,5-dihydro-2H-naphth[1,2-d]imidazole (**5.18**) rather than the anticipated mesomeric betaine (**5.16**) is formed when 2-chloro-3-aminonaphtho-1,4-quinone is allowed to react with phenyl isocyanante in the presence of triethylamine.[86] A rearrangement involving a

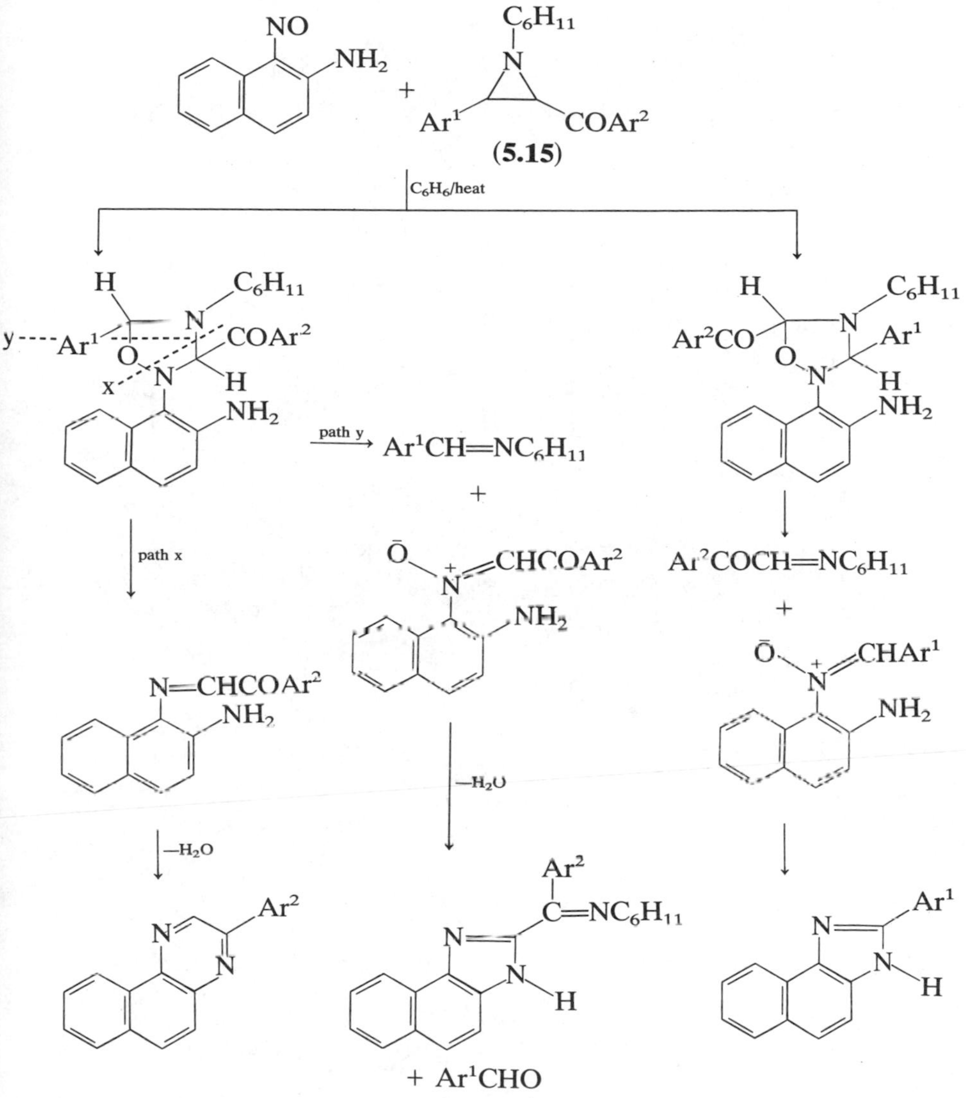

Scheme 5.2

spiro intermediate (**5.17**) is invoked to rationalize the formation of these unusual naphthimidazole derivatives (see Scheme 5.3). The 4-chloro-substituent of **5.18** can be displaced by nucleophiles such as azide ion, and the ensuing azido derivative has been used to prepare 4-imino-, 4-oxo-, and 4-triphenylphosphoranylidenamino analogs.

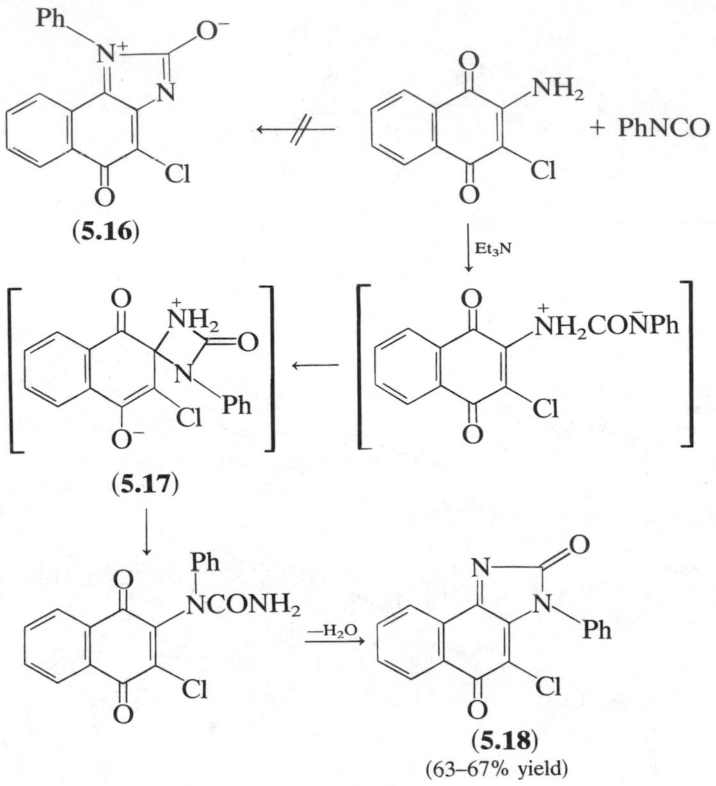
(5.16)
+ PhNCO
Et3N
(5.17)
−H2O
(5.18)
(63–67% yield)

Scheme 5.3

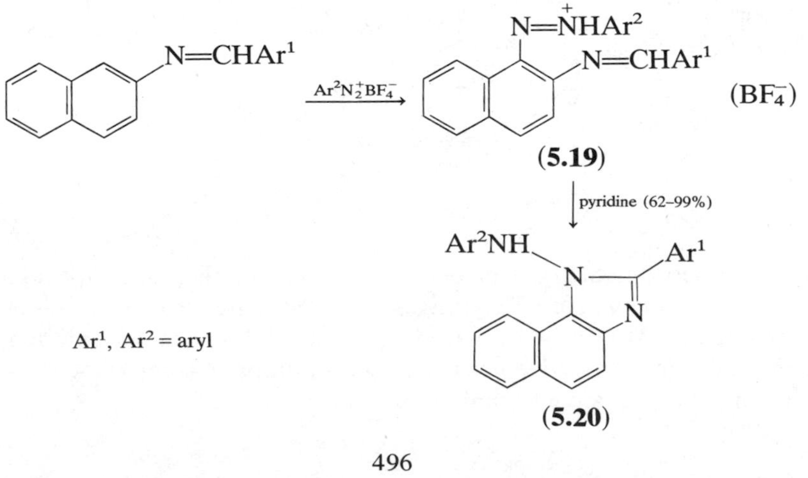
N=CHAr1
Ar2N2+BF4−
(BF4−)
(5.19)
pyridine (62–99%)
(5.20)
Ar1, Ar2 = aryl

Synthesis from N-benzylidenenaphthylamine Derivatives

N-Arylidene-2-naphthylamines react with aryl diazonium tetrafluoroborates to give tetrafluoroborate salts (5.19).[88] Decomposition of the latter by
pyridine provides an efficient procedure for the synthesis of 1-arylamino
derivatives in the naphth[1,2-d]imidazole series (5.20); 3-arylamino analogs
are also accessible by a similar method from N-arylidene-1-naphthylamines
but yields are lower.

Synthesis from Amidines and Guanidines

The synthesis of benzimidazoles by cyclization of amidines and related
compounds has been described in section 1.2.7 of Chapter 1. This type of
reaction has been used to prepare 2-phenylnaphth[1,2-d]imidazole
(5.21).[89,90]

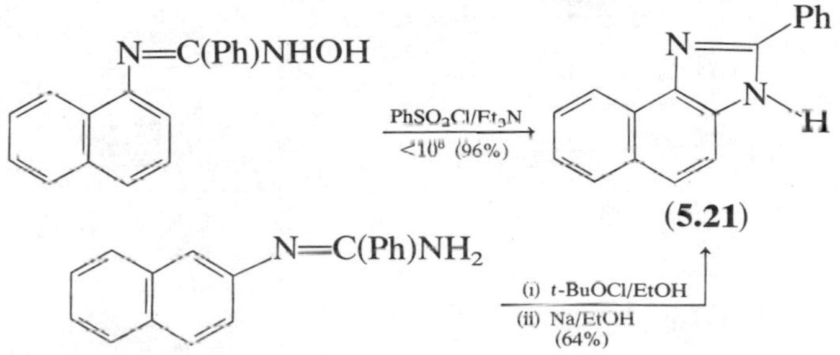

N=C(Ph)NHOH

PhSO₂Cl/Ft₃N
<10° (96%)

(5.21)

N=C(Ph)NH₂

(i) t-BuOCl/EtOH
(ii) Na/EtOH
(64%)

The guanidine derivatives (5.22) are converted by oxidative cyclization
into further examples of compounds in the uncommon 3,5-dihydro-2H-
naphth[1,2-d]imidazol-5-one series (5.23)[91] (cf. Ref. 86 in section 5.1.1).

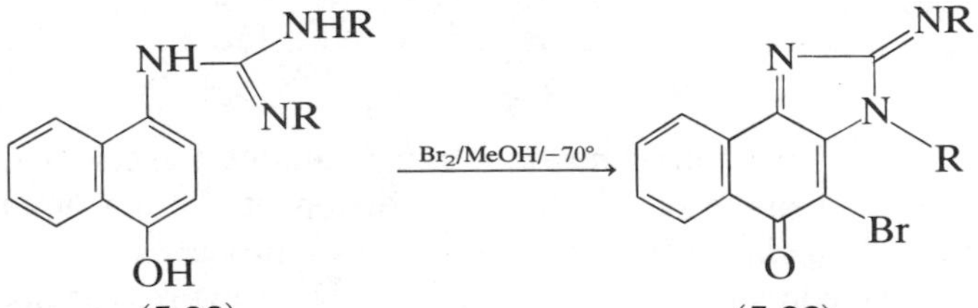

NH—C(NHR)(NR)

OH
(5.22)

Br₂/MeOH/−70°

(5.23)
R = t-Bu (72%)
R = cyclohexyl (73%)

Synthesis from Heterocyclic Compounds

A variety of naphthimidazole derivatives has been synthesized by methods
in which heterocyclic compounds are used as starting materials; such procedures involve substrates which may or may not contain an intact imidazole

ring. Two advantages accrue from this approach: first, it is possible to design the synthesis of naphthimidazole derivatives with unusual substitution patterns in the carbocyclic ring (cf. the aryne method described below); and second, the use of strongly acidic or basic reaction media is usually circumvented.

SYNTHESIS FROM CONDENSED IMIDAZOLE DERIVATIVES. The Diels–Alder reaction has been successfully applied to the synthesis of tetrahydro derivatives in the naphth[2,3-d]imidazol-4,9-dione series (5.24)[92,93] and also to provide an alternative route to partially reduced compounds in the uncommon 3,5-dihydro-2H-naphth[1,2-d]imidazol-5-one group[91] (5.25; cf. also preceding discussion in this section).

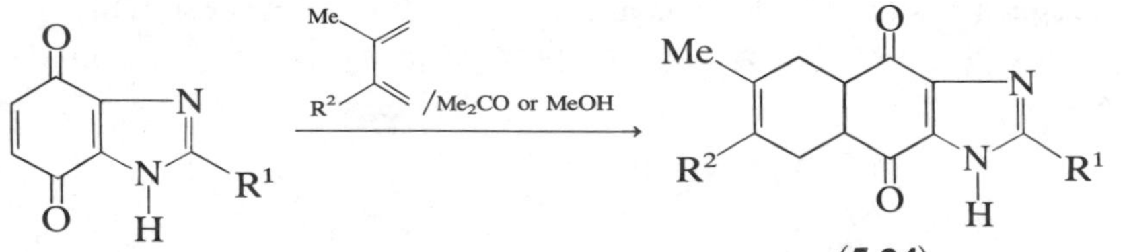

(5.24)

R[1] = OH, R[2] = H (96%)[92]
R[1] = H, R[2] = H (55%)[93]
R[1] = H; R[2] = Me (48%)[93]

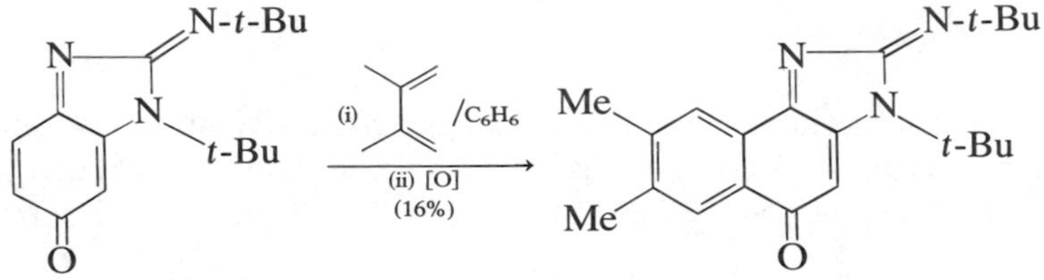

(5.25)

Substituted 1,2,3-benzotriazines undergo reactions typical of diazonium compounds, one example of which is conversion of the diazonium moiety into hydroxyl by acidic hydrolysis. This type of process is exemplified by conversion of the condensed system (5.26) into 2-(o-hydroxyphenyl)naphth-[1,2-d]imidazole (5.27).[94]

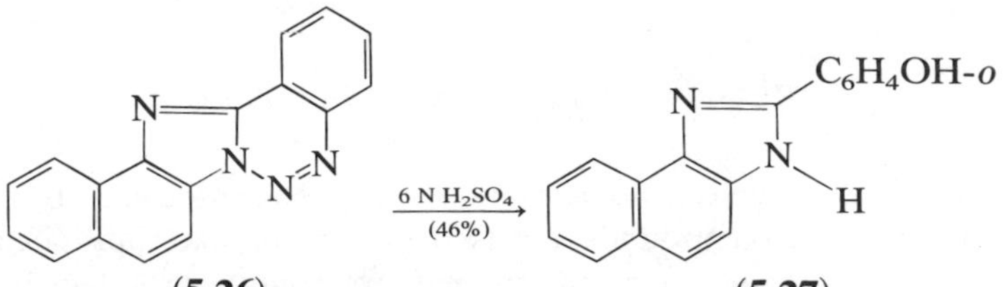

(5.26)　　　　　　　　　　**(5.27)**

Scission of a condensed heterocyclic ring has also been used to prepare a naphth[2,3-*d*]imidazole derivative (**5.28**).[95] A particularly useful feature of this benzyne route is that it can be used to provide compounds with an unusual substitution pattern in the carbocyclic ring.

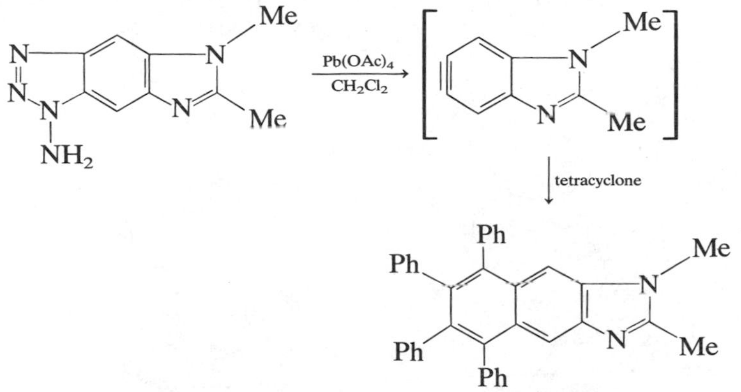

(**5.28**)

82% yield

SYNTHESIS FROM OTHER HETEROCYCLIC COMPOUNDS. Pyrolysis of 3,4-disubstituted 1,2,4-oxadiazol-5-ones and of 1,5-diaryl tetrazoles have been discussed in section 1.2.9 of Chapter 1 in the context of benzimidazole synthesis. Both of these approaches, which presumably occur via imino nitrenes, have been used for the preparation of 2-phenylnaphth[1,2-*d*]-imidazole (see Scheme 5.4).[96,97]

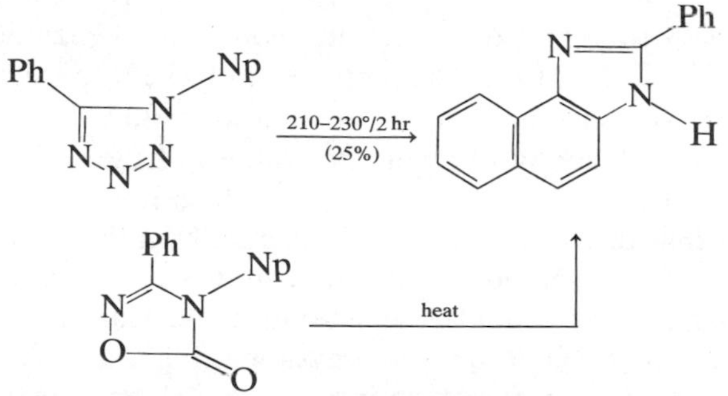

Np = 1- or 2-naphthyl

Scheme 5.4

2-Alkyl- and 1-alkenylnaphthimidazoles have been obtained from the acid-promoted rearrangement of condensed 1,2,5-oxadiazines (**5.29**)[98] and also by thermolysis of naphtho[1,4]diazepines (**5.30**[99] and **5.31**[100]).

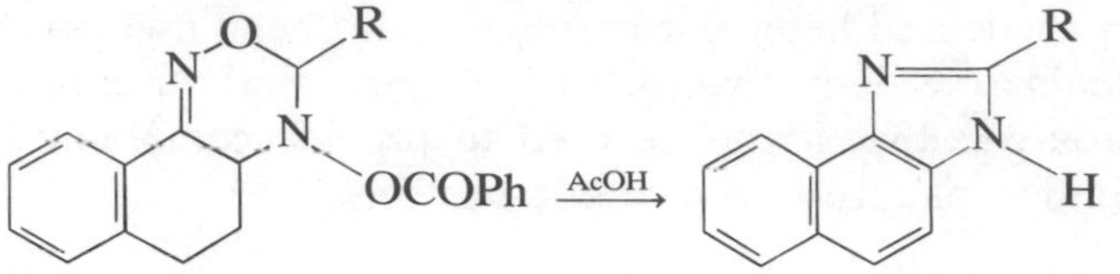

(5.29)

R = H, Me
(65% yield for R = Me)

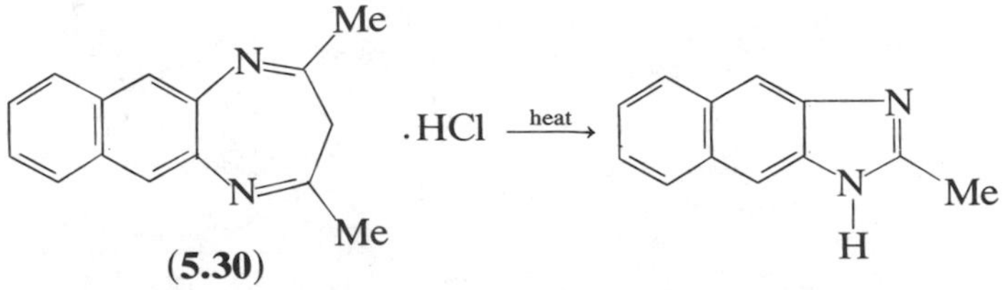

(5.30)

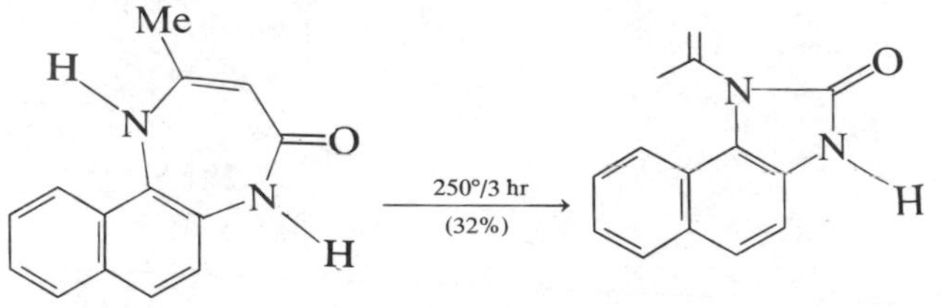

(5.31)

5.1.2. Physicochemical Studies

Spectroscopic Studies

INFRARED SPECTRA. Unlike the situation for benzimidazoles (see sections 1.3.3 and 3.2.3 in Chapters 1 and 3), the infrared spectra of naphthimidazoles have not been systematically studied.

Infrared data for 1-methyl-2-amino derivatives of the [1,2-*d*] (ν_{max} 3395, 3310, 1655 cm^{-1}) and [2,3-*d*] series (ν_{max} 3450, 3320, 1665 cm^{-1}) lead to the conclusion that these derivatives exist primarily in the amino form in the solid phase and also in solution.[101] Infrared studies have also illustrated that in 9-hydroxynaphth[1,2-*d*]imidazole an intramolecular hydrogen bond is formed from the hydroxyl group to the pyridine-like 1-nitrogen atom.[102] An evaluation of ring-chain tautomerism in S-(acyl)alkyl derivatives (cf. equation 5.1) has been made by use of NMR and infrared spectral analysis.[103] The tautomeric equilibrium is dependent upon the nature of side-chain sub-stituents and also upon the medium. In the example shown, the compound exists in the cyclic carbinolamine form in the crystalline state but predomin-antly in the open-chain form in chloroform solution [ν_{max} 1712 (C=O), 3460 (N–H), 3570, 3600 (O–H weak)].

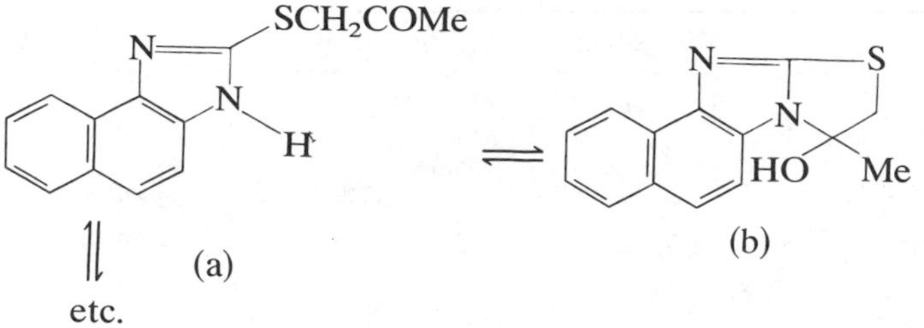

$$SCH_2COMe \quad (a) \rightleftharpoons (b) \quad (5.1)$$

etc.

The position of the C=O stretching frequency in naphthimidazolones is well documented (cf. the variations in C=O stretch in benzimidazolones described in section 3.2.3 of Chapter 3). Examples indicating the position of this stretching mode are illustrated within structures **5.32** to **5.35**. It is interesting to note that the C=O stretching frequency of 2-alkyl naphth-[2,3-d]imidazol-4,9-diones occurs as a single absorption at 1675 cm^{-1}, but in 1,2-dialkyl analogs this is split into a doublet [e.g., ν_{max} 1661, 1678 cm^{-1} for a 1-benzyl-2-(3'-cyclohexyl)propyl derivative];[66] a similar doublet splitting is apparent in the spectra of compounds **5.33** and **5.34**.

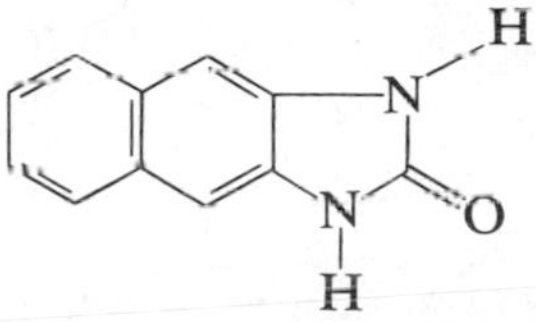

(5.32)

1751 cm^{-1} [83]

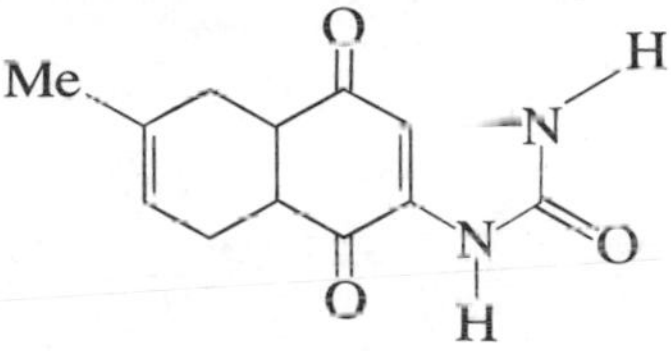

(5.33)

1737 cm^{-1} [92]

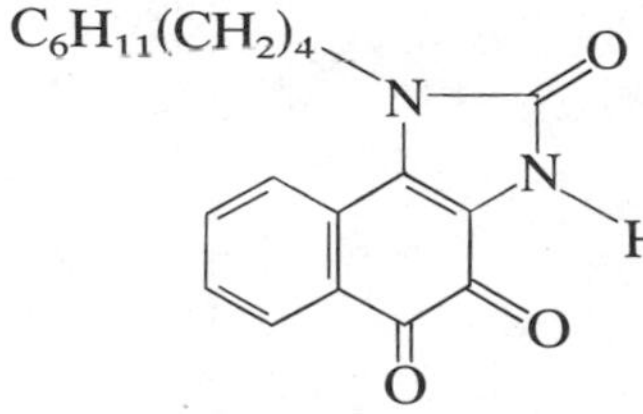

(5.34)

1712 cm^{-1} [67]

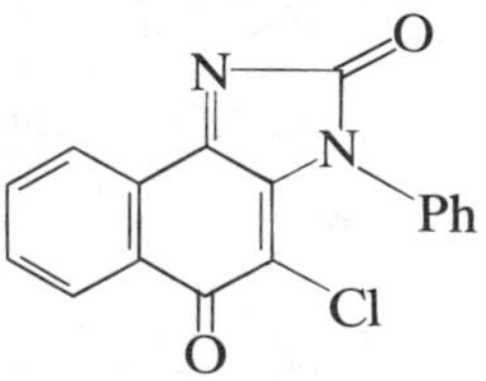

(5.35)

1736 cm^{-1} [86]

ELECTRONIC ABSORPTION AND LUMINESCENCE SPECTRA. Systematic studies of the electronic absorption of naphthimidazoles have been carried out,[44] and selected data are collected in Table 5.10. There is a considerable bathochromic shift of the long-wavelength band (neutral molecule and cation) in passing from benzimidazole to naphth[1,2-d]imidazole and in turn to naphth[2,3-d]imidazole (cf. also Table 1.19 in Chapter 1). Fine structure

TABLE 5.10. CHARACTERISTIC BANDS IN THE ELECTRONIC
ABSORPTION SPECTRA OF NAPHTHIMIDAZOLES

Substituent	Ring system	λ_{max} (nm) (Solvent)	log ε	Ref.
None	[2,3-*d*]	235	4.79	
(neutral molecule)		317	3.83	44
		327	3.88	
		342 (H_2O)	3.74	
None (cation)	[2,3-*d*]	235	4.75	
		318	3.84	44
		325 (H_2O)	3.90	
2-Et	[2,3-*d*]	239	4.84	
(neutral molecule)		321	3.92	44
		338 (H_2O)	3.76	
2-Et (cation)	[2,3-*d*]	237	4.82	
		318	3.92	44
		326 (H_2O)	3.95	
2-OEt	[2,3-*d*]	222 (sh)	4.68	
		235	4.82	
		238 (sh)	4.80	
		276 (sh)	3.55	
		285 (sh)	3.70	52
		290	3.71	
		296 (sh)	3.78	
		300	3.81	
		309	3.79	
		314	3.76	
		323	3.80	
2-Ph	[2,3-*d*]	264	4.54	
		273	4.56	
		327 (sh)	4.11	24
		338 (95% EtOH)	4.23	
None	[1,2-*d*]	222	4.54	
(neutral molecule)		240	4.57	
		273	3.63	
		279	3.65	44
		313	3.46	
		319	3.38	
		326 (H_2O)	3.54	
None (cation)	[1,2-*d*]	240	4.37	
		273	3.78	
		284 (sh)	3.57	
		295	3.32	44
		308	3.58	
		315	3.46	
		322 (H_2O)	3.76	
2-$C_6H_4NH_2$-*o*	[1,2-*d*]	241	4.66	
		268 (sh)	4.39	
		275	4.41	104
		352 (EtOH)	4.27	
2-C_6H_4OH-*o*	[1,2-*d*]	231	4.68	
		264 (sh)	4.33	
		273	4.41	
		306	4.10	94
		320	4.21	
		334	4.51	
		350	4.62	

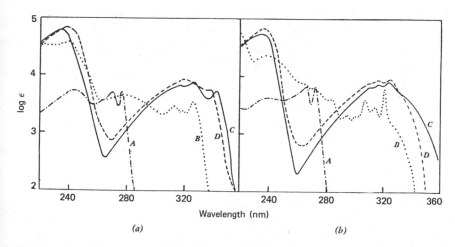

Fig. 5.1. Electronic absorption spectra of A, benzimidazole; B, naphth[1,2-d]imidazole; C, naphth[2,3-d]imidazole; and D, 2-ethylnaphth[2,3-d]imidazole as neutral molecules (a) and cations (b).

also increases with the complexity of the system and is thus more in evidence in the naphth[1,2-d] rather than the naphth[2,3-d]imidazole system (see Figs. 5.1a and b). The effect of substitution of an aryl group at the 2-position is to cause a bathochromic shift in the ultraviolet absorption and also to increase the intensity of the spectra.

The ultraviolet spectra of 5,6,7,8-tetrahydronaphthimidazoles (Table 5.11)[1,21] resemble the spectra of benzimidazoles rather than naphthimidazoles. Spectra of the compounds in the [1,2-d] series are very similar to their [2,3-d] analogs, but small hypsochromic shifts are observed.

The spectra of naphth[2,3-d]imidazole-4,9-diones (**5.36**)[66] closely resemble the spectra of unreduced derivatives (Table 5.10), whereas naphth[1,2-d]imidazole-4,5-diones (**5.37**)[67] show a weak absorption in the visible region. The absorption spectra of 1-alkyl analogs of **5.37** are very similar to **5.37**, which might indicate that the phenyl group is skewed from coplanarity as a result of the steric influence of the 9-hydrogen.

(**5.36**)

λ_{max}	ϵ
244	38,000
277 (broad)	15,200
330	2,900

(**5.37**)

λ_{max}	ϵ
251	sh
261	27,200
268	26,900
444	1,600

TABLE 5.11.   CHARACTERISTIC BANDS IN THE ELECTRONIC ABSORPTION
SPECTRA OF 5,6,7,8- AND 6,7,8,9-TETRAHYDRONAPHTHIMI-
DAZOLES[a]

Substituent	Ring system	λ_{max} (nm) (Solvent)	log ϵ	Ref.
None	[2,3-d]	252	3.62	
(neutral molecule)		282	3.78	21
		291 (H$_2$O)	3.78	
None (cation)	[2,3-d]	279	3.87	21
		289 (H$_2$O)	3.86	
None	[1,2-d]	248	3.77	
(neutral molecule)		273	3.55	1
		281	3.55	
None (cation)	[1,2-d]	272	3.71	1
		280	3.64	
5,6-Dimethylbenzimidazole	—	246	3.59	
(neutral molecule)		277[b]	3.66	
		279	3.66	21
		286	3.67	
5,6-Dimethylbenzimidazole	—	246	3.41	
(cation)		274	3.77	
		276[b]	3.77	21
		283	3.79	

[a] The spectra of 5,6-dimethylbenzimidazole are included for comparison.
[b] Inflexion.

An attempt has been made to establish the nature of amino-imino
tautomerism in 1-methyl-2-aminonaphthimidazoles by ultraviolet spectroscopy,[101] but this problem is more amenable to investigation by infrared
methods [see first part of this section].

Luminescence spectra have been recorded for 2-arylnaphth[2,3-
d]imidazoles (see Table 5.12).[24] These data indicate a marked dependence
of luminescence wavelength on the structure of the aryl group.

NUCLEAR MAGNETIC RESONANCE SPECTRA. Unlike the situation for benzimidazoles (cf. section 1.3.3 in Chapter 1), there have been no systematic

TABLE 5.12.   LUMINESCENCE SPECTRA OF 2-ARYL-
NAPHTH[2,3-d]IMIDAZOLES[24]

Substituent	λ_{emiss}^{solid} (nm)	$\lambda_{emiss}^{CH_2Cl_2}$ (nm)	Emission band width (nm)[a]
2-C$_6$H$_4$OH-o	468	485	62
2-(2-Hydroxy-1-naphthyl)	485	465	62
2-(4-Hydroxy-3-biphenylyl)	490	512	62
2-(2,5-Dihydroxyphenyl)	515	530	65
2-(3-Hydroxy-2-naphthyl)	610	480	95

[a] Width at half height of emission band measured in methylene chloride.

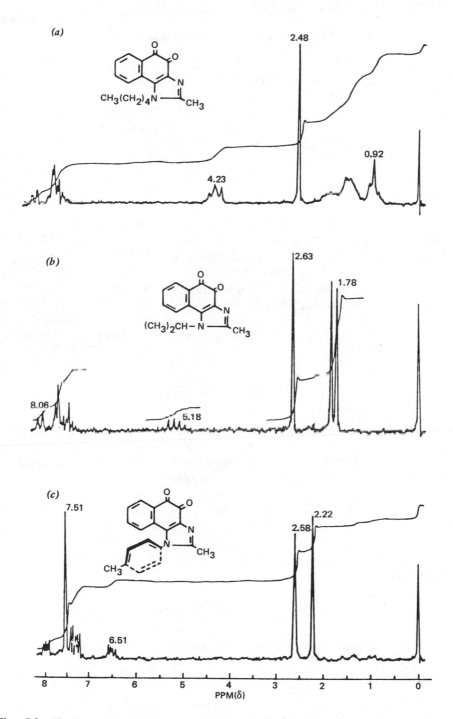

Fig. 5.2. Nuclear magnetic resonance spectrum of (*a*) 1-pentyl-2-methylnaphth[1,2-*d*]imidazole-4,5-dione, (*b*) 1-isopropyl-2-methylnaphth[1,2-*d*]imidazole-4,5-dione, and (*c*) 1-(*p*-tolyl)2-methylnaphth[1,2-*d*]imidazole-4,5-dione. [Reprinted with permission from F. I. Carroll and J. T. Blackwell, *J. Heterocycl. Chem.*, **7**, 297 (1970).]

studies of the NMR spectra of naphthimidazoles; spectra from nuclei other than hydrogen have not been recorded.

In 6,7-dimethylnaphth[2,3-d]imidazolc, H-2, H-4 and H-9, and H-5 and H-8 appear at δ 8.4, δ 8.0, and δ 7.7, respectively,[17] while in the 2-ethoxy derivative, the aromatic protons appear as two multiplets at δ 7.86 and 7.38.[52] In naphth[2,3-d]imidazol-2-one the aromatic protons appear in the region δ 7.29 to 7.95 and at δ 7.27 to 8.23 for 3-isopropylnaphth[1,2-d]-imidazole.[100] The NMR spectrum of the reduced derivative (**5.33**)[92] shows resonances at δ 11.42 (H-1, H-3), 5.33 (H-7), 2.22 (4H at C-5 and C-8), and 1.62 (Me).

Evidence for steric crowding of groups in the 1-position of 1,2-disubstituted naphth[1,2-d]imidazole-4,5-diones has been adduced from ^{1}H NMR spectra (see Fig. 5.2).[67] For example, the 2-Me resonance of the 1-isopropyl-2-methyl derivative appears at δ 2.63 (Fig. 5.2b), whereas the same group appears at δ 2.48 in the 1-pentyl analog (Fig. 5.2a); this effect is ascribed to increased shielding of the 2-methyl group by the methyl groups of the isopropyl function. The chemical shifts of the 9-H and 2-Me (δ 6.51 and 2.58; see Fig. 5.2c) of the 1-p-tolyl derivative lead to the conclusion that the 1-aryl group is skewed and causes an anistropic effect.

General Studies

ACIDITY CONSTANTS. Progressive annulation of imidazole weakens the basic properties: the basic pK_a of imidazole (7.03) may be compared with that of benzimidazole (~5.5; cf. Table 1.27 in Chapter 1) and in turn with naphth-[1,2-d]imidazole (5.28) and naphth[2,3-d]imidazole (5.24).

The acidity of naphthimidazole and its derivatives is manifested by the formation of metal salts (e.g., Pb,[105] Sn and Ti,[106] Hf,[107] Ag and Hg[108]). As expected, the weak acidity of imidazole (pK_a = 14.2) is increased by annulation (cf. approximately 13.2 in benzimidazole) and the naphthimidazoles are even more acidic (approximately 12.5). A selection of pK_a values is given in Table 5.13.

TABLE 5.13.   ACIDITY CONSTANTS OF NAPHTHIMIDA-
ZOLES

Substituent	Ring system	pK_a as Base	pK_a as Acid	Ref.
None	[2,3-d]	5.24[a]	12.52[b]	109
2-Me	[2,3-d]	6.11[a]	12.9[b]	109
2-NH$_2$	[2,3-d]	7.01[a]	12.8[b]	109
2-OEt	[2,3-d]	4.07[b]	11.15[b]	52
None	[1,2-d]	5.28[a]	12.54[b]	109
2-OH[c]	[1,2-d]	−1.7[b]	11.95[b]	109

[a] From potentiometric titration.
[b] From spectroscopic data.
[c] Tautomeric with naphth[1,2-d]imidazol-2-one.

The pK_a values of 5,6,7,8-tetrahydronaphth[2,3-d]- and 6,7,8,9-tetrahydronaphth[1,2-d]imidazoles are very similar.[1] The basicity (e.g., $pK_a = 5.99$ for 6,7,8,9-tetrahydronaphth[1,2-d]imidazole) is higher in these derivatives than in the parent naphthimidazoles and approximates to the value for 5,6-dimethylbenzimidazole ($pK_a = 6.1$).

MISCELLANEOUS STUDIES. Dipole moments of naphth[1,2-d]imidazole derivatives have been estimated from dielectric data measured in dioxan at 25°.[110] Values (in Debyes) for unsubstituted, 3-methyl, and 1-ethyl derivatives are 4.12, 3.86, and 4.04 and are thus similar to values for benzimidazole derivatives (see Table 1.16 in Chapter 1).

The polarographic characteristics of naphth[2,3-d]imidazole-4,7-dione derivatives are similar to those of anthraquinones.[111]

A number of molecular orbital calculations relating to naphthimidazoles have been carried out.[112–114] These have been concerned in part with providing estimates of pK_a values[113] and with ascertaining the charge at the 2-position.[114]

5.1.3. Reactions of Naphthimidazoles

Reactions with Electrophilic Reagents

SUBSTITUTION IN THE CARBOCYCLIC RINGS. There has been no systematic investigation of the behavior of naphthimidazoles in regard to electrophilic aromatic substitution. Naphth[1,2-d]imidazole forms a 5-bromo derivative when treated with bromine in acetic acid at 0°;[115] an addition product is also formed under these conditions. It may be noted that the reaction is carried out in bright sunlight and a process of radical substitution cannot be ruled out. Other examples of electrophilic aromatic substitution in the [1,2-d] series are azo coupling at the 4-position of a 2-methyl-5-hydroxy derivative[116] and nitration at the 4-position in a 5-acetylamino-3-N-oxide derivative.[61]

The pattern of electrophilic sulfonation of naphth[2,3-d]imidazole-4,9-dione has analogy in anthraquinone chemistry: in the presence of mercuric oxide as a catalyst the 5-sulfonic acid is formed, but in its absence, substitution occurs at the 6-position (**5.38 → 5.39** or **5.40**).[117]

ELECTROPHILIC ATTACK AT THE IMIDAZOLE NITROGEN ATOMS: ALKYLATION AND RELATED REACTIONS. Some but not all of the alkylation methods used in benzimidazole chemistry have been employed to alkylate naphthimidazoles (see Table 5.14 and cf. section 1.4.1 in Chapter 1). Compounds in the [1,2-d] system are alkylated by all reagents at the 3-position usually in good yields with the exception of the formation of 1-alkenyl derivatives (3.6% for the naphth[1,2-d]imidazol-2-one derivative[124] listed in Table 5.14).

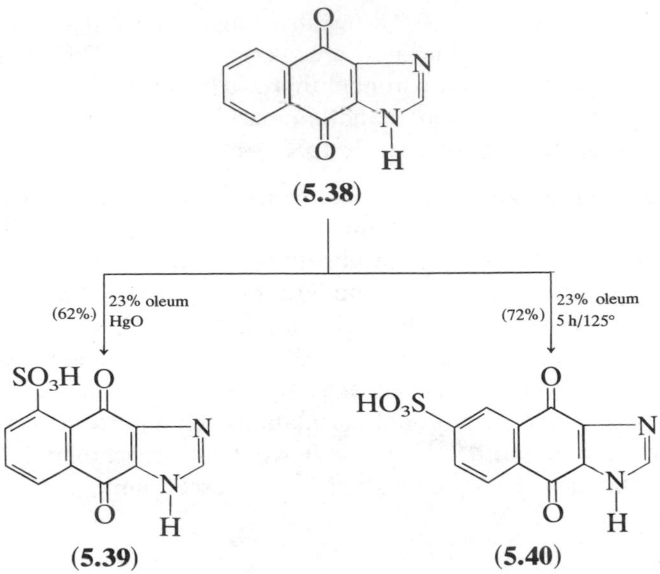
O
N
N
O
H
(5.38)
(62%)
23% oleum
HgO
(72%)
23% oleum
5 h/125°
SO3H
O
N
N
O
H
(5.39)
HO3S
O
N
N
O
H
(5.40)

The literature on alkylation of compounds in the [2,3-*d*] system is sparse. A special case of alkylation is illustrated by conversion of the 1-trimethylsilyl derivative (**5.41**) into 1-*β*-D-ribofuranosylnaphth[2,3-*d*]imidazol-4,9-dione (**5.41**).[128] Compounds of this type are interesting in view of the involvement of many quinones in biological redox processes.[129]

TABLE 5.14. ALKYLATION OF NAPHTHIMIDAZOLES

Substituents in the naphthimidazole	Ring system	Alkylating agent	Product	Ref.
None	[1,2-*d*]	MeI/KOH	3-Me	5
2-Cl	[1,2-*d*]	$BrCH_2COR$/NaOEt (R = alkyl, aryl)	2-Cl-3-CH_2COR	118
2-SMe	[1,2-*d*]	$BrCH_2COMe$/NaOEt	2-SMe-3-CH_2COR	119
2-(2-Furyl)	[1,2-*d*]	MeI/base	2-(2-Furyl)-3-Me	37
None	[1,2-*d*]	Me_2SO_4/80°	3-Me	5
2-Me	[1,2-*d*]	Me_2SO_4/NaOH	2,3-Me_2	120
None	[1,2-*d*]	$Me_2(Ph)\overset{+}{N}CH_2Ph(Cl^-)$	3-CH_2Ph	121
None	[1,2-*d*]	$CH_2O/HN[(CH_2)_2Cl]_2$	3-$CH_2N[(CH_2)_2Cl]_2$	122
None	[1,2-*d*]	C_2H_2/KOH/ZnO pyridine	3-$CH{=}CH_2$	123
2-OH	[1,2-*d*]	C_2H_2/KOH	1,3-$(CH{=}CH_2)_2$-2-C=O	124
None	[2,3-*d*]	$CF_2{=}CFCl$	1-CF_2CHFCl	125
2-SH	[2,3-*d*]	CH_2O	1,3-$(CH_2OH)_2$-2-C=S	126
4,9-Dioxo	[2,3-*d*]	Me_2SO_4	1-Me	127

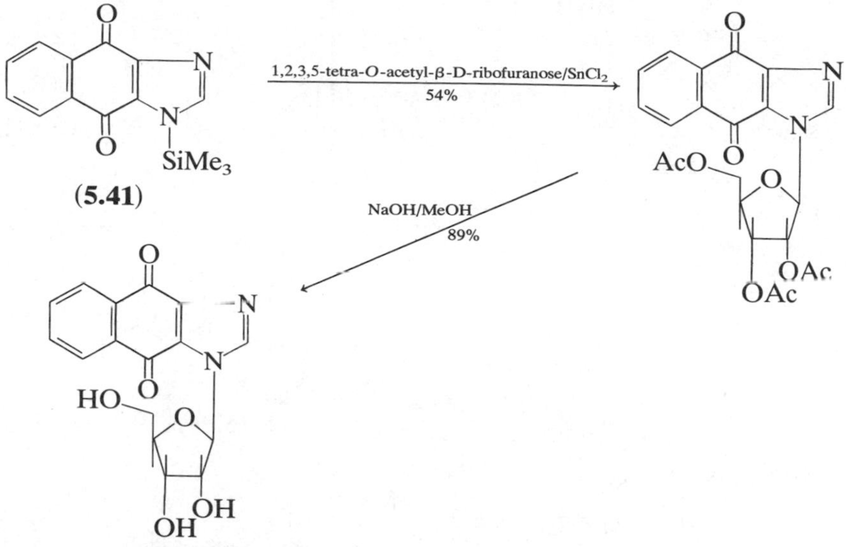

(5.41)

1,2,3,5-tetra-*O*-acetyl-β-D-ribofuranose/SnCl$_2$
54%

NaOH/MeOH
89%

(5.42)

INTRAMOLECULAR ELECTROPHILIC ATTACK AT THE IMIDAZOLE NITROGEN ATOMS: FORMATION OF CONDENSED NAPHTHIMIDAZOLES. Intramolecular electrophilic attack at the imidazole nitrogen in naphth[1,2-*d*]imidazoles occurs preferentially at the 3-position (e.g., **5.43 → 5.44**),[130] but both cyclization modes occur in cyclization of the 2-chloroethyl derivative (**5.45 → 5.46 + 5.47**).[131] Further examples of intramolecular cyclizations are illustrated by the conversions **5.48 → 5.49**,[132] **5.50 → 5.51**,[27] **5.52 → 5.53**,[45] and **5.54 → 5.55**.[133]

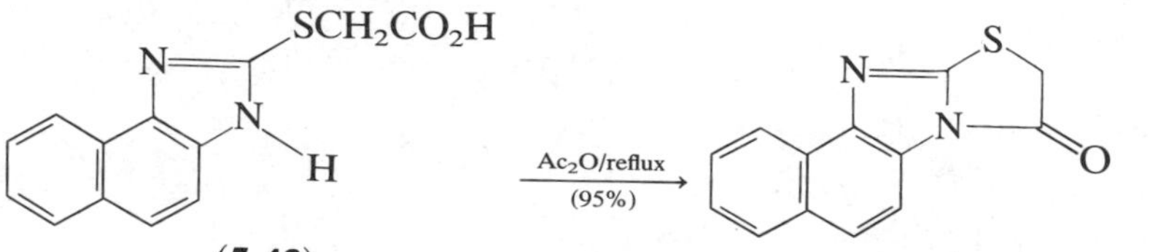

SCH$_2$CO$_2$H

(5.43)

Ac$_2$O/reflux
(95%)

(5.44)

S(CH$_2$)$_2$OH

(5.45)

SOCl$_2$, neat
or in DMF
(>80%)

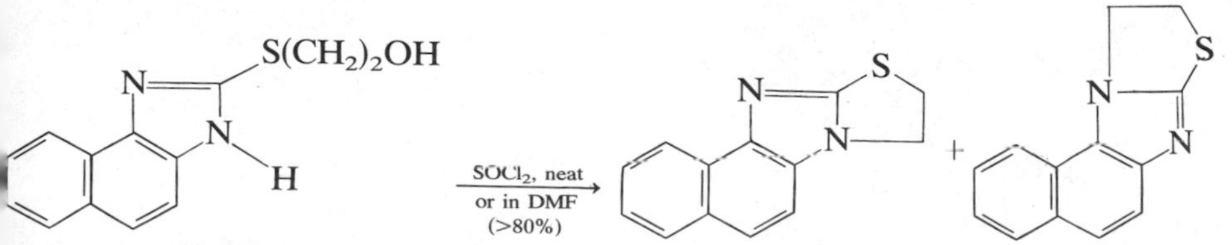

(5.46)

+

(5.47)

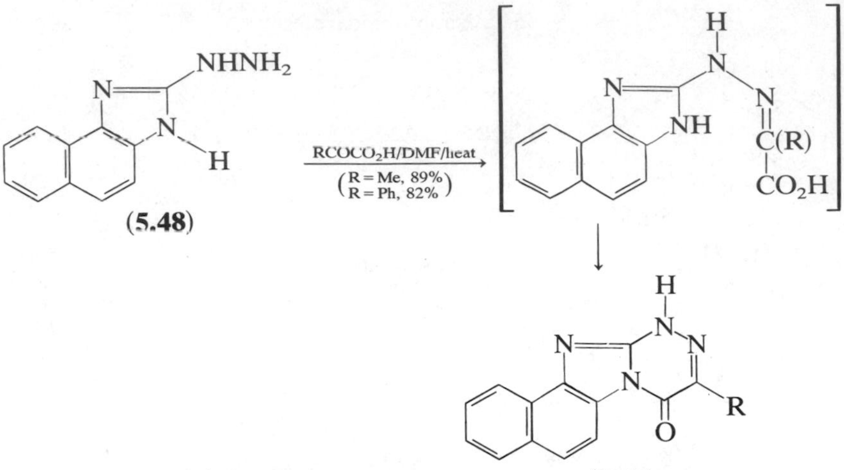

(5.48)

RCOCO$_2$H/DMF/heat
$\left(\begin{array}{l}R = Me, 89\% \\ R = Ph, 82\%\end{array}\right)$

(5.49)

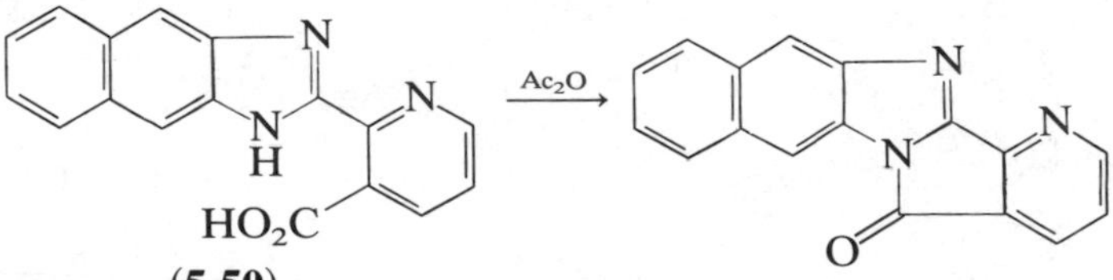

(5.50)

Ac$_2$O

(5.51)

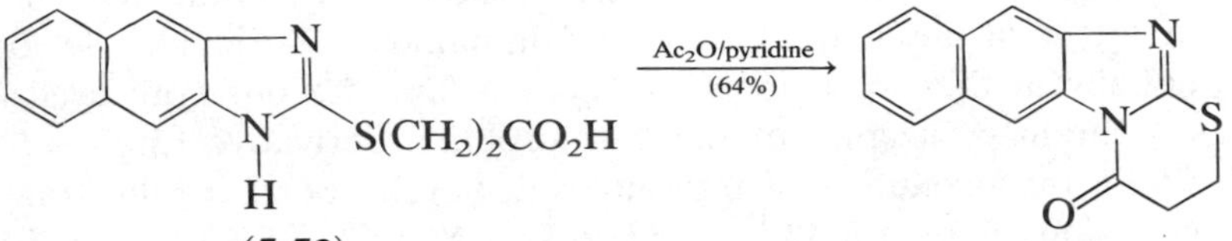

(5.52)

Ac$_2$O/pyridine
(64%)

(5.53)

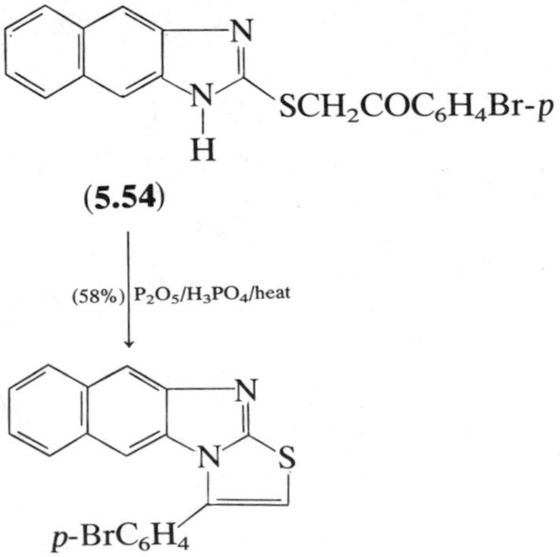

(5.54)

(58%) P$_2$O$_5$/H$_3$PO$_4$/heat

p-BrC$_6$H$_4$

(5.55)

TABLE 5.15.   REACTIONS INVOLVING ELECTROPHILIC ATTACK AT
              NAPHTHIMIDAZOLE SUBSTITUENTS

Substituent undergoing electrophilic attack	Ring system	Electrophilic reagent	Reaction product	Ref.
2-SH	[2,3-d]	p-BrC$_6$H$_4$COCH$_2$Br	2-SCH$_2$COC$_6$H$_4$Br-p^a	133
2-SH	[2,3-d]	Br(CH$_2$)$_2$CO$_2$H	2-S(CH$_2$)$_2$CO$_2$H^a	45
2-NH$_2$	[2,3-d]	RNCO	2-NHCONHR	49
	[1,2 d]	(R = aryl, C$_6$H$_{11}$)		
2-(CH$_2$)$_n$NR^2(NH)$_m$H (n = m − 0, 1; R^2 = H, Me)	[2,3-d]	RNCY (Y = O, S; R = aryl; cycloalkyl)	2-(CH$_2$)$_n$N(NH)$_m$CY (with R^2 and NHR substituents)	134
2-NH$_2$	[1,2-d]	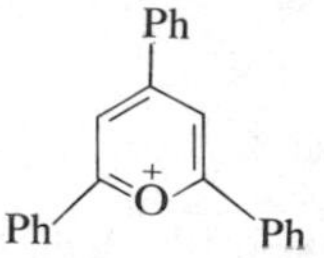 2,4,6-triphenylpyrylium (ClO$_4^-$)	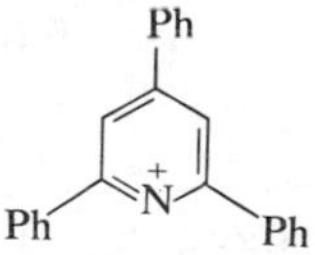 pyridinium product (ClO$_4^-$) (R = 3-methylnaphth-[1,2-d]imidazol-2-yl)	

^{a} See section 5.1.3 for examples of the use of these derivatives for the synthesis of tetracyclic heterocycles.

Intramolecular cyclization in relation to the ring-chain tautomerism of 2-S-(acyl)alkyl-substituted derivatives in the [1,2-d] series has been described in section 5.1.2.[103]

ELECTROPHILIC ATTACK AT SIDE-CHAIN SUBSTITUENTS. Routine procedures involving electrophilic attack at substituent positions in naphthimidazoles are shown in Table 5.15. In some cases, the initial electrophilic reaction is followed by cyclization, as depicted in transformations **5.56 → 5.57**[50] and **5.58 → 5.59**.[136]

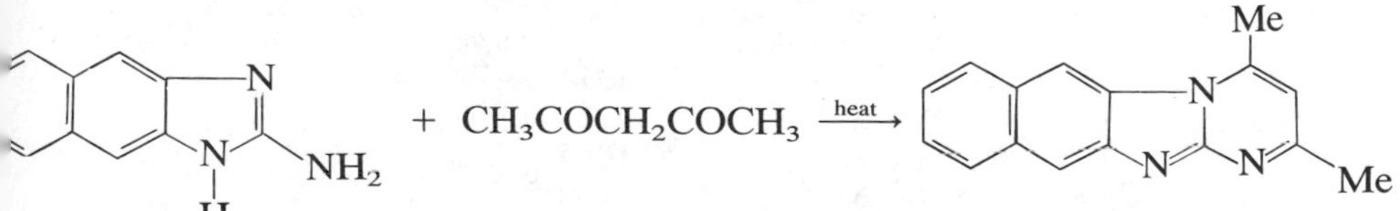

(5.56)   + CH$_3$COCH$_2$COCH$_3$ $\xrightarrow{\text{heat}}$ (5.57)

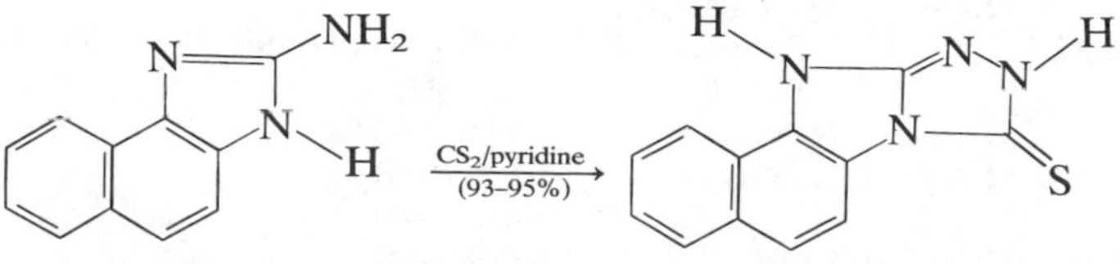

(5.58) $\xrightarrow[\text{(93–95%)}]{\text{CS}_2/\text{pyridine}}$ (5.59)

Reactions with Nucleophilic Reagents

NUCLEOPHILIC SUBSTITUTION IN THE CARBOCYCLIC RING. There is only one example of a nucleophilic aromatic substitution in this category: the 3-chloro substituent in the naphth[1,2-d]imidazole-2,5-dione derivative (5.60) is displaced by azide ion to give the azido compound (5.61).[86]

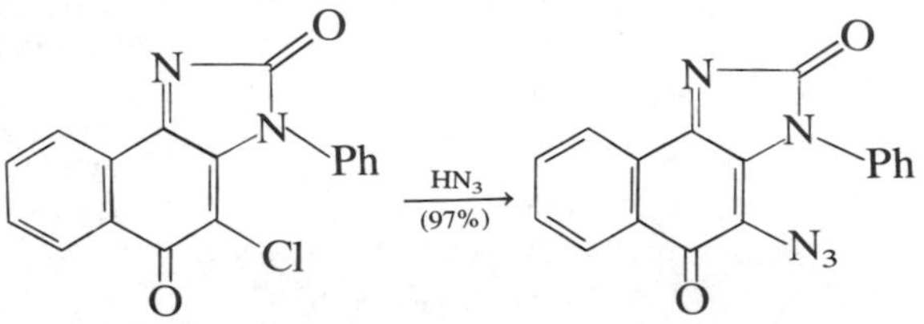

(5.60)   **(5.61)**

NUCLEOPHILIC SUBSTITUTION IN THE IMIDAZOLE RING. Examples of the nucleophilic displacement of substituents in the imidazole ring of naphthimidazoles are shown in Table 5.16. These involve displacement of 2-chloro and 2-methylthio substituents by amines, and two examples of the Chichibabin procedure in which a 2-amino substituent is introduced directly into the imidazole ring. In the latter examples, it may be noted that one nitrogen of the imidazole ring is alkylated (cf. also the analogous requirement in benzimidazoles described in section 1.4.3 of Chapter 1). In a related

TABLE 5.16.   NUCLEOPHILIC SUBSTITUTION REACTIONS IN THE IMIDAZOLE RING OF NAPHTHIMIDAZOLES

Substituent displaced	Ring system	Nucleophile	Naphthimidazole product	Ref.
2-Cl	[1,2-d]	Pyrrolidine, piperidine, morpholine,	2-Pyrrolidinyl, 2-piperidinyl, 2-morpholinyl, (including 1-Me and 3-CH$_2$Ph analogs)	137
2-Cl	[1,2-d]	PhNH$_2$ Me$_2$NH	2-NHPh} -6,7,8,9-tetrahydro 2-NMe$_2$}	1
2-Cl	[1,2-d]	PhCH$_2$NH$_2$	2-NHCH$_2$Ph	138
2-Cl	[2,3-d]	N$_2$H$_4$	2-NHNH$_2$	139
2-Cl	[1,2-d]	(H$_2$N)$_2$CS	2-SH	140
2-SMe	[2,3-d]	PhCH$_2$NH$_2$, PhCH$_2$NHMe, PhNH$_2$	2-NHCH$_2$Ph, N(Me)CH$_2$Ph, NHPh	44
2-H	[1,2-d]	NaNH$_2$	2-NH$_2$-3-CH$_2$Ph	121
2-H	[1,2-d]	NaNH$_2$	1-Me-2-NH$_2$ 3-Me-2-NH$_2$	5
2-H	[1,2-d]	KOH	2-OH-3-Me	141

procedure, 3-methylnaphth[1,2-*d*]imidazole is transformed by heating it with potassium hydroxide into the imidazol-2-one derivative in 90% yield, thus providing a very effective alternative route to such derivatives[141] (cf. methods based on diamines described in Table 5.5).

Nucleophilic displacement of 2-methylthio and 2-chloro substituents by amines has been used to generate intermediates that undergo cyclization in reactions leading to condensed naphthimidazole derivatives (e.g., **5.62** → **5.63**,[119,142] **5.64** → **5.65**,[143] **5.66** → **5.67**,[142,144] **5.68** → **5.69**).[145]

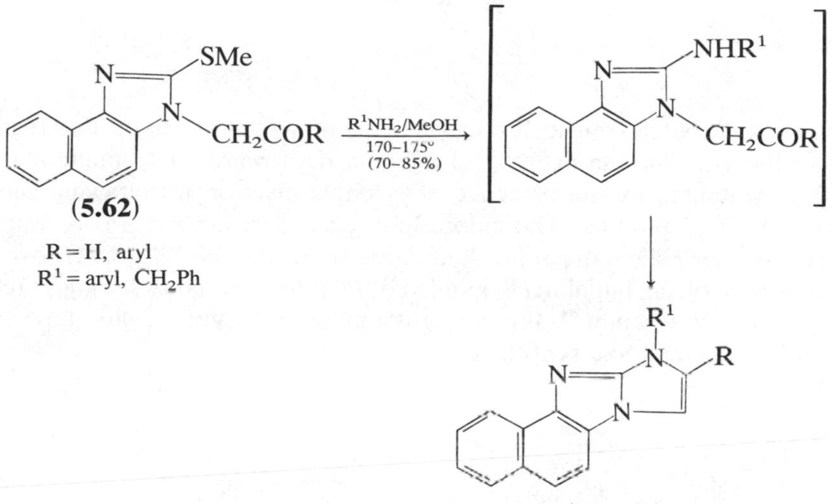

(5.62)

R = H, aryl
R¹ = aryl, CH₂Ph

(5.63)

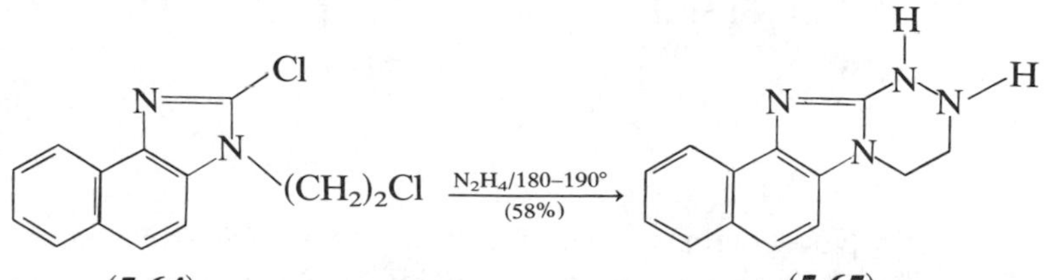

(5.64)

(5.65)

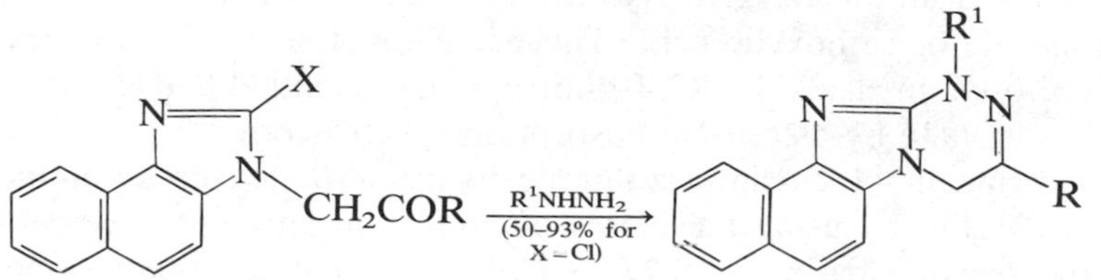

(5.66)

(5.67)

R,R¹ = alkyl, aryl, hetaryl
X = Cl,[144] SMe[142]

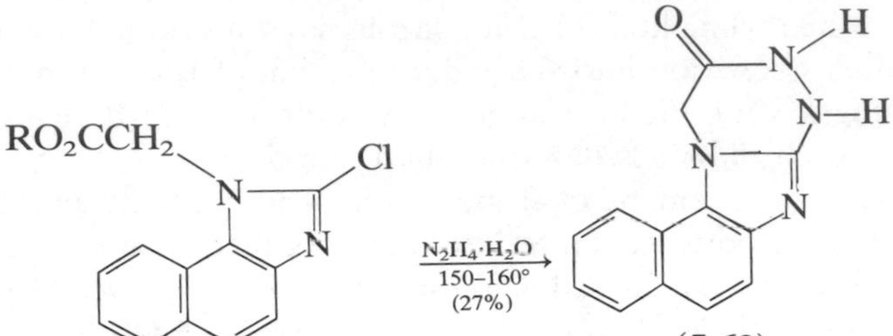

RO$_2$CCH$_2$—N⟨...⟩N, Cl

$\xrightarrow[\substack{150-160° \\ (27\%)}]{N_2H_4\cdot H_2O}$

(5.69)

(5.68)

R = Me, Et

ADDITION OF NUCLEOPHILIC REAGENTS TO THE IMIDAZOLE RING. The reactivity of the 2-position in naphth[2,3-d]imidazole toward nucleophilic reagents is accentuated by the presence of strongly electron-withdrawing substituents in the 1-position. The chlorofluoroalkyl derivative (5.70) is transformed by ethanolic potassium hydroxide into the acetal (5.71) by a mechanism involving initial nucleophilic addition to the imidazole ring[125] (cf. section 1.4.3 of Chapter 1 for a discussion of the value of this type of process in benzimidazole synthesis).

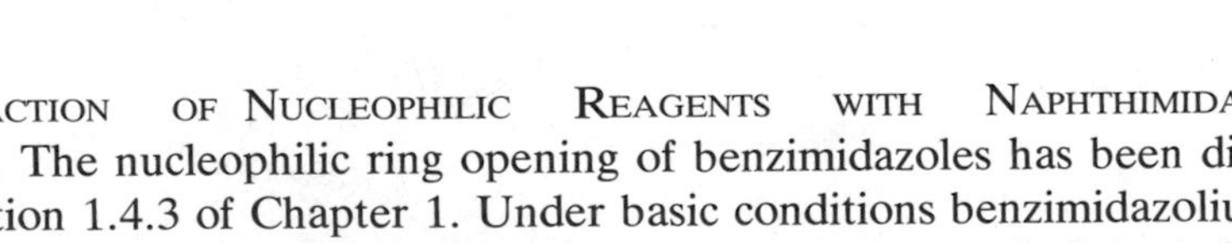

$\xrightarrow[(59\%)]{\text{KOH/EtOH/heat}}$

CF$_2$CFHCl   CH(OEt)$_2$

(5.70)   **(5.71)**

REACTION OF NUCLEOPHILIC REAGENTS WITH NAPHTHIMIDAZOLIUM SALTS. The nucleophilic ring opening of benzimidazoles has been discussed in section 1.4.3 of Chapter 1. Under basic conditions benzimidazolium salts are cleaved and ultimately give rise to N,N'-dialkyl-o-phenylene diamines and a molecule of carboxylic acid. The behavior of such ring opening has been monitored in 1-alkyl-3-(2,4-dinitrophenyl)naphth[1,2-d]imidazolium p-toluene sulfonate by electronic absorption spectroscopy.[146]

1,3-Dialkylnaphth[1,2-d]imidazolium salts are also cleaved by amide ion (5.72 → 5.73), but by using an excess of potassium amide it is possible to modify the reaction pathway (5.72 → 5.74);[147] a similar oxidative method using dimsyl potassium has been described in section 3.2.1 of Chapter 3 in relation to the synthesis of benzimidazolones.

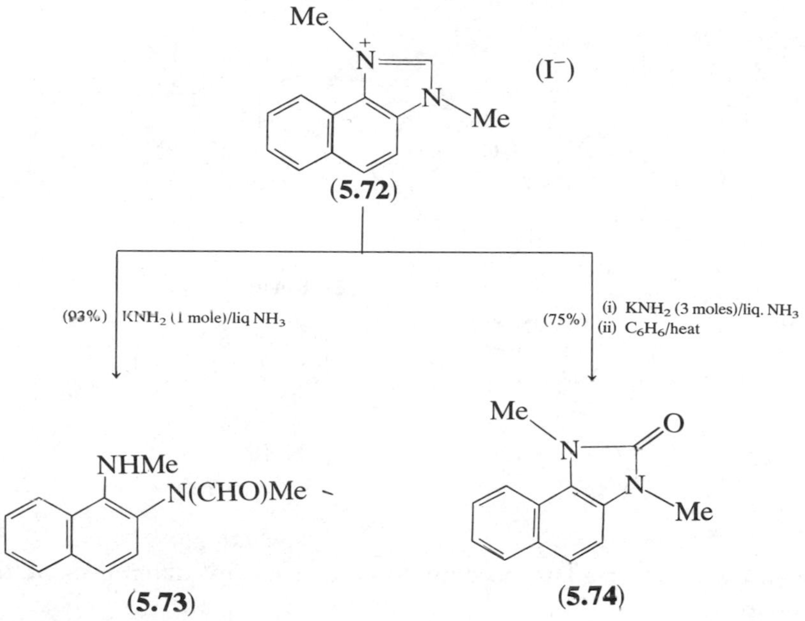

(5.72)

(93%) | KNH$_2$ (1 mole)/liq NH$_3$

(75%) | (i) KNH$_2$ (3 moles)/liq. NH$_3$
(ii) C$_6$H$_6$/heat

(5.73)

(5.74)

Thermal and Photochemical Reactions

THERMAL REACTIONS. 2-Methoxy-1-methylnaphth[2,3-d]imidazole-4,9-dione (5.75) is transformed thermally into a 2,4,9-trione derivative (5.76).[127] This type of reaction has analogy in the behavior of 2-alkoxybenzimidazoles (see eq. 38 in Chapter 3).

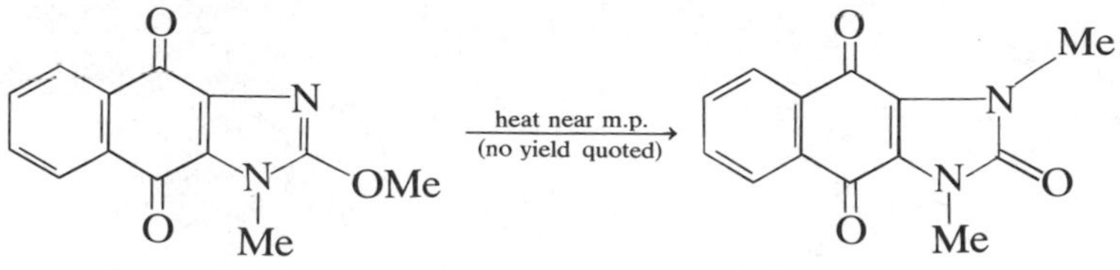

(5.75)

heat near m.p.
(no yield quoted)

(5.76)

PHOTOCHEMICAL REACTIONS. Ultraviolet irradiation of the 4,9-dione derivative (5.77) with amines causes substitution of the methoxy substituent (cf. 5.78a and b); in both cases the product of substitution at the 5-position predominates over that at the 8-position.[148] Reactions of this type may have synthetic potential since analogous methods of thermal substitution are very inefficient at moderate temperatures. Evaluation of π-electron densities by the PPP SCF-MO procedure indicates values of 0.808 and 0.817 for the 5- and 8-positions, respectively, of 5.77 in the excited state, and these small

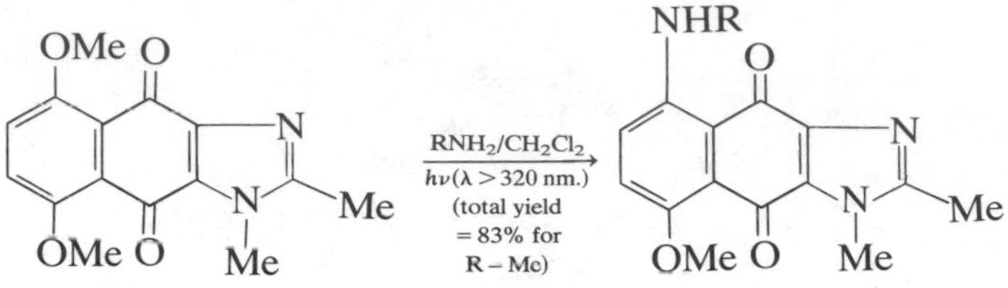

(5.77)   (5.78a)

Product ratios:

R	(5.78a)	(5.78b)
Me	1.75	1.00
cyclo-C$_6$H$_{11}$	1.80	1.00

+

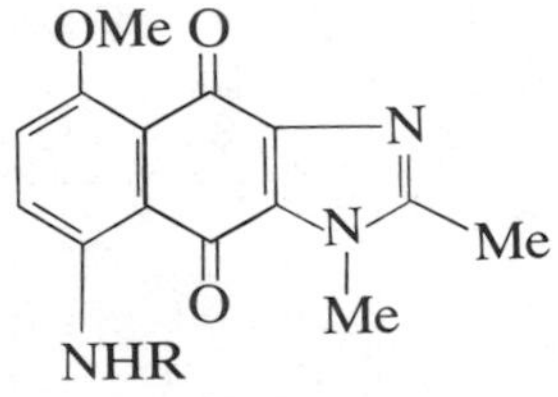

(5.78b)

differences are deemed to account for the reactivity differences at these positions.

Oxidation

Reaction conditions for effecting the conversion of naphthimidazoles into dione derivatives are illustrated by reactions $5.79 \rightarrow 5.80$[125] and $5.81 \rightarrow 5.82$.[149] Oxidation of naphth[1,2-d]imidazole by potassium dichromate causes cleavage of a carbocyclic ring, but the product is an imidazole derivative (5.83)[149] and not a benzimidazole derivative (5.84) as had been suggested earlier.[150]

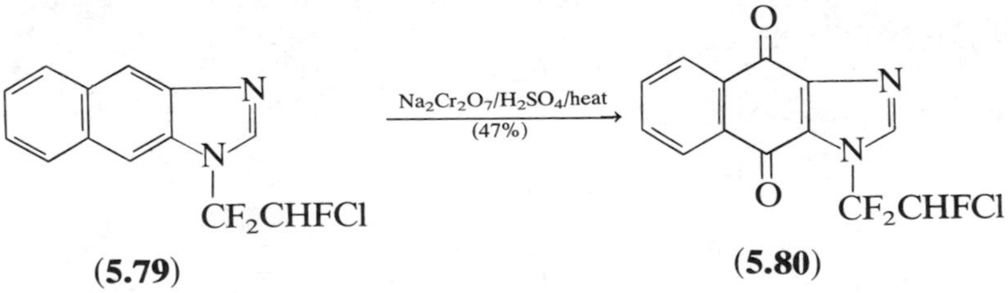

(5.79)   (5.80)

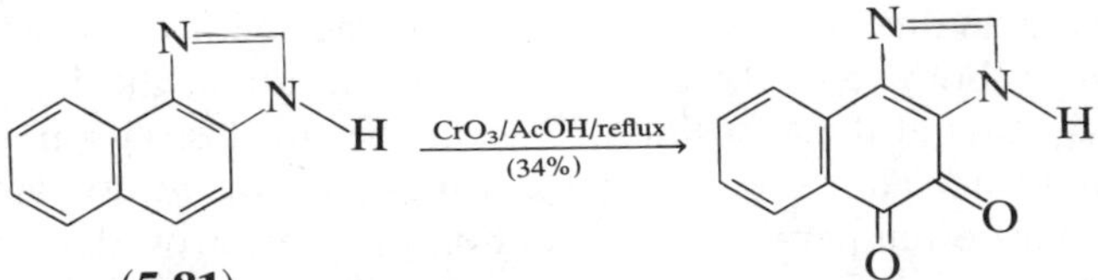

(5.81)   (5.82)

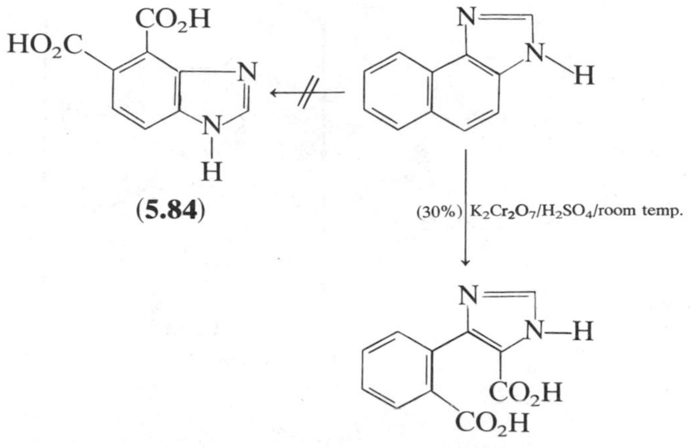

(5.84) ←//← (30%) $K_2Cr_2O_7/H_2SO_4$/room temp.

(5.83)

Procedures for oxidation of naphthimidazole substituent groups are collected in Table 5.17. The methods illustrated have precedent in the chemistry of benzimidazoles, and the interesting conversion of 1-benzyl 2-amino-napth[1,2-*d*]imidazole into an azo and a 2-nitro derivative has been discussed in detail in section 1.4.6 of Chapter 1 (cf. also Table 1.44).

TABLE 5.17. OXIDATION OF NAPHTHIMIDAZOLE SUBSTITUENTS

Substituent(s) oxidized	Ring system	Reaction conditions	Naphthimidazole product	Ref.
2-Me	[1,2-*d*]	SeO_2/Ac_2O/reflux	2-CHO	151
4,5-$(OH)_2$[a]	[1,2-*d*]	HNO_3	4,5-Dione	152
2-CH(OH)Me	[1,2-*d*] [2,3-*d*]	CrO_3/AcOH/100°	2-COMe	11
2-$NHNH_2$[b]	[1,2-*d*]	Stand in EtOH one week	2-NHN=C(Me)N=N-naphth-[1,2-*d*]imidazol-2-yl	153
2-NH_2[c]	[1,2-*d*]	Na/air	RN=NR (75%) RNO_2 (21%) (R = naphth[1,2-*d*]imidazol-2-yl)	154

[a] The dihydroxy derivative is prepared from the appropriate diamine and formic acid, and is oxidized *in situ*.
[b] 1- and 3-Methyl analogs are also converted to formazan derivatives by this method.
[c] A 1-benzyl substituent is cleaved during this reaction.

Oxidation of the hydrazine derivative **(5.85)** with lead dioxide in chloroform gives rise to a red diamagnetic crystalline product.[155] A structure involving a condensed naphthimidazolium cation **(5.86)** is speculatively suggested,[155] and it would be of interest to investigate this material by X-ray crystallographic analysis.

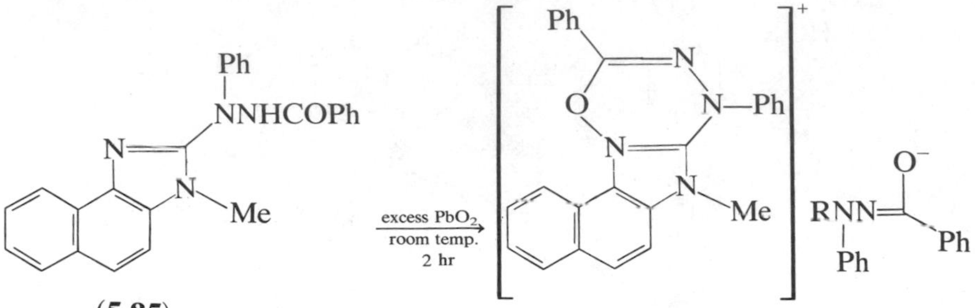

(5.85)

excess PbO₂
room temp.
2 hr

(5.86)

R = 3-methylnaphth[2,3-d]imidazol-2-yl

Reduction

Controlled catalytic hydrogenation of 9-aminonaphth[1,2-d]imidazole causes reduction of the ring remote from the imidazole moiety; using the same catalyst and more forcing conditions an octahydro derivative is produced (5.87 → 5.88 → 5.89).[156] Reductions of type 5.87 → 5.88 can also be achieved by catalytic hydrogenation using platinum oxide in acetic acid,[157]

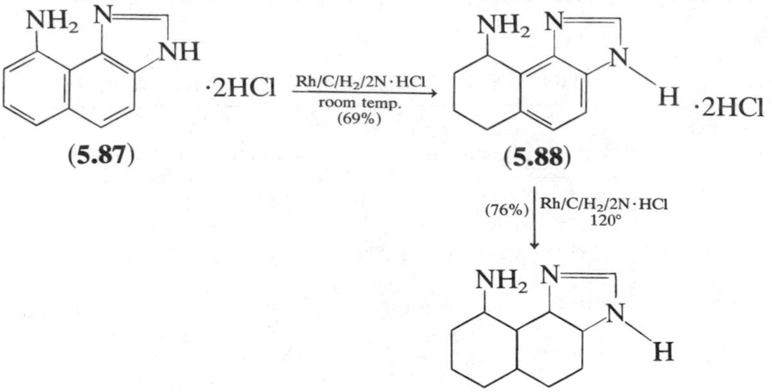

(5.87)   ·2HCl   Rh/C/H₂/2N·HCl
                          room temp.
                          (69%)

(5.88)   ·2HCl

(76%) | Rh/C/H₂/2N·HCl
        120°

(5.89)

TABLE 5.18.   REDUCTION OF NAPHTH[1,2-d]IMIDAZOLE SUBSTITUENTS

Substituent reduced	Reaction conditions	Reaction product	Ref.
1-OH[a]	H₂/Raney Ni/MeOH	Naphth[1,2-d]imidazole	87
4-NO₂	H₂/Pd on C/EtOH	2-Me-3-OH-4-NH₂-5-NHAc[a]	61
3-C(Me)=CH₂	H₂/Pt/EtOH	2-OH-3-i-Pr[a]	100
2-COC₆H₄NO₂-p	N₂H₄/Raney Ni	2-COC₆H₄NH₂-p	78
3-CH₂C(=NNH₂)Me	NaOH/aq. EtOH/180°	2-OH-3-C₃H₇[a,b]	118

[a] Designated for convenience as hydroxy derivatives rather than N-oxide or naphth[1,2-d]imidazol-2-one derivatives.
[b] The starting material contains a 2-chloro substituent.

and this method has also been used to prepare 5,6,7,8-tetrahydronaphth-[2,3-d]imidazole.[157]

Routine methods that have been used for reduction of naphth[1,2-d]-imidazole substituents are collected in Table 5.18.

Systematic Survey of Naphthimidazole Derivatives

TABLE 5.19.   UNSUBSTITUTED- AND ALKYLNAPHTH[1,2-d]imidazoles[a]

Substituent(s)	Melting point (°C)	Yield (%)	Method of preparation	Ref.
Unsubstituted	174	89	D(H₂SO₄)/HCO₂H/10% HCl/heat	149
Unsubstituted	183.0–183.5	—	Naphth[1,2-d]imidazole-1-oxide/ H₂/Raney Ni	87
1-Et	Picrate, 205–208	—	Desulfurization of naphth[1,2-d]-imidazo[1,2-b]thiazoline by Raney Ni	131
2-Me	168	—	D/McC(OEt)₃/160°/4 hr	51
2-Me	170.5–171.5	91	1-NH₂-2-NHCOMe-C₁₀H₆/225°/20 min	73a
2-Me	>320	—	1,2-(NHAc)₂–C₁₀H₆/1 N HCl/heat	60
2-Me	Picrate, 253–254	—	Ring contraction of a naphtho-diazepin derivative	99
2-Me	168–169	94	Ring contraction of a naphtho-diazepin derivative	100
2-Me-4-Br	233–234	59	1-NO₂-2-NH—COCH₃-3-Br—C₁₀H₅/ H₂/Raney Ni	75
2-Me-7-Br	216–218	—	1-NH₂-2NHAc-6-Br—C₁₀H₅/heat	75
2-Me-7-NH₂	126.5–128	87	1,6-(NH₂)₂-2-NHAc—C₁₀H₅/heat	75
2-Me-9-NH₂	205–207	45	1,8-(NO₂)₂-2-NHAc—C₁₀H₅; (i) Fe/EtOH, (ii) SnCl₂/HCl	75
2-Me-7-Cl	235.5–236.5	43	2-Me-7-NH₂ derivative/Sandmeyer reaction	75
2-Me-6-Br	242	—	1-NH₂-2-NO₂-5-Br—C₁₀H₅/ SnCl₂/AcOH	59
2,3-Me₂	142–143	—	D/Ac₂O/4N·HCl	10
2-Me-5-OH	>360	95	1-NHAc-2-NH₂-4-OAc—C₁₀H₅/ Ac₂O/HCl/reflux	62
2-Me-5-NHAc	258	42	1,4-(NHAc)₂-2-NH₂—C₁₀H₅/ AcOH/reflux	61
2-i-Pr(HCl)	239–240	—	D/butyraldehyde/(Weidenhagen procedure)	33
2-Hexyl (HCl)	199–202	—	D/heptaldehyde (Weidenhagen procedure)	33
3-Me	137–138	70[b]	(i) Alkylation by Me₂SO₄/80°, or (ii) Alkylation by MeI/KOH/EtOH, or (iii) D/HCO₂H/reflux	5
1-CH₂Ph-2-CH₂Cl	121–122	60	D/carboxylic acid	8
2-CH(Ph)C₆H₄OH-o	294–295	—	D/lactone derivative	55
2-CH₂-ferrocenyl	126–128	—	D/carboxylic acid	16
3-CH₂Ph	170	—	Alkylation by Me₂PhN⁺(CH₂Ph)Cl⁻	121

[a] D refers to the appropriate naphthylene diamine derivative.
[b] Yield by method (ii).

TABLE 5.20. UNSUBSTITUTED- AND ALKYLNAPHTH[2,3-*d*]IMIDAZOLES[a]

Substituent(s)	Melting point (°C)	Yield (%)	Method of preparation	Ref.
Unsubstituted	218	—	D/HCO$_2$H	18
Unsubstituted	221; mono Ac, 172	—	D/HCO$_2$H/reflux 1 hr	19
1-CF$_2$CHFCl	104–105	57	1-Unsubstituted derivative/ CF$_2$=CFCl/THF/14 hr/150°	125
2-Me	285	—	D/AcOH	18
2-Me	286	—	D/AcOH	19
2-Me	283–284	—	Ring contraction of a naphtho-diazepin derivative	99
2-Me	286	83	2-N=C(Me)CH$_2$COPh-3-NH$_2$—C$_{10}$H$_6$/ heat at 220°	56
2-Et	250	—	D/EtCO$_2$H	44
5,6-Me$_2$	263–265	90	D/HCO$_2$H	17

[a] D refers to the appropriate naphthylene diamine derivative.

TABLE 5.21. ALKENYL DERIVATIVES OF NAPHTH[1,2-*d*]IMIDAZOLE[a]

Substituent(s)	Melting point (°C)	Yield(%)	Method of preparation	Ref.
1-CH=CH$_2$	30–38; Picrate 210–211 (b.p. 224–227/5 mm)	—	1-vinylation by C$_2$H$_2$/THF/ KOH/ZnO/pyridine/130°	123
2-CH=CHPh	Sulfate, 221–223; free base, 104–106	—	D(sulfate)/PhCH=CHCHO/ EtOH/heat	35
2-CH=CH-(5-X-2-furyl)			2-Me derivative/furfural derivative/boric acid/	37
(X = H)	195–196	31	195–200°/10 h	
(X = Br)	208–209	26	2-Me derivative/furfural derivative/boric acid/ 195–200°/10 hr	
(X = NO$_2$)	226–227	36	2-Me derivative/furfural derivative/boric acid/ 195–200°/10 hr	
3-C(Me)=CH$_2$-2-OH	198	—	D/CH$_3$COCH$_2$CO$_2$Et	100

[a] D refers to the appropriate naphthylene diamine derivative.

TABLE 5.22. ARYL DERIVATIVES OF NAPHTH[1,2-d]IMIDAZOLES[a]

Substituent(s)	Melting point (°C)	Yield (%)	Method of preparation	Ref.
1-Ph-5-OMe	138	—	D/HCO$_2$H	6
2-Ph	Sulfate 190; free base, 217–218		D(sulfate)/PhCHO/EtOH/heat	35
2-Ph	218	57	D/PhCH=$\overset{+}{N}(\overset{-}{O})C_6H_4NMe_2$-$p$/AcOH/heat	57
2-Ph	217–218	80–90	1-NH$_2$-2-NHCOPh-C$_{10}$H$_6$/225°/20 min	73a
2-Ph	217	—	1-NH$_2$-2-NHCOPh-C$_{10}$H$_6$/4 NHCl/ reflux	69
2-Ph	218	75	2-N=C(Ph)NHOH-C$_{10}$H$_7$/dry C$_6$H$_6$/ PhSO$_2$Cl/pyridine <10°	89
2-Ph	215–216	24, 64	N-(1- or 2-naphthyl)benzamidine/ t-BuOCl then Na/EtOH	90
2-Ph	122(hydrate), 218(anhydrous), 310(HCl)	—	1-NO$_2$-2-NHCOPh-C$_{10}$H$_6$/Zn/HCl	73c
2-Ph	215 hydrochloride, 314–315, hydrate of free base, 118–120	25	Pyrolysis of naphthyl tetrazoles (see section 5.1.1)	96
2-Ph	296	—	1-NO$_2$-2-NHCOPh-C$_{10}$H$_6$/ Zn/HCl/EtOH/heat	73b
2-C$_6$H$_4$CO$_2$H-o	>300 304–305	—	D/phthalic anhydride	29
2-C$_6$H$_4$OH	281.5–283	46	Acid-catalyzed ring opening of a condensed benzotriazine derivative	94
2-C$_6$H$_4$NH$_2$-o	210–211	42	D/anthranilic acid	104
2-Ph-3-CH$_2$Ph	120	—	1-NH$_2$-2-N=CHPh-C$_{10}$H$_6$/ (i) PhCHO, (ii) concn. HCl	85
2,3-Ph$_2$-5-OBz	181	90	1-NHCOPh-2-NHPh-3-OBz-C$_{10}$H$_5$/AcOH/heat	70

[a] D refers to the appropriate naphthylamine derivative.

TABLE 5.23. ARYL DERIVATIVES OF NAPHTH[2,3-d]IMIDAZOLE[a]

Substituent(s)	Melting point (°C)	Yield (%)	Method of preparation	Ref.
2-Ph	122–124 (hydrate) 260 (picrate)	—	—	161
2-$C_6H_4N(CH_2CH_2Cl)_2$-p	140	78	D/aldehyde	36
2-$C_6H_4SO_3H$-o	436–437	—	D/(o-$HO_3SC_6H_4CO)_2$O/heat	32
2-[2-OH-5-R-C_6H_3]	R = H; 340–346	51	D/aldehyde bisulfite addition compound/DMF	24
	R = Ph; 328–330	13		
	R = OH; >380	33		
	R = OMe; 338–342	54		
2-(2-OH-1-naphthyl)	337–339	28	D/aldehyde bisulfite addition compound/DMF	24
2-(2-OH-3-naphthyl)	340–342	12	D/carboxylic acid	24
5,6,7,8-Ph_4-1,2-Me_2	340	82	5,6-dehydrobenzimidazole derivative/tetracyclone	95

[a] D refers to the appropriate arylamine diamine derivative.

TABLE 5.24. HETARYL DERIVATIVES OF NAPHTH[1,2-d]IMIDAZOLE[a,b]

Substituent(s)	Melting point (°C)	Yield (%)	Method of preparation	Ref.
2-(5-X-2-Furyl)				
(X = H)	121–122	49	D/aldehyde derivative/$PhNO_2$/ 170–180°	37
(X = Br)	145–146	52	D/aldehyde derivative/$PhNO_2$/ 170–180°	52
(X = NO_2)	253–254	43	D/aldehyde derivative/$PhNO_2$/ 170–180°	
2-(2,3-Dimethyl-1-phenyl-5-oxo-3-pyrazolin-4-yl)	270	—	D/aldehyde	38
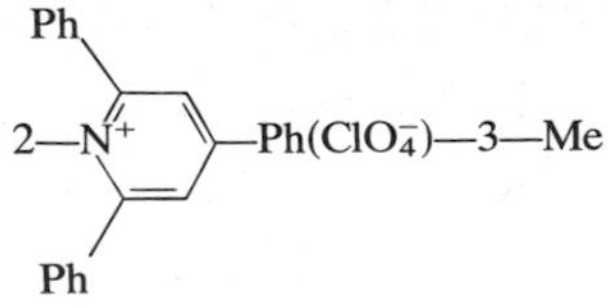 2—N$^+$(Ph)—Ph(ClO$_4^-$)—3—Me (Ph substituents)	305	85	2-NH_2-3-Me derivative/2,4,6-triphenylpyrilium perchlorate/DMF/reflux	135
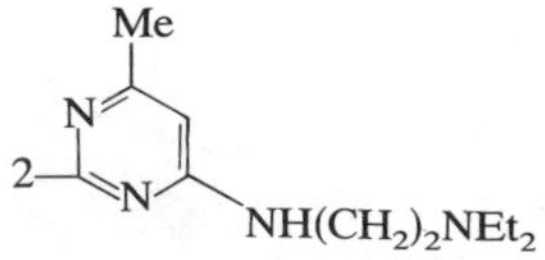 2- (4-Me-pyrimidinyl)-NH(CH$_2$)$_2$NEt$_2$	226	—	4-Chloropyrimidine derivative/ Et$_2$N(CH$_2$)$_2$NH$_2$	159
2-(2-Phenyl-4-quinolinyl)	—	—	D/carboxylic acid	15

[a] 2-(5-Phenyl-2-thienyl)naphth[2,3-d]imidazole has been prepared by reaction of naphthalene-2,3-diamine with the appropriate carboxylic acid derivative (see Ref. 26).
[b] D refers to the appropriate naphthylamine derivative.

TABLE 5.25. AMINO DERIVATIVES OF NAPHTH[1,2-d]IMIDAZOLE[a]

Substituent(s)		Melting point (°C)	Yield (%)	Method of preparation	Ref.
1-NHPh-2-Me		237	—	1-N$_2$Ph-2-NH$_2$-C$_{10}$H$_6$/CH$_3$COCO$_2$H	77
1-NHR-2-R^1					
R	R^1			2-N=CHR^1-C$_{10}$H$_7$/RN$_2^+$BF$_4^-$	88
C$_6$H$_4$OMe-p	C$_6$H$_4$OMe-p	177	80		
C$_6$H$_4$OMe-p	Ph	185	88		
Ph	Ph	199	62		
C$_6$H$_4$Br-p	C$_6$H$_4$NO$_2$-p	249	95		
C$_6$H$_4$NO$_2$-p	C$_6$H$_4$NO$_2$-p	304	99		
2-NH$_2$		212 5(dec.)	—	D/BrCN/H$_2$O	48
2-NH$_2$		218-221	—	D/CNBr	49
2-NH$_2$-3-Me		252-253	40	3-Me derivative/NaNH$_2$/PhNMe$_2$/140°	5
		p-Nitrobenzylidene derivative, 197			
2-NH$_2$-1-Et		202-203;	—	1-Et derivative/NaNH$_2$/PhNMe$_2$	5
		p-Nitrobenzylidene derivative, 211-212			
2-NH$_2$-3-CH$_2$Ph		256;	—	3-CH$_2$Ph derivative/NaNH$_2$/PhNMe$_2$/110°	121
		p-Nitrobenzylidene derivative, 271			

523

TABLE 5.25 (Continued)

Substituent(s) R	R¹	Melting point (°C)	Yield (%)	Method of preparation	Ref.
2-NHR-3-CH₂CH(OH)R¹					
H	H	240–242 (dec.)	65	2-Cl derivative/primary amine/MeOH/heat	199
CH₃	H	216–218 (dec.)	48	2-Cl derivative/primary amine/MeOH/heat	199
C₂H₅	C₆H₅	167–168 (dec.)	70	2-Cl derivative/primary amine/MeOH/heat	199
C₃H₇	C₆H₅	178–179 (dec.)	40	2-Cl derivative/primary amine/MeOH/heat	199
C₄H₉	C₆H₅	200–201 (dec.)	50	2-Cl derivative/primary amine/MeOH/heat	199
iso-C₄H₉	C₆H₅	203–205 (dec.)	65	2-Cl derivative/primary amine/MeOH/heat	199
C₆H₅	H	255–257 (dec.)	70–73	2-Cl derivative/primary amine/MeOH/heat	199
C₆H₅	C₆H₅	273–275 (dec.)	28	2-Cl derivative/primary amine/MeOH/heat	199
m-CH₃C₆H₄	H	220–221 (dec.)	62	2-Cl derivative/primary amine/MeOH/heat	199
p-CH₃C₆H₄	H	234–236 (dec.)	60	2-Cl derivative/primary amine/MeOH/heat	199
p-CH₃OC₆H₄	H	195–196 (dec.)	64	2-Cl derivative/primary amine/MeOH/heat	199
p-CH₃OC₆H₄	C₆H₅	248–250 (dec.)	30	2-Cl derivative/primary amine/MeOH/heat	199
p-C₂H₅OC₆H₄	H	231–232 (dec.)	60	2-Cl derivative/primary amine/MeOH/heat	199
m-ClC₆H₄	H	226–227 (dec.)	55	2-Cl derivative/primary amine/MeOH/heat	199
p-ClC₆H₄	H	198–199 (dec.)	55–60	2-Cl derivative/primary amine/MeOH/heat	199
m-BrC₆H₄	H	230–232 (dec.)	53	2-Cl derivative/primary amine/MeOH/heat	199
p-BrC₆H₄	H	237–238 (dec.)	50	2-Cl derivative/primary amine/MeOH/heat	199
C₆H₅CH₂	H	173–175 (dec.)	30	2-Cl derivative/primary amine/MeOH/heat	199
C₁₀H₇	H	288–290 (dec.)	50–85	2-Cl derivative/primary amine/MeOH/heat	199
2-Pyrrolidinyl		256–258; HCl salt, 294–296	—	2-Cl derivative/amine	137
2-Piperidino		230–232; HCl salt, 247–249	—	2-Cl derivative/amine	137
2-Morpholino		119–122; HCl salt 291–293	—	2-Cl derivative/amine	137

Compound	mp	Yield (%)	Method/derivative	Ref
2-Pyrrolidino-3-Me	140–141; HCl salt, 275–276	—	2-Cl derivative/amine	137
2-Pyrrolidino-1-Me	HCl salt, 230–232	—	2-Cl derivative/amine	137
2-Pyrrolidino-3-CH$_2$Ph	187–189; HCl salt, 273–277	—	2-Cl derivative/amine	137
2-Piperidino-3-Me	131–133; HCl salt, 216–217	—	2-Cl derivative/amine	137
2-Piperidino-1-Me	HCl salt, 223–225	—	2-Cl derivative/amine	137
2-Piperidino-3-CH$_2$Ph	128–130; HCl salt 216–220	—	2-Cl derivative/amine	137
2-Morpholino-1-Me	HCl salt, 291–292	—	2-Cl derivative/amine	137
2-Morpholino-3-CH$_2$Ph	156–158; HCl salt, 243–246	—	2-Cl derivative/amine	137
2-CH$_2$NMe$_2$-1-CH$_2$Ph	108	70	2-CH$_2$Cl-1-CH$_2$Ph derivative/Me$_2$NH/50°	8
2-CH$_2$(1-morpholinyl)	241–243	10	N-Acyl o-naphthylene diamine derivative/reflux in xylene	10
2-CH$_2$(-1-piperidyl)-3-Me	134.8–135.0	97	2-CH$_2$Cl-3-Me derivative/piperidine	10
2-CH$_2$(1-morpholinyl)-3-Me	134.0–134.4	—	2-CH$_2$Cl-3-Me derivative/morpholine	10
2-CH(OH)CH$_2$(1-piperidyl)-3-Me	149.2–149.6	95	2-CH(OH)CH$_2$Cl-3-Me derivative/piperidine	10
2-CH(OH)CH$_2$(1-morpholinyl)-3-Me	168.4–169.0	—	2-CH(OH)CH$_2$Cl-3-Me derivative/morpholine	10
2-(CH$_2$)$_2$NH$_2$	di HCl, 256–258; free base, 88–89	71	D/β-alanine	9
3-NHC$_6$H$_4$OMe-p-2-C$_6$H$_4$OMe-p	195	30	1-N=CHC$_6$H$_4$OMe-p-C$_{10}$H$_7$/p-MeOC$_6$H$_4$N$_2^+$BF$_4^-$	88
3-CH$_2$N(CH$_2$CH$_2$Cl)$_2$	165	34	Naphth[1,2-d]imidazole/CH$_2$O/HN-(CH$_2$CH$_2$Cl)$_2$/EtOH	122
9-NH$_2$	di HCl, 299–300; picrate, 209–210	27	1,3-(NO$_2$)$_2$-2-NHCHO-C$_{10}$H$_5$/i) H$_2$/Raney Ni, ii) HCl (di HCl)	156

[a] D refers to the appropriate naphthylene diamine derivative.

TABLE 5.26. AMINO DERIVATIVES OF NAPHTH[2,3-*d*]IMIDAZOLE[a]

Substituent(s)	Melting point (°C)	Yield (%)	Method of preparation	Ref.
2-NH$_2$	301	—	D/CNBr	44
2-NH$_2$	297–299	—	D/CNBr	49
2-NH$_2$	291–293	81	D/BrCN/H$_2$O/O-5°	50
2-NHCH$_2$Ph	246	—	2-SMe derivative/PhCH$_2$NH$_2$/180°	44
2-N(Me)CH$_2$Ph	235	—	2-SMe derivative/PhCH$_2$NIIMe/heat	44
2-NHPh	284–285	—	2-SMe derivative/PhNH$_2$/heat	44
2-(CH$_2$)$_2$NH$_2$	di HCl, 292–293; free base, 197(dec.)	68 (di HCl)	D/β-alanine	9

[a] D refers to the appropriate naphthylene diamine derivative.

TABLE 5.27. NAPHTH[1,2-*d*]- AND [2,3-*d*]IMIDAZOLES WITH NITROGEN-CONTAINING
SUBSTITUENTS: UREAS, HYDRAZINES, HYDRAZONES, HYDRAZIDES, NIT
AND IMINO COMPOUNDS[a]

Substituent	Ring system	Melting point (°C)	Yield (%)	Method of preparation	R
2-NHCONHPh	[2,3-*d*]	362–364	—	2-NH$_2$/PhNCO/THF	4
2-NHCONH-1-C$_{10}$H$_7$	[2,3-*d*]	330 (dec.)	—	2-NH$_2$/1-C$_{10}$H$_7$NCO/THF	4
2-NHCONHC$_6$H$_4$F-*p*	[2,3-*d*]	364–365	—	2-NH$_2$/*p*-FC$_6$H$_4$NCO/THF	4
2-NHCONHC$_6$H$_{11}$	[2,3-*d*]	252–254	—	2-NH$_2$/C$_6$H$_{11}$NCO/THF	4
2-NHCONHPh	[1,2-*d*]	348–349	—	D/PhNCO/THF	4
2-NHCONH-1-C$_{10}$H$_7$	[1,2-*d*]	331–332	—	D/1-C$_{10}$H$_7$NCO/THF	4
2-NHCONHC$_6$H$_4$Cl-*p*	[1,2-*d*]	354–355	—	D/*p*-Cl-C$_6$H$_4$NCO/THF	4
2-NHCONHC$_6$H$_4$F-*p*	[1,2-*d*]	349–350	—	D/*p*-FC$_6$H$_4$NCO/THF	4
2-NHNH$_2$	[2,3-*d*]	245	—	3-Cl derivative/N$_2$H$_4$/< 80°	13
2-NHN=C(Ph)CO$_2$H	[1,2-*d*]	375 (dec.)	78	2-NHNH$_2$ derivative/ PhCOCO$_2$H/heat	13
2-N(Ph)NHCOPh	[1,2-*d*]	255–257	70	2-N(Ph)NH$_2$ derivative/PhCOCl	15.
1-CH$_2$CONHNH$_2$-2-Cl	[1,2-*d*]	255–256	85	1-CH$_2$CO$_2$-alk-2-Cl derivative/ N$_2$H$_4$·H$_2$O/20–78°	14:
2-NO$_2$	[1,2-*d*]	170	21	1-CH$_2$Ph-2-NH$_2$ derivative/ Na(4 g atom)	15-
2-C(Ar)=NC$_6$H$_{11}$ (Ar = C$_6$H$_4$OMe-*p*)	[1,2-*d*]	130–131	—	1-NO-2-NH$_2$ naphthalene/ Ar¹CH——CHAr/C$_6$H$_6$ reflux	
(Ar = C$_6$H$_4$Me-*p*)		134–135	—		8
(Ar = C$_6$H$_5$)		110–112	—		
(Ar = C$_6$H$_4$Br-*p*)		149–150.5	—	(Ar¹ = aryl)	

[a] D refers to the appropriate naphthylamine derivative.

526

TABLE 5.28. HALOGENO DERIVATIVES OF NAPHTH[1,2-d]IMIDAZOLE[a,b]

Substituent(s)	Melting point (°C)	Yield (%)	Method of preparation	Ref.
2-Cl	210–212; HCl salt, >350	—	2-OH derivative/POCl₃	42
2-Cl-3-Me[c]	154–155	65	2-Cl derivative/MeI/NaOH	42
2-Cl-3-CH₂Ph	134–135	70	2-Cl derivative/PhCH₂Cl/NaOH	42
2-Cl-3-CH₂COMe	133–134	60	2-Cl derivative/BrCH₂COMe/ NaOEt/EtOH	118
2-Cl-3-CH₂COt-Bu	155–156	80	Alkylation as above	118
2-Cl-3-CH₂COPh	200–201	80	Alkylation as above	118
2-Cl-3-CH₂COC₆H₄Me-p	205–206	65	Alkylation as above	118
2-Cl-3-CH₂COC₆H₄OMe-p	223–224	78	Alkylation as above	118
2-Cl-3-CH₂COC₆H₄Cl-p	238–239	70	Alkylation as above	118
2-Cl-3-CH₂COC₆H₄Br-p	231–232	75	Alkylation as above	118
2-Cl-3CH₂CH₂OH	186–187	85	2-Cl derivative/NaOEt/BrCH₂CH₂OH	118
2-Cl-3-CH₂CH(OH)Ph	195–196	62	Alkylation as above	118
2-Cl-3-CH₂CH(OH)C₆H₄- NO₂-p	226–227	71	Alkylation as above	118
2-Cl-3-CH₂CH₂Cl	115–116	35	2-Cl derivative/ClCH₂CH₂Cl/aq. NaOH	118
2-Cl-3-CH₂Br	106–107	40	Alkylation as above	118
2-Cl-3-CH₂CH(Cl)Ph	141–142	60	Alkylation as above	118
2-Cl-3-CH₂CO₂Et	105–106	75	2-Cl derivative/NaOEt/BrCH₂CO₂Et	118
2-CH₂Cl 3-Me	160–162 (dec.)	34	D/ClCH₂CO₂H	10
2-Br	272 [by method (i)] 274 [by method (ii)]	— —	(i) Naphth[1,2-d]imidazole/Br₂/KOH or (ii) D/HCO₂H/reflux	115

2-Trifluoromethylnaphth[2,3-d]imidazole has been prepared from naphthalene 2,3-diamine and CF₃CO₂H (see Ref. 23).

D refers to the appropriate naphthylene diamine derivative.

The 1-Me-2-Cl isomer (3%) and 1,3-dimethylnaphth[1,2-d]imidazol-2-one (20%) are also formed.

TABLE 5.29. HYDROXY DERIVATIVES OF NAPHTH[1,2-d]IMIDAZOLES[a]

Substituent(s)	Melting point (°C)	Yield (%)	Method of preparation	Ref.
2-CH$_2$OH	253–255	45	D/HOCH$_2$CO$_2$H/150°/6 N HCl	10
2-CH$_2$OH	253–255	40	D/HOCH$_2$CO$_2$H/4 N HCl/heat	167
2-CH(OH)Me	221	—	D/MeCH(OH)CO$_2$H	11
2-CH(OH)CH$_3$-3-Me	184.1–184.6	96	D/lactic acid/HCl	10
2-CH(OH)CH$_2$Cl-3-Me	168.9–169.0	8–10	D/β-chlorolactic acid	10
2-CH(OH)C$_6$H$_4$R				
(R = H)	229–230	50		
(R = 2-Cl)	212	35		
(R = 3-Cl)	91	39	D/carboxylic acid/4 N HCl/	170
(R = 4-Cl)	96–98	31	heat	
2-CH(OH)C$_6$H$_3$Cl$_2$-2′,4′	315	46	D/carboxylic acid/4 N HCl/ heat	170
2-(L-arabo-tetrahydroxybutyl)	244–245	62	D/Ca arabonate/HCl/H$_3$PO$_4$/ EtOH/heat	165
9-OH	230–233 (dec.)	71	9-OMe derivative/HBr/room temperature	160

[a] D refers to the appropriate naphthylamine diamine derivative.

TABLE 5.30. HYDROXY DERIVATIVES OF NAPHTH[2,3-d]IMIDAZOLE[a]

Substituent(s)	Melting point (°C)	Yield (%)	Method of preparation	Ref.
2-CH$_2$OH	224 (dec.)	47	D/HOCH$_2$CO$_2$H/4 N HCl/heat	167
2-CH(OH)Me	244–246	—	D/CH$_3$CH(OH)CO$_2$Et/concn. HCl/reflux	11
2-CH(OH)C$_6$H$_4$R				
(R = H)	221–222	65		
(R = 2-Cl)	205	75	D/carboxylic acid/4 N HCl/heat	170
(R = 3-Cl)	213–215	74		
(R = 4-Cl)	205	62		
2-CH(OH)C$_6$H$_3$Cl$_2$-2′,4′	180	65	D/carboxylic acid/4 N HCl/heat	170
2-CH$_2$CH(OH)Me	192	87	2-CH$_2$COMe derivative/H$_2$/ Raney Ni	56
2-CH$_2$CH(OH)Ph	204–206	80	2-CH$_2$COPh derivative/H$_2$/ Raney Ni	56
2-(γ-Hydroxypropyl)	216	—	D/γ-butyrolactone/HCl/heat	54

[a] D refers to the appropriate naphthylamine derivative.

528

TABLE 5.31. NAPHTH[1,2-d]IMIDAZOLE 1- AND 3-N-OXIDE DERIVATIVES[a]

Substituent(s)	Melting point (°C)	Yield (%)	Method of preparation	Ref.
1-OH	214–215	77	1-Nitroso-2-naphthol/CH$_2$O/ MeOH/NH$_4$OH/45°/6 hr	87
2-Me-3-OH-5-NHAc	292	96	1,4-(NHAc)$_2$-2-NO$_2$-C$_{10}$H$_5$/Pd/ C/H$_2$/AcOH	61
2-Me-3-OH-4-NO$_2$-5-NHAc	287 (dec.)	60	2-Me-3-OH-5-NHAc derivative/ concn. H$_2$SO$_4$/concn. HNO$_3$/0–5°	61
2-Me-3-OH-4-NH$_2$-5-NHAc	332 (indef.)	92	2-Me-3-OH-4-NO$_2$-5-NHAc derivative/Pd/C/H$_2$/EtOH	61

[a] Designated for convenience as N-hydroxy compounds.

Table 5.32. NAPHTH[1,2-d]- AND [2,3-d]IMIDAZOLES WITH MISCELLANEOUS OXYGEN-CONTAINING SUBSTITUENTS: CARBOXYLIC ACIDS, ESTERS, ETHERS, KETONES AND ALDEHYDES[a]

Substituent	Ring system	Melting point (°C)	Yield (%)	Method of preparation	Ref.
2-(CO$_2$H structure)	[1,2-d]	263–265	—	D/carboxylic acid	31
2-SCH$_2$CO$_2$H	[2,3-d]	208 (dec.)	—	2-SH derivative/1 N NaOH/ ClCH$_2$CO$_2$H	44
2-CO$_2$Et	[1,2-d]	186	—	1-N=NC$_6$H$_4$NO$_2$-p-2-NHCH$_2$ CO$_2$Et-C$_{10}$H$_6$/AcOH/heat	76
2-OEt	[2,3-d]	241–242	>90	D/(EtO)$_4$C	52
2-OPr	[2,3-d]	167–168	>90	D/(PrO)$_4$C	52
9-OMe	[1,2-d]	180.5–181.5	49	D/HCO$_2$H/heat	160
2-CH(OEt)$_2$	[2,3-d]	196–197	59	1-CF$_2$CHFCl derivative/KOH/ EtOH/heat	125
2-COMe	[1,2-d]	181–190	—	2-CH(OH)Me derivative/CrO$_3$/ AcOH/100°	11
2-COMe	[2,3-d]	225	—	2-CH(OH)Me derivative/CrO$_3$/ AcOH/100°	11
2-COAr	[1,2-d]			1-N$_2$Ph-2-NH$_2$-C$_{10}$H$_6$/BrCH$_2$COAr/ heat	
Ar = Ph		210	58		
C$_6$H$_4$Br-p		245	54		
C$_6$H$_4$OMe-p		224	41		78
C$_6$H$_4$NO$_2$-p		268	61		
2-COC$_6$H$_4$NH$_2$	[1,2-d]	249	67	2-COC$_6$H$_4$NO$_2$ derivative/N$_2$H$_4$/ Raney Ni	78
2-CH$_2$COMe	[2,3-d]	210	81	D/MeCOCH$_2$CO$_2$Et	56
2-CH$_2$COPh	[2,3-d]	213	85	D/PhCOCH$_2$CO$_2$Et	56
2-CHO	[1,2-d]	225–228 (dec.) dinitrophenylhydrazone, 300–310	54	2-Me derivative/SeO$_2$/Ac$_2$O/reflux	151
2-CHO	[2,3-d]	Phenylhydrazone, 265	—	2-CH(OEt)$_2$ derivative/ concn. HCl/heat	125

[a] D refers to the appropriate naphthylene diamine derivative.

529

TABLE 5.33. NAPHTH[1,2-*d*]IMIDAZOLES WITH SULFUR-CONTAINING
SUBSTITUENTS: ALKYL- AND (SUBSTITUTED)ALKYLTHIO
DERIVATIVES AND SULFONIC ACIDS[a]

Substituent(s)	Melting point (°C)	Yield (%)	Method of preparation	Ref.
2-SMe	220–222	86	2-SH derivative/NaOEt/MeI	119
2-SCH$_2$CO$_2$H	225–227 (dec.)	81	2-SH derivative/ClCH$_2$CO$_2$H/ DMF/60–65°	130
2-S(CH$_2$)$_2$CO$_2$H	220–221 (dec.)	96	Alkylation as above	130
2-SCH$_2$CO$_2$Me	114–115 (dec.)	70	2-SH derivative/ClCH$_2$CO$_2$Me/ MeOH/reflux	130
2-S(CH$_2$)$_2$OH	159–160	98	2-SH derivative/Cl(CH$_2$)$_2$OH/ NaOEt/EtOH	131
2-SMe-3-CH$_2$COR			2-SMe derivative/BrCH$_2$COR/ NaOH/EtOH	119
Ph	185–186 (dec.)	20		
p-MeC$_6$H$_4$	187–188 (dec.)	20		
p-MeOC$_6$H$_4$	190–191 (dec.)	20		
p-ClC$_6$H$_4$	188–190 (dec.)	20		
p-BrC$_6$H$_4$	196–198 (dec.)	33		
2-SMe-3-CH$_2$CH(OH)R			2-SMe derivative/chlorohydrin	119
(R = HO(CH$_2$)$_2$)	143–144 (dec.)	22–30	derivative or epoxide	
(R = PhCH(OH)CH$_2$)	164–165 (dec.)	20–50		
(R = *p*-O$_2$NC$_6$H$_4$CH (OH)CH$_2$)	200–202 (dec.)	47–53		
4-SO$_3$H-2-Me-7-OH			D/Ac$_2$O/90–95°	28
7-SO$_3$H		60	1-NO-2-NHMe-6-SO$_3$H-C$_{10}$H$_5$/ AcOH/ZnCl$_2$/heat	80

[a] D refers to the appropriate naphthylene diamine derivative.

TABLE 5.34. ALKYLTHIO AND (SUBSTITUTED)ALKYLTHIO DERIVATIVES
OF NAPHTH[2,3-*d*]IMIDAZOLE

Substituent(s)	Melting point (°C)	Yield (%)	Method of preparation	Ref.
2-SMe	241	—	2-SH derivative/MeI/KOH/50°	44
2-SCH$_2$COC$_6$H$_4$Br-*p*	2·HBr, 250; free base, 190–191	84	2-SH derivative/*p*-BrC$_6$H$_4$COCH$_2$ Br/EtOH	133
2-S(CH$_2$)$_2$CO$_2$H	215	41	2-SH derivative/Br(CH$_2$)$_2$CO$_2$H/ NaAc/dry EtOH/heat	45

naphth[1,2-d]imidazole derivative	Melting point (°C)	Yield (%)	Method of preparation	Ref.
$(CH_2)_nR$				
(n = 1)	290	26	D/dicarboxylic acid/polyphosphoric acid/	171
(n = 2)	221–223	70	150–170°/7–8 hr	
(n = 3)	253–255	66		
(n = 4)	172 (dec.)	65		
(n = 5)	133–134	72		
(n = 6)	218–219	70		
(n = 7)	182–183	68		
CH=CHR	>325	28	D/dicarboxylic acid/polyphosphoric acid/ 150–170°/7–8 hr	171
$[CH(OH)]_nR$				
n = 2)	314–315	63		
n = 4)	274–275	46	D/dicarboxylic acid/polyphosphoric acid/ 150–170°/7–8 hr	171
C_6H_4R-o	304; diOAc, 236	—	Naphthalene-1,2-$(NH_2)_2$/phthalic anhydride	30
C_6H_4R-o	248–250	45	D/dicarboxylic acid/polyphosphoric acid/ 150–170°/7–8 hr	171
C_6H_4R-p	>320	64	D/dicarboxylic acid/polyphosphoric acid/ 150–170°/7–8 hr	171
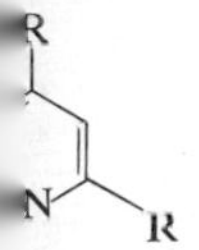 (structure)	>310	42	D/dicarboxylic acid/polyphosphoric acid/ 150–170°/7–8 hr	171
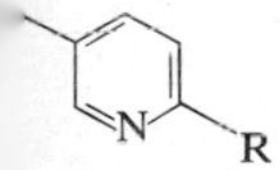 (pyridinyl structure)	245 (dec.)	68	D/dicarboxylic acid/polyphosphoric acid/ 150–170°/7–8 hr	171
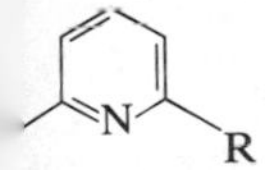 (pyridinyl structure)	230 (dec.)	42	D/dicarboxylic acid/polyphosphoric acid/ 150–170°/7–8 hr	171
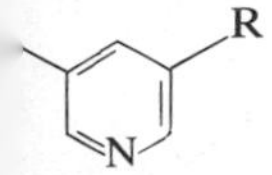 (pyridinyl structure)	>320	65	D/dicarboxylic acid/polyphosphoric acid/ 150–170°/7–8 hr	171
$(SCH_2)_2$	222–223	40	2-SH derivative/$X(CH_2)_2X$/NaOEt/EtOH (X = Cl or Br)	131
N=NR	340	74.6	1-CH_2Ph-2-NH_2 naphth[1,2-d]imidazole/ Na(4 g atom)	154

D refers to the appropriate naphthylene diamine derivative.

R = naphth[1,2-d]imidazol-2-yl.

Bis naphth[1,2-d]imidazoles of the formazan type have also been reported.[153] Structure of R in NHN=C(Me)N=NR, melting points and yield are: naphth[1,2-d]imidazol-2-yl, 214–217 (dec.), 26; 1-methylnaphth[1,2-d]imidazol-2-yl, 191–193 (dec.); 3-methylnaphth[1,2-d]imidazol-2-yl, 169–172 (dec.), 40.

TABLE 5.36. BIS NAPHTH[2,3-d]IMIDAZOLES[a,b]

Bis naphth[2,3-d]-imidazole derivative	Melting point (°C)	Yield (%)	Method of preparation	Ref.
R(CH$_2$)$_n$R				
(n = 1)	260–264	30	D/dicarboxylic acid/polyphosphoric acid/ acid/150–170°/7–8 hr	171
(n = 2)	296–297	34		
(n = 3)	259–260	35		
(n = 4)	150	30		
(n = 5)	163–164	48		
(n = 6)	236–238	42		
(n = 7)	203–204	48		
RCH=CHR	263–265	22	D/dicarboxylic acid/polyphosphoric acid/ 150–170°/7–8 hr	171
R[CH(OH)]$_2$R	255–256	26	D/dicarboxylic acid/polyphosphoric acid/ 150–170°/7–8 hr	171
RC$_6$H$_4$R-o	295	54	D/dicarboxylic acid/polyphosphoric acid/ 150–170°/7–8 hr	171
RC$_6$H$_4$R-p	>320	42	D/dicarboxylic acid/polyphosphoric acid/ 150–170°/7–8 hr	171
RC$_6$H$_4$R-p	>490	90	D/HO$_2$CC$_6$H$_4$CO$_2$H-p/polyphosphoric acid	25
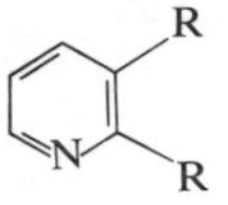 (pyridine structure, R at 3, R at 2)	302–303	62	D/dicarboxylic acid/polyphosphoric acid	171
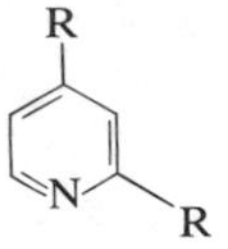 (pyridine structure, R at 4, R at 2)	>320	49	D/dicarboxylic acid/polyphosphoric acid	171
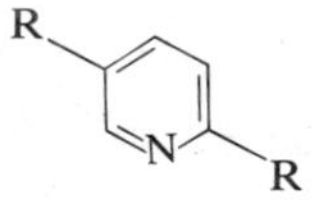 (pyridine structure, R at 4, R at 2)	>320	40	D/dicarboxylic acid/polyphosphoric acid	171
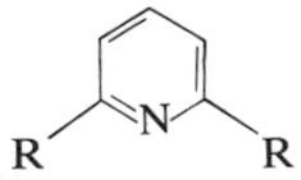 (pyridine structure, R at 2, R at 6)	212(dec.)	48	D/dicarboxylic acid/polyphosphoric acid	171
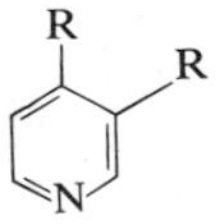 (pyridine structure, R at 4, R at 3)	>320	55	D/dicarboxylic acid/polyphosphoric acid	171
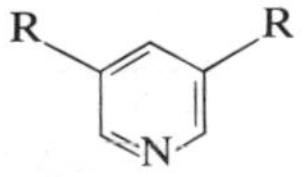 (pyridine structure, R at 3, R at 5)	>310	51	D/dicarboxylic acid/polyphosphoric acid	171

[a] D refers to the appropriate naphthylene diamine derivative.
[b] R = naphth[2,3-d]imidazol-2-yl.

TABLE 5.37. NAPHTH[1,2-d]- AND NAPHTH[2,3-d]IMIDAZOLIUM COMPOUNDS

Ring system	1-Substituent	3-Substituent	Gegenion	Other substituents	Melting point (°C)	Yield (%)	Ref.
[2,3-d]	Me	Me	ClO_4^-	2-Me-4,9-dioxo	300		34
[2,3-d]	Me	Me	ClO_4^-	2-Ph-4,9-dioxo	300	73	34
[2,3-d]	Me	Me	ClO_4^-	2-p-$Me_2NC_6H_4$-4,9-dioxo	308–310	90	34
[2,3-d]	Me	Me	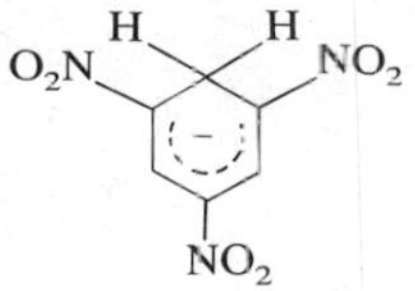 (dinitrophenyl structure: O_2N, NO_2, NO_2, H, H)	None	134–135	90	168
[2,3-d]	Me	Me	—	2-Me-4,9-dioxo	220(dec.)		162
[2,3-d]	Me	Ph	ClO_4^-	2-CH=CHR-4,9-dioxo (R = 1-Me-3-Ph-4,9-dioxo-2(3H)-naphth[2,3-d]imidazolinylidenemethyl)	281		164,166
[2,3-d]	Me	Ph	ClO_4^-	(R = 2-Me-2-benzothiazolinylidenemethyl)	148–150		
[2,3-d]	Me	Ph	ClO_4^-	(R = -Me-2(1H)quinolinylidenemethyl)	286–287		
[2,3-d]	Me	Ph	ClO_4^-	(R = p-$Me_2NC_6H_4$)	314		
[1,2-d]	CH$_2$COMe	Me	Br^-	2-Me	230		7
[1,2-d]	CH$_2$COEt	Me	Br^-	2-Me	259		7
[1,2-d]	CH$_2$COPh	Me	Br^-	2-Me	222		7
[1,2-d]	CH$_2$COMe	Me	Br^-	2-Et	243		7
[1,2-d]	CH$_2$COPh	Me	Br^-	2-Et	210		7

533

TABLE 5.38. NAPHTH[1,2-d]- AND NAPHTH[2,3-d]IMIDAZOLE 2-THIONES[a]

Ring System	1-Substituent	3-Substituent	Melting point (°C)	Yield (%)	Method of preparation	R
[1,2-d]	H	H	290(dec.)	35	D/CS$_2$/alcoholic NaOH/reflux	4
[1,2-d]	H	H	>300	—	D/CS$_2$	4
[1,2-d]	II	H	>300	20	1,2-(NHCSNHPh)$_2$-C$_{10}$H$_6$/15% KOH/heat	8
[2,3-d]	H	H	305	—	D/thiourea/195°	4
[2,3-d]	H	H	299	55	D/CS$_2$	4
[1,2-d]	H	(CH$_2$)$_2$OH	250–252 (dec.)	90	1-NH$_2$-2-NH(CH$_2$)$_2$OH-C$_{10}$H$_6$/K ethyl xanthate	1
[1,2-d]	H	t-Bu	305–307	83	2-Cl derivative/(H$_2$N)$_2$CS	1
[1,2-d]	H	Ph	290–292	83	2-Cl derivative/(H$_2$N)$_2$CS	1
[1,2-d]	H	C$_6$H$_4$OMe-p	295–297	83	2-Cl derivative/(H$_2$N)$_2$CS	1
[1,2-d]	H	C$_6$H$_4$Br-p	295–297	96	2-Cl derivative/(H$_2$N)$_2$CS	1
[1,2-d]	H	2-thienyl	308–310	71	2-Cl derivative/(H$_2$N)$_2$CS	1

[a] D refers to the appropriate naphthylamine diamine derivative.

TABLE 5.39. NAPHTH[1,2-d]IMIDAZOL-2-ONES[a]

1-Substituent	3-Substituent	Melting point (°C)	Yield (%)	Method of preparation	Ref.
H	H	345–347	90	D/urea/170–180°	42
H	H	>345	75–95	D/COCl$_2$/HCl/room temp.	40
H	Me	293–294	90	3-methylnaphth[1,2-d]imidazole/ melt with KOH	141
H	Me	295	—	D(2.HCl)/urea/160–170°	42
H	Me	302–305	60	2-Cl derivative/KOH/150–165°	118
H	Pr	221–222	62	2-Cl derivative/KOH/150–165°	118
H	i-Pr	200–201	—	(i) 2-OH-3-C(Me)=CH$_2$ deriva-tive[b]/H$_2$/Pt or (ii) 1-NH$_2$-2-NH-i-Pr-C$_{10}$H$_6$/ COCl$_2$/xylene/60°	100
H	(CH$_2$)$_2$OH	194–196	57	2-Cl derivative/KOH/150–165°	118
H	CH$_2$Ph	310–320	—	D/urea	42
H	CH$_2$COPh	242–244	83	2-Cl derivative/KOH/150–165°	118
Me	Me	175–176	75	1,3-Dimethylnaphth[1,2-d]-imidazolium iodide/KNH$_2$ (3 moles)/liq. NH$_3$	147
CH=CH$_2$	CH=CH$_2$	122.5–123.8	3.6	Unsubstituted imidazolone derivative/C$_2$H$_2$/KOH	124

[a] D refers to the appropriate naphthylene diamine derivative.
[b] Designated for convenience as a 2-hydroxy derivative.

TABLE 5.40. NAPHTH[2,3-d]IMIDAZOL-2-ONES[a]

1-Substituent	3-Substituent	Melting point (°C)	Yield (%)	Method of preparation	Ref.
H	H	320	85	D/N,N'-carbonyl diimidazole	39
H	H	320 (dec.)	—	2,3-(NHCO$_2$Et)$_2$-C$_{10}$H$_6$/ NaOH/100°	83
H	H	≮320	—	D/urea/180°	44
Ac	H	238	—	2-NHAc-3-CON$_3$-C$_{10}$H$_6$/heat in C$_6$H$_6$	19
Ac	Ac	258	—	2-NHAc-3-CON$_3$-C$_{10}$H$_6$/heat in Ac$_2$O	19

[a] D refers to the appropriate naphthylene diamine.

TABLE 5.41. DERIVATIVES OF NAPHTH[1,2-d]IMIDAZOLE-2,5-DIONES AND RELATED COMPOUNDS

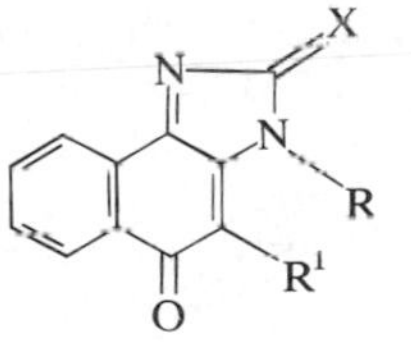

(5.90)

Compound 5.90

X	R	R^1	Melting point (°C)	Yield (%)	Method of preparation	Ref.
O	Ph	Cl	255	63–67	2-NH$_2$-3-Cl naphtho-1,4-quinone/PhNCO/Et$_3$N	86
O	Ph	N$_3$	150–160 (dec.)	97	2-Cl derivative/HN$_3$	86
O	Ph	N=PPh$_3$	265	89	2-N$_3$ derivative/Ph$_3$P/C$_6$H$_6$/heat	86
N-t-Bu	t-Bu	H	165	16	Oxidative cyclization of 1-naphthyl guanidinium salts (see section 5.1)[a]	91[a]
N-Cyclo-C$_6$H$_{11}$	Cyclo-C$_6$H$_{11}$	H	178	22	Oxidative cyclization of 1-naphthyl guanidinium salts (see section 5.1)[a]	91

[a] The 7,8-dimethyl analog of this material has been synthesized in a Diels-Alder reaction on a benzimidazole-5-one derivative m.p., 188; yield, 16%.[91]

Substituent(s)	Melting point (°C)	Yield (%)	Method of preparation	Re
Unsubstituted	318–320(34) darkens, 300	—	Naphth[1,2-*d*]imidazole/AcOH/ CrO$_3$/room temp.	14
Unsubstituted	Darkens, 210; unmelted at 250		(i) 1,2-(NH$_2$)$_2$-3,4-(OH)$_2$-C$_{10}$H$_4$/ HCO$_2$H/NaAc or (ii) HNO$_3$	15
1-*i*-Pr-2-Me	250–253	64	4-Amino-3-acylaminonaphtho-1,2-quinones/AcOH/heat	
1-(CH$_2$)$_4$CH$_3$-2-Me	158–159	99	4-Amino-3-acylaminoaphtho-1,2-quinones/AcOH/heat	67
1-cyclopentyl-2-Me	242–245	44	4-Amino-3-acylaminonaphtho-1,2-quinones/AcOH/heat	67
1-*n*-C$_5$H$_{11}$-2-CH$_2$Cl	180–184 (dec.)	88	4-Amino-3-acylaminonaphtho-1,2-quinones/AcOH/heat	67
1-(CH$_2$)$_4$C$_6$H$_{11}$-2-[3,4,5-tri-(OMe)$_3$C$_6$H$_2$]CH$_2$	212–216	74	4-Amino-3-acylaminonaphtho-1,2-quinones/AcOH/heat	67
1-(CH$_2$)$_4$C$_6$H$_{11}$-2-CH=CHPh	252–253	57	4-Amino-3-acylaminonaphtho-1,2-quinones/AcOH/heat	67
1-(CH$_2$)$_4$C$_6$H$_{11}$-2-(CH$_2$)$_3$C$_6$H$_{11}$	150–152	95	4-Amino-3-acylaminonaphtho-1,2-quinones/AcOH/heat	67
1-(CH$_2$)$_4$C$_6$H$_{11}$-2-*t*-Bu	233–234	37	4-Amino-3-acylaminonaphtho-1,2-quinones/AcOH/heat	67
1-CH$_2$Ph-2-Me	248–249	90	4-Amino-3-acylaminonaphtho-1,2-quinones/AcOH/heat	67
1-CH$_2$Ph-2-CH$_2$Cl	222–224	85	4-Amino-3-acylaminonaphtho-1,2-quinones/AcOH/heat	67
1-(CH$_2$)$_3$NEt$_2$-2-(CH$_2$)$_4$CH$_3$	156–158	32	4-Amino-3-acylaminonaphtho-1,2-quinones/AcOH/heat	68
1-(CH$_2$)$_3$NEt$_2$-2-CH$_2$Ph	174–176 (dec.)	37	4-Amino-3-acylaminonaphtho-1,2-quinones/AcOH/heat	68
1-(CH$_2$)$_3$NEt$_2$-2-C$_6$H$_{11}$	153–156	16	4-Amino-3-acylaminonaphtho-1,2-quinones/AcOH/heat	68
1-Ph-2-Me	314–315	99	4-Amino-3-acylaminonaphtho-1,2-quinones/AcOH/heat	67
1-Ph-2-Ph	312	83	4-Amino-3-acylaminonaphtho-1,2-quinones/AcOH/heat	72
1-C$_6$H$_4$Me-*p*-2-Me	301–303	99	4-Amino-3-acylaminonaphtho-1,2-quinones/AcOH/heat	67
1-C$_6$H$_4$Cl-*p*-2-Me	315–317	100	4-Amino-3-acylaminonaphtho-1,2-quinones/AcOH/heat	67
1-[3,4,5-(OMe)$_3$C$_6$H$_2$]-2-Me	290–293	96	4-Amino-3-acylaminonaphtho-1,2-quinones/AcOH/heat	67

TABLE 5.43A. NAPTH[2,3-d]IMIDAZOLE-4,9-DIONES: UNSUBSTITUTED AND MONO-SUBSTITUTED DERIVATIVES[a]

Substituent(s)	Melting point (°C)	Yield (%)	Method of preparation	Ref.
Unsubstituted	370	95	2-NH$_2$-3-NHCOMe-naphtho-1,4-quinone/H$_2$SO$_4$/ethyl formate	63
Unsubstituted	369	70	D/HCO$_2$H	162
1-Me	286–287	82	Unsubstituted derivative/Me$_2$SO$_4$	127
1-CF$_2$CHFCl	171–172	47	Oxidation of parent 1-CF$_2$CHFCl derivative with Na$_2$Cr$_2$O$_7$/H$_2$SO$_4$	125
1-β-D-Ribofuranosyl	209–210; triacetate, 135–136	89	1-SiMe$_3$ derivative/1,2,3,5-tetra-O-acetyl β-D-ribofuranose/SnCl$_2$	128
1-Ph	239	70	2-NHPh-3-NHCHO-naphtho-1,4-quinone/alcoholic KOH/heat	169
2-CF$_3$	250.5	67	2-NH$_2$-3-NHCOCF$_3$-naphtho-1,4-quinone/NaOH-EtOH	162
2-Me	>350	90[b]	2-NH$_2$-3-NHAc-naphtho-1,4-quinone/2 N NaOH or HCO$_2$H	66
2-Me	368	63	2-NH$_2$-3-NHCOR-naphtho-1,4-quinone/NaOH/EtOH/heat	63
2-Et	304.4–305	70	2-NH$_2$-3-NHCOR-naphtho-1,4-quinone/NaOH/EtOH/heat	63
2-n-Pr	221.8–222.2	65	2-NH$_2$-3-NHCOR-naphtho-1,4-quinone/NaOH/EtOH/heat	63
2-i-Pr	260.1–261.3	83	2-NH$_2$-3-NHCOR-naphtho-1,4-quinone/NaOH/EtOH/heat	63
2-n-Bu	232–232.8	74	2-NH$_2$-3-NHCOR-naphtho-1,4-quinone/NaOH/EtOH/heat	63
2-i-Bu	250–251.4	57	2-NH$_2$-3-NHCOR-naphtho-1,4-quinone/NaOH/EtOH/heat	63
2-n-C$_5$H$_{11}$	182.3–183.5	68	2-NH$_2$-3-NHCOR-naphtho-1,4-quinone/NaOH/EtOH/heat	63
2-n-C$_5$H$_{11}$	191–192	68[c]	2-NH$_2$-3-NHCOR-naphtho-1,4-quinone/2 N NaOH or HCO$_2$H	66
2-(CH$_2$)$_3$C$_6$H$_{11}$	201–203	72[c]	2-NH$_2$-3-NHCOR-naphtho-1,4-quinone/2 N NaOH or HCO$_2$H	66
2-C$_6$H$_{11}$	274–276	55[b]	2-NH$_2$-3-NHCOR-naphtho-1,4-quinone/2 N NaOH or HCO$_2$H	66
2-(CH$_2$)$_4$Cl	170–171	39[b]	2-NH$_2$-3-NHCOR-naphtho-1,4-quinone/2 N NaOH or HCO$_2$H	66
2-CHEt$_2$	240.7–242	81	2-NH$_2$-3-NHCOR-naphtho-1,4-quinone/NaOH/EtOH/heat	63
2-CH$_2$Ph	279–280.3	62	2-NH$_2$-3-NHCOR-naphtho-1,4-quinone/NaOH/EtOH/heat	63
2-CH$_2$C$_6$H$_4$Cl-p	287–291	67	2-NH$_2$-3-NHCOR-naphtho-1,4-quinone/HCO$_2$H	66
2CH$_2$NEt$_2$(HCl)	264.2–265.4	76	2-NH$_2$-3-NHCOR-naphtho-1,4-quinone/NaOH/EtOH/heat	63
2-CH$_2$(1-piperidyl)HCl	310(dec.)	42	2-NH$_2$-3-NHCOR-naphtho-1,4-quinone/NaOH/EtOH/heat	63
2-CH$_2$(1-morpholinyl)HCl	317(dec.)	80	2-NH$_2$-3-NHCOR-naphtho-1,4-quinone/NaOH/EtOH/heat	63
2-undecyl	127.5–129	71	2-NH$_2$-3-NHCOR-naphtho-1,4-quinone/NaOH/EtOH/heat	64
2-pentadecyl	107–108	64	2-NH$_2$-3-NHCOR-naphtho-1,4-quinone/NaOH/EtOH/heat	64
2-[12-(3-cyclopentenyl)dodecyl]	110–111	57	2-NH$_2$-3-NHCOR-naphtho-1,4-quinone/NaOH/EtOH/heat	64

537

TABLE 5.43A. (*Continued*)

Substituent(s)	Melting point (°C)	Yield (%)	Method of preparation	Ref.
2-CH$_2$OH	276–278(dec.)	66	2-NH$_2$-3-NHCOR-naphtho-1,4-quinone/NaOH/EtOH/heat	64
2-phenylvinyl	340–342(dec.)	20	2-NH$_2$-3-NHCOR-naphtho-1,4-quinone/NaOH/EtOH/heat	64
2-(4-carboxylbutyl)	267–268(dec.)	74	2-NH$_2$-3-NHCOR-naphtho-1,4-quinone/NaOH/EtOH/heat	64
2-Ph	371	100	2,3-(NHCOPh)$_2$-naphtho1,4-quinone/aq. alcoholic KOH	162
2-C$_6$H$_4$Cl-p	344-346	30	2-NH$_2$-3-NHCOC$_6$H$_4$Cl-p-naphtho-1,4-quinone/HCO$_2$H/heat	66
5-SO$_3$H	>380	62	4,9-Dione derivative/23% oleum/HgO	117
6-SO$_3$H	>380	72	4,9-Dione derivative/23% oleum/125°	117
6-SO$_2$Cl-p	270–272	47	6-SO$_3$H derivative/ClSO$_3$H/100°	117
6-SO$_2$NMe$_2$	301–302 ·	—	6-SO$_2$Cl derivative/Me$_2$NH	117
6-OH	No definite m.p.	100	6-SO$_3$H derivative/NaOH/180°	117
5-Cl	312–313(dec.)	—	D/HCO$_2$H	117
6-Cl	355(dec.)	74 by method (ii)	(i) 6-SO$_3$H derivative/1 N HCl/KClO$_3$ or (ii) D/HCO$_2$H	117
5-NO$_2$	315–330(dec.)	56	Nitration of unsubstituted derivative	64

[a] D refers to the appropriate naphthylene diamine derivative.

[b] Acid-catalyzed procedure.

[c] Base-catalyzed procedure.

TABLE 5.43B. NAPHTH[2,3-d]IMIDAZOLE-4,9-DIONES: DISUBSTITUTED DERIVATIVES

Substituent(s)	Melting point (°C)	Yield (%)	Method of preparation	Ref.
1-Me-2-Me	247–248	—	2-NH$_2$-3-NHMe naphtho-1,4-quinone/MeOH/CH$_3$CO$_2$Et/room temp.	163
1-Me-2-Me	248 305–308	—	2-NHMe-3-NHCOMe-naphtho-1,4-quinone/HCl/heat or alcoholic KOH/heat	162, 169
1-Me-2-Me	247–249	60	2-NHAc-3-NHMe-naphtho-1,4-quinone/2 N NaOH/EtOH/heat	65
1-Me-2-CF$_3$	170–171	—	2-NHMe-3-NHCOCF$_3$-naphtho-1,4-quinone/alcoholic KOH/heat	169
1-Me-2-Cl	230–231	52	1-Me derivative/1 N HCl/KClO$_3$	127
1-Me-2-Br	244–245	—	1-Me derivative/Br$_2$/heat 120°/10 hr	127
1-Me-2-OMe	Rearranges near m.p. to 1,3-Me$_2$ derivative		1-Me-2-Cl derivative/MeONa/dry MeOH/heat	127
1-Me-2-OPh	184	65	1-Me-2-Cl derivative/PhONa/dioxan/heat	127
1-Me-2-SPh	183–184		1-Me-2-Cl derivative/PhSNa/dioxan/heat	127
1-Me-2-NH$_2$	314–315	76	1-Me-2-Cl derivative/NH$_3$/EtOH/150°	127
1-Me-2-NHMe	304–305	31	Amination as above	127
1-Me-2-NHPh	305.5–306.5	32.5	Amination as above	127
1-Me-2-NEt$_2$	95–96	35	Amination as above	127
1-Me-2-(1-morpholino)	220	100	Amination as above	127
1-Me-2-Ph	194–195	—	2-NHMe-3-NHCOPh-naphtho-1,4-quinone/alcoholic KOH/heat	169
1-Me-2-C$_6$H$_4$NMe$_2$-p	246–247	—	Cyclization as above	169
1-Et-2-Me	181–183	—	2-NHEt-3-NHCOMe-naphtho-1,4-quinone/HCl/heat	162
1-Et-2-Me	182–183 168–170(53)	—	2-NHEt-3-NHCOMe-naphtho-1,4-quinone/alcoholic NaOH	164, 165
1-n-Pr-2-Me	145–147	78	Cyclization as above	65
1-i-Pr-2-Me	175	—	2-NH-i-Pr-3-NHAc-naphtho-1,4-quinone/HCl/heat	162
1-i-Pr-2-Me	175–176	—	2-NH-i-Pr-3-NHAc-naphtho-1,4-quinone/2 N NaOH/EtOH/heat	65
1-i-Pr-2-Ph	185–186	90	Cyclization as above	65
1-n-Bu-2-Me	115–117	75	Cyclization as above	65
1-n-Bu-2-Et	144–145	66	Cyclization as above	65
1-n-Bu-2-Ph	178–179	70	Cyclization as above	65
1-sec-Bu-2-Me	161–163	62	Cyclization as above	65
1-t-Bu-2-Me	158–159	—	2-NH-t-Bu-3-NHAc-naphtho-1,4-quinone/HCl/heat	162
1-C$_6$H$_{11}$-2-Me	221	—	Cyclization as above	162

TABLE 5.43B. (*Continued*)

Substituent(s)	Melting point (°C)	Yield (%)	Method of preparation	Ref.
1-n-C_6H_{13}-2-Me	128–129	38	2-NHC$_6$H$_{13}$-3-NHAc-naphtho-1,4-quinone/2 N NaOH/EtOH/heat	65
1-$C_{14}H_{29}$-2-Me	72–73	72	Cyclization as above	65
1-CH$_2$CH=CH$_2$-2-Ph	156–158	60	Cyclization as above	65
1-CH$_2$Ph-2-Me	195–196	85	Cyclization as above	65
1-CH$_2$Ph-2-Et	129–130	74	Cyclization as above	65
1-CH$_2$Ph-2-Ph	164–166	65	Cyclization as above	65
1-Ph-2-Me	239,[71] 240–241[65]	—	Cyclization as above	65, 7
1-Ph-2-Me	239	—	2-NHPh-3-NHAc-naphtho-1,4-quinone/HCl/heat	162
1-Ph-2-CF$_3$	238–239	—	2-NHPh-3-NHCOCF$_3$-naphtho-1,4-quinone/alcoholic KOH/heat	169
1-Ph-2-Et	174–175	84	2-NHPh-3-NHCOEt-naphtho-1,4-quinone/2 N NaOH/EtOH/heat	65
1-Ph-2-Ph	177–180	50	Cyclization as above	65
1-Ph-2-$C_6H_4NMe_2$-p	310–313	—	2-NHPh-3-NHCOC$_6$H$_4$NMe$_2$-p-naphtho-1,4-quinone/alcoholic KOH/heat	169
1-C_6H_4Me-o-2-Me	189–191	40	2-NHC$_6$H$_4$Me-o-3-NHAc-naphtho-1,4-quinone/2 N NaOH/EtOH/heat	65
1-C_6H_4Me-p-2-Me	260–263	40	Cyclization as above	65
1-C_6H_4OMe-o-2-Me	193–195	68	Cyclization as above	65
1-C_6H_4OMe-o-2-Ph	224–226	85	Cyclization as above	65
1-C_6H_4Cl-p-2-Me	236–238	73	Cyclization as above	65
1-C_6H_4Br-p-2-Me	242–244	82	Cyclization as above	65

TABLE 5.43C. NAPHTH[2,3-d]IMIDAZOLE-4,9-DIONES: TRISUBSTITUTED DERIVATIVES

Substituent(s)	Melting point (°C)	Method of preparation	Ref.
1,2-Me$_2$-5-NHMe-8-OMe	206–208[a]	1,2-Me$_2$-5,8-(OMe)$_2$ derivative/MeNH$_2$/CH$_2$Cl/hν at $\lambda > 320$ nm	148
1,2-Me$_2$-5-OMe-8-NHMe	198–200[a]	1,2-Me$_2$-5,8-(OMe)$_2$ derivative/MeNH$_2$/CH$_2$Cl/hν at $\lambda > 320$ nm	148
1,2-Me$_2$-5-NHC$_6$H$_{11}$-8-OMe	226–228	As above, using cyclohexylamine	148
1,2-Me$_2$-5-OMe-8-NHC$_6$H$_{11}$	210–211	As above, using cyclohexylamine	148

[a] Combined yield of 5- and 8-NHMe derivatives is 83%

TABLE 5.43D.   NAPHTH[2,3-d]IMIDAZOLIUM-4,9-DIONE DERIVATIVES

1-Substituent	3-Substituent	Other Substituent	Anion	Melting point (°C)	Yield (%)	Ref.
Et	Me	Me	I	247–250	85	65
Et	n-Pr	Me	I	177–181	60	65
i-Pr	Me	Me	I	249–251	92	65
i-Pr	n-Pr	Me	I	211–214	81	65
n-Bu	Me	Me	I	224–226	91	65
n-Bu	Et	Me	I	127–129	35	65
n-Bu	PhCH$_2$	Me	Cl	179–181	30	65
n-Bu	4-O$_2$NC$_6$H$_4$CH$_2$	Me	Br	235–236	83	65
n-Bu	PhCH$_2$CH$_2$	Me	Br	203–206	75	65
PhCH$_2$	Me	Me	I	210 (dec.)	87	65
Ph	Me	Me	I	273–275	60	65
2-MeOC$_6$H$_4$	Me	Me	I	277–280	50	65
Ph	4-O$_2$NC$_6$H$_4$CH$_2$	Me	Br	234–236	74	65
i-Bu	Me	Me	I	226–227	62	65
i-Bu	4-O$_2$NC$_6$H$_4$COCH$_2$	Me	Br	230–231	86	65
n-Bu	Me	Ph	I	135–136	71	65
Me	i-Pr	CH=CHPh	I	218–219.5	90	65
Me	i-Pr	CH=CHC$_6$H$_4$Cl-o	I	214–215	13	65
Me	i-Pr	CH=CHC$_6$H$_4$NMe$_2$-p	I	212–214	31	65
Me	i-Pr	CH=CHC$_6$H$_4$OMe-p	I	210–211	30	65
Me	i-Pr	CH=CHC$_6$H$_4$OH-p	I	245–247	36	65

ubstituent or ompound	Melting point (°C)	Yield (%)	Method of preparation	Ref.
nsubstituted	364	—	D(HCl)/urea/165°	43
-Me	296–298	76	1-Me-2-Cl derivative of 4,9-dione/10% KOH/heat	127
-n-Bu	223–224.5	62	2-NHCONHCOPh-3-NHnBu-naphtho-1,4-quinone/2 N NaOH/heat	84
-i-Bu	253–254	59	2-NHCONHCOPh-3-NHiBu-naphtho-1,4-quinone/2 N NaOH/heat	84
-Ph	>360	86	2-NHCONHCOPh-3-NHPh-naphtho-1,4-quinone/2 N NaOH/heat	84
	346–350 (dec.)	96	Diels-Alder reaction on a benzimidazoline trione derivative	92

TABLE 5.45. 6,7,8,9-TETRAHYDRO DERIVATIVES OF NAPHTH[1,2-d]IMIDAZOLES[a]

Substituent(s)	Melting point (°C)	Yield (%)	Method of preparation	Ref.
None	200–201; 204–205	90	D/HCO$_2$H	1,5
None	202–204 [by method (i)]; 201–203 [by method (ii)]		(i) Naphth[1,2-d]imidazole/AcOH/25% H$_2$SO$_4$/PtO$_2$/H$_2$ or (ii) D/HCO$_2$H/4 N HCl/reflux	157
1-Me	69–70	60 —	Alkylation of unsubstituted derivative by Me$_2$SO$_4$/100°	5
1-Me-2-NH$_2$	228–229		1-Me derivative/NaNH$_2$/PhNMe$_2$	5
2-Me	192–193; hydrate, 70	— — —	(i) 2-Methylnaphth[1,2-d]imidazole derivative/AcOH/PtO$_2$/H$_2$ or (ii) D/CH$_3$CO$_2$H/4 N HCl/reflux	157
2-Me	188	—	D/CH$_3$CO$_2$H/HCl	1
2-cyclo-C$_6$H$_{11}$	218–219, hydrate, 120	— —	2-Phenylnaphth[1,2-d]imidazole/ AcOH/PtO$_2$/H$_2$	157
2-OH	307–308	—	D/urea	1
2-SH	375	—	D/thiourea	1
2-SMe	196–198	—	2-SH derivative/1 N NaOH/MeI	1
2-SCH$_2$CO$_2$H	219	—	2-SH derivative ClCH$_2$CO$_2$H/1 N NaOH	1
2-Cl	175–178	—	2-OH derivative/POCl$_3$/reflux	1
2-NHPh	198–199	—	2-Cl derivative/PhNH$_2$/reflux	1
2-NMe$_2$	154–155	—	2-Cl derivative/Me$_2$NH/Cu/175°	1
2-NH$_2$	178 (dec.); sinter, 170	—	D/CNBr	1
2-Me-3-cyclo-hexyl	275–278	45 —	1-NO$_2$-2-NHAc-3-cyclo-C$_6$H$_{11}$-C$_{10}$H$_5$/ H$_2$, Raney Ni/dioxan/room temp.	74
2-NH$_2$	diHCl, 298–302 picrate 223–226	69(di HCl)	9-Aminonaphth[1,2-d]imidazole/Rh/C/ H$_2$/2 N HCl/room temp.	156

[a] D refers to the appropriate naphthylene diamine derivative.
[b] For examples of dihydro- and octahydronaphth[1,2-d]imidazoles, see compounds **5.8**[58] and **5.89**[15] respectively.

TABLE 5.46. 5,6,7,8-TETRAHYDRO DERIVATIVES OF NAPHTH[2,3-d]
IMIDAZOLES[a,t]

Substituent(s)	Melting point (°C)	Method of preparation	Ref.
None	178–180 picrate, 231-235	Naphth[2,3-d]imidazole/AcOH/25% H_2SO_4 PtO_2/H_2	157
None	139–141[20] 134–135[21]	D/HCO_2H	20, 21
2-Me	251–252	D/AcOH/heat	22
2-Et	242–243	D/$EtCO_2H$	21
2-OH[c]	355	D/urea	21
2-SH[c]	310	D/thiourea	21
1-Me-2-SH[c]	250–252	D/thiourea	21
2-SMe	195	2-SH derivative/NaOH/Me_2SO_4	21
2-Cl	211	2-OH derivative/$POCl_3$	21
2-NHPh	247–249	2-Cl derivative/$PhNH_2$/reflux	21
2-NMe_2	210–211	2-Cl derivative/Me_2NH/reflux	21
2-SCH_2CO_2H	194–195	2-SH derivative/$ClCH_2CO_2H$/1 N NaOH/heat	21

[a] D refers to the appropriate naphthylene diamine derivative.
[b] For examples of 1,3-dihydro- and 5,8-dihydronaphth[2,3-d]imidazoles, see Refs. 58 and 93, respectively.
[c] Designated for convenience as the 2-hydroxy or 2-mercapto tautomer

TABLE 5.47. MISCELLANEOUS NAPHTH[1,2-d]IMIDAZOLE DERIVATIVES OF
POTENTIAL COMMERCIAL INTEREST

Substituents	Commercial area of interest	Ref.
2-Ph-4,5-dihydro	Contraceptives	172
Imidazolium salts of 4,9-dione derivatives	Antimicrobial and fungicidal agents	173, 174
2-Me-5-N_2R (R = 2, 1-HOC$_{10}H_6$) 3-Me-1-Ph-5-oxo-4-pyrazolinyl	Dyestuffs	175
6-$N_2C_6H_4NEt_2$ p-7-OH	Dyestuffs	176
2-$C_6H_4NH_2$-m-6-OH-7-N_2Ar-8-SO_3H[a]	Dyestuffs (cellulose)	177
2-Me-6-OH-7-R-8-SO_3H	Dyestuffs (cellulose)	178

| 2-(3-NH_2-4-OMe-C_6H_3)-6-OH-7-R-8-SO_3H | Dyestuffs (polyamides) | 179 |

TABLE 5.47. (*Continued*)

Substituents	Commercial area of interest	Ref.
1-NHAr-3-Me-7-SO$_2$R^1-8-N$_2$Ar-9-OH (Ar = aryl; R^1 = OH, N(Me)Ph, NEt$_2$ etc.)	Fiber-reactive dyes	180
2-CH=CHR-3-Ph (R = aryl, hetaryl)	Luminophores	181

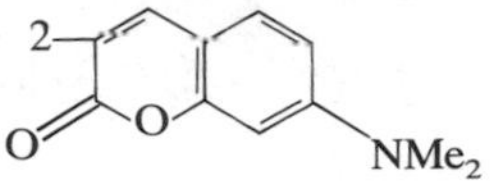	Fluorescent agent	182

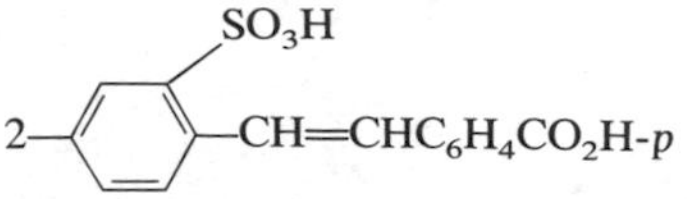	Fluorescent agent	183
None	Optical brightener for textile fibers	12
2-CH$_2$OH, (CH$_2$)$_2$OH; OH at positions 5, 6, 7 or 8	Additives for two-component diazo-type paper	184
2-(3-pyridyl)	Additive in the preparation of photographic emulsions	13
2-NHCOPh	Photographic sensitizer	185
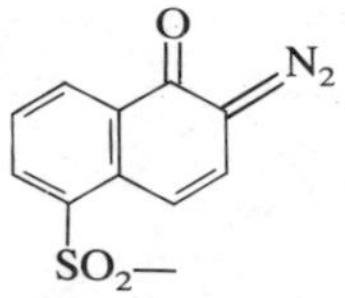 3-R-8-OH-9-N$_2$R^1	Photosensitive compounds for printing processes	186

(R^1 = 2-OH-1-naphthyl)

1-R-2-R-8-OCOR^1 (R = alkyl) 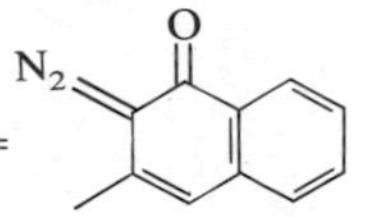	Photosensitive compounds for printing processes	187

1-(or 3-)Pr-2-Et-8-R 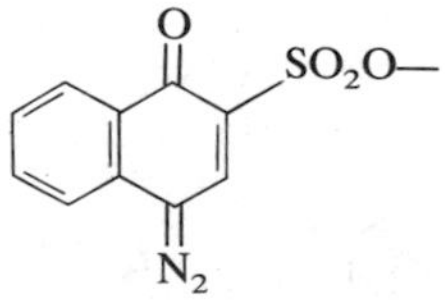	Photosensitive compounds for printing processes	188

2-C$_6$H$_4$N$_3$ with various substituents in the carbocyclic rings	Photosensitive compounds for printing processes	189
1,3-Et$_2$-2-R 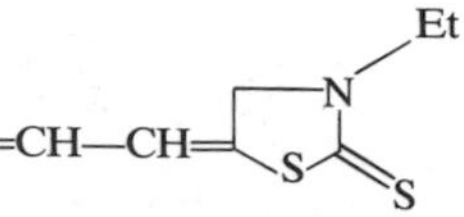	Dyes for increasing spectral sensitivity of photographic silver halide emulsions	190

TABLE 5.47. (*Continued*)

Substituents	Commercial area of interest	Ref.
2-C$_6$H$_4$NMe$_2$-p, 2-(2-pyridyl), 1-Me-2-(2-hydroxyphenyl), 1-Et-2-(4-Me$_2$NC$_6$H$_4$)-7-OMe	Materials for electrographic reproduction	191

[a] These derivatives are diazotized and coupled with 8-hydroxyquinoline to provide metallizable dyes for cellulose fibers.

TABLE 5.48.   MISCELLANEOUS NAPHTH[2,3-d]IMIDAZOLE DERIVATIVES OF POTENTIAL COMMERCIAL INTEREST[a]

Substituent or compound	Commercial area of interest	Ref.
2-(CH$_2$)$_n$N(R^2)(NH)$_m$C(=Y)NRR^1 (Y = O, S; R^2 – H, Me; $m = n = 0, 1$; R = cycloalkyl, aryl; R^1 = H, Ph)	Antiviral and immunization reaction- suppressing agents	134
Imidazolium salts of 4,9-dione derivatives	Antimicrobial and fungicidal agents	173, 192
2-R-4,9-dioxo (R = H, alkyl, cycloalkyl)	Fungicides	193
N-Vinyl[b]	Monomers for the preparation of polymers and copolymers	194
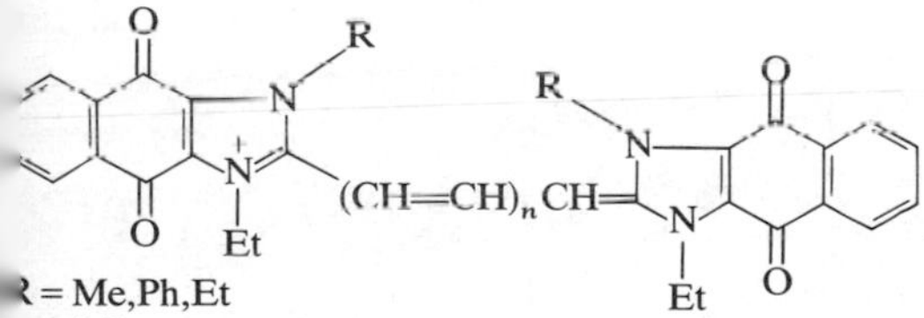 R = Me,Ph,Et n = 1,2	Dyestuffs	195
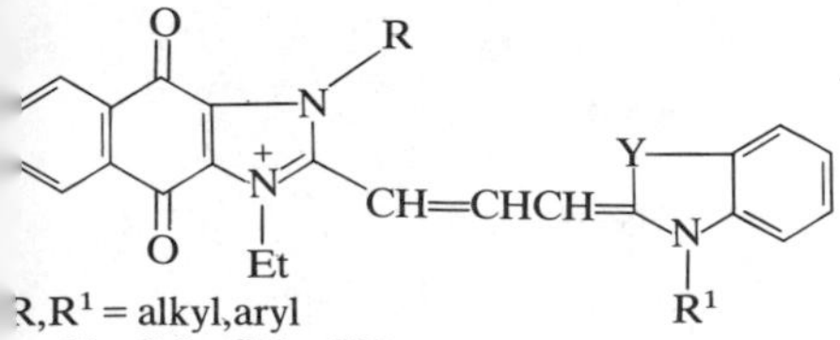 R,R^1 = alkyl,aryl Y = S,Se,CH=CH	Dyestuffs	195
1-R-2-Me-4,9-dioxo (R = Et, allyl)	Catalysts for acceleration of the dye bleaching in Ag dye bleach process	196
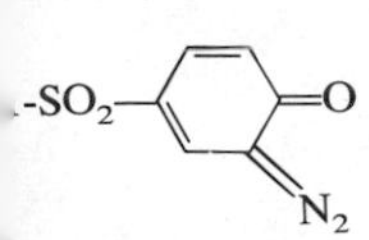 2-SO$_2$-	Photographic developer	197

[a] See Ref. 198 for an evaluation of the inhibitory activity of a variety of naphth[2,3-d]imidazoles against influenza B virus.

[b] The type of naphthimidazole ring system used is not stated.

545

5.2. TRICYCLIC 6-6-5 FUSED BENZIMIDAZOLES WITH ONE ADDITIONAL HETERO ATOM

A large number of ring systems based on a tricyclic, 6-6-5 fused benzimidazole framework having one hetero atom in addition to the nitrogen atoms of the imidazole ring are theoretically possible. However, only derivatives of the imidazo[4,5-g]benzopyran (5.91), imidazo[4,5-h]benzopyran (5.92), imidazo[4,5-f]benzothiopyran (5.93), imidazo[4,5-g]quinoline (5.94), imidazo[4,5-f]quinoline (5.95), imidazo[4,5-h]quinoline (5.96), and imidazo[4,5-h]isoquinoline (5.97) ring systems appear to have been reported to date (Scheme 5.5 and Table 5.49). Moreover, of these only the chemistry of the fused quinoline ring systems (5.94), (5.95), and (5.96) (Scheme 5.5) have been investigated to any extent.

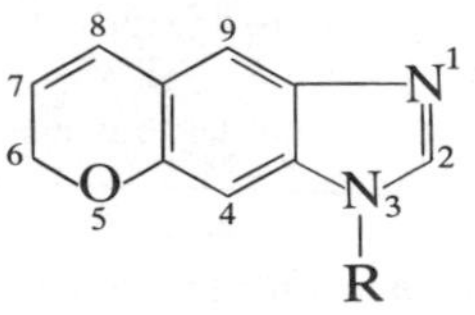

(5.91)

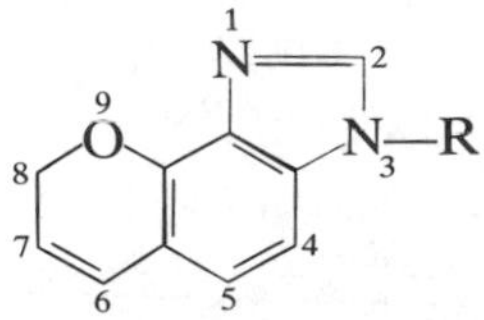

(5.92)

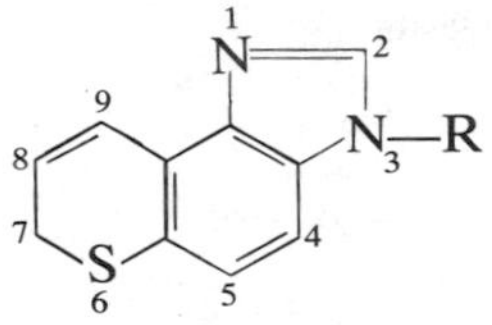

(5.93)

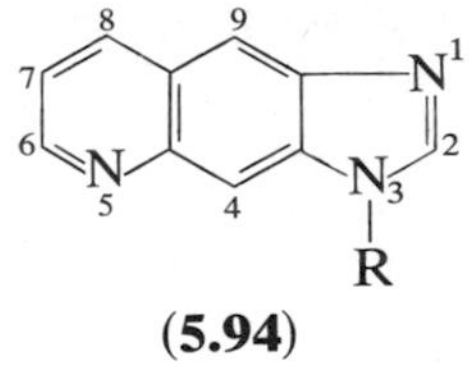

(5.94)

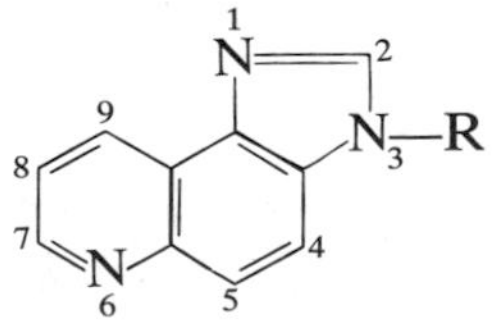

(5.95)

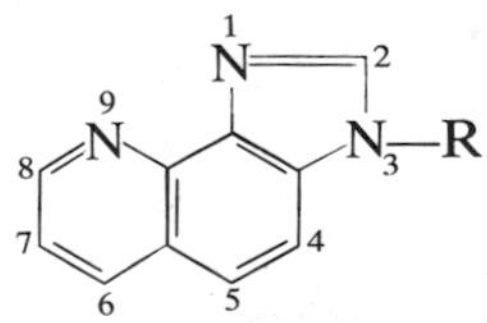

(5.96)

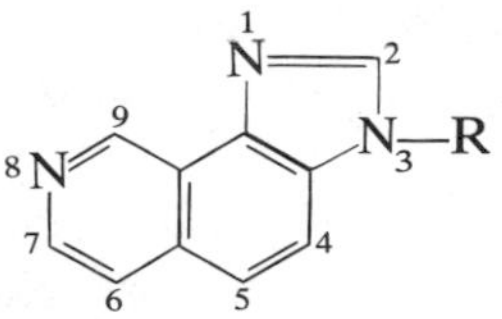

(5.97)

Scheme 5.5

TABLE 5.49. TRICYCLIC 6-6-5 FUSED
BENZIMIDAZOLES WITH ONE
ADDITIONAL HETERO ATOM

Structure[a]	Name[b]
(5.91)	6H-imidazo[4,5-g]benzopyran[c]
(5.92)	8H-imidazo[4,5-h]benzopyran[d]
(5.93)	7H-imidazo[4,5-f]benzothiopyran[e]
(5.94)	Imidazo[4,5-g]quinoline
(5.95)	Imidazo[4,5-f]quinoline
(5.96)	Imidazo[4,5-h]quinoline
(5.97)	Imidazo[4,5-h]isoquinoline

[a] See Scheme 5.5.
[b] Based on the Ring Index.
[c] 8H tautomer is also possible.
[d] 6H tautomer is also possible.
[e] 9H tautomer is also possible.

5.2.1. Synthesis

Ring-Closure Reactions of Benzimidazole Derivatives

Derivatives of the 8H-imidazo[4,5-g][5,6]benzopyran ring system (5.91) have been synthesized by closure of the pyran ring in suitable 5-acyl-6-hydroxybenzimidazole precursors. For example, the sodium-ethoxide-catalyzed condensation of the benzimidazole derivative (5.98) with diethyl oxalate followed by acid-catalyzed cyclization of the open-chain condensation product, furnishes the imidazo[4,5-g][5,6]benzopyran derivative (5.99) (Scheme 5.6), albeit in low yield (Table 5.50).[200]

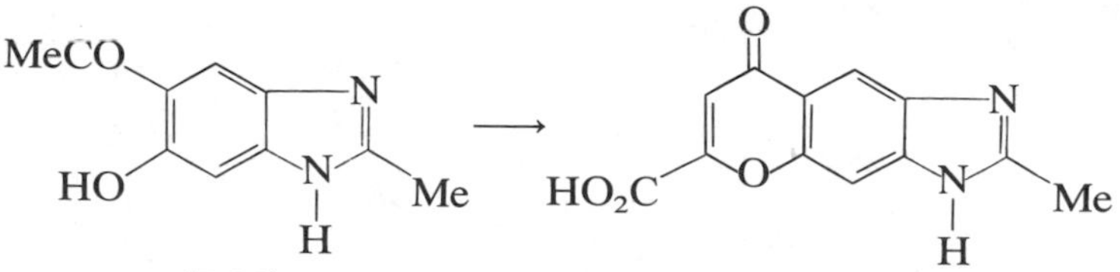

(5.98)   (5.99)

Scheme 5.6

Typical quinoline syntheses (Skraup, Doebner–von Miller, Conrad-Limpach) utilizing 5-aminobenzimidazole derivatives as starting materials can, depending on the direction of cyclization, lead to linear imidazo[4,5-g]-quinolines (Scheme 5.7, 5.100 → 5.101) or angular imidazol[4,5-f]quinolines (Scheme 5.7, 5.100 → 5.102). In practice, application of the Skraup[201–202] or Doebner–von Miller[204–208] procedures to 4-unsubstituted 5-amino-benzimidazoles results in preferential ring closure at the free 4-position of the aminobenzimidazole [Scheme 5.7, 5.100 (R³ = H) → 5.102], thus providing flexible, general, and usually efficient methods for the synthesis of

TABLE 5.50.　SYNTHESES OF 8H-IMIDAZO[4,5-g][5,6]BENZOPYRAN-8-ONES (**5.99**), AND (**5.112**), 6H-IMIDAZO[4,5-h][5,6]BENZOPYRAN-6-ONES (**5.114**) AND (**5.115**), AN 7H-IMIDAZO[4,5-f][5,6]BENZOTHIOPYRAN-7-ONES (**5.117**)

Starting material	Reaction conditions[a]	Product	Yield (%)	Melting point (°C)	Solvent of crystallization	Crystal form	Ref.
(**5.98**)	A	(**5.99**)	20	>320	—[b]	Pale green solid	200
(**5.111**)	B	(**5.112**)	82	274–5 (decomp.)	—[b]	—[c]	211
(**5.113**)	C	(**5.114**)	—[d]	—[e]	—[b]	—[c]	200
(**5.114**)	D	(**5.115**)	—[d]	232 (decomp.)	—[b]	Buff powder	200
(**5.116**) (R = Ac)	E	(**5.117**) (R = Me)	—[d]	294–295	Ethanol	Prisms	212
(**5.116**) (R = H)	F	(**5.117**) (R = H)	—[d]	257–258	Methanol	Colorless crystals	212

[a] A, (i) diethyl oxalate, NaOEt–EtOH/100°/2 hr. (ii) HCl–AcOH/reflux/1.5 hr. B, Ac$_2$O–pyridine/heat/ briefly. C, Ac$_2$O. D, HCl, AcOH. E, dilute HCl/reflux/30 min. F, HCO$_2$H/reflux/2 hr.
[b] Not specified.
[c] Crystal form not specified.
[d] Yield not specified.
[e] Melting point not specified.

variously substituted imidazo[4,5-f]quinoline derivatives (Table 5.52). These procedures only lead to alternative ring closure at the 6-position of the 5-aminobenzimidazole when the 4-position is blocked by a substituent. Despite this limitation, however, the application of the Skraup and Doebner–von Miller quinoline syntheses to 4-substituted 5-aminobenz-imidazoles again provides reasonably general and moderately efficient routes [Scheme 5.7, **5.100** (R^3 ≠ H) → **5.101**] to a variety of imidazo[4,5-g]-quinoline derivatives (Table 5.51).[201,203,204]

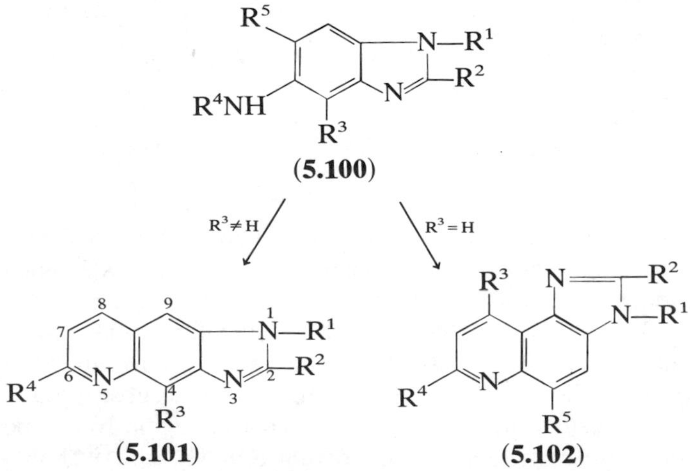

Scheme 5.7

TABLE 5.51. SYNTHESES OF IMIDAZO[4,5-g]QUINOLINE DERIVATIVES (5.101) AND (5.107) BY RING-CLOSURE REACTIONS OF 5-AMINOBENZIMIDAZOLE DERIVATIVES (5.100) AND (5.105)

Starting material	Reaction conditions[a]	Product	Yield (%)	Melting point (°C)	Solvent of crystallization	Crystal form	Ref.
(5.100) (R^1 = R^4 = R^5 = H, R^2 = Ph, R^3 = Br)	A	(5.101) (R^1 = R^4 = H, R^2 = Ph, R^3 = Br)	—[b]	165	Ethanol–water	Colorless needles	201
(5.100) (R^1 = Ph, R^2 = R^4 = R^5 = H, R^3 = Cl)	A	(5.101) (R^1 = Ph, R^2 = R^4 = H, R^3 = Cl)	—[b]	199–201	Ethanol–water	—[c]	201
(5.100) (R^1 = p-MeC$_6$H$_4$, R^2 = R^4 = R^5 = H, R^3 = Cl)	A	(5.101) (R^1 = p-MeC$_6$H$_4$, R^2 = R^3 = R^4 = R^5 = H)	—[b]	235	Benzene–ethanol	Yellow prisms	201
(5.100) (R^1 = R^2 = R^5 = H, R^3 = Cl, R^4 = CHO)	B	(5.101) (R^1 = R^2 = R^4 = H, R^3 = Cl)^{d,e}	31	304	Ethanol	—[c]	203
(5.100) (R^1 = R^4 = R^5 = H, R^2 = R^3 = Me)	C	(5.101) (R^1 = H, R^2 = R^3 = R^4 = Me)	39	275	Ethanol–water	Colorless crystals	204
(5.105) (R^1 = R^2 = R^3 = Me, R^4 = H)	D	(5.107) (R^1 = R^2 = R^3 = Me, R^4 = H)^f	79^g	—^{h}	—^{i}	—[c]	205
(5.105) (R^1 = Et, R^2 = R^3 = Me, R^4 = H)	D	(5.107) (R^1 = Et, R^2 = R^3 = Me, R^4 = H)^j	85^g	—^{h}	—^{i}	—[c]	205

^{a} A, As$_2$O$_5$, glycerine, conc. H$_2$SO$_4$/reflux/7 hr. B, As$_2$O$_5$, glycerine, conc. H$_2$SO$_4$/reflux/10 hr. C, crotonaldehyde, conc. HCl/reflux/2 hr. D, diphenyl ether/250–260°/15 min.
^{b} Yield not specified.
^{c} Crystal form not specified.
^{d} Hydrochloride; free base has m.p. 304° (from ethanol).
^{e} Forms a picrate, m.p. >230° (decomp.) (from water).
^{f} Mixture with (5.106) (R^1 = R^2 = R^3 = Me, R^4 = H).
^{g} Total yield of mixture.
^{h} Not recorded.
^{i} Not purified.
^{j} Mixture with (5.106) (R^1 = Et, R^2 = R^3 = Me, R^4 = H).

The acid-catalyzed condensation of β-keto esters **(5.104)** with 4-unsubstituted 5-aminobenzimidazoles **(5.103)** to afford crotonates **(5.105)** susceptible to preferential thermal cyclization at the 4-position exemplifies the Conrad–Limpach procedure (Scheme 5.8) used extensively[208–210] for the general synthesis (Table 5.52) of imidazo[4,5-f]quinolin-9(6H)-ones **(5.106)**. In a variant of this type of synthesis (Table 5.52), 5-aminobenzimidazoles are condensed with the ethoxymethylene derivatives of β-keto esters to afford intermediates that undergo smooth, base-catalyzed cyclization to imidazo[4,5-f]quinolin-9(6H)-one-8-carboxylic acids (e.g., Scheme 5.9, **5.108 → 5.109**).[210] The latter products are readily decarboxylated (e.g., Scheme 5.9, **5.109 → 5.110**)[210] to give the corresponding imidazo[4,5-f]-quinolin-9(6H)-ones in high yield (Table 5.52). Imidazo[4,5-f]quinoline derivatives are usually the exclusive products of cyclization reactions of the Conrad–Limpach type (cf. Schemes 5.8 and 5.9). However, the recent demonstration[205] of the concomitant formation (Table 5.51) of the linear imidazo[4,5-g]quinolin-8(5H)-one isomers **(5.107)** in some instances suggests that ring closure to the 6-position may compete with cyclization to the 4-position under favorable circumstances. This demonstration may be significant in relation to possible structural ambiguity in the products of such imidazoquinoline syntheses.[208–210]

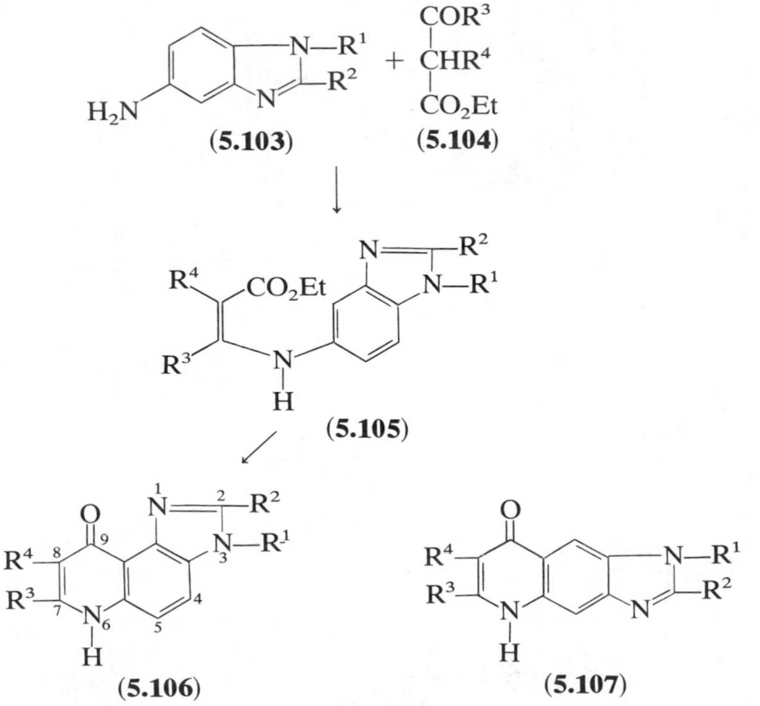

Scheme 5.8

TABLE 5.52. SYNTHESES OF IMIDAZO[4,5-f]QUINOLINE DERIVATIVES (5.102) AND (5.106) BY RING-CLOSURE REACTIONS OF 5-AMINOBENZIMIDAZOLE DERIVATIVES (5.100) AND (5.105)

Starting materials	Reaction conditions	Product	Yield (%)	Melting point (°C)	Solvent of crystallization	Crystal form	Ref.
(5.100) ($R^1 = R^3 = R^4 = R^5 = H$, $R^2 = Ph$)	A	(5.102) ($R^1 = R^3 = R^4 = R^5 = H$, $R^2 = Ph$)	36	270	Ethanol	Yellow needles	201
(5.100) ($R^1 = Ph$, $R^2 = R^3 = R^4 = R^5 = H$)	B	(5.102) ($R^1 = Ph$, $R^2 = R^3 = R^4 = R^5 = H$)	—[b]	165	Ethanol–water	Colorless needles	201
(5.100) ($R^1 = p\text{-Me.C}_6\text{H}_4$, $R^2 = R^3 = R^4 = R^5 = H$)	A	(5.102) ($R^1 = p\text{-Me.C}_6\text{H}_4$, $R^2 = R^3 = R^4 = R^5 = H$)	—[b]	180	Benzene	Yellow plates	201
(5.100) ($R^1 = R^2 = R^3 = R^4 = R^5 = H$)[c]	C	(5.102) ($R^1 = R^2 = R^3 = R^4 = R^5 = H$)[d]	47	78 / 214	Water / Benzene	—[e] / —[e]	202
(5.100) ($R^1 = R^3 = R^4 = R^5 = H$, $R^2 = Me$)	D	(5.102) ($R^1 = R^3 = R^5 = H$, $R^2 = R^4 = Me$)	56	103–105	Water	Colorless needles	204
(5.100) ($R^1 = R^3 = R^4 = H$, $R^2 = R^5 = Me$)	D	(5.102) ($R^1 = R^3 = H$, $R^2 = R^4 = R^5 = Me$)[f]	51	95	Ethanol–water	Colorless needles	204
(5.100) ($R^1 = R^3 = R^4 = H$, $R^2 = Me$, $R^5 = MeO$)	D	(5.102) ($R^1 = R^3 = H$, $R^2 = R^4 = Me$, $R^5 = MeO$)	49	115–117	—[g]	—[e]	204
(5.100) ($R^1 = R^2 = Me$, $R^3 = R^4 = R^5 = H$)	D	(5.102) ($R^1 = R^2 = R^4 = Me$, $R^3 = R^5 = H$)	75	201.5–202.5	Water	Colorless needles	205
(5.100) ($R^1 = R^2 = Me$, $R^3 = R^4 = R^5 = H$)	E	(5.102) ($R^1 = R^3 = R^4 = Me$, $R^2 = R^5 = H$)	75	171–172	Ethanol–water	Colorless needles	204
(5.100) ($R^1 = Et$, $R^2 = Me$, $R^3 = R^4 = R^5 = H$)	D	(5.102) ($R^1 = Et$, $R^2 = R^3 = R^5 = H$, $R^4 = Me$)	81	149.5–150.5	Ethanol–water	Colorless needles	205
(5.100) ($R^1 = Et$, $R^2 = Me$, $R^3 = R^4 = R^5 = H$)	E	(5.102) ($R^1 = Et$, $R^2 = R^5 = H$, $R^3 = R^4 = Me$)	87	141.5–142.5	Ethanol–water	Colorless needles	205
(5.100) ($R^1 = Et$, $R^2 = Me$, $R^3 = R^4 = R^5 = H$)	F	(5.102) ($R^1 = Et$, $R^2 = R^3 = Me$, $R^4 = R^5 = H$)	18	176–177	Ligroin	Prisms	206
(5.100) ($R^1 = Ph$, $R^2 = R^3 = R^4 = R^5 = H$)	G	(5.102) ($R^1 = R^4 = Ph$, $R^2 = R^3 = R^5 = H$)	[e]	198–199	Benzene	Yellow plates	207

551

TABLE 5.52 (Continued)

Starting material	Reaction conditions	Product	Yield (%)	Melting point (°C)	Solvent of crystallization	Crystal form	Ref.
(5.100) (R^1 = CH$_2$Ph, R^2 = R^4 = R^5 = H)	G	(5.102) (R^1 = CH$_2$Ph, R^2 = R^3 = R^5 = H, R^4 = Ph)	36	165–167	—^{h}	Yellow prisms	207
(5.105) (R^1 = R^2 = R^3 = R^4 = H)	H	(5.106) (R^1 = R^2 = R^3 = R^4 = H)	89	366–368	Ethanol	—^{e}	208
(5.105) (R^1 = R^2 = R^4 = H, R^3 = Me)	H	(5.106) (R^1 = R^2 = R^4 = H, R^3 = Me)	91	345–347	Dimethylformamide	—^{e}	208, 210
(5.105) (R^1 = R^2 = R^4 = H, R^3 = Et)	H	(5.106) (R^1 = R^2 = R^4 = H, R^3 = Et)	84	334–335	Dimethylformamide	—^{e}	208
(5.105) (R^1 = R^2 = R^4 = H, R^3 = Ph)	H	(5.106) (R^1 = R^2 = R^4 = H, R^3 = Ph)	45	320–322	Dimethylformamide	—^{e}	208
(5.105) (R^1 = R^4 = H, R^2 = R^3 = Me)	H	(5.106) (R^1 = R^4 = H, R^2 = R^3 = Me)	81	>400	Dimethylformamide	—^{e}	208
(5.105) (R^1 = R^4 = H, R^2 = Ph, R^3 = Me)	H	(5.106) (R^1 = R^4 = H, R^2 = Ph, R^3 = Me)	70	332–336 (decomp.)	Dimethylformamide	—^{e}	208, 209
(5.105) (R^1 = R^4 = H, R^2 = 2-furyl, R^3 = Me)	I	(5.106) (R^1 = R^4 = H, R^2 = 2-furyl, R^3 = Me)	74	280–282	Nitromethane	Tan solid	209
(5.105) (R^1 = R^4 = H, R^2 = p-MeC$_6$H$_4$, R^3 = Me)	J	(5.106) (R^1 = R^4 = H, R^2 = p-Me.C$_6$H$_4$, R^3 = Me)^i	70	280	Nitromethane	Brown solid	209
(5.105) (R^1 = H, R^2 = Ph, R^3 = R^4 = Me)^i	K	(5.106) (R^1 = H, R^2 = Ph, R^3 = R^4 = Me)	49	328–338	Dimethylformamide	—^{e}	209
(5.105) (R^1 = R^4 = H, R^2 = R^3 = Ph)	K	(5.106) (R^1 = R^4 = H, R^2 = R^3 = Ph)	58	304–307	Dimethylformamide	—^{e}	209

552

Reactant	Product	Method	Yield (%)	m.p. (°C)	Solvent of crystallization		Ref.
(5.105) (R^1=R^4=H, R^2=o-ClC$_6$H$_4$, R^3=Me)	(5.106) (R^1=R^4=H, R^2=o-ClC$_6$H$_4$, R^3=Me)	L	32	366–367	Dimethylformamide	—^{e}	209
(5.108)	(5.109)	M	86	358–360	Dimethylformamide	—^{i} —^{e}	210
(5.109)	(5.110)	N	89	366–368	—^{n}	—^{e}	210
(5.105) (R^1=R^2=R^3=Me, R^4=H)	(5.106) (R^1=R^2=R^3=Me, R^4=H)^k	O	79^l	—^{m}	—^{n}	—^{e}	205
(5.105) (R^1=Et, R^2=R^3=Me, R^2=H)	(5.106) (R^1=Et, R^2=R^3=Me, R^4=H)^o	O	85^i	—^{m}	—^{n}	—^{e}	205

^{a} A, As$_2$O$_5$, glycerine, concn. H$_2$SO$_4$/reflux/7 hr. B, As$_2$O$_5$, glycerine, concn. H$_2$SO$_4$/reflux/14 hr. C, glycerine, concn. H$_2$SO$_4$/150–155°/1.5 hr. D, crotonaldehyde, concn. HCl/reflux/2 hr. E, propenyl methyl ketone, concn. HCl/reflux/2 hr. F, methyl vinyl ketone, FeCl$_3$, ZnCl$_2$, ethanol/65°/2 hr then reflux/4 hr. G, cinnamaldehyde, H$_3$PO$_4$/110°/4–20 hr. H, Dowtherm/reflux/6 min. I, Dowtherm/230°/15 min. J, Dowtherm/reflux/1 hr. K, Dowtherm/reflux/0.5 hr. L, Dowtherm/reflux/15 min. M, 2 M NaOH, reflux/3 hr. N, quinaldine/reflux/9 hr. O, diphenyl ether/250–260°/15 min.

^{b} Yield not specified.

^{c} As the dihydrochloride.

^{d} Forms a dihydrochloride, m.p. 284–286° (decomp.) (from ethanol–ether).

^{e} Crystal form not specified.

^{f} Forms a picrate m.p. 240° (decomp.) (from ethanol).

^{g} Could not be crystallized.

^{h} Purified by molecular distillation.

^{i} Hydrate.

^{j} Purified by precipitation from HCl solution by NH$_4$OH.

^{k} Mixture with (5.107) (R^1=R^2=R^3=Me, R^4=H).

^{l} Total yield of mixture.

^{m} Not recorded.

^{n} Not purified.

^{o} Mixture with (5.107) (R^1=Et, R^2=R^3=Me, R^4=H).

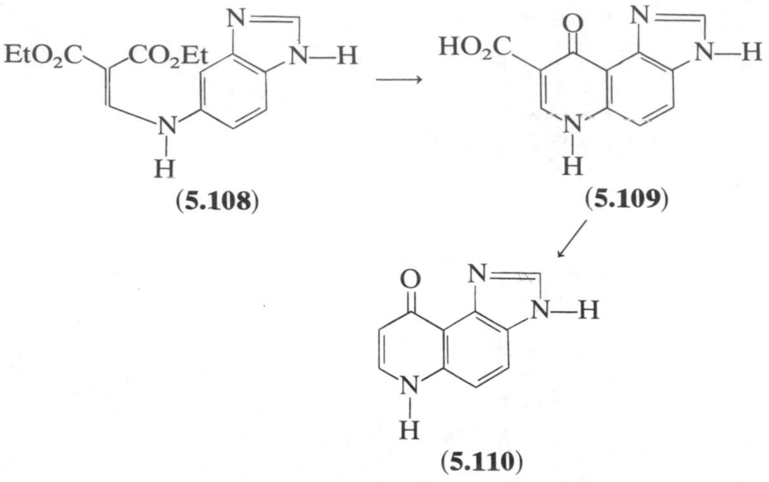

(5.108)   (5.109)

(5.110)

Scheme 5.9

Ring-Closure Reactions of Other Heterocycles

Ring-closure reactions of *ortho*-diamino derivatives of benzopyrans, ben-zothiopyrans, quinolines, and isoquinolines provide perhaps the most general basis for the synthesis of tricyclic 6-6-5 fused benzimidazoles containing one additional hetero atom. In one version of this synthetic approach, the 6- and 8-amino derivatives of 7-acetylamino[5,6]benzopyran-4-ones are cyclodehydrated by treatment with acetic anhydride to afford imidazo[4,5-g]-[5,6]benzopyran-8-ones (e.g., Scheme 5.10, **5.111 → 5.112**),[211] and imidazo-[4,5-h][5,6]benzopyran-6-ones (e.g., Scheme 5.10, **5.113 → 5.114**)[200] in ac-ceptable yields (Table 5.50). Subsequent hydrolysis of the esters produced in the latter synthesis also provides access (Table 5.50) to imidazo[4,5-h][5,6]-benzopyran-6-one-8-carboxylic acids (e.g., Scheme 5.10, **5.114 → 5.115**).[200] Imidazo[4,5-f][5,6]benzothiopyran-7-ones are likewise readily prepared (Table 5.50) by the cyclodehydration of 6-acetylamino-5-amino[5,6]benzo-thiopyran-2-ones [Scheme 5.11, **5.116** (R = Ac) → **5.117** (R = Me)],[212] or by the ring closure of a 5,6-diamino[5,6]benzothiopyran-2-one with a carbox-ylic acid such as formic acid [Scheme 5.11, **5.116** (R = H) → **5.117** (R = H)].[212]

Acylative ring closure of 6,7-diaminoquinoline derivatives, either pre-formed or prepared *in situ* by reduction of the more accessible dinitro or nitro–amino precursors, provides a widely used synthetic route to the imidazo[4,5-g]quinoline ring system. This method, which complements that through benzimidazole derivatives, is exemplified by the condensation[203] of 6,7-diaminoquinoline (**5.119**) [prepared *in situ* by catalytic reduction of 7-amino-6-nitroquinoline (**5.118**)] with formic acid to afford 1(3)H-imidazo-[4,5-g]quinoline (**5.120**) in high yield (Table 5.53).

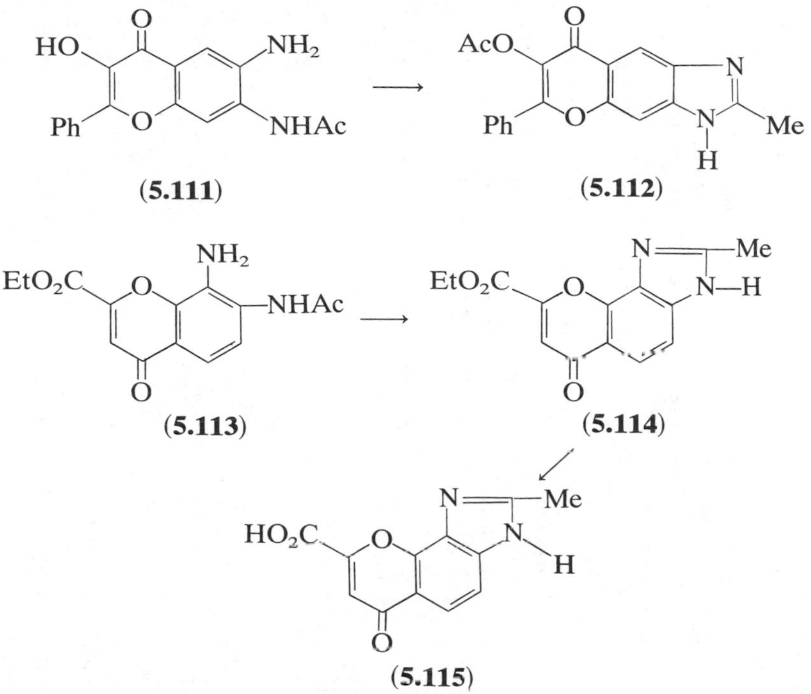

(5.111)   **(5.112)**

(5.113)   **(5.114)**

(5.115)

Scheme 5.10

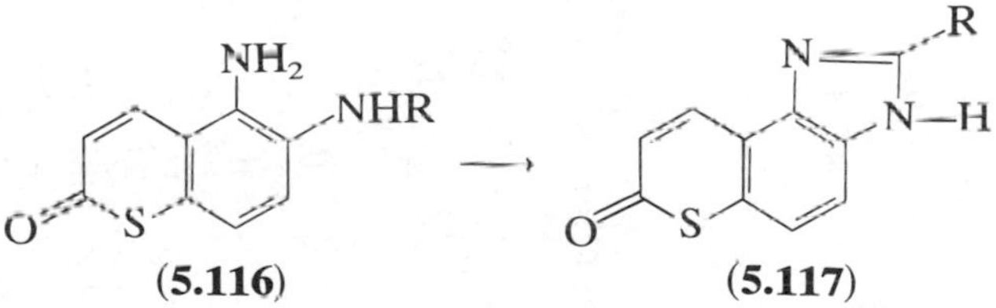

(5.116)   **(5.117)**

Scheme 5.11

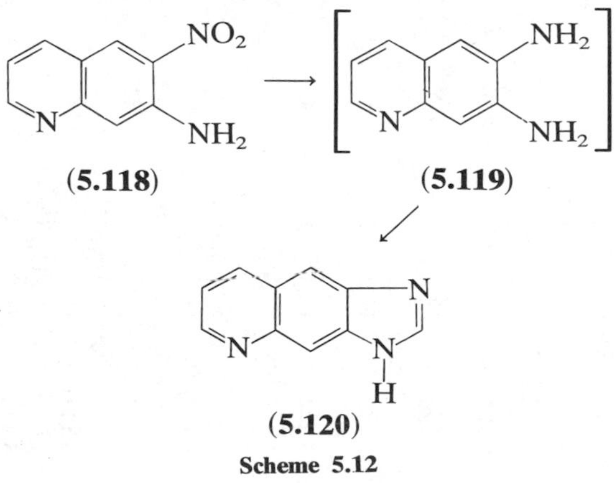

(5.118)   **(5.119)**

(5.120)

Scheme 5.12

TABLE 5.53. SYNTHESES OF IMIDAZO[4,5-g]QUINOLINE DERIVATIVES (5.120), (5.123), AND (5.127) BY RING-CLOSURE REACTIONS OF 6,7-DIAMINOQUINOLINE DERIVATIVES

Starting material	Reaction conditions[a]	Product	Yield (%)	Melting point (°C)	Solvent of crystallization	Crystal form	Ref.
(5.118)	A	(5.120)[a,b]	77	222	Ethanol–ligroin	—[c]	203
(5.121) (R = H)	B	(5.123) (R^1 = R^2 = H, R^3 = 2-furyl)	56	>340	Nitrobenzene	Orange red crystals	213
(5.121) (R = H)	B	(5.123) (R^1 = R^2 = H, R^3 = 4-pyridyl)	60	360 (decomp.)	Dimethylformamide	Yellow crystals	213
(5.121) (R = H)	B	(5.123) (R^1 = R^2 = H, R^3 = Ph)	52	>360 (decomp.)	Dimethylformamide	Yellow crystals	213
(5.121) (R = Me)	B	(5.123) (R^1 = Me, R^2 = H, R^3 = 2-vanillinyl)	45	327 (decomp.)	Dimethylformamide	Orange crystals	213
(5.122) (R^1 = R^2 = H, R^3 = Me)	C	(5.123; R^1 = R^2 = H, R^3 = Me)	69	>360 (decomp.)	Dimethylformamide	Yellow brown crystals	213
(5.122) (R^1 = H, R^2 = Ph, R^3 = Me)	C	(5.123; R^1 = H, R^2 = Ph, R^3 = Me)	93	275	1,2,4-trichlorobenzene	Yellow crystals	213
(5.122) (R^1 = H, R^2 = Ph, R^3 = Et)	C	(5.123) (R^1 = H, R^2 = Ph, R^3 = Et)	92	282 (decomp.)	1,2-dichlorobenzene	Yellow crystals	213
(5.122) (R^1 = R^3 = Me, R^2 = o-MeOC$_6$H$_4$)	C	(5.123) (R^1 = R^3 = Me, R^2 = o-MeOC$_6$H$_4$)	—[e]	270	Pyridine	—[c]	213
(5.122) (R^1 = H, R^2 = p-Me$_2$NC$_6$H$_4$, R^3 = Me)	C	(5.123) (R^1 = H, R^2 = p-Me$_2$NC$_6$H$_4$, R^3 = Me)	—[e]	306 (decomp.)	Pyridine	—[c]	213
(5.122) (R^1 = H, R^2 = o-MeOC$_6$H$_4$, R^3 = Me)	C	(5.123) (R^1 = H, R^2 = o-MeOC$_6$H$_4$, R^3 = Me)	—[e]	270 (decomp.)	Chlorobenzene	—[c]	213
(5.122) (R^1 = H, R^2 = p-Me$_2$NC$_6$H$_4$, R^3 = Et)	C	(5.123) (R^1 = H, R^2 = p-Me$_2$NC$_6$H$_4$, R^3 = Et)	—[e]	329–331	Chlorobenzene	Red crystals	213
(5.124) (R = H)	D	(5.127) (R^1 = R^2 = H)	97	126–128	Ethanol	—[c]	214
(5.124) (R = Ac)	D	(5.127) (R^1 = Ac, R^2 = H)	95	>270	Ethanol	—[c]	214
(5.126)	D	(5.127) (R^1 = H, R^2 = Bu^n)	100	150–152	Ethanol	—[c]	214

[a] A, (i) Raney Ni, H$_2$, ethanol/room temp./atm. press. (ii) HCO$_2$H/reflux/1 hr. B, R^{3}CHO, AcOH, H$_2$O/reflux/2 hr. C, 10% aq. NaOH, ethanol/reflux/20 min. D, (i) Raney Ni—N$_2$H$_4$. H$_2$O, ethanol, H$_2$O/100°. (ii) CS$_2$, pyridine/reflux/3 hr.

[b] Forms a dihydrochloride, m.p. >270° (decomp.) (from methanol–ether).

[c] Crystal form not specified.

[d] Forms a picrate, m.p. 266 (from water).

In closely related cyclizations,[213] readily accessible 6,7-diaminoquinoline-5,8-quinones (**5.121**) condense with aromatic and heterocyclic aldehydes in aqueous acetic acid to afford moderate to good yields (Table 5.53) of the corresponding imidazo[4,5-g]quinoline-4,9-quinones (**5.123**). An alternative general route to the latter is also provided by the high yield (Table 5.53) alkali-catalyzed cyclization of 6-acylamino-7-aminoquinoline-5,8-quinones (**5.122** → **5.123**).[213]

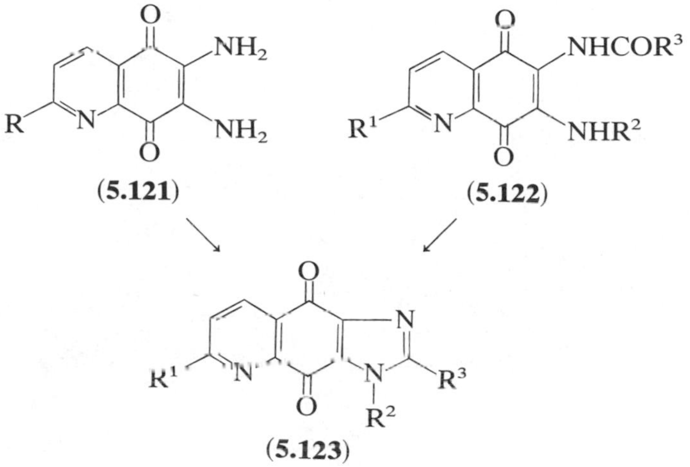

(5.121)   (5.122)

(5.123)

Scheme 5.13

Thioacylative ring-closure occurs when 6,7-diamino-1,2,3,4-tetrahydroquinoline derivatives (prepared *in situ* by reduction of the corresponding 6,7-dinitro- or 7-amino-6-nitro-1,2,3,4-tetrahydroquinolines with hydrazine in the presence of Raney-nickel) are allowed to react with carbon disulphide in refluxing pyridine, affording excellent yields (Table 5.53) of 5,6,7,8-tetrahydroimidazo[4,5-g]quinoline-2(1H,3H)-thiones (Scheme 5.14, **5.124** or **5.126** → **5.125** → **5.127**).[214]

Acylative ring-closure reactions of 5,6-diaminoquinolines are also a fertile source of a wide variety of imidazo[4,5-f]quinoline derivatives (Scheme 5.15) (Table 5.54). For example, 5,6-diaminoquinoline (**5.129**, $R^1 = R^2 =$ H)[215] and its N(6)-alkyl (**5.129**; $R^1 =$ alkyl)[216] and N(6)-aryl (**5.129**, $R^1 =$ aryl)[217,218] derivatives react smoothly with formic or acetic acids, or acetic anhydride, alone or in the presence of hydrochloric acid under reflux to afford consistently high yields (Table 5.54) of the corresponding imidazo-[4,5-f]quinolines (**5.131**; $R^1 =$ H, alkyl, or aryl, $R^2 =$ H, $R^3 =$ H or methyl). Analogous reactions (Scheme 5.16) of 5,6-diamino-1-methylquinolinium benzenesulfonates (**5.134**) with hot formic acid in the presence of concentrated hydrochloric acid, followed by treatment with hydriodic acid, lead to 6-methylimidazo[4,5-f]quinolinium iodides (**5.135**), albeit in unspecified

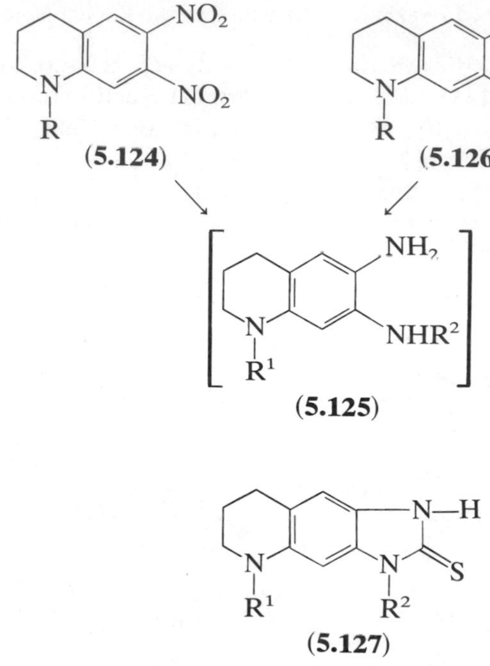

Scheme 5.14

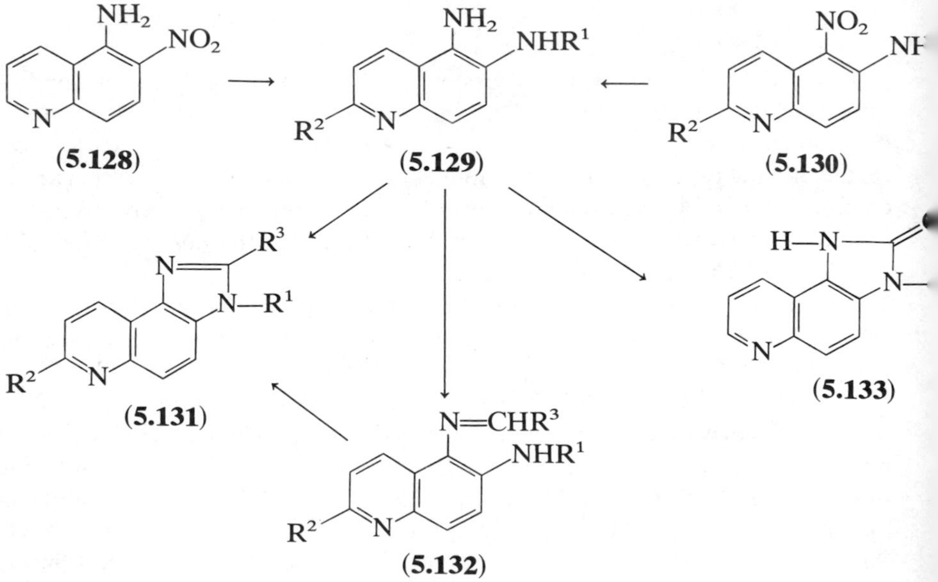

Scheme 5.15

TABLE 5.54. SYNTHESES OF IMIDAZO[4,5-f]QUINOLINE DERIVATIVES BY RING-CLOSURE REACTIONS OF 5,6-DIAMINO QUINOLINE DERIVATIVES

Starting materials	Reaction conditions[a]	Product	Yield (%)	Melting point (°C)	Solvent of crystallization	Crystal form	Ref.
(5.129) ($R^1 = R^2 = H$)	A	(5.131) ($R^1 = R^2 = R^3 = H$)	100	215–217	Water	Colorless prisms	215
(5.129) ($R^1 = R^2 = H$)	B	(5.131) ($R^1 = R^2 = H$, $R^3 = Me$)[b]	77	313	Acetone–water	Colorless prisms	215
(5.129) ($R^1 = Me$, $R^2 = H$)	C	(5.131) ($R^1 = Me$, $R^2 = R^3 = H$)	100	189–190	Benzene	Needles	216
(5.129) ($R^1 = CH_2Ph$, $R^2 = H$)[c]	D	(5.131) ($R^1 = CH_2Ph$, $R^2 = R^3 = H$)	94	158	Ethanol–water	Colourless needles	216
(5.129) ($R^1 = Ph$, $R^2 = H$)[d]	E	(5.131) ($R^1 = Ph$, $R^2 = R^3 = H$)	73	164.5	Benzene	Colorless plates	217
(5.129) ($R^1 = $ 2,4-dinitrophenyl, $R^2 = H$)	F	(5.131) ($R^1 = $ 2,4-dinitrophenyl, $R^2 = R^3 = H$)	—[e]	232	—[f]	—[g]	218
(5.129) ($R^1 = $ 2,4-dinitrophenyl, $R^2 = H$)	G	(5.131) ($R^1 = $ 2,4-dinitrophenyl, $R^2 = H$, $R^3 = Me$)	—[g]	319	—[f]	—[g]	218
(5.134) ($R = H$)	H	(5.135) ($R = H$)	—[e]	234–235	Ethanol	Yellow crystals	217
(5.134) ($R = Me$)	H	(5.135) ($R = Me$)	—[e]	282	Ethanol	Yellow crystals	217
(5.134) ($R = Ph$)	H	(5.135) ($R = Ph$)	—[e]	269–270	Ethanol	Yellow crystals	217
(5.134) ($R = CH_2Ph$)	H	(5.135) ($R = CH_2Ph$)	—[e]	264–265	Ethanol	Yellow crystals	217
(5.128)	I	(5.131) ($R^1 = R^2 = R^3 = H$)[h]	—[e]	214	Benzene	—[g]	219
(5.128)	J	(5.131) ($R^1 = R^2 = H$, $R^3 = Me$)[i]	—[e]	200	Water[i]	—[g]	219
(5.130) ($R^1 = R^2 = Me$)	K	(5.131) ($R^1 = R^2 = R^3 = Me$)	79	201.5–202.5	Water	Colorless needles	205
(5.130) ($R^1 = Et$, $R^2 = Me$)	K	(5.131) ($R^1 = Et$, $R^2 = R^3 = Me$)	82	149.5–150.5	Ethanol–water	Colorless needles	205
(5.130) ($R^1 = CH_2Ph$, $R^2 = Me$)	L	(5.131) ($R^1 = CH_2Ph$, $R^2 = R^3 = Me$)[j]	51	103–105	Water	Colorless needles	204
(5.129) ($R^1 = CH_2Ph$, $R^2 = H$)[c]	M	(5.133)	79	280	Ethanol	Colorless needles	216

559

TABLE 5.54 (*Continued*)

Starting materials	Reaction conditions	Product	Yield (%)	Melting point (°C)	Solvent of crystallization	Crystal form	Ref.
(5.129) (R¹ = R² = H)	N	(5.131) (R¹ = R² = R³ = H)	—[e]	215–217	Water	Colorless prisms	215
(5.136) (R = H) + (5.137) (n = 0, X = H)	O	(5.138; n = 0, R = X = H)	25	—	—	—[q]	220
(5.136) (R = H)[c] + (5.137) (n = 0, X = Br)	P	(5.138; n = 0, R = H, X = Br) + (5.139, n = 0, X = Br)[r]	13	230–231	[f]	—[g]	220
		(5.139, n = 0, X = Br)[r]	14	170–174	[f]	—[g]	220
(5.136) (R = Me)[c] + (5.137) (n = 0, X = H)	P	(5.138; n = 0, R = Me, X = H)	10	139–142	[f]	—[g]	220
(5.136) (R = Me) + (5.137) (n = 0, X = Br)	O	(5.138; n = 0, R = Me, X = Br)	27	175–176	[f]	—[g]	220
(5.136) (R = Me) + (5.137) (n = 0, X = NO₂)	O	(5.138) (n = 0, R = Me, X = NO₂)	25	234–235	[f]	—[g]	220
(5.136) (R = H)[c] + (5.137) (n = 1, X = H)	P	(5.138) (n = 0, R = X = H) + (5.139) (n = 1, X = H)[r]	67	204–205	[f]	—[g]	220
		(5.139) (n = 1, X = H)[r]	12	192–198	[f]	—[q]	220
(5.136) (R = Me) + (5.137) (n = 1, X = H)	O	(5.138) (n = 1, R = Me, X = H)	23	—	—	—[g]	220
(5.136) (R = Me) + (5.137) (n = 1, X = Br)	O	(5.138) (n = 1, R = Me, X = Br)	18	143–145	[f]	—[g]	220
(5.136) (R = Me) + (5.137) (n = 1, X = NO₂)	O	(5.138) (n = 1, R = Me, X = NO₂)	24	248–250	[f]	—[q]	220
(5.136) (R = H)[c] + (5.140) (n = 0, X = Br)	Q	(5.138) (n = 0, R = H, X = Br)	—	—	—	—[q]	220
(5.136) (R = Me)[c] + (5.140) (n = 0, X = NO₂)	Q	(5.138) (n = 0, R = Me, X = NO₂)	75	—	—	—	220
(5.136) (R = H)[c] + (5.140) (n = 1, X = H)	Q	(5.138) (n = 1, R = X = H)	40	—	—	—	220

560

(5.136) (R = Me)^c + (5.140) (n = 1, X = NO₂)	Q		62	—	—	220
(5.138) (n = 1, R = Me, X = NO₂)						
(5.132) (R¹ = R² = H, R³ = p-NO₂C₆H₄)^s	N		—^e	334.5	HCl aq.	Yellow crystals	215
(5.131) (R¹ = R² = H, R³ = p-NO₂C₆H₄)^s						
(5.132) (R¹ = R² = H, R³ = CH=CHPh)	N		79	258	Benzere	Yellow plates	215
(5.131) (R¹ = R² = H, R³ = CH=CHPh)^t						

^{a} A, HCO_2H/reflux/4 hr. B, $MeCO_2H$/reflux/3 hr. C, HCO_2H, concn. HCl/reflux/4 hr. D, HCO_2H/125-135°/10 hr. E, HCO_2H/reflux/6 hr. F, HCO_2H, con. HCl/reflux/5 hr. G, Ac_2O, AcOH, concn. HCl/reflux/4 hr. H, 80% HCO_2H, concn. HCl/reflux/4 hr, then treat with HI. I. $SnCl_2$, concn. HCl, HCO_2H/reflux/several hr. J, $SnCl_2$, concn. HCl. $MeCC_2H$/reflux/several hr. K, () PdC/H_2, room temp. atm press; (ii) 4 M HCl, Ac_2O/reflux/0.5 hr. L, (i) PdC/H_2, room temp., atm press; (ii) 15% HCl, Ac_2O/100°/0.5 hr. M, urea; 140°. N. R^3CHO, Cu(II) acetate, EtOH, H_2O/100°/15 min. O, piperidine, DMF/reflux/1 hr. P, Cu(II) acetate, H_2O, EtOH/reflux/2 hr. Q, EtOH/reflux/1 hr.

^{b} Hydrochloride; anhydrous free base has no definite m.p. and forms a hydrate.
^{c} Dihydrochloride.
^{d} Monohydrochloride.
^{e} Yield not specified.
^{f} Solvent not specified.
^{g} Crystal form not specified
^{h} Forms a hydrate, m.p. 78°.
^{i} Forms a hydrate, m.p. 70°.
^{j} Forms a picrate, m.p. 230° (decomp.) (from ethanol).
^{k} Forms a trihydrate.
^{l} Forms a hydrochloride, prisms, m.p. 282–284° (from EtOH·HCl).
^{m} Forms a picrate, m.p. 269° (from water).
^{n} Free base, m.p. 184° (dihydrate).
^{o} Free base, m.p. 100–105° (anhydrous); forms a crystalline monohydrate.
^{p} Nitrate; free base, colorless cubes (1.5 hydrate), m.p. 270° (ethanol-water).
^{q} Oil.
^{r} Isomer mixture.
^{s} Hydrochloride; free base, yellow brown prisms, m.p. 356°.
^{t} Forms a hydrochloride, m.p. 280° (decomp.) (from HCl aq.).

561

yield (Table 5.54).[217] In acylative condensations of type (**5.129**→ **5.131**), the 5,6-diaminoquinoline substrate (**5.129**) need not be preformed but can be generated by the reduction of the corresponding 6-amino-5-nitro- or 5-amino-6-nitroquinoline and subjected to ring closure without isolation. This variant is exemplified (Table 5.54) by the reductive conversion[219] of 5-amino-6-nitroquinoline (**5.128**) by stannous chloride in the presence of formic or acetic acids into the imidazo[4,5-*f*]quinolines (**5.131**; $R^1 = R^2 = H$, $R^3 = H$ or Me) via the presumed intermediacy of 5,6-diaminoquinoline or an acyl derivative (**5.129**; $R^1 = H$ or acyl, $R^2 = H$) (Scheme 5.15). Correspondingly, catalytic reduction of 6-amino-5-nitroquinoline derivatives (**5.130**) and reaction of the resulting 5,6-diaminoquinoline (**5.129**), without isolation, with acetic anhydride in the presence of hydrochloric acid affords the anticipated imidazo[4,5-*f*]quinolines (**5.131**) (Scheme 5.15)[204,205] in moderate to excellent yield (Table 5.54).

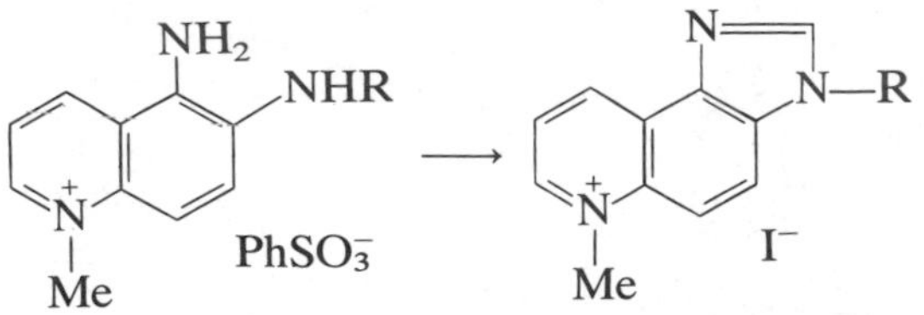

(**5.134**)   (**5.135**)

Scheme 5.16

Reagents suitable for effecting the ring closure of 5,6-diaminoquinoline derivatives to imidazo[4,5-*f*]quinolines are not confined to carboxylic acids or their anhydrides. Processes of this type can also be accomplished directly by fusing with urea or perhaps more important in a synthetic sense, by oxidative condensation with aldehydes, both aliphatic and aromatic. The former type of ring closure provides a potentially general, if relatively unexploited (Table 5.54), synthetic route to imidazo[4,5-*f*]quinolin-2(1*H*,3*H*)-ones [e.g., Scheme 5.15, **5.129** ($R^1 = CH_2Ph$, $R^2 = H$) → **5.133**].[216] Ring closure of 5,6-diaminoquinolines with aldehydes to give imidazo[4,5-*f*]quinolines is accomplished oxidatively in the presence of Cu(II) acetate.[215,220] Since yields are generally high (Table 5.54) and a wide range of alkyl, aryl, and hetaryl aldehydes can be employed, this method is probably the single most flexible process for the synthesis of imidazo[4,5-*f*]-quinolines from 5,6-diaminoquinoline precursors. However, one complicating feature of such imidazo[4,5-*f*]quinoline syntheses, particularly apparent in the oxidative condensation of 5,6-diaminoquinoline itself with furan carboxaldehydes,[220] is the co-formation of bis-condensates (Table 5.54) derived by concurrent attack at both amino centers. For example, the bis-condensates (**5.139**; $n = 0$, X = Br) and (**5.139**; $n = 1$, X = H) accompany the orthodox products (**5.138**; $n = 0$, R = H, X = Br) and (**5.138**; $n = 1$,

R = X = H) in the oxidative condensation of 5,6-diaminoquinoline (**5.136**; R = H) with 5-bromofuran-2-carboxaldehyde (**5.137**; $n = 0$, X = Br) and 2-furylacrolein (**5.137**; $n = 1$, X = H), respectively (Scheme 5.17 and Table 5.54).[220] However, this complication is readily circumvented and at the same time the yields are improved (Table 5.54) by the use of the imidate esters of furan-2-carboxylic acids (**5.140**; $n = 0$) or 2-(2-furyl)acrylic acids (**5.140**; $n = 1$) as reagents.[220]

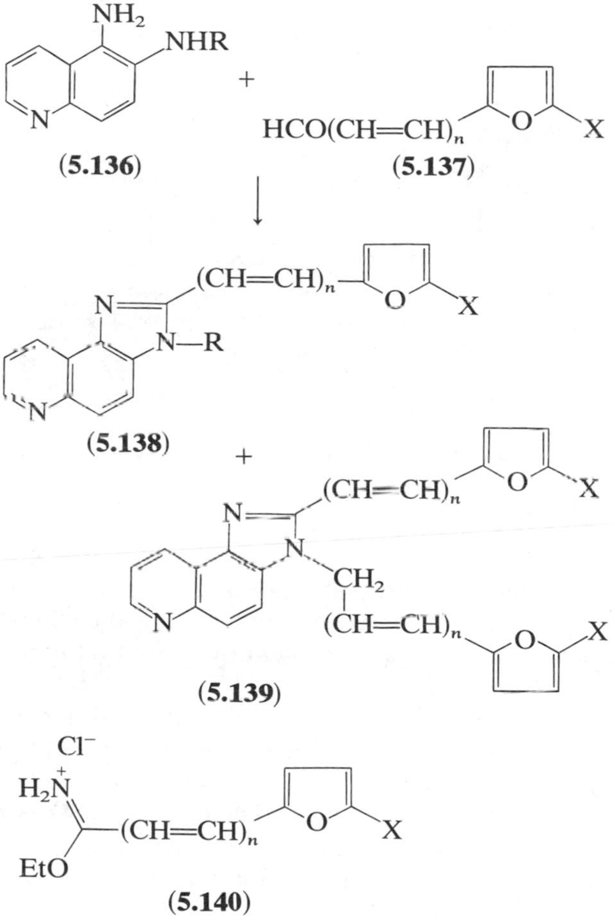

Scheme 5.17

The demonstration that preformed Schiff bases of the type (**5.132**) (Scheme 5.15) can be oxidatively cyclized to the corresponding imidazo[4,5-f]-quinolines using Cu(II) acetate (Scheme 5.15, **5.132** → **5.131**)[215] supports a mechanism (Scheme 5.18) for the direct aldehyde condensations involving initial Schiff base formation (**5.141** → **5.142**) followed by cyclization to an

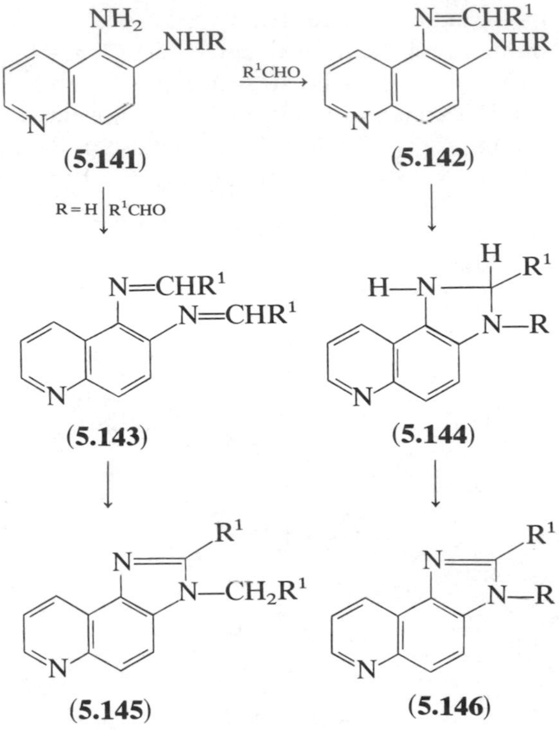

Scheme 5.18

imidazoline intermediate and subsequent oxidation (**5.142 → 5.144 → 5.146**). Formation of bis-condensation products may then be rationalized by the intermediate formation and subsequent oxidative cyclization of bis Schiff bases (**5.141 → 5.143 → 5.145**).

Imidazo[4,5-*h*]quinolines have principally attracted attention[221–224] as chelating agents, and as such, have in general been prepared (Table 5.55) by the polyphosphoric-acid-catalyzed condensation of nitriles with the relatively inaccessible 7,8-diaminoquinoline (Scheme 5.19, **5.147 → 5.148**),[221,222] or as the 5-nitro derivatives (Table 5.55) by polyphosphoric-acid-catalyzed ring-closure of the more accessible 7,8-diamino-5-nitroquinoline with

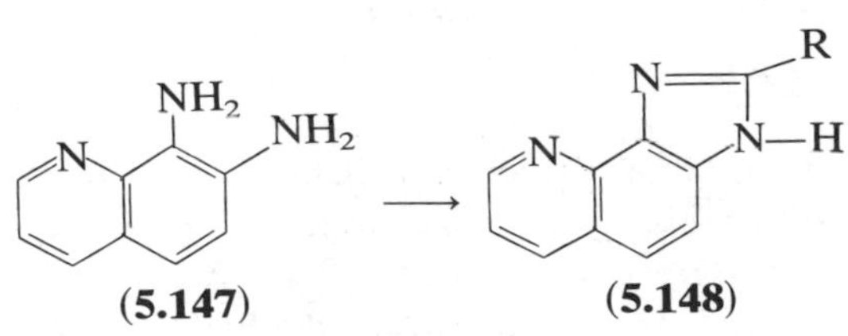

Scheme 5.19

TABLE 5.55. SYNTHESES OF IMIDAZO[4,5-h]QUINOLINE DERIVATIVES

Starting material	Reaction conditions	Product	Yield (%)	Melting point (°C)	Solvent of crystallization	Crystal form	Ref.
(5.147)	A	(5.148) (R 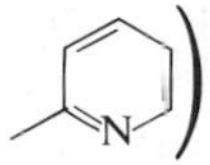 =)	79	235	Ethanol	—[b]	221
(5.147)	A	(5.148) (R 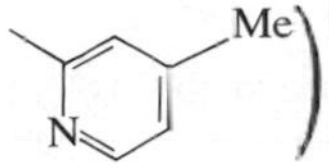 =)	23	174	Benzene	—[b]	221
(5.147)	A	(5.148) (R 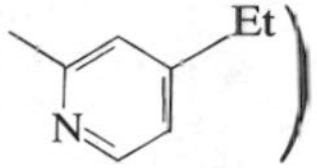 =)	39	104	Benzene–light petroleum	—[b]	221
(5.147)	A	(5.148) (R 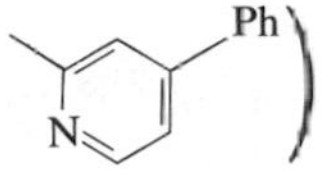 =)	44	111	Benzene	—[b]	221
(5.147)	A	(5.148) (R 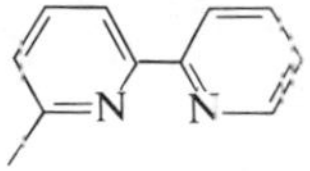 =)	56	180	Benzene	—[b]	221
(5.147)	A	(5.148) (R 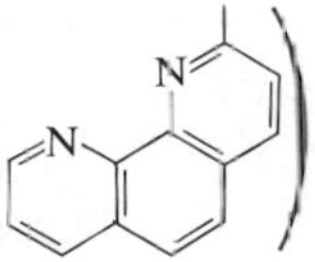 =)	20	270	Benzene	—[b]	222
(5.147)	A	(5.148) (R 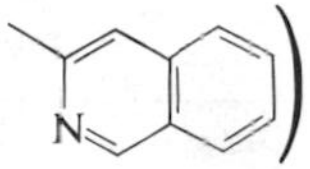 =)	37	236	Benzene	—[b]	222
(5.147)	A	(5.148) (R 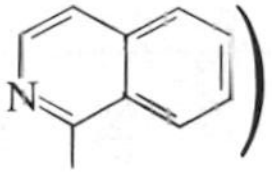 =)	53	220	Benzene	—[b]	222
(5.149)	A	(5.148) (R = H)	55	341–342 (decomp.)	Ethanol	Yellow crystals	223

565

TABLE 5.55. (*Continued*)

Starting material	Reaction conditions	Product	Yield (%)	Melting point (°C)	Solvent of crystallization	Crystal form	Ref.
(**5.149**)	B	(**5.152**) (R = Me)	50	307	Ethanol	Yellow crystals	223
(**5.149**)	B	(**5.152**) (R = Ph)	—[c]	—[d]	—[e]	—[b]	224
(**5.149**)	B	(**5.152**) (R = CH$_2$Ph)	—[c]	—[d]	—[e]	—[b]	224
(**5.149**)	B	(**5.152**) (R = CF$_3$)	—[c]	—[d]	—[e]	—[b]	224
(**5.150**) (R = H)	C	(**5.153**) (R^1 = R^2 = H)[f]	70	282–283	Methanol	Yellow crystals	224
(**5.150**) (R = H)	C	(**5.153**) (R^1 = H, R^2 = Me)[g]	15	154	Methanol-dichloromethane	Yellow needles	224
(**5.150**) (R = H)	C	(**5.154**) (R = CF$_3$)	7	312	Methanol-ethanol	Yellow needles	224
(**5.150**) (R = H)	C	(**5.154**) (R = Ph)[i]	9	252 (decomp.)	Methanol	Yellow crystals	224
(**5.150**) (R = H)	C	(**5.154**) (R = CH$_2$Ph)[j]	17	263–264	Ethanol	Colorless needles	224
(**5.150**) (R = Me)	C	(**5.153**) (R^1 = Me, R^2 = H)[k]	20	320.5 (decomp.)	—[e]	Colorless needles	224
(**5.150**) (R = Me)	C	(**5.153**) (R^1 = R^2 = Me)[l]	12	298	—[e]	Colorless needles	224

[a] A, RCN, polyphosphoric acid/250°/4 hr. B, RCO$_2$H, polyphosphoric acid/250°/2 hr. C, RCO$_2$H, SnCl$_2$, concn. HCl/reflux/8 hr.
[b] Crystal form not specified.
[c] Yield not reported.
[d] Melting point not reported.
[e] Solvent not specified.
[f] Forms a dihydrochloride, orange needles, m.p. 265–6° (decomp.) (from methanol).
[g] Forms a dihydrochloride, red crystals, m.p. 293° (decomp.) (from acetone).
[h] Forms a monohydrochloride, yellow needles, m.p. 214° (from acetone).
[i] Forms a dihydrochloride, yellow crystals, m.p. 252° (decomp.) (from methanol).
[j] Forms a dihydrochloride, colorless needles, m.p. 239° (from acetone).
[k] Forms a dihydrochloride, colorless needles, m.p. 317° (decomp.) (from methanol).
[l] Forms a dihydrochloride, colorless needles, m.p. 312° (from methanol).

carboxylic acids (Scheme 5.20, **5.149 → 5.152**).[223,224] In a closely related procedure, carboxylic acids are condensed with 5,7,8-triaminoquinolines (produced *in situ* by stannous-chloride-catalyzed reduction of the requisite 5,7-dinitro-8-aminoquinolines) to afford 5-aminoimidazo[4,5-*h*]-quinolines (Scheme 5.20, **5.150 → 5.151 → 5.153**), albeit in low overall yield (Table 5.55). In some instances (specifically using trifluoroacetic, phenylacetic, or benzoic acids as reagents),[224] the latter synthetic approach results in subsequent hydrolysis of the 5-amino group, the products then being the corresponding 5-hydroxyimidazo[4,5-*h*]quinolines (Scheme 5.20, **5.150 → 5.151 → 5.154**) (Table 5.55). In contrast to related syntheses of imidazo[4,5-*h*]quinolines (cf. Table 5.54), all of these ring closures to imidazo[4,5-*h*]quinolines appear to proceed in low yield (Table 5.55), although reaction conditions may well not have been optimized.

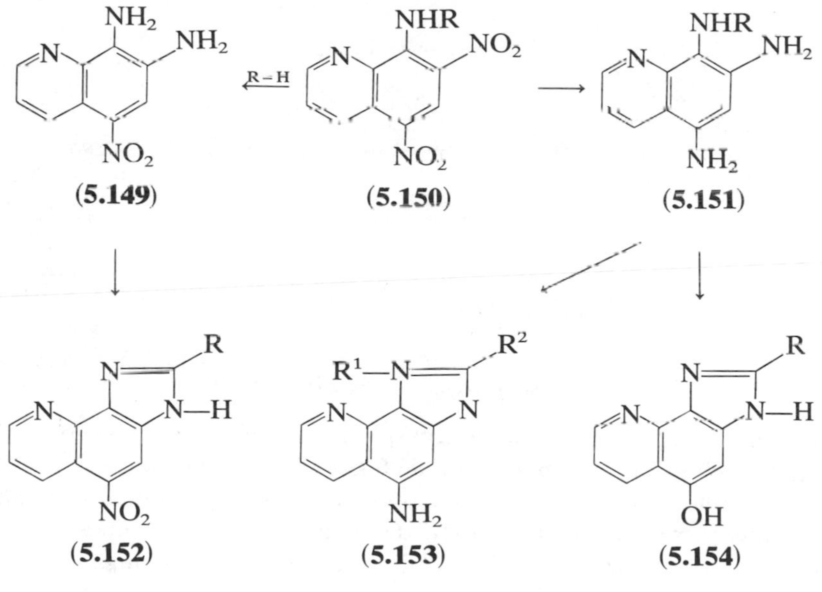

Scheme 5.20

The parent member (**5.158**) of the imidazo[4,5-*h*]isoquinoline ring system is synthesized in moderate-to-excellent yield by the condensation of 7,8-diaminoisoquinoline [readily available by the stannous chloride or catalytic reduction of 8-amino-7-nitroisoquinoline (**5.155**) or 7-amino-8-nitroquin-oline (**5.156**)] with formic acid (Scheme 5.21, **5.157 → 5.158**).[225]

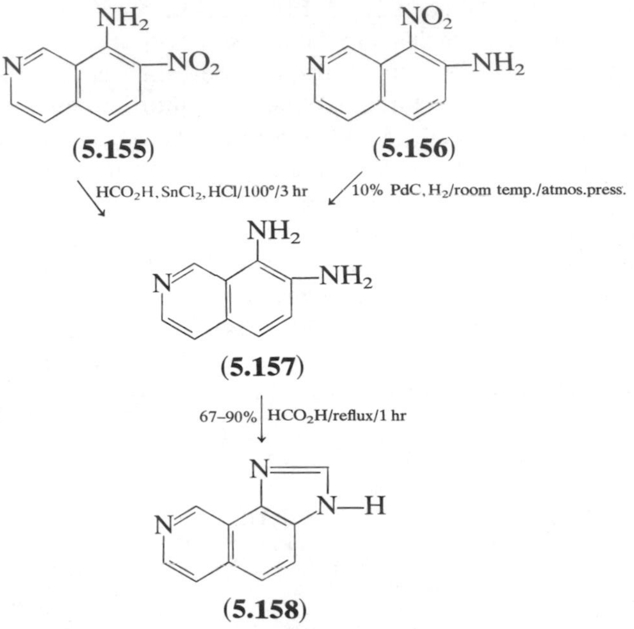

(5.155)   **(5.156)**

HCO$_2$H, SnCl$_2$, HCl/100°/3 hr   10% PdC, H$_2$/room temp./atmos.press.

(5.157)

67–90% | HCO$_2$H/reflux/1 hr

(5.158)

[colourless needles, m.p. 221° (from benzene); forms a dihydrochloride, m.p. 298–301° (decomp.) (from methanol-ether)]

Scheme 5.21

5.2.2. Physicochemical Properties

Spectroscopic Studies

INFRARED SPECTRA. Routine detection of functional groups apart, the infrared spectra (Table 5.56) of tricyclic 6-6-5 fused benzimidazoles containing one additional hetero atom have only been studied in detail in relation to the tautomeric character of N-1(3)H-imidazo[4,5-g]quinolines (Scheme 5.22a, **5.159 ⇌ 5.160**) and N-1(3)H-imidazo[4,5-h]quinolines (Scheme 5.22a, **5.161 ⇌ 5.162**). In the latter case, the position of tautomeric equilibria of type (**5.161 ⇌ 5.162**) is crucial in relation to the role of imidazo[4,5-h]-quinolines as metal complexing agents (see later in this section). The infrared spectra of both N-1(3)H-imidazo[4,5-g]quinolines (**5.159 ⇌ 5.160**)[204] and N-1(3)H-imidazo[4,5-h]quinolines (**5.161 ⇌ 5.162**)[223,224] exhibit broad absorption bands in the 3500 to 3100 cm^{-1} region originating in the stretching vibrations of associated imidazole-NH groups. Broad NH absorption in the regions 3240 to 3120 cm^{-1} and 2490 cm^{-1} is taken as evidence for the existence of potentially tautomeric 2-thioimidazo[4,5-g]-quinolines (Scheme 5.22a, **5.163 ⇌ 5.164**) in the thione form (**5.163**) rather than the mercapto form (**5.164**).[214]

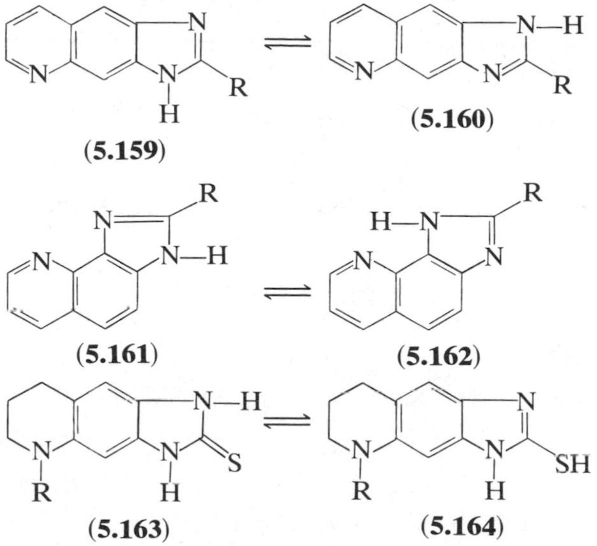

(5.159)   **(5.160)**

(5.161)   **(5.162)**

(5.163)   **(5.164)**

Scheme 5.22a

ULTRAVIOLET, VISIBLE, AND LUMINESCENCE SPECTRA. Not unexpectedly, in view of their condensed aromatic structures, the ultraviolet-visible absorption characteristics of the imidazo[4,5-g]quinoline,[205,214] imidazo[4,5-f]-quinoline,[205,218,220,226] and imidazo[4,5-h]quinoline[223,224,227] ring systems have been extensively investigated (Table 5.57).

The ultraviolet spectra of imidazo[4,5-g]quinolines[205] are typified by the presence of a single intense absorption band around 250 nm accompanied by a lower intensity band at longer wavelength (ca. 330 nm) (Table 5.57). Hexahydroimidazo[4,5-g]quinolines show the expected decrease in the intensity of ultraviolet absorption (Table 5.57) consequent on the presence of a less extensive chromophore.[214]

The ultraviolet spectra of imidazo[4,5-f]quinolines[205,218,220,226] (Table 5.57) differ markedly from those of their imidazo[4,5-g]quinoline counterparts (a fact that has been used[205] to distinguish the linear [4,5-g] ring system from the angular [4,5-f] ring system) and are characterized[226] by the presence of two intense maxima at approximately 250 and 320 nm, respectively. The ultraviolet absorption of imidazo[4,5-f]quinolines generally resembles that of naphth[1,2-d]imidazoles but with the difference that the long-wavelength band in molecules of the former type is more intense. In acid solution the long-wavelength ultraviolet band of imidazo[4,5-f]quin-olines undergoes a strong bathochromic shift of approximately 50 nm,[226] which is also induced when imidazo[4,5-f]quinolines are converted into the corresponding $N(6)$-substituted quaternary salts. These observations in conjunction with the knowledge (see later in this section) that imidazo[4,5-f]-quinolines protonate preferentially at $N(6)$ implies the involvement of this

TABLE 5.56. INFRARED SPECTRA OF IMIDAZO[4,5-g]QUINOLINES AND IMIDAZO[4,5-h]QUINOLINES

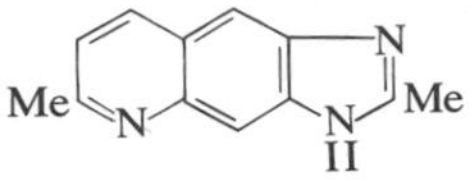

(5.165) 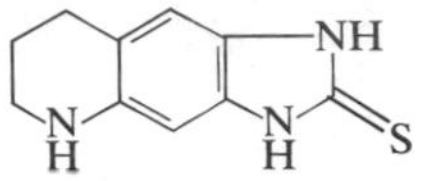   (5.166) 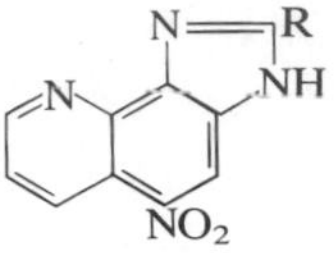   (5.167)

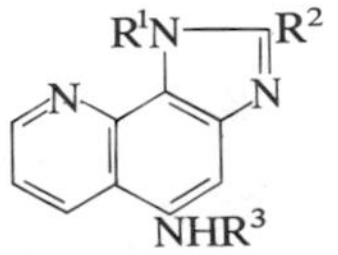

(5.168) 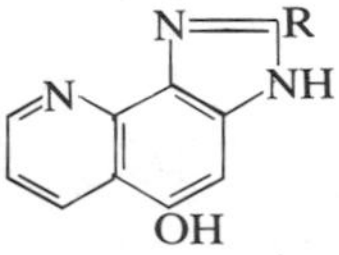   (5.169) 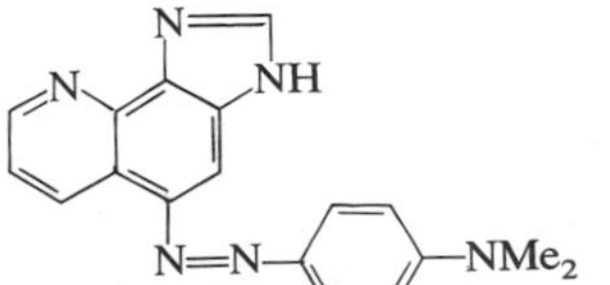   (5.170)

Compound	R	R^1	R^2	R^3	ν_{max}, cm^{-1}	Ref.
(5.165)	—	—	—	—	3280–2250	204
(5.166)	—	—	—	—	3240–3120, 2490	214
(5.167)	H	—	—	—	3220w, 3100–2900br, 1625m	223
(5.167)	Me	—	—	—	3180–2700br, 1620m	223
(5.168)	—	H	H	H	3450, 3380, 3100	224
(5.168)	—	H	H	H^a	3370, 3315, 3200	224
(5.168)	—	H	Me	H	3400, 3370, 3280	224
(5.168)	—	H	Me	H^a	3375, 3340, 3230	224
(5.168)	—	H	Me	SO$_2$T^b	3350, 3185, 1175^c	224
(5.170)	—	—	—	—	3400, 1600^d	224
(5.168)	—	H	H	NH$_2$	3200–2400br	224
(5.168)	—	Me	Me	H	3375, 3210	224
(5.168)	—	Me	H	H	3325, 3200	224
(5.168)	—	Me	H	H^a	2540–2270br	224
(5.169)	CF$_3$	—	—	—	3200–3100br, 1190^e	224
(5.169)	CF$_3$	—	—	—^{f}	1135^e	224
(5.169)	Ph	—	—	—	3830br	224
(5.169)	Ph	—	—	—^{a}	3400, 3240–3000	224
(5.169)	CH$_2$Ph	—	—	—	3400–3320	224
(5.169)	CH$_2$Ph	—	—	—^{a}	3050–2400br	224

^{a} Dihydrochloride.
^{b} T = p-tolyl.
^{c} SO$_2$.
^{d} N = N.
^{e} CF$_3$.
^{f} Monohydrochloride.

TABLE 5.57. ULTRAVIOLET SPECTRA OF IMIDAZO[4,5-g]QUINOLINES (5.171) AND (5.172), IMIDAZO[4,5-f]QUINOLINES (5.173), AND IMIDAZO[4,5-h]QUINOLINES (5.174)

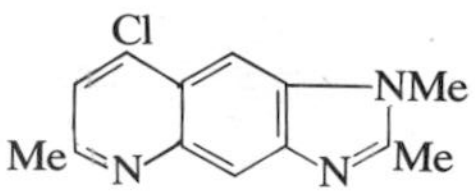

Structures: (5.171), 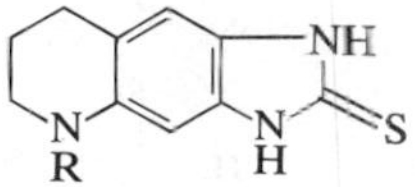 (5.172), 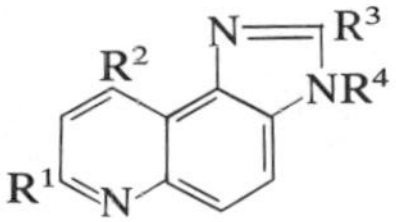 (5.173), 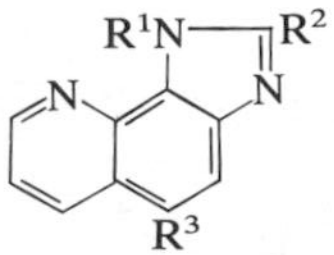 (5.174)

Compound	R	R^1	R^2	R^3	R^4	Solvent^a	λ_{max} nm (log ε)	Ref.
(5.171)	—	—	—	—	—	A	247.5(4.76), 332(4.00)	205
(5.172)	H	—	—	—	—	A	325(3.81)	214
(5.172)	n-Bu	—	—	—	—	A	260(3.94)	214
(5.172)	Ac	—	—	—	—	A	320(4.34), 4?5(2.72)	214
(5.172)	Ac	H	H	H	H	B	267(4.18), 289(4.18), 322(4.17)	214
(5.172)	—	H	H	H	4-nitrophenyl (NO$_2$)	A	250(4.38), 253.5(4.41)	218
(5.173)	—	H	H	H	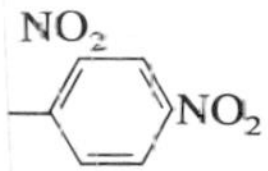 H	A	250(4.79), 300(4.02), 328(3.93)	218
(5.173)	—	Me	H	Me	Me	A	255sh(4.57), 258.5(4.59), 310(3.60)	205
(5.173)	—	Me	H	Me	Et	A	255sh(4.60), 258.5(4.62), 310(4.62)	205
(5.173)	—	Me	Me	Me	Me	A	253sh(4.59), 256.5(4.61), 310(3.65)	205
(5.173)	—	Me	Me	Me	Et	A	253sh(4.61), 256.5(4.63), 310(3.67)	205
(5.173)	—	Me	Cl	Me	Me	A	247.5(4.76), 332(4.00)	205
(5.173)	—	H	H	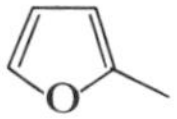 5-methylfuran-2-yl	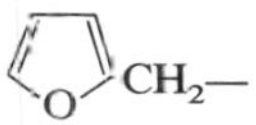 furan-2-ylmethyl (—CH$_2$—)	C	242(4.37), 285(4.50), 331(4.18)	220
(5.173)	—	H	H	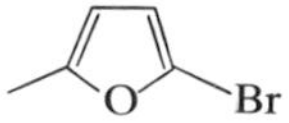 5-bromofuran-2-yl	H	C	242(4.05), 294(4.31), 335(4.05)	220
(5.173)	—	H	H	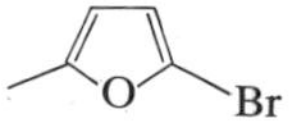 5-bromofuran-2-yl	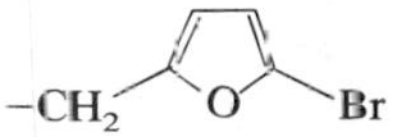 (5-bromofuran-2-yl)methyl (—CH$_2$—)	C	241(4.61), 291(4.63), 333(4.42)	220

571

TABLE 5.57 (Continued)

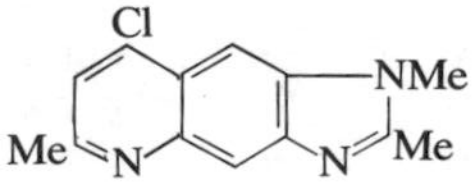

(5.171)

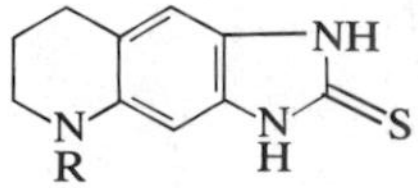

(5.172)

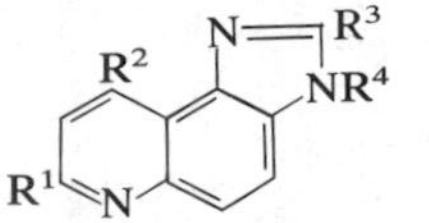

(5.173)

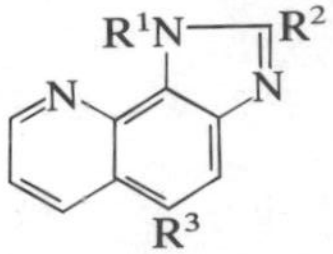

(5.174)

Compound	R	R^1	R^2	R^3	R^4	Solvent^a	λ$_{max}$ nm (log ε)	Ref.
(5.173)	—	H	H	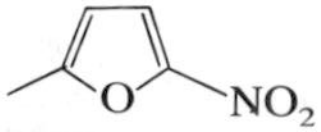 [furan-Me, NO$_2$]	Me	C	235(4.26), 273(4.23), 377(4.10)	220
(5.173)	—	H	H	 CH=CH—[furan]	H	C	251(4.28), 311(4.38), 357(4.72)	220
(5.173)	—	H	H	 CH=CH—[furan]	 [furan]CH=CHCH$_2$—	C	247(4.35), 294(4.39), 350(4.32),	220
(5.173)	—	H	H	—CH=CH—[furan-Br] 	Me	C	247(4.35), 294(4.39), 350(4.32)	220
(5.173)	—	H	H	—CH=CH—[furan-NO$_2$] 	Me	C	246(4.26), 294(4.10), 404(4.16)	220
(5.173)	—	H	H	—CH=CH—[furan-COMe] 	H	C	251(4.02), 280(3.81), 385(4.23)	220
(5.174)	—	H	H	NO$_2$	—	B	244(4.60), 343(3.80)	223
(5.174)	—	H	Me	NO$_2$	—	B	247(4.60), 356(3.90)	223

572

						Solvent	λ_{max} (log ε)	Ref.
(5.174)	—	H	H	NH$_2$		C	228(4.05), 266(4.45), 340(3.40)	224
(5.174)	—	H	H	NH$_2$		D	227(4.30), 342(3.40)	224
(5.174)	—	H	H	NH$_2$		E	257.5(4.35), 325(3.60)	224
(5.174)	—	H	Me	NH$_2$		C	270(4.60), 345(3.60)	224
(5.174)	—	H	Me	NH$_2$		D	271(4.40), 347.5(3.30)	224
(5.174)	—	H	Me	NH$_2$		E	265(4.40), 338(3.70)	224
(5.174)	—	Me	H	NH$_2$		C	268(4.35), 343(3.50)	224
(5.174)	—	Me	H	NH$_2$		D	265(4.40), 348(3.30)	224
(5.174)	—	Me	H	NH$_2$		E	264(4.40), 340(3.70)	224
(5.174)	—	Me	Me	NH$_2$		C	268(4.40), 340(3.50)	224
(5.174)	—	Me	Me	NH$_2$		D	270(4.40), 350(3.30)	224
(5.174)	—	Me	Me	NH$_2$		E	265(4.35), 340(3.70)	224
(5.174)	—	H	CF$_3$	OH		C	233.5(4.1), 261(4.50), 323(3.50)	224
(5.174)	—	H	CF$_3$	OH		D	244.5(4.10), 288.5(4.45), 351(3.50)	224
(5.174)	—	H	CH$_2$Ph	OH		C	231(4.20), 267(4.70), 332(3.50)	224
(5.174)	—	H	CH$_2$Ph	OH		D	280(4.60), 343.5(3.70)	224
(5.174)	—	H	CH$_2$Ph	OH		E	231(4.25), 259(4.50), 324(3.75)	224
(5.174)	—	H	Ph	OH		C	256(4.30), 294(4.50), 342(4.10)	224
(5.174)	—	H	Ph	OH		D	273.5(4.4), 283(4.35), 307(4.40), 364(4.20)	224
(5.174)	—	Ph	Ph	OH		E	245(4.40), 293.5(4.40), 340(4.10)	224
(5.174)	—	Me		(naphthalene–CONH– / OH)		C	348(4.12), 402(3.79)	227

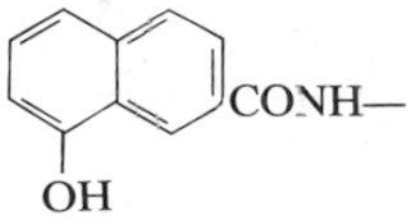

^{a} Solvents: A, Ethanol; B, Dimethyl sulfoxide; C, Methanol; D, 0.1 M KOH, Methanol; E, 0.1 M HCl, Methanol.

center in the long-wavelength absorption of imidazo[4,5-f]quinolines and hence a $\pi \rightarrow \pi^*$ origin for the latter. The marked bathochromic shift[226] also brought about in the short-wavelength band of imidazo[4,5-f]quinolines by protonation or $N(6)$-quaternization is inconsistent with the origin of this band in a transition involving the imidazole ring. Quaternization and protonation at $N(6)$ in imidazo[4,5-f]quinolines are known to reduce the basicity of the imidazole ring[226] and hence should result in a hypsochromic rather than a bathochromic shift in any associated absorption band. It follows that the short-wavelength band in the ultraviolet spectra of imidazo-[4,5-f]quinolines is also probably the result of a $\pi \rightarrow \pi^*$ transition.[226]

The $N(6)$-quaternary salts of imidazo[4,5-f]quinolines, but not the parent bases, exhibit fluorescence in methanol solution, the associated fluorescence spectra being characterized by the presence of a single, high-intensity absorption band at 458 to 496 nm. This property of $N(6)$-substituted imidazo[4,5-f]quinolinium salts has been attributed[226] to an inability to absorb energy (due to the similar hybridization of $N(6)$ in the quaternary salt in the ground and excited states) and consequently to undergo intersystem crossover followed by radiationless decay to the ground state.

The ultraviolet absorption spectra of imidazo[4,5-h]quinolines[223,224,227] like those of their imidazo[4,5-f]quinoline counterparts are also characterized by the presence of two intense maxima at approximately 250 and 350 nm (Table 5.57). However, unlike the bands in the ultraviolet spectra of imidazo[4,5-f]quinolines, those associated with imidazo[4,5-h]quinolines undergo a small but non-the-less significant hypsochromic shift in acid solution (Table 5.57).[224] Correspondingly, the ultraviolet spectra of N-1(3)H-imidazo[4,5-h]quinolines taken in alkaline solution exhibit the anticipated bathochromic shift in the absorption maxima (Table 5.57) consequent on deprotonation of the acidic imidazole NH group.[224]

NUCLEAR MAGNETIC RESONANCE SPECTRA. Perhaps the most important general aspect of the ^{1}H NMR absorption of tricyclic 6-6-5 fused benzimidazoles containing one additional hetero atom is its utility for the differentiation of linear structures from their angular counterparts and vice versa.[204,205,208,223,224,228] For example, the lack of observable splitting in the benzenoid proton [H(4),H(9)] resonances (Table 5.58) of the imidazobenzopyran (**5.175**)[200] (Scheme 5.22b) is clearly inconsistent with the alternative angular structure (**5.176**) in which the benzenoid protons [H(4),H(5)] would exhibit *ortho*-coupling and hence well-defined splitting. This latter situation pertains in the case[228] of the angular imidazobenzothiopyrans (**5.177**; R = H or Me) wherein the proton resonances due to H(4) and H(5) (Table 5.58) appear as well-defined doublets with a coupling constant of 9 Hz.

The ^{1}H NMR absorption of imidazo[4,5-g]quinolines[204,205] and imidazo-[4,5-f]quinolines[204,205,208,218] (Table 5.59) also permits a clear-cut distinction between the linear [4,5-g] isomers on the one hand and the angular [4,5-f] isomers on the other. For example, the lack of coupling between the

TABLE 5.58. ^{1}H N.M.R. SPECTRA^{a,b} OF THE IMIDAZO[4,5-a][5,6]-BENZOPYRAN-8-ONE (**5.175**) AND THE IMIDAZO[4,5-f] [5,6]BENZOTHIOPYRAN-7-ONES (**5.177**)

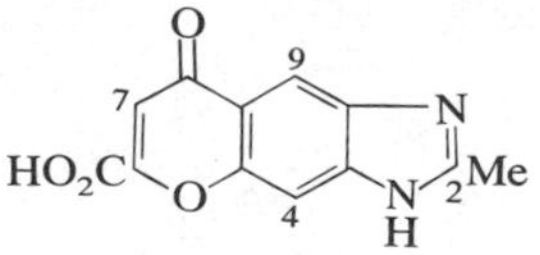

(**5.175**) 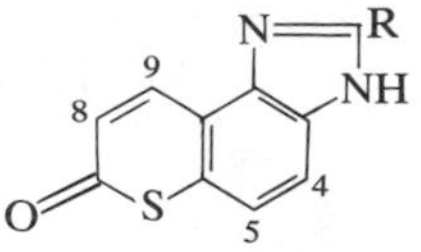   (**5.177**)

Compound	Solvent^c	H(2)	H(4)	H(5)	H(7)	H(8)	H(9)	Me(2)	Ref.
(**5.175**)	A	—	7.47	—	7.02	—	7.20	2.12	200
(**5.177**) (R = H)	B	9.37	7.98d^{d,e}	8.12d^{d,e}	—	←8.03q^{f,g} →		—	228
(**5.177**) (R = Me)	B	—	7.91d^{d,e}	8.12d^{d,e}	—	←7.98q^{f,g} →		3.15	228

^{a} δ values in ppm. measured from tetramethylsilane.
^{b} Signals are sharp singlets unless specified as: d = doublet; q = quartet.
^{c} Solvents: A, (CD$_3$)$_2$SO; B, CF$_3$CO$_2$H.
^{d} These signal assignments may be reversed.
^{e} J = 9 Hz.
^{f} Centre of a quartet.
^{g} J = 10 Hz.

benzenoid protons [H(4) and H(9)] in imidazo[4,5-g]quinolines (**5.178**) (Table 5.59) is clearly only in accord with a linear formulation. Conversely, the consistent coupling (J = 9 Hz) between H(4) and H(5) observed in a wide range of imidazo[4,5-f]quinoline derivatives (**5.179**) and (**5.180**) (Table 5.59) is only reconcilable with the angular structures of such compounds. A further feature of the ^{1}H NMR absorption of imidazo[4,5-f]quinolines (Table 5.59)

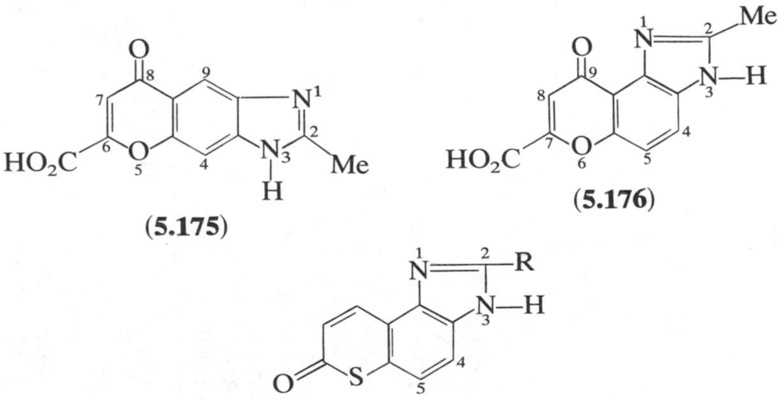

(**5.175**)   (**5.176**)

(**5.177**)

Scheme 5.22b

TABLE 5.59. ¹H NMR. SPECTRA^{a,b} OF IMIDAZO[4,5-g]QUINOLINES (5.178), IMIDAZO[4,5-f]QUINOLINES (5.179) AND (5.180) AND IMIDAZO[4,5-h]QUINOLINES (5.181)

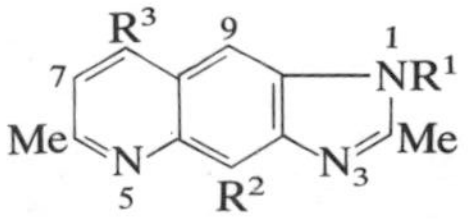

(5.178) 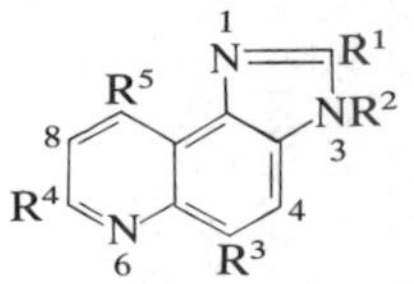 (5.179) 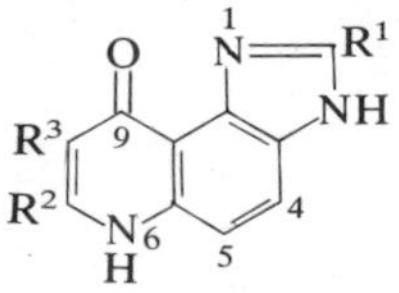 (5.180) 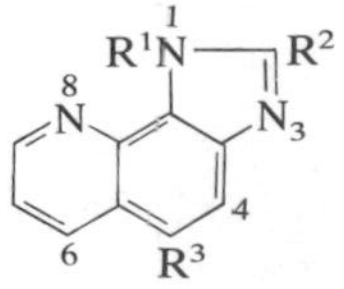 (5.181)

Compound	Solvent^c	H(2)	H(4)	H(5)	H(6)	H(7)	H(8)	H(9)	Me	Others	Ref.
(5.178) (R¹=R³=H, R²=Me)	A	—	—	—	—	8.03d^d	9.25d^d	8.60	3.25, 3.19, 3.19	—	204
(5.178) (R¹=Me, R²=H, R³=Cl)	B	—	8.29	—	—	7.20	—	7.78	3.73^e, 2.62^f 2.71^f	—	205
(5.178) (R¹=Et, R²=H, R³=Cl)	B	—	8.32	—	—	7.30	—	7.87	1.44^h, 2.63^f 2.69^g	4.22^i	205
(5.179) (R¹=R⁴=Me, R²=R³=R⁵=H)	A	—	8.45^i	8.48^i	—	—	8.21^d	9.59d^d	6.75, 6.79	—	204
(5.179) (R¹=R³=R⁴=Me, R²=R⁵=H)	A	—	8.35	—	—	—	8.22d^d	9.60d^d	6.94, 6.75, 6.86	—	204
(5.179) (R¹=R⁴=Me, R²=R⁵=H, R³=OMe)	A	—	7.84	—	—	—	8.22d^k	9.52d^k	4.31^l, 6.78, 6.90	—	204
(5.179) (R¹=R²=R⁴=Me, R³=R⁵=H)	B	—	7.56d^d	7.85d^d	—	—	7.36d	8.78d	3.72^e, 2.63^f 2.75^m	—	205
(5.179) (R¹=R⁴=Me, R²=Et, R³=R⁵=H)	B	—	7.62d^d	7.88d^d	—	—	7.38d	8.82d	1.44^h, 2.68^f 2.76^m	4.22^i	205
(5.179) (R¹=R²=R⁴=Me, R³=H, R⁵=Cl)	B	—	7.58d^d	7.86d^d	—	—	7.42	—	3.75^d, 2.68^f 2.71^m	—	205
(5.179) (R¹=R⁴=Me, R²=Et, R³=H, R⁵=Cl)	B	—	7.66d^d	7.89d^d	—	—	7.44	—	1.44^h, 2.73^f 2.73^m	4.25^i	205

Compound	Method									Ref
(5.179) (R^1 = R^2 = R^4 = R^5 = Me, R^3 = H)	B	—	7.51d^a	7.86d^d	—	7.14	—	2.59,^{f} 2.68,^{m} 3.68,^{e} 3.14^n	—	205
(5.179) (R^1 = R^4 = R^5 = Me, R^2 = Et, R^3 = H)	B	—	7.56d^d	7.86d^d	—	7.14	—	2.64,^{f} 2.69,^{m} 3.16,^{n} 1.41^g	4.17^i	205
(5.179) (R^1 = R^3 = R^4 = R^5 = H, R^2 = 2,4-dinitrophenyl)	—	(7.68, 7.90, 8.24, 8.70, 8.78, 9.06)^o							—	218
(5.180) (R^1 = R^3 = H, R^2 = Me)	C	8.23	7.45d^c	8.01d^d	—	6.15	—	—^{p}	—^{p}	208
(5.180) (R^1 = R^3 = H, R^2 = Et)	C	8.26	7.48d^d	8.05d^d	—	6.20	—	—^{p}	—^{p}	208
(5.180) (R^1 = R^3 = H, R^2 = Ph)	C	8.30	7.71d^d	8.10d^d	—	6.61	—	—^{p}	—^{p}	208
(5.180) (R^1 = R^2 = Me, R^3 = H)	C	—	7.43d^d	7.75d^d	—	6.15	—	—^{p}	—^{p}	208
(5.180) (R^1 = Ph, R^2 = Me, R^3 = H)	C	—	7.46d^d	8.06d^d	—	5.15	—	—^{p}	—^{p}	208
(5.180) (R^1 = R^2 = R^3 = H)	C	8.33	7.53d^d	8.13d^d	6.37d^q	8.07d^q	—	—	—^{p}	208
(5.180) (R^1 = R^2 = H, R^3 = CO$_2$H)	D	8.25	7.57d^d	7.90d^d	8.79	—	—	—	—^{p}	208
(5.179) (R^1 = R^2 = R^3 = H, R^4 = Me, R^5 = Cl)	C	8.46	7.86d^d	8.16d^d	—	7.66	—	—^{p}	—	208
(5.179) (R^1 = R^2 = R^3 = H, R^4 = Et, R^5 = Cl)	C	8.46	7.86d^d	8.06d^d	—	7.68	—	—^{p}	—	208
(5.179) (R^1 = R^2 = R^3 = H, R^4 = Ph, R^5 = Cl)	C	8.51	8.00d^a	8.21d^d	—	7.51	—	—^{p}	—	208
(5.179) (R^1 = R^4 = Me, R^2 = R^3 = H, R^5 = Cl)	C	—	7.70d^d	8.01d^d	—	7.60	—	—^{p}	—	208
(5.179) (R^1 = Ph, R^2 = R^3 = H, R^4 = Me, R^5 = Cl)	C	—	7.85d^d	3.33d^d	—	—	—	—^{p}	—	208
(5.179) (R^1 = R^2 = R^3 = R^4 = H, R^5 = Cl)	C	8.55	7.96d^r	8.23d^d	7.80d^r	8.85d^r	—	—	—	208
(5.181) (R^1 = R^2 = H, R^3 = NO$_2$)	C	8.84	8.62	—	9.03^a	7.79^s	9.10^s	—	3.30^t	223

TABLE 5.59 *(Continued)*

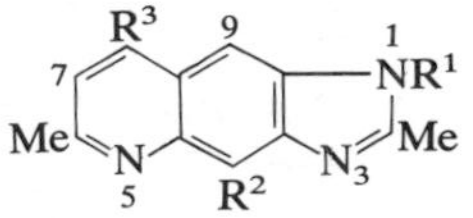

(5.178)

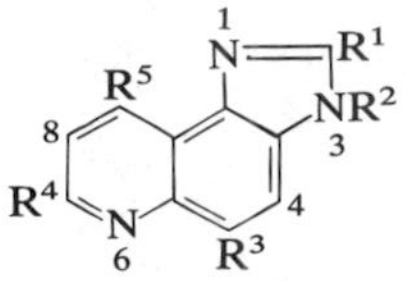

(5.179)

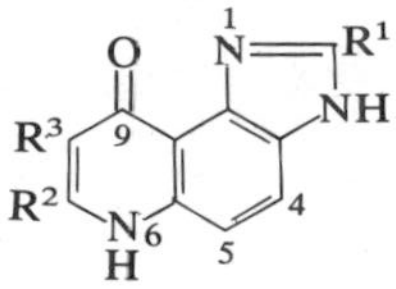

(5.180)

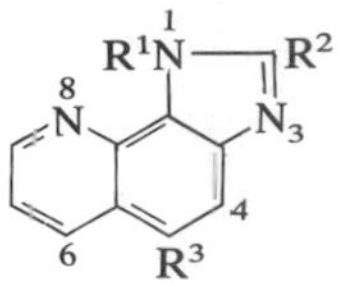

(5.181)

Compound	Solvent[c]	H(2)	H(4)	H(5)	H(6)	H(7)	H(8)	H(9)	Me	Others	Ref.
(5.181) (R^1=H, R^2=Me, R^3=NO_2)	C	—	8.64	—	8.98[s]	7.70[s]	9.00[s]	—	2.65	3.38[t]	223
5.181 (R^1=R^2=H, R^3=NH_2)	E	8.08[u] 8.18[v]	7.02[u] 7.20[v]	—	8.60[s]	7.44[s]	8.88[s]	—	—	—[p]	224
(5.181) (R^1=H, R^2=Me, R^3=NH_2)	E	—	6.94	—	8.35[s]	7.35[s]	8.79[s]	—	—[w]	—[p]	224
(5.181) (R^1=R^2=Me, R^3=NH_2)	E	—	6.93	—	8.64[s]	7.46[s]	8.87[s]	—	4.30,[e] 2.65[z]	4.10[x]	224
(5.181) (R^1=H, R^2=CF_3, R^3=OH)	E	—	7.20	—	8.66[s]	7.58[s]	8.97[s]	—	—	—	224
(5.181) (R^1=H, R^2=CH_2Ph, R^3=OH)	E	—	7.11	—	8.56	—[p]	8.85	—	—	7.36,[y] 4.20,[z] 3.56[t]	224

[a] δ in ppm. measured from tetramethylsilane.
[b] Signals are sharp singlets unless otherwise specified as: d = doublet.
[c] Solvents: A, CF_3CO_2H. B, $CDCl_3$. C, $(CD_3)_2SO$. D, D_2O-NaOD. E, $(CH_3)_2SO$.
[d] J = 9 Hz.
[e] NMe.
[f] Me(2).
[g] Me(6).
[h] Me of Et group.
[i] CH_2 of Et group.
[j] These signals may be interchanged.
[k] J = 10 Hz.
[l] OMe.

[n] Me(9).
[o] Not assigned.
[p] δ value not quoted.
[q] J = 7 Hz.
[r] J = 5 Hz.
[s] J not quoted.
[t] NH.
[u] N(3) tautomer.
[v] N(1) tautomer.
[w] Me masked by $(CH_3)_2SO$.
[x] NH_2.
[y] ArH.

is the marked deshielding of H(9), presumably due to the anisotropic effect of the proximate $N(1)$ nitrogen atom.

The ¹H NMR spectra of imidazo[4,5-h]quinolines (**5.181**) (Table 5.59) also reveal marked *ortho*-coupling between the benzenoid protons [H(4) and H(5)], and again this feature is in accord with the angular character of the imidazo[4,5-h]quinoline ring system. ¹H NMR studies of N-1(3)H-imidazo-[4,5-h]quinolines (**5.181**; $R^1 = H$) (Table 5.59) also provide useful information on the position of the tautomeric equilibrium in such molecules and hence on their ability to function as metal-complexing agents (see the following). The situation in the case of 5-amino-1(3)H-imidazo[4,5-h]quinoline (**5.182**) is illustrative.[224] The ¹H NMR spectrum of this molecule taken in dimethyl sulfoxide (Table 5.59) exhibits signals attributable to the presence of both possible tautomeric forms (Scheme 5.23, **5.182a** ⇌ **5.182b**) in equilibrium. On the basis of the anticipated[224] higher shielding of H(2) and H(4) (Table 5.59) in the N-(3)H-tautomer (**5.182b**), an equilibrium ratio of 1:1.5 for the tautomers **a**:**b** may be adduced.[224]

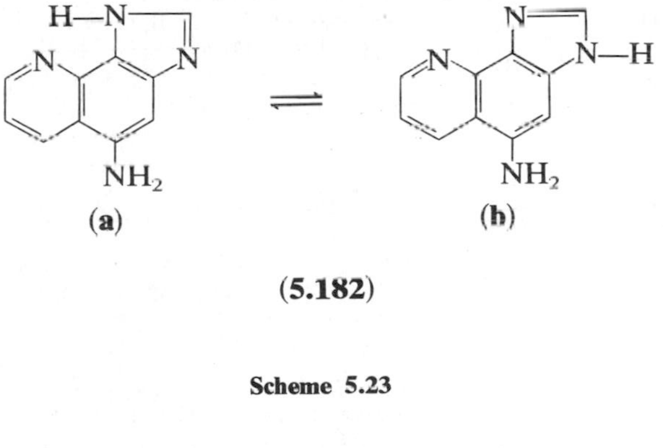

(5.182)

Scheme 5.23

MASS SPECTRA. Only the electron-impact-induced fragmentation of imidazo[4,5-h]quinolines has been subjected to detailed scrutiny (Table 5.60).[224] Analysis of variously labeled ¹⁵N derivatives of 5-amino-1(3)H-imidazo[4,5-h]quinoline has revealed the fragmentation pattern outlined in Scheme 5.24.[224] Breakdown is initiated at the imidazole ring and occurs by loss of HCN before or after loss of a hydrogen atom to afford fragment ions that break down by further sequential loss of HCN.[224] The mass spectral fragmentation patterns of other 5-aminoimidazo[4,5-h]quinolines are broadly similar (Table 5.60). On the other hand, the mass spectral fragmentation of 5-nitroimidazo[4,5-h]quinolines (Table 5.60) is initiated by loss of the nitro group,[224] whereas in the case of 5-hydroxy-2-trifluoromethyl-1(3)H-imidazo[4,5-h]quinoline (Table 5.60), initial loss of HF is followed by extrusion of a fluorine atom or CO.[224]

TABLE 5.60. MASS SPECTRA OF 5-AMINO-, 5-HYDROXY-, AND 5-NITRO-IMIDAZO[4,5-*h*]QUINOLINES (**5.183**) (X = NHR, OR, NO$_2$)[224,227]

(**5.183**)

R^1	R^2	X	m/e (%)
H	H	NH$_2$	184(100),[a] 183, 157, 156, 130, 129, 103, 102, 79, 76, 52
H	Me	NH$_2$	198(100),[a] 197, 183, 170, 157, 156, 143, 130, 129, 103, 177^{2+}, 79, 76, 52
H	CH$_2$Ph	NH$_2$	275(30), 274(100), 273(51), 198(8), 197(12), 170(2), 169(2), 158(4), 156(7), 137(11), 136.5(8), 130(4), 129(6), 103(5), 102(3), 91(11), 79(3), 78(5), 77(4), 76(3), 65(4), 58(8), 57(9), 55(9), 52(2), 41(8), 39(3)
H	Me	NHSO$_2$T^b	353(4), 352(14), 351(9), 320(2), 287(5), 272(1), 246(2), 219(0.5), 213(1), 198(15), 197(100), 170(5), 156(4), 155(3), 142(4), 130(55), 129(4), 104(4), 103(2), 102(2), 91(9), 79(2), 78(3), 77(2), 65(4), 52(2), 39(2)
Me	H	NH$_2$	198(100),[a] 197, 183, 180, 170, 156, 143, 129, 117, 116, 103, 99, (M^{2+}), 89, 72
Me	Me	NH$_2$	212(100),[a] 211, 197, 184, 170, 156, 144, 143, 116, 106, 89, 71.5, 58, 28
H	Me	NHNH$_2$	213(100),[a] 198, 197, 183, 181, 170, 156, 147, 143, 130, 123.5, 102, 78, 63, 52
H	CF$_3$	OH	253(100),[a] 252, 233, 214, 205, 185, 157, 130, 103, 102
H	CF$_3$	OSO$_2$T^b	408(7), 407(17), 388(3), 342(1), 315(1), 253(16), 252(100), 233(2), 232(2), 224(4), 204(8), 174(4), 159(1), 155(8), 130(1), 129(2), 102(5), 91(30), 65(9), 51(3), 39(4)
H	CH$_2$Ph	OH	276(20), 275(99), 274(100), 273(12), 272(4), 199(3), 198(13), 170(3), 157(3), 137.5(15), 137(10), 130(4), 103(5), 102(4), 91(20), 89(4), 79(4), 78(5), 77(5), 76(5), 75(3), 65(7), 63(3), 57(3), 55(2), 52(2), 51(6), 50(3), 43(4), 41(3), 39(4)
H	Ph	OH	263(2), 262(20), 261(100), 260(9), 233(2), 232(2), 205(2), 158(2), 157(7), 131(4), 130.5(10), 130(13), 129(5), 104(6), 103(9), 102(6), 79(8), 78(4), 77(6), 76(6), 75(3), 69(3), 57(6), 55(2), 52(2), 51(5), 50(2), 43(4), 41(2), 39(2)
H	Me	NO$_2$	228(100),[a] 198, 182, 170, 167, 155, 149, 141, 128, 114, 102, 87, 71, 63, 57, 43
H	Me	[naphthalenyl with OH]CONH—	369(7), 368(28), 225(2), 199(2), 198(100), 197(33), 171(37), 170(12), 156(14), 143(19), 130(16), 115(55)

[a] Accurate % abundances not quoted.
[b] T = p-MeC$_6$H$_4$.

580

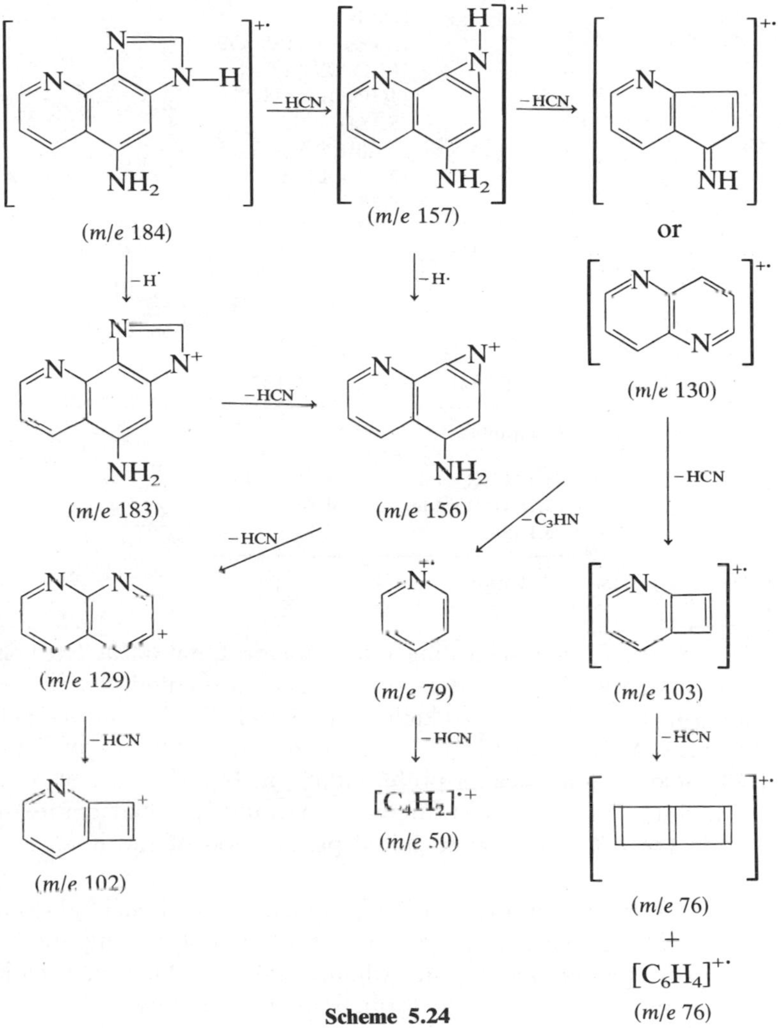

Scheme 5.24

General Studies

IONIZATION CONSTANTS. The basicity of imidazo[4,5-g]quinolines, imidazo-[4,5-f]quinolines, and imidazo[4,5-h]quinolines conferred by the presence of both a quinoline ring and an imidazole ring is amply demonstrated by the ability of derivatives of these ring systems to form well defined hydrochloride and picrate salts (cf. Tables 5.51 to 5.55). However, the pK_a values (Table 5.61)[229] of 3-methylimidazo[4,5-f]quinoline (**5.184**; R = H) and its 2-methyl derivative (**5.184**; R = Me) demonstrate that such molecules are less basic then [5,6]benzoquinoline (**5.185**) (Table 5.61). Moreover, the pK_a for protonation of the imidazole ring in the 3,6-dimethylimidazo[4,5-f]quinolinium

TABLE 5.61. BASICITY CONSTANTS[a] OF IMIDAZO[4,5-f]-QUINOLINES (**5.184**) AND [5,6]BENZO-QUINOLINE (**5.185**)[229]

(**5.184**)   (**5.185**)

Compound	pKa
(**5.184**) (R = H)	4.55 ± 0.06
(**5.184**) (R = Me)	4.84 ± 0.06
(**5.185**)	5.15

[a] Measured in aqueous solution.

cation is 1.86 ± 0.07, demonstrating that prior protonation at N(6) sharply reduces the basicity of the imidazole ring. The close similarity in the pK_a values for 3-methylimidazo[4,5-f]quinoline (**5.184**; R = H) and its 2-methyl derivative (**5.184**; R = Me) (Table 5.61) demonstrates that protonation of imidazo[4,5-f]quinolines occurs preferentially at N(6).[229] The introduction of a 2-methyl substituent (base strengthening) would have had a more dramatic effect on the pK_a value were initial protonation of the imidazole ring involved.

In contrast to the apparent situation in imidazo[4,5-f]benzimidazoles, the nature of the C(2) substituent plays a major role in determining the basicity of imidazo[4,5-h]quinolines.[224] Thus whereas electron-donating substituents at C(2) promote dihydrochloride formation, the presence of an electron-withdrawing group such as trifluoromethyl at this site reduces the basicity of the molecule to the extent that only monohydrochloride formation is observed.[224] The implication of these observations, namely, that in contrast to imidazo[4,5-f]quinolines, the imidazole ring in an imidazo[4,5-h]-quinoline is more basic than the quinoline ring, is supported by the results of molecular orbital calculations.[224] These indicate that N(1) in the 3H-tautomer of an imidazo[4,5-h]quinoline is more basic than N(9) and further that +M and +I substituents at C(2) by increasing the electron density at N(1) and N(9) will increase the basicity of these positions while lowering the acidity of the imidazole NH group. Predictably, −I and −M substituents have the reverse effect.[224] Molecular orbital studies[224] also reveal the not unexpected (see section 5.2.3) acidic character of the imidazole NH group in N-1(3)-unsubstituted imidazo[4,5-h]quinolines. A more unexpected revelation of such studies,[224] however, is the prediction of the weakly acidic nature of a 5-amino substituent in an imidazo[4,5-h]quinoline. This prediction is supported in practice by the apparent inability of a 5-amino group in an imidazo[4,5-h]quinoline to form salts with mineral acids and by the ready hydrolytic conversion of 5-aminoimidazo[4,5-h]quinolines into the corresponding hydroxy compounds.[224]

METAL COMPLEX FORMATION. N-1(3)H-Imidazo[4,5-h]quinolines contain both an electron-donor center [N(9)] and a salt-forming center [N(1)H] in suitable juxtaposition and consequently have attracted considerable attention[224,230-232] as ligands for the formation of both charged (5.186) and neutral (5.187) complexes with metal ions [Cu(I), Cu(II), Fe(II), Co(II)] (Scheme 5.25). The spectral properties of such metal complexes have been investigated,[230,231] and their utility as components of new color-coupling agents has been described.[232]

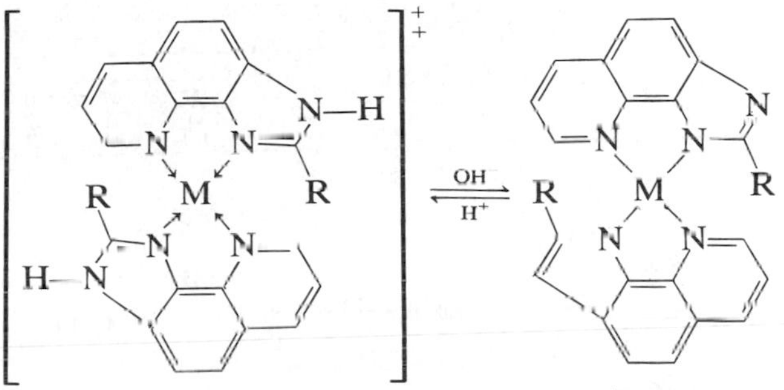

(5.186)   (5.187)

Scheme 5.25

5.2.3. Reactions

Reactions with Electrophiles

Most information concerning the behavior of tricyclic 6-6-5 fused benzimidazoles containing one additional hetero atom toward electrophilic attack has come from the study of imidazo[4,5-f]quinolines and imidazo-[4,5-h]quinolines.

PROTONATION. As already discussed (section 5.2.2), pK$_a$ data indicate that protonation of imidazo[4,5-f]quinolines occurs preferentially at N(6) (i.e., in the pyridine ring),[220] whereas the electronic effect of C(2) substituents on basicity and the results of molecular orbital studies suggest that imidazo[4,5-h]quinolines are protonated initially in the imidazole ring.[224] The reason for this contrasting behavior of two structurally very similar ring systems toward protonation has not been established.

ALKYLATION AND ARYLATION. The mode of alkylation of imidazo[4,5-f]-quinolines is markedly dependent on the reaction conditions.[201,206,217,233] Under effectively basic conditions attack is mediated by initial deprotonation of the acidic imidazole NH group and occurs at N(1) and/or N(3) in the imidazole ring. For example, heating N-1(3)H-imidazo[4,5-f]quinoline (**5.188**; $R^1 = R^2 = R^3 = H$) with phenyltrimethylammonium hydroxide results in the formation, in good yield (Table 5.62), of a readily separated mixture of the N(3) and N(1) methyl derivatives (**5.189**) and (**5.190**) (Scheme 5.26).[217]

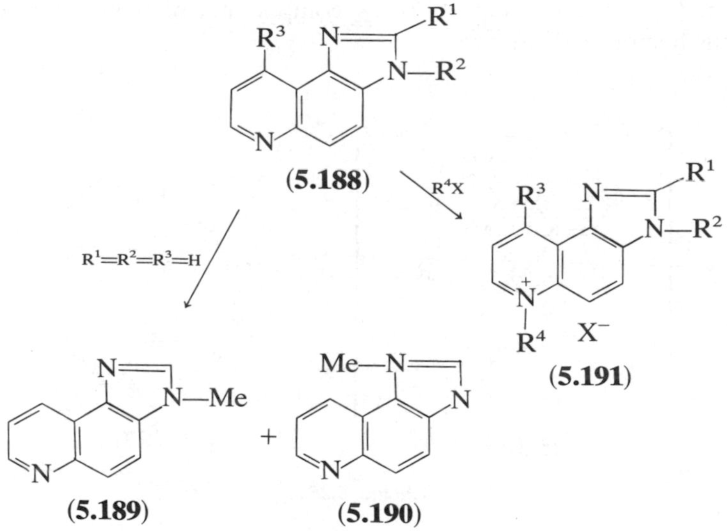

Scheme 5.26

Under effectively neutral conditions, alkylation of imidazo[4,5-f]quinolines occurs at N(6) in the quinoline ring and results in quaternary salt formation. Thus heating under reflux in ethanol or, more efficiently, in benzene with methyl iodide, or fission with methyl benzenesulfonate affords excellent yields (Table 5.62) of the corresponding N(6)-methylimidazo[4,5-f]-quinolinium salts [Scheme 5.26, **5.188→5.191** ($R^4 = Me$, $X = I$ or $PhSO_3$)].[217] Heating imidazo[4,5-f]quinolines with alkyl chlorides or iodides under autoclave conditions (Table 5.62) achieves the same result if less conveniently.[201,206,217,233]

Cyclic lactams in the imidazo[4,5-f]quinoline series undergo alkylation under basic conditions in an orthodox fashion at nitrogen. The example (Scheme 5.27, **5.192 → 5.193**) (Table 5.62) is illustrative.[229]

TABLE 5.62. ALKYLATION AND ARYLATION REACTIONS OF IMIDAZO[4,5-f]QUINOLINES (5.188) AND (5.194)

Starting materials	Reaction conditions	Product	Yield (%)	Melting point (°C)	Solvent of crystallization	Crystal form	Ref.
(5.188) (R^1 = R^2 = R^3 = H)	A	(5.189)	63	190	—^{b}	—^{c}	217
		(5.190)	15	187	—^{b}	Colorless needles	217
(5.188) (R^1 = R^2 = R^3 = H)	B	(5.191) (R^1 = R^2 = R^3 = H, R^4 = Me, X = I)	73	234–235	Ethanol	Yellow crystals	217
(5.188) (R^1 = R^3 = H, R^2 = Me)	B	(5.191) (R^1 = R^3 = H, R^2 = R^4 = Me, X = I)	65	282	Ethanol	Yellow crystals	217
(5.188) (R^1 = R^3 = H, R^2 = Me)	C	(5.191) (R^1 = R^3 = H, R^2 = R^4 = Me, X = I)	91	—	—	—	217
(5.188) (R^1 = R^3 = H, R^2 = Ph)	B	(5.191) (R^1 = R^3 = H, R^2 = Ph, R^4 = Me, X = I)	70	262–270	Ethanol	Yellow crystals	217
(5.188) (R^1 = R^3 = H, R^2 = Ph)	C	(5.191) (R^1 = R^3 = H, R^2 = Ph, R^4 = Me, X = I)	95	—	—	—	217
(5.188) (R^1 = R^3 = H, R^2 = CH$_2$Ph)	B	(5.191) (R^1 = R^3 = H, R^2 = CH$_2$Ph, R^4 = Me, X = I)	68	264–265	Ethanol	Yellow crystals	217
(5.188) (R^1 = R^3 = H, R^2 = CH$_2$Ph)	C	(5.188) (R^1 = R^3 = H, R^2 = CH$_2$Ph, R^4 = Me, X = I)	95	—	—	—	217
(5.188) (R^1 = R^2 = R^3 = H)	D	(5.191) (R^1 = R^2 = R^3 = H, R^4 = Me, X = Cl)^d	—^{e}	234–235	Ethanol-acetone	Colorless prisms	217
(5.188) (R^1 = R^3 = H, R^2 = Ph)	E	(5.191) (R^1 = R^3 = H, R^2 = Ph, R^4 = Me, X = PhSO$_3$)	90	269–270	Ethanol	Colorless needles	217
(5.188) (R^1 = R^3 = H, R^2 = CH$_2$Ph)	E	(5.191) (R^1 = R^3 = H, R^2 = CH$_2$Ph, R^4 = Me, X = PhSO$_3$)	88	261–262	Ethanol	Colorless needles	217
(5.188) (R^1 = R^3 = H, R^2 = Ph)	F	(5.191) (R^1 = R^3 = H, R^2 = Ph, R^4 = Me)	—^{e}	280 (decomp.)	Water	Yellow needles	201

585

TABLE 5.62 (Continued)

Starting material	Reaction conditions	Product	Yield (%)	Melting point (°C)	Solvent of crystallization	Crystal form	Ref.
(5.188) (R¹ = Ph, R² = R³ = H)	F	(5.191) (R¹ = Ph, R² = R³ = H, R⁴ = Me)	—[e]	295	Water	—[c]	201
(5.188) (R¹ = R³ = Me, R² = Et)	G	(5.191) (R¹ = R³ = Me, R² = R⁴ = Et)	—[e]	286	Pyridine	—[c]	206, 233
(5.192)	H	(5.193)	20	182	—[f]	Yellow prisms	229
(5.194) + (5.195)	I	(5.196)	53	232	—[b]	—[c]	218

[a] A, PhN̊Me₃OH⁻/122°/1.5 hr. B, MeI, EtOH/60–65°/4 hr. C, MeI, benzene/reflux/4 hr. D, MeCl, EtOH/autoclave, 70°/14 hr. E, PhSO₃Me/50–55°/few minutes. F, MeI/autoclave, 100°/12 hr. G, EtI/sealed tube, 100°/14 hr. H, MeI, 40% aq. NaOH/room temp./few minutes I, NaOAc/120–130°/3 hr.

[b] Solvent of crystallization not specified.

[c] Crystal form not specified.

[d] Forms the iodide with KI, m.p. 234–235°.

[e] Yield not quoted.

[f] Purified by molecular distillation.

586

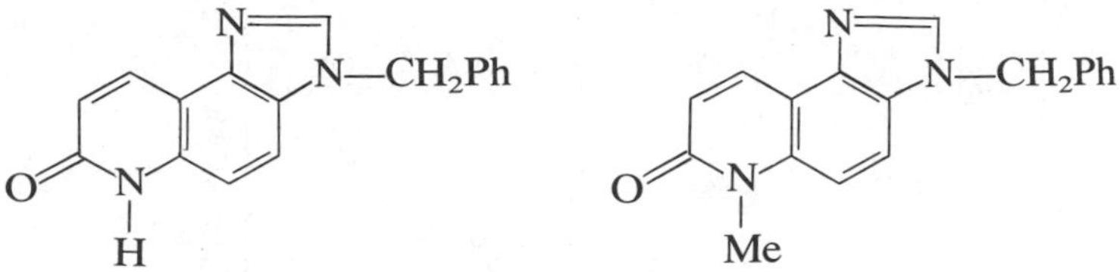

(5.192)   **(5.193)**

Scheme 5.27

N-1(3)H-Imidazo[4,5-f]quinoline derivatives are also arylated by activated aryl halides under basic conditions, attack occurring preferentially at the N(3)-position of the imidazole ring (Scheme 5.28, **5.194** + **5.195** → **5.196**).[218]

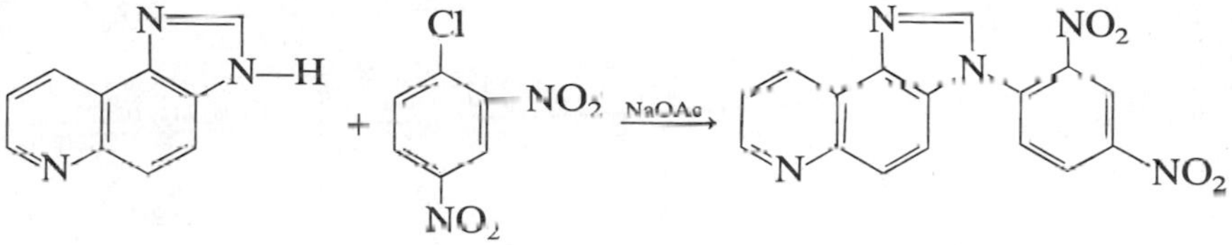

(5.194)   **(5.195)**   **(5.196)**

Scheme 5.28

ACYLATION. Friedel–Crafts type reactions of tricyclic 6-6-5 fused benzimidazoles containing one additional hetero atom have not been studied to any extent, although one report records the inertness of the imidazo[4,5-f]-quinoline ring to such conditions.[220] On the other hand, N-1(3)H-imidazo-[4,5-f]quinolines are readily acylated by acid chlorides at the N(3)-position of the imidazole ring under basic conditions (e.g., Scheme 5.29, **5.197** → **5.198**) (Table 5.63).[215]

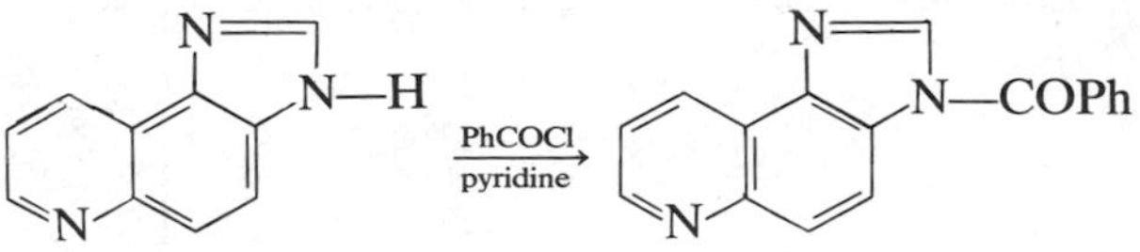

(5.197)   **(5.198)**

Scheme 5.29

TABLE 5.63. ACYLATION REACTIONS OF IMIDAZO[4,5-f]QUINOLINES AND IMIDAZO[4,5-h]QUINOLINES

Starting materials	Reaction conditions[a]	Product	Yield (%)	Melting point (°C)	Solvent of crystallization	Crystal form	Ref.
(5.197)	A	(5.198)	68	166	Ethanol–water	Colorless prisms	215
(5.199)	B	(5.200) (R = CH$_2$COMe)	—[b]	230	—[c]	—[d]	232
(5.199)	B	(5.200) (R = CH$_2$COPh)	—[b]	241	—[c]	—[d]	232
(5.199)	B	(5.200) (R = CH$_2$CO–C$_6$H$_4$–NO$_2$) 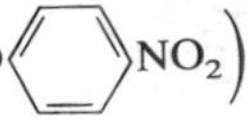	—[b]	275	—[c]	—[d]	232
(5.199)	B	(5.200) (R = CH$_2$CO–C$_6$H$_4$–NO$_2$) 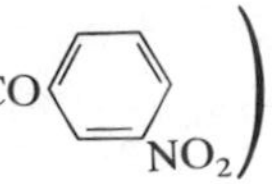	—[b]	235	—[c]	—[d]	232
(5.199)	C	(5.200) (R 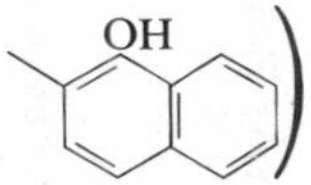 = 2-methyl-naphthol group)	60	227	Pyridine	Yellow crystals	227
(5.201) (R = Me)	D	(5.202)	—[b]	255.5	Ethanol	Yellow needles	224
(5.201) (R = H)	D	(5.203)	—[b]	206.5	Acetone–water	Colorless needles	224
(5.204)	D	(5.205)	—[b]	124	Acetone or water	Colorless needles	224
(5.206)	E	(5.207)	—[b]	262.3	Pyridine	Orange prisms	205
(5.208 + 5.209)	F	(5.210)	—[b]	—[e]	—[c]	—[d]	206, 233

[a] A, PhCOCl, pyridine/room temp./few min. B, RCOCH$_2$CO$_2$Et, pyridine/heat. C, RCO$_2$Ph/180°. D, TSO$_2$Cl, NaHCO$_3$, acetone/reflux/4 hr. E, p-NO$_2$C$_6$H$_4$CHO pyridine/heat/1 hr. F, Et$_3$N, EtOH/reflux/1.5 hr.
[b] Yield not specified.
[c] Solvent of crystallization not specified.
[d] Crystal form not specified.
[e] Melting point not quoted.

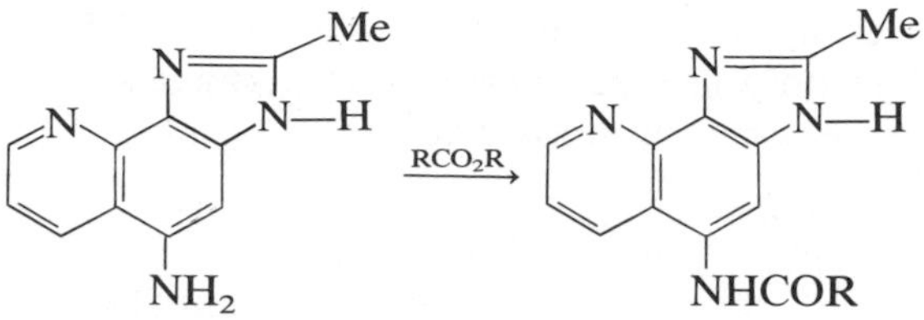

(5.199) (5.200)

Scheme 5.30

The acylation[227,232] of 5-amino-2-methylimidazo[4,5-*h*]quinoline (**5.199**) by esters (Table 5.63) on the other hand occurs at the amino-group and not at an imidazole ring nitrogen atom (Scheme 5.30, **5.199** → **5.200**). In contrast, the reaction of 5-amino-*N*1(3)H-imidazo[4,5-*h*]quinolines with toluene-*p*-sulphonyl chloride in the presence of sodium carbonate can result in both tosylation at N(3) or the 5-amino group with or without concomitant solvolysis of the latter (Scheme 5.31) (Table 5.63).[224] Conversely, tosylation of 5-hydroxy-2-trifluoromethylimidazo[4,5-*h*]quinoline (**5.204**) results in exclusive attack at the hydroxy-group (Scheme 5.31, **5.204** → **5.205**) (Table 5.63).[224]

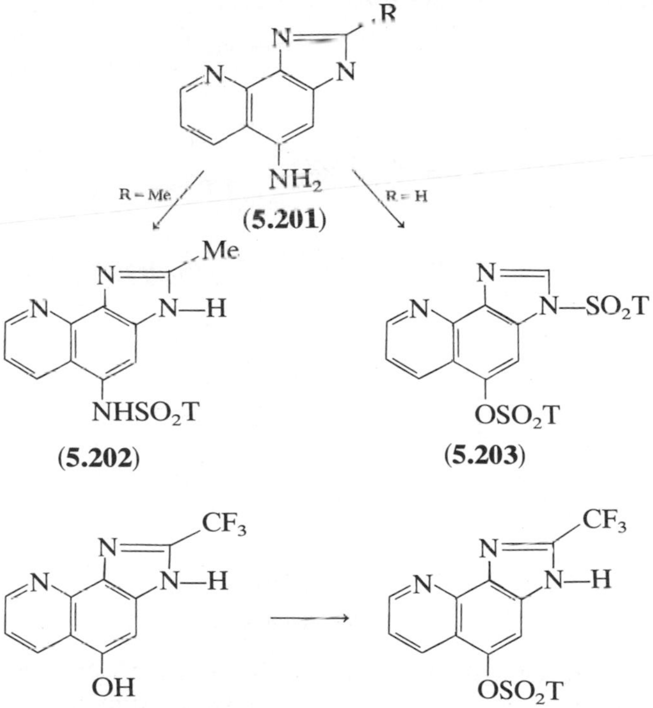

(5.201) (5.202) (5.203) (5.204) (5.205)

[T = *p*-tolyl]

Scheme 5.31

The amino group in 2-amino-N-1(3)H-imidazo[4,5-f]quinoline condenses smoothly with p-nitrobenzaldehyde in pyridine solution to afford the corresponding azomethine in unspecified yield (Table 5.63) (Scheme 5.32, **5.206 → 5.207**).[205]

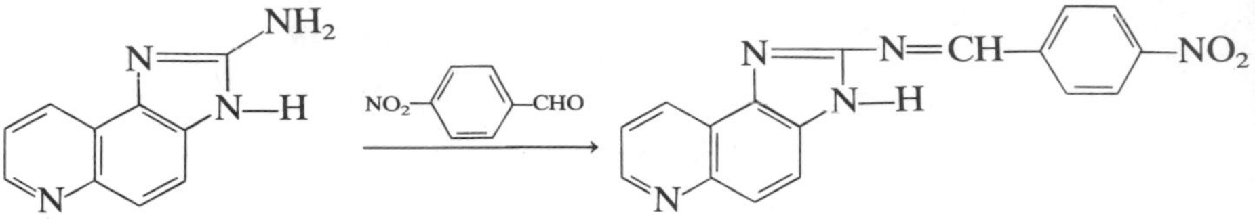

Scheme 5.32

Alkyl groups α or γ to a quaternary center in an imidazo[4,5-f]quinoline exhibit the expected methylene reactivity manifested by ready base-catalyzed deprotonation to carbanions capable of undergoing aldol-type condensations with a variety of electrophilic carbon reagents. Processes of this type provide the basis for the general synthesis of cyanine dyes containing an imidazo[4,5-f]quinoline chromophore, as illustrated by the example shown in Scheme 5.33 (**5.208 + 5.209 → 5.210**) (Table 5.63).[206]

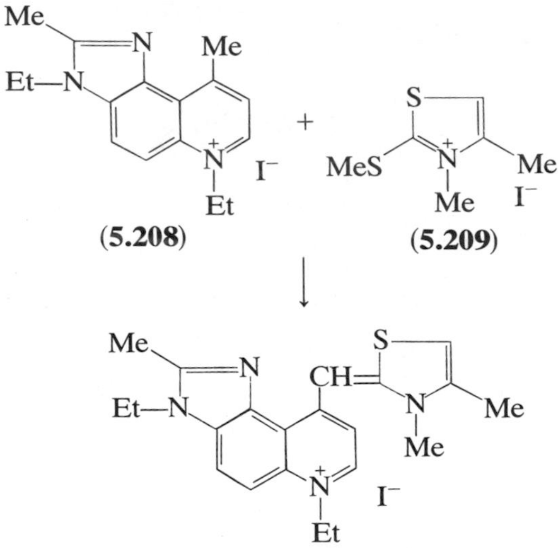

Scheme 5.33

NITRATION AND DIAZOTIZATION. As with Friedel–Crafts acylation (see before) no systematic studies of the electrophilic substitution (e.g., nitration, halogenation) of tricyclic 6-6-5 fused benzimidazoles containing one additional hetero atom appear to have been carried out, though an isolated report[220] records the failure of the imidazo[4,5-f]quinoline ring to undergo nitration. On the other hand, C(5)-amino substituents attached to imidazo[4,5-h]quinoline rings behave in an orthodox manner and may be converted into diazonium salts which can be coupled with activated arenes and reduced to the corresponding hydrazines (cf. Scheme 5.34).[224]

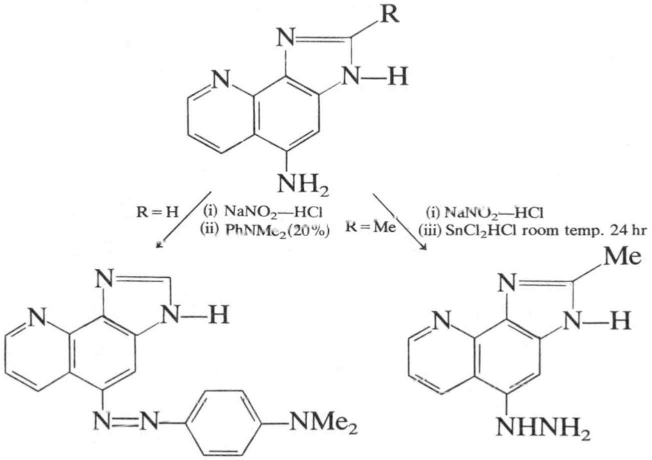

[red needles, m.p. 292° (from ethanol)]   [yellow crystals, m.p. 238° (from methanol)]

Scheme 5.34

Reactions with Nucleophiles

Theoretical studies suggest that the C(2) and C(7) positions of unionised imidazo[4,5-f]quinolines should be susceptible to nucleophilic attack,[229] the C(2) position being the more reactive in this respect. On the other hand, the C(7) position in an N(6)-substituted imidazo[4,5-f]quinolinium cation is predicted to have the highest positive charge and the lowest anion localization energy[229] and hence is the position most prone to undergo nucleophilic attack. It is known (see earlier in this section) that N(6) is the site of preferred electrophilic coordination in imidazo[4,5-f]quinolines. Consequently, if prior coordination is a prerequisite of successful nucleophilic attack, this will be expected to occur at the C(7)-position rather than at the C(2)-position of imidazo[4,5-f]quinolines. It will be seen later in this section that this prediction is born out experimentally. Molecular orbital studies[224] also predict that −M and −I substituents at C(2) in an imidazo[4,5-h]quinoline will promote nucleophilic attack at C(5), and again this prediction has been verified experimentally.

ARYLATION. In accord with the predictions of theoretical studies (see earlier in this section), imidazo[4,5-f]quinolines are phenylated by phenyl-lithium exclusively at C(7), albeit in low yield (Scheme 5.35, **5.211** → **5.212**) (Table 5.64).[229]

AMINATION. Despite an earlier report to the contrary,[216] 3-benzylimidazo-[4,5-f]quinoline is aminated in low yield at C(7) (Table 5.64) by heating

TABLE 5.64. NUCLEOPHILIC SUBSTITUTION REACTIONS OF IMIDAZO[4,5-g]QUINOLINES AND IMIDAZO[4,5-f]QUINOLINES

Starting materials				Reaction conditions[a]	Product					Yield (%)	Melting point (°C)	Solvent of crystallization	Crystal form	Ref.
R	R¹	R²	R³			R	R¹	R²	R³					
(5.211)	Ph	—	—	—	A	(5.212)	Ph	—	—	—	—[b]	198–199	Benzene	Yellow plates	229
(5.211)	CH₂Ph	—	—	—	A	(5.212)	CH₂Ph	—	—	—	30	165–167	—[c]	Yellow prisms	229
(5.211)	CH₂Ph	—	—	—	B	(5.213)	—	—	—	—	17	—[d]	—	—[h]	229
(5.215)	—	H	H	—	C	(5.216)	—	H	H	OMe[e]	58	—[f]	—[g]	—[h]	210
(5.215)	—	H	Me	—	D	(5.216)	—	H	Me	H[e]	40	368–370	Ethanol	Tan needles	210, 234
(5.215)	—	H	Me	—	D	(5.216)	—	H	Me	Br[e]	81	343 (decomp.)	Methanol	—[h]	234
(5.215)	—	H	Me	—	D	(5.216)	—	H	Me	Cl[e]	29	Gradual decomp.	Methanol	—[h]	235
(5.215)	—	H	Me	—	D	(5.216)	—	H	Me	F[e]	57	341–345	Ethanol	—[h]	235
(5.215)	—	H	Me	—	D	(5.216)	—	H	Me	I[e]	98	313–315	Ethanol	—[h]	234
(5.215)	—	H	Me	—	D	(5.216)	—	H	Me	OH[e]	48	393–395	Methanol	—[h]	235
(5.215)	—	H	Me	—	D	(5.216)	—	H	Me	OMe[e]	73	300–305	Ethanol-ether	—[h]	234
(5.215)	—	H	Me	—	D	(5.216)	—	H	Me	NMe₂[e]	42	313–317	Methanol	—[h]	234
(5.215)	—	H	Me	—	D	(5.216)	—	H	Me	COPh[e]	90	348–350	Methanol	—[h]	235
(5.215)	—	H	Me	—	D	(5.216)	—	H	Me	CO₂Et[e]	60	275–280	Ethanol	—[h]	210, 234
(5.215)	—	H	Et	—	D	(5.216)	—	H	Et	OMe[e]	63	308–310	Ethanol	—[h]	210, 234
(5.215)	—	H	Ph	—	D	(5.216)	—	H	Ph	OMe[e]	48	294 (decomp.)	Dimethyl-formamide	—[h]	210, 234
(5.215)	—	Me	Me	—	D	(5.216)	—	Me	Me	OMe[e]	96	315–317	Methanol	—[h]	210, 234
(5.215)	—	Ph	Me	—	D	(5.216)	—	Ph	Me	OMe[e]	68	258–274	Ethanol	—[h]	210, 234
(5.217)	—	—	—	—	E	(5.218)	—	—	—	—	49	273 (decomp.)	Pyridine	Colorless needles	216
(5.219)	—	—	—	—	F	(5.221)	—	—	—	—	62	182	Ethanol	—[h]	229
(5.213)	—	—	—	—	G	(5.214)	—	—	—	—	—[b]	—[d]	—	—[h]	229
(5.222) + (5.225)	Me	—	—	—	H	(5.224) + (5.227)	Me	—	—	—	6	203–204	Benzene	—[h]	205
(5.225)	—	Me	Me	Me			—	Me	Me	Me	35	220.5–221.5	Benzene	—[h]	205

592

Reactant	Product	R^1	R^2	R^3	Method	Yield (%)[b]	mp (°C)[f]	Solvent[g]	Crystal form[h]	Ref.
(5.222) + (5.225)	(5.224) + (5.227)	Et	—	—	—	6	175–176	Benzene-n-hexane	—[h]	205
(5.225)	(5.227)	Me	Et	Me	—	39	182.5–183	Benzene	Colorless crystals	208, 210
(5.225)	(5.227)	H	H	H	I	84	<400 (decomp.)	Methanol	—[h]	208, 210
(5.225)	(5.227)	H	H	Me	I	86	>300	Ethanol–water	Colorless needles	208, 210
(5.225)	(5.227)	H	H	Me	I	86	>300	Ethanol–water	Colorless needles	208, 210
(5.225)	(5.227)	H	H	Et	I	85	189–193	Ethanol	—[h]	208, 210
(5.225)	(5.227)	H	H	Ph	I	51	>230 (decomp.)	Ethanol	—[h]	208, 210
(5.225)	(5.227)	Me	H	Me	H	57	>400 (decomp.)	Methanol	—[h]	208, 210
(5.225)	(5.227)	Ph	H	Me	H	44	175–179	Methanol	—[h]	208, 210
(5.235)	(5.237)	—	—	—	H	83	>400	Dimethylformamide	—[h]	208, 210
(5.233)	(5.234)	—	—	—	I	92	170–171	Ethanol	Colorless needles	216

[a] A, (i) PhLi, ether/room temp./2 hr; (ii) H_2O; (iii) $PhNO_2$/reflux/5 min. B, $NaNH_2$, dimethylaniline/160°/4 hr. C, $p\text{-}R^3C_6H_4NH_2$, ethanol/reflux/10 hr. D, $p\text{-}R^3C_6H_4NH_2$, ethanol or dimethylformamide/reflux/14 hr. E, $CuSO_4$, CuCl, NH_4OH, EtOH, H_2O/autoclave/30 hr. F, KOH, $K_3Fe(CN)_6$, benzene/100°/5 hr. G, 40% NaOH aq. ethanol/reflux/1.5 hr. H, $POCl_3$/reflux/3 hr. I, $POCl_3$, dimethylformamide/85°/2 hr. then room temp./17 hr.

[b] Yield not specified.

[c] Purified by molecular distillation.

[d] Not obtained pure.

[e] Hydrochloride.

[f] Melting point not specified.

[g] Solvent not specified.

[h] Crystal form not specified.

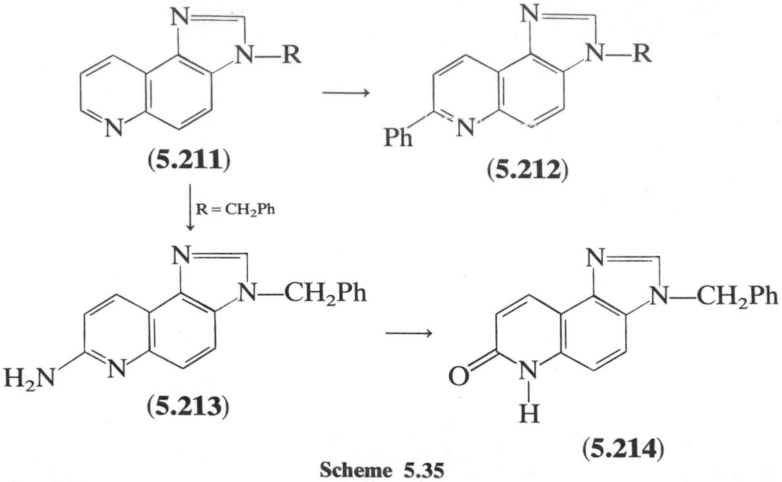

(5.211) (5.212) (5.213) (5.214)

Scheme 5.35

with sodamide in dimethylaniline [Scheme 5.35, **5.211** ($R=CH_2Ph$)→ **5.213**].[229]

By analogy with the well-established[233] reactivity of the halogen atom in γ-halogenopyridines towards nucleophilic displacement, the chlorine atom in readily accessible (see later in this section) C(9)-chlorinated imidazo[4,5-*f*]quinolines is smoothyl substituted by arylamines giving the corresponding 9-arylaminoimidazo[4,5-*f*]quinolines. Reactions of this type,[210,234,235] which occur with a wide range of nuclear substituted arylamines but are exemplified in the present discussion for *para*-substituted arylamines only (Scheme 5.36, **5.215**→**5.216**), proceed smoothly and usually in excellent yield (Table 5.64) simply by heating the chloroimidazo[4,5-*f*]quinoline with the

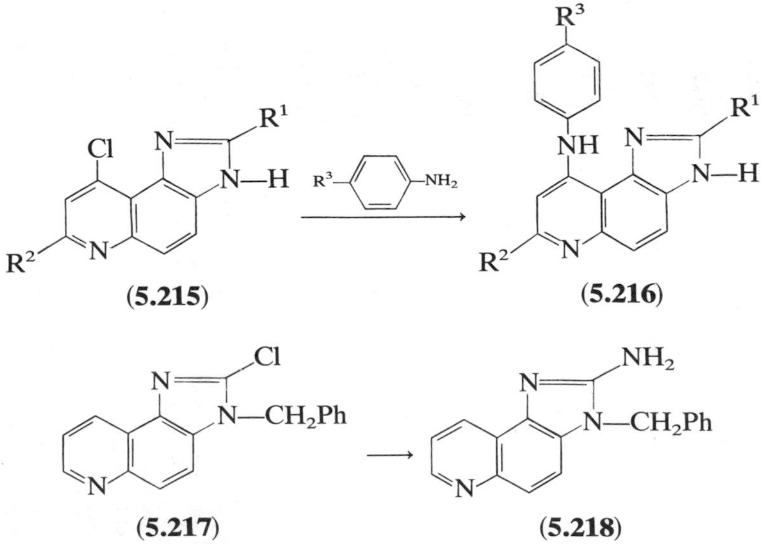

(5.215) (5.216) (5.217) (5.218)

Scheme 5.36

TABLE 5.65.   CATALYTIC REDUCTION OF IMIDAZO[4,5-g]QUINOLINES, IMIDAZO[4,5-f]QUINOLINES, AND IMIDAZO[4,5-h]-QUINOLINES

Starting material	Reaction conditions[a]	Product	Yield (%)	Melting point (°C)	Solvent of crystallization	Crystal form	Ref.
(5.235)	A	(5.236)[b,c] ($R^1 = R^2 = Me$)	74	222	Ethanol–ligroin	—[d]	203
(5.239) ($R^1 = R^2 = Me$)	B	(5.240) ($R^1 = R^2 = Me$)	87	201.5–202.5	Water	Colorless needles	205
(5.239) ($R^1 = Me$, $R^2 = Et$)	B	(5.240) ($R^1 = Me$, $R^2 = Et$)	81	149.5–150.5	Ethanol–water	Colorless needles	205
(5.236)	C	(5.237)[e,f]	35	268–272	Methanol–ether	—[d]	203
(5.235)	C	(5.237)[e,f]	35	—	—	—	203
(5.237)	D	(5.238)[e,g]	78	>300	Ethanol–ether	—[d]	203
(5.236)	D	(5.238)[e,g]	78	—	—	—	203
(5.235)	D	(5.238)[e,g]	78	—	—	—	203
(5.240) ($R^1 = R^2 = H$, H for Me)	C	(5.241)[e]	85	274–278 (decomp.)	Methanol–ether	—[d]	202
(5.240) ($R^1 = R^2 = H$, H for Me)	E	(5.242)[e,h]	76	>300	Ethanol–ether	—[d]	202
(5.242)	C	(5.244)[e,i]	85	>300 (decomp.)	Methanol–ether	—[d]	225
(5.243)	F	(5.245)[e,j]	80	286–291 (decomp.)	Ethanol–ether	—[d]	225

[a] A, Raney-Ni, H$_2$, NaOH, EtOH. B, PdC, H$_2$/room temp./atm. pressure. C, 5% RhC, 2 M HCl, H$_2$/room temp.,atm. pressure. D, 5% RhC, 2 M HCl H$_2$/room temp./90 atm. E, 5% RhC, 2 M HCl, H$_2$/80°/130 atm./72 hr. F, 5% RhC, 2 M HCl, H$_2$/80°/50 atm.

[b] Forms a dihydrochloride, m.p. >270° (decomp.) (from methanol-ether).

[c] Forms a picrate, m.p. 266° (decomp.).

[d] Crystal form not specified.

[e] Dihydrochloride.

[f] Forms a dipicrate, m.p. 170–173° (from water).

[g] Forms a dipicrate, m.p. 155–164° (from water).

[h] Forms a dipicrate, m.p. 176–186°.

[i] Forms a dipicrate, m.p. 260–261° (from water).

[j] Forms a dipicrate, m.p. 228–230° (from water).

arylamine in a suitable solvent (e.g., ethanol or dimethylformamide). Chlorine atoms at the C(2)-position in imidazo[4,5-f]quinolines are also activated towards nucleophilic replacement, as demonstrated[216] by the copper-catalyzed reaction of 2-chloro-3-benzylimidazo[4,5-f]quinoline (**5.217**) with ammonia to afford 2-amino-3-benzylimidazo[4,5-f]quinoline (**5.218**) (Scheme 5.36) (Table 5.64). The latter type of reaction has synthetic potential since, as noted previously (see section 5.2.3) 2-aminoimidazo[4,5-f]quinolines are not accessible by direct amination, amination in imidazo-[4,5-f]quinolines occurring preferentially at the C(7)-position.[229]

HYDROXYLATION. Unlike amination (as discussed previously), direct hydroxylation of simple imidazo[4,5-f]quinoline derivatives does not appear to have been reported to date. On the other hand, N(6)-substituted imidazo-[4,5-f]quinolinium salts react readily at C(7) with hydroxide ion to afford pseudo bases that are readily oxidized *in situ* by ferricyanide to give the anticipated imidazo[4,5-f]quinolin-7(6H)-ones (e.g., Scheme 5.37, **5.219** → **5.220** → **5.221**) (Table 5.64).[229] Hydroxylative replacement of a substituent at C(7) in an imidazo[4,5-f]quinoline framework is exemplified by the conversion of the amine (**5.213**) into the imidazo[4,5-f]quinolinone (**5.214**) on heating with 40% ethanolic sodium hydroxide (Scheme 5.35) (Table 5.64).[229] The tosylation of 5-aminoimidazo[4,5-h]quinolines in the presence of sodium hydrogen carbonate is also accompanied by the hydroxylative replacement of the 5-amino substituent (cf. section 5.2.3, Scheme 5.31, and Table 5.63).[224]

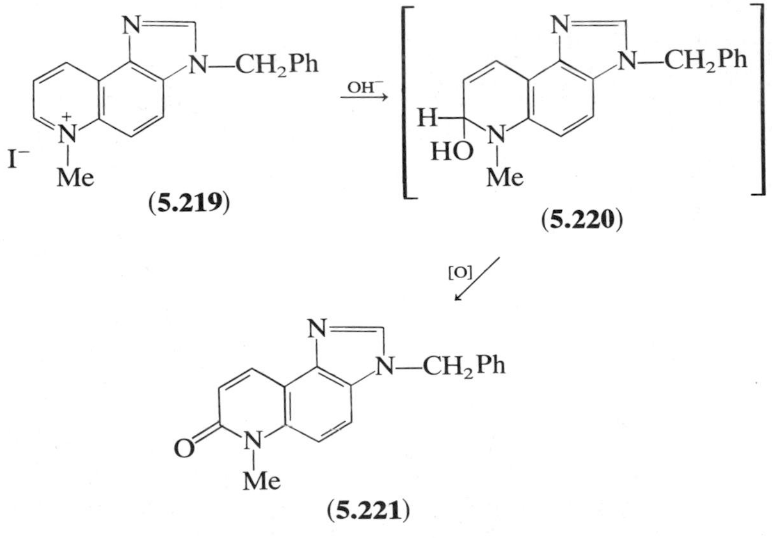

Scheme 5.37

HALOGENATION. N-1(3)H-Imidazo[4,5-g]quinolin-8(5H)-ones (**5.222**)[205] and N-1(3)H-imidazo[4,5-f]quinolin-9(6H)-ones (**5.225**)[205,208–210] are tautomeric with the corresponding hydroxy-compounds (**5.223** and **5.226**) (Scheme 5.38). As such, they behave as orthodox quinolin-4-ones and undergo halogenation with phosphorus oxychloride to afford moderate-to-excellent yields (Table 5.64) of the corresponding 8-chloroimidazo[4,5-g]-quinolines (**5.224**)[205] and 9-chloroimidazo[4,5-f]quinolines (**5.227**),[205,208,210] respectively. Imidazo[4,5-f]quinolin-1(3)H-ones are also chlorinated in high yield (Table 5.64) by heating under reflux with phosphorus oxychloride (e.g., Scheme 5.39, **5.228 → 5.229**).[216] In the context of this reaction, it is interesting to note that the imidazo[4,5-f]quinoline-2,9-dione (**5.230 ⇌ 5.231**) is reported[210] to react with phosphorus oxychloride with exclusive

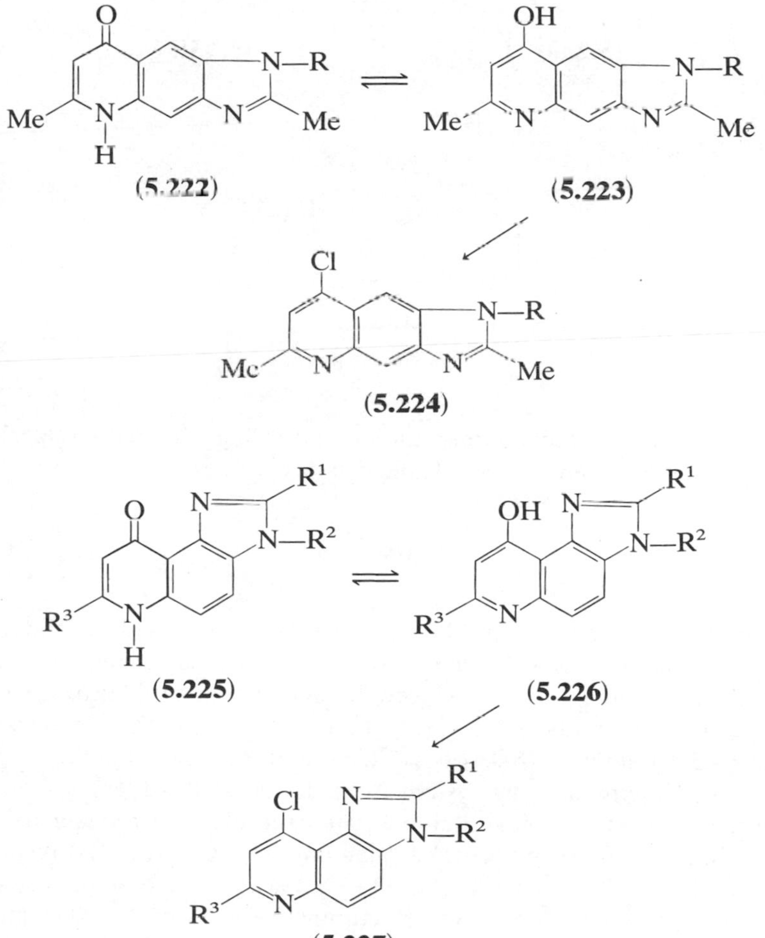

(**5.222**)   (**5.223**)

(**5.224**)

(**5.225**)   (**5.226**)

(**5.227**)

Scheme 5.38

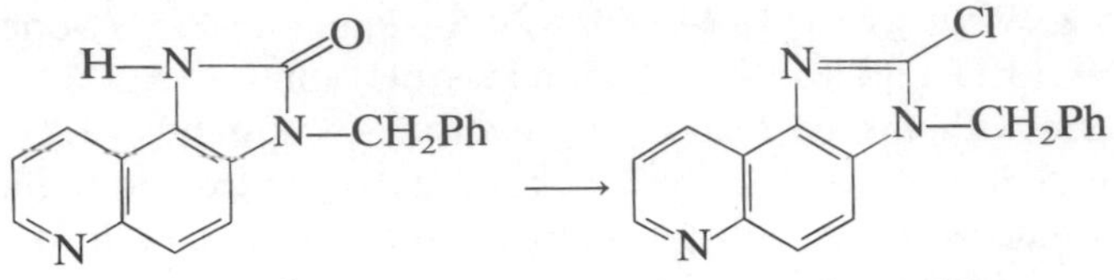

(5.228)   (5.229)

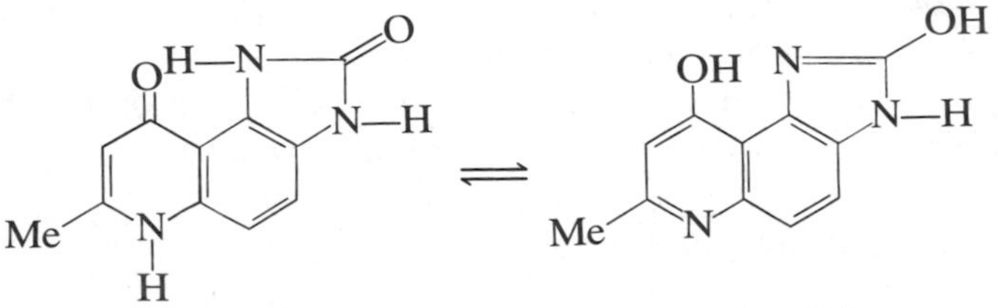

(5.230)   (5.231)

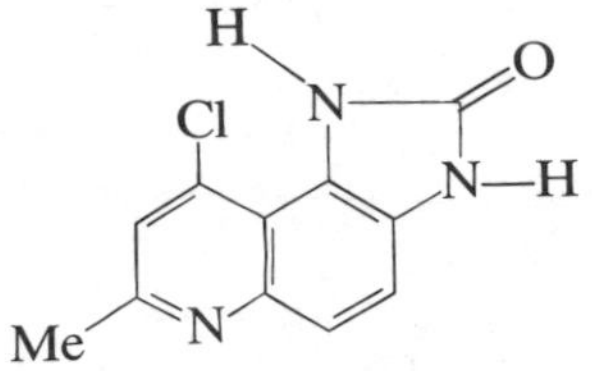

(5.232)

Scheme 5.39

chlorination of the quinolinone moiety affording the mono-chlorinated product (**5.232**) (Scheme 5.39) (Table 5.64).

Oxidation

Information on the susceptibility of tricyclic 6-6-5 fused benzimidazoles containing one additional hetero atom to oxidation is sparse. Ferricyanide oxidation of pseudo bases produced by the action of hydroxide ion on $N(9)$-substituted imidazo[4,5-f]quinolinium salts to the corresponding imidazo[4,5-f]quinolin-7(6H)-ones demonstrates the stability of the imidazo[4,5-f]quinoline ring system to oxidation of this type.

Polarographic studies have demonstrated the electron-transfer oxidation of 5-amino-N-1(3)H-imidazo[4,5-h]quinolines to the corresponding quinone diimines (Scheme 5.40, **5.233** → **5.234**).[236] The ease of such oxidation is influenced markedly by the electronic character of the $N(1)$ and $C(2)$ substituents and also by the nature of the solvent. Ease of oxidation increases with increase in the electron-donating capacity of both the $N(1)$

and the C(2) substituent and also occurs more readily in dimethyl sulfoxide than in acetonitrile. The latter effect is attributed to the greater basicity of dimethyl sulfoxide and hence its facilitation of N–H dissociation and thus oxidation.[236] In comparison with 5-amino-N-1(3)H-imidazo[4,5-h]quinolines, the redox behavior of the 5-nitro compounds [Scheme 5.40; **5.233** (NO$_2$ for NH$_2$)] is quite different, such compounds being destroyed on attempted polarographic oxidation.

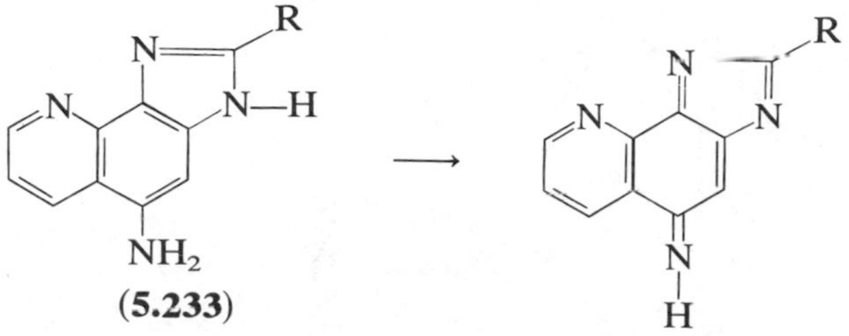

(5.233) → **(5.234)**

Scheme 5.40

Reduction

A number of studies have been concerned with the behavior of the imidazo[4,5-g]quinoline,[203] imidazo[4,5-f]quinoline,[202,205] and imidazo[4,5-h]isoquinoline[225] ring systems to catalytic hydrogenation. In all cases, the pattern is the same (Schemes 5.41 to 5.43). The imidazo[4,5-g]quinoline[203] and imidazo[4,5-f]quinoline[205] ring systems are stable to hydrogenation at normal temperature and pressure over palladium or nickel catalysts, and under these conditions substituents such as chlorine suffer hydrogenolysis

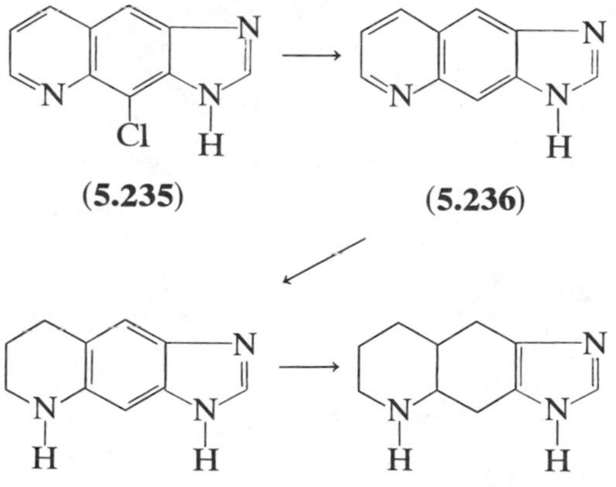

(5.235) → **(5.236)**

(5.237) → **(5.238)**

Scheme 5.41

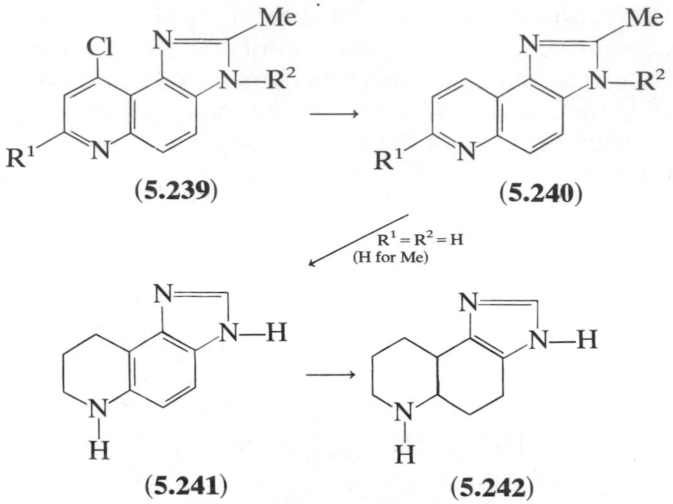

Scheme 5.42

(Table 5.65) without reduction of the ring system itself (Scheme 5.41, **5.235 → 5.236**;[203] and Scheme 5.42, **5.239 → 5.240**).[205] On the other hand, the use of a rhodium catalyst at normal temperature and pressure promotes selective reduction (Table 5.65) of the pyridine ring in imidazo[4,5-g]quinolines (Scheme 5.41, **5.236 → 5.237**),[203] imidazo[4,5-f]quinolines (Scheme 5.42, **5.240 → 5.241**),[202] and imidazo[4,5-h]isoquinolines (Scheme 5.43, **5.243 → 5.244**),[225] whereas under forcing conditions (Table 5.65)[202,203,205] reduction of both the pyridine and the benzene moieties is observed, the imidazole ring in all cases remaining inviolate (Scheme 5.41, **5.237 → 5.238**; Scheme 5.42, **5.241 → 5.242**; and Scheme 5.43, **5.244 → 5.245**).

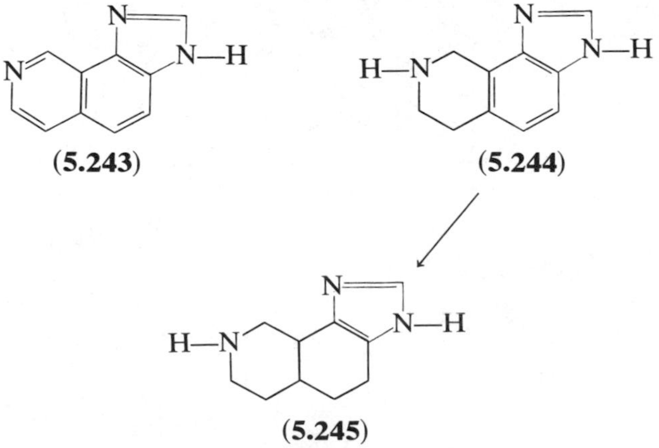

Scheme 5.43

5.2.4. Practical Applications

Biological Properties

Quinones of the imidazo[4,5-g]quinoline series have been patented as tuberculostatic agents,[213] while imidazo[4,5-f]quinoline derivatives are potent anthelmintics.[209,210,234,235] The imidazo[4,5-h]quinoline nucleus has also been incorporated in analogues of streptonigrin, a useful anticancer agent.[237]

Dyestuffs

The imidazo[4,5-f]quinoline chromophore has been incorporated in a large number of cyanine dyes,[206] while imidazo[4,5-h]quinoline derivatives have attracted considerable interest as chelating agents,[221,222,224,230] notably for incorporation in color-coupling agents.[227,232]

5.3. TRICYCLIC 6-6-5 FUSED BENZIMIDAZOLES WITH TWO ADDITIONAL HETERO ATOMS

The title category includes a large number of possible ring systems derived by the fusion of a six-membered, two-hetero-atom ring across the 4:5, 5:6,

TABLE 5.66. TRICYCLIC 6-6-5 FUSED BEN-
ZIMIDAZOLES WITH TWO ADDI-
TIONAL HETERO ATOMS

Structure[a]	Name[b]
(5.246)	Imidazo[4,5-g][1,4]benzothiin
(5.247)	Imidazo[4,5-g][1,4]benzoxazine
(5.248)	Imidazo[4,5-f][1,4]benzothiazine
(5.249)	Imidazo[4,5-g]quinazoline (*lin*-benzopurine)[c]
(5.250)	Imidazo[4,5-f]quinazoline (*prox*-benzopurine)[d]
(5.251)	Imidazo[4,5-h]quinazoline (*dist*-benzopurine)[d]
(5.252)	Imidazo[4,5-g]quinoxaline
(5.253)	Imidazo[4,5-f]quinoxaline

[a] See Scheme 5.44.
[b] Based on the Ring Index.
[c] Alternative name denoting the linear "stretched purine" framework of this ring system.
[d] Alternative names denoting the angular "stretched purine" frameworks of these ring systems, the prefixes *prox* and *dist* referring to the proximate and remote spatial relationships, respectively, of C(9) and C(6) in the pyrimidine ring to N(1) in the imidazole ring.

or 6:7 positions in benzimidazole. Of these, the only examples that appear to have been studied to any extent are those based on [1,4]dithiin, [1,4]oxazine, or [1,4]thiazine rings (**5.246** to **5.248**), the three possible imidazoquinazoline structures (**5.249** to **5.251**), and the two possible imidazoquinoxaline structures (**5.252** and **5.253**). The systematic nomenclature for each of these ring systems is summarized in Table 5.66. The alternative name, lin-benzoadenine, has been coined[238] for the imidazo[4,5-g]quinazoline ring system (**5.249**) and emphasizes its linear "stretched purine" structure. The alternative names,[238,239] *prox*-benzoadenine and *dist*-benzoadenine, for the imidazo[4,5-*f*]quinazoline and imidazo[4,5-*h*]quinazoline ring systems likewise emphasize their angular "stretched purine" structures, the prefixes *prox* and *dist* denoting the remote and proximate spatial relationship, respectively, between $N(1)$ in the imidazole ring and the amino group at $C(9)$ or $C(6)$ in the pyrimidine ring.

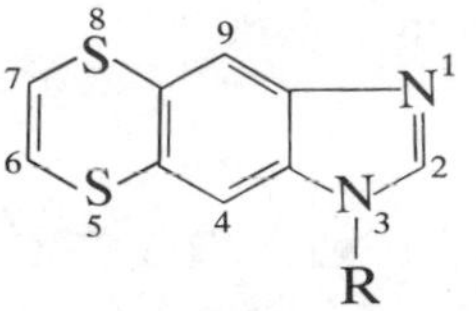

(5.246) 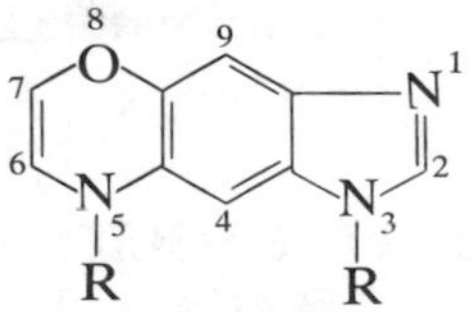   (5.247)

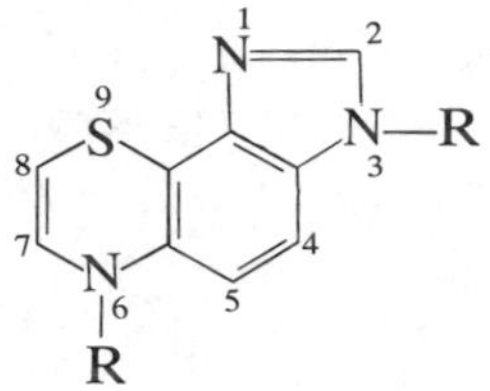

(5.248) 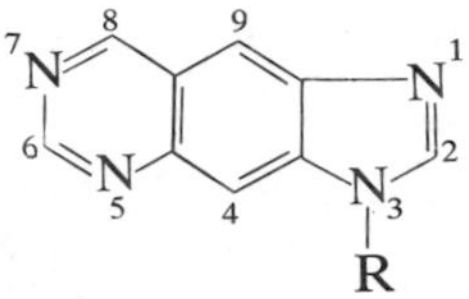   (5.249)

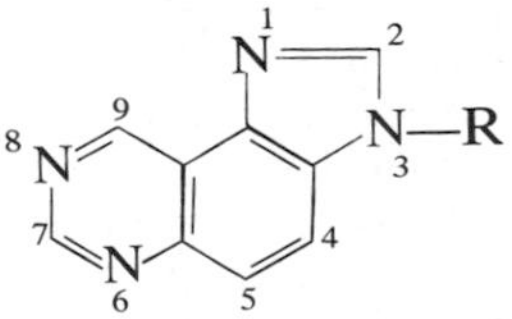

(5.250) 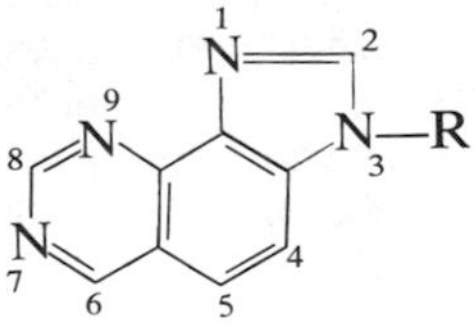   (5.251)

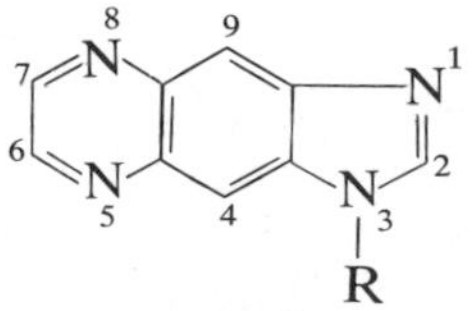

(5.252) 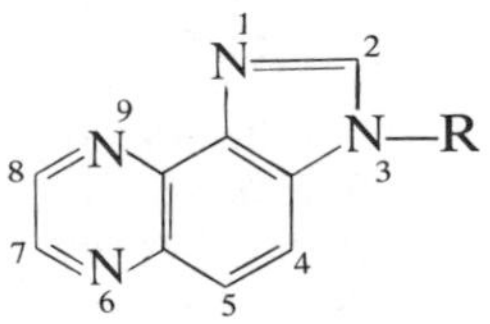   (5.253)

Scheme 5.44

5.3.1. Synthesis

Ring-Closure Reactions of Benzimidazole Derivatives

The addition of ethanedithiol to benzimidazole-4,7-quinone (**5.254**) provides a method (Scheme 5.45) (Table 5.67) for the synthesis of one of the few reported[240] derivatives (**5.255**) of the imidazo[4,5-g][1,4]benzothiin ring system.

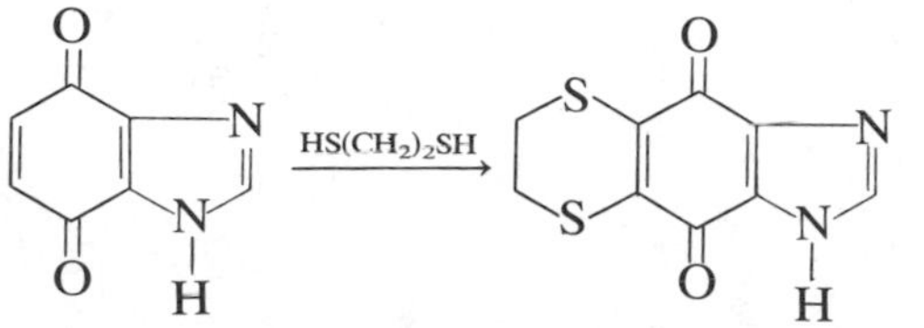

(**2.254**)   (**2.255**)

Scheme 5.45

What appears to be the sole representative (**5.257**) of the imidazo[4,5-g][1,4]benzoxazine ring system has been prepared[241] by the reductive bis-ring closure of (4-acetylamino-2,5-dinitro)phenoxyacetic acid (Scheme 5.46, **5.256** → **5.257**) (Table 5.67). In closely related processes, high yields (Table

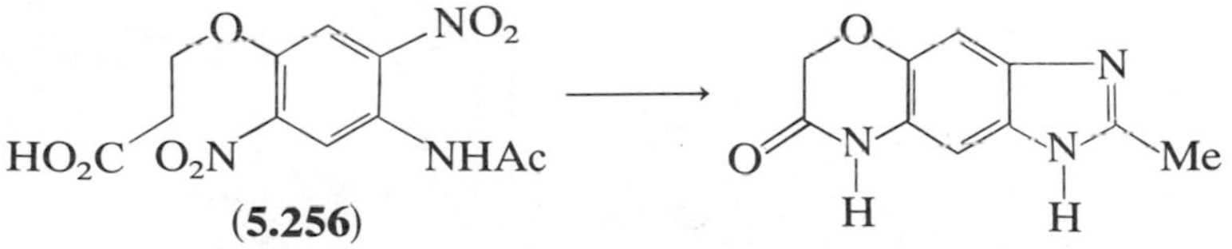

(**5.256**)   (**5.257**)

Scheme 5.46

5.67) of imidazo[4,5-h][1,4]benzothiazin-7-ones (**5.261**) are produced by the acid- or base-catalyzed cyclization of S-(5-aminobenzimidazol-4-yl)thio-acetic acids (**5.260**), which can be preformed or prepared in situ by the reaction of readily accessible 4-mercapto-5-aminobenzimidazole alkali metal salts (**5.258**) with chloroacetic acid (Scheme 5.47).[242,243] In a logical extension of the latter synthetic mode, reaction of 5-amino-4-mercapto-2-methyl-1-phenylbenzimidazole potassium salt (**5.258**; $R^1 = Me$, $R^2 = Ph$, $M = K$) with phenacyl bromide affords the 8H-imidazo[4,5-g][1,4]benzo-thiazine (**5.259**) in moderate yield (Table 5.67).[243]

TABLE 5.67. SYNTHESES OF IMIDAZO[4,5-g][1,4]BENZOTHIINS, IMIDAZO[4,5-g][1,4]BENZOXAZINES, AND IMIDAZO[4,5-h][1,4]BENZOTHIAZINES

Starting material	Reaction conditions[a]	Product	Yield (%)	Melting point (°C)	Solvent of crystallization	Crystal form	Ref.
(5.254)	A	(5.255)	53	191–193 (decomp.)	—[b]	—[c]	240
(5.256)	B	(5.257)	—[d]	243	—[b]	Colorless plates	241
(5.260) (R^1 = H, R^2 = CH$_2$Ph)	C	(5.261) (R^1 = H, R^2 = CH$_2$Ph)[e]	72	234–235	Isopropanol	—[c]	242
(5.258) (R^1 = H, R^2 = CH$_2$CH$_2$OH, M = Na)	D	(5.261) (R^1 = H, R^2 = CH$_2$CH$_2$OH)[f]	81	250–251	Ethanol–water	—[c]	242
(5.258) (R^1 = Me, R^2 = Ph, M = K)	E	(5.261) (R^1 = Me, R^2 = Ph)	65	288	Ethanol	—[c]	243
(5.259) (R^1 = Me, R^2 = Ph, M = K)	F	(5.259)	56	190	Ethanol	—[c]	243

[a] A, HSCH$_2$CH$_2$SH, methanol/room temp./17 hr. B, Zn, HCl. C, polyphosphoric acid/160°/40 min. D, ClCH$_2$CO$_2$H, NaOH, 80% ethanol/60°/1 hr. E, ClCH$_2$CO$_2$H, KOH aq./100°/2 hr. F, PhCOCH$_2$Br, ether/room temp./20 hr.

[b] Solvent not specified.

[c] Crystal form not specified.

[d] Yield not quoted.

[e] Forms a hydrochloride, m.p. 278–284° (from ethanol–water).

[f] Mono-hydrochloride, free base has m.p. 301–302° (from water).

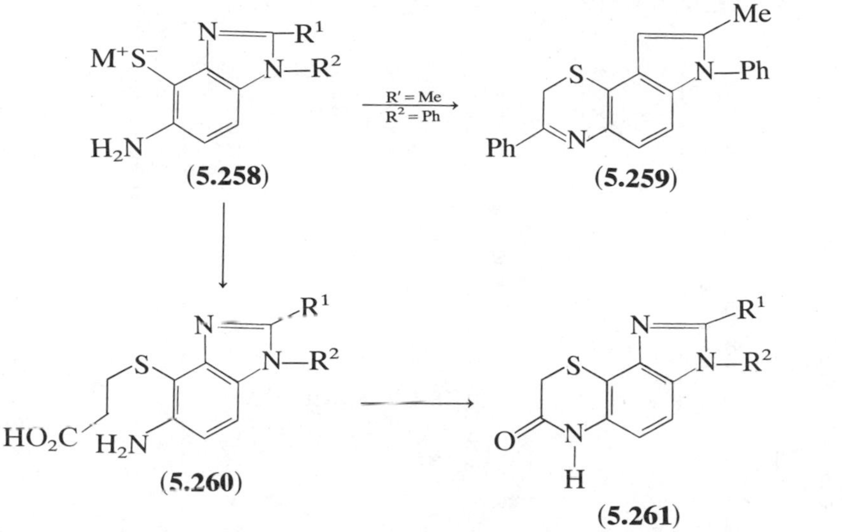

Scheme 5.47

The formation in excellent yield (Table 5.68) of 6-amino-N-1(3)H-imidazo[4,5-g]quinazolin-8(7H)-one (*lin*-benzoguanine) (**5.263**) by the condensation of 5-amino-6-ethoxycarbonylbenzimidazole hydrochloride (**5.262**) with cyanamide in ethanol[244] exemplifies (Scheme 5.48) one potentially flexible synthetic approach to imidazo[4,5-g]quinazolines of biological interest as "stretched purines."[244] However, although useful for the synthesis of imidazo[4,5-g]quinazolines with a variety of C(6) and C(7) substituents, this approach is not practical for the unequivocal synthesis of N(3)-substituted derivatives.[244]

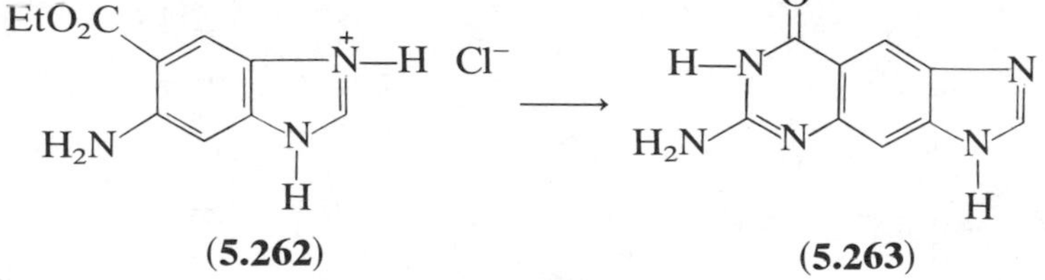

Scheme 5.48

Imidazo[4,5-g]quinoxaline derivatives (**5.266**) are generally accessible[245] in very good yield (Table 5.69) by the condensation of 5,6-diaminobenzimidazoles (**5.265**) with α-dicarbonyl compounds (**5.264**) (Scheme 5.49). The alternative, if less conventional, approach to imidazo[4,5-g]quinoxalines involving the reaction of a benzimidazole-5,6-quinone with a 1,2-diamino

TABLE 5.68. SYNTHESES OF IMIDAZO[4,5-g]QUINAZOLINES, IMIDAZO[4,5-f]QUINAZOLINES, AND IMIDAZO[4,5-h]-QUINAZOLINES

Starting material	R¹	R²	R³	Reaction conditions	Product	R¹	R²	R³	Yield (%)	Melting point (°C)	Solvent of crystallization	Crystal form	Ref.
(5.262)	—	—	—	A	(5.263)	—	—	—	98	>300	—e	—d	244
(5.269)	NH₂	H	H	B	(5.271)	NH₂	H	—	86	—	—	—	244
(5.269)	H	H	H	C	(5.271)	H	H	—	54	>320	Water	Colorless crystals	238
(5.270)	H	H	H	D	(5.271)	H	H	—	94	—	—	—	238
(5.270)	H	H	CH₂Ph	E	(5.271)	H	H	CH₂Ph	93	>320	Dimethylformamide–water	Yellow prisms	238
(5.270)	H	—	H	F	(5.271)	H	—	—	85	268–270	Ethanol	Colorless prisms	238
(5.270)	H	CH₂Ph	H	D	(5.271)	H	CH₂Ph	—	75	248–249	Ethanol	Colorless prisms	238
(5.272)	H	H	—	B	(5.274)	H	H	—	69	>320	Water	Colorless prisms	239
(5.273)	n-Bu	H	—	G	(5.274)	n-Bu	H	—	—b	103–104	—c	—d	247
(5.273)	n-Bu	Ac	—	H	(5.274)	n-Bu	Me	—	—b	126–127	—c	—d	247
(5.273)	n-Bu	H	—	I	(5.275)	—	—	—	—b	231–232	—c	—d	247
(5.276)	—	—	—	B	(5.278)	—	—	—	69	>320	Water	Colorless crystals	239

a A, NH₂CN, ethanol/reflux/18 hr. B, 10% PdC, H₂/3 atm./30 min, then HCO₂H/reflux/2 hr. C, PdC, H₂, 98% HCO₂H/3 atm./1–2 hr, then reflux. D, HCO₂H/reflux/2 hr. E, HCO₂H/reflux/90 min. F, 98% HCO₂H/reflux/1 hr. G, HCO₂H/reflux. H, CH₃CO₂H/reflux. I, potassium ethyl xanthate ethanol/reflux.
b Yield not specified.
c Solvent not specified.
d Crystal form not specified.
e Purified by precipitation from 1M HCl solution using ammonia.

TABLE 5.69. SYNTHESES OF IMIDAZO[4,5-*g*]QUINOXALINES AND IMIDAZO[4,5-*f*]-QUINOXALINES

Starting materials	Reaction conditions[a]	Product	Yield (%)	Melting point (°C)	Solvent of crystallization	Crystal form	Ref.
(**5.264**) (R^2 = H)[b] + (**5.265**) (R^1 = Me)	A	(**5.266**) (R^1 = Me, R^2 = H)	57	214	Toluene	Yellow needles	245
(**5.264**) (R^2 = Me) + (**5.265**) (R^1 = Me)	B	(**5.266**) (R^1 = R^2 = Me)	75	318	*n*-Butanol	Yellow needles	245
(**5.264**) (R^2 = Ph) + (**5.265**) (R^1 = Me)	C	(**5.266**) (R^1 = Me, R^2 = Ph)	75	221	Methanol	—[c]	245
(**5.264**) (R^2 = Ph) + (**5.265**) (R^1 = H)[d]	C	(**5.266**) (R^1 = H, R^2 = Ph)	60	285–286	Ethanol	Yellow needles	245
(**5.267**)	D	(**5.268**)	91	300 (decomp.)	Acetic acid	Cream solid	246
(**5.279**)	E	(**5.281**)	31	249–249.5	—[f]	—[c]	248

[a] A, 4M aq. NaOAc, 2M AcOH/60°/1 hr. B, methanol/100°/3 hr. C, ethanol/reflux/10 min. D, ethanol/reflux/104 min. E, Fe, Ac_2O, AcOH/reflux/0.5 hr.
[b] Bisulfite complex.
[c] Crystal form not specified.
[d] Dihydrochloride.
[e] Forms a hydrochloride, m.p. >450° (decomp.).
[f] Purified by sublimation.

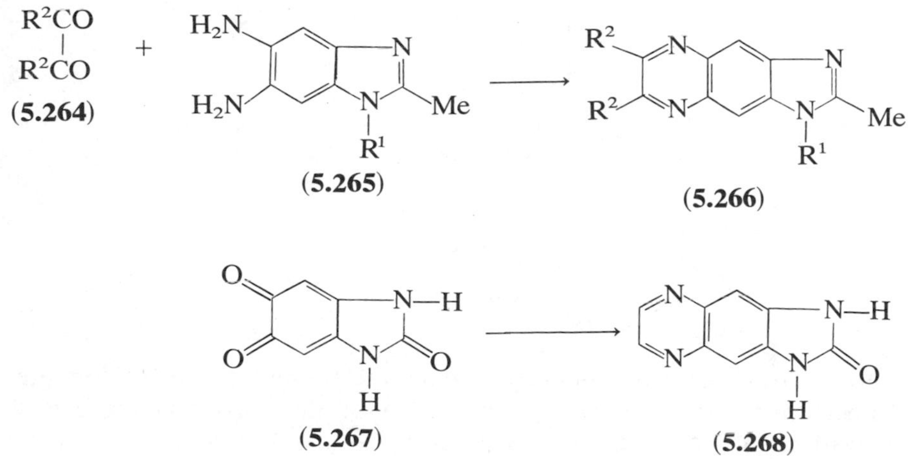

Scheme 5.49

reagent has been demonstrated in one instance (Scheme 5.49).[246] Thus the benzimidazoletrione (**5.267**) reacts smoothly with ethylenediamine to afford imidazo[4,5-g]quinoxalin-2(1H,3H)-one (**5.268**) in excellent yield (Table 5.69).[246]

Ring-Closure Reactions of Other Heterocycles

Condensation[244] of 2,6,7-triaminoquinazolin-4(3H)-one [Scheme 5.50, (**5.270**; $R^1 = NH_2$, $R^2 = R^3 = H$)] [prepared *in situ* by catalytic reduction of the 6-nitro derivative (**5.269**; $R^1 = NH_2$, $R^2 = R^3 = H$) with formic acid affords a very good yield (Table 5.68) of the "stretched purine," *lin*-benzoguanine (**5.271**; $R^1 = NH_2$, $R^2 = R^3 = H$) and exemplifies a general route (Scheme 5.50) to imidazo[4,5-g]quinazolines that complements that involving ring-closure in 5-amino-6-acylbenzimidazole derivatives (cf. Scheme 5.48). The route through 6,7-diaminoquinazoline intermediates (Scheme 5.50) has the advantage over the benzimidazole approach (Scheme 5.50) of permitting the unequivocal synthesis of N(3)-substituted imidazo[4,5-g]-quinazolines (Table 5.68) useful as starting materials for the preparation of biologically interesting *lin*-benzoadenine derivatives (see later in this section).[238]

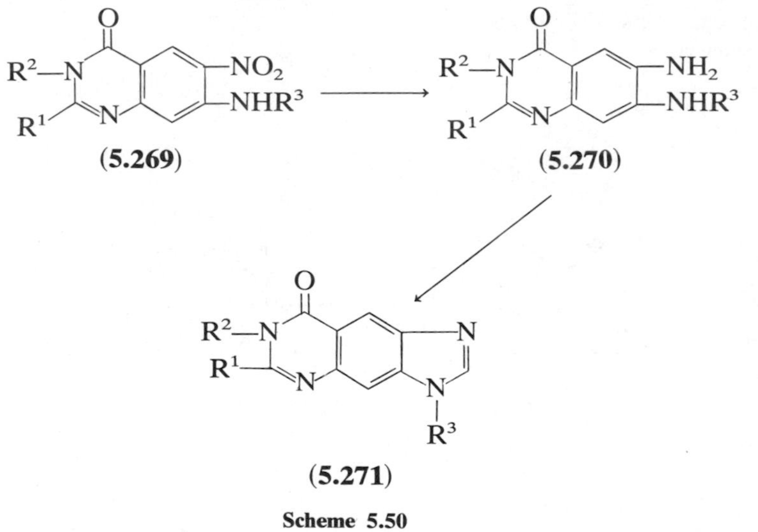

Scheme 5.50

Ring closure of 5,6-diaminoquinazolin-4(3H)-ones (**5.273**) [either preformed or prepared *in situ* by reduction of the corresponding 5-nitro derivatives (**5.272**)] with carboxylic acids also provides an efficient (Table 5.68) and flexible general route (Scheme 5.51) to imidazo[4,5-f]quinazolinones (**5.274**).[239,247] The latter are of interest as "stretched purines" having

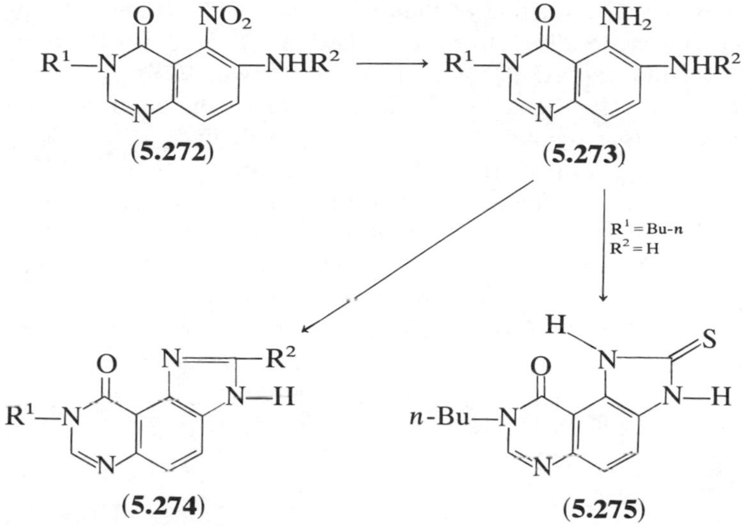

Scheme 5.51

an angular structure suitable as synthetic precursors of *prox*-benzo-adenines,[239] the imidazo[4,5-*f*]quinazoline analogs of *lin*-benzoadenines. A variant of this type of imidazo[4,5-*f*]quinazoline synthesis[247] is illustrated by the reaction of 3-*n*-butyl-5,6-diaminoquinazolin-4(3*H*)-one with potassium ethyl xanthate to give the corresponding 2-thione [Scheme 5.51, **5.273** (R¹ = Bu-*n*, R² = H)→**5.275**] (Table 5.68). The formic-acid-mediated ring closure (Scheme 5.52)[239] of 7,8-diaminoquinazolin-4(3*H*)-one (**5.277**)

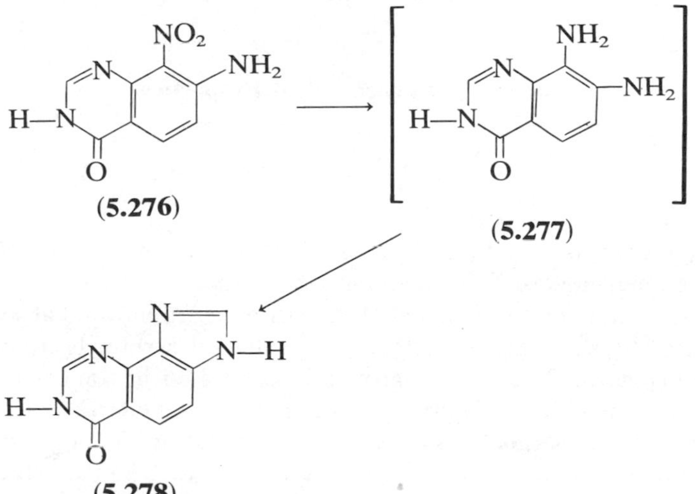

Scheme 5.52

[derived by *in situ* reduction of the readily accessible 7-amino-8-nitroquin-azolin-4(3*H*)-one (**5.276**)] affords a method (Table 5.68) for the preparation of *N*-1(3)*H*-imidazo[4,5-*h*]quinazolin-6(7*H*)-one (**5.278**), the key starting material for the synthesis of *dist*-benzoadenine, the imidazo[4,5-*h*]quin-azoline analog (see section 5.3.3) of *lin* benzoadenine.

The acetylative ring closure of 5,6-diaminoquinoxaline (**5.280**; R = H) [generated *in situ* by reduction of 5,6-dinitroquinoxaline (**5.279**)] affords the anticipated fused benzimidazole (**5.281**), a derivative of the rare imidazo-[4,5-*f*]quinoxaline ring system.[248] Surprisingly, attempts[248] to effect the acid- or base-catalyzed cyclization of 5,6-diacetylaminoquinoxaline (**5.280**; R = Ac) resulted in the formation of tars or the recovery of the unreacted starting material.

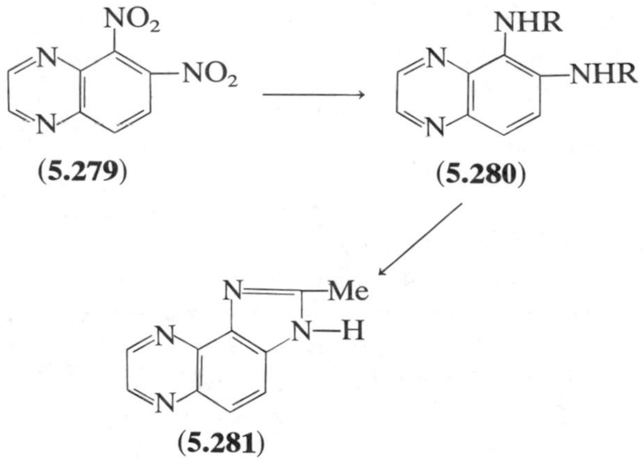

Scheme 5.53

5.3.2. Physicochemical Properties

Spectroscopic Studies

INFRARED SPECTRA. The paucity of systematic infrared studies of tricyclic 6-6-5 fused benzimidazoles containing two additional hetero atom precludes any general comment on the effect of structure on the infrared absorption of such ring systems. Not unexpectedly, the infrared carbonyl stretching frequencies (Table 5.70) of lactam groups incorporated in frameworks of the type in question vary markedly with the linear or angular character of the structure and the particular site within the molecule. Though insufficient information is available to allow generalizations to be made, the data in Table 5.70 clearly demonstrate the significantly higher infrared carbonyl frequency of a lactam group when incorporated in the imidazole ring as

TABLE 5.70. INFRARED CARBONYL STRETCHING FREQUENCIES OF
LACTAM DERIVATIVES OF THE IMIDAZO[4,5-h]-
[1,4]BENZOTHIAZINE, IMIDAZO[4,5-g]QUINAZOLINE,
IMIDAZO[4,5-f]QUINAZOLINE, AND IMIDAZO[4,5-g]-
QUINOXALINE RING SYSTEMS

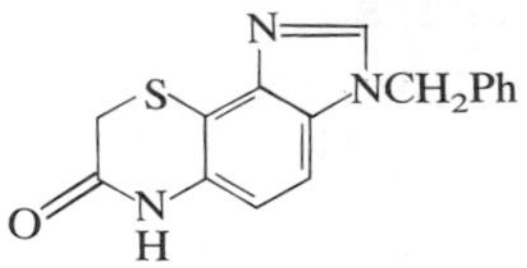

(5.282) 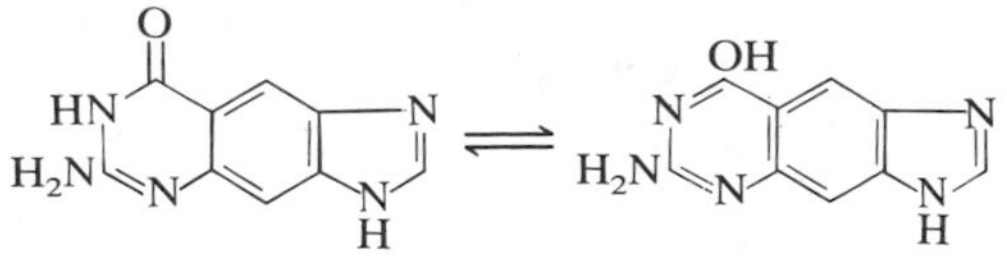   (5.283)   (5.284)

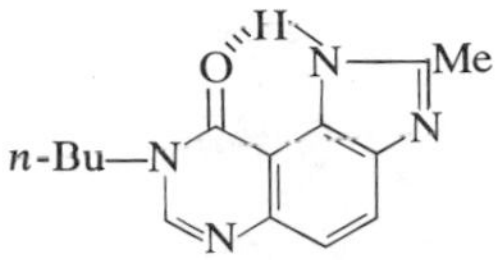

(5.285) 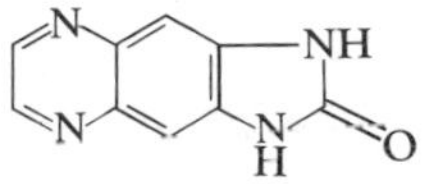   (5.286)

Compound	ν_{max} (C=O) (cm^{-1})	Medium	Ref.
(5.282)	1670	Nujol	242
(5.283)	1700	KBr	244
(5.285)	1600	—^{a}	247
(5.286)	1721	KBr	246

^{a} Not specified

compared with the heterocyclic six-membered ring. The low-infrared frequency of the nonenolizable carbonyl group in the imidazo[4,5-f]quinazolinone (5.285) is noteworthy when compared with the infrared carbonyl frequency reported for the enolisable imidazo[4,5-g]quinazolinone (5.283) (Table 5.70) and may be attributed to strong intramolecular hydrogen bonding (see 5.285).[247]

ULTRAVIOLET AND FLUORESCENCE SPECTRA. The ultraviolet absorption (Tables 5.71, 5.72, and 5.73) of imidazoquinazolines of structural types (5.287 to 5.292) (Scheme 5.54)[238,239,244,247,249,250] has been extensively investigated, most notably because of its utility for establishing unequivocally the site of alkylation and more specifically ribosidation in "stretched purines." In comparison, investigations of the ultraviolet absorption characteristics of the other known tricyclic 6-6-5 fused benzimidazole ring systems with two additional hetero atoms are sparse. For example, the ultraviolet absorption[246] of only a single derivative (5.294) of the imidazo[4,5-g]quinoxaline ring system has been recorded.

The ultraviolet spectrum of 5-amino-N-1(3)H-imidazo[4,5-g]quinazoline (*lin*-benzoadenine) (5.287; R^1 = NH$_2$, R^2 = H) (Table 5.71) broadly resembles that of anthracene, whereas the spectra of the angular isomers *prox*-benzoadenine (5.290; R = NH$_2$) and *dist*-benzoadenine (5.292 R = NH$_2$)

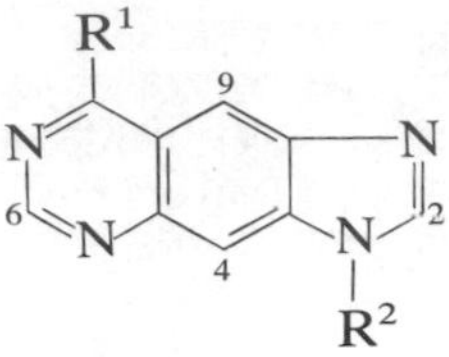

(5.287)

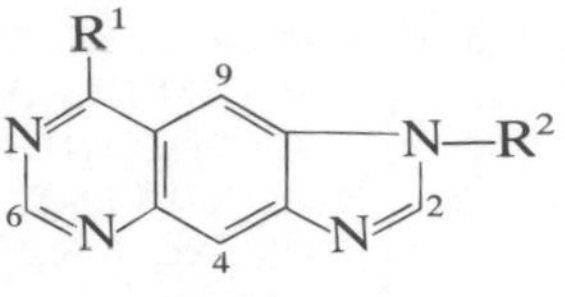

(5.288)

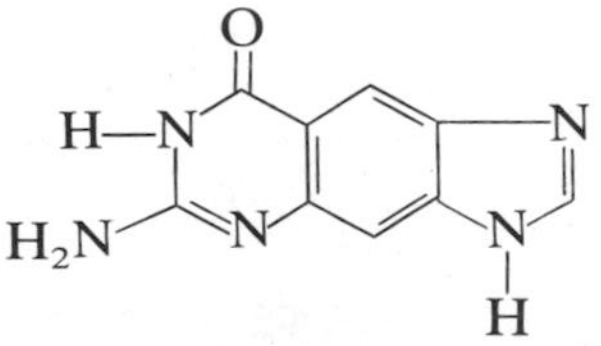

(5.289)

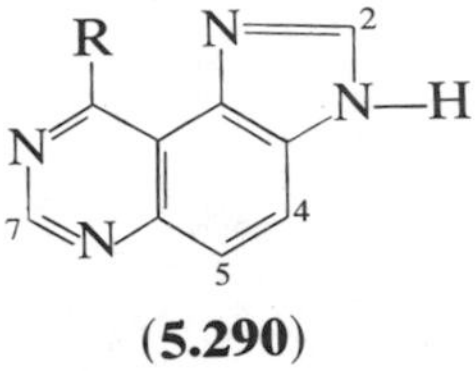

(5.290)

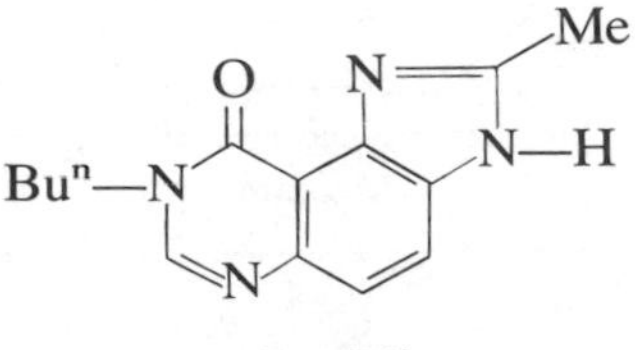

(5.291)

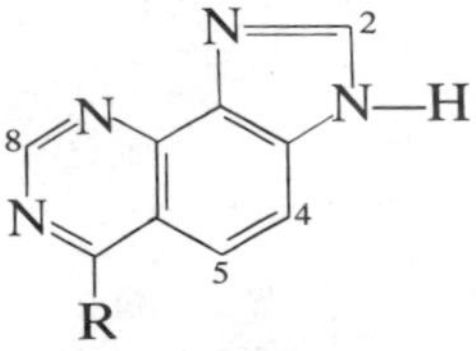

(5.292)

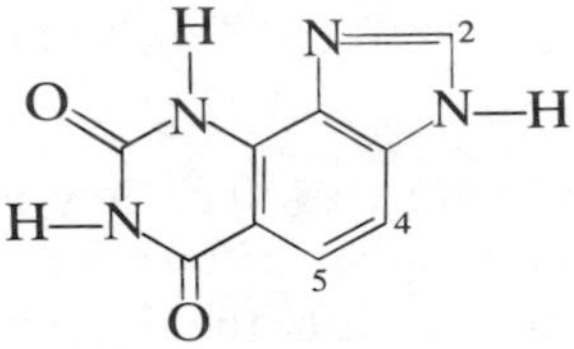

(5.293)

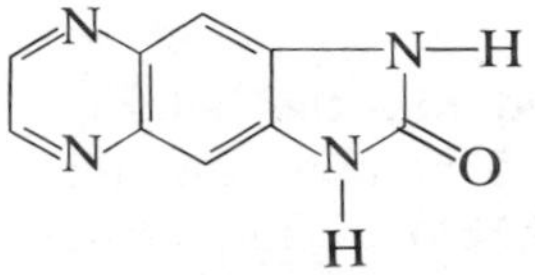

(5.294)

Scheme 5.54

resemble the ultraviolet spectrum of phenanthrene.[239] In accord with the generalization that the electronic transitions in linear aromatic systems are of lower energy than those of their angular counterparts, the low-energy bands in the ultraviolet spectrum of *dist*-benzoadenine (**5.293**; R = NH$_2$) are shifted approximately 30 nm to shorter wavelength (Table 5.73) compared with the same bands in *lin*-benzoadenine (**5.287**; R^1 = NH$_2$, R^2 = H). In contrast, the hypsochromic shift of the low-energy bands in the ultraviolet spectrum of *prox*-benzoadenine (**5.290**; R = NH$_2$) compared with the same bands in the spectrum of *lin*-benzoadenine is only 10 nm. The smaller hypsochromic shift in this case may be explained in terms of a countering bathochromic shift in the *prox* isomer resulting from the extension of the π-system due to conjugative interaction with the C(9)-amino group, which

TABLE 5.71. ULTRAVIOLET SPECTRA OF IMIDAZO[4,5-g]QUINAZOLINE DERIVATIVES (5.287), (5.288) AND (5.289)

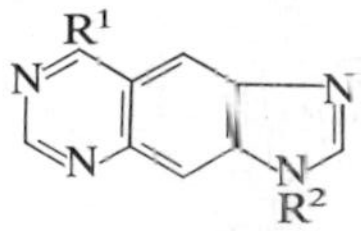

(5.287) 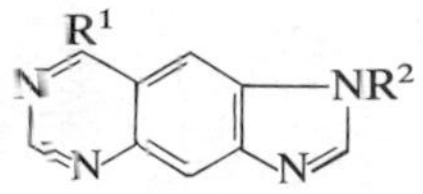   (5.288) 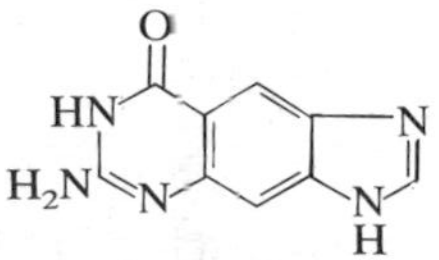   (5.289)

Compound	R¹	R²	Solvent[a]	λ_{max} nm (log ε)	Ref.
(5.287)	NH_2	H	A	237(4.42), 242(4.43), 258(4.17), 261 sh, 298 sh, 318(3.8), 332(3.96), 348(3.85)	238
(5.287)	NH_2	H	E	224(4.38), 234(4.36), 258(4.10), 265 sh, 289 sh, 302 sh, 322 sh, 333(4.16), 349(4.14)	238
(5.287)	NH_2	H	C	248(4.57), 262(4.51), 270(4.53), 310 sh, 322(3.90), 350(3.77), 365 sh	238
(5.287)	NH_2	CH_2Ph	A	231(4.54), 242 sh, 260(4 11), 266(4.11), 360 sh, 319(3.83), 333(3.94), 349(3.83)	238
(5.287)	NH_2	CH_2Ph	E	228(4.37), 236(4.38), 263(4.09), 271(4.00), 321(3.90), 335(3.98), 351(3.93)	238
(5.287)	NH_2	CH_2Ph	C	242(4.42), 260(4.13), 265(4.13), 306 sh, 318(3.84), 338(3.85), 349(3.85)	238
(5.288)	NH_2	CH_2Ph	A	223(4.43), 242(4.53), 264(4.20), 295(3.75), 306(3.64), 327(3.79), 343(3.97), 259(3.91)	238
(5.288)	NH_2	CH_2Ph	B	236(4.38), 267(4.10), 275(4.01), 299(3.81), 317(3.89), 328 sh, 336(4 00), 352(4.00), 363 sh	238
(5.288)	NH_2	CH_2Ph	C	237(4.58), 264(4.28), 294(3.89), 306(3.84), 328(3.88), 343(4.01), 358(3.93)	238
(5.288)	NH_2	CH_2Ph	D	237(4.43), 267(4.17), 275(4.22), 313(4.01), 326.5(4.00), 343(3.99), 359(3.96)	238
(5.287)	$NHCH_2Ph$	H	A	227(4.58), 236 sh, 264(4.11), 272(4.09), 322 sh, 334(4.15), 351(4.C5)	238

TABLE 5.71 (*Continued*)

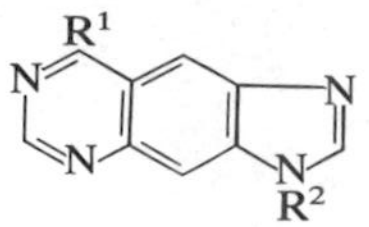

(5.287) 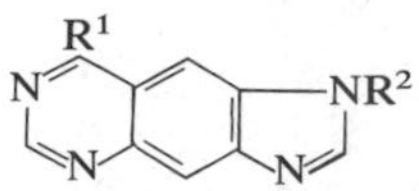   (5.288) 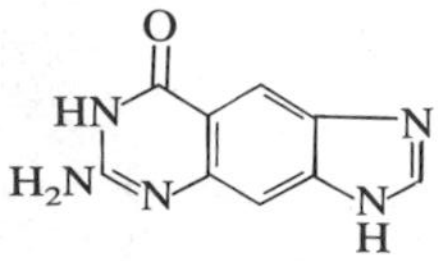   (5.289)

Compound	R¹	R²	Solvent[a]	λ_{max} nm (log ε)	Ref.
(5.287)	NHCH₂Ph	H	B	217(4.54), 237 sh, 265(3.89), 324 sh, 338(4.30), 354(4.32)	238
(5.287)	NHCH₂Ph	H	C	244(4.63), 248 sh, 270 sh, 277(4.50), 324(4.10), 349(3.98), 365(3.90)	238
(5.287)	NH₂	c-C₆H₁₁	A	231(4.67), 242 sh, 260 sh, 266(4.23), 307 sh, 319(3.96), 334(4.07), 351(3.96)	238
(5.287)	NH₂	c-C₆H₁₁	B	225(4.51), 233 sh, 260(4.13), 266 sh, 307 sh, 322(4.01), 336(4.13), 352(4.10)	238
(5.287)	NH₂	c-C₆H₁₁	C	242 sh, 260(4.26), 266 sh, 307 sh, 319(3.98), 334(4.08), 351(3.98)	238
(5.287)	NHCH₂CH=CMe₂	H	A	228(4.60), 239 sh, 245(4.08), 261 sh, 297(3.76), 310 sh, 335(4.15), 350(4.06)	250
(5.287)	NHCH₂CH=CMe₂	H	B	218(4.59), 237 sh, 263(3.95), 288 sh, 326 sh, 338(3.59), 354(3.61)	250
(5.287)	NHCH₂CH=CMe₂	H	C	244(4.64), 249(4.61), 270(4.41), 278(4.46), 315(3.98), 326(4.10), 348(4.00), 365(3.92)	250
(5.287)	NH₂	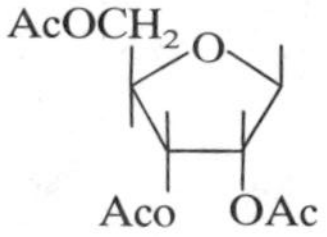 AcOCH₂ O OAc (Aco)	E	231(4.63), 259(4.25), 316(3.89), 330(3.99), 345(3.86)	249

614

(5.287)	NH_2	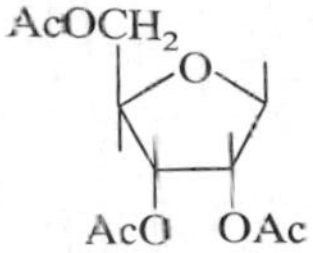 AcOCH2 / AcO OAc	F	226(4.50), 234 sh, 259(4.14), 267(4.06), 290 sh, 308 sh, 32C sh, 332(4.09), 347(4.04)	249
(5.287)	NH_2	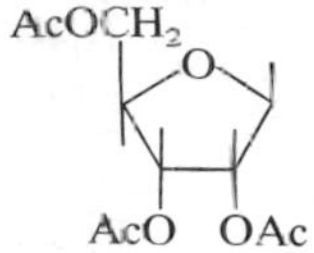 AcOCH2 / AcO OAc	G	231(4.63), 306 sh, 317(3.88), 330(3.99), 346(3.85)	249
(5.288)	NH_2	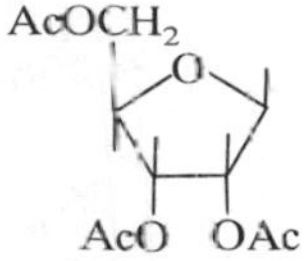 AcOCH2 / AcO OAc	E	223(4.49), 237(4.49), 2€3(4.23), 292(3.75), 304 sh, 324(3.80), 339(3.94), 354(3.84)	249
(5.288)	NH_2	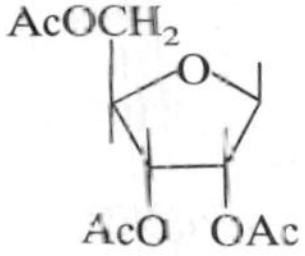 AcOCH2 / AcO OAc	F	221(4.51), 232 sh, 259(4.08), 267(4.06), 300(3.80), 320 sh, 332(4.05), 347(4.01)	249
(5.288)	NH_2	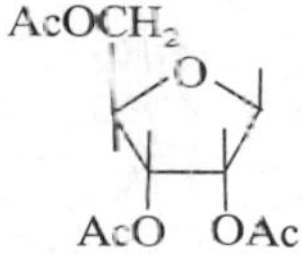 AcOCH2 / AcO OAc	G	239(4.52), 260(4.24), 292(3.75), 305(3.67), 326(3.78), 339(3.93), 346(3.83)	249
(5.287)	SMe	H	A	240(4.65), 284(3.88), 327(4.06), 341(4.02), 358(3.88)	238
(5.287)	SMe	H	B	244(4.33), 290 sh, 299(3.72), 344(4.13), 355 sh	238
(5.287)	SMe	H	C	255(4.61), 279(4.33), 323 sh, 337(4.20), 364 sh	238
(5.287)	SMe	CH2Ph	A	243(4.75), 284(3.94), 329(4.13), 342(4.10), 359(3.94)	238

TABLE 5.71. (Continued)

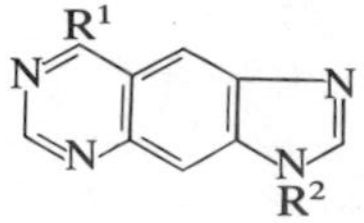

(5.287) 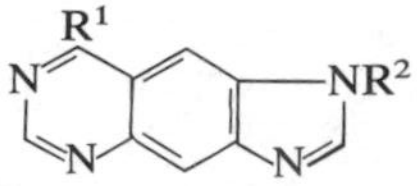 (5.288) 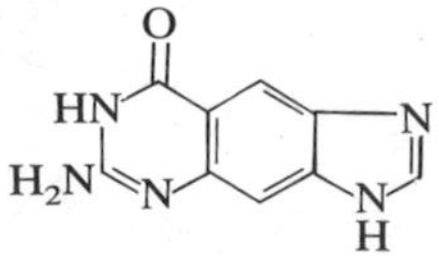 (5.289)

Compound	R¹	R²	Solvent[a]	λ_{max} nm (log ε)	Ref.
(5.287)	SMe	CH$_2$Ph	B	252(4.47), 298(3.79), 352(4.18)	238
(5.287)	SMe	CH$_2$Ph	C	243(4.78), 284(3.99), 329(4.11), 338 sh, 359(391)	238
(5.288)	SMe	CH$_2$Ph	A	243(4.61), 248 sh, 284(3.88), 323(3.98), 348(4.01), 366(3.92)	238
(5.288)	SMe	CH$_2$Ph	B	248(4.35), 298(3.74), 348(4.15)	238
(5.288)	SMe	CH$_2$Ph	C	243(4.64), 248 sh, 284(3.87), 327(4.00), 348(4.00), 366(3.89)	238
(5.289)	—	—	H	238(4.58), 254(4.60), 280 sh, 296 sh, 310(3.81), 345(3.75), 374 sh	244
(5.289)	—	—	I	234(4.61), 270(3.86), 283(3.85), 324(3.76)	244
(5.289)	—	—	J	233(4.64), 254(4.35), 270(3.84), 285(3.86), 324(3.77)	244
(5.289)	—	—	K	235 sh, 240(4.64), 266 sh, 277 sh, 290 sh, 317(3.76), 329 sh	244
(5.289)	—	—	L	228 sh, 234(4.63), 240(4.66), 256(3.95), 267 sh, 278(3.56), 316(3.79), 325 sh	244

[a] A, 95% ethanol. B, 0.1 M HCl, 95% ethanol. C, 0.1 M NaOH, 95% ethanol. D, 0.01 M HCl, 95% ethanol. E, pH = 7. F, pH = 1. G, pH = 12. H, 0.1 M NaOH. I, 0.1 M Na$_2$CO$_3$ buffer, pH = 10.3. J, 0.05 M phosphate, pH = 6.8. K, 0.01 M NH$_4$HCO$_3$. L, 0.1 M HCl.

TABLE 5.72. ULTRAVIOLET SPECTRA OF IMIDAZO[4,5-*f*]QUINAZOLINE DERIVATIVES (**5.290**) AND (**5.291**)

(**5.290**)   (**5.291**)

Compound	R	Solvent[a]	λ_{max} nm (log ε)	Ref.
(**5.290**)	NH_2	A	238(4.28), 251(4.24), 258(4.26), 278 sh, 312(3.77), 324(3.97), 335 sh, 339(3.93)	239
(**5.290**)	NH_2	B	229(4.04), 237(4.03), 270(4.31), 290 sh, 328(3.93), 335 sh	239
(**5.290**)	NH_2	C	269(4.44), 296(3.69), 333(3.88), 346 sh	239
(**5.290**	$NHCH_2Ph$	D	238(4.09), 265(4.25), 277 sh, 311 sh, 323(3.98), 336(3.93)	250
(**5.290**)	$NHCH_2Ph$	E	232(4.12), 276(4.31), 307 sh, 312 sh, 323(4.02), 329 sh	250
(**5.290**)	$NHCH_2Ph$	F	272(4.44), 296 sh, 332(3.93), 343 sh	250
(**5.290**)	$NHCH_2CH=CMe_2$	D	230 sh, 238(4.10), 262(4.12), 311 sh, 322(3.98), 337(3.92)	250
(**5.290**)	$NHCH_2CH=CMe_2$	E	231(4.13), 276(4.24), 309 sh, 321(4.02), 332 sh	250
(**5.290**)	$NHCH_2CH=CMe_2$	F	273(4.41), 297 sh, 331(3.93), 341 sh	250
(**5.291**)	—	A	213(4.19), 247(4.46), 272(3.71), 282(3.71), 294(3.78), 322(3.99), 335(3.95)	247

[a] A, 95% ethanol. B, 0.1 M HCl, 95% ethanol. C, 0.1 M NaOH, 95% ethanol. D, pH = 7. E, pH = 1. F, pH = 13.

TABLE 5.73. ULTRAVIOLET SPECTRA OF IMIDAZO[4,5-*h*]QUINAZOLINES (**5.292**)

(**5.292**)

Compound	R	Solvent[a]	λ_{max} nm (log ε)	Ref.
(**5.292**)	NH_2	A	251(4.61), 280 sh, 295(3.95), 303 sh, 309 sh, 315 sh	239
(**5.292**)	NH_2	B	255(4.55), 295 sh, 307(3.89), 315 sh	239
(**5.292**)	NH_2	C	262(4.79), 320(3.84)	239
(**5.292**)	$NHCH_2Ph$	A	228(4.42), 253(4.62), 300(4.15), 314(4.10)	250
(**5.292**)	$NHCH_2Ph$	B	222(4.31), 261(4.53), 314(4.17)	250
(**5.292**)	$NHCH_2Ph$	C	266(4.81), 314(4.05)	250
(**5.292**)	$NHCH_2CH=CMe_2$	A	227(4.45), 253(4.56), 301(4.17), 314(4.13), 316 sh	250
(**5.292**)	$NHCH_2CH=CMe_2$	B	222(4.32), 233 sh, 261(4.53), 315(4.18), 328 sh	250
(**5.292**)	$NHCH_2CH=CMe_2$	C	266(4.92), 312(4.09)	250

[a] A, 95% ethanol. B, 0.1 M HCl, 95% ethanol. C, 0.1 M NaOH, 95% ethanol.

is held coplanar with the ring-system by hydrogen-bonding to $N(1)$ and $N(8)$. Such coplanarity (and hence n-π conjugative interaction) is precluded in the *lin* and *dist* isomers due to steric interaction between the amino group and the *peri*-hydrogen atom [$H(9)$ in *lin*-benzoadenine; $H(5)$ in *dist*-benzo-adenine].

The low-energy absorption bands in the ultraviolet spectra (Table 5.71) of *lin*-benzoadenine and its $N(1)$ and $N(3)$-alkyl derivatives (**5.288**; $R^1 = NH_2$, $R^2 = CH_2Ph$) and (**5.287**; $R^1 = NH_2$, $R^2 = H$, CH_2Ph, c-C_6H_{11}) (Scheme 5.54) undergo only negligible shifts in changing from neutral to acidic solution.[238] In contrast, the low-energy absorption bands in the ultraviolet spectra (Table 5.71) of the thiomethyl analogs, (**5.287**; $R^1 = SMe$, $R^2 = H$ or CH_2Ph) and (**5.288**; $R^1 = SMe$, $R^2 = CH_2Ph$) (Scheme 5.54), which show no change in position in going from neutral to basic solution, collapse to a single broad band and undergo a simultaneous hypsochromic shift in acidic media.[238] The essential lack of change in the ultraviolet absorption (Table 5.71) of the aminoimidazo[4,5-g]quinazolines (**5.287**; $R^1 = NH_2$, $R^2 = H$, CH_2Ph) with change in pH indicates that the imidazo[4.5-g]quinazoline ring system protonates preferentially at $N(5)$ and $N(7)$ in the pyrimidine ring.[238,249] On the other hand, the absorption maxima in *lin*-benzoadenine (**5.287**; $R^1 = NH_2$, $R^2 = H$) undergo the expected bathochromic shift (Table 5.71) in changing from neutral to basic solution consequent on the deprotonation of the acidic imidazole NH group.[238] The marked shifts in the absorption bands (Table 5.71) in the ultraviolet spectrum of *lin*-benzo-guanine (**5.289**) (Scheme 5.54) with change in pH are consistent with the pK_a assignments for this molecule.[244]

Comparison of the ultraviolet spectra (Table 5.71) of the $N(1)$- and $N(3)$-benzyl derivatives of *lin*-benzoadenine of unambiguous structure, (**5.288**; $R^1 = NH_2$, $R^2 = CH_2Ph$) and (**5.287**; $R^1 = NH_2$, $R^2 = CH_2Ph$) (Scheme 5.54), with those of the products of ribosylation of *lin*-benzoadenine (see later in this section) has allowed the sites of ribosylation in the latter molecule to be established with some degree of certainty as $N(1)$ and $N(3)$ (i.e., at the imidazole ring nitrogen atoms).[249] Ultraviolet studies have also proved valuable for pinpointing the sites of enzymic oxidation in imidazo-[4,5-g]quinazolin-8(7H)-one (*lin*-benzohypoxanthine), imidazo[4,5-f]quin-azolin-9(8H)-one (*prox*-benzohypoxanthine), and imidazo[4,5-h]quinazolin-6(7H)-one (*dist*-benzohypoxanthine) (see section 5.33).[249]

The presence of two strong maxima at relatively long wavelength in the ultraviolet spectrum of the derivative (**5.294**) [λ_{max}(MeOH), nm (log ε), 245 (4.31), 247 sh, 356 (4.23)][246] demonstrates the delocalized character of the imidazo[4,5-g]quinoxaline ring system.

Derivatives of the imidazo[4,5-g]quinazoline, imidazo[4,5-f]quinazoline, and imidazo[4,5-h]quinazoline ring systems are fluorescent in both polar and nonpolar solvents (Table 5.74).[244,249,250] The fluorescence of the *lin*-benzoadenine derivatives, (**5.287**; $R^1 = NH_2$, $R^2 = 1$-β-D-ribofuranosyl) and (**5.287**; $R^1 = NHCH_2Ph$ or $NHCH_2CH=CMe_2$, $R^2 = H$) is particularly

TABLE 5.74. FLUORESCENCE SPECTRA[244,249,250] OF IMIDAZO[4,5-g]QUINAZOL-
INES (5.287), (5.288), AND (5.289), IMIDAZO[4,5-f]QUINAZOLINES
(5.290), AND IMIDAZO[4,5-h]QUINAZOLINES (5.292)

(5.287)   (5.288)   (5.289)

(5.290)   (5.291)

Compound	Solvent[a]	Excitation (nm)[b,c]	Emission (nm)[c]	τ (nsec)[d]	ϕ[e]
(5.287) ($R^1 = NH_2$, $R^2 =$ 1-β-D-ribofuranosyl)	A	320 sh, 332, 348	358, 372, 385	3.7	0.44[f]
(5.287) ($R^1 = NH_2$, $R^2 =$ 1-β-D-ribofuranosyl)	B	322 sh, 334, 350	358, 374, 390 sh	3.7	0.44
(5.287) ($R^1 = NH_2$, $R^2 =$ 1-β-D-ribofuranosyl)	C	322 sh, 336, 352	360, 376, 392	3.7	0.44
(5.288) ($R^1 = NH_2$, $R^2 =$ 1-β-D-ribofuranosyl)	A	328 sh, 340, 356	365, 382, 396	3.0	0.18
(5.287) ($R^1 = NHCH_2Ph$, $R^2 = H$)	A	320 sh, 335, 350	364 sh, 380, 395 sh	1.3	0.14
(5.287) ($R^1 = NHCH_2Ph$, $R^2 = H$)	B	320 sh, 335, 350	365 sh, 380, 395 sh	1.8	0.26
(5.287) ($R^1 = NHCH_2Ph$, $R^2 = H$)	C	320 sh, 335, 350	365, 380, 395 sh	2.2	0.29
(5.287) ($R^1 = NHCH_2CH= CMe_2$, $R^2 = H$)	A	320 sh, 335, 350	365 sh, 380, 395 sh	0.9	0.07
(5.287) ($R^1 = NHCH_2CH= CMe_2$, $R^2 = H$)	B	320 sh, 335, 350	365 sh, 380, 395 sh	1.7	0.29
(5.287) ($R^1 = NHCH_2CH= CMe_2$, $R^2 = H$)	C	320 sh 335, 350	365, 380, 395 sh	2.0	0.33
(5.289)	D	248, 257, 270, 283, 332[g]	426[h]	—	—
(5.289)	E	256, 290, 340[i]	380 sh, 403[h]	—	—
(5.289)	F	256, 268, 316, 356[j]	410[h]	—	—
(5.290) ($R = NHCH_2Ph$)	A	325, 338	350, 362, 378 sh	2.0	0.12
(5.290) ($R = NHCH_2CH= CMe_2$)	A	325, 340	350, 362, 378 sh	1.0	0.09
(5.292) ($R = NHCH_2Ph$)	A	302 sh, 312	328 sh, 340, 355 sh	1.7	0.007
(5.292) ($R = NHCH_2CH= CMe_2$)	A	302 sh, 312	330 sh, 340, 355 sh	1.5	0.003

[a] A, water. B, ethanol. C, dioxane. D, 0.1 M HCl. E, 0.05 M phosphate. F, 0.1 M NaOH.
[b] >300 nm.
[c] Uncorrected.
[d] Fluorescence lifetimes.
[e] Absolute quantum yields (corrected) determined relative to lin-benzoadenosine (5.287) ($R^1 = NH_2$, $R^2 = $ 1-β-D-ribofuranosyl).
[f] Determined relative to quinine in 1N H_2SO_4 ($\phi = 0.70$).
[g] Emission fixed at 426 nm.   [i] Emission fixed at 403 nm.
[h] Excitation at 332 nm.   [j] Emission fixed at 410 nm.

619

TABLE 5.75 ^{1}H NMR SPECTRA[a,b] OF IMIDAZO[4,5-g]QUINAZOLINE DERIVATIVES (**5.287**), (**5.288**), (**5.289**), (**5.295**), (**5.296**), AND (**5.297**)

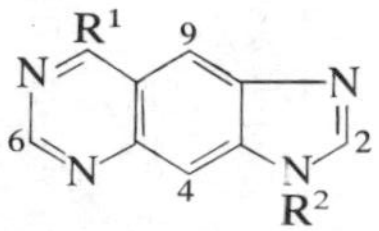

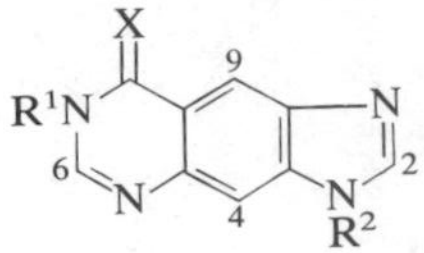

(**5.287**)

(**5.288**)

(**5.289**)

(**5.295**)

(**5.296**)

(**5.297**)

Compound	R	R^1	R^2	Solvent[c]	H(2)	H(4)	H(6)	H(9)[d]	PhH	CH_3	CH_2	Others	Ref.
(**5.287**)	—	NH_2	H[e]	A	8.71	8.96	9.26	9.65	—	—	—	—	238
(**5.287**)	—	NH_2	CH_2Ph	A	8.43	8.95	9.22	9.43	7.50	—	5.83	—	238
(**5.288**)	—	NH_2	CH_2Ph	A	8.77	9.15	9.23	9.28	7.52	—	5.88	—	238
(**5.287**)	—	NH_2	C_6H_{11}	B	8.62	9.04	9.44	9.56	—	—	—	1.22–2.20 br	238
(**5.295**)	—	NH_2	OH	B	8.04	8.40	8.71	8.77	—	—	3.10	3.96–4.26 m 4.35–4.55 m 5.12–5.83 br 6.01 d[f] 7.82 br[g]	249
(**5.296**)	—	NH_2	OH	B	7.96	8.33	8.64	8.82	—	—	3.64	3.90–4.26 m 4.56 t[f] 5.2–5.7 br 6.02 d[h] 7.64 br[g]	249
(**5.287**)	—	$NHCH_2Ph$	—	B	7.86	8.37	8.48	8.65	7.08–7.50 m	—	4.83	—	238, 250
(**5.287**)	—	$NHCH_2$— CH=CMe_2	—	B	7.86	8.46	8.55	8.67	—	1.75	4.0–	3.00–4.00 m 5.20–6.60 m 8.20–8.70	250

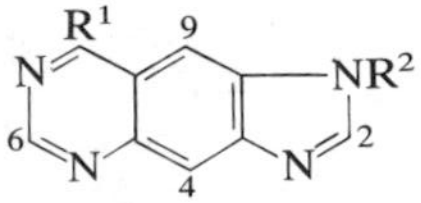

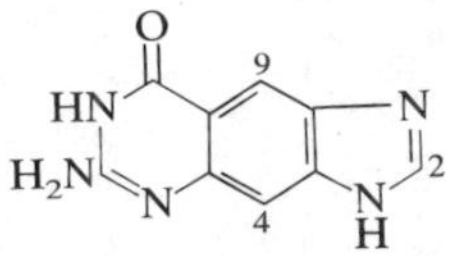

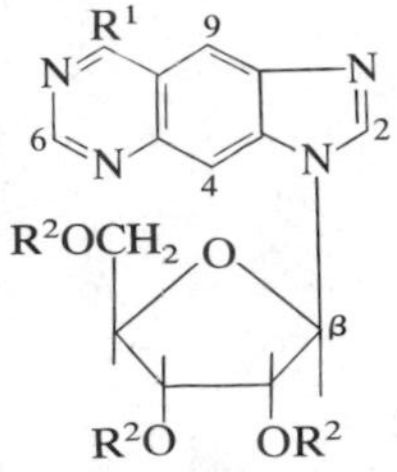

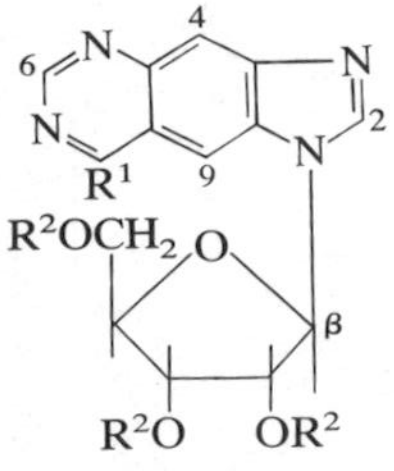

Compound				Solvent									
(5.287)	—	SMe	H	A	8.83	9.13	9.22	9.70	—	3.10[i]	—	—	238
(5.287)	—	SMe	CH$_2$Ph	A	8.53	9.12	9.22	9.50	7.50	3.08[i]	5.87	—	238
(5.288)	—	SMe	CH$_2$Ph	A	8.00	8.83	9.17	9.48	7.52	3.05[i]	5.90	—	238
(5.295)	—	SMe	Ac	C	8.00	8.24	8.28	8.73	—	2.01; 2.09; 2.22; 2.67[i]	4.37	4.06 m; 4.55 m; 5.33–5.55 m; 5.56–5.75 m; 6.08 d[h]	249
(5.296)	—	SMe	Ac	D	8.19	8.32	8.42	8.87	—	2.03–2.20 m; 2.70[i]	4.46 br	5.35–5.70 m; 6.19 d[f]	249
(5.297) (X=O)	—	H	H	A	8.82	9.20	9.45	9.66	—	—	—	—	238
(5.297) (X=S)	—	H	H	A	8.54	9.04	9.34	9.49	—	—	—	—	238
(5.297) (X=O)	—	H	CH$_2$Ph	A	8.55	9.18	9.48	—	7.57	—	5.88	—	230
(5.297) (X=S)	—	H	CH$_2$Ph	A	8.28	9.02	9.22	9.33	7.52	—	5.82	—	238
(5.297) (X=O)	—	H	C$_6$H$_{11}$	B	7.90	8.01	8.40	8.55	—	—	—	1.20–2.20 br	238
(5.297) (X=S)	—	H	C$_6$H$_{11}$	B	8.08	8.13	8.80	8.90	—	—	—	1.20–2.20 br	238
(5.297) (X=O)	—	CH$_2$Ph	H	A	8.72	9.20	9.28	9.63	7.53	—	5.55	—	238
(5.297) (X=S)	—	CH$_2$Ph	H	B	7.98	8.64	8.78	9.30	7.34	—	5.98	—	238
(5.289)[j]	—	—	—	E	7.48	8.18	8.25	—	—	—	—	—	244

[a] δ Values in ppm measured from tetramethylsilane.
[b] Signals are either sharp singlets unless otherwise specified as:
br = broad; d = doublet; m = multiplet.
[c] Solvents: A, CF$_3$CO$_2$H; B, (CD$_3$)$_2$SO; C, CCl$_4$; D, CDCl$_3$; E, 1M NaOH, D$_2$O.
[d] Signals not assigned.
[e] Hydrochloride.
[f] J = 5 Hz.
[g] NH$_2$.
[h] J = 6 Hz.
[i] SMe.
[j] Dihydrochloride.

strong, being associated with relatively high quantum yields (Table 5.74) and giving rise to fluorescence spectra having a strong emission maximum at approximately 370 to 380 nm (Table 5.74).[249,250] The corresponding *prox*-benzoadenine derivatives (**5.290**; R = NHCH$_2$Ph or NHCH$_2$CH=CMe) exhibit analogous fluorescence spectra, but are not so strongly fluorescent as their *lin*-benzoadenine counterparts (Table 5.74).[249,250] The *dist*-benzoadenine analogs (**5.292**; R = NHCH$_2$Ph or NHCH$_2$CH=CMe$_2$) are only weakly fluorescent, the low quantum yields in these cases (Table 5.74) being attributed[249,250] to fluorescence quenching by hydrogen transfer between $N(1)$ and $N(9)$. The biological activity[250] of "stretched purines" in conjunction with their fluorescence property make them attractive probes for the study of binding at active sites in biological systems.[249,250]

NUCLEAR MAGNETIC RESONANCE SPECTRA. With the exception of a single derivative of the imidazo[4,5-g]quinoxaline ring system,[246] detailed NMR studies of tricyclic 6-6-5 fused benzimidazoles with two additional hetero atoms have been confined to imidazo[4,5-g]quinazolines (Table 5.75),[238,249,250] imidazo[4,5-f]quinazolines (Table 5.76),[239,247,250] and imidazo[4,5-h]quinazolines (Table 5.77).[239,249,250] In most cases, precise assignments for the benzenoid, quinazoline, and imidazole ring protons in such structures are precluded by their closely similar chemical shifts. On the other hand, the benzenoid protons at $C(4)$ and $C(5)$ in imidazo[4,5-f]quinazolines (Table 5.76) and imidazo[4,5-h]quinazolines (Table 5.77) are clearly differentiated by well-defined *ortho* coupling. Moreover, the imidazole proton [$H(2)$] and the quinazoline protons [$H(7)$ or $H(8)$] in such molecules may be assigned (at least tentatively) (Tables 5.76 and 5.77) on the basis of the greater shielding of the $C(2)$ proton in benzimidazoles[251] compared with the protons at $C(2)$ and $C(4)$ in quinazolines.[252] In accord with their linear structures, imidazo[4,5-g]quinazoline derivatives show no coupling in the *para*-orientated $C(4)$ and $C(9)$ protons, thus making assignments of the benzenoid, quinazoline, and imidazole ring protons in such molecules ambiguous in the absence of further information. The demonstration[249] of *ortho* coupling in the benzenoid protons of the product (*dist*-benzoxanthine) (**5.293**) (Scheme 5.54 and Table 5.77) of the enzymic oxidation (see section 5.3.3) of *dist*-benzohypoxanthine pinpoints the site of oxidative attack as the quinazoline ring rather than the benzene ring.

The β-configuration has been assigned to the "stretched purine" nucleoside *lin*-benzoadenosine on the basis of the chemical shift of the anomeric proton [C-1$'$(H)] taken in conjunction with the magnitude of the C-1$'$(H)/C-2$'$(H) coupling constant in its isopropylidene derivative.[249]

^{13}C–^{15}N Coupling in specifically labeled substrates (Scheme 5.55, **5.300** and **5.301**) has been used to establish the sites of benzylation in thiomethylimidazo[4,5-g]quinazoline derivatives, thereby providing ultraviolet models with which to identify the specific site of ribosylation in *lin*-benzoadenine derivatives.[253] Thus the ^{13}C NMR spectrum of the $N(3)$-benzyl derivative

TABLE 5.76 ¹H NMR SPECTRA[a,b] OF IMIDAZO[4,5-f]QUINAZOLINE DERIVATIVES (5.290) and (5.298)

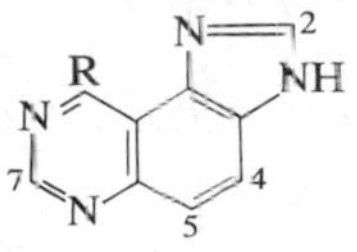

(5.290)

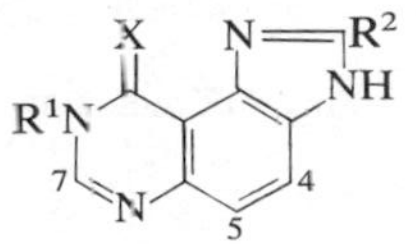

(5.298)

Compound	R	R¹	R²	Solvent[c]	H(2)	H(4)	H(5)	H(7)	PhH	CH₃	CH₂	Others	Ref.
(5.290)	NH₂	—	—	A	8.72^d	7.73^{e,f}	8.25d^e	8.82^d	—	—	—	—	250
(5.290)	NHCH₂Ph	—	—	B	8.93^d	8.13d^{e,f}	8.60d^{e,f}	9.45^d	7.25–7.55 m	—	5.32	—	250
(5.290)	NHCH₂CH=CMe₂	—	—	A	8.53^d	7.58d^{e,f}	8.02d^{e,f}	8.56^d	—	1.78	4.12–4.47 m	5.29–5.68 m	250
(5.298) (X=O)	—	H	H	B	10.92^d	8.22d^{e,f}	8.63d^{e,f}	10.97^d	—	—	—	—	239
(5.298) (X=S)	—	H	H	B	10.63^d	8.25d^{e,f}	8.56d^{e,f}	10.91^d	—	—	—	—	239
(5.298) (X=O)	—	n-Bu	Me	—[g]	—	7.56d^d	8.08c^d	—[h]	—	—[h]	—[h]	—	247

[a] δ Values in ppm measured from tetramethylsilane.
[b] Signals are either sharp singlets unless otherwise specified as: d = doublet; m = multiplet.
[c] Solvents: A,(CD₃)₂SO. B, CF₃CO₂H.
[d] These signals may be interchanged.
[e] These signals may be interchanged.
[f] AB doublet, J = 9 Hz.
[g] Solvent not specified.
[h] Chemical shifts of other signals not specified.

623

TABLE 5.77 ^{1}H NMR SPECTRA^{a,b} OF IMIDAZO[4,5-h]QUINAZOLINES (5.292), (5.293), AND (5.299)

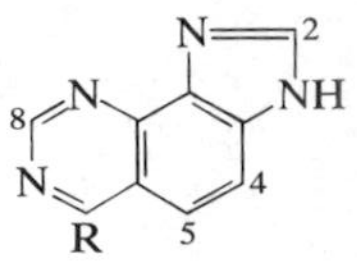

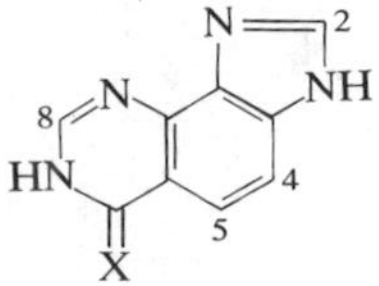

Compound	R	Solvent^c	H(2)	H(4)	H(5)	H(8)	ArH	CH$_3$	CH$_2$	Others	Ref.
(5.292)	NH$_2$	A	8.28^d	7.80d^{e,f}	8.42d^{e,f}	8.42^d	—	—	—	—	239
(5.292)	NHCH$_2$Ph	A	8.37^d	7.79c^{e,f}	8.09d^{e,f}	8.58^d	7.18–7.48m	—	4.82d^g	8.77^g	250
(5.292)	NHCH$_2$CH=CMe$_2$	A	8.37^d	7.73d^{e,f}	8.01d^{e,f}	8.57^d	—	1.72 1.75	5.38m	4.17t^g	250
(5.293)	—	A	8.32^h	7.30d^{e,i}	7.72d^{e,i}	—	—	—	—	—	249
(5.299) (X=O)	—	B	9.37^d	8.14d^{e,f}	8.74d^{e,f}	9.60^d	—	—	—	—	239
(5.299) X=S	—	B	9.00^d	8.17d^{e,f}	8.92d^{e,f}	9.47^d	—	—	—	—	239

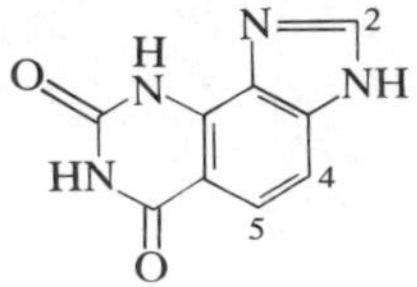

^{a} δ Values in ppm measured from tetramethylsilane.

^{b} Signals are either sharp singlets unless otherwise specified as: d = doublet; t = triplet; m = multiplet.

^{c} Solvents: A, (CD$_3$)$_2$SO. B, CF$_3$CO$_2$H.

^{d} These signals may be interchanged.

^{e} These signals may be interchanged.

^{f} AB doublet, $J = 9$ Hz.

^{g} $J = 6$ Hz.

^{h} Treatment with NaOD/100°/4 hr results in a shift to δ 7.85 and loss of 80% of signal intensity.

^{i} AB doublet, $J = 8$ Hz.

624

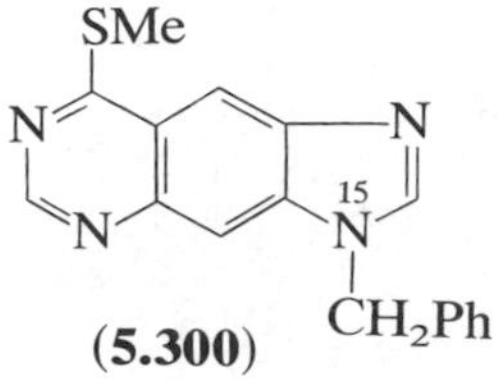

(5.300)   CH₂Ph

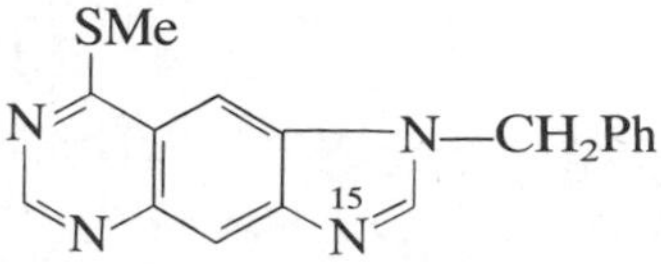

(5.301)

Scheme 5.55

(5.300) contains a doublet ($J_{^{15}N-C} = 8.6$ Hz) at δ 50.66 ppm due to the benzylic carbon atom clearly differentiating it from the $N(1)$-benzyl isomer (5.301), the benzylic carbon atom in which resonates as a singlet at δ 52.3 ppm. The two isomers are further distinguished by the fact that the benzylic protons of the $N(1)$-benzyl compound (5.301) resonate as a singlet, whereas the corresponding protons in the $N(3)$-benzyl compound (5.300) appear as a doublet due to $^{15}N-C_{\alpha}(H)$ coupling.[253]

General Studies

IONIZATION CONSTANTS. The presence of a pyrimidine ring and an imidazole ring in N-1(3)H-imidazo[4,5-g]quinazolines, N-1(3)H imidazo[4,5-f]quin azolines, and N-1(3)H-imidazo[4,5-h]quinazolines not unexpectedly confers both basic character (pyrimidine and imidazole nitrogen atoms) and acidic character (imidazole NH group) on all three types. pK_a values have been measured[239,244,249] in 66% aqueous dimethylformamide for derivatives of all three ring systems (Table 5.78) and show that of the three "stretched" benzoadenine derivatives (Scheme 5.56), $prox$-benzoadenine [9-amino-N-1(3)H-imidazo[4,5-f]quinazoline] (5.303) is the most acidic ($pK_a = 11.4$), while $dist$-benzoadenine [9-amino-N-1(3)H-imidazo[4,5-h]quinazoline] (5.304) is the least acidic ($pK_a = 12.25$).[239] This order of acidity is consistent with stabilization (by intramolecular hydrogen bonding) of the anion of $prox$-benzoadenine (5.303) and the free base of $dist$-benzoadenine (5.304). Conversely, the latter compound is the least readily protonated ($pK_a = 4.9$) (Table 5.78).[239]

TABLE 5.78   IONIZATION CONSTANTS^a OF IMIDAZOQUINAZOLINE DE-RIVATIVES

Compound	pK_a Values	Ref.
(5.302)	5.6, 11.7	239
(5.305)	5.3, (5.6)^b	249
(5.303)	5.2, 11.4	239
(5.304)	4.9, 12.2	239
(5.306)	5.3	249
(5.289)	<2.0, 4.5, 9.5, >11.0	244

^{a} Measured in 66% dimethylformamide-water.
^{b} Measured in water at 20°.

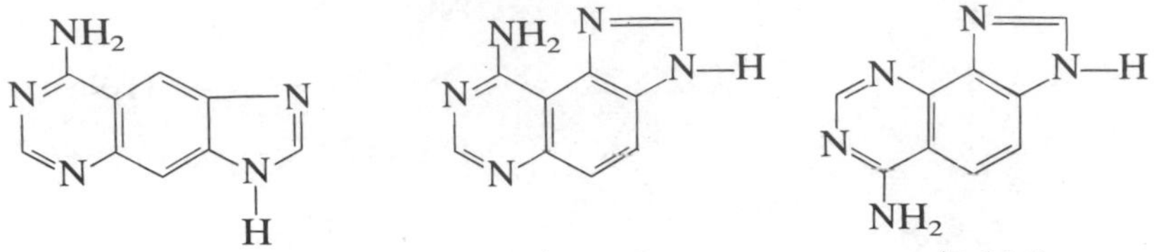

(5.302)   **(5.303)**   **(5.304)**

Scheme 5.56

The fact that the "stretched purine" riboside, *lin*-benzoadenosine (**5.305**) is a stronger base ($pK_a = 5.6$ in water) (Table 5.78) than adenosine ($pK_a = 3.5$ in water) may account (together with the 3.5 Å stretch in linear dimension) for the differing behavior in enzymic systems of the two ribosides.[249] The pK_a values for *lin*-benzoadenosine (**5.305**) and its $N(1)$ isomer (**5.306**) in 66% aqueous dimethylformamide are essentially the same (Table 5.78).[249]

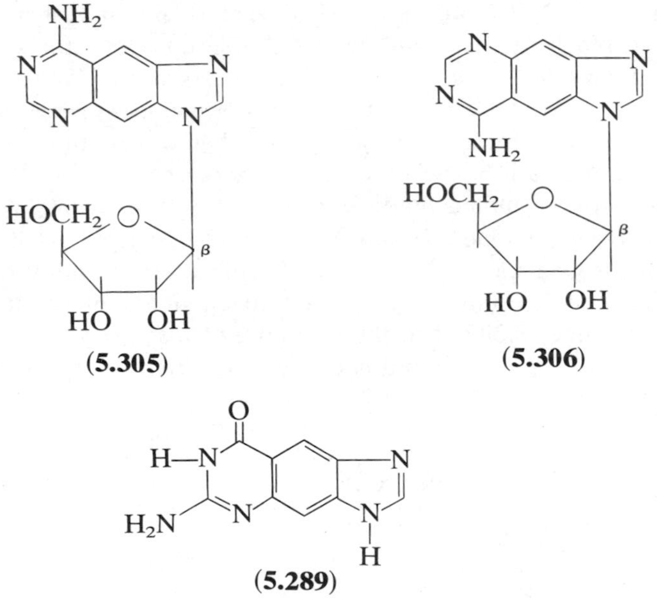

(5.305)   **(5.306)**

(5.289)

Scheme 5.57

The pK_a data for *lin*-benzoguanine (**5.289**) (Table 5.78) reveals that the molecule exhibits quinazoline and benzimidazole, rather than "stretched purine" character.[244]

5.3.3. Reactions

Reactions with Electrophiles

PROTONATION. The basicity of the imidazo[4,5-g][1,4]benzoxazine,[241] imidazo[4,5-h][1,4]benzothiazine,[242] and imidazo[4,5-a]quinoxaline[246] ring systems is illustrated by the ability of various derivatives to form well-defined salts with mineral acids (cf. Tables 5.67 and 5.69). Correspondingly, the change in the ultraviolet absorption (cf. Tables 5.71, 5.72, and 5.73) of imidazo[4,5-g]quinazolines,[238,244,249,250] imidazo[4,5-f]quinazolines,[239,250] and imidazo[4,5-h]quinazolines[239,250] in going from neutral to acidic solution, clearly demonstrates proton uptake by these ring systems and moreover is interpretable in each case in terms of preferential protonation at the nitrogen atoms of the pyrimidine ring.

ALKYLATION. Methylation[243] of the imidazo[4,5-h][1,4]benzothiazine derivative (**5.307**) by heating with methyl sulfate in nitrobenzene results in quaternization at N(1) to give (after treatment with potassium iodide) the salt (**5.308**) in very good yield (Table 5.79). Analogous methylation[243] of the imidazo[4,5-h][1,4]benzothiazin-7-one (**5.309**) under these essentially non-basic conditions also results in preferential methylation at N(1) to afford the quaternary salt (**5.310**) in high yield (Table 5.79) without any apparent competing attack at N(6) or at the C(7) oxygen atom.

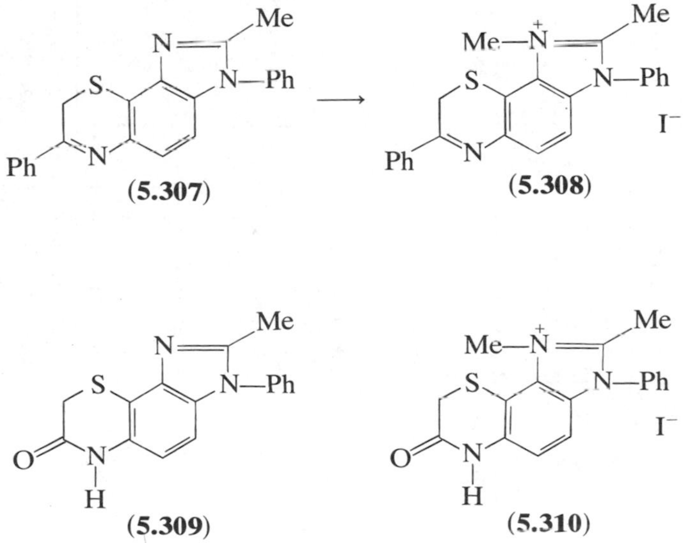

Scheme 5.58

TABLE 5.79   ALKYLATION REACTIONS OF IMIDAZO[4,5-h][1,4]BENZOTHIAZINES, IMIDAZO[4,5-g]-QUINAZOLINES, IMIDAZO[4,5-f]QUINAZOLINES, AND IMIDAZO[4,5-g]QUINOXALINES

Starting materials	Reaction conditions[a]	Product	Yield (%)	Melting point (°C)	Solvent of crystallization	Crystal form	Ref.
(5.307)	A	(5.308)	85	242	Ethanol	—[b]	243
(5.309)	A	(5.310)	90	261	Ethanol	—[b]	243
(5.311) (R = H)	B	(5.312) (R = H)	79	311–314	Ethanol–water	Yellow needles	238
(5.311) (R = CH₂Ph)	B	(5.312) (R = CH₂Ph)	86	224–225	Ethanol	Colorless needles	238
(5.312) (R = H)	C	(5.313) +	41	211–212	Ethanol–water	Colorless needles	
		(5.314)	42	223–225	Ethanol–water	Colorless needles	238
(5.312) (R = H) + (5.315)	D	(5.316) + (5.317)	—[c] —[c]	— —	— —	—[d] —[d]	249
(5.318)	E	(5.319)	—[c]	115–116	—[e]	—[d]	247
(5.320) (R¹ = Ph, R² = H)	F	(5.321)	76	221	Methanol	—[b]	245
(5.320) (R¹ = H, R² = Me) G		(5.322) (R¹ = H, R² = Me, X = ClO₄)	—[c]	—[f]	Ethanol	—[b]	245
(5.320) (R¹ = R² = Me)	H	(5.322) (R¹ = R² = Me, X = I)	—[c]	275	Ethanol	Yellow needles	245
(5.320) (R¹ = Ph, R² = Me)	I	(5.322) (R¹ = Ph, R² = Et, X = I)	—[c]	291–292	Ethanol	Yellow needles	245

[a] A, Me₂SO₄, PhNO₂/115–120°/15 min, then treat with KI. B, MeI, KOH, H₂O/25°/15 min. C, PhCH₂Br, K₂CO₃, dimethylformamide/25°/10 hr. D, Hg(CN)₂, MeNO₂/reflux/4 hr, then 100°/10 hr. E, EtI, NaOEt, F, Me₂SO₄, NaOH. G, Me₂SO₄, PhNO₂/40°, then treat with HClO₄. H, MeI/100° (sealed tube)/5 hr. I, TSO₃Et/heat, then treat with KI (T = p-tolyl).

[b] Crystal form not specified.
[c] Yield not specified.
[d] Glass.
[e] Solvent not specified.
[f] Melting point not specified.

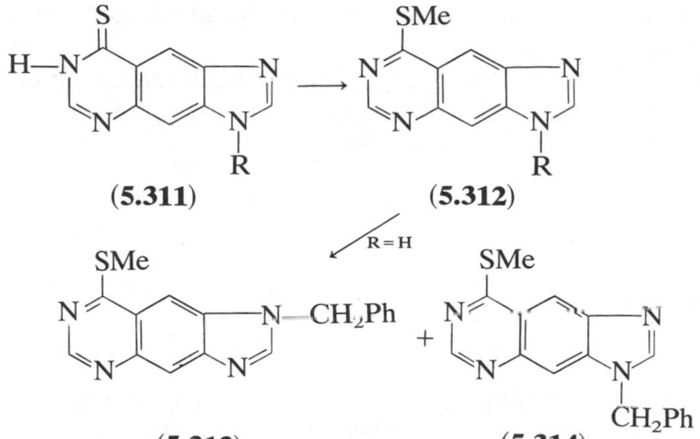

(5.311)   **(5.312)**

R=H

(5.313)   **(5.314)**

Scheme 5.59

In contrast, methylation of N-1(3)H-imidazo[4,5-g]quinazoline-8(7H)-thione (**5.311**, R = H) using methyl iodide under basic conditions[238] results in selective methylation at sulfur to afford the S-methyl derivative (**5.312**; R = H) in high yield (Table 5.79). Benzylation of the latter under basic conditions (Table 5.79) leads to a readily separable mixture of the N(1)- and N(3)-benzyl derivatives (**5.313**) and (**5.314**),[238] the latter also being the product of the S-methylation (Table 5.79) of the 3-benzylimidazo[4,5-g]-quinazolinethione (**5.311**; R = CH$_2$Ph).[238] Ribosylation[249] of the methylthio-imidazo[4,5-g]quinazoline (**5.312**; R = H) also occurs smoothly both at N(1) and N(3) (Table 5.79) to afford an easily separated mixture of the respective ribosides (**5.316**) and (**5.317**) (Scheme 5.60), the latter compound being the

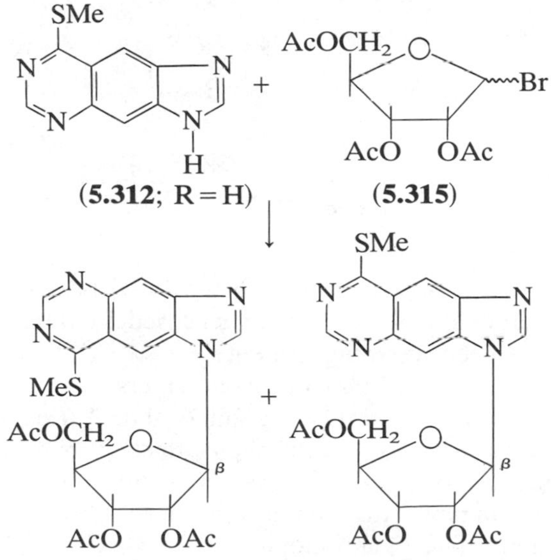

(5.312; R = H)   **(5.315)**

(5.316)   **(5.317)**

Scheme 5.60

key intermediate in the synthesis of the "stretched purine" nucleoside, *lin*-benzoadenosine.

Ethylation of the imidazo[4,5-*f*]quinazoline-2(1*H*,3*H*)-thione (**5.318**) under basic conditions[247] occurs at sulfur rather than at nitrogen giving the *S*-ethyl derivative (**5.319**) (Table 5.79).

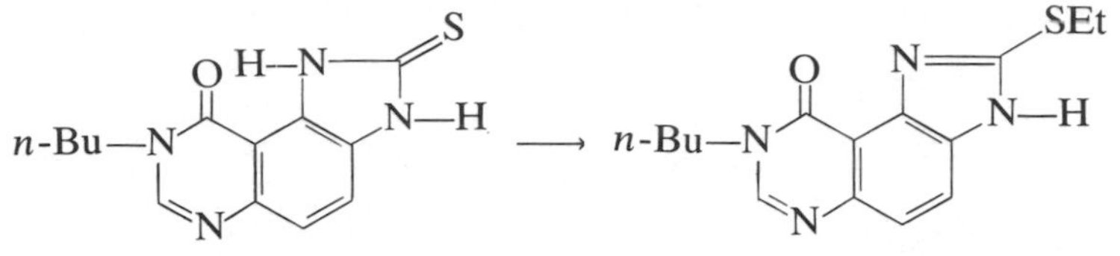

(**5.318**)   (**5.319**)

Scheme 5.61

The methylation of *N*-1(3)*H*-imidazo[4,5-*g*]quinoxalines under basic conditions occurs in high yield (Table 5.79) with replacement of the acidic imidazole *NH*-group [e.g., Scheme 5.62, **5.320** (R^1 = Ph, R^2 = H) → **5.321**].[245] Conversely, alkylation of *N*(3)-substituted imidazo[4,5-*g*]quinoxaline derivatives leads to the formation of the corresponding *N*(1)-quaternary salt [Scheme 5.62, **5.320** (R^2 = Me) → **5.322**].[245]

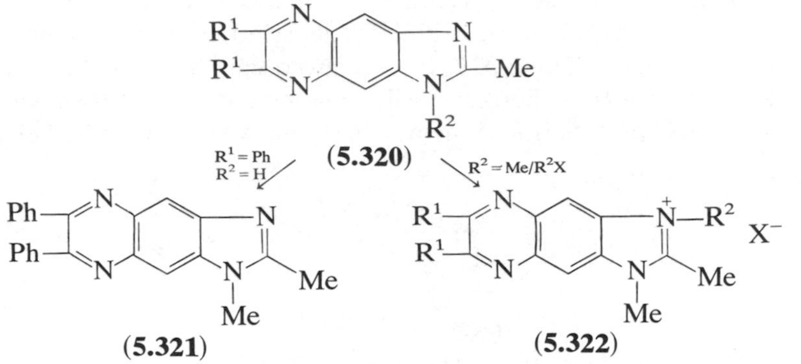

Scheme 5.62

ACYLATION. The acylation of tricyclic 6-6-5 fused bezimidazole ring systems with two additional hetero atoms under Friedel-Crafts conditions does not appear to have been attempted to date.

Methyl groups α to quaternary nitrogen centers in the imidazo[4,5-*h*]-[1,4]benzothiazine[243] and imidazo[4,5-*g*]quinoxaline[245] ring systems are, as might be expected, activated to electrophilic attack. This reactivity has been exploited in the form of acylation type procedures for the synthesis of cyanine dyes containing the imidazo[4,5-*h*][1,4]benzothiazine[243] and imidazo[4,5-*g*]quinoxaline[245] chromophores.

DEAMINATION. Amino groups at the C(8) position in imidazo[4,5-g]-quinazolines are susceptible to enzyme-mediated deamination in processes that for the present purposes are assumed (though without justification) to be electrophilic in character. For example, the enzyme adenosine deaminase converts *lin*-benzoadenine (**5.323**; R = H) into *lin*-benzohypoxanthine (**5.324**, R = H), as well as the riboside, *lin*-benzoadenosine (**5.323**; R = 1-β-D-ribofuranosyl) into *lin*-benzoinosine (**5.324**; R = 1-β-D-ribofuranosyl).[249]

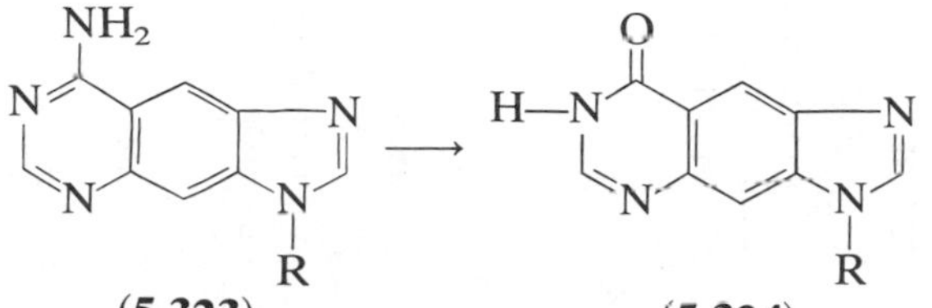

(**5.323**)   (**5.324**)

Scheme 5.63

Reactions with Nucleophiles

AMINATION. Despite the presence of component rings (i.e., imidazole, quinazoline, and quinoxaline) with high potential reactivity toward nucleophilic attack, no example of the direct amination (e.g., under Chichibabin conditions) of a tricyclic 6-6-5 fused benzimidazole ring system having two additional hetero atoms appears to have been described so far. However, amination of the imidazo[4,5-g]quinazoline[238,249,250] the imidazo[4,5-f]-quinazoline,[239,250] and the imidazo[4,5-h]quinazoline[239,250] ring systems is readily achieved in high yield (Table 5.80) by the action of ammonia or amines on the corresponding and readily accessible mercapto or methylmercapto derivatives (Scheme 5.64). Reactions of this type have, most notably, been used to synthesize linear and angular benzoadenines, "stretched purines" of considerable biological interest (Table 5.80).

HYDROXYLATION. Examples of the direct hydroxylation of tricyclic 6-6-5 fused benzimidazole ring systems having two additional hetero atoms are lacking. On the other hand, thiohydroxylation of synthetically readily accessible imidazo[4,5-g]quinazolin-8(7H)-ones (**5.331**),[238] imidazo[4,5-f]quinazolin-9(8H)-ones (**5.335**),[239] and imidazo[4,5-h]quinazolin-6(7H)-ones (**5.339**)[239] to give the corresponding thiones (**5.332**, **5.336**, and **5.340**) is a crucial step in the synthesis of benzoadenine derivatives, as discussed earlier in this section. Thiohydroxylation of this type is readily effected in high yield (Table 5.80) by heating the substrate with phosphorus pentasulfide in pyridine under reflux.

TABLE 5.80 NUCLEOPHILIC SUBSTITUTION REACTIONS OF IMIDAZOQUINAZOLINES

Starting material	Reaction conditions[a]	Product	Yield (%)	Melting point (°C)	Solvent of crystallization	Crystal form	Ref.
(5.325) ($R^1 = R^2 = H$)	A	(5.326) ($R^1 = R^2 = H$)[b]	85	>320	Ethanol–ether–water	Beige crystals	238
(5.325) ($R^1 = Me$, $R^2 = H$)	A	(5.326) ($R^1 = R^2 = H$)	75	—	—	—	238
(5.325) ($R^1 = R^2 = H$)	B	(5.326) ($R^1 = CH_2Ph$, $R^2 = H$)	92	>320	Ethanol	Colorless needles	238
(5.325) ($R^1 = Me$, $R^2 = H$)	B	(5.326) ($R^1 = CH_2Ph$, $R^2 = H$)	—[c]	—	—	—	250
(5.325) ($R^1 = Me$, $R^2 = H$)	C	(5.326) ($R^1 = CH_2CH{=}CMe_2$, $R^2 = H$)	70	287–289	Ethanol	—[d]	250
(5.325) ($R^1 = H$, $R^2 = 3\text{-}CH_2Ph$)	D	(5.326) ($R^1 = H$, $R^2 = 3\text{-}CH_2Ph$)	90	>320	Dimethylformamide	Tan needles	238
(5.325) ($R^1 = Me$, $R^2 = 3\text{-}CH_2Ph$)	D	(5.326) ($R^1 = H$, $R^2 = 3\text{-}CH_2Ph$)	75	—	—	—	238
(5.325) ($R^1 = Me$, $R^2 = 1\text{-}CH_2Ph$)	D	(5.36) ($R^1 = H$, $R^2 = 1\text{-}CH_2Ph$)	78	295–296	Ethanol	—[d]	238
(5.325) ($R^1 = H$, $R^2 = 3\text{-}C_6H_{11}$)	D	(5.326) ($R^1 = H$, $R^2 = 3\text{-}C_6H_{11}$)	85	>320	Dimethylformamide	Colorless needles	238
(5.325) ($R^1 = Me$, $R^2 = 3$-(1-β-D-ribofuranosyl)	D	(5.326) ($R^1 = H$, $R^2 = 3$-(1-β-D-ribofuranosyl)	25	294–296	Water	Colorless needles	249
(5.325) ($R^1 = Me$, $R^2 = 1$-(1-β-D-riboruranosyl)	D	(5.326) ($R^1 = H$, $R^2 = 1$-(1-β-D-riboruranosyl)	9	277–280	Water	Colorless needles	249

	Method[a]		Yield (%)[c]	mp (°C)	Solvent	Form	Ref.
(5.327)	A	(5.328) (R = H)	67	>320	Ethanol	Colorless crystals	239
(5.327)	B	(5.328) (R = CH$_2$Ph)	89	270 (decomp.)	Ethanol	Dark crystals	250
(5.327)	C	(5.328) (R = CH$_2$CH=CMe$_2$)	88	217–220	Ethanol	—[d]	250
(5.327)	E	(5.328) (R = NH$_2$)	100	225 (decomp.)	—[e]	Tan crystals	239
(5.329)	A	(5.330) (R = H)	50	>320	Acetic acid–ethanol	Colorless powder	239
(5.329)	B	(5.330) (R = CH$_2$Ph)	88	290–291	Ethanol	Dark crystals	250
(5.329)	C	(5.330) (R = CH$_2$CH=CMe$_2$)	70	253–256	Ethanol	—[d]	250
(5.331) (R^1 = R^2 = H)	F	(5.332) (R^1 = R^2 = H)	79	>320	Acetic acid	Yellow crystals	238
(5.331) (R^1 = H, R^2 = CH$_2$Ph)	F	(5.332) (R^1 = H, R^2 = CH$_2$Ph)	78	>320	Dimethylforma-mide–water	Bronze needles	238
(5.331) (R^1 = CH$_2$Ph, R^2 = H)	F	(5.332) (R^1 = CH$_2$Ph, R^2 = H)	79	290–291	Dimethylformamide–water	Yellow solid	238
(5.331) (R^1 = H, R^2 = C$_6$H$_{11}$)	F	(5.332) (R^1 = H, R^2 = C$_6$H$_{11}$)	69	>320	Acetic acid–methanol	Yellow prisms	238
(5.335)	F	(5.336)	78	>320	Dimethyl sulf-oxide–water	Yellow crystals	239
(5.339)	F	(5.340)	70	>320	Acetic acid–dimethylforma-mide	Yellow crystals	239

[a] A, NH$_3$, n-butanol/200–220° (sealed tube)/24–28 hr. B, PhCH$_2$NH$_2$, ethanol/20)° (sealed tube)/48 hr. C, NH$_2$CH$_2$CH=CMe$_2$, ethanol/200° (sealed tube)/48 hr. D, NH$_3$, ethanol/150–200° (sealed tube)/24–48 hr. E, N$_2$H$_4$, methyl cellosolve/100°/2 hr. F, P$_2$S$_5$, pyridine/reflux/12–24 hr.

[b] Hydrochloride.

[c] Yield not specified.

[d] Crystal form not specified.

[e] Solvent of crystallization not specified.

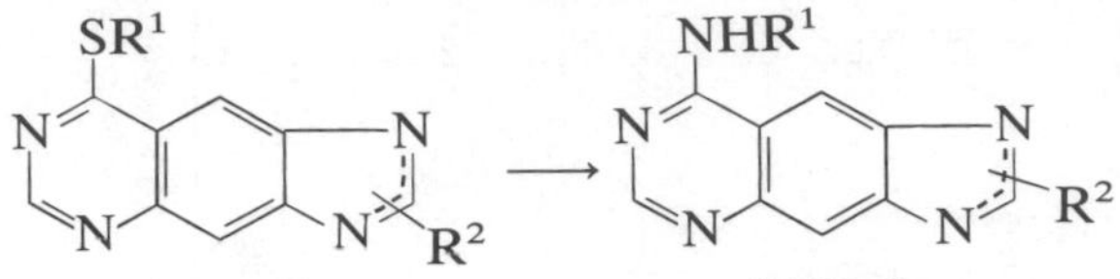

(5.325) → (5.326)

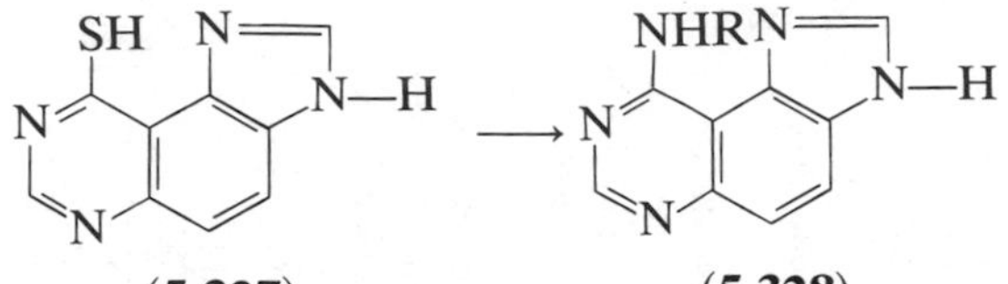

(5.327) → (5.328)

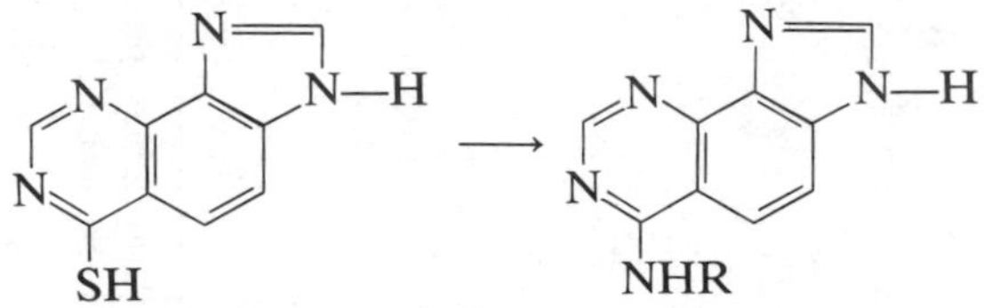

(5.329) → (5.330)

Scheme 5.64

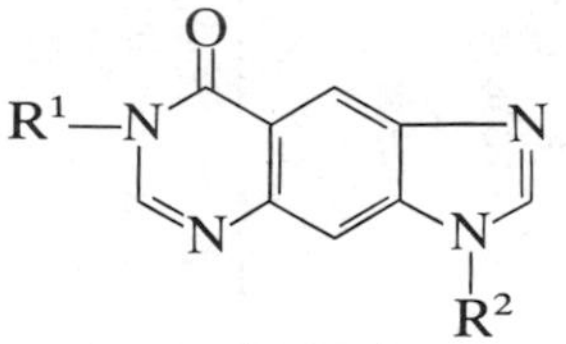

(5.331)

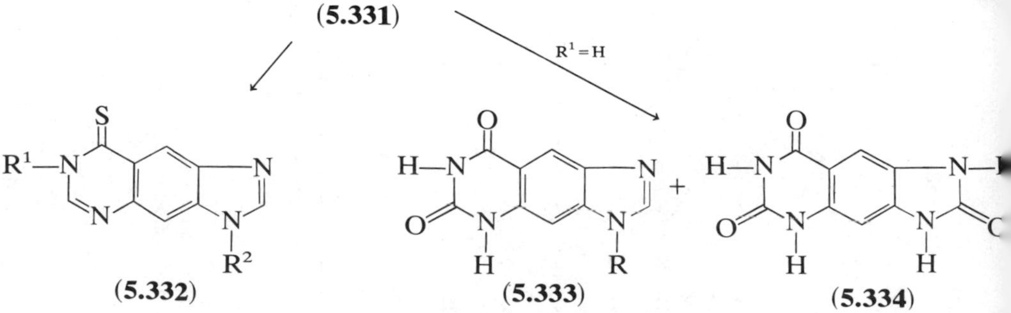

R¹ = H

(5.332)   (5.333)   +   (5.334)

Scheme 5.65

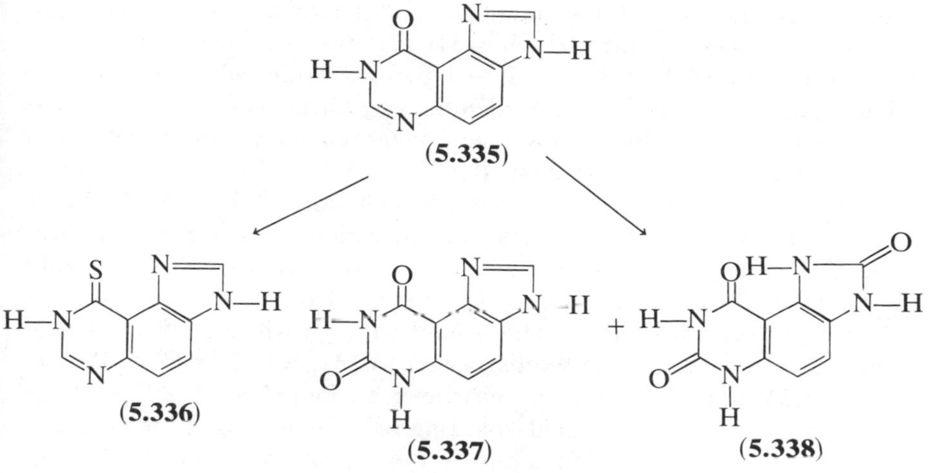

Scheme 5.66

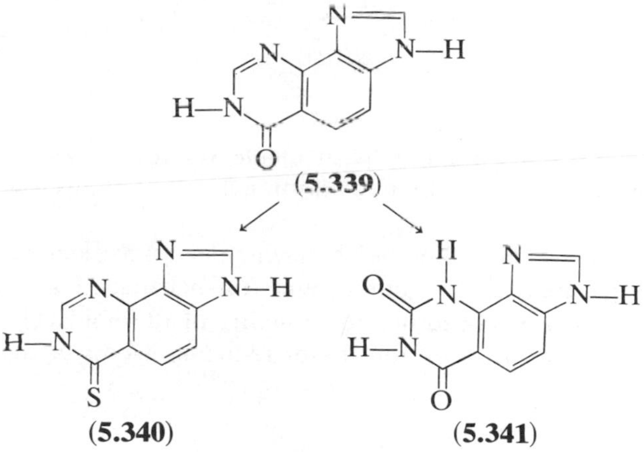

Scheme 5.67

Oxidation

Information on the susceptibility of tricyclic 6-6-5 fused benzimidazoles containing two additional hetero atoms to oxidation is almost totally lacking. However, the enzyme (xanthine oxidase) mediated air oxidation of imidazo-[4,5-g]quinazolin-8(7H)-ones (**5.331**; R = II), imidazo[4,5-f]quinazolin-9(8H)-ones (**5.335**), and imidazo[4,5-h]quinazolin-6(7H)-ones (**5.340**) has been investigated in some detail[249] and interestingly has been shown to result in oxidation at both the pyrimidine and the imidazole rings. Thus *lin*-benzohypoxanthine (**5.331**; $R^1 = R^2 = H$) and the derived ribonucleoside, *lin*-benzoinosine (**5.331**; $R^1 = H$, $R^2 = 1$-β-D-ribofuranosyl) are oxidized to mixtures of two types of product, identified by ultraviolet, NMR, and mass spectroscopy as the mono-oxidized and dioxidized species (**5.333**; $R = H$ or 1-β-D-ribofuranosyl) and (**5.334**; $R = H$ or 1-β-D-ribofuranosyl). *Prox*-benzohypoxanthine (**5.335**) likewise affords compounds (**5.337**) and (**5.338**). In contrast, the xanthine-oxidase-catalyzed oxidation of *dist*-benzohypoxanthine (**5.339**) does not proceed beyond the mono-oxidised *dist*-benzoxanthine stage (**5.341**). Comparison of the rates[249] of these enzymic oxidations reveals that the *prox* isomer (**5.335**) is oxidized approximately two orders of magnitude more slowly than the *lin* isomer (**5.331**; $R^1 = R^2 = H$) or the *dist* isomer (**5.339**). This sluggishness has been attributed to shielding of the $C(9)$ carbonyl group by the imidazole ring or to its hydrogen bonding with the $N(1)H$ group. The termination of the oxidation of the *dist* isomer (**5.339**) at the mono-oxidation stage (**5.341**) indicates that the substrate cannot be accommodated at the requisite oxidation site on the enzyme, possibly because of steric interference by the imidazole ring or preferential intramolecular hydrogen-bonding between the $N(9)H$ group and $N(1)$.[249]

Reduction

Only very limited information is available on the behavior of tricyclic 6-6-5 fused benzimidazoles with two additional hetero atoms towards reduction.

The debenzylation of 8-amino-3-benzylimidazo[4,5-g]quinazoline to 8-amino-N-1(3)H-imidazo[4,5-g]quinazoline (*lin*-benzoadenine) using sodium and liquid ammonia demonstrates the stability of the imidazo[4,5-g]quinazoline ring system to metal-proton donor reducing agents of this type.[238]

5.3.4. Practical Applications

Biological Properties

Certain 8-aminated imidazo[4,5-g]quinazoline derivatives represent the first examples of active cytokinins having a triheterocyclic structure.[250] Derivatives of the imidazo[4,5-h][1,4]benzothiazine ring exhibit a stimulating effect on the central nervous system,[242] and the hypoglycaemic activity of imidazo[4,5-f]quinazoline derivatives has been reported.[247]

Dyestuffs

Cyanine dyes containing the imidazo[4,5-*h*][1,4]benzothiazine[242] and imidazo[4,5-*g*]quinoxaline[245] chromophores have been described.

REFERENCES

1. D. J. Brown and R. J. Harrison, *J. Chem. Soc.* **1960,** 1837.
2. H. Dellweg, E. Becher, and K. Bernhauer, *Biochem. Z,* **328,** 96 (1956); *Chem. Abstr.,* **51,** 18043b.
3. B. A. Porai-Koshits, L. N. Kononova, and L. S. Efros, *Zh. Obshch. Khim.,* **24,** 507 (1954); *Chem. Abstr.,* **49,** 6235.
4. Z. Yoshida, Y. Shimada, and R. Oda, *J. Chem. Soc. Japan,* **55,** 354 (1952); *Chem. Abstr.,* **48,** 11398c.
5. A. D. Gornovskii and A. M. Simonov, *Zh. Obshch. Khim.,* **31,** 1941 (1961); *Chem. Abstr.,* **55,** 27278.
6. G. Leandri and I. Maioli, *Ann. Chim. (Rome),* **45,** 3 (1955); *Chem. Abstr.,* **49,** 15785h.
7. G. P. Kutrov, E. P. Kozlovskaya, and F. S. Babichev, *Ukr. Khim. Zh.,* **35,** 738 (1969); *Chem. Abstr.,* **71,** 124325.
8. D. Jerchel, H. Fischer, and M. Kracht, *Annalen,* **575,** 162 (1952).
9. G. R. Revankar, S. Siddappa, and D. C. Vonarani, *Indian J. Chem.,* **2,** 489 (1964).
10. E. W. Malmberg and C. S. Hamilton, *J. Am. Chem. Soc.,* **70,** 2415 (1948).
11. H. Zellner, G. Zellner, F. Koeppl, and J. Dirnberger, *Monat,* **98,** 643 (1967).
12. M. Duennenberger, A. E. Siegrist, E. Maeder, and P. Liechti, Swiss Patent 374,361; *Chem. Abstr.,* **62,** 679e. See also British Patent 941,048; *Chem. Abstr.,* **61,** 5828.
13. R. J. Pollet, H. A. Philippaerts, J. F. Willems, and F. H. Claes, German Patent 2,053,023; *Chem. Abstr.,* **75,** 103658.
14. P. M. Dzadzic, B. L. Bastic, and M. V. Polatic, *Glas. Hem. Drus., Beograd,* **36** (3–4), 137 (1971); *Chem. Abstr.,* **78,** 16096.
15. P. M. Dzadzic, M. V. Pilatic, and B. Bastic, *Glas. Hem. Drus., Beograd,* **37,** 257 (1972); *Chem. Abstr.,* **79,** 53215.
16. D. Heydenhauss and H. Schubert, *Z. Chem.,* **4,** 459 (1964); *Chem. Abstr.,* **62,** 9172.
17. P. Renz and A. J. Bauer-David, *Z. Naturforsch. B,* **27,** 539 (1972).
18. H. Goldstein and M. Streuli, *Helv. Chim. Acta,* **20,** 520 (1937).
19. K. Fries, R. Walter, and K. Schilling, *Annalen,* **516,** 248 (1935).
20. G. Boretti, D. Caltapan, A. Minghetti, M. Reggiani, U. Valcavi, and L. Valentini, *Chem. Ber.,* **92,** 3023 (1959).
21. D. J. Brown and R. J. Harrison, *J. Chem. Soc.,* **1959,** 3332.
22. G. Schroeter, E. Kindermann, C. Dietrich, C. Beyschlag, C. Fleischauer, E. Riebensahm, and C. Oesterlin, *Annalen,* **426,** 17 (1922).
23. K. H. Büchel, *Z. Naturforsch. B,* **25,** 934 (1970).
24. C. M. Orlando, J. G. Wirth, and D. R. Heath, *J. Heterocycl. Chem.,* **7,** 1385 (1970).
25. S. Chatterjee and J. Wolski, *J. Indian Chem. Soc.,* **43,** 660 (1966).
26. Ciba Ltd., Belgian Patent 620,372; *Chem. Abstr.,* **59,** 14145. See also Swiss Patent 443,664; *Chem. Abstr.,* **69,** 37123.
27. B. L. Bastic and V. B. Golubovic, *Glasnik. Khem. Drushtava, Beograd,* **18,** No. 4, 235 (1953); *Chem. Abstr.,* **52,** 2005f.
28. L. Blangey, *Helv. Chim. Acta,* **38,** 744 (1955).
29. H. Lieb, *Monatshaft,* **39,** 873 (1918); *J. Chem. Soc.,* **116,** I, 174. G. C. Chakravarti, *Q. J. Indian Chem. Soc.,* **1,** 19 (1924).
30. H. Lieb and G. Schwartzer, *Monatshaft,* **41,** 573 (1921).
31. M. V. Betrabet and G. C. Chakravarti, *J. Indian Chem. Soc.,* **7,** 191 (1930).

32. H. Moehrke, H. Koch, and H. V. Freyberg, German Patent 865,305; *Chem. Abstr.*, **52,** 20201f.
33. I. G. Farbenindustrie A.-G, German Patent 676,196 (1939); *Chem. Abstr.*, **33,** 6343. See also R. Weidenhagen, *Chem. Ber.*, **69B,** 2263 (1936).
34. A. V. El'tsov and V. S. Kuznetsov, *Zh. Org. Khim.*, **2,** 1482 (1966); *Chem. Abstr.*, **66,** 46368.
35. K. Zaitsu and Y. Ohkura, *Chem. Pharm. Bull.*, **23,** 1057 (1975).
36. F. D. Popp and A. Catala, *J. Heterocycl. Chem.*, **1,** 108 (1964).
37. L. Y. Oleinikova and F. T. Pozharskii, *Khim., Geterotsikl. Soedin.*, **1972,** 1555; *Chem. Abstr.*, **78,** 58315.
38. M. Ridi, L. Lazzi, and P. Corti, *Bull. Chim. Farm.*, **107,** 667 (1968); *Chem. Abstr.*, **70,** 87668.
39. H. A. Staab, *Annalen*, **609,** 75 (1957).
40. R. L. Clark and A. A. Pessolano, *J. Am. Chem. Soc.*, **80,** 1657 (1958).
41. M. V. Povstyanoi and P. M. Kochergin, *Khim. Issled. Farm.*, **1970,** 47; *Chem. Abstr.*, **76,** 46136.
42. N. P. Bednyagina, I. N. Getsova, and I. Y. Povstovskii, *Zh. Obshch. Khim.*, **32,** 3011 (1962); *Chem. Abstr.*, **58,** 9050f.
43. A. V. El'tsov, V. S. Kuznetsov, and M. B. Kolesova, *Zh. Org. Khim.*, **1,** 1117 (1965); *Chem. Abstr.*, **63,** 13254e.
44. D. J. Brown, *J. Chem. Soc.*, **1958,** 1974.
45. V. K. Chadha and H. K. Pujari, *Indian J. Chem.*, **8,** 1039 (1970).
46. W. G. Bywater, D. A. McGinty, and N. D. Jenesel, *J. Pharmacol.*, **85,** 14 (1945).
47. I. G. Farbenindustrie A.-G., German Patent 557,138 (1931); *Chem. Abstr.*, **27,** 1233.
48. G. B. Crippa and S. Maffei, *Gazz Chim. Ital.*, **71,** 418 (1941).
49. C. J. Paget, K. Kisner, R. L. Stone, and D. C. Delong, *J. Med. Chem.*, **12,** 1010 (1969).
50. W. Reid and W. Müller, *J. Prakt. Chem.*, **8,** 132 (1959).
51. E. Mutschler and M. Rummel, *Arch. Pharm.*, **300,** 695 (1967).
52. D. J. Brown and R. K. Lynn, *J. Chem. Soc. Perkin I*, **1974,** 349.
53. N. Ishikawa and T. Muranatsu, *Nippon Kagaku Kaishi*, **1973,** 563; *Chem. Abstr.*, **78,** 147873.
54. W. Reppe, *Annalen*, **596,** 158 (1955).
55. A. Bistrzycki and W. Schmutz, *Annalen*, **415,** 1 (1918); *J. Chem. Soc.*, **114,** I, 452.
56. W. Ried and E. Torinus, *Chem. Ber.*, **92,** 2902 (1959).
57. F. Kröhnke and H. Leister, *Chem. Ber.*, **91,** 1479 (1958).
58. F. D. Popp, *J. Heterocycl. Chem.*, **6,** 125 (1969).
59. K. Fries and E. Köhler, *Chem. Ber.*, **57B,** 496 (1924).
60. N. J. Leonard and A. M. Hyson, *J. Am. Chem. Soc.*, **71,** 1961 (1949).
61. M. Kamel, M. A. Allam, and N. Y. Abou-Zeid, *Tetrahedron*, **23,** 1863 (1967).
62. M. Kamel and H. I. Nasr, *Kolor. Ert.*, **10,** 177 (1968); *Chem. Abstr.*, **69,** 96582.
63. J. R. E. Hoover and A. R. Day, *J. Am. Chem. Soc.*, **76,** 4148 (1954).
64. J. M. Wilbur and A. R. Day, *J. Org. Chem.*, **25,** 753 (1960).
65. P. Truitt, D. Hayes, and L. T. Creagh, *J. Med. Chem.*, **7,** 362 (1964).
66. F. I. Carroll and J. T. Blackwell, *J. Heterocycl. Chem.*, **6,** 909 (1969).
67. F. I. Carroll and J. T. Blackwell, *J. Heterocycl. Chem.*, **7,** 297 (1970).
68. F. I. Carroll and J. T. Blackwell, *J. Med. Chem.*, **13,** 312 (1970).
69. Z. Yoshida and R. Oda, *J. Chem. Soc. Japan*, **55,** 447 (1952); *Chem. Abstr.*, **49,** 301.
70. H. Goldstein and P. Grandjeau, *Helv. Chim. Acta*, **26,** 468 (1943).
71. K. Fries and K. Billig, *Chem. Ber.*, **58B,** 1128 (1925).
72. H. Goldstein and G. Genton, *Helv. Chim. Acta*, **21,** 56 (1938).
73. (a) C. F. Kelly and A. R. Day, *J. Am. Chem. Soc.*, **67,** 1074 (1945); (b) P. Galimberti, *Gazz. Chim. Ital.*, **63,** 96 (1933); (c) L. Hunter, *J. Chem. Soc.*, **1945,** 806.
74. R. Delavigne, *Bull. Soc. Chim. France*, **1959,** 1200.
75. J. A. Stephens and C. S. Hamilton, *J. Am. Chem. Soc.*, **73,** 4297 (1951).

76. H. E. Fierz and R. Sallmann, *Helv. Chim. Acta.*, **5**, 560 (1922).
77. G. B. Crippa, T. Cessi, and G. Perroncito, *Gazz. Chim. Ital.*, **63**, 251 (1933).
78. M. Guarneri and F. Fiorini, *Gazzetta*, **92**, 1262 (1962); *Chem. Abstr.*, **59**, 600e.
79. O. Fischer, C. Dietrich, and F. Weiss, *J. Prakt. Chem.*, **100**, 167 (1920); *J. Chem. Soc.*, **120**, I, 57.
80. G. T. Morgan and H. S. Rooke, *J. Soc. Chem. Ind.*, **41**, 1–3T (1922).
81. J. W. Lown and M. H. Akhtar, *Can. J. Chem.*, **49**, 1610 (1971).
82. T. N. Ghosh, *J. Indian Chem. Soc.*, **10**, 583 (1933).
83. M. S. Chauhan and R. G. Cooke, *Aust. J. Chem.*, **23**, 2133 (1970).
84. P. Truitt and J. T. Witkowski, *Can. J. Chem.*, **45**, 997 (1967).
85. S. Maffei, *Gazzetta*, **76**, 345 (1946); *Chem. Abstr.*, **42**, 1266.
86. L. Capuano and W. Ebner, *Chem. Ber.*, **103**, 3104 (1970).
87. K. J. Hayes, U.S. Patent 3,915,985 (1974); *Chem. Abstr.*, **84**, 32604.
88. A. Messner and O. Sziman, *Angew. Chem.*, **6**, 250 (1967).
89. M. W. Partridge and H. A. Turner, *J. Chem. Soc.*, **1958**, 2086.
90. E. Harnki, T. Inaiki, and E. Imoto, *Bull. Chem. Soc. Japan*, **38**, 1805 (1965); ibid., **41**, 1361 (1968).
91. A. Heesing, V. Kappler, and W. Rauh, *Annalen*, **1976**, 2222.
92. L. C. March and M. M. Joullié, *J. Heterocycl. Chem.*, **7**, 39 (1970).
93. L. C. March and M. M. Joullié, *J. Heterocycl. Chem.*, **7**, 425 (1970).
94. L. L. Zaika and M. M. Joullié, *J. Heterocycl. Chem.*, **3**, 444 (1966).
95. R. K. Smalley and R. C. Perera, *Chem. Commun.*, **1970**, 1458.
96. P. A. S. Smith and E. Leon, *J. Am. Chem. Soc.*, **80**, 4647 (1958).
97. T. Bacchetti and A. Alcmagna, *Atti Accad. Nazl. Lincei. Rend. Classe Sci. Fis Mat e Nat.*, **28**, 824 (1960); *Chem. Abstr.*, **56**, 7304.
98. L. B. Volodarskii and V. A. Koptyug, *Zh. Obshch. Khim.*, **34**, 227 (1964); *Chem. Abstr.*, **60**, 12014.
99. D. Lloyd, R. H. McDougall, and D. R. Marshall, *J. Chem. Soc.*, **1965**, 3785.
100. M. Israel and E. C. Zoll, *J. Org. Chem.*, **37**, 3566 (1972); see also W. Ried and W. Höhne, *Chem. Ber.*, **87**, 1801 (1954).
101. A. F. Pozharskii, I. S. Kashparov, Yu P. Andreichikov, A. I. Buryak, A. A. Konstantinchenko, and A. M. Simonov, *Khim. Geterotsikl. Soedin*, **7**, 807 (1972); *Chem. Abstr.*, **76**, 24436.
102. M. V. Gorelik, T. K. Glasysheva, N. N. Sharetko, and T. A. Mikhailova, *Khim. Geterotsikl, Soedin.*, **7**, 238 (1971); *Chem. Abstr.*, **75**, 34977.
103. L. M. Alekseeva, E. M. Peresleni, Yu N. Sheinker, P. M. Kochergin, A. N. Krasovskii, and B. V. Kurmaz, *Khim. Geterotsikl. Soedin.*, **1972**, 1125; *Chem. Abstr.*, **77**, 139322, also E. G. Knysh, A. N. Krasovskii, and P. M. Kochergin, *Khim. Geterotsikl. Soedin.*, **7**, 1128 (1971); *Chem. Abstr.*, **76**, 59529.
104. L. L. Zaika and M. M. Joullié, *J. Heterocycl., Chem.*, **3**, 289 (1966).
105. N. Biradar and N. N. Sirmokadam, *J. Inorg. Nuc. Chem.*, **35**, 3639 (1973).
106. A. D. Garnovskii, O. A. Osipov, V. T. Panyushkin, and A. F. Pozharskii, *Zh. Obshch. Khim.*, **36**, 1063 (1966); *Chem. Abstr.*, **65**, 11742.
107. V. T. Panyushkin, A. D. Garnovskii, O. A. Osipov, A. L. Sinyavin, and A. F. Pozharskii, *Zh. Obshch. Khim.*, **37**, 312 (1967); *Chem. Abstr.*, **67**, 28852.
108. F. Feigl and H. Gleich, *Monatshaft*, **49**, 385 (1928).
109. D. J. Brown, *J. Chem. Soc.*, **1958**, 1974.
110. O. A. Osipov, A. M. Simonov, V. I. Minkin, and A. D. Garnovskii, *Dokl. Akad. Nauk. SSSR*, **137**, 1374 (1961); *Chem. Abstr.*, **55**, 24173. See also *Chem. Abstr.*, **57**, 13262 and **62**, 3494g.
111. V. S. Kuznetsov, N. A. Sobina, L. Y. Kheifets, and L. S. Efros, *Zh. Obshch. Khim.* **37**, 1802 (1967); *Chem. Abstr.*, **68**, 8764. See also L. S. Efros and G. N. Kulbitskii, *Zh. Obshch. Khim.*, **38**, 981 (1968); *Chem. Abstr.*, **69**, 76345.
112. A. F. Pozharskii and E. N. Malysheva, *Khim. Geterotsikl. Soedin.*, **1970**, 103; *Chem. Abstr.*, **72**, 120864.

113. V. I. Minkin, V. A. Bren, A. D. Garnovskii, and R. I. Nikitina, *Khim. Geterotsikl. Soedin.*, **1972,** 552; *Chem. Abstr.*, **77,** 61151.

114. A. D. Garnovskii, A. M. Simonov, and V. I. Minkin, *Khim. Geterotsikl. Soedin.*, **1973,** 99; *Chem. Abstr.*, **80,** 2941.

115. L. S. Efros, L. N. Kononkova, and Y. Eded, *Zh. Obshch. Khim.*, **24,** 488 (1954); *Chem. Abstr.*, **49,** 6236.

116. M. Kamel and H. I. Nasr, *Kolor Ert.*, **10,** 216 (1968); *Chem. Abstr.*, **70,** 12627.

117. G. N. Kulbitskii and L. S. Efros, *Zh. Org. Khim.*, **2,** 1305 (1966); *Chem. Abstr.*, **66,** 85725.

118. M. V. Povstyanoi and P. M. Kochergin, *Khim. Geterotsikl. Soedin.*, **7,** 1115 (1971); *Chem. Abstr.*, **76,** 140648. See also ref. 41.

119. M. V. Povstyanoi and P. M. Kochergin, *Khim. Geterotsikl. Soedin.*, **1972,** 816; *Chem. Abstr.*, **77,** 88395.

120. A. V. Stetsenko and V. G. Motornyi, *Ukr. Khim. Zh.*, **33,** 722 (1967); *Chem. Abstr.*, **68,** 60517.

121. B. I. Khristich and A. M. Simonov, *Khim. Geterotsikl. Soedin.*, **1966,** 611; *Chem. Abstr.*, **66,** 37830.

122. C. R. Revankar and S. Siddappa, *Monatshaft,* **98,** 169 (1967).

123. W. Reppe, *Annalen,* **601,** 134 (1956).

124. B. I. Mikhantev and V. V. Kalmykov, *Tr. Probl. Lab. Khim., Vyoskomol, Soedin., Voronezh. Gos. Univ.,* **1966,** No. 4, 51; *Chem. Abstr.*, **69,** 27336.

125. W. Reid and H. Lohwasser, *Annalen,* **699,** 88 (1966).

126. H. Zellner, Austrian Patent 176,561; *Chem. Abstr.*, **48,** 10778b. For an alternative preparation, See H. Zellner, Austrian Patent 176,560; *Chem. Abstr.*, **48,** 10778e.

127. G. N. Kulbitski and L. S. Efros, *Zh. Org. Khim.*, **3,** 575 (1967); *Chem. Abstr.*, **67,** 11456.

128. M. V. Pickering, P. Dea, D. G. Streeter, and J. T. Witkowski, *J. Med. Chem.*, **20,** 818 (1977). The Lewis acid procedure is reported by V. Niedballa and H. Vorbrüggen, *Angew. Chem. Int. Ed. Engl.*, **9,** 461 (1970).

129. "Biochemistry of Quinones," R. A. Morton, Ed., Academic Press, New York 1965.

130. E. G. Knysh, A. N. Krasovskii, P. M. Kochergin, and P. M. Shabelnik, *Khim. Geterotsikl. Soedin.*, **1972,** 399; *Chem. Abstr.*, **77,** 88398.

131. E. G. Knysh, A. N. Krasovskii, and P. M. Kochergin, *Khim. Geterotsikl. Soedin.*, **1972,** 257; *Chem. Abstr.*, **77,** 48331.

132. M. V. Povstyanoi, E. A. Yakubovskii, and P. M. Kochergin, *Zh. Org. Khim.*, **12,** 2044 (1976); *Chem. Abstr.*, **86,** 16643.

133. V. K. Chadha, K. S. Sharma, and H. K. Pujari, *Indian J. Chem.*, **9,** 913 (1971); *Chem. Abstr.*, **75,** 151730.

134. C. J. Paget and J. L. Sands, German Patent 2,003,841; *Chem. Abstr.*, **73,** 87920.

135. E. A. Zvezdina, M. P. Zhdanova, V. A. Bren, and G. N. Dorofeenko, *Khim. Geterotsikl. Soedin.*, **1974,** 1461; *Chem. Abstr.*, **82,** 97303.

136. I. N. Getsova, L. L. Sribnaya, and N. P. Bednyagina, *Khim. Geterotsikl. Soedin. Akad Nauk. Latv. SSR,* **1965,** 129; *Chem. Abstr.*, **63,** 4282b.

137. I. N. Getsova, I. V. Panov, and N. P. Bednyagina, *Zh. Obshch. Khim.*, **34,** 2026 (1964); *Chem. Abstr.*, **61,** 8297c.

138. P. M. Kochergin, M. V. Povstyanoi, B. A. Priimenko, and V. S. Ponomar, *Khim. Geterotsikl. Soedin.*, **1970,** 129; *Chem. Abstr.*, **72,** 90370.

139. H. C. Koppel, R. H. Springer, C. D. Danes, and C. C. Cheng, *J. Pharm. Sci.*, **52,** 81 (1963); *Chem. Abstr.*, **62,** 11808d.

140. E. G. Knysh and A. N. Krasovskii, *Khim. Issled. Farm.*, **1970,** 30; *Chem. Abstr.*, **75,** 140763.

141. I. S. Kashparov and A. F. Pozharskii, *Khim. Geterotsikl. Soedin.*, **7,** 124 (1971); *Chem. Abstr.*, **75,** 35922.

142. M. V. Povstyanoi, P. M. Kochergin, and E. A. Yakubovskii, *Tezisy. Dokl.-Nauchno-Tekh. Konf. "Khim. Primen. Formazanov 2nd,* 1974 (1975), 26; *Chem. Abstr.*, **87,** 53159.

143. M. V. Povstyanoi, P. M. Kochergin, E. V. Logachev, and E. A. Yakubovskii, *Khim. Geterotsikl. Soedin.*, **1975**, 422; *Chem. Abstr.*, **83**, 28197.

144. M. V. Povstyanoi, E. A. Yakubovskii, and P. M. Kochergin, *Ukr. Khim. Zh.*, **42**, 64 (1976); *Chem. Abstr.*, **84**, 150602.

145. M. V. Povstyanoi, P. M. Kochergin, E. V. Logachev, E. A. Yakubovskii, A. V. Akimov, and V. P. Kruglenko, *Khim. Geterotsikl. Soedin.*, **1976**, 1424; *Chem. Abstr.*, **86**, 55350.

146. A. D. Garnovskii, A. M. Simonov, V. I. Minkin, and V. D. Dionisev, *Zh. Obshch. Khim.*, **34**, 272 (1964); *Chem. Abstr.*, **60**, 10508e.

147. A. F. Pozharskii, V. V. Kuzmenko, I. S. Kashparov, V. I. Sokolov, and M. M. Medvedeva, *Khim. Geterotsikl. Soedin.*, **1976**, 356; *Chem. Abstr.*, **85**, 78070.

148. G. Green-Buckley and J. Griffiths, *Chem. Commun.* **1977**, 396.

149. B. A. Porai-Koshits, L. N. Kononova, and L. S. Efros, *J. Gen. Chem. USSR*, **24**, 517 (1954); *Chem. Abstr.*, **49**, 9630.

150. O. Fischer, *Ber.*, **32**, 1314 (1899).

151. H. Schubert and G. Bohme, *Wiss. Z. Martin-Luther Univ.*, **8**, 1037 (1959); *Chem. Abstr.*, **55**, 12389b.

152. L. F. Fieser and M. A. Ames, *J. Am. Chem. Soc.*, **49**, 2604 (1967).

153. I. N. Getsova and N. P. Bednyagina, *Khim. Geterotsikl. Soedin.*, **1965**, 284; *Chem. Abstr.*, **63**, 9935.

154. A. F. Pozharskii, E. A. Zvezdina, Y. P. Andreichikov, A. M. Simonov, V. A. Anisimova, and S. F. Popova, *Khim. Geterotsikl. Soedin.*, **1970**, 1267; *Chem. Abstr.*, **75**, 5804.

155. L. A. Petrov, R. O. Matevosyan, V. D. Galyaminskikh, and N. I. Abramova, *Zh. Org. Khim.*, **2**, 501 (1966); *Chem. Abstr.*, **65**, 18573.

156. E. Lebenstedt and W. Schunack, *Arch. Pharm.*, **310**, 455 (1977).

157. H. Schubert and H. Fritsche, *J. Prakt. Chem.*, **7**, 207 (1958).

158. E. G. Knysh, A. N Krasovskii, and P. M. Kochergin, *Khim. Geterotsikl. Soedin.*, **1972**, 33; *Chem. Abstr.*, **76**, 153679.

159. F. R. Basford, F. H. S. Curd, and F. L. Rose, U.S. Patent 2,460,409; *Chem. Abstr.*, **43**, 3854.

160. T. K. Gladychova and M. V. Gorelik, *Khim. Geterotsikl. Soedin.*, **1970**, 554; *Chem. Abstr.*, **73**, 87858.

161. K. Dziewonski and L. Sternbach, *Bull. Int. Acad. Polonaise, Classe Scu. Math. Nat.*, **1935A**, 327.

162. V. S. Kuznetsov and L. S. Efros, *Zh. Org. Khim.*, **1**, 1458 (1965); *Chem. Abstr.*, **64**, 727.

163. G. A. Efimova and L. S. Efros, *Zh. Org. Khim.*, **3**, 162 (1967); *Chem. Abstr.*, **66**, 104957.

164. A. V. Stetsenko and V. K. Lishko, *Ukr. Khim. Zh.*, **28**, 218 (1962); *Chem. Abstr.*, **58**, 6818e.

165. G. N. Dorofeenko and Y. A. Zhdanov, *Zh. Obshch. Khim.*, **30**, 3451 (1960); *Chem. Abstr.*, **55**, 18712b.

166. A. V. Stetsenko and V. K. Lishko, *Ukr. Khim. Zh.*, **28**, 218 (1962); *Chem. Abstr.*, **58**, 6818e.

167. T. J. Lane and K. P. Quinlan, *J. Am. Chem. Soc.*, **82**, 2997 (1960).

168. A. F. Pozharskii, A. N. Suslov, and V. V. Kitaev, *Dokl. Akad. Nauk SSSR*, **234**, 841 (1977); *Chem. Abstr.*, **87**, 183573.

169. V. S. Kuznetsov and L. S. Efros, *Zh. Org. Khim.*, **3**, 393 (1967); *Chem. Abstr.*, **67**, 3069.

170. G. R. Revankar, S. B. Nerali, and S. Siddappa, *J. Karnatek Univ.*, **9–10**, 44 (1964–5); *Chem. Abstr.*, **64**, 6641.

171. G. R. Revankar and S. Siddappa, *J. Karnatek Univ.*, **9–10**, 36 (1964–1965); *Chem. Abstr.*, **64**, 6642a.

172. A. Omodei-sale, E. Toia, G. Galliani, and L. J. Lerner, German Patent 2,551,868; *Chem. Abstr.*, **85**, 192731.

173. N. D. Mikhnovska and O. V. Stetsenko, *Microbiol. Zh.*, **29**, 242 (1967); *Chem. Abstr.*, **67**, 97905.

174. N. M. Stetsenko, L. A. Sirenko, A. V. Stetskenko, and V. V. Arendarchuk, "*Tsvetenie*" *Vody*, **1969**, No. 2, 186; *Chem. Abstr.*, **73**, 13406.

175. M. A. Allan and N. Y. Abou-Zeid, *Egypt J. Chem.*, **15**, 339 (1972); *Chem. Abstr.*, **81**, 51085.

176. S. C. Slifkin, U.S. Patent 2,542,715; *Chem. Abstr.*, **45**, 6521.

177. Ciba Ltd., Swiss Patent 239,325; *Chem. Abstr.*, **43**, 5961. See also Geigy A.-G., Swiss Patent 228,821, *Chem. Abstr.*, **43**, 4485 and W. Anderan, US Patent 2,411,646; *Chem. Abstr.*, **41**, 1455.

178. R. Fleischhauer and F. Aldebert, German Patent 1,081,990; *Chem. Abstr.*, **55**, 19255.

179. C. E. Müller and W. Kirst, German Patent 883,284; *Chem. Abstr.*, **51**, 2300.

180. A. Wohlkoenig, P. Hindermann, F. Beffa, and G. Hegàr, German Patent 2,247,838; *Chem. Abstr.*, **78**, 160903.

181. A. M. Simonov, B. I. Khristich, O. A. Osipov, M. I. Knyazhanskii, and O. T. Asmaev, USSR Patent 202,155; *Chem. Abstr.*, **68**, 96808.

182. H. Haeusermann and J. Volz, German Patent, 1,098,125; *Chem. Abstr.*, **56**, 10158.

183. J. Pirkl, Czechoslovakian Patent 161,347; *Chem. Abstr.*, **85**, 79699.

184. Chemische Fabriek L. van der Grinten N. V. , British Patent 953,908; *Chem. Abstr.*, **61**, 3852b.

185. A. E. van Dormael, J. Nys, and A. de Cat. U.S. Patent 2,680,686; *Chem. Abstr.*, **49**, 759.

186. M. P. Schmidt and O. Sues, U.S. Patent 3,046,115; *Chem. Abstr.*, **58**, 12484. See also U.S. Patent 3,046,116; *Chem. Abstr.*, **58**, 12573h. U.S. Patent 3,046,111; *Chem. Abstr.*, **58**,12573e. U.S. Patent 3,046,124; *Chem. Abstr.*, **59**, 1649e German Patent 872,154; *Chem. Abstr.*, **53**, 12899d.

187. O. Sues, U.S. Patent 3,046,114; *Chem. Abstr.*, **59**, 636.

188. M. P. Schmidt and O. Sues, U.S. Patent 2,754,209; *Chem. Abstr.*, **51**, 918. See also German Patent 876,202; *Chem. Abstr.*, **53**, 9870. *German Patent* 957,392; *Chem. Abstr.*, **53**, 8095b. British Patent 784,672; *Chem. Abstr.*, **52**, 8213b. U.S. Patent 3,046,131; *Chem. Abstr.*, **60**, 477.

189. T. Tsunoda, S. Maeda, and M. Ozutsumi, Japanese Patent 75 40,626; *Chem. Abstr.*, **83**, 115921.

190. E. J. Van Lare and L. G. S. Brooker, U.S. Patent 2,548,571; *Chem. Abstr.*, **45**, 6100.

191. O. Sues, M. Tomanek, and E. Lind, German Patent 1,137,625; *Chem. Abstr.*, **59**, 11515. See also U.S. Patent 3,257,204; *Chem. Abstr.*, **65**, 20130. British Patent 895,001; *Chem. Abstr.*, **57**, 14577.

192. N. M. Stetsenko, L. A. Sirenko, A. V. Stetsenko, and V. V. Arendarchuk, *Tsvetenie Vody*., **1969**, No. 2, 186., *Chem. Abstr.*, **73**, 13406.

193. Oesterreichesche Stichkstoffwerke A.-G., Belgian Patent 610,311; *Chem. Abstr.*, **57**, 17138i.

194. C. Schuster and R. Gehm, German Patent 941,999; *Chem. Abstr.*, **52**, 19256c.

195. A. V. Stetsenko and M. M. Kul'chitskii, *Ukr. Khim. Zh.*, **35**, 288 (1969); *Chem. Abstr.*, **71**, 22892.

196. A. Ogawa and T. Ishido, German Patent 2,600,966; *Chem. Abstr.*, **86**, 36316.

197. O. Sues, U.S. Patent 3,046,122; *Chem. Abstr.*, **59**, 1648.

198. I. Tamm, R. Bablannan, M. M. Nemes, C. H. Shunk, F. M. Robinson, and K. A. Folkers, *J. Exp. Med.*, **113**, 625 (1961).

199. M. V. Povstyanoi, P. M. Kochergin, *Khim. Geterotsikl. Soedin.*, **7**, 1125 (1971); *Chem. Abstr.*, **76**, 172457.

200. A. O. Fitton and B. T. Hatton, *J. Chem. Soc. C*, **1970**, 2518.

201. K. Fries, E. Modrow, B. Raeke, and K. Weber, *Annalen*, **454**, 191 (1927).

202. E. Lebenstedt and W. Schunack, *Arch. Pharm.*, **307**, 894 (1974); *Chem. Abstr.*, **82**, 72877 (1975).

203. E. Lebenstedt and W. Schunack, *Arch. Pharm.*, **308**, 977 (1975); *Chem. Abstr.*, **84**, 90072 (1976).

204. S. Ishiwata and Y. Shiokawa, *Chem. Pharm. Bull. (Japan)*, **17,** 1153 (1969); *Chem. Abstr.*, **71,** 61293 (1969).
205. S. Ishiwata and Y. Shiokawa, *Chem. Pharm. Bull. (Japan)*, **17,** 2455 (1969); *Chem. Abstr.*, **72,** 100603 (1970).
206. M. Gandino and A. R. Katritzky *Ann. Chim. (Rome)*, **60,** 462 (1970); *Chem. Abstr.*, **73,** 110855 (1970); M. Gandino and L. Magnani, U.S. Patent 3,909,275; *Chem. Abstr.*, **82,** 195252 (1975); German Patent 2,002,409; *Chem. Abstr.*, **73,** 121570 (1970).
207. B. I. Khristich, V. A. Kruchinin, A. F. Pozharskii, and A. M. Simonov, *Chem. Heterocyc. Compounds*, **1971,** 759; *Khim. Geterotsikl. Soedin.*, **1971,** 814; *Chem. Abstr.*, **76,** 24349 (1972).
208. C. F. Spencer, H. R. Snyder, and R. J. Alaimo, *J. Heterocyc. Chem.*, **12,** 1319 (1975)
209. C. F. Spencer and R. J. Alaimo, U.S. Patent 3,868,378; *Chem. Abstr.*, **83,** 58818 (1975).
210. C. F. Spencer and H. R. Snyder, U.S. Patent 3,947,434; *Chem. Abstr.*, **85,** 33014 (1976); German Patent 2,427,409; *Chem. Abstr.*, **82,** 170927 (1975); German Patent 2,427,410; *Chem. Abstr.*, **82,** 170926 (1975).
211. L. Reichel and G. Hempel, *Annalen*, **693,** 216 (1966).
212. A. Ricci, M. Negri, and C. Rossi, *Ann. Chim. (Rome)*, **53,** 1507 (1963); *Chem. Abstr.*, **60,** 14464 (1964).
213. C. W. Schellhammer, S. Petersen, and G. Domagk, U.S. Patent 3,084,165; *Chem. Abstr.*, **59,** 13956 (1963); German Patent 1,095,837; *Chem. Abstr.*, **57,** 7279 (1962); German Patent 1,102,161; *Chem. Abstr.*, **57,** 836 (1962).
214. I. G. Ilina, N. B. Kazennova, V. G. Bakhmut-Skaya, and A. P. Terent'ev, *Chem. Heterocyc. Compounds*, **1973,** 1028; *Khim. Geterotsikl. Soedin.*, **1973,** 1112; *Chem. Abstr.*, **79,** 126396 (1973).
215. R. Weidenhagen and U. Weeden, *Chem. Ber.*, **71,** 2347 (1938).
216. B. I. Khristich and A. M. Simonov, *Chem. Heterocyc. Compounds*, **1966,** 465; *Khim. Geterotsikl. Soedin.*, **1966,** 611; *Chem. Abstr.*, **66,** 37830 (1967).
217. A. M. Simonov, B. I. Khristich, and V. G. Poludnenko, *Chem. Heterocyc. Compounds*, **1967,** 731; *Khim. Geterotsikl. Soedin.*, **1967,** 927; *Chem. Abstr.*, **68,** 114499 (1968).
218. M. S. Kamel, S. Sherif, R. M. Issa, and F. I. Abd-El-Hay, *Tetrahedron*, **29,** 221 (1973).
219. R. Huisgen, *Annalen*, **559,** 101 (1948).
220. F. T. Pozharskii and L. Y. Oleinikova, *Chem. Heterocyc. Compounds*, **1973,** 406; *Khim. Geterotsikl. Soedin.*, **1973,** 440; *Chem. Abstr.*, **79,** 18635 (1973).
221. F. H. Case, *J. Heterocyc. Chem.*, **4,** 157 (1967).
222. F. H. Case and L. Kennon, *J. Heterocyc. Chem.*, **4,** 483 (1967).
223. H. Hennig and J. Tauchnitz, *J. Prakt. Chem.*, **312,** 1191 (1970).
224. H. Hennig, J. Tauchnitz, H. J. Hofmann, M. Herrmann, and W. Schindler, *J. Prakt. Chem.*, **313,** 646 (1971).
225. E. Lebenstedt and W. Schunack, *Arch. Pharm.*, **308,** 413 (1975); *Chem. Abstr.*, **83,** 114292 (1975).
226. B. I. Khristich, M. I. Knyazhanskii, O. A. Osipov, O. T. Asmaev, and A. M. Simonov, *Chem. Heterocyc. Compounds*, **1970,** 220; *Khim. Geterotsikl. Soedin.*, **1970,** 234; *Chem. Abstr.*, **72,** 120645 (1970).
227. H. Hennig, J. Tauchnitz, G. Kohler, and W. Schindler, *J. Prakt. Chem.*, **317,** 853 (1975).
228. G. Grandoline, A. Ricci, A. Martani, and F. Delle Monach, *J. Heterocycl. Chem.*, **3,** 302 (1966).
229. B. I. Khristich, V. A. Kruchinin, A. F. Pozharskii, and A. M. Simonov, *Chem. Heterocy. Compounds*, 759 (1971); *Khim. Geterotsikl. Soedin.*, 814 (1971); *Chem. Abstr.*, **76,** 24349 (1972).
230. A. A. Schilt and K. R. Kluge, *Talanta*, **15,** 1055 (1968); *Chem. Abstr.*, **70,** 53515 (1969); A. A. Schilt, W. E. Dunbar, B. W. Gandrud, and S. E. Warren, *Talanta*, **17,** 649 (1970); *Chem. Abstr.*, **73,** 76351 (1970).
231. H. Hennig, R. Kirmse, J. Tauchnitz, and W. Windsch., *Z. Chem.*, **11,** 115 (1971); *Chem. Abstr.*, **75,** 13278 (1971).

232. H. Hennig, J. Tauchnitz, B. Wittmann, and W. Schindler, *Z. Chem.*, **15,** 113 (1975); *Chem. Abstr.*, **83,** 44708 (1975); H. Hennig, J. Tauchnitz, J. Koerner, and W. Schindler, *J. Signalaufzeichnungs-materialen*, **4,** 125 (1976); *Chem. Abstr.*, **85,** 64778 (1976); H. Hennig, W. Schindler, and J. Tauchnitz, East German Patent 116,040; *Chem. Abstr.*, **84,** 181,622 (1976); East German Patent 116,039; *Chem. Abstr.*, **84,** 181,623 (1976).

233. H. E. Mertel, in "Pyridine and Derivatives," Part 2, E. Klingsberg, Ed., Wiley-Interscience, New York, **1961,** Chap. 6, p. 345 *et seq.*

234. C. F. Spencer, H. R. Snyder, H. A. Burch, and C. J. Hatton, *J. Med. Chem.*, **20,** 829 (1977); *Chem. Abstr.*, **87,** 33429 (1977).

235. H. R. Snyder, C. F. Spencer, and R. Freedman, *J. Pharm. Sci.*, **66,** 1204 (1977); *Chem. Abstr.*, **88,** 22744 (1978).

236. H. Hennig, J. Tauchnitz, O. Gurtler, H. Hofmann, and W. Schindler, *J. Prakt. Chem.*, **313,** 781 (1971).

237. W. B. Kremer and J. Laszlo, *Cancer Chemother. Rep.*, **51,** 19 (1967); *Chem. Abstr.*, **66,** 84303 (1967); N. S. Mizuno, *Biochem. Pharmacol.*, **16,** 933 (1967); *Chem. Abstr.*, **67,** 31137 (1967); M. A. Chirigos, J. W. Pearson, T. S. Papas, W. A. Woods, H. B. Wood, and G. Spahn, *Cancer Chemother. Rep.*, **57,** 305 (1973); *Chem. Abstr.*, **80,** 75 (1974).

238. N. J. Leonard, A. G. Morrice, and M. A. Sprecker, *J. Org. Chem.*, **40,** 356 (1975).

239. A. G. Morrice, M. A. Sprecker, and N. J. Leonard, *J. Org. Chem.*, **40,** 363 (1975).

240. L. C. March and M. M. Joullié, *J. Heterocyc. Chem.*, **7,** 249 (1970).

241. C. C. Howard, *Chem. Ber.*, **30,** 2103 (1897).

242. E. A. Kuznetsova, N. T. Pryanishnikova, L. I. Gaidukova, I. V. Fedina, and S. V. Zhuravlev, *Chem. Pharm. J.*, **9,** 759 (1975); *Khim. Farm. Zh.*, **9,** 11 (1975); *Chem. Abstr.*, **84,** 130139 (1976); USSR Patent 419,524; *Chem. Abstr.*, **80,** 146180 (1974).

243. S. G. Fridman and L. I. Kotova, *Ukr. J. of Chem.*, **37,** 60 (1971); *Ukr. Khim. Zh.*, **37,** 920 (1971); *Chem. Abstr.*, **76,** 101195 (1972).

244. G. E. Keyser and N. J. Leonard, *J. Org. Chem.*, **41,** 3529 (1976).

245. S. G. Fridman and L. I. Kotova, *J. Gen. Chem. USSR*, **32,** 2829 (1962); *Zh. Obshch. Khim.*, **32,** 2871 (1962); *Chem. Abstr.*, **58,** 11495 (1963).

246. L. C. March and M. M. Joullié, *J. Heterocyc. Chem.*, **7,** 39 (1970).

247. C. M. Gupta, A. P. Bhaduri, and N. M. Khanna, *Indian J. Chem.*, **7,** 1166 (1969); *Chem. Abstr.*, **72,** 31752 (1970).

248. M. J. S. Dewar and P. M. Maitlis, *J. Chem. Soc.*, **1957,** 2518.

249. N. J. Leonard, M. A. Sprecker, and A. G. Morrice, *J. Am. Chem. Soc.*, **98,** 3987 (1976).

250. M. A. Sprecker, A. G. Morrice, B. A. Gruber, N. J. Leonard, R. Y. Schmitz, and F. Skoog, *Phytochemistry*, **15,** 609 (1976); *Chem. Abstr.*, **85,** 46584 (1976).

251. T. J. Batterham, in "NMR Spectra of Simple Heterocycles," Wiley-Interscience, New York, **1973,** p. 267.

252. T. J. Batterham, in "NMR Spectra of Simple Heterocycles," Wiley-Interscience, New York, **1973,** p. 329.

253. D. F. Wiemer, D. I. C. Scopes, and N. J. Leonard, *J. Org. Chem.*, **41,** 3051 (1976).

Author Index

Numbers in *italics* indicate pages where reference appear.

Subject Index

Italicized page numbers refer to tables.